# OVARIAN FOLLICULAR AND CORPUS LUTEUM FUNCTION

# ADVANCES IN EXPERIMENTAL MEDICINE AND BIOLOGY



## Recent Volumes in this Series

Volume 105
NUTRITIONAL IMPROVEMENT OF FOOD AND FEED PROTEINS
Edited by Mendel Friedman

Volume 106
GASTROINTESTINAL HORMONES AND PATHOLOGY OF THE DIGESTIVE SYSTEM
Edited by Morton Grossman, V. Speranza, N. Basso, and E. Lezoche

Volume 107
SECRETORY IMMUNITY AND INFECTION
Edited by Jerry R. McGhee, Jiri Mestecky, and James L. Babb

Volume 108
AGING AND BIOLOGICAL RHYTHMS
Edited by Harvey V. Samis, Jr. and Salvatore Capobianco

Volume 109
DRUGS, LIPID METABOLISM, AND ATHEROSCLEROSIS
Edited by David Kritchevsky, Rodolfo Paoletti, and William L. Holmes

Volume 110
HUMAN INTERFERON: Production and Clinical Use
Edited by Warren R. Stinebring and Paul J. Chapple

Volume 111
HORMONES AND ENERGY METABOLISM
Edited by David M. Klachko, Ralph R. Anderson, and Murray Heimberg

Volume 112
OVARIAN FOLLICULAR AND CORPUS LUTEUM FUNCTION
Edited by Cornelia P. Channing, John M. Marsh, and William A. Sadler

Volume 113
PARKINSON'S DISEASE – II: Aging and Neuroendocrine Relationships
Edited by Caleb E. Finch, David E. Potter, and Alexander D. Kenny

Volume 114
FUNCTION AND STRUCTURE OF THE IMMUNE SYSTEM
Edited by Wolfgang Müller-Ruchholtz and Hans Konrad Müller-Hermelink

Volume 115
DYNAMICS OF ARTERIAL FLOW
Edited by Stewart Wolf and Nicholas T. Werthessen

# OVARIAN FOLLICULAR AND CORPUS LUTEUM FUNCTION

Edited by

**Cornelia P. Channing**
Department of Physiology
University of Maryland School of Medicine
Baltimore, Maryland

**John M. Marsh**
Department of Biochemistry
University of Miami School of Medicine
Miami, Florida

and

**William A. Sadler**
Center for Population Research
National Institute of Child Health and Human Development
Bethesda, Maryland

PLENUM PRESS • NEW YORK AND LONDON

Library of Congress Cataloging in Publication Data

Workshop on Ovarian Follicular and Corpus Luteum Function, 2d, Miami Beach, Fla., 1978.
Ovarian follicular and corpus luteum function.

(Advances in experimental medicine and biology; v. 112)
"Proceedings of the second workshop . . . sponsored by the Institute of Child Health and Human Development, National Institutes of Health, Bethesda, Maryland."
1. Ovulation – Congresses. 2. Graafian follicle – Congresses. 3. Corpus luteum – Congresses. I. Channing, Cornelia P. II. Marsh, John M. III. United States. National Institute of Child Health and Human Development. IV. Title. V. Series.
[DNLM: 1. Corpus luteum – Physiology – Congresses. 2. Graafian follicle – Physiology – Congresses. W1 AD559 v. 112/WP320 W9266 1978o]
QP261.W66 1978 599'.01'66 79-48
ISBN 978-1-4684-3476-7 ISBN 978-1-4684-3474-3 (eBook)
DOI 10.1007/978-1-4684-3474-3

Proceedings of the Second Workshop on Ovarian Follicular and Corpus Luteum Function held in Miami Beach, Florida, June 11–13, 1978. The workshop was sponsored by the Institute of Child Health and Human Development, National Institutes of Health, Bethesda, Maryland.

Published in 1979
Softcover reprint of the hardcover 1st edition 1979

Plenum Press, New York
A Division of Plenum Publishing Corporation
227 West 17th Street, New York, N.Y. 10011

## Preface

The purpose of this volume, the proceedings of the Second Ovarian Workshop, is to review the state of the art in the field of ovarian follicular and corpus luteum function. This workshop was sponsored by the Center of Population Research, Institute of Child Health and Human Development, National Institutes of Health, Bethesda, Maryland. We also acknowledge the support of The Channing-Weinberg Company, Mr. Henry Burckhardt, The Merck Institute for Experimental Research, The Ortho Pharmaceutical Company, The Expanded Program for Human Reproduction of the World Health Organization, and The University of Miami.

We thank the organizing committee consisting of Drs. Hans Lindner, Georgianna Jagiello, Neena Schwartz, Rees Midgley, Gil Greenwald, Andrew Nalbandov, Sheldon Segal, Julia Labotsky, and Griff Ross for their generous assistance in setting up the program. We also thank the chairpersons of each session for their skillful work and for using their expertise in keeping the discussions to the point. We also are grateful to them for editing the written discussions. Chairpersons included Drs. Robert J. Ryan, Roy O. Greep, Allen W. Schuetz, Neena B. Schwartz, John M. Marsh, John W. Wilks, Hans R. Lindner, and Scheldon J. Segal. Special thanks go to all the invited speakers, discussants, participants and particularly Dr. Irving Rothchild for his summing up remarks.

The above organization of Ms. Betty Howard and her staff is gratefully acknowledged. The above secretarial assistance of Ms. Gail Williams, Ms. Evelyn Wisowaty, Ms. Theda Smith and Ms. Dorothy Tengood is acknowledged with thanks. The expert editorial assistance of Ms. Alison Schwartz-Kripner is to be given special recognition.

## Contents

SESSION NUMBER: I

TITLE: INTRODUCTORY REMARKS

PRINCIPAL SPEAKER: Gilbert Greenwald

# INTRODUCTORY REMARKS: RUMINATIONS ON OVARIAN FUNCTION

Gilbert S. Greenwald

Department of Physiology, University of Kansas Medical Center
Kansas City, Kansas 66103

The opportunity to serve as the gadfly in residence for this meeting is an irresistible temptation. Without the usual editorial constraints, one can bare one's prejudices, fancies and even follies. I therefore propose to discuss selected aspects of follicular development in a series of mini-essays.

1. Development of primary follicles: LIFO, FIFO or chance:

Several years ago, when the specter of inflation first appeared, a number of corporations adopted new accounting procedures to handle their inventories. The acronyms then in vogue were FIFO and LIFO, standing for "first in, first out" and "last in, first out". These terms aptly apply to living as well as to corporate entities in describing the possible sequence of release of the products of endocrine and exocrine glands. For example, thyroglobulin fits the pattern of LIFO secretion since the most recently formed product is released preferentially on stimulation of the thyroid gland.

In considering the cues converting primordial to primary follicles, a hypothesis proposed by Edwards fits the FIFO model; i.e., the first formed oocytes are the first ones to be mobilized in the postnatal period. This is an interesting speculation but with no direct evidence - either pro or con. Pulse labelling of the prenatal ovary and subsequent autoradiographic studies at various times postnatally might indicate whether the maturation of oocytes is temporally programmed.

Alternatively, the recruitment of primordial follicles may represent a random process with chance proximity to blood vessels, nerves, more advanced follicles or corpora lutea (CL) acting as the signal for passage out of the resting pool.

The state of the art has at least advanced to the point where the problem is open to experimental attack. Trypsin-collagenase digestion and subsequent culture of primordial follicles with peptide and/or steroid hormones, catecholamines, CL or more advanced follicles might yield interesting results.

2. What stages of folliculogenesis are pituitary independent?

The usual textbook account suggests that follicles can develop through the latest preantral stage before they fall under the influence of FSH and LH. In fact, a number of studies indicate that pituitary dependence begins at a much earlier stage in follicular development. For example, after hypophysectomy of the rat, mouse or hamster, the most advanced follicles fall far short of the normal 8-12 layered preantral stage. Moreover, in hypogonadotropic women, follicles are rarely encountered beyond the formation of 1 to 2 layers of granulosa cells.

In hypophysectomized hamsters and rats, replacement therapy with steroids and FSH and LH affects the number and histological appearance of small follicles. Similarly, in immature mice, deprivation of pituitary hormones also disrupts follicular development. Hence, there are quantitative and qualitative differences in preantral follicles in the absence of pituitary hormones. The concept of pituitary independence in early stages of folliculogenesis, therefore, seems to have little basis in fact.

3. When does the follicle begin secreting estrogen?

It is surprising that there is so little direct evidence on this point. It is axiomatic that the antral follicle is the principal source of the hormone - at least in non-primate species - but whether earlier preantral stages secrete estrogens for either local or distant consumption is unknown. At least in the hamster none of the enzymes involved in steroidogenesis can be localized by histochemical techniques in preantral follicles. In the rat, granulosa cells harvested from large preantral follicles and maintained in vitro, when supplied with androgen precursors, begin secreting estrogens after a lag period of 1-2 days. However, by this time these cells really may be physiologically comparable to antral granulosal cells. The methods are now available to resolve directly whether estrogen secretion is limited to the antral follicle. If this turns out to be true the interesting possibility then arises that younger follicles somehow tap the estrogen required for their further differentiation from more advanced follicles.

4. An antral follicle is not an antral follicle.

Detailed analysis of the antral follicle is now under way and for obvious reasons attention has focused on ripe follicles which are on the verge of ovulating. The obvious caveat is that the conclusions reached in analyzing mature antral follicles may not necessarily apply to younger stages. In its life history the tertiary follicle passes from infancy to maturity to senescence. Failure to keep this sequence in mind may account for some of the confusion on the steroidogenic capabilities of thecal and granulosa cells.

Morphological, biochemical and physiological differences exist between immature and mature antral follicles, which must be considered in the interpretation of results. Moreover, antral follicles of the same size and age may differ in their steroidal profiles and responses depending on the local hormonal milieu (e.g., the influence of the corpus luteum).

Thus, until the life cycle of the antral follicle is worked out in much greater detail we have only a fragmentary understanding of follicular function.

5. "Pharmacological" doses of steroids are physiological in the ovary.

It is now evident that the various ovarian compartments are bathed with extraordinarily high levels of steroids produced either *in situ* or transferred from one compartment to another. The pioneering research of Williams and Bradbury showed that in hypophysectomized rats massive amounts of stilbestrol (1 mg/daily) were needed to stimulate granulosal development in preantral follicles and for the facilitatory interactions with gonadotropins. Similarly, several laboratories have shown recently that in the pregnant rat large amounts of exogenous steroids are needed to reverse the local luteolytic effects of anti-LH serum.

This is an interesting turn of events. Amounts of steroids which not so long ago were dismissed as unphysiological are now known to be necessary to restore the normal ovarian environment of hypophysectomized animals.

Intuitively, one feels that these high local levels of steroids are involved in various negative and positive feedback interactions at the ovarian level. However, one of our arch skeptics has raised the possibility that the elevated concentrations of estrogen in follicular fluid merely represents a storage phenomenon and that the hormone may be of greater consequence after its postovulatory release and subsequent actions on the oviduct. At present, one has to concede that there really is no direct *in vivo* proof that follicular fluid steroids have the profound effects on the ovary that have been postulated.

Two worthwhile experiments come to mind. First, when labelled estrogen is instilled directly into an antral follicle, what is its distribution, with time, throughout the rest of the ovary? In other words, how diffusible is the steroid and where is it found? Secondly, species such as the sheep and guinea pig develop many more follicles than the numbers normally ovulating. If suitably marked antral follicles are injected with estradiol, what is their subsequent fate? Will peripheral levels of estrogen increase above normal values and how many CL will be formed?

6. Intraovarian interactions of steroids and gonadotropins: the new frontier.

It was apparent several years ago that the most vigorously pursued area in the 70s would be direct analysis of ovarian interactions. Numerous situations exist in which peripheral levels of hormones do not reflect profound changes in ovarian function. For example, depending on the species, the emergence of a single or group of estrogen secreting follicles is often not correlated with significant changes in circulating levels of gonadotropins. The role of LH as a luteotropic hormone in several species is not associated with increased titers of the hormone; in fact, in pregnant rodents serum LH levels may be actually lower than in cyclic animals. As a final example, the onset of functional luteolysis is usually not paralleled by alterations in gonadotropins or prolactin. Hence, changes in tissue level responsiveness emerge as the focal point for research in the next decade.

I do not want to belabor this point because it is bound to be the theme of many of the presentations at this meeting. Attention so far, has been directed largely towards the positive intraovarian interplay of steroids and gonadotropins, such as the build up of receptors in the favored follicle(s). But negative feedback is most likely an equally important factor in a number of situations. For example, what effect does progesterone have on follicular development? One can confidently predict that increasing attention will be paid to the interactions of steroid hormones in which rate limiting enzymes in steroidogenesis may be directly affected. To cite one possible example: peak preovulatory levels of estrogen are followed by a drastic decline as a consequence of the LH surge. Is this a direct effect of LH or is the collapse in estrogen mediated by steroids produced as a result of LH activation?

7. Species differences in ovarian function.

The underlying mechanisms regulating ovarian function in mammals probably represent a common system, e.g., all species have a luteotropic hormone or complex of hormones which may vary at different stages of the cycle or pregnancy. In this context, luteotropin is used in a generic sense.

However, this obviously does not mean that species differences can be disregarded. In fact, they should be exploited to take advantage of the diversity of ways that different mammals have solved the same problem. Several years ago Drs. Geschwind and Michaels published an abstract entitled "The mouse is not a small rat". In a similar vein, A. S. Parkes said that if elephants were the size of rats and were used as extensively in the laboratory, then our concepts of reproductive physiology would be drastically different from present views.

As one concrete example of species differences, we have recently compared the *in vitro* production rate of progesterone, testosterone and estradiol by newly formed CL and the non-luteal ovary of the guinea pig, hamster, mouse, rabbit and rat (Terranova, Saidapur and Greenwald, unpublished). The rat, 10-12 hours after ovulation, is in a class by itself in terms of the amounts of progesterone produced *in vitro* and *in vivo*. On the other hand, the non-luteal ovary of the guinea pig is a very rich source of testosterone, presumably synthesized by the great numbers of atretic antral follicles.

I believe that more attention has to be paid to species variations and to the fact that the rat is not the touchstone of reproductive physiology.

8. Models of follicular control: Short versus long cycled species.

Like many other investigators, I have dabbled with models of follicular development, based on various manipulations of cyclic hamsters, mice and rats. Two features have struck me as characteristic of the short cycled species: firstly, the rapidity -within even hours - with which the follicular population responds to changes in exogenous gonadotropins or to endogenous gonadotropins following unilateral ovariectomy. Secondly, the periovulatory changes in gonadotropins - especially, but not exclusively, FSH -recruit the next crop of preantral follicles which will ovulate 4 to 5 days later.

I wonder how applicable these findings are to species with long cycles, such as the rhesus monkey, sheep and pig. For the latter species, there are several features which may pose problems in fitting them into the same mold as the hamster or rat. First, in the long-cycled species, several sets of antral follicles develop before the group ultimately destined to ovulate is differentiated. These changes in follicular activity do not correlate with corresponding cyclic variations in gonadotropins. Secondly, after the extirpation of the dominant follicle from the rhesus ovary, the next ovulation occurs some 10-12 days later. The maturation of the new follicle is not associated with an immediate increase in gonadotropins which would launch the follicle on the path to ovulation.

For the long cycled species, does this mean that there is continued entry and exit of antral follicles into different size categories unrelated to appreciable fluctuations in gonadotropins? Alternatively, it is possible that the lack of parallelism between follicular and pituitary events results from inadequate sampling frequencies for determination of gonadotropin levels.

In the case of the hamster and rat, only 15 to 20% of the amount of LH normally released at proestrus is needed for ovulation of the full complement of ova. It is equally feasible that a similar redundancy exists for FSH; only a fraction of the normal periovulatory quota may be needed to mobilize the next set of preantral follicles.

9. The luteal-follicular shift.

The most dramatic changes in sex steroids, gonadotropins and follicular development occur during the periovulatory period. For this reason, attention has focused on the prime mover initiating these events and in all species investigated this appears to be the positive feedback effects of estrogen. For non-primate species, the next logical question is: what ovarian compartment acts to restrain follicular maturation and estrogen secretion until a few days before ovulation? It appears likely that the functioning CL serve in this role; only on their demise can the final phases of follicular development proceed. It is well documented that the life span of the CL determines the length of the estrous cycle and thus the consequent importance of the luteal-follicular shift.

The temporal pattern of progesterone and estrogen during the cycle of most species fits this scheme - full blown estrogen secretion usually does not occur until after functional luteolysis. The rat has a 4 or 5 day cycle, the duration depending on whether the CL are functional for 2 or 3 days. On the other hand, the unusually stable 4 day cycle of the hamster (in which the CL secrete progesterone for only two days) can be routinely turned into a 5 day cycle by administering as little as 25 μg progesterone on day 3 of the cycle.

Conversely, premature destruction of the CL accelerates follicular development. For example, luteal extirpation leads to premature ovulation in guinea pig, sheep, cow, and pig. Luteal regression induced by prostaglandins or antiserum to LH similarly promotes follicular growth and estrogen secretion.

How does decreased luteal secretion affect subsequent events in the cycle? The best established effects are via the hypothalamic-pituitary axis, by increasing gonadotropin levels. This is well shown in the ewe by work from Dr. Karsch's laboratory: luteal regression is followed by higher levels of serum LH. As sheer heresy, I propose that under some circumstances progesterone affects the ovary directly, possibly by suppressing receptor populations in antral follicles or directly interfering with later steps in steroidogenesis. Whatever mechanisms are involved, a luteal-follicular shift appears to be an important regulatory mechanism in a number of species.

The preparation of this manuscript has turned out to be an enjoyable exercise. The half-life of some of the more outrageous speculations is, of course, unknown. Hopefully, these remarks at least touch on some of the issues to be considered in the following papers.

SESSION NUMBER: II

TITLE: FOLLICULAR FUNCTION - HORMONE RECEPTORS AND ADENYLATE CYCLASE

PRINCIPAL SPEAKER: JoAnne S. Richards

INVITED DISCUSSANT: Mary Hunzicker-Dunn

PRESENTED ABSTRACTS:

1.) M.L. Dufau, K. Hayashi, G. Sala, A. Baukal and K.J. Catt. Gonadal LH Receptors: Resolution from adenylate cyclase and transfer to heterologous cells.

2.) G. Gibori, J.S. Richards and P.L. Keyes. Corpora Lutea of Pregnant Rats.

CHAIRPERSON: Robert Ryan

# FOLLICULAR GROWTH IN THE RAT: A REEVALUATION OF THE ROLES OF FSH AND LH.

J.S. Richards, J. Th.J. Uilenbroek and J.A. Jonassen

Reproductive Endocrinology Program, Depart. Pathology,

University of Michigan, Ann Arbor, MI 48109

Although it has been well established that there are two structurally distinct pituitary gonadotropins, the question of why there are two remains an intriguing, unresolved question to those studying the hormonal control of follicular development. The purpose of this communication is to describe some mechanisms by which luteinizing hormone (LH) as well as follicle-stimulating hormone (FSH) may act to stimulate follicular growth (1) and to examine how the effects of LH appear dependent on the differentiation of theca cells and granulosa cells. To provide a conceptual framework for this review, first we will describe studies concerning the interactions of estradiol and FSH on granulosa cell differentiation in the hypophysectomized immature rat. This will be followed by a description of studies demonstrating a role for LH in follicular growth in the estradiol primed, hypophysectomized rats and in adult cycling and pregnant rats. These studies suggest that LH acts not only to increase follicular estradiol production via stimulation of theca cell androgen synthesis (2), but may also act directly on granulosa cells of developing preovulatory follicles to increase the content of LH receptor.

## 1. EFFECTS OF ESTRADIOL AND FSH ON GRANULOSA CELL FUNCTION USING THE IMMATURE HYPOPHYSECTOMIZED RAT.

We have shown that the induction of LH receptor in granulosa cells is dependent on steroid and protein hormone regulation of granulosa cell differentiation (Figure 1). Specifically estradiol administered to hypophysectomized immature rats stimulates granulosa cell proliferation (3-4), causes the growth of large preantral follicles which are exquisitely sensitive to gonadotropins (4-10), and enhances the ability of FSH to act _in vivo_ or _in vitro_ to

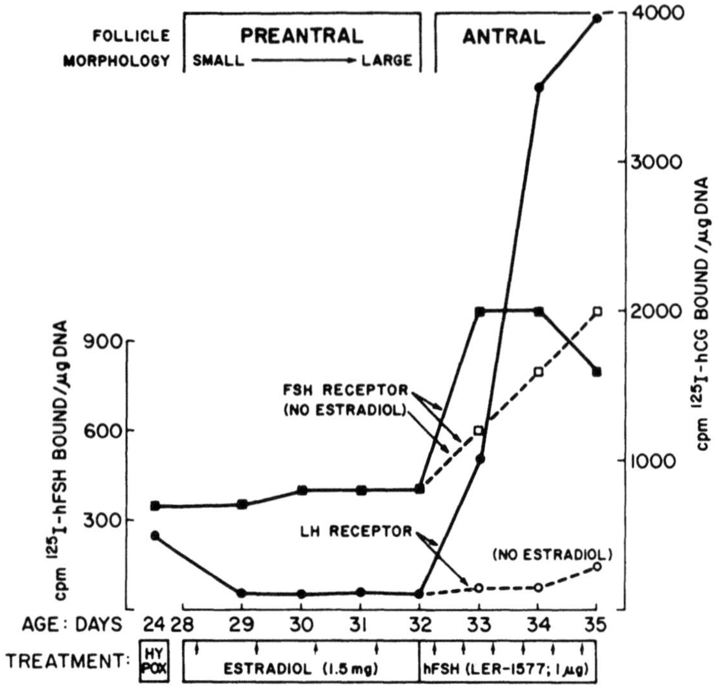


Figure 1. Summary of changes in follicular morphology and in granulosa cell content of receptors for FSH and LH in hypophysectomized, immature rats primed with estradiol and highly purified hFSH. Receptor content has been measured by determining the cpm of $^{125}$I-gonadotropin bound per µg DNA granulosa cell (11,12).

stimulate the appearance of granulosa cell LH receptor in developing antral follicles (11-13).

Because estradiol alone enhances the responsiveness of granulosa cells to FSH without altering the number of FSH binding sites per granulosa cell (11,14; Fig. 1), we have proposed that estradiol must act on some other component(s) of the FSH response system (see Fig. 2): to increase the ability of FSH to stimulate cAMP production (1); to increase cAMP binding proteins (cAMP-BP) which are functionally distinct from (2) or identical to (3) the regulatory subunits of protein kinases ($R_2C_2$); or to induce synthesis of a specific protein (4) which is a substrate for a cAMP dependent protein kinase (12,15-17).

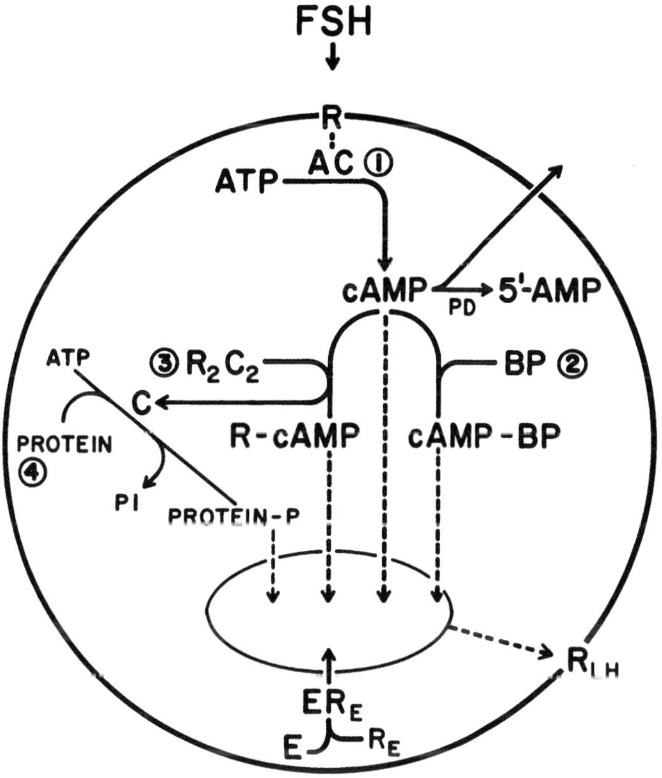


Figure 2. Possible sites of estradiol action on components of the FSH-cAMP effector system in granulosa cells.

To examine the effects of estradiol on the ability of FSH to stimulate granulosa cell cAMP accumulation, granulosa cells from hypophysectomized rats primed with saline, estradiol ($HE_3$) or estradiol and hFSH ($HE_3F_2$) were isolated and incubated in 0.3 ml of Medium 199 containing 0.5 mM isobutylmethyl xanthine, an inhibition of phosphodiesterase (Fig. 3; 17). As shown, parallel dose-response relationships were observed for hFSH stimulation of cAMP in granulosa cells from small preantral, large preantral and antral follicles. However, the total amount of cAMP accumulated was 2.5 and 4.0 fold greater in the cells from estradiol and estradiol-FSH treated rats respectively than from hypophysectomized rats (17).

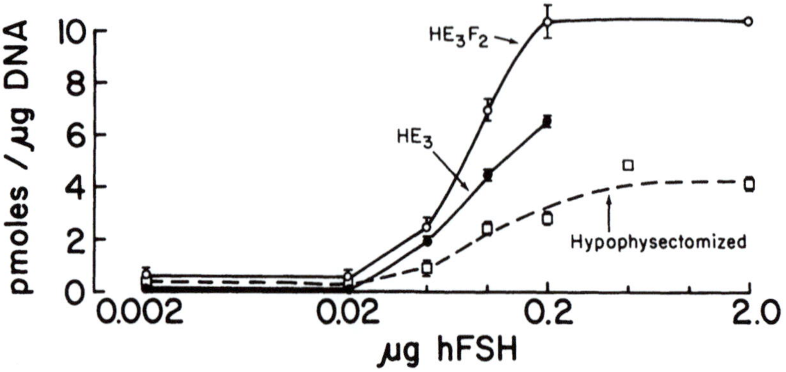


Figure 3. Accumulation of cAMP in granulosa cells _in vitro_. Isolated granulosa cells were incubated 10 min at 35C in the presence of increasing concentrations of highly purified hFSH (LER-1577). CAMP was measured by radioimmunoassay and all values corrected for the percent of viable cells in each assay (17).

A similar enhancement of FSH stimulation of cAMP in granulosa cells of hypophysectomized rats primed with estradiol was observed when hFSH was administered _in vivo_ (Figure 4; 17). As shown in the insert, little cAMP accumulation was observed in granulosa cells of hypophysectomized rats 30 min following an intravenous (tail vein) injection of 2 μg of highly purified hFSH (LER-1577) whereas demonstrable dose-dependent increases were observed in the estradiol primed rats following injections of 1, 2 and 10 μg of hFSH (17). The lower portion of Figure 4 illustrates the dose-dependent induction of LH receptor in granulosa cells of estradiol primed rats by 0.25-1.0 μg of hFSH given twice daily for 2 days. The progressive increase in LH receptor in granulosa cells of estradiol primed rats contrasts with the absence of a response in the hypophysectomized rats.

These _in vitro_ and _in vivo_ studies suggest that estradiol does act to enhance FSH stimulation of cAMP in granulosa cells (15,17). Further, since estradiol did not appear to alter the rapid disappearance rate of cAMP _in vivo_ (Fig. 4) and since cAMP accumulation _in vitro_ was measured in the presence of an inhibitor of phosphodiesterase (Fig. 3), estradiol does not appear to be acting to prevent or change the rate of cAMP degradation. The data further suggest that there is a dose-dependent relationship between FSH stimulation of cAMP and FSH stimulation of LH receptor in the estradiol primed rats but not in hypophysectomized rats.

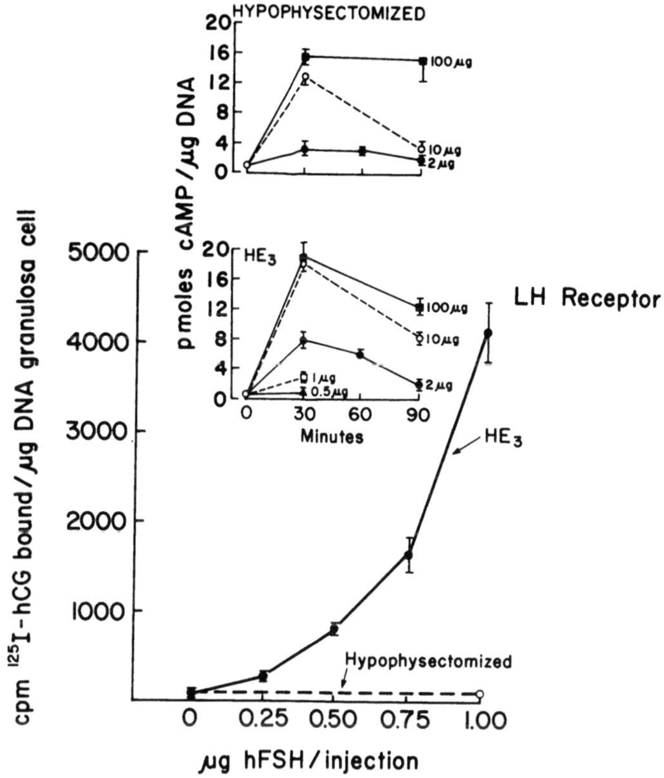


Figure 4. Dose-response stimulation of cAMP and induction of LH receptor in granulosa cells _in vivo_ (17).

Therefore, it would appear that if cAMP is involved in LH receptor induction, cAMP may not be the only requirement and that estradiol may be acting at other sites within the granulosa cell as well.

Therefore, the effects of estradiol on the number of cAMP binding sites in granulosa cell cytosol (30,000 x g supernatant) of hypophysectomized rats was examined (Fig. 5; 15). As shown, estradiol alone has no effect on cAMP binding sites. However, estradiol does enhance hFSH stimulation of cAMP binding sites. This is demonstrated further by incubating equal amounts of cytosol protein with $10^{-8}$M $^{3}$H-cAMP in the presence or absence of $10^{-6}$M unlabeled cAMP for 16 hr at 4° with subsequent resolution of the specifically bound cAMP binding peak by polyacrylamide gel electrophoresis. Only the cytosol of estradiol-FSH primed rats exhibited a specific peak of radioactivity. The extent of migration suggests that the binding protein is acidic, one characteristic of the regulatory subunits of cAMP dependent protein kinases.

Although estradiol alters FSH stimulation of cAMP it is clear that a specific site(s) of estradiol action in the granulosa cell remains to be defined.

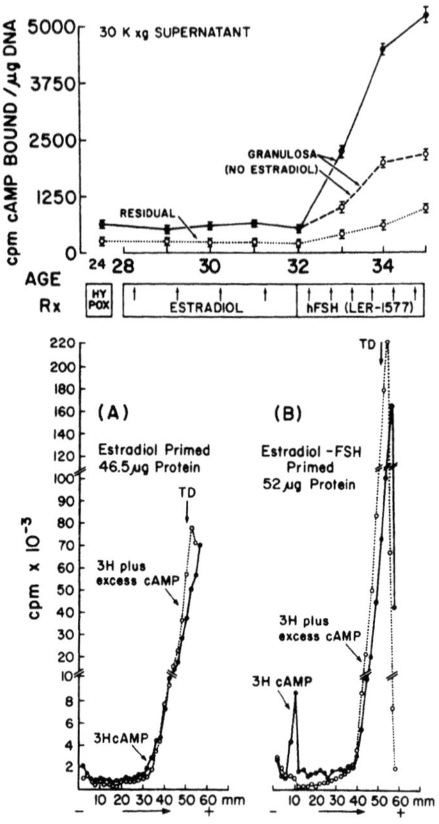


Figure 5. Induction of cAMP binding protein by estradiol and FSH (15).

2. EFFECTS OF LH ON GRANULOSA CELL FUNCTION.

It is well known that LH induces ovulation and luteinization of granulosa cells in large preovulatory follicles, that luteinization is associated with a marked decrease in the number of receptors for both FSH and LH (11,18), and that this decrease in gonadotropin receptor content occurs by mechanisms in addition to the occupation of binding sites (18). LH (hCG) has also been shown to cause a decrease in its own receptor in other target cells suggesting that LH might act generally on gonadal cells to reduce its own receptor (19,20). Therefore, it seemed reasonable to postulate that LH might act on preantral or small antral follicles to antagonize FSH mediated induction of LH receptor. To examine this hypothesis an ovulatory dose of 5 IU of hCG was administered to estradiol primed hypophysectomized rats 12 h prior to a single, subminimal dose of 2 μg of hFSH or saline (Fig. 6; 21). As shown, hCG alone has little or no effect on LH receptor in granulosa cells while hFSH alone caused a small, transient increase. By 48 h follicles in both groups were undergoing atresia. However, when hCG preceded the hFSH, large antral follicles developed and LH receptor in granulosa cells was markedly increased (21). Thus, hCG acted to increase rather than decrease its own receptor. We have postulated that the small increase in LH receptor occurring within 24 h after FSH (or 36 h after hCG) allowed hCG to be "seen" by the granulosa cells as a low dose of LH activity and to evoke "FSH-like" responses (21).

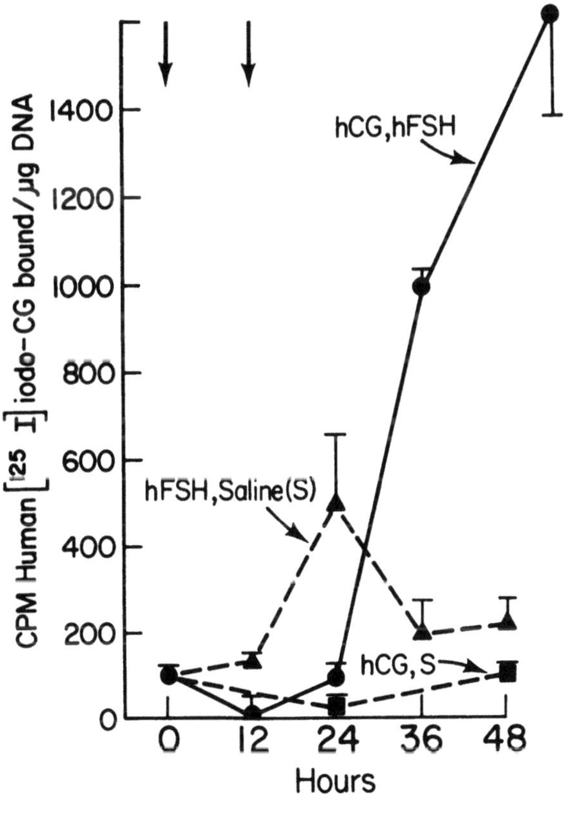


Figure 6. Induction of LH receptor by hCG and hFSH in estradiol primed hypophysectomized rats (21).

If hCG is capable of acting directly on granulosa cells, it would be anticipated, according to the current theories of gonadotropin action, that hCG would stimulate cAMP accumulation only in granulosa cells possessing LH receptor, that is in granulosa cells from ovaries of estradiol, FSH primed rats but not in cells isolated from ovaries of hypophysectomized or estradiol primed hypophysectomized rats. Dose-response studies illustrated in the lower panel of Figure 7 support this hypothesis. These studies were done on the same pools of cells used in the FSH dose-response study (Figure 2 and repeated in the upper panel of Figure 7) and indicate that stimulation of cells from estradiol-FSH primed rats with maximal doses of FSH or hCG leads to similar increases in cAMP concentrations.

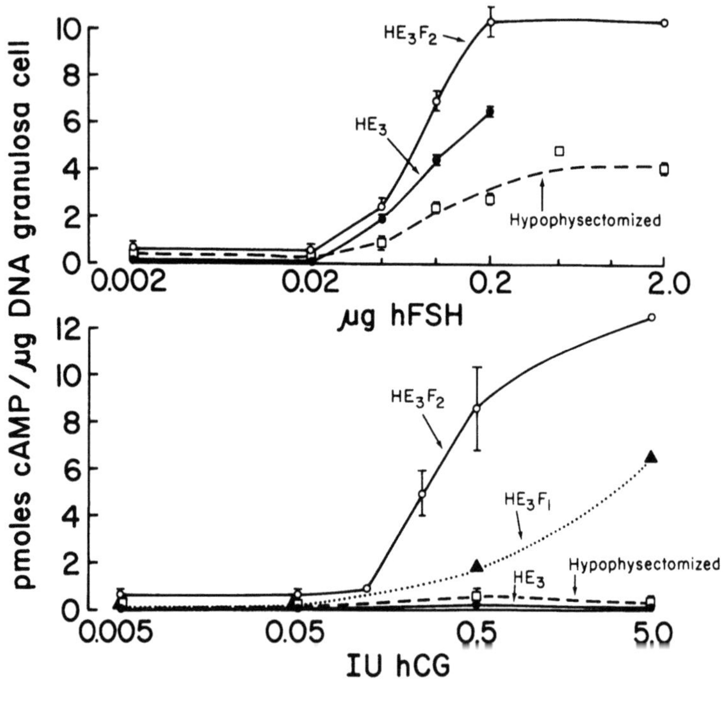


Figure 7. Accumulation of cAMP in granulosa cells _in vitro_ (see legend of Figure 3; 17).

These results suggested that if cells possessed receptor sites for both FSH and LH, the receptors might be coupled to the same adenylate cyclase system. Previous studies revealed that cAMP accumulation _in vivo_ following acute administration of hFSH or hCG was transient: peak concentrations at 30 min returned to basal values within 90-120 min (Fig. 4; 18). Therefore, we decided to use this system to determine if hFSH and hCG shared an adenylate cyclase system. It was reasoned that if the rapid decrease in cAMP accumulation occurring between 30-120 min following hormone administration reflected a loss of hormone stimulatable cAMP accumulation, then stimulation of granulosa cells by one hormone would prevent a response to a subsequent exposure of cells to the heterologous hormone (22). Conversely, in granulosa cells possessing only receptor for FSH, hCG given prior to hFSH would not be expected to inhibit the response to hFSH. Data presented in Figure 8 support this hypothesis since hCG given intravenously to estradiol FSH primed rats at t=0 prevented FSH from stimulating an increase in cAMP accumulation when given at 120 min. In the reverse experiment (lower panel, Figure 8; 22) FSH given to estradiol, FSH primed rats at t=0 prevented hCG stimulation of cAMP when hCG was given at 120 min. Neither hormone given to estradiol primed rats altered the response of ovarian cells to the heterologous hormone (Figure 8).

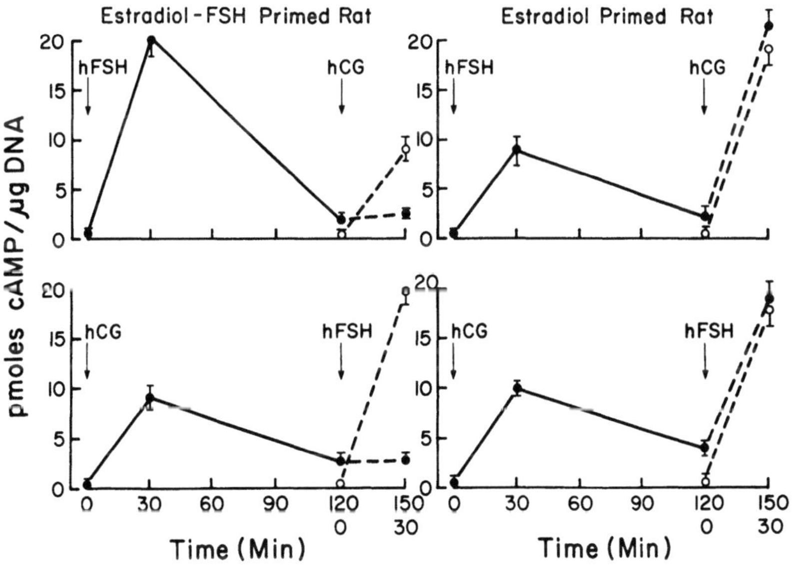


Figure 8. Effects of hCG (or hFSH) on subsequent hFSH (or hCG) stimulation of cAMP accumulation *in vivo* (22).

These results suggest that in granulosa cells possessing receptors for both gonadotropins, the receptors may be functionally coupled to the same adenylate cyclase system. Alternatively, the loss of hormone stimulatable cAMP accumulation may occur via intracellular mechanisms inactivating all adenylate cyclase systems equally (22).

## 3. FOLLICULAR GROWTH IN ADULT CYCLING AND PREGNANT RATS.

Kinetic studies of follicular growth in the mouse (23) and rat (24) suggest that once a follicle begins to grow it takes at least 19 days for it to reach the preovulatory stage (12). This period is roughly equivalent to 4 consecutive estrous cycles or the entire period of pregnancy. Thus, follicles ovulating in any given estrous cycle have been exposed to at least 3-4 prior surges of gonadotropins whereas those ovulating in response to the postpartum gonadotropic hormone surge actually developed under conditions of sustained tonic concentrations of gonadotropins. Thus the functional significance of gonadotropin surges in regulating follicular growth prior to ovulation is unclear. Therefore, we compared changes in follicular cell function during the estrous cycle and pregnancy by examining the content of receptors for FSH and LH in both granulosa cells and theca cells of large follicles and also by determining the ability of individual isolated follicles to synthesize estradiol (Fig. 9). The number of receptor sites for FSH in granulosa cells remained remarkably constant during the estrous cycle and pregnancy. However, a slight increase in FSH binding sites did occur in follicles isolated on days 20-22 of pregnancy. This increase was followed by a marked decline in FSH binding sites in postovulatory follicles isolated on the first day postpartum.

Unlike receptor content for FSH, receptor sites for LH were extremely low in granulosa cells isolated from the largest follicles on estrus, metestrus and diestrus 1 but increased dramatically on diestrus 2 and proestrus. A similar increase in LH receptor occurred between days 4-6 of pregnancy (day 5 of pregnancy being equivalent to proestrus of the cycle) suggesting that the development of Graafian, preovulatory follicles resumes during early pregnancy just as it would in a subsequent cycle. Although ovulation does not normally occur, 20-50 IU of hCG induced ovulation on days 4-6 but not on day 15 of pregnancy. Thus, follicles of early pregnancy possess the mechanisms (including LH receptor) necessary for ovulation and luteinization but apparently do not receive the appropriate stimulus (LH surge).

Following day 6 of pregnancy, LH receptor content decreased and remained low until day 21 of pregnancy. The progressive increase in LH receptor at the end of pregnancy (days 21-23) coincides with the growth of preovulatory follicles which ovulate in response to the postpartum LH surge. As with FSH receptor, LH receptor was low in postovulatory, luteinizing follicles still exhibiting an ovulatory pore (i.e. approximately 12-18 h after the endogenous LH surge). This decline in gonadotropin receptors has been observed in association with hCG-induced luteinization of

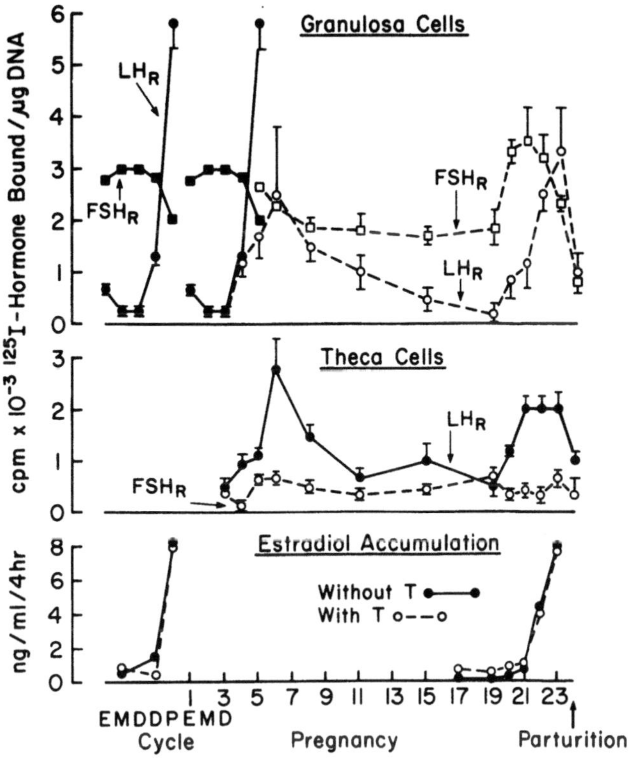


Figure 9. Receptors for FSH and LH in granulosa cells and theca cells of large follicles obtained during the estrous cycle and pregnancy: relationships with follicular production of estradiol in vitro.

preovulatory follicles from estradiol-FSH primed hypophysectomized rat (11,18).

LH receptor content in theca cells of developing Graafian follicles (middle panel; Fig. 9) increased on days 4-6 and days 19-22 of pregnancy suggesting that the theca cells as well as the granulosa cells may become more responsive to LH at these times. Although the functional significance of the increased content of LH receptor in the theca cell layer remains to be determined, it may serve to increase the amount of LH bound to theca cells and thus increase theca cell synthesis of androgen in the absence of increasing peripheral concentrations of LH. An increase in androgen might then be utilized to increase follicular production of estradiol by aromatization within the membrana granulosa (25).

To assess the ability of whole follicles to produce estradiol, follicles were isolated and incubated individually without or with 50 ng/ml testosterone. As shown in the lower panel of Figure 9, estradiol synthesis (accumulation per 4 h incubation) was greatest during the estrous cycle on proestrus and at the end of pregnancy on days 22-23. It can be seen that the increased ability of follicles to produce estradiol in the presence or absence of exogenous substrate was associated temporally with increased LH receptor in granulosa and theca cells. These relationships support the data obtained in hypophysectomized rats (Figs. 1 and 6) which suggest that estradiol is required for FSH/LH induction of LH receptor. Importantly, the increases in granulosa cell LH receptor and follicular estradiol synthesis occur without major changes in either serum concentrations of gonadotropins (26,27) or FSH receptor content in granulosa cells, but do occur with an increase in FSH responsive adenylate cyclase activity (28). Thus, as in the hypophysectomized rat (Figs. 3 and 4), estradiol may act to enhance the FSH response system via affecting the hormone responsive adenylate cyclase system.

When estradiol synthesis was depicted on a logarithmic scale (Fig. 10), the data revealed further that the content of estradiol

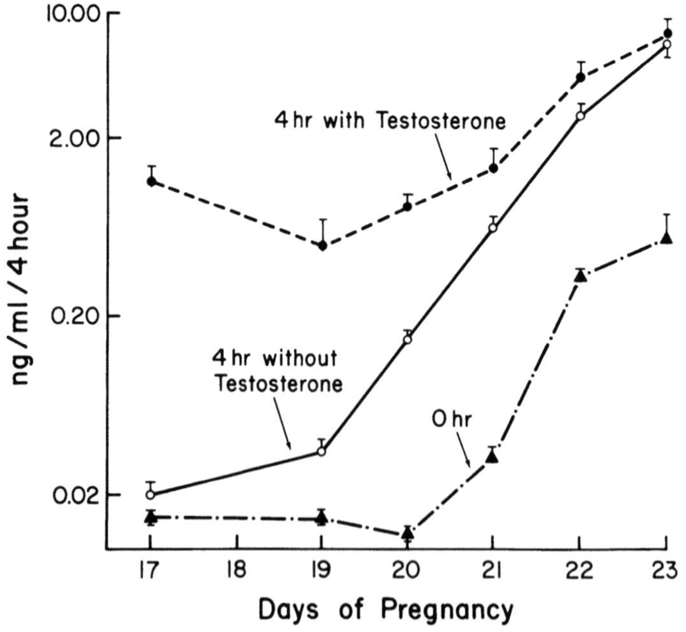


Figure 10. Estradiol accumulation in individual follicles obtained from rat ovaries at the end of pregnancy.

in follicles at the time of isolation (t=0) increased progressively between days 21-23 of pregnancy. When follicles were incubated for 4 h in the absence of exogenous testosterone substrate, estradiol accumulation increased steadily from days 20-23. When testosterone was added to the incubations estradiol synthesis on days 17-20 was markedly increased. However, by days 21-23 the addition of exogenous substrate became progressively less effective in promoting additional estradiol synthesis suggesting that the follicles were producing sufficient endogenous androgen to maintain maximal rates of estradiol production. Thus, these data suggest that follicles on days 17-20 have some aromatase activity but produce little or no androgen substrate *in vitro*. Beginning on day 21 and continuing through day 23, production of androgen substrate as well as aromatase activity increases.

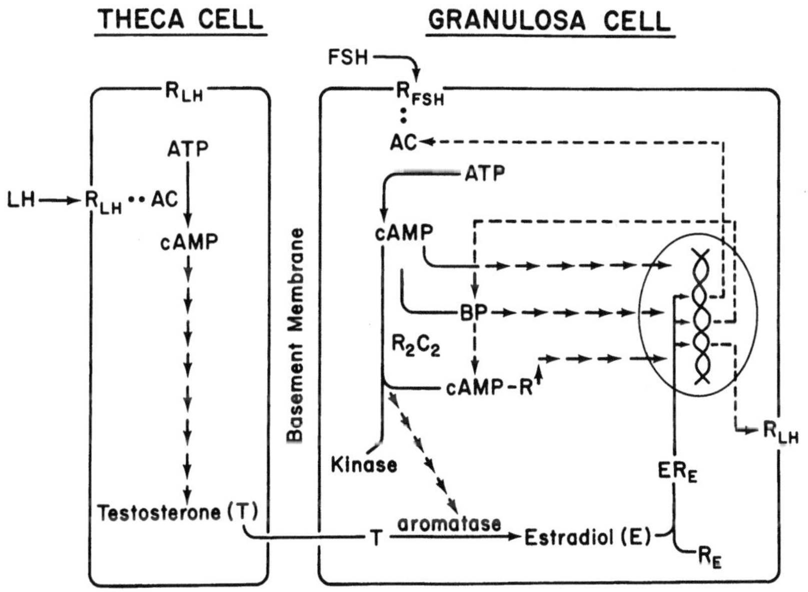


Figure 11. Schematic integration of effects of FSH, estradiol and LH during final states of follicular growth.

The summary depicted in Figure 11 suggests that granulosa cells of preantral and small antral follicles (<400 μm) would appear to be predominantly under the regulatory control of FSH. FSH binding to its receptor ($R_{FSH}$) stimulates cAMP accumulation, activation of protein kinase ($R_2C_2$) and maintenance of some basal aromatase enzyme activity. The growth of the Graafian, preovulatory follicle appears to be dependent on LH and an enhanced capacity of theca cells to respond to LH (possibly involving increased

LH receptor $R_{LH}$) and to produce androgens. Increasing androgen production would lead to increased estradiol synthesis by aromatization within the membrana granulosa. Estradiol acting via its receptor ($R_E$) would then act on the granulosa cell genome to stimulate events enhancing FSH stimulation of cAMP via adenylate cyclase (AC), FSH stimulation of cAMP binding sites (likely the regulatory subunits of a particular protein kinase) and FSH stimulation of aromatase activity. Thus, FSH and estradiol interact within the granulosa cell to comprise an intracellular positive feedback system which ultimately results in the appearance of LH receptor. Once some LH receptor appears on granulosa cells of the developing Graafian follicle, LH can then act directly in a manner similar to FSH and appears to become the hormone regulating not only theca cell function but also granulosa cell differentiation. Thus, during the final stages of follicular growth LH may play a predominant role.

## SUMMARY

The growth of preovulatory ovarian follicles involves hormonally induced proliferation and differentiation of theca cells and granulosa cells resulting ultimately in an increased ability of follicles to produce estradiol and to respond to the pituitary gonadotropins. The increased ability of follicles to produce estradiol appears to depend on an increased ability of theca cells to produce androgen as well as an increased ability of granulosa cells to aromatize androgens to estradiol. Estradiol in turn, appears to be required for FSH or FSH and LH to stimulate the appearance of functional receptors for LH in granulosa cells. Thus, the intrafollicular hormone estradiol enhances the response of granulosa cells to the gonadotropin. Therefore production of estradiol appears to determine which follicles will gain the mechanisms, including LH receptor, necessary for ovulation and luteinization. Since LH can act to increase its own receptor in the presence of estradiol and low amounts of FSH, it is possible that LH plays a predominant role in the final stages of preovulatory follicular growth both to promote estradiol production stimulation of theca cell androgen production as well as by facilitating an increase in its receptor by acting directly on granulosa cells.

## Acknowledgements

Supported in part by NIH-HD-09110, a Program Project grant NIH-HD-08333 and a Research Career Development Award (JSR).

## REFERENCES

(1) A.J. Lostroh and R.E. Johnson. Endocrinology 79(1966)991.

(2) J.E. Fortune and D.T. Armstrong. Endocrinology 100(1977)1341.

(3) R.L. Goldenberg, J.L. Vaitukaitis and G.T. Ross. Endocrinology 90(1972)1492.

(4) M.C. Rao, A.R. Midgley, Jr. and J.S. Richards. Cell 14(1978) 71.

(5) R.I. Pencharz. Science 91(1940)554.

(6) P.C. Williams. Nature 145(1940)388.

(7) M.E. Simpson, H.M. Evans, H.L. Fraenkel-Conrat and C.H. Li. Endocrinology 28(1941)37.

(8) J.S. Richards. Endocrinology 97(1975)1174.

(9) J.E. Meyer and J.T. Bradbury. Endocrinology 66(1960)121.

(10) R.W. Payne and R.H. Runser. Endocrinology 62(1958)313.

(11) J.S. Richards, J.J. Ireland, M.C. Rao, G.A. Bernath, A.R. Midgley, Jr. and L.E. Reichert, Jr. Endocrinology 99(1976)1562.

(12) J.S. Richards and A.R. Midgley, Jr. Biol. Reprod. 14(1976)82.

(13) A. Nimrod, A. Tsafriri and H.R. Lindner. Nature 267(1977)632.

(14) J.-P. Louvet and J.L. Vaitukaitis. Endocrinology 99(1976)758.

(15) J.S. Richards, M.C. Rao and J.J. Ireland, in: Control of Ovulation, ed. D.B. Crighton (Butterworths, Sevenoaks, G.B., 1978) p. 197.

(16) J.S. Richards, in: Evolution of the Vertebrate Ovary, ed. R.E. Jones (Plenum Press, New York, 1978) in press.

(17) J.S. Richards, J.A. Jonassen, A.I. Rolfes, K. Kersey and L.E. Reichert, Jr. submitted.

(18) M.C. Rao, J.S. Richards and A.R. Midgley, Jr. Endocrinology 101(1977)512.

(19) A.J.W. Hsueh, M.L. Dufau and K.J. Catt. Proc. Nat. Acad. Sci. 74(1977)592.

(20) M. Conti, J.P. Harwood, A.J. Hsueh, M.L. Dufau and K.J. Catt. J. Biol. Chem. 251(1976)7729.

(21) J.J. Ireland and J.S. Richards. Endocrinology 102(1978)1458.

(22) J.A. Jonassen and J.S. Richards. 11th Annual Meeting of the Society for the Study of Reproduction, Carbondale, 1978, Abstract #129.

(23) T. Pedersen. Acta. Endocrinology (Kbh) 64(1970)304.

(24) A.N. Hirshfield and A.R. Midgley, Jr. Biol. Reprod. 18 (in press).

(25) Y.S. Moon, J.H. Dorrington and D.T. Armstrong. Endocrinology 97(1973)244.

(26) K.W. Cheng. J. Reprod. Fert. 48(1976)129.

(27) D.M. Linkie and G.D. Niswender. Endocrinology 90(1972)632.

(28) M. Hunzicker-Dunn and L. Birnbaumer. Endocrinology 99(1976) 198.

# LH-INDUCED DESENSITIZATION OF THE ADENYLYL CYCLASE SYSTEM IN OVARIAN FOLLICLES

Mary Hunzicker-Dunn*, Richard Jungmann*, Dennis Derda*, and Lutz Birnbaumer°

*Department of Beochemistry, Northwestern University Medical School, Chicago, Illinois 60611
°Department of Cell Biology, Baylor College of Medicine, Houston, Texas 77035

## INTRODUCTION

The preovulatory surge of LH promotes stimulation, closely followed by desensitization of the ovarian follicular adenylyl cyclase (AC) system (1-6). Desensitization of the AC system in estrous rabbit follicles is characterized by an LH-specific, dose-dependent decline in LH-stimulated AC activity and results, by definition, in a retarded rate of cAMP synthesis (2). Although a coincident decline in LH- and FSH-stimulated AC activities occurs, there is no concurrent decline in prostaglandin $(PG)E_1$- or NaF- stimulated AC activities (2). LH refractoriness is demonstrable within min following mating of the iv injection of an ovulatory dose of hCG and is complete prior to ovulation. The follicular AC system does not acquire LH responsiveness again until 48-72 h after the induction of ovulation, when the new CL has formed. Desensitization of the AC system, then, is clearly one of the earliest events prompted by the surge of LH in follicles. However, little is known concerning the mechanism underlying the desensitizing process.

The studies reported herein extend our previous findings on LH-induced desensitization of the AC system in estrous follicles and further characterize some of the molecular requirements of desensitization and resensitization of the follicular AC system.

## STUDIES USING RABBIT OVARIAN FOLLICLES

In our previous studies, the AC system of follicles obtained from untreated estrous rabbits was highly responsive both to LH and FSH, being only slightly more responsive to LH (2). HCG(iv)- or LH (endogenous via coitus)-induced desensitization in these follicles was characterized by a seemingly coincident decline in the responsiveness of the AC system both to LH and FSH (2). This result suggested that the receptors for both LH and FSH were coupled to the same catalytic unit of the AC system. However, in view of the significant contamination of the FSH preparation with LH, we have reinvestigated the effect of hCG on the responsiveness of the follicular AC system to highly purified preparations of LH and FSH, kindly made available to us by Dr. Harold Papkoff (H.P.), Hormone Research Laboratory, University of California Medical Center, San Francisco.

We initially determined the responsiveness of the AC system to stimulation by H.P.-FSH vs. NIH-FSH in homogenates of popped follicles obtained from untreated, estrous rabbits (Fig. 1). The responsiveness of the AC to the two fractions of FSH was also compared with that to the two fractions of LH. Both LH preparations maximally and equally stimulated AC at 0.75 μg/ml. At least a 30-fold larger concentration of the NIH-FSH was required to stimulate AC to the same degree as the LH fractions. Even at the highest concentration used, the H.P.-FSH did not achieve the maximal stimulation of AC exhibited by either fraction of LH or by NIH-FSH. The 3-fold difference in sensitivity exhibited by the two FSH fractions probably reflects the degree of LH contamination in the NIH-FSH preparation. These results show that the AC in popped estrous rabbit follicles is at least 100-fold less sensitive to stimulation by highly purified FSH compared to LH.

Utilizing H.P.-LH (5 μg/ml) and FSH (10 μg/ml) to measure the gonadotrophin responsiveness of the AC system, the specificity of hCG-induced desensitization of AC was determined in follicular homogenates at various times after the injection of an ovulatory dose of hCG into estrous rabbits (Fig. 2). One h after the injection of hCG, only the LH-stimulated AC activity had declined (by 50%); FSH-stimulated AC activity was not changed. However, by 3 h FSH-stimulated AC activity had dropped by 40%; and by 5 h, it was not significantly different from basal activity. NaF-stimulated AC activities (not shown) remained unchanged during this time period (19.1 ± 1.1 pmoles cAMP/min/mg protein, n=16), while epinephrine-stimulated AC activities began to increase as ovulation approached (Fig. 2; 10.6 ± 1.1 pmoles cAMP/min/mg protein, n=2, at 9 h post hCG).

Thus, while the initial period of the hCG-induced desensitizing process is specific for LH, by 3 h a degree of nonspecificity is demonstrable in the desensitizing process. In rabbits, the coital-induced surge of LH, which peaks at 90 min, is followed at 150 min

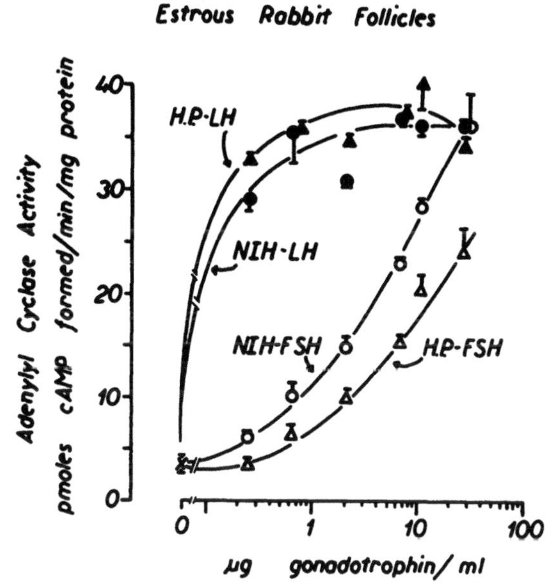


Fig. 1. Effect of LH and FSH obtained from NIH and Harold Papkoff (H.P.) on AC activities in estrous rabbit follicles. Graafian follicles were dissected from ovaries of estrous rabbits, homogenized and filtered, and 20 µl aliquots were assayed for basal and hormonally-responsive AC activity (2). AC incubations were for 10 min at 37 C in 50 µl of a medium containing 3 mM [$\alpha$-$^{32}$P]ATP, 5 mM $MgCl_2$, 1 mM EDTA, 1 mM [$^3$H]cAMP, 25 mM TRIS-HCl, 20 mM creatine phosphate, and 0.2 mg/ml creatine kinase, pH 7.0, homogenate protein, and BSA, LH (NIH-LH-B9, 0.70 x NIH-LH-S1, <0.05 x NIH-FSH-S1, ref. [7]; H.P.-oLH, G3-222B, 2.75 x NIH-LH-S14, 0.014 x NIH-FSH-S1) or FSH (NIH-FSH-P1, 0.76 x NIH-FSH-S1, 0.0075 x NIH-LH-S1, ref. [8]; H.P.-oFSH, G4-150C, 50 x NIH-FSH-S1, 0.001-0.01 x NIH-LH-S1) in concentrations indicated on the figure. The reaction was terminated and [$^{32}$P]cAMP was isolated as previously described (2). Mean ± SEM is shown where 2-6 such assays were performed.

by only a small increase in serum FSH levels (approximately 2-fold greater than tonic levels; ref. 9,10). The effect of this small physiological increase in serum FSH on the gonadotrophin responsiveness of the follicular AC system in rabbits is not known and is currently being investigated.

Incubation in vitro of estrous follicles with BSA or LH (2 h) also results in an LH-induced, concentration-dependent desensitization of the LH-sensitive AC system (2). Utilizing this in vitro incubation system, we investigated some of the requirements of desensitization. (a) Preincubation of follicles 1 h with BSA

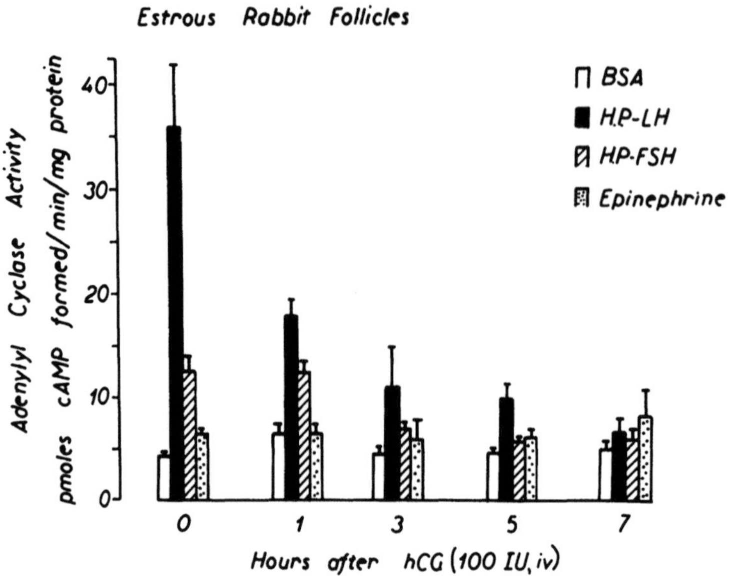


Fig. 2. Effect of the administration of an ovulatory dose of hCG (100 IU, iv/rabbit) on the responsiveness of the AC system in estrous follicles to H.P.-LH, H.P.-FSH, and epinephrine. Rabbits were sacrificed at the times indicated on the figure. AC activity was determined in follicular homogenates in the presence of BSA, 5 μg/ml H.P.-LH, 10 μg/ml H.P.-FSH, or 20 μg/ml epinephrine. For rest of conditions, see legend to Fig. 1.

followed by a second h with 10 μg/ml LH promoted a 62% decline in LH-stimulated AC activity and a 3-fold increase in basal activity (Fig. 3, compare panels 1 and 5). Preincubation of follicles with 11 μM puromycin, 30 μM cycloheximide, or 8 μM actinomycin D followed by a second h with LH did not in any case retard the LH-induced decline in LH-sensitive AC activity. The enhanced reduction in LH- and NaF-stimulated AC activity following preincubation with puromycin and cycloheximide is probably attributable to a partial inhibition or poisoning of the AC system which is demonstrable in equivalent BSA incubations. Thus, the RNA and protein synthetic events prevented by these drugs are not initially required for LH to induce desensitization of the AC system in estrous follicles. Although Lamprecht et al. (6) found that actinomycin D (8 μM) prevented LH-induced refractoriness in cAMP accumulation in rat follicles, the mechanism of desensitization may be different in the two species since desensitization of the AC to LH-stimulation in rat follicles in vivo (2, 3) and in vitro (6) requires a much longer time than in rabbit follicles. (b) To determine if desensitization was mediated by cAMP, estrous rabbit follicles were

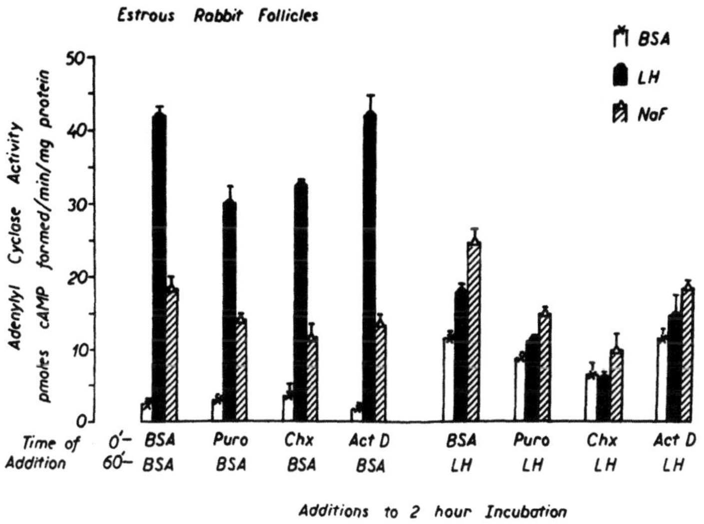


Fig. 3. Effect of in vitro exposure of rabbit follicles to 11 μM puromycin, 30 μM cycloheximide, or 8 μM actinomycin D on LH-induced desensitization of the AC system. Follicles from estrous rabbits were incubated (4-5 follicles/flask) at 37 C for 1 h with continuous gassing (95% $O_2$/5% $CO_2$) in 2.5 ml Krebs-Ringer bicarbonate buffer (KRB) and 1 mg/ml glucose containing BSA, puromycin (Puro), cycloheximide (Chx), or actinomycin D (Act D). BSA or 10 μg/ml LH was then added to each flask and incubations were continued for 1 h. Flasks were chilled, follicles were homogenized, and AC activity was determined in the presence of BSA, 10 μg/ml LH(NIH), or 10 mM NaF. At these concentrations, and under identical incubations, puromycin and cycloheximide inhibited [$^3$H]amino acid (added at time 0 to 2 h incubation) incorporation into trichloroacetic acid precipitates by 90% and actinomycin D inhibited [$^3$H]uridine incorporation by 93%. For rest of conditions, see legend to Fig. 1. Mean ± SD of duplicate determinations is shown.

incubated for 2 h in the presence of BSA, LH or various concentrations of dibutryl(d)-cAMP (Fig. 4). Incubations with LH rendered the AC system unresponsive to LH, due to a 70% decline in LH-stimulated AC activity and an 8-fold increase in basal AC activity. Incubations with increasing concentrations of d-cAMP (in the absence of LH) promoted an apparent concentration-dependent decline in LH-stimulated AC and no change in NaF-stimulated AC. Although the highest concentration of d-cAMP promoted a 52% decline in LH-stimulated AC activity, the cyclase still retained a 5-fold responsiveness to stimulation by LH. These results indicate that d-cAMP, in concentrations which have previously been shown to promote

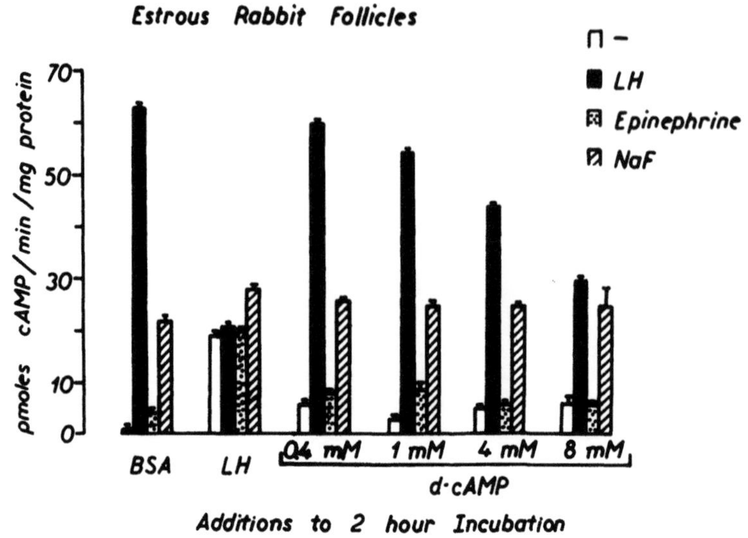


Fig. 4. Effect of in vitro exposure of rabbit follicles to d-cAMP (Sigma Chemical Co.) on AC activity. Follicles from estrous rabbits were incubated for 2 h at 37 C in KRB-glucose containing BSA, 10 μg/ml LH, or d-cAMP in the concentrations indicated on the figure. For rest of details, see legends to Figs. 1, 2, and 3. Mean ± SD of duplicate determinations is shown.

luteinization of estrous follicles (11, 12), causes a partial reduction in LH-stimulated AC activity. However, we do not know if these experiments are physiologically significant since it is not possible to ascertain the quantity of d-cAMP which entered or if degredation products of d-cAMP entered the cells (13).

## STUDIES USING PORCINE OVARIAN FOLLICLES

LH-induced desensitization of a highly LH-responsive AC system can also be obtained in an in vitro incubation system utilizing a cell-free membrane preparation obtained from porcine follicles (14). Partial desensitization of the LH-responsive AC occurs within 20 min upon incubation of the membranes in the presence of LH and the complete AC assay medium (14). Analysis of the dependence of desensitization upon the individual components of the medium demonstrated critical, concentration-dependent requirements for LH, Mg, and ATP (14). Substitution of ATP with AMP-P(NH)P (an analog of ATP which serves as a substrate for the AC but cannot donate its terminal phosphate in a phosphorylation reaction [15]) did not permit LH-induced desensitization to obtain. Based upon the critical Mg and ATP dependence of the in vitro desensitization reaction, we investigated the possibility that LH-induced desensitization of the AC system was dependent upon a phosphorylation reaction (16-21).

Porcine follicular membranes which exhibit AC activity were found to contain both cAMP-dependent and cAMP-independent protein kinase (PK). DEAE-cellulose chromatography of the membranes revealed the predominance of a cAMP-dependent isoenzyme eluting at the location of the type II variety (22-25) as well as the presence of a cAMP-independent PK, verified by PK inhibitor studies (26, 27), which eluted at a lower salt concentration (Fig. 5). The elution profile of PK obtained from pig follicle cytosol was qualitatively similar to that of PK from membranes, however the low salt peak of PK activity derived from the cytosol was not exclusively a cAMP-independent enzyme as was that from membranes.

Porcine follicular membranes also contain endogenous phosphatase activity. Phosphatases associated both with the membranes and cytosol exhibited pH optima of 6.8 (not shown). To determine the type of phosphatase present, we investigated the effects of various divalent metal cations on as well as potential substrates for the endogenous membrane- and cytosol-associated phosphatase activity, since acid, alkaline and protein phosphatases each characteristically prefer different cations and substrates (29-32). $Mn^{2+}$ strongly activated enzyme activities from both sources, being more effective on the cytosol preparation by a factor of 3, while $Zn^{2+}$ markedly inhibited both (Table I). The other divalent metal ions were without effect. Neither α-glycerol phosphate nor phosphoserine competed with protamine as a substrate for the phosphatases (Table II). However, phosphatase activities from both preparations were severely inhibited by ATP and, to a lesser extent, by ADP. These characteristics, along with the neutral pH preference, suggest that the membrane- and cytosol-phosphatases are phosphoprotein as opposed to nonspecific acid or alkaline phosphatases.

Since $Mn^{2+}$ (Table I; ref. 31) and dithiothreitol (DTT), routinely included in phosphatases assays (32, 33), strongly activate follicular membrane phosphatases, we investigated the effect of these substances on the LH-sensitive AC system. $Mn^{2+}$ promoted a concentration-dependent resensitization of the LH-sensitive AC system (Fig. 6). Neither basal (measured in the complete absence of LH) nor LH-stimulated AC activity (measured with LH present only during the 6 min AC assay) were significantly altered, but LH-stimulated desensitization (measured with LH present throughout 20 min incubation and 6 min AC assay) was reversed. DTT, in concentrations less than 1 mM, also promoted a concentration-dependent reversal of desensitization, affecting only LH-stimulated desensitization and not basal or LH-stimulated AC activity (Fig. 7). Although the effects of $Mn^{2+}$ and DTT that we observed on resensitization of the LH-sensitive AC system in porcine follicular membranes may well be due to

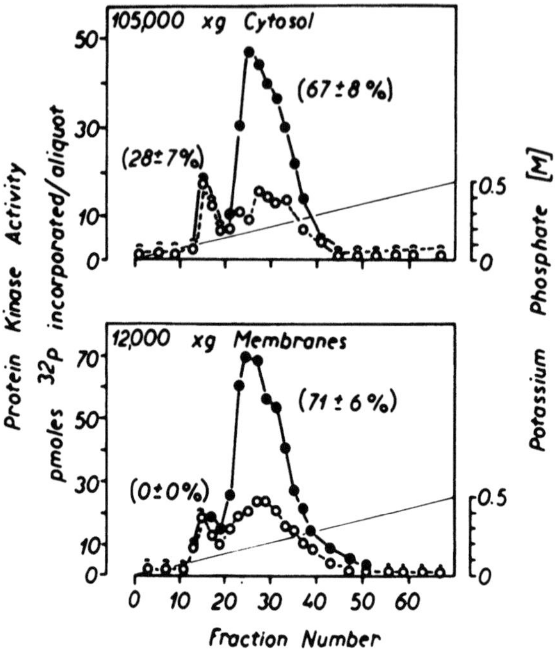


Fig. 5. DEAE-cellulose elution profiles of porcine follicular PK. Follicles exceeding 6mm in diameter were dissected from ovaries of slaughterhouse pigs and homogenized, filtered and centrifuged at 12,000 x g for 15 min (14). The pellet was washed 2 times by repelleting. The original supernatant was centrifuged at 105,000 x g for 60 min. The 105,000 x g supernatant and the washed 12,000 x g pellet, resuspended by homogenization in 10 mM TRIS-HCl, 1 mM EDTA, pH 7.5, were loaded onto separate 2 x 5 cm DEAE-cellulose columns. PK activity was determined (22, 26) in the presence and absence (O) of $4.5 \times 10^{-7}$ M cAMP in 50 µl aliquots of 1 ml fractions eluted with a 10 to 500 mM potassium phosphate buffer. Peak fractions were subsequently assayed in the presence of cAMP and saturating concentrations of the heat-stable PK inhibitor, isolated from rabbit skeletal muscle (26, 28). Percent inhibition of cAMP-sensitive PK activity determined in the presence of inhibitor are shown in parenthesis (± SD of duplicate determinations, n=3) adjacent to respective peaks of PK activity.

activation of endogenous phosphatases, we cannot exclude the possibility that $Mn^{2+}$ and DTT are directly modulating the AC system.

In view of the indirect evidence for a role of phosphatases in resensitization of the AC system, provided by $Mn^{2+}$- and DTT-induced reversal of desensitization, we investigated the effects of various phosphatases on LH-induced desensitization in porcine fol-

TABLE I

Effect of Divalent Metal Ions on Phosphatase Activity in Pig Follicular Cytosol and Membrane Preparations*

| Divalent Metal Ion | Relative % Activity | |
|---|---|---|
| | Pig Follicular Cytosol | Pig Follicular Membranes |
| None | 100 | 100 |
| $Mg^{2+}$ | 110 ± 13 | 101 ± 2 |
| $Ca^{2+}$ | 96 ± 5 | 106 ± 0 |
| $Mn^{2+}$ | 929 ± 75 | 284 ± 14 |
| $Co^{2+}$ | 82 ± 4 | N.T.† |
| $Zn^{2+}$ | 7 ± 0 | 7 ± 1 |

* Phosphatase activity was determined by measuring the release of $^{32}$P-orthophosphate from [$^{32}$P]protamine chloride according to the technique of Meisler and Langan (33). Phosphatase activity was determined in a washed 12,000 x g pellet obtained from pig follicles (see legend to Fig. 5 for details of tissue preparation) and in a phosphatase preparation partially purified, according to the technique of Nakai and Thomas (34), from a 105,000 x g supernatant derived from pig follicles (see legend to Fig. 5 for tissue preparation and DEAE-cellulose chromatography procedures). Final concentrations of the divalent metal ions were 2.5 mM. Results are expressed as mean ± SD of duplicate determinations.

† Not tested.

TABLE II

Effect of Various Substrates on Phosphatase Activity in Pig Follicular Cytosol and Membrane Preparations*

| Additions | Relative % Activity | |
|---|---|---|
| | Pig Follicular Cytosol | Pig Follicular Membranes |
| None | 100 | 100 |
| β-Glycerol phosphate | 91 ± 1 | 91 ± 5 |
| Phosphoserine | 117 ± 8 | 89 ± 2 |
| ATP | 42 ± 0 | 54 ± 1 |
| ADP | 56 ± 3 | 71 ± 7 |
| AMP | 126 ± 11 | 99 ± 11 |

* Results are expressed as mean ± SD of duplicate determinations. Final concentrations of all additions are 2.5 mM. All procedures are as described in the legend to Table I.

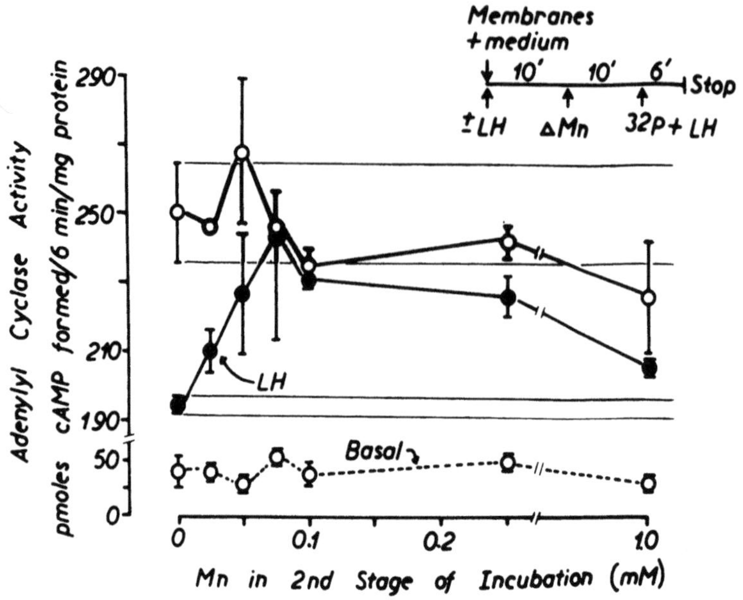


Fig. 6. Effect of $Mn^{2+}$ on the AC activity of porcine follicular membranes. Graafian follicle membranes (18 μg protein; see legend to Fig. 5 for details and ref. 14) were incubated in 3 stages at 30 C as diagrammed on the figure. (a) In the first stage (10 min), 20 μl of membranes were incubated in a volume of 30 μl in a medium consisting of 41.75 mM bis-TRIS-propane-HCl (BTP) buffer, 10.02 mM $MgCl_2$, 1.67 mM EDTA, 1.67 mM EGTA, 1.67 mM [$^3$H]cAMP, 2.92 mM ATP, 0.334 mg/ml creatine kinase, 33.4 mM creatine phosphate, in the presence of BSA or 20.0 μg/ml LH (all of which are at concentrations 1.67 times those desired in the third stage of the incubation). (b) In the second stage of the incubation (10 min), $Mn^{2+}$ (in concentrations indicated on the figure) was added to the first incubation mixture in a volume of 10 μl. (c) In the third stage of incubation (AC assay), 10 μl of a medium consisting of [α-$^{32}$P]ATP and BSA or 12.5 μg/ml LH were added to the incubation mixture. Incubations were terminated after 6 min (26 min total incubation time) and [$^{32}$P] cAMP was isolated as previously described (2). Incubations were thus performed in the complete absence of LH (--O--), with LH present in the 6 min AC assay only (—O—), or with LH present throughout the 26 min incubation time (—●—). Mean ± SD of duplicate determinations is shown.

licular membranes. Incubation of the membranes with a phosphatase partially purified from pig follicle cytosol (Tables I and II) reversed desensitization of the LH-sensitive AC system in a concentration dependent manner (Fig. 8). Although LH-stimulated activity

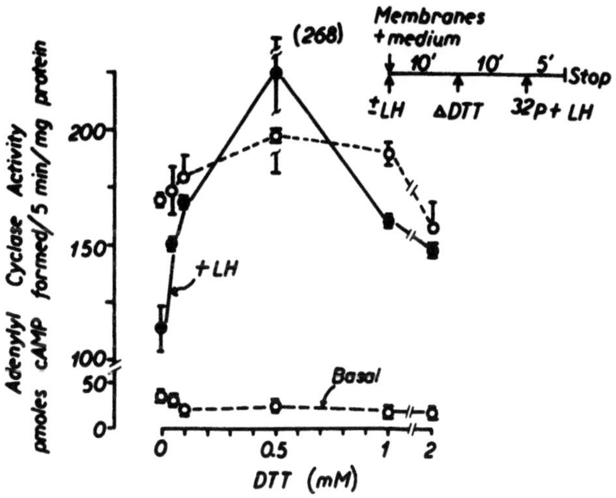


Fig. 7. Effect of DTT on AC activity of porcine follicular membranes (18 μg protein). Experimental protocol is as described in the legend to Fig. 6. DTT, in concentrations indicated on the figure, was added to the second stage of the incubation. Mean ± SD of duplicate determinations is shown.

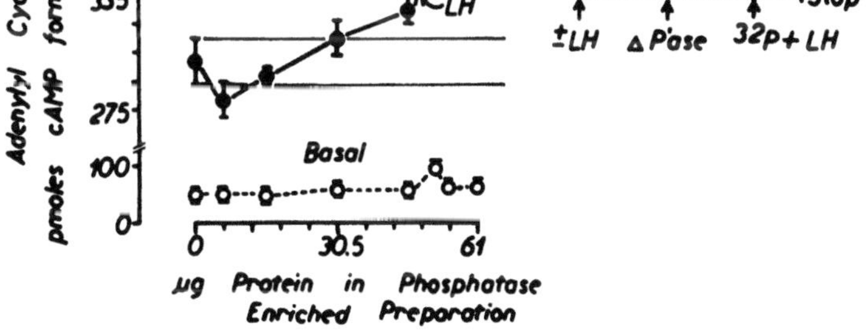


Fig. 8. Effect of phosphatase-enriched preparation on AC activity of porcine follicular membranes (28.5 μg protein). Experimental protocol is as described in the legend to Fig. 6. Protein phosphatase, partially purified from pig follicular cytosol and free of PK activity (Table I), was added to the second stage of the incubation in concentrations indicated on the figure. Mean ± SD of duplicate determinations is shown. *, $P<0.05$.

(measured with LH present only in the 5 min AC assay) was slightly increased, the major effect of the phosphatase-enriched preparation was on LH-stimulated desensitization (measured with LH present throughout 25 min incubation). Boiling of the phosphatase-containing preparation at each concentration tested resulted in maximal desensitization (Fig. 9). In a separate experiment using a different batch of membranes and cytosol-derived phosphatase, as well as slightly different incubation times, a phosphatase preparation (containing 35 μg protein and exhibiting 33 pmoles/min/mg protein of specific activity) reversed LH-stimulated desensitization by 57% (not shown).

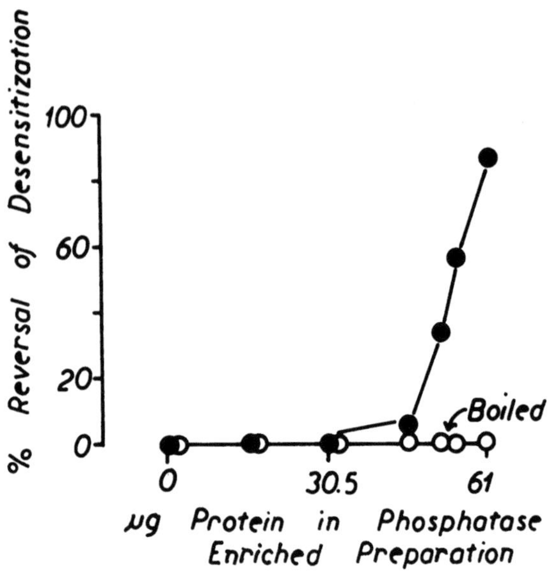


Fig. 9. Effect of intact and boiled phosphatase-enriched preparation on the reversal of desensitization. AC values using intact phosphatase are derived from the data in Fig. 8 (●). Equivalent concentrations of the phosphatase were boiled at 100 C for 20 min, and then centrifuged at 10,000 x g for 20 min, and the supernatants were added to the second stage of the incubation (O). Percent desensitization is calculated as the decline in LH-stimulated AC activity measured with LH present in all 3 stages of the incubation vs. that with LH present only in the AC assay (stage 3). Zero percent reversal of desensitization corresponds to values obtained in the absence of added phosphatase (BSA control; 21% desensitization).

Complete reversal of LH-induced desensitization was obtained using a phosphatase-enriched preparation (containing 15 μg protein and exhibiting 110 pmoles/min/mg protein specific activity) obtained from calf thymus nuclei (not shown). Experiments using various concentrations of commercially available alkaline phosphatases derived from human placenta or calf intestine modified the AC system but did not reverse desensitization by specifically increasing the LH-stimulated desensitization (not shown).

The ability of phosphoprotein phosphatase-rich preparations as well as $Mn^{2+}$ and DTT, both of which can activate endogenous membrane protein phosphatases, to reverse LH-stimulated desensitization suggested that the resensitization of the AC system to LH stimulation may be mediated by a dephosphorylation reaction. It follows then that desensitization may be mediated by a phosphorylation reaction.

To this end we investigated whether beef heart PK (Sigma Chemical Co.), a type II cAMP-dependent PK (25), could enhance LH-induced desensitization. Concentrations of 0.5 and 1 μg (Fig. 10) and 0.05 μg (not shown) beef heart PK did not enhance or reduce the extent of LH-induced desensitization. Also, incubations of the pig follicular membranes with varying concentrations of heat-stable PK inhibitor did not alter LH-induced desensitization (Table III). These latter two experiments suggest that desensitization is not mediated by a cAMP-dependent PK, at least one of heterologous origin. This suggestion is further substantiated by the ability of LH to promote desensitization of the porcine follicular AC system in the absence of exogenous cAMP (14).

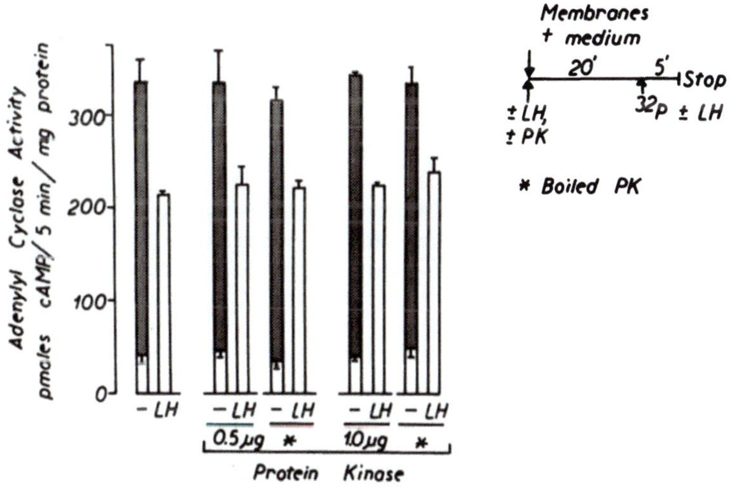


Fig. 10. Effect of beef heart PK on AC activity of porcine follicular membranes (37 μg protein). Protocol is as described in the legend to Fig. 6. PK, in concentrations indicated on the figure, was added to the second stage of the incubation. Equivalent concentrations of PK were boiled and treated as described in the legend to Fig. 9. "-, LH" on abscissa indicates BSA (-) or LH additions to first stage of incubation. Hatched bars indicate LH added only to last 5 min AC assay. Mean ± SD of duplicate determinations is shown.

TABLE III

Lack of Effect of Heat-Stable PK Inhibitor on Desensitization of the AC System*

| μg Protein of Inhibitor | % Desensitization |
|---|---|
| None | 43.7 ± 3.5 |
| 6.0 | 41.5 ± 2.9 |
| 3.0 | 38.5 ± 4.9 |
| 1.5 | 39.6 ± 2.5 |
| 0.6 | 41.1 ± 1.4 |

* PK inhibitor was added to the first stage of a 2 stage incubation in which follicular membranes (48 μg protein) were incubated at 30 C for 20 min in a medium consisting of buffer, EDTA, EGTA, cAMP, ATP, and an ATP-regenerating system in the absence and presence of 12.5 μg/ml (in concentrations 1.25 times final AC assay concentrations). The second stage of the incubation (AC assay) was started with the addition of 10 μl of [$\alpha$-$^{32}$P]ATP in the presence of BSA or 12.5 μg/ml LH and was terminated after 5 min (2). For rest of conditions, see legend to Fig. 6. Six μg of protein of the inhibitor, isolated from rabbit skeletal muscle (26, 28), inhibited beef heart PK by 96%. Results are expressed as mean ± SD of triplicate determinations.

## CONCLUSIONS

Agonist-induced desensitization of the AC system has been described in a great many other tissues as well as in the ovarian follicular system, as recently reviewed by Hunzicker-Dunn et al. (35). The specific characteristics of desensitization vary greatly. For example, hormone-induced refractoriness of the AC system may reach completion within a few h or it may require as many as 24-48 h to reach completion. Desensitization of AC may be specific for the agonist (homologous) or may be nonspecific (heterologous), with the AC in the latter case exhibiting refractoriness not only to the stimulating agent but also to other agonists (36). Heterologous desensitization usually occurs more slowly than agonist-specific, is mediated by cAMP, and may occur secondarily to agonist-specific desensitization. Although receptor loss often, but not always, accompanies desensitization, it is not known if receptor loss is the cause or the result of desensitization of the AC system. Lastly, the molecular mechanism by which desensitization occurs is not known, even in terms of the involvement of a protein synthetic step, in which results in the literature are conflicting (35, 37, 38).

The rabbit ovarian follicle is unique among the various systems in which agonist-induced desensitization of the AC system has been described in that desensitization is a physiological event: it is normally induced in vivo in response to coitus and known physiological responses follow which ultimately culminate in ovulation and CL formation. Furthermore, desensitization of the follicular AC system can also be induced in vivo by hCG in concentrations which promote ovulation and CL formation as well as in vitro by exposing intact follicles or in a cell-free follicular membrane system to LH.

Studies using hCG administered in vivo demonstrate that the initial phase of desensitization, which occurs in the first h subsequent to LH-receptor interaction, is of the homologous type. It is agonist-specific and, based upon the experiments in which porcine follicular membranes were incubated 20 min in the absence and presence of 1 mM cAMP, is not mediated by cAMP (14). This initial, homologous phase of desensitization of the AC system in follicles appears to be followed by a secondary, heterologous phase, as indicated by the following. (a) By 3 h after hCG administration, the follicular AC becomes refractory to FSH as well as LH. (b) Incubation of follicles for 2 h with d-cAMP promotes a dose-dependent reduction in LH-sensitive AC activity. Heterologous desensitization characteristically is not agonist specific and is mediated by cAMP (35).

The initial homologous phase of desensitization does not appear to require RNA or protein synthetic steps. Furthermore, the following lines of indirect evidence obtained in the in vitro porcine follicular membrane system indicate that the activity of the AC system may be partially regulated by the level of phosphorylation of one or more of the membrane-associated components of the AC system. (a) AC-rich membrane preparations contain both cAMP-dependent (type II) and cAMP-independent PK as well as phosphoprotein phosphatases. (b) LH-induced desensitization exhibits an absolute dependence on Mg and ATP (14). (c) AMP-P(NH)P, a nonphosphorylating analog of ATP, cannot be substituted for ATP (14). (d) DTT and $Mn^{2+}$, shown capable of activating endogenous membrane protein phosphatases, specifically reverse LH-dependent desensitization in a concentration-dependent maner, rendering the AC system responsive to LH. (e) Protein phosphatase-enriched preparations also specifically reverse LH-dependent desensitization. (f) Nonspecific alkaline phosphatases do not specifically resensitize the desensitized AC system. Although these evidences indicate that resensitization is mediated by a dephosphorylation reaction, final proof rests with experiments using a purified phosphoprotein phosphatase.

The initial phase of desensitization does not appear to be mediated by a cAMP-dependent PK. Neither beef heart PK nor an inhibitor of PK which complexes with endogenous catalytic subunits modifies desensitization in any detectable manner. These conclusions are substantiated by results obtained using variants of the S 49 mouse lymphoma cell line which lack cAMP-dependent PK (39). These investigators demonstrated equivalent isoproterenol-induced desensitization of the AC system in the cells which contained cAMP-dependent PK as in the variants which lacked the PK.

In summary, the AC desensitizing process in follicles appears to consist of two phases: an initial, LH-specific phase and a second phase which is not specific for LH. The initial phase occurs during the first h after hCG-induced activation of the AC system, does not require RNA or protein synthetic events, and even though it is demonstrable in a cell-free system only in the presence of Mg and ATP, the desensitizing process does not appear to be mediated by a cAMP-dependent PK. Furthermore, our results suggest that reversal of this first phase of AC desensitization involves a dephosphorylation reaction. The second phase of AC desensitization is demonstrable within 3 h after hCG-induced activation of the AC system and may be mediated by cAMP.

Although the nature of the participation of gonadotrophin-induced AC desensitization in follicular events which culminate in CL formation are not currently understood, a more thorough understanding of the desensitizing process may well provide vital information on the cellular mechanisms by which LH promotes ovulation and CL formation. Thus, further investigations are necessary to elucidate the physiological significance and the molecular mechanism of desensitization of the follicular AC system.

## SUMMARY

Studies on the gonadotrophin-responsive adenylyl cyclase (AC) system of rabbit and porcine ovarian follicles reveal that hCG or LH-induced desensitization of the AC system can be divided into two phases: an initial, LH-specific phase and a second phase which is not specific for LH. The first phase occurs within the first hour after LH-hCG-receptor interaction, is agonist specific, and is not mediated by protein synthetic events or by cAMP. In view of our previous demonstration of the critical dependence of the LH-induced desensitizing process in cell-free membrane preparations of porcine follicles upon $Mg^{2+}$ and ATP, we investigated the role of a phosphorylation reaction in the first phase of the AC desensitizing process. Porcine follicular membranes rich in LH-sensitive AC activity were found to contain the molecular requirements necessary for a phosphorylation reaction: namely, cAMP-dependent and cAMP-independent protein kinases as well as phosphoprotein phosphatases. The following lines of indirect evidence indicated that reversal or resensitization of the desenzitized AC system to LH was mediated by a dephosphorylation reaction. Activators of endogenous phosphoprotein phosphatases--$Mn^{2+}$ and dithiothreitol--promoted a specific resensitization of the follicular

AC system to LH. Likewise, a partially purified phosphoprotein phosphatase also resensitized the desensitized, LH unresponsive AC to LH, and boiling of the phosphatase prevented its effect. LH-induced desensitization of the AC system, on the other hand, did not appear to be mediated by a cAMP-dependent protein kinase, as evidenced both by the inability of beef heart protein to promote desensitization of AC and by the inability of an inhibitor of cAMP-dependent protein kinase to prevent LH-induced desensitization. These results indicate that the level of phosphorylation of membrane-associated components may, in part, regulate the activity of the AC system during this first phase of homologous desensitization.

The second phase of desensitization, which occurs after the first hour following hCG-LH-receptor interaction, is characterized by a loss of responsiveness to FSH as well as to LH and can be promoted by dibutryl cAMP (in the absence of LH).

These results provide new evidence on the characteristics and molecular mechanism of LH-induced desensitization of the follicular AC system.

## ACKNOWLEDGMENTS

The expert technical assistance of Ming-ann Wang is gratefully acknowledged. This work was supported by Grant HD 111356 from the National Institutes of Health.

## REFERENCES

1. L. Birnbaumer, P.-C. Yang, M. Hunzicker-Dunn, J. Bockaert, and J.M. Duran. Endocrinology 99(1976)163.
2. M. Hunzicker-Dunn and L. Birnbaumer. Endocrinology 99(1976)185.
3. M. Hunzicker-Dunn and L. Birnbaumer. Endocrinology 99(1976)198.
4. J.M. Marsh, T.M. Mills and W.J. LeMaire. Biochim.Biophys.Acta 304(1973)197.
5. S.A. Lamprecht, U. Zor, A. Tsafriri, and H.R. Lindner. J. Endocrinol. 57(1973)217.
6. S.A. Lamprecht, U. Zor, Y. Salomon, Y. Koch, K. Ahren, and H.R. Lindner. J.Cyc.Nuc.Res. 3(1977)69.
7. L.E. Reichert, Jr. and A.E. Wilhelmi. Endocrinology 102(1978) 982.
8. L.E. Reichert, Jr. and A.E. Wilhelmi. Endocrinology 92(1973) 1301.
9. L. Dufy-Barbe, P. Franchimont, and J.M.A. Faure. Endocrinology 92(1973)1318.
10. A.L. Goodman and J.D. Neill. Endocrinology 99(1976)852.
11. L.R. Ellsworth and D.T. Armstrong. Endocrinology 92(1976)840.
12. J.B. Miller and P.L. Keyes. Endocrinology 95(1973)253.

13. F. J. Chlapowski, L.A. Kelly, and R.W. Butcher, in: Advances in Cyclic Nucleotide Research, Vol. 6, ed. P. Greengard and G.A. Robison (Raven Press, New York, 1975) p. 245.
14. J. Bockaert, M. Hunzicker-Dunn and L. Birnbaumer. J. Biol. Chem. 251(1976)2653.
15. M. Rodbell, L. Birnbaumer, S.L. Pohl, and K.M.J. Krans. J. Biol. Chem. 246(1971)1877.
16. P. Layne, A. Constantopoulos, J.F.X. Judge, R. Rauner, and V.A. Najjar. Biochem. Biophys. Res. Comm. 53(1973)800.
17. A. Constantopoulos, and V.A. Najjar. Biochem. Biophys. Res. Comm. 53(1973)794.
18. V.A. Najjar and A. Constantopoulos. Molec. and Cell. Biochem 2(1973)87.
19. A.S. Gordon, C.G. Davis, D. Melfay and I. Diamond. Nature 267 (1977)539.
20. V.I. Teichberg, A. Sobel, and J.-P. Changeux. Nature 267 (1977)540.
21. H. Lo and G.S. Levey. Endocrinology 98(1976)251.
22. M. Hunzicker-Dunn and R.A. Jungmann. Endocrinology 103(1978) 431.
23. M. Hunzicker-Dunn and R.A. Jungmann. Endocrinology 103(1978) 441.
24. J.D. Corbin, S.L. Keely, and C.R. Park. J. Biol. Chem. 250 (1975)218.
25. J.D. Corbin, and S.L. Keely. J. Biol. Chem. 252(1977)910.
26. M. Hunzicker-Dunn and R.A. Jungmann. Endocrinology 103(1978) 420.
27. C.D. Ashby and D.A. Walsh. J. Biol. Chem. 248(1973)1255.
28. D.A. Walsh, C.D. Ashby, C. Gonzalez, D. Calkins, E.H. Fischer, and E.G. Krebs. J. Biol. Chem. 246(1971)1977.
29. T.W. Reid, and I.B. Wilson, in: The Enzymes, Vol. 4, ed. P.D. Boyer (Academic Press, New York, 1971) p. 373.
30. H.N. Fernley, in: The Enzymes, Vol. 4, ed. P.D. Boyer (Academic Press, New York, 1971) p. 417.
31. H. Maeno and P. Greengard. J. Biol. Chem. 247(1972)3269.
32. S. Azhar and K.M.J. Menon. Eur. J. Biochem. 55(1975)487.
33. M.H. Meisler and T.A. Langan. J. Biol. Chem. 244(1969)4961.
34. C. Nakai and J.A. Thomas. J. Biol. Chem. 249(1974)6459.
35. M. Hunzicker-Dunn, J. Bockaert, and L. Birnbaumer, in: Receptors and Hormone Action, Vol. 3, ed. B.W. O'Malley and L. Birnbaumer (Academic Press, New York, 1978) p. 393.
36. Y.-F. Su, L. Cubeddu, and J.P. Perkins. J. Cyc. Nuc. Res. 2 (1976)257.
37. G.J. Lauzon, S. Kulshrestha, L. Starr, and H.-P. Bar. J. Cyc. Nuc. Res. 2(1976)99.
38. J.W. Kebabian, M. Zate, J.A. Romero, and J. Axelrod, Proc. Natl. Acad. Sci. USA 72(1975)3735.
39. M. Shear, P.A. Insel, K.L. Melmon, and P. Coffino. J. Biol. Chem. 25(1976)7572.

# GONADAL LH RECEPTORS: RESOLUTION FROM ADENYLATE CYCLASE AND TRANSFER TO HETEROLOGOUS CELLS

M. L. Dufau, K. Hayashi, G. Sala, A. Baukal
and K. J. Catt

Endocrinology and Reproduction Research Branch, NICHD
National Institutes of Health, Bethesda, Md. 20014

The majority of particulate high-affinity receptors for LH in the steroidogenic cells of the testis and ovary are membrane-bound and capable of functional coupling to adenylate cyclase (1,2). The gonadal receptors for LH can be solubilized together with adenylate cyclase by a variety of non-ionic detergents (3). In addition to the membrane-associated receptors that require detergent for solubilization, the luteinized ovary has been found to contain a small proportion of spontaneously "soluble" receptor sites (up to 10%) that can be recovered in the floating lipid fraction of the 360,000 x g supernate prepared from ovarian homogenates (4).

This report will describe the resolution of LH receptors and adenylate cyclase in detergent extracts of testis and ovary, and the results of studies on the transfer of lipid-associated ovarian LH receptors to adrenal fasciculata cells.

Chromatographic analysis of soluble LH/hCG receptors and adenylate cyclase.

To examine the physical and functional relationships between solubilized gonadotropin receptors and adenylate cyclase, detergent extracts from the rat testis and luteinized ovaries of PMSG/hCG-treated rats were analyzed by gel filtration (3), anion exchange chromatography, and lectin adsorption chromatography (5).

Gel filtration analysis of Lubrol-solubilized ovarian LH receptors and adenylate cyclase was performed in the presence of 5 mM sodium fluoride and 5 mM magnesium chloride to preserve enzyme activity. Chromatography of the soluble preparation on Sepharose 6B showed two peaks of receptor binding with coincident peaks of adenylate cyclase activity. The major peak was eluted with Kav 0.28, and there was a minor peak at the void volume. In these studies, we did not observe the small molecular weight species of receptor binding activity (Kav 0.58) previously noted in experiments where solubilization and fractionation was carried out in the absence of sodium fluoride and magnesium chloride (3). This finding suggests that these agents may stabilize the receptor molecule, preventing dissociation of a portion or subunit of the LH receptor containing the binding site.

Further studies by group-specific affinity chromatography on Sepharose-concanavalin A clearly showed that testicular LH receptor sites were eluted separately from the corresponding adenylate cyclase activity. The enzyme was not adsorbed by the lectin affinity column, and eluted with the majority of the protein in the soluble preparation, while all the LH/hCG receptor binding activity was adsorbed to the column and could be subsequently eluted by 0.2 M methylmannoside. In contrast, resolution of detergent-soluble ovarian LH receptors by lectin-affinity chromatographic procedure (Fig. 1) showed two populations of LH receptors, a major fraction that bound to concanavalin A and did not possess adenylate cyclase activity, and a smaller population that was not adsorbed to the lectin, similar to the adenylate cyclase activity. The binding characteristics and physical properties (sedimentation coefficient, resolution in Sephadex or Sepharose chromatography) were identical for both classes of receptors. Ion-exchange chromatography revealed that adenylate cyclase was not coincident with either of the receptor peaks, indicating that as for the Leydig cells, adenylate cyclase activity and LH receptor sites reside in separate molecular entities (Fig. 2).

In contrast, the receptor activity of the ovarian lipid-associated sites did not adsorb to the Sepharose-concanavalin A and eluted with most of the protein of the preparation (Fig. 3). Unlike the detergent-solubilized material, this preparation did not contain detectable adenylate cyclase activity.

The preceding studies indicated that gonadotropin receptor and adenylate cyclase are physically separate entities that can be resolved during appropriate fractionation procedures. However, these findings did not exclude the possibility that the two species are associated in some manner in the lipid bilayer of the plasma membrane. Recent studies employing cell fusion (6,7) and reconstitution studies of detergent-extracted adenylate cyclase components (8) have also indicated that hormone receptors and adenylate cyclase exist as separate components within the cell membrane.

It was therefore of interest to follow the functional properties of LH receptors from gonadal tissue after incorporation into heterologous steroidogenic cells. For this purpose,

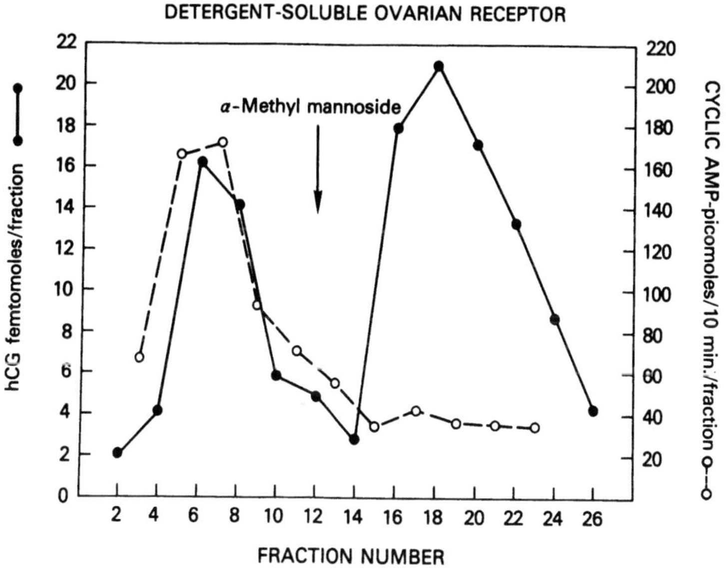


Figure 1. Elution pattern of LH/hCG receptors and adenylate cyclase from ovarian membrane-rich particulate fraction with Lubrol PX, during Sepharose-concanavalin A chromatography. The first 12 ml fractions were eluted by 0.5% Lubrol in Phosphate-buffered saline and fractions 12-26 with 0.2 M methyl-Mannoside.

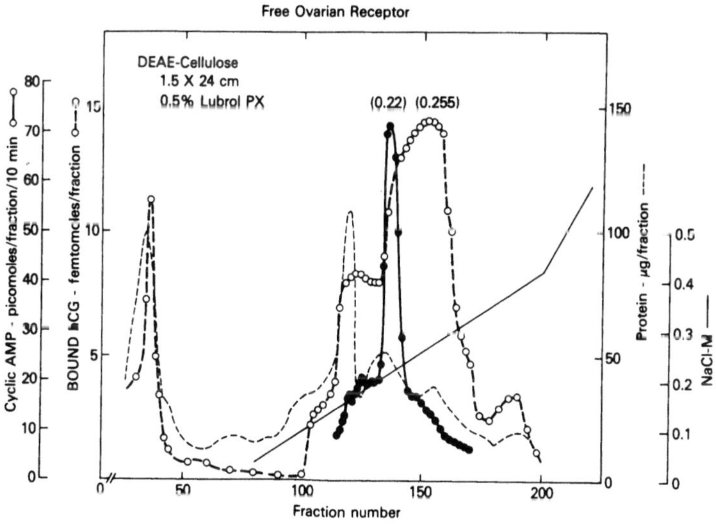


Figure 2. Ion exchange chromatography of detergent-soluble ovarian LH/hCG receptors and adenylate cyclase performed in a DEAE cellulose with elution by linear gradient NaCl (0-500 mM).

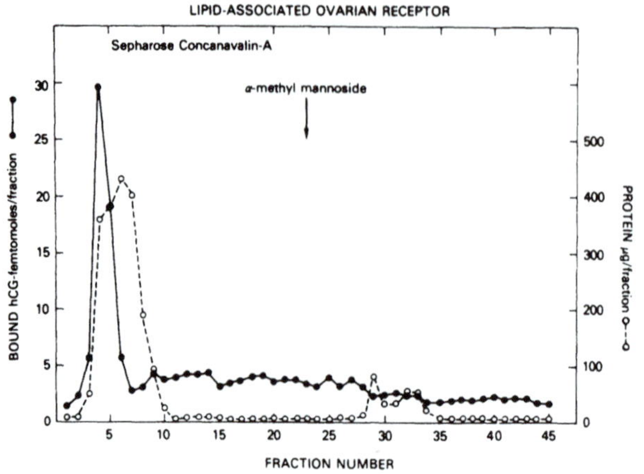


Figure 3. Elution profile of lipid-associated ovarian LH/hCG receptors from Sepharose-concanavalin A chromatography.

the lipid-associate LH receptors were incorporated into fasciculata cells prepared from the rat adrenal gland.

Transfer of lipid associated ovarian LH receptor to adrenal cells. After collagenase dispersion of adrenal fasciculata-reticularis zones (9) the cell suspension was incubated with lipid-associated LH receptors (4) at 16° for 4 hrs. The cells were then washed and incubated at 37°C for 3 hrs: (1) with $^{125}$I-hCG for binding studies to assess the number of LH receptors incorporated in the adrenal cells; (2) with ACTH or hCG to evaluate responses mediated by the natural (ACTH) and transferred (LH) receptors. The cell responses to hormonal stimuli were monitored by radioimmunoassay of cAMP (10) and corticosterone (11) in the incubation medium and cell extracts. Electron microscopic analysis of negatively stained preparations of the dispersed lipid receptor samples used as a source of receptor to transfer showed fat globules with a size range of 85 to 350 nanometers. The lipid composition of the preparation was: triglyceride 78%; esterified cholesterol 17%; free cholesterol 1.1% and phospholipid 3.5%; phospholipid analysis showed major contents of phosphatidyl choline (54%) and phosphatidyl ethanolamine (24%).

Studies on the uptake of ovarian LH receptors by adrenal cells (Fig. 4) showed that these receptors were incorporated into the cell membrane and were subsequently available to bind $^{125}$I-hCG. The number of LH receptors increased with time, and at 4 hrs the cell bound an average of 200 femtomoles of hCG per $10^6$ cells, or about 120 receptors per cell. The efficiency of uptake of the LH receptors was relatively high (80-90%). Non-specific binding of $^{125}$I-hCG to control adrenal cells was always very low ( 1%).

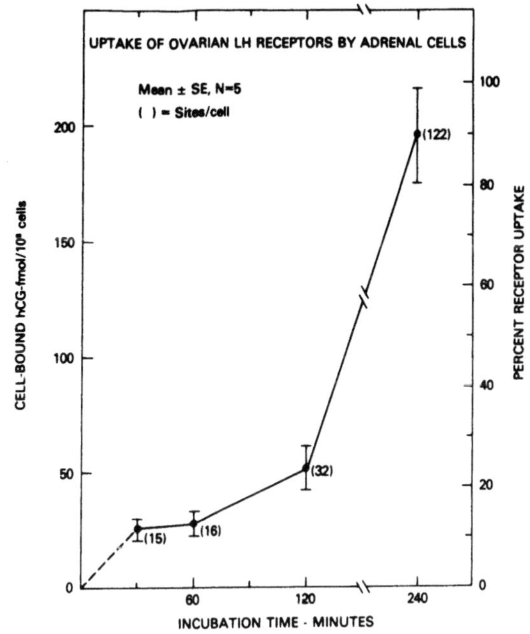


Figure 4. Specific uptake of $^{125}$I-hCG by adrenal cells bearing LH/hCG receptors. The number of receptors per cell are shown in parentheses. Values are mean $\pm$ SE of quadruplicate incubations.

Stimulation of control adrenal cells without LH receptors by a maximum ACTH concentration for steroidogenesis (1 nM) provoked a seven-fold increase of cyclic AMP production over control values (Fig. 5, left). In contrast, no stimulation of cAMP or steroid production was elicited by 10 to 100 nM hCG in control cells. In adrenal cells bearing LH receptors, significant stimulation of extracellular cAMP production was observed during incubation with 10 and 100 nM hCG, with increases of 25 and 50% respectively above control values (Fig. 5, right). The corresponding steroidogenic responses in normal adrenal cells showed that only ACTH increased corticosterone production, and no stimulation was observed with hCG. In adrenal cells bearing transferred LH receptors, hCG elicited significant increments in corticosterone production. In parallel to the rise in extracellular cAMP, increased levels of intracellular cAMP were also observed during hormone stimulation of LH receptor-bearing cells (Fig. 6). Incubation with a supramaximal dose of ACTH stimulated intracellular and extracellular cAMP and corticosterone maximally, while 1 nM hCG increased intracellular cAMP by 60%, extracellular cAMP by 100%, and corticosterone production by 60%. With 10 nM hCG, all responses were again significantly increased.

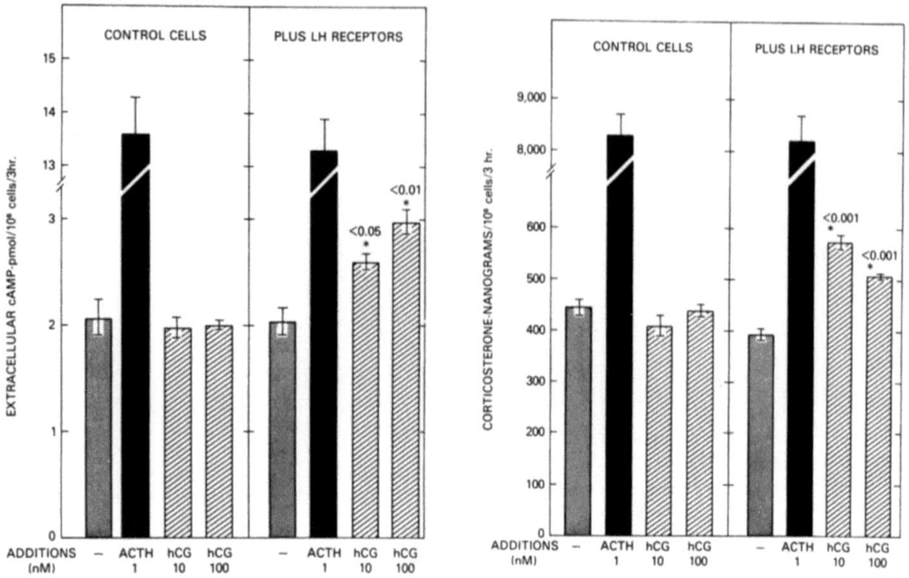


Fig. 5. Extracellular cAMP (Left) and corticosterone production (Right) by control adrenal cells and by cells bearing LH receptors in the presence and absence of ACTH or hCG. Values were mean $\pm$ SD, n = 4.

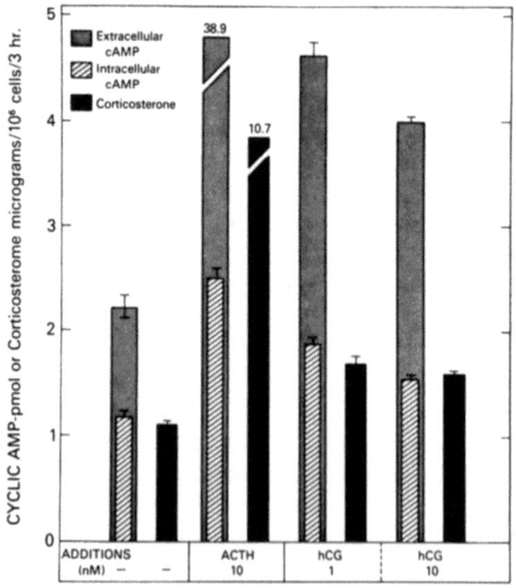


Fig. 6. Extracellular and intracellular cAMP and corticosterone production during stimulation of adrenal cells with LH receptors by ACTH or hCG.

These studies have demonstrated that testicular and ovarian LH receptors can be resolved from adenylate cyclase by Sepharose-concanavalin A and/or DEAE cellulose chromatography. Lipid-associated ovarian LH receptors can be incorporated into adrenal cells, and such receptors have been shown to be functionally coupled to adenylate cyclase as demonstrated by cAMP and corticosterone responses to gonadotropin. It can be concluded from these studies that gonadotropin receptors and adenylate cyclase exist in the cell membrane as separate entities that become associated to form an active complex during receptor occupancy by the homologous hormone.

## References

1. Catt, K.J., T. Tsuruhara, C. Mendelson, J.-M. Ketelslegers and M.L. Dufau. In: Hormone Binding Activation in the Testis (Dufau, M.L. and A.R. Means, eds.), Plenum Press, New York, 1-30, 1974.

2. Seong, S.H., H.J. Rajaniemi, M.O. Cho, A.N. Hirshfield and A.R. Midgley. Endocrinology 95(1974)589.

3. Dufau, M.L., A.J. Baukal, D. Ryan and K.J. Catt. Mol. Cell Endocrinol. 6(1977)253.

4. Conti, M., M.L. Dufau and K.J. Catt. Biochim. et Biophys. Acta 541(1978)35.

5. Dufau, M.L., E.H. Charreau and K.J. Catt. J. Biol. Chem. 248(1973)6973.

6. Schramm, M., J. Orly, S. Eimerl, J. Korner. Nature 268 (1977)268.

7. Schulster, D., J. Orly, G. Seidel and M. Schramm. J. Biol. Chem. 253(1978)1201.

8. Ross, E.M. and A.G. Gillman. J. Biol. Chem. 252(1977)6966.

9. Douglas, J., G. Aguilera, T. Kondo and K.J. Catt. Endocrinology 102(1978)685.

10. Dufau, M.L., T. Tsuruhara, A.K. Horner, E. Podesta and K.J. Catt. Proc. Natl. Acad. Sci. USA 74(1977)3419.

11. Gross, H.A., H.J. Ruder, K.S. Brown and M.B. Lipsett. Steroids 20(1972)681.

# PROLACTIN CONTROL OF RECEPTOR FOR ESTRADIOL IN CORPORA LUTEA OF PREGNANT RATS

G. Gibori, J.S. Richards and P.L. Keyes

Dept. Physiol., Univ. Illinois, Chicago, Illinois

Repro. Endo. Prog., Univ. Mich., Ann Arbor, Michigan

Both prolactin and a prolactin-like hormone from the placenta, rat placental lactogen, appear to be required for the luteotropic effect of estrogen in corpora lutea of pregnant rats (1-3). However, the specific synergistic role of these two hormones with estrogen has not been defined. Because receptors for both estradiol and prolactin have been identified and characterized in corpora lutea of pregnant rats, the luteotropic effects of these two hormones are presumed to be direct (4,5). The following studies were designed to determine if the synergistic role of prolactin in maintaining luteal cell progesterone production might involve, at least in part, maintenance of luteal cell receptors for estradiol. To examine the roles of pituitary and placental hormones during early pregnancy two approaches were taken. In the first, pituitary prolactin surges (6) were selectively inhibited by the administration of 2-bromo-α-ergocryptine (ECO; CB-154; 7-10). In the second, both the pituitary and placenta were surgically removed.

Administration of ECO on day 6 of pregnancy caused serum concentrations of progesterone to fall rapidly and remain low (6.0 $\pm$ 1.0 ng/ml) through day 8 (Fig. 1). However, despite the reduction in serum progesterone intrauterine hemorrhages were not observed at the time of autopsy. Administration of prolactin in the ECO treated rats prevented but did not completely reverse the marked decline in serum progesterone (Fig. 1). Further, although ECO caused serum progesterone to decline, cytosolic and nuclear concentrations of estradiol receptor were not markedly different from those observed in the vehicle treated rats. Only on day 7 was nuclear content of estradiol receptor in ECO treated rats demonstrably lower than that in vehicle treated rats (Fig. 1).

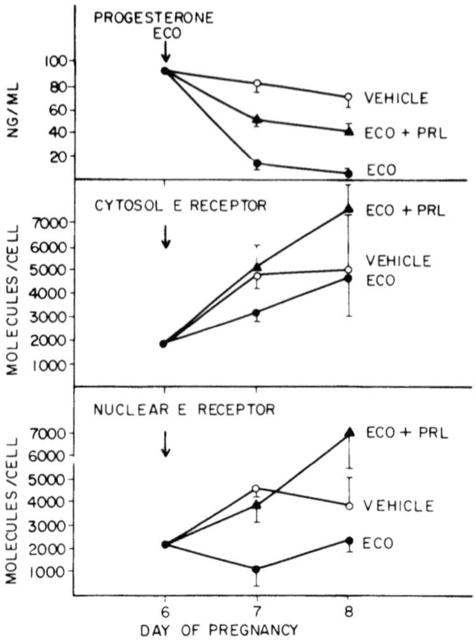


Figure 1: Effect of 2 Bromo-α-ergocryptine (ECO) and prolactin (PRL) replacement on luteal cell content of cytosol and nuclear estrogen receptor and on serum progesterone. A 0.4% (w/v) solution of ECO in 70% aqueous ethanol was injected subcutaneously (s.c) in 0.25 ml. Ovine prolactin (PRL) LER-860-2, 25 IU/mg and NIH-P-S12, 35 IU/mg) was administered in 50% polyvinylpyrrolidone, pH 9.0, twice daily (125 μg in 0.25 ml/injection) (7,10). Serum progesterone was quantified as reported previously (11). To measure estradiol receptors, corpora lutea were separated from non-luteal ovarian tissue and homogenized (250-300 mg/3.5 ml) in 0.01 M Tris, 0.0015 M EDTA, and 12 mM monothioglycerol. The homogenate was centrifuged at 800 x g for 10 min and the supernatant centrifuged at 105,000 x g for 1 hr at 4°C. Aliquots of cytosol (500 μl) were incubated with 10 nM of $^3$H-estradiol with or without 1 μM diethylstilbesterol (DES) for 2 hr at 0°C. The $^3$H-estradiol bound to cytosol was separated from that which was free by dextran charcoal adsorption: exposure to 0.5 ml of dextran (Pharmacia -T70) charcoal (Norit A) (0.065% - 0.65%) for 15 min at 4°C. The charcoal was removed by centrifugation at 3000 x g for 10 min and the supernatant containing $^3$H-estradiol bound to cytosol was poured directly into scintillation vials containing 10 ml of scintillation cocktail. Estradiol receptor content in luteal cell nuclei was measured by nuclear exchange assay as described previously (4).

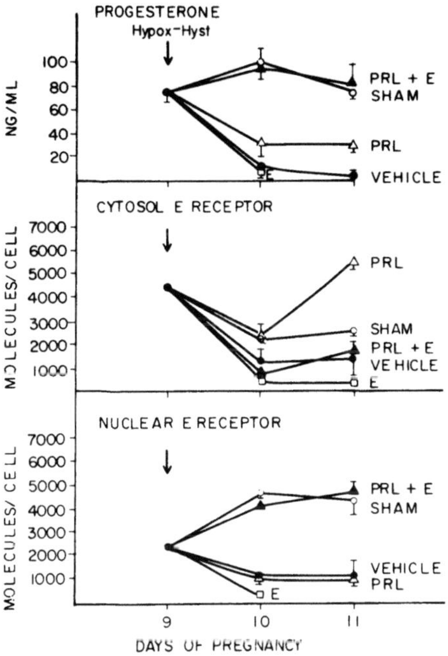


Figure 2: Luteal content of cytosol and nuclear estrogen receptor and serum concentrations of progesterone in rats hypophysectomized-hysterectomized on day 9 of pregnancy and treated daily with prolactin (PRL), estrogen (E) or PRL with E. A clean but not aseptic technique was used during all surgical procedures, with ether as anesthesia. Hypophysectomy was done by the transauricular approach using a stereotaxic instrument (13). Completeness of removal of the pituitary was judged by the recovery of the pituitary at operation and by the absence of any fragments in the fossa at autopsy. Hysterectomy was performed through a midline abdominal incision. Sham operations for hysterectomy and hypophysectomy consisted of laparotomy and penetration of the ear drum by the aspirating syringe without removal of the pituitary. Estradiol 17β was dissolved in sesame oil and 100 μg were injected s.c. daily in a volume of 0.25 ml. Control pregnant rats were injected with the same volume of sesame oil. Daily estradiol or oil treatment was maintained from the day of operation until autopsy. The last injection was 2 hr before sacrifice. Serum from day 12 pregnant rats (PRS) was used as the source of rat placental lactogen (14,15). In order to strip the steroids from serum, 100 ml of serum were incubated twice with 0.8 gm of charcoal at 45°C in a water bath for 20 min. After centrifugation at 3000 x g for 10 min, the supernatant was filtered through Celite and stored at -20°C in small aliquots. Once thawed, the serum was used no longer than 2 days. One ml serum was injected intraperitoneally (i.p.) twice daily. See legend of Figure 1 for other methods. Progesterone concentrations obtained from Gibori and Richards (2).

However, the administration of prolactin to ECO treated rats did cause nuclear estradiol receptor content to increase on day 8 suggesting that prolactin was capable of effecting changes in luteal cell content of estradiol receptor.

When pregnant rats were hypophysectomized and hysterectomized on day 9, serum progesterone concentrations fell rapidly and remained low for 48 h (Fig. 2; 2). Further, as reported, prolactin and estradiol given together, but neither hormone alone, reversed the effects of ablation (2). As shown in this investigation hypophysectomy and hysterectomy also caused luteal cell content of nuclear estradiol receptor to decrease and remain low for 48 h (Fig. 2). Administration of prolactin alone had no effect on nuclear content of estradiol receptor but did cause cytosolic receptor to increase markedly by 48 h (Fig. 2). Similarly, if serum of day 12 pregnant rats was given as a source of rat placental lactogen, cytosolic estradiol receptor content increased (Table 1). Since luteal cell content of estradiol is low following hypophysectomy and hysterectomy (12), translocation of the cytosolic receptor to the nucleus did not occur (Fig. 2).

Administration of estradiol alone to hypophysectomized-hysterectomized rats caused a further decrease in both cytosolic and nuclear receptors (Fig. 2). However, when prolactin treatments were combined with exogenous estradiol, nuclear estradiol receptor was maintained at elevated concentrations observed in sham operated pregnant rats (Fig. 2).

TABLE 1

| Treatment | Serum Progesterone ng/ml | Specific cytosol bound $^3$H-E molecules/cell |
|---|---|---|
| Sham operated | 75 $\pm$ 4<br>6 | 2555 $\pm$ 21<br>2 |
| Hypophysectomized-hysterectomized | 3 $\pm$ 0.4<br>4 | 1424 $\pm$ 670<br>3 |
| Hypophysectomized-hysterectomized + PRS | 35 $\pm$ 10<br>4 | 4104 $\pm$ 544<br>2 |

Table 1. Effect of day 12 pregnant rat serum (PRS) on luteal cell content of cytoplasmic estradiol receptor and serum progesterone concentration in rats hypophysectomized and hysterectomized on day 9 of pregnancy. The results represent the values found on day 11 of pregnancy. One ml of PRS was given i.p. twice daily on days 9-11. Values shown for estradiol receptor are the mean $\pm$ SE with the number of replicates indicated below each mean; each observation represents receptor determination from a pool of corpora lutea obtained from 5 to 8 rats. For serum progesterone, each point represents the mean $\pm$ SE, with the number of rats indicated below each mean.

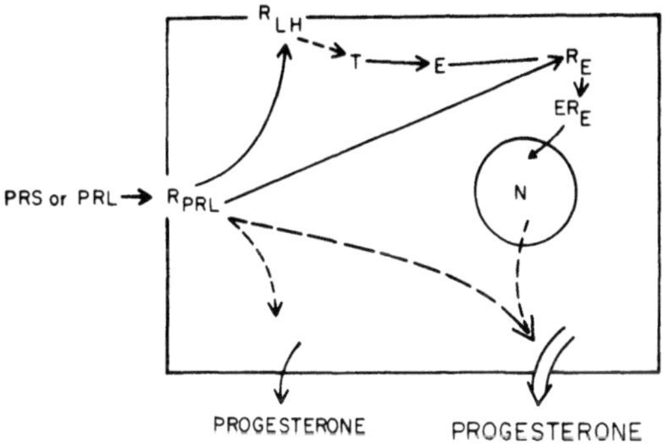


Figure 3: Regulation of luteal cell function. See text for explanation.

Results of these studies combined with those already reported revealed that there are distinct and separable luteotropic effects of prolactin on the corpus luteum of the pregnant rat. One action of prolactin appears to be its ability to stimulate and maintain luteal cell LH receptor (Fig. 3, 2). Because LH is required to maintain luteal cell estradiol content, presumably by providing aromatizable androgen substrate (11,16) prolactin, by maintaining LH receptor, is indirectly supporting luteal cell production of estradiol synthesis (Fig. 3). Second, prolactin also appears to maintain luteal cell content of estradiol receptor, allowing estrogen to bind to its receptor, to be translocated to the nucleus and to increase progesterone secretion. In addition, however, prolactin seems to be required at some other site(s) to maintain luteal cell production of progesterone.

Results of these studies suggest that prolactin or rat placental lactogen can increase the concentrations of cytosolic estradiol receptor in corpora lutea of pregnant rats. Further, in the absence of pituitary and placental hormones, prolactin in the presence of exogenous estradiol maintained elevated concentrations of estradiol receptor in luteal cell nuclei and elevated concentrations of serum progesterone. However, because the selective inhibition of prolactin surges from the pituitary by ECO on days 6-8 of pregnancy decreased serum progesterone without causing similar changes in luteal cell estradiol receptor, these results suggest that prolactin acts to maintain progesterone synthesis by some mechanism(s) in addition to maintaining estradiol receptor.

Because estradiol alone in hypophysectomized-hysterectomized rats caused estradiol receptor to decrease below the values observed following ablation, estradiol alone does not appear capable of maintaining its own receptor.

## REFERENCES

1. G. Gibori, E. Antczak and I Rothchild. Endocrinology 100(1977) 1483.

2. G. Gibori and J.S. Richards. Endocrinology 102(1978)767.

3. J.J. Alloiteau and J. Bouhours. C.R.Acad.Sci. 259(1964)4141.

4. J.S. Richards. Endocrinology 95(1974)1046.

5. J.S. Richards and J.J. Williams. Endocrinology 99(1976)1571.

6. M.S. Smith and J.D. Neill. Endocrinology 98(1976)696.

7. W.K. Morishige and I. Rothchild. Endocrinology 95(1974)260.

8. W. Wuttke, E. Cassel and J. Meites. Endocrinology 88(1971)737.

9. E.S. Kisch and M.C. Shelesnyak. J.Reprod.Fertil. 15(1968)401.

10. J.A. Holt, J.S. Richards, A.R. Midgley, Jr. and L.E. Reichert, Jr. Endocrinology 98(1976)1005.

11. D.J. Elbaum, E.M. Bender, J.M. Brown and P.L. Keyes. Biol.Reprod. 13(1975)541.

12. G. Gibori and P.L. Keyes. Endocrinology 102(1978)1176.

13. V.L. Gay. Endocrinology 81(1967)1177.

14. D.M. Linkie and G.D. Niswender. Biol.Reprod. 8(1973)48.

15. P.A. Kelly, R.P.C. Shiu, M.C. Robertson and H.G. Friesen. Endocrinology 96(1975)1187.

16. G. Gibori, P.L. Keyes and J.S. Richards. Endocrinology 103 (1978) in press.

## ACKNOWLEDGEMENTS

Supported in part by NIH-HD-09110, a Program Project, NIH-HD-08333 and PHS-HD-11119.

DISCUSSION - CHAIRPERSON - R.J. RYAN

L. BIRNBAUMER: Joanne, that was very nice. I would like to make a small comment. In your experiments on desensitization with hCG and FSH, it appears that you did not find what you looked for, but found what I looked for some years ago and did not find. In 1973, I looked at John Marsh's data on the desensitization of the cyclic AMP system in preovulatory follicles, and thought that his findings measuring follicular cAMP accumulation were not due to cyclase desensitization but rather to general metabolic changes including changes in phosphodiesterase and alterations in compartmentalization of cyclic AMP. With Mary Dunn, we then measured adenylyl cyclase and found that Marsh's interpretations were right and my speculation was not. However, in fat cells there is a well-known phenomenon that leads to results similar to yours. Exposure to one hormone results in loss of responsiveness not only to itself but also to other hormones--i.e., it results in heterologous desensitization. Thus, if fat cells are incubated with epinephrine, and cAMP accumulation by cells is evaluated, a transient increase is observed. If, after 15 minutes or more, ACTH is added to these cells, no effect on cAMP levels can be seen; they remain low. This resembles what happened to you with rat granulosa cells, where FSH caused refractoriness to hCG and vice versa (which was the type of response I thought I would find in rabbit preovulatory follicles and did not). I think that cross-talk of the type seen by you in granulosa cells and by others before in fat cells (cf. Manganiello et al. [1971] J. Biol. Chem. 246:2195-2202) is probably the result of a second phase response that either develops independently or is subsequent to a first phase response associated with homologous desensitization.

J. S. RICHARDS: In terms of the ability of FSH and hCG to turn off the adenylate cyclase system, I believe we're observing an early process of desensitization occurring in association with luteinization. Loss of receptors for FSH and LH occur later and thus appear to be secondary steps in the desensitization process associated with luteinization. In contrast, in preantral follicles, FSH does not cause loss of FSH receptors nor do the subcutaneous injections of FSH (1 μg/injection) to estradiol primed rats reduce subsequent intravenous FSH stimulation of cyclic AMP production in vivo. Thus, I think if FSH inhibits FSH responsive adenylate cyclase in the estradiol primed follicle, the inhibition will be transient. Thus, I think we need to distinguish the events occurring in association with terminal differentiation such as luteinization from those occurring during the progressive stages of differentiation in the developing antral follicle.

A. L. GOODMAN: I would like to ask Dr. Richards if she could explain in greater detail the difference between the regulation of the recruitment of follicles destined to ovulate after the postpartum estrus in the rat, and the recruitment of follicles during the estrus cycle. Several lines of evidence, for example--studies by Schwartz and by Hirschfield and Midgley, indicate that the proestrus surge of FSH release is required to recruit follicles for ovulation in the next cycle. Since no surge of FSH occurs during pregnancy in the rat, what is responsible for the presence of ovulable follicles postpartum?

J. S. RICHARDS: I think that's a good question, because I don't think we know. During the cycle, it has been proposed that the surge of gonadotropins selects a group of follicles which grow and ovulate at the subsequent proestrus. During pregnancy, Graafian follicles develop in the absence of surge concentrations of gonadotropins. Thus, surge concentrations of FSH or LH may not be required to stimulate the development of preovulatory follicles. Rather, very small changes in circulating levels of FSH or LH may regulate which follicles gain the mechanisms necessary for luteinization, including increased estrogen production and LH receptor in granulosa cells. Small changes, but nevertheless changes, in FSH and LH that do occur at the end of pregnancy, more or less coincident with the fall in progesterone are, therefore, probably significant.

A. L. GOODMAN: Are you saying that actually very little of the proestrus surge of FSH is required to recruit new follicles?

J. S. RICHARDS: Yes, that's what I'm saying.

A. L. GOODMAN: I was also uncertain about another aspect of your presentation. Are you saying the decrease in progesterone levels is having a direct enhancing effect on follicle growth, or is it permitting an estrogen effect on recruitment?

J. S. RICHARDS: I don't think it's been clearly determined yet. I would propose that the decrease in progesterone that is occurring at the end of pregnancy is removing the negative feedback of progesterone. Once progesterone declines, gonadotropin levels will increase, to a level that is sufficient to promote follicular growth.

A. L. GOODMAN: I was thinking in terms of Clark's (in Houston) findings that progesterone decreases the number of cytoplasmic estradiol receptors, and whether the action of progesterone on follicle recruitment was a blockade of estradiol receptors.

J. S. RICHARDS: There has been no evidence obtained to indicate that progesterone acts on the granulosa cells to antagonize estradiol action in the hypophysectomized rat.

H.R. LINDNER: I have a comment for Dr. Richards. You may be

aware that we have obtained support in an in vitro system for your hypothesis implicating FSH in the induction of the LH receptor: granulosa cells in preantral follicles explanted from hypophysectomized estrogen-treated immature rats responded to addition of FSH to the culture medium with a dramatic increase in the number of LH receptors per cell [A. Nimrod, A. Tsafriri and H. R. Lindner, Nature 267:632 (1977)]. The inducing effect was specific for FSH, in that this hormone could not be replaced by androgen, estrogen or agents stimulating cyclic AMP production in granulosa cells, such as prostaglandin $E_2$, or even by 8-bromo-cyclic AMP. However, we were unable to demonstrate the FSH effect on the generation of LH-receptors in cultures of isolated granulosa cells, as distinct from cultured follicles, suggesting that some other cellular component of the follicle is essential for this induction; or that the mode of association of the follicle cells is crucial for the response.

R. J. RYAN: At this point, I would like to call on Dr. Lindner to make a brief presentation that is pertinent to Dr. Hunzicker-Dunn's topic.

(Dr. Lindner's data can be found in the back of the book under Ezra, Lindner and Salomon).

H. R. LINDNER: In the light of these data, I would like to ask Dr. Hunzicker-Dunn if GTP was invariably present in her experiments?

M. HUNZICKER-DUNN: Dr. Birnbaumer, would you like to comment on the GTP studies?

L. BIRNBAUMER: To answer Dr. Lindner's question: Yes, we had GTP present throughout. I'd like to add a comment on the nucleotide requirement of two systems that show desensitization in isolated membranes: the LH-stimulated pig Graafian follicle adenylyl cyclase and the glucagon-stimulated rat liver adenylyl cyclase. With pig Graafian follicle membranes, if reactions are carried out at very low ATP levels (0.1 mM) and then supplemented with GTP, desensitization does not occur. Thus, in this system GTP does not support desensitization by itself. This points to the need of ATP. Since the ATP used in the experiments I am referring to (Bockaert et al., [1976] J. Biol. Chem. 251:2653-2663) may have been contaminated with GTP (up to 0.1-0.5%), these experiments do not indicate whether GTP, while not being sufficient, is necessary. More recently, Dr. Iyengar and I have been studying glucagon-induced desensitization of the adenylyl cyclase system present in purified liver membranes. In general, this system behaves similarly to the pig Graafian follicle membrane system--i.e., (1) desensitization does occur; (2) desensitization requires ATP; (3) GTP does not support desensitization at low ATP; and (4) desensitization is not supported by AMP-P(NHP)P, a nonphosphorylating analog of ATP. As with pig Graafian follicles, we have not yet investigated whether GTP, while not sufficient, might not be necessary for glucagon's desensitization. Yet, seeing Dr. Salomon's results, we shall test specifically for this possibility.

L. NILSSON: Under certain circumstances you can get different responses between whole intact follicles and granulosa cells in short-term incubation studies. We have worked with PMS-treated rats and isolated either follicles or granulosa cells from rats before and after the endogenous gonadotropin peak, and before and after the injection of gonadotropin in vivo. The endogenous gonadotropin peak occurs about 4 p.m. on the day before ovulation. Before the gonadotropin peak, we have a rather marked effect of LH on cAMP production in whole follicles, but after the endogenous peak, the response is reduced. It is not completely abolished, but we think this is a partial desensitization. The LH response lasts up to ovulation, which would take place around 3 a.m. The effect on granulosa cells is different. After the endogenous peak, the response to LH in vitro is not diminished but increased in comparison with the data obtained in the morning before the LH surge. We obtained the same response if exogenous gonadotropins are administered in vivo prior to follicle harvest.
Could you give me some explanation for that?

M. HUNZICKER-DUNN: No, I can't. In our rabbit follicular system, we've measured both the popped follicle, in which we lose a percentage of granulosa cells when we pop them, as well as intact follicles, and the responsiveness of the adenylyl cyclase to LH declines in both. I believe that Dr. Lindner's lab measured adenylyl cyclase in intact follicles from rats, if I'm not mistaken, and he gets desensitization.

J. BAHR: We obtain follicles from rabbits which are definitely in estrus, and we put them into a perifusion system in which medium flows through the chambers constantly for 36 hours. We inject a maximum dose of LH, 1 microgram of NIH/LH, plus 50 nanograms of highly purified FSH, which was obtained from Dr. Harold Papkoff. We inject these hormones directly into the chambers at 0/6/12/18/24 and 30 hours.

We monitored steroid production by the follicles. After the initial injection at 0 time, a large amount of testosterone and estradiol were secreted and a small amount of progesterone. After the second challenge with LH, there was a minor peak of testosterone and estradiol, and a significant progesterone peak. With subsequent challenges of LH and FSH, the follicles continue to produce progesterone only. Now, if the follicles are desensitized after exposure to a maximum dose of LH which we are using as determined by dose-response curves, then why are we continually getting progesterone production?

One of the questions I raised recently with Dr. Hunzicker-Dunn was that probably when she popped the follicles, she was losing some of the granulosa cells. It appears from our results that what we're seeing is that new receptors are being induced in the granulosa cells which are responsive to gonadotropins. These follicles in our perifusion system do luteinize.

M. HUNZICKER-DUNN: I still don't have an explanation. Possibly in both popped and intact follicles we are measuring primarily thecal adenylyl cyclase, due either to a low absolute adenylyl cyclase activity in granulosa cells and/or to proportionately fewer granulosa cells compared to theca cells. We haven't separated the two cell types as yet.

H. G. M. RAJ: I want to add a few comments to the conversation that took place between Dr. Richards and Dr. Goodman. We have repeatedly observed that if you give anti-LH serum and block LH in pregnant rats, or pregnant hamsters, the ovarian picture is one of great degeneration. You do see large numbers of follicles, which can be classified as both primary and secondary follicles, but you do not see any Graffian follicles. This agrees with the data that Dr. Richards presented, that tonic LH modulates and perhaps has a role in the development of the follicles.

R. J. RYAN: Dr. Marsh is next. We will have to make this the last comment.

J. M. MARSH: My comment relates to Dr. Nilsson's discussion. If I interpret him correctly, he is suggesting that the desensitization that we have seen before in the follicle is desensitization of the thecal cell component and not the granulosa cell component. We have some data on prostaglandin synthesis which indirectly supports that viewpoint. A Graafian follicle taken from a rabbit 9 hours after hCG injection is insensitive to LH, and does not make much cyclic AMP. Recently, however, we have found that if one removes the granulosa cells from those follicles, they will make a great deal of prostaglandins in an in vitro incubation system. We also believe that these isolated granulosa cells can produce elevated levels of cyclic AMP since we have other evidence to indicate that increased prostaglandin synthesis involves cyclic AMP. So I think that the desensitization seen in whole follicles is probably due to thecal desensitization.

R. J. RYAN: Thank you, Dr. Marsh. I am now supposed to summarize this discussion but, before doing so, I would like to add some additional information.

If you examine electron microscopic autoradiographs of luteal cells obtained from rats 7 days after PMSG-hCG priming and 6 to 24 hours after intravenous injection of 5 μg of $^{125}$I-hCG, you will note several interesting phenomena. First, the luteal cell has two faces. Along the capillary surface, there are numerous microvilli which contain microfilaments. The basolateral surfaces are smooth and regular and frequently have gap junctions as part of their abutment with adjacent luteal cells. The second phenomenon is that the majority (80-90%) of plasma membrane-associated silver grains, presumed to be $^{125}$I-hCG bound to its receptor, are localized on the microvillous surface. Only a small minority are present on the basolateral surface (Anderson, Kang, Perotti, Bramley and Ryan, Biology of Reproduction, in press).

Two years ago, Tony Bramley from our lab reported at The Endocrine Society's meeting that, following homogenization and sucrose density gradient centrifugation, two plasma membrane fractions could be obtained from rat luteal tissue. The first, with a density of 1.13 contained the majority of hCG binding sites (85%) and no adenylate cyclase activity. The second, with a density of 1.18, contained the minority of binding sites (15%) but all of the cyclase enzyme and both basal and that stimulated by NaF, hCG and $PGE_2$. These data are now in press (Endocr., 103, 1978).

Recently, these membrane fractions have been proven to be of luteal cell origin (Bramley and Ryan, submitted). After further purification and examination by electron microscopy, the light microscopic (d 1.13) membranes were found to be composed of small plasma membrane vesicles, of rather uniform size, that contain microfilaments. Further, disc gels prepared from light membrane extracts have a protein band with the same mobility as actin. The heavy (d 1.18) membrane fraction is composed of large irregularly shaped vesicles of varying size, often including smaller vesicles or with smaller vesicles attached on the outside. Junctional complexes can frequently be noted where vesicles are attached to each other.

We conclude from these studies that there are two classes of hCG receptors. The majority are on the microvillous surface of the cell and are unassociated with the cyclase enzyme and the minority are on the basolateral surface where they may be coupled to the cyclase enzyme. We cannot, however, show differences between these two classes with respect to hormone binding specificity, equilibrium constants, association or dissociation rate constants or Hill coefficients.

These data, and the discussions that we have heard this evening, lead me to conclude that we still have much to learn about the mechanism of action of gonadotropins. We see receptors come and go but we know nothing of their synthesis and precious little about their degradation. We see cyclase activation and desensitization as phenomena, but we still do not know the intervening steps. I think the situation is nicely summarized in the Figure which was prepared by my talented friend and colleague, Tony Bramley.

MECHANISM OF ACTION OF LH-HCG or THE 'SIGNAL-CALLER'S' DILEMMA

R. J. RYAN: During the initial studies on purification, you showed cyclase activity; was that hormonally stimulated or fluoride stimulated cyclase?

M. L. DUFAU: The adenylate cyclase activity shown in the fractionation experiments was maximally stimulated enzyme activity. In order to preserve adenylate cyclase activity, solubilization of the enzyme was carried out with Lubrol PX in the presence of 5 mM sodium fluoride and 5 mM magnesium chloride. When these were omitted, the detergent-solubilized enzyme could be activated by sodium fluoride, but was not responsive to hCG or LH (M. L. Dufau, A. J. Baukal, D. Ryan and K. J. Catt, Mol. and Cell. Endocrinol. 6:253 [1977]).

R. J. RYAN: Are there other questions or comments?

QUESTION: I wasn't sure whether the cyclase was the cyclase transferred along with the LH receptor or the LH receptor was coupled with a different cyclase in the adrenal cell.

M. L. DUFAU: The lipid-associated receptor used in the transfer studies did not possess adenylate cyclase activity. We incorporated the spontaneously soluble ovarian receptors into adrenal cells containing adenylate cyclase, and coupled responses occurred during binding of gonadotropin to the transferred LH receptors in the adrenal cells.

L. E. REICHERT: Have you been able to re-insert the receptor into the ovarian membrane and re-establish hormone synthesis, after you remove the receptor by solubilization?

M. L. DUFAU: This was not the purpose of our experiments. To return receptors to cells and look for restoration of hormonal stimuli and consequent metabolic events, it will be necessary to use cells that have lost receptors. Selective loss of receptors from target cells can be accomplished by controlled proteolytic enzyme treatment, as reported in the fat cell. However, luteal and Leydig cells are very susceptible to proteolytic enzymes, which cause impairment of metabolic functions and cell death. Another way to carry out such studies would be to use desensitized cells. However, even if it would be possible to restore receptors, one would not necessarily expect to restore the post-receptor activities due to the steroidogenic lesion that occur in desensitized Leydig and luteal cells (S. B. Cigorraga, M. L. Dufau and K. J. Catt, J. Biol. Chem. 253:4297 [1978]).

H. G. M. RAJ: We initially demonstrated that LH is luteotropic in the pregnant rat (Raj and Moudgal, Endocrinology 86:874, 1970). In these studies, anti-LH, given at any time between days 8 and 11 of pregnancy, caused abortion in the rat. We have been intrigued as to the role of LH and prolactin in luteal maintenance since our initial demonstration. In our recent investigations, we have examined the effects of maintaining the feto-placental unit by dydrogesterone (a synthetic progestin), on luteal functionality. Rats were given LH antiserum or LH antiserum + dydrogesterone from day 8 through day 14 and autopsied. LH antiserum treatment leads to abortion and lowering of ovarian weight. Luteal weights, progesterone content, as well as concentration, and plasma progesterone levels, were lowered drastically with this treatment. All these effects of the LH antiserum were fully reversed when the feto-placental unit was maintained by the synthetic progestin. (See our detailed article in this book). The decline in luteal progesterone production in the antiserum-treated group was associated with a decline in luteal estrogen content and concentration.

Furthermore, the maintenance of feto-placental unit and luteal progesterone production was associated with elevated intraluteal estradiol levels. Thus, if you maintain the feto-placental unit, the estrogen also is maintained, so this indeed supports the involvement of estrogen in the luteal regulation and production of progesterone in the pregnant corpus luteum.

G. GIBORI: The decidual tissue has been shown to secrete some luteotropic hormones in pseudopregnant animals. We still don't know what it's doing in the pregnant animals; however, its role has not been carefully examined.

H. G. M. RAJ: Yes, this could be. We cannot exclude that hCG-like material coming out of this feto-placental unit could in turn be involved in luteal maintenance.

G. GIBORI: I don't know, because the evidence shows that on the eighth day of pregnancy there is no hCG-like activity in the placenta of the rat; it starts around Day 10. So I tend more to think that the decidual tissue is involved.

H. G. M. RAJ: I would still not exclude the possibility. Also, the observations of Morishige and Rothchild that ergocornine is not effective in terminating pregnancy on or after Day 8 suggests that there could be a placental prolactin-like material being put out at that time even though you may not detect it in serum.

G. GIBORI: Let me answer the question differently. To maintain progesterone synthesis between Days 8-12 of pregnancy, estrogen is needed. Treatment with estrogen or with an androgen which is converted to estrogen by the luteal tissue, will maintain progesterone synthesis and pregnancy in rats hypophysectomized or treated with LH antiserum between Days 8-12. It seems that the important factor is the availability of androgen which could be produced not only by the ovaries but also by the adrenals.

H. G. M. RAJ: I think you are getting away from the prolactin end of the story. What I am trying to say is that, if you maintain the feto-placental unit in the LH antiserum-treated rat with a synthetic progesterone, the placenta has the potential (as opposed to what happens in the absence of the progestin, where the placenta is aborted with LH antiserum) of maintaining luteal function in the face of LH deprivation by LH antiserum .

G. GIBORI: Yes, it has been shown by Alloiteau (1956) that, after hypophysectomy on Day 6, treatment with progesterone from Days 6-12 allowed pregnancy to continue normally. Because corpora lutea regress in the absence of prolactin, the placenta, probably the decidual tissue of the placenta, keeps the CL sensitive to reactivation by rat placental lactogen.

R. J. RYAN: Dr. Rothchild wants to make a concluding comment, so we can go to cocktails.

I. ROTHCHILD: If you will permit me an interpretation of what Dr. Gibori said, in response to Dr. Raj: -- Even if the rat was not pregnant, if the LH antiserum effect was overcome by the maintenance of the uterine decidual tissue, the decidual tissue alone (in the presence of the prolactin that would still be secreted) would be enough to maintain the corpus luteum.

G. GIBORI: Thank you for the nice translation.

I. ROTHCHILD: It's not necessary for several hormones at this point to be involved. It may be, but it is not necessary.

R. J. RYAN: Thank you all very much.

# LATERAL MOBILITY AND INTERNALIZATION OF HORMONE RECEPTORS TO HUMAN CHORIONIC GONADOTROPIN IN CULTURED RAT GRANULOSA CELLS

A. Amsterdam, F. Kohen, A. Nimrod and H. R. Lindner

Department of Hormone Research, The Weizmann Institute of Science, Rehovot, Israel

Since the introduction by Singer and Nicholson (1) of the concept of membrane fluidity and lateral mobility of membrane proteins, such mobility has been demonstrated for several membrane proteins, including receptors to immunoglobulins (2), lectins (3), catecholamines (4) and acetylcholine (5,6). Moreover, it was recently suggested that receptors on the plasma membrane can be internalized after binding specific ligands (7,8). In this study we have shown by high resolution autoradiography that clusters of receptor-bound human chorionic gonadotropin (hCG) are formed on the cell membrane of cultured granulosa cells, and that the bound hormone is subsequently internalized into lysosomes where degradation may occur. Lateral movement of receptor-bound hormone to form clusters, patches and caps following association with antibody to the hormone was demonstrated using the indirect immunofluorescence technique.

## LOCALIZATION OF RECEPTOR-BOUND $^{125}$I-hCG BY HIGH RESOLUTION AUTORADIOGRAPHY

When cultured rat granulosa cells, obtained from preovulatory follicles on the day of proestrus, were incubated for 2-3 h with $^{125}$I-hCG (2 IU/ml; 30 μCi/μg) at 37$^{o}$C, 70-80% of the bound label was found to be associated with the cell membrane (Fig. 1). Most of the label was found to be randomly distributed on the cell surface, but a significant portion (20-30%) was aggregated in clusters (Figs. 1,2). These clusters were often associated with invaginations of the cell membrane, and filamentous material was often observed in close vicinity to such invaginations. Such areas were devoid of ribosomes (Fig. 2). At this time, 6-7% of the radioactivity was found in intracellular vesicles (Fig. 3) and in lysosome-like structures (11-19%). Internalization was more extensive in cells that had been incubated for additonal 5 h either in hormone-free medium or in the presence of 20 IU/ml of unlabeled hormone. At this stage, only 26% of the label was found associated with the cell membrane while the rest was associated with intracellular vesicles beneath the cell membranes (14%) and with lysosome-like structures (50%, Fig. 4). Similar internalization of $^{125}$I-hCG was observed by Chen *et al.* (9) in corpora lutea cells *in vivo*. A fraction of the label was found within nuclei (6%; Fig. 5) and in the Golgi complex (2%; Fig. 6).

It has previously been shown (10) that prolonged exposure of ovarian follicular tissue to LH induces refractoriness, so that the cells do not respond to further hormonal challenge with cAMP accumulation. In our cell cultures, it was found that almost complete desensitization to the hormone was achieved within the first 3 h (data not shown), suggesting that desensitization precedes internalization. This suggestion is supported by the findings of Lamprecht *et al.* (10) that desensitization to LH in cultured rat Graafian follicles is initially not accompanied by loss of binding sites.

The internalized hormone in granulosa cells is probably degraded in lysosomes, since after 24 h 80% of the radioactivity initially bound is released into the medium and most of this radioactivity is recoverable as free mono- and diiodotyrosine. The fate of the receptor molecule is not yet clear. It was recently reported that shortly after internalization of the bound hormone in corpus luteum hormone-receptor complex can be recovered from cell homogenates (11). Whether the receptor molecules are finally degraded or can be reutilized remains to be seen. Our observation that some of the internalized radioactivity is associated with the Golgi complex points to the latter possibility, since the Golgi apparatus has been implicated in membrane recirculation (12). The internalized label found in the nuclei at late stages (8 h of incubation) may represent products of hormone catabolism rather than biologically active hormone.

## MOBILITY OF hCG RECEPTORS VISUALIZED BY IMMUNOFLUORESCENCE

When cultured granulosa cells were incubated for 90-120 min with unlabeled hCG (20 IU/ml) and subsequently exposed to specific anti-hCG IgG raised in rabbit (1 mg/ml) followed by goat anti-rabbit IgG conjugated to fluorescein, lateral mobility of receptors on the cell membrane could be visualized by fluorescence microscopy (13). Cells fixed with 2% formaldehyde (30 min at 25$^{\circ}$C) after initial incubation with the hormone for 90 min showed fluorescence distributed along their entire circumference (Fig. 7), though this fluorescent band was often broken up into fine, discrete dots, and not all cells

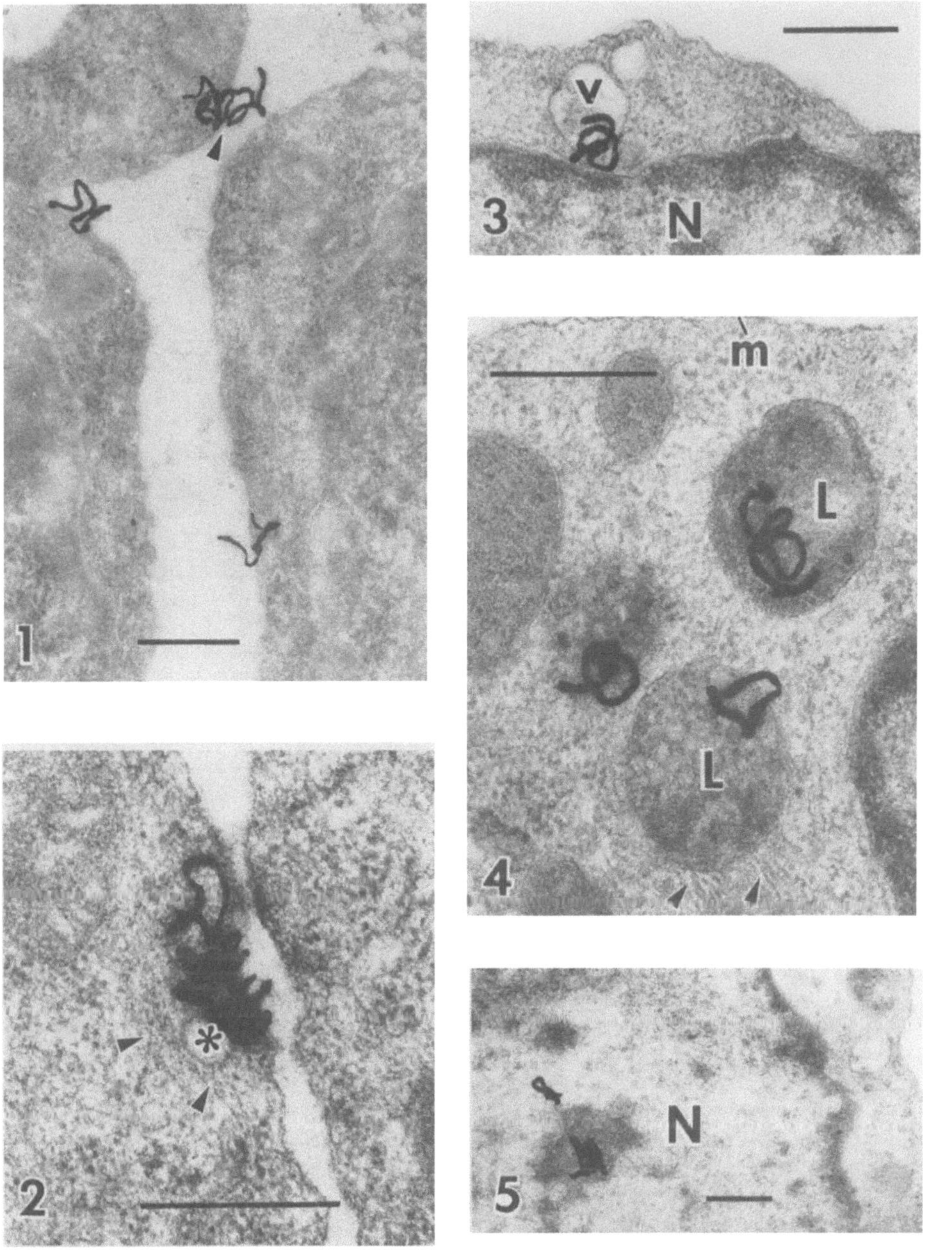


Figs. 1-6. Electron microscope autoradiographs of thin sections of cultured granulosa cells incubated for 3 h with $^{125}$I-hCG (2 IU/ml; 30 µCi/µg; Figs. 1-3), and subsequently for 5 h with 20 IU/ml unlabeled hormone (Figs. 4-6). Cells were fixed with 1% formaldehyde-3% glutaraldehyde followed by 1% $OsO_4$ in 0.1 M

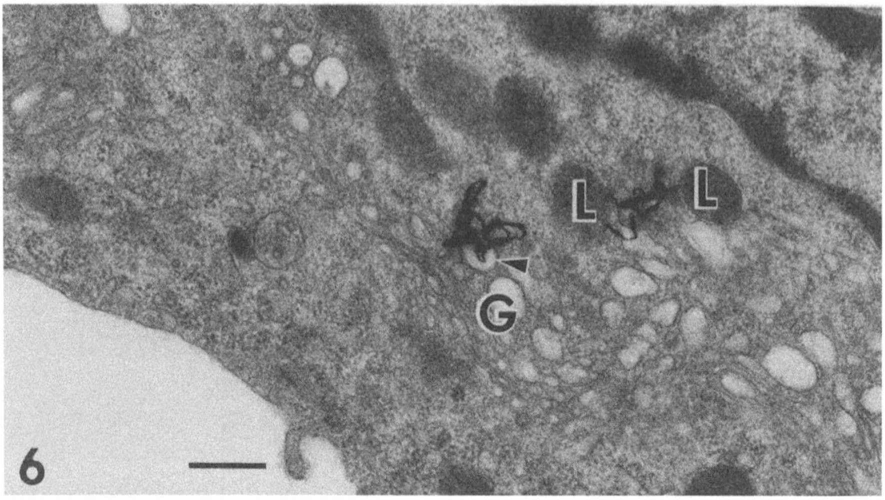


cacodylate buffer, pH 7.4, and processed for autoradiography (14). The photographic emulsion was exposed for 60 days. Calibration = 0.5 µm (Amsterdam, Nimrod, Lamprecht, Burstein and Lindner, in preparation).

Fig. 1. Distribution of silver grains on the cell membrane of granulosa cells. Arrowhead indicates cluster.

Fig. 2. Cluster of grains associated with cell membrane invagination (star). Note filamentous material in close vicinity (arrowheads).

Fig. 3. Part of granulosa cell. Label over intracellular vesicle (v) beneath the cell membrane. N, nucleus.

Fig. 4. Silver grains overlying 3 lysosome-like structures (L): the cell membrane (m) is devoid of labeling. Filamentous material is associated with one of the lysosomes (arrowheads).

Fig. 5. Silver grains over the nucleus of a granulosa cell.

Fig. 6. Golgi complex (G) of granulosa cell showing silver grains over electron-translucent vesicle (arrowhead). Another cluster of grains seems to be associated with lysosome-like structures (L).

were labeled. When the fixation step was omitted and the cells incubated with antibody, clustering of receptors was evident within 30 min - 2 h and large patches as well as caps were observed within 2-3 h (Figs. 8,9). Internalization of the fluorescent probes was evident after 4 h incubation (data not shown). Failure of some of the cells to bind the fluorescent label is in line with our observation that the density of receptors for hCG varies between different granulosa cell layers *in vivo* (15).

The immunofluorescence data suggest that the hormone receptor can move in the plane of the membrane and aggregate to form patches

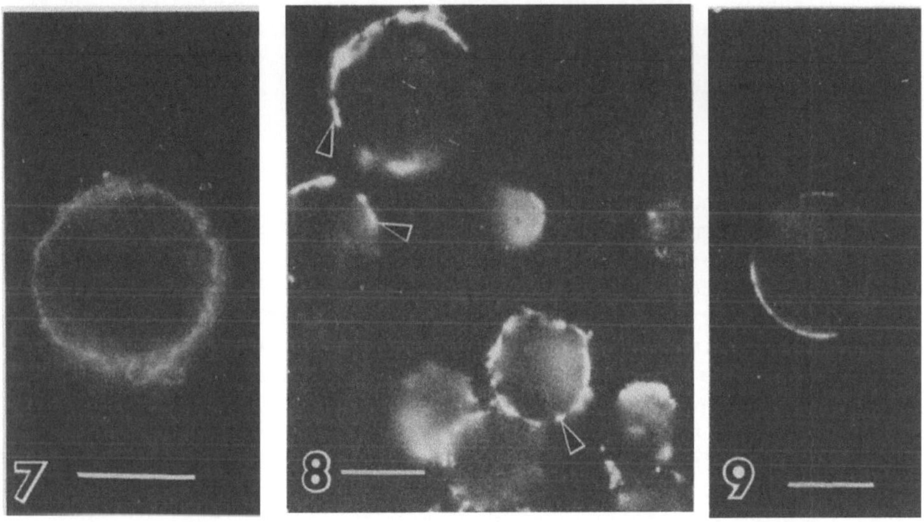


Figs. 7-9. Localization of receptor-bound hCG in granulosa cell cultures by indirect immunofluorescence. Cells were incubated with 20 IU/ml hCG for 90-120 min and subsequently with antibodies to hCG raised in rabbits (IgG fraction, 1.0 mg/ml) for 30 min and with fluorescein-conjugated goat anti-rabbit IgG (1.0 mg/ml) for 30 min at $37^{o}C$. Calibration = 10 μm (Amsterdam, Kohen, Lamprecht, Nimrod and Lindner, in preparation).

Fig. 7. Cells were fixed (2% formaldehyde for 30 min) after 90 min incubation with the hormone and subsequently incubated with the antibodies. A fluorescent ring is seen at the circumference of the cell.
Fig. 8. Unfixed cultures were treated with both antibodies and then incubated in antibody-free medium for 1 h. Discrete patches of fluorescence are evident at the circumference of the cells (arrowheads).
Fig. 9. Treatment as for Figure 8, but incubation in antibody-free medium continued for 3 h. Granulosa cell shows capping of the fluorescent probe.

and caps in the presence of antibodies to the receptor-bound hormone. However, cluster formation, though not unequivocal capping, was evident also in the absence of antibodies, as shown in our autoradiographic studies (Figs. 1,2).

Within 30 min of adding anti-hCG IgG to cells treated with the hormone, no further accumulation of cyclic AMP occurred, even in response to fresh hormone. Cells treated with nonspecific IgG continued to accumulate this nucleotide. Thus, aggregation of receptor-hormone complexes and consequent restriction of their mobility, possibly enhanced by the antibody, may interfere with receptor function, which requires coupling to and activation of hormone-sensitive adenylate cyclase.

## CONCLUDING REMARKS

Mobility of hCG receptors on granulosa cells was visualized by the indirect immunofluorescence technique. With the aid of specific antibodies to the bound hormone, patching and capping of hormone receptor complexes was demonstrated. High resolution autoradiography revealed the existence of clusters of receptor to hCG which can be internalized, resulting in hormone degradation within lysosomes. The fate of the receptor molecule is not yet clear, particularly the question whether the receptor molecule can be used for only one cycle of stimulation or be reused. Desensitization of the cell to the hormone seems to precede hormone and receptor internalization. Therefore, the primary mechanism of the desensitization phenomenon should be sought at the level of the cell membrane, whereas the down-regulation of receptor density may be due to a later, intracellular process. Restriction of receptor mobility within the membrane should be considered as one possible mechanism underlying the induction of desensitization.

## ACKNOWLEDGMENT

We are grateful to Mrs. A. Azrad, Mrs. E. Epstein, Mr. S. Gordon and Mr. Z. Hollander for skilful technical assistance. Supported by grants from the United States-Israel Binational Foundation, Jerusalem, Israel (to A.A.) and from the Ford Foundation and Population Council N.Y. (to H.R.L.). A.A. is the incumbent of the Gestetner Career Development Chair. H.R.L. is the Adlai E. Stevenson 111, Professor of Endocrinology and Reproductive Biology at the Weizmann Institute of Science.

## REFERENCES

(1) S.J. Singer and G.L. Nicolson. Science 7(1972)720.

(2) S. de Petris and M. Raff. Eur.J.Immunol. 2(1972)523.

(3) G.M. Edelman. Science 192(1976)318.

(4) J. Orly and M. Schramm. Proc.Nat.Acad.Sci.U.S.A. 73(1976)4410.

(5) J. Prives, I. Silman and A. Amsterdam. Cell 7(1976)543.

(6) J. Prives, L. Hoffman, R. Tarab-Hazdai, S. Fuchs and A. Amsterdam, in preparation.

(7) G. Carpenter and S. Cohen. J.Cell.Biol. 71(1976)159.

(8) M.S. Brown and J.L. Goldstein. Science 191(1976)150.

(9) T.T. Chen, J.H. Abel, Jr., M.I. McCellan, H.R. Sawyer, M.A. Diekman and G.D. Niswender. Cytobiologie. 14(1977)412.

(10) S.A. Lamprecht, U. Zor, Y. Salomon, Y. Koch, K. Ahrén and H.R. Lindner. J.Cycl.Nucl.Res. 3(1977)69.

(11) P.M. Conn, M. Conti, J.P. Harwood, M.L. Dufau and K.J. Catt. J.Cell.Biol. 75(1977)192a.

(12) G.E. Palade. In: "Subcellular Particles" ed. T. Hayashi (Donald Press, New York). 1959, pp. 64-80.

(13) A. Amsterdam, Z. Hollander, A. Nimrod, R. Reisel and F. Kohen. J.Cell.Biol. 75(1977)222a.

(14) A. Amsterdam and J.D. Jamieson. J.Cell.Biol. 63(1974)1057.

(15) A. Amsterdam, Y. Koch, M.E. Lieberman and H.R. Lindner. J.Cell. Biol. 67(1975)894.

FSH STIMULATION OF MUCOPOLYSACCHARIDE SYNTHESIS IN RAT AND PORCINE OVARY

R. L. Ax, A. R. LaBarbera and R. J. Ryan

Mayo Clinic and Foundation

Rochester, Minnesota 55901

## INTRODUCTION

Incorporation of mucopolysaccharides (MPS) into follicular fluid occurs during antrum formation in follicular development. Staining of follicular fluid with toluidine blue or autoradiographic localization of $^{35}S$ showed that concentrations of MPS varied during the ovulatory cycle (3,9,10,11). FSH stimulated *in vivo* uptake of $^{35}S$ into ovarian follicular MPS in hypophysectomized immature female rats and *in vitro* uptake of $^{35}S$ by rat granulosa cells (8). Conversely, a decrease of *in vitro* incorporation of $^{35}S$ by rat ovarian slices was evoked by either LH or progesterone (4), implying antagonistic functions of FSH and LH on MPS accumulation. The physiological significance of MPS on follicular and oocyte maturation and on ovulation remains to be resolved. This report confirms FSH modulation of incorporation of MPS into follicular fluid *in vivo* and *in vitro* and contains data showing quantitative changes in the concentrations of MPS in porcine follicular fluid.

## MATERIALS AND METHODS

### Incorporation of $^{35}S$ into Rat Ovaries in vivo:

At 26 days of age female rats were injected with 1.0 mg DES/rat/day for four days, followed by injections of either graded doses of FSH for three days or PBS-BSA vehicle (controls). In the first experiment, half the rats in each treatment regimen were injected with 10 µCi $Na_2{}^{35}SO_4$ concurrent with the last injection of FSH. The remaining rats were injected with 5 µCi $Na_2{}^{35}SO_4$ with both the second and third injections of FSH (dual pulse). All rats were killed 18 hours after the last injection of FSH. Ovaries were excised, weighed, digested with NCS tissue solubilizer (Amersham), and $^{35}S$ was measured.

In the second *in vivo* experiment, the procedure was identical to the first except that (a) all rats received a dual pulse of $^{35}S$ and (b) only one ovary from each rat was digested for measurement of $^{35}S$. Contralateral ovaries were homogenized, centrifuged, and the supernatants chromatographed on a 22 X 0.7 cm Bio-Gel p-10 gel filtration column to separate $^{35}S$ and $^{35}S$ incorporated into macromolecules. Protein was determined according to Lowry et al. (7).

Incorporation of $^{35}S$ into Macromolecules by Porcine Granulosa Cells *in vitro*:

Ovaries were removed from 3-6 month old pigs at slaughter (Hormel Co., Austin, MN) and transported to the laboratory within 2 hrs after removal. Granulosa cells were aspirated aseptically from small follicles and cultured for 9 days in Medium 199 + 10% fetal calf serum + $10^{-7}$M dexamethasone + 10 mIU/ml insulin + 50 μg/ml gentamicin. Cells ($10^6$/ml) were cultured with no additional hormones or with 100 nM DES, 1 μg/ml NIH-FSH-S9 or DES + FSH at the same concentrations. Hormones were added daily. On day seven, 500 μCi of $Na_2{}^{35}SO_4$ were added to each culture. Cultures were terminated 48 hours later, centrifuged, thrice washed, and the cells sonicated. Cytosol was prepared by centrifuging the sonicates at 105,000 x g for 1 hr. Culture media and cytosol were chromatographed as described above for the *in vivo* experiments. DNA from each culture was measured (6).

High Pressure Liquid Chromatography (HPLC) of MPS in Porcine Follicular Fluid:

Follicles on porcine ovaries were classified as small (1-2 mm dia), medium (3-5 mm dia), or large (5+ mm dia) and aspirated. Cells and cellular debris were (a) removed by centrifugation, or (b) not removed. Samples were flash frozen and stored at -20°C. Hyaluronic acid, chondroitin sulfates A, B and C and heparin (all Sigma) were chromatographed as standards, and the areas of the peaks were integrated. The methodology for isolation of MPS from follicular fluid and subsequent HPLC will appear elsewhere (1).

## RESULTS

Incorporation of $^{35}S$ into rat ovaries increased in response to FSH, and the increase was significant in rats which received a dual pulse of isotope (Table 1).

When rats were injected with graded doses of FSH, all doses of FSH increased the uptake of $^{35}S$ by rat ovaries, and the increases were significant with the 100 μg and 200 μg doses of NIH-FSH-B1 (Table 2).

Table 1

dpm $^{35}S$/mg Ovary ($\bar{X} \pm$ SEM)

| Treatment | Single Pulse $^{35}S$ (10 μCi Total) | Dual Pulse $^{35}S$ (10 μCi Total) |
|---|---|---|
| Control | 48.6 ± 4.1 | 36.6 ± 2.9 |
| 100 μg NIH-FSH-S9 X 3 days | 55.9 ± 7.1 | 46.9 ± 3.7 ($p < 0.01$) |

Table 2

dpm $^{35}S$/mg Ovary ($\bar{X} \pm$ SEM)

| Treatment | Dual Pulse $^{35}S$ (10 μCi Total) |
|---|---|
| Control | 59.7 ± 5.48 |
| 50 μg NIH-FSH-B1 X 3 days | 67.7 ± 8.16 |
| 100 μg NIH-FSH-B1 X 3 days | 69.6 ± 4.42 ($p < 0.02$) |
| 200 μg NIH-FSH-B1 X 3 days | 71.2 ± 5.27 ($p < 0.01$) |

The increased total uptake of $^{35}S$ noted with increasing doses of FSH was not linear. However, comparisons of void volume fractions eluted from the gel filtration columns showed a linear log-dose change in $^{35}S$ incorporated into macromolecules in the ovaries over the range of doses injected (Figure 1).

Results of the <u>in vitro</u> cultures of porcine granulosa cells showed that radioactivity incorporated into macromolecules was located in the cytosol as well as the culture media, suggesting secretory products from the granulosa cells. Using the specific activity of 1 Ci/mmole $^{35}S$, a MW of 50,000 for the MPS, and assuming molar ratios of one sulfur:one hexosamine residue, concentrations of MPS in the culture media were calculated (Figure 2). DES markedly suppressed the production of MPS by granulosa cells compared to other culture conditions.

HPLC was used to quantitate MPS in porcine follicular fluid. For all MPS standards there was a linear relationship between mass applied to the HPLC column and the mass eluted. Retention times for each standard did not vary by more than 2% for all experiments.

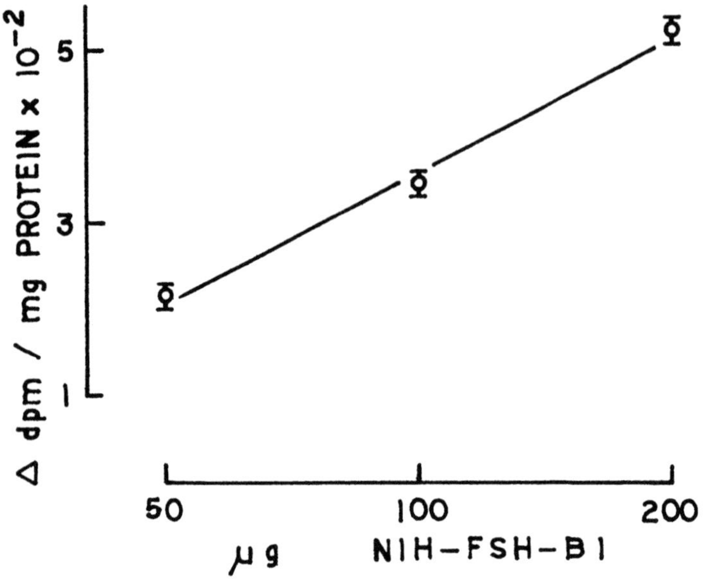


Figure 1: $^{35}S$ incorporated into macromolecules (void volume on Bio-Gel P-10) in rat ovarian supernatants. Difference between control dpm and FSH-treated dpm.

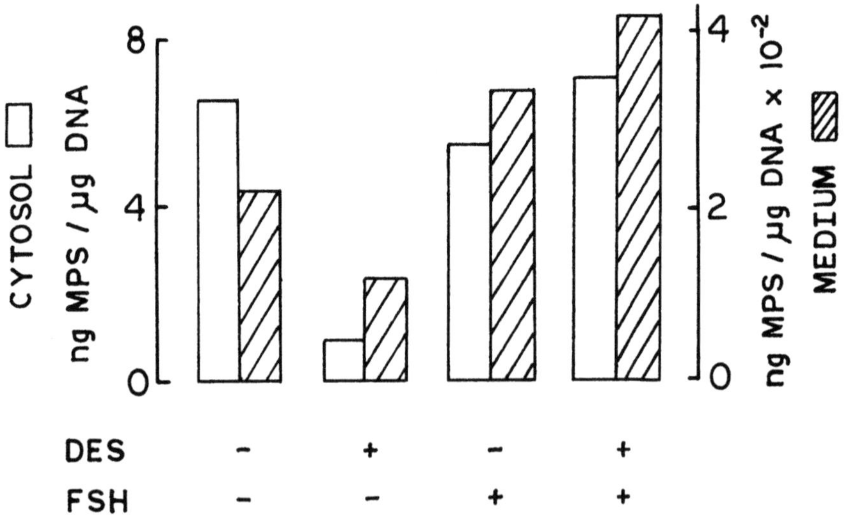


Figure 2: $^{35}S$-labeled mucopolysaccharides present in cultures of porcine granulosa cells.

The MPS in porcine follicular fluid detected using HPLC were hyaluronic acid and chondroitin sulfates, but not heparin (Table 3), in agreement with constituents of bovine follicular fluid (5), and rat ovarian homogenates (8), but in discordance with MPS produced by rat ovarian slices *in vitro* (4).

Table 3

Amounts of MPS Present as Detected by HPLC of 100 μl of Follicular Fluid from Porcine Ovaries

| Size of Follicle | Area eluted at RT's 3.2 - 3.3 Mean ± SEM, n = 3 (Hyaluronic acid - Chondroitin sulfates) | Concentration (μg/ml) |
|---|---|---|
| Small, fluid + cells | 50434 ± 859 | 2179 |
| Medium, fluid + cells | 10909 ± 256 | 471 |
| Large, fluid + cells | 8470 ± 120 | 366 |
| Small, fluid only | 13420 ± 208 | 580 |
| Large, fluid only | 8811 ± 95 | 381 |
| Normal pig serum | Undetectable | |

A greater concentration of MPS was observed in follicular fluid from small follicles containing granulosa cells when compared to fluid not containing granulosa. Contrastingly, little difference was noted in concentrations of MPS in follicular fluid from large follicles with or without granulosa, suggesting changes in synthesis and/or secretion of MPS by granulosa cells during follicular maturation.

## DISCUSSION

Our results substantiate accumulation in follicular fluid of MPS which appear to be secretory products from granulosa cells. Incorporation of $^{35}S$ into macromolecular substances in rat ovaries was linear with respect to the log-dose of FSH in agreement with previous *in vivo* experiments (8). Cytosol and medium from cultured porcine granulosa cells also contained macromolecules which had incorporated $^{35}S$ and FSH or FSH plus DES increased the amount of MPS secreted compared to cells cultured in the absence of hormones.

Incubations with DES alone caused a marked reduction in amounts of MPS found in cytosol of medium from porcine granulosa cells, but the concentration of MPS secreted in relation to the concentration in cytosol was greater with DES than controls, FSH, or DES + FSH administered to cells.

The changes in concentrations of MPS in porcine follicular fluid decreased with follicular maturation, in contrast to most other constituents assayed in porcine follicular fluid (2). If granulosa cells and follicular fluid were assayed for MPS, concentrations of MPS were much greater if granulosa cells remained in fluid from small follicles. This phenomenon was not observed in large follicles.

It was reported that LH or progesterone decreased the uptake of $^{35}S$ into heparin-like substances by rat ovarian slices *in vitro* (4). Heparin has not been identified as a major component of MPS in response to FSH by rat ovaries, but differences in gonadotropic action may account for changes in composition, synthesis, and secretion of MPS by granulosa cells in response to FSH or LH. Since follicular concentrations of MPS do not parallel concentrations of other constituents during follicular growth and appear to be regulated by gonadotropins and gonadal steroids, MPS present a new approach to study follicular maturation, ovulation, and mechanisms of gonadotropic action.

## REFERENCES

(1) Ax, R. L. and R. J. Ryan. Manuscript in preparation.

(2) Chang, S. C. S., J. D. Jones, R. D. Ellefson, and R. J. Ryan. Biol. Reprod. 15(1976)321.

(3) Dziewiatkowski, D. D. Int. Rev. Cyt. 7(1958)159.

(4) Gebauer, H., H. R. Lindner, and A. Amsterdam. Biol. Reprod. 18(1978)350.

(5) Jensen, C. E. and F. Zachariae. Acta Endocr. 27(1958)356.

(6) Kissane, J. M. and E. Robins. J. Biol. Chem. 233(1958)184.

(7) Lowry, O. H., N. J. Rosebrough, A. L. Farr, and R. J. Randall. J. Biol. Chem. 193(1951)265.

(8) Mueller, P. L., J. R. Schreiber, A. W. Lucky, J. D. Schulman, D. Rodbard, and G. T. Ross. Endocrinology 102(1978)824.

(9) Oldeblad, E. Acta Endocr. 11(1952)269.

(10) Oldeblad, E. Acta Endocr. 15(1954)313.

(11) Zachariae, F. and C. E. Jensen. Acta Endocr. 27(1958)343.

# IMMUNOLOGIC IMPAIRMENT OF $\Delta^5$-3B HYDROXYSTEROID DEHYDROGENASE AND ITS EFFECT ON OVARIAN FUNCTION: THE USE OF A PURIFIED ENZYME

Shawky Badawy, M.D., M.ch., F.A.C.O.G.*

William M. O'Leary, Ph.D.**

Kenneth A. Laurence, Ph.D.***

*Department of Obstetrics and Gynecology, Upstate Medical Center, Syracuse, New York 13210.

**Department of Microbiology, Cornell Medical Center, New York, New York 10021.

***Department of Biological Sciences, University of Idaho, Moocow, Idaho 83843

This work was done at the Biomedical Division of the Population Council, Rockefeller University, New York, New York 10021.

## ABSTRACT

The effects of actively immunizing rats with a purified preparation of $\Delta^5$-3B-OHSDH are reported. The following parameters were examined: (1) vaginal cytology; (2) pregnancy; (3) antibody titres; (4) $\Delta^5$-3B-OHSDH activity in the ovary by (a) histochemical, (b) biochemical methods; (5) 20$\alpha$-OHSDH activity in the ovary determined histochemically; (6) progesterone levels in plasma.

The enzyme was prepared by the growth of pseudomonas testosteroni ATCC 11996 in medium containing testosterone, and purified according to the methods described by Talalay(1962),( 2 ) and Dixon and Webb(1964),( 3 ).

By six-seven weeks following the completion of immunization, 80% of the rats were anestrous and 20% showed irregularity of the cycles. The $\Delta^5$-3B-OHSDH activity, as demonstrated histochemically, showed marked reduction in both the interstitial tissue and the corpora lutea in the ovaries of immunized rats. This was confirmed by the biochemical assay. The enzyme activity was found to be 13.9 $\pm$ 1.6 and 7.4 $\pm$ 1.3 ug androstenedione/mgm ovarian weight/minute in the control and immunized group respectively. The difference is highly significant ($P < 0.005$). The plasma progesterone levels dropped in the immunized group in comparison with the controls. The drop was statistically significant during the estrus and the first diestrus day ($P < 0.05$). Although mating occurred, the majority showed failure of implantation or various degrees of resorption in those animals that demonstrated high levels of antibody titers.

This work supports our previous findings using the crude enzyme and, in addition, it indicates that the effects are not due to impurities as $\alpha$-hydroxysteroid dehydrogenase or cell debris, both of which have been eliminated by the purification process. It also suggests that purification of the enzyme did not interfere with its immunologic activity.

Active immunization using a purified $\Delta^5$-3B-OHSDH demonstrated that the interstitial tissue and corpus luteum are important sites for the enzyme activity and also for the process of steroidogenesis in the ovary. By immunologic suppression of the enzyme activity in the ovary, there resulted interference with the normal cycles and interference with implantations. This was shown to be due to the reduction in the progesterone levels but it might also suggest a reduction in the estrogen surge necessary for implantation. During normal pregnancy there is an increase in the $\Delta^5$-3B-OHSDH activity with a reduction in 20 $\alpha$-OHSDH activity in the corpora lutea. In the animals that showed resorption or absence of implantations there was a reduction in $\Delta^5$-3B-OHSDH activity and an increase in 20$\alpha$-OHSDH. This is in support of the work of others that proved the association of 20$\alpha$-OHSDH with luteal regression. Since 20 $\alpha$-OHSDH mediates catabolism of progesterone to 20$\alpha$-dihydroprogesterone, these changes in the ovarian enzyme activities following immunization with $\Delta^5$-3B-OHSDH might suggest an upset of the ratio of progesterone: 20$\alpha$-dihydroprogesterone with failure to maintain the pregnancy.

## INTRODUCTION

Badawy and Laurence(1975),( 1 ), reported on the interruption of pregnancy in the female rat following immunization with a crude type of $\Delta^5$-3B-hydroxysteroid dehydrogenase. The relationship between LH,

$\Delta$ 5-3B hydroxysteroid dehydrogenase, 20$\alpha$-hydroxysteroid dehydrogenase, progesterone production and pregnancy maintenance in the female rat, led us to examine the effect of active immunization with this steroidogenic enzyme, on these various biologic phenomena. The present work reports on the purification of $\Delta^5$-3B hydroxysteroid dehydrogenase and its use for immunologic suppression of the ovarian enzyme.

## MATERIALS AND METHODS

The purified enzyme was prepared by the growth of pseudomonas testosteroni ATCC 11996 in medium containing testosterone. The enzyme was recovered by modification of the methods described by Talalay(1962),(2) and Dixon and Webb(1964),(3). Each animal was immunized intradermally three times at weekly intervals with 1.0 ml of the adjuvant antigen emulsion or with the saline adjuvant emulsion in the experimental and controls respectively. One hundred normal adult female rats were selected on the basis of four to five day estrous cycle for a period of four weeks prior to the study. During and following immunization, daily vaginal smears were examined. Six weeks following the completion of immunization some of the animals were laparotomized under ether anesthesia. The ovaries were dissected and quick frozen at -20$^o$C. The animals were also bled by cardiac puncture and the plasma was separated and stored at -20$^o$C. until processed for the antibody titres and progesterone levels. Some of the animals were mated six-seven weeks following the last injection of the immunogen. Laparotomy was performed on the controls and experimental animals on day 9 of the pregnancy. The ovaries were dissected and frozen at -20$^o$C. until processed for the enzyme assay. The antibody titer was determined by the hemagglutination test of Boyden(1951),(4). The histochemical studies of $\Delta^5$-3B-OHSDH and 20$\alpha$-OHSDH in the ovaries were carried out by the method of Levy et al(1959),(5), and Balogh(1964),(6),respectively, with modifications (Loewit and Laurence, 1969),(7). The quantitative assay of $\Delta^5$-3B-OHSDH in the ovaries followed the method of Rubin et al(1961) (8). Plasma progesterone levels were determined by the competitive protein binding analysis of Murphy(1967),(9), and modifications by Neil et al(1967),(10).

## RESULTS

Vaginal cytology. By six-seven weeks almost 80 per cent of the immunized female rats became anestrous, the remaining 20 per cent continued with irregular cycles.

Antibody titer. The titers were 1:10240 to 20480 and 1:640 to 1280 at antigen concentrations of 0.02 and 0.02 mgm. respectively.

$\Delta^5$-3B-OHSDH activity. In the ovaries of immunized animals, the enzyme activity was markedly suppressed in both the corpora lutea and interstitial tissue. This was verified by the quantitative assay where the enzyme activity was 13.9 $\pm$ S.E. 1.9 (N=22), and 7.4 $\pm$ S.E. 1.3 (N=29) ug androstenedione/mgm ovarian weight/minute in controls and experimental animals respectively (P 0.005).

Plasma progesterone levels. There was a decrease in progesterone levels in the immunized group in comparison with the controls. The difference was significant ($P < 0.05$) during the estrus and first day of diestrus.

Pregnancy. The immunized group showed various results varying from absent implantations to various degrees of resorption. Histochemical studies of the ovarian enzymes revealed a low concentration of $\Delta^5$-3B-OHSDH activity and a marked increase in the activity of 20$\alpha$-OHSDH, compared with the controls.

## DISCUSSION

The purified enzyme used in this study was produced by growth of pseudomonas testosteroni and was tested to be free from $\alpha$- hydroxysteroid dehydrogenase. Immunization of female rats with the purified bacterial enzyme produced higher levels of antibody titres than those previously reported with the crude enzyme (Badawy and Laurence,1975), ( 1 ), and that these antibodies successfully suppressed $\Delta^5$-3B-OHSDH activity in the rat ovaries. It appears that the purification process improved the antigenic potency of this enzyme. In addition, by removal of the cell debris from the medium, we eliminated any nonspecific antigenic reaction. Although there are differences between mammalian (Ford and Engel,1974)(11), and bacterial (Batzold et al, 1975) (12), $\Delta^5$-3B-OHSDH, it is possible that these enzymes are structurally similar at a catalytic site, and antibodies directed against this region of a bacterial protein could inactivate the mammalian enzyme.

By six-seven weeks following the completion of immunization, there was significant reduction in $\Delta^5$-3B-OHSDH activity in both the interstitial tissue and corpora lutea. Quantitatively, the reduction was found to be almost 50 per cent of the control values ($P < 0.005$) as demonstrated by the biochemical studies of the ovaries. This was also associated with a reduction in plasma progesterone levels, thus suggesting a cause-relationship.

During normal pregnancy there is an increase in the activity of $\Delta^5$-3B-OHSDH in the rat ovaries. At the same time the activity of 20$\alpha$-OHSDH decreases rapidly during early pregnancy to remain almost undetectable during most of the gestation period (Loewit, Badawy and Laurence, 1969),(13). The result of such changes in enzyme activity is a decrease in 20$\alpha$-hydroxyprogesterone and an increase in the amount of progesterone (Fajer and Barraclaugh, 1967),(14); Hashimoto and Malampy,1967),(15). In the group of immunized female rats that mated but failed to show implantations or demonstrated a resorption effect, the ovaries demonstrated a marked suppression of $\Delta^5$-3B-OHSDH and an increase in 20$\alpha$-OHSDH activity in comparison with the controls; this would increase 20$\alpha$-hydroxyprogesterone with very low progestational activity.

Since Goldman et al,(1971),(16), have shown that there is an increase in both pituitary content of LH and serum levels of LH following inhibition of the enzyme $\Delta^5$-3B-OHSDH activity with cyanoketone; therefore, the immunologic suppression of $\Delta^5$-3B-OHSDH in the rat ovary, and the subsequent decrease in progesterone levels will result in interruption of the negative feedback mechanism on the hypothalamo-pituitary axis and this will lead to increase in circulating LH levels. These high LH levels will lead to an increase in 20$\alpha$-OHSDH activity in the rat ovary. This might seem to be contradictory to the previous work of Loewit, Badawy and Laurence (1969),(13), in which neutralization of circulating LH with anti-LH serum in pregnant rats produced an increase in the ovarian 20$\alpha$-OHSDH activity. However, it seems that 20$\alpha$-OHSDH activity might be due to very low or very high levels of LH. Rothchild,(1965),(17), suggested that luteolysis due to LH is a dose-dependent process and that a luteotrophic process might involve a synergism between specific ratios of LH and prolactin in the rat. Also, Browning et al,(1962),(18), reported that LH synergizes with prolactin in prolonging the functional life of corpora lutea on intraocular ovarian isografts. In addition, Guillemin et al,(1963),(19), have shown that the effectiveness of LH in depletion of ascorbic acid from lutein tissue, can be restored by exogenous prolactin administration in hypophysectomized rats.

Another aspect of the immunologic suppression of $\Delta^5$-3B-OHSDH is the possible effect on the embryo as the enzyme has been demonstrated in day 4, 5 and 6 embryos by Dey and Dickmann,(1974),(20). In addition, resorption of implantation might be partially related to a direct effect of antibodies on the trophoblastic $\Delta^5$-3B-OHSDH, as the enzyme was demonstrated in the trophoblast by Marcal et al (1975),(21).

REFERENCES

(1) S. Badawy and K.A. Laurence. Fertil. Steril. 26(1975)1198.

(2) P. Talalay, in: Methods of Enzymology, Vol. V, Ed. S.P. Colowick and N.O. Kaplan (Academic Press, New York, 1962) p. 512ff.

(3) M. Dixon and E.C. Webb, in: Enzymes, 2nd Ed. (Academic Press, New York, 1964) p. 39ff.

(4) S.W. Boyden. J. Exp. Med. 93(1951)107.

(5) H. Levy, H.W. Deane and B.L. Rubin. Endocrinology 65(1959)932.

(6) K. Balogh. J. Histochem. Cytochem. 12(1964)670.

(7) K.K. Loewit, K.A. Laurence. Fertil. Steril. 20(1969)679.

(8) B.L. Rubin, G. Leipsner and H.W. Deane. Endocrinology 69 (1961)619.

(9) B.E.P. Murphy. J. Clin. Endocrinol. Metab. 27(1967)973.

(10) J.D. Neill, E.D.B. Johansson, J.K. Datta and E. Knobil. J. Clin. Endocrinol. 27(1967)1167.

(11) H.C. Ford and L.L. Engle. J. Biol. Chem. 249(1974)1363.

(12) F.S. Batzold, A.M. Benson, D.F. Covey, C.H. Robinson and P. Talalay. Adv. Enzyme Regul. 14(1976)243.

(13) K.K. Loewit, S. Badawy and K.A. Laurence. Endocrinology 84 (1969)244.

(14) A.B. Fajer and E.A. Barraclough. Endocrinology 81(1967)617.

(15) I. Hashimoto and R.M. Malampy. Fed. Proc. 26(1967)485.

(16) A.S. Goldman, A.W. Root, D.R. Russ and B.T. Rudd. Proc. Soc. Exp. Biol. Med. 136(1971)672.

(17) I. Rothchild. Acta Endocrinol. 49(1965)107.

(18) H.C. Browning, G.A. Larke and W.D. White. Proc. Soc. Exp. Biol. Med. 111(1962)686.

(19) R. Guillemin and E. Sakiz. Endocrinology 72(1963)813.

(20) S.K. Dey and Z. Dickmann. Endocrinology 95(1974)321.

(21) J.M. Marcal, N.J. Chew, D.S. Solomon and M.I. Sherman. Endocrinology 96(1975)1270.

# RECEPTORS FOR GONADOTROPHINS IN THE OVINE FOLLICLE DURING GROWTH AND ATRESIA.

R.S. Carson, J.K. Findlay and H.G. Burger*

Reproduction Research Section, University of Melbourne, Parkville, 3052 and *Medical Research Centre, Prince Henry's Hospital, Melbourne, 3004, Australia.

## INTRODUCTION

Studies in rats (1,2) have shown that granulosa cells of small follicles acquire the capacity to bind FSH and, in the presence of oestrogen, further follicular development is accompanied by formation of antral fluid and the appearance of receptors for LH on the granulosa. LH binding to these receptors results in a loss of receptor and induces differentiation of granulosa to luteal cells which in the presence of prolactin and LH, will secrete progesterone. In the sheep, follicles may undergo degeneration i.e. atresia at any stage of the oestrous cycle (3) and in all sizes of follicles (4). The relationship between gonadotrophin binding and atresia is not clear, although atresia is known to be associated with loss of LH receptor in rat granulosa cells (5).

A study of individual sheep follicles has enabled us to relate changes in the binding of $^{125}$I-labelled gonadotrophins to follicular growth and atresia.

## MATERIALS AND METHODS

*Tissues:* Ovaries were harvested from adult ewes within 15 minutes of death and placed in 0.15M NaCl (saline) containing 50 U/ml Kanamycin at ambient temperature until dissected. During dissection of follicles from surrounding stromal tissue (6), ovaries and isolated follicles were maintained in incubation medium (5mM HEPES-buffered Medium 199, pH 7.4, containing 4% (v/v) calf serum and 50 U/ml Kanamycin) equilibrated with 95% $O_2$ : 5% $CO_2$.

*Classification:* Follicles were grouped according to size (to the nearest mm) and stage of atresia as assessed morphologically by examination of each follicle using transmitted light at 8 x magnification. The classes of atresia were based upon continuity of the membrana granulosa, the degree of thecal vascularization and the presence of a cumulus-oocyte unit (7,8). Class I represented non-atretic follicles and class V represented follicles in the most advanced stage of atresia. The morphological assessments were confirmed by histological examination of groups of dissected follicles.

*Preparation of $^{125}$I-labelled gonadotrophins:* Human FSH (NIH-FSH-HSI, 4990 IU/mg) and human chorionic gonadotrophin (hCG, CR119, 11600 IU/mg) were iodinated using lactoperoxidase/$H_2O_2$ (9) and chloramine-T (10) respectively, purified on Biogel P-60 (25 x 1 cm), and the specific activity calculated (25-35 μCi/μg). The respective labelled preparations exhibiting maximal specific binding in a bovine testicular membrane receptor assay for FSH (11) and a rat ovarian homogenate receptor assay for LH-hCG (12) were subsequently used in binding studies.

*Incubation:* Groups of follicles (4-20 if possible) were placed in incubation medium (see *Tissues*) to which $^{125}$I-labelled gonadotrophin ($10^{-10}$M) had been added and incubated at 37° under 95% $O_2$ : 5% $CO_2$ with constant shaking. Non-specific binding was determined by the addition of Primantron (FSH) and Pregnyl (hCG) to a final concentration of 40 IU/ml. After incubation (27-36 h) follicles were washed in ice-cold saline and separated into theca, granulosa and follicular fluid fractions by hemisection, centrifugation, and washing at < 4°C. Binding was expressed as cpm specifically bound/mg protein (13).

## RESULTS

*Uptake of $^{125}$I-labelled gonadotrophin by whole follicles:* Equilibration of $^{125}$I-hFSH between incubation medium and follicular fluid required 27 hours whereas 36 hours was required in the case of $^{125}$I-hCG. For both $^{125}$I-labelled hormones, the partition co-efficient at equilibrium was 0.8. Neither the time required for equilibration nor the partition co-efficient were altered in the presence of excess, unlabelled gonadotrophin.

*Binding as a function of follicular diameter:* Specific binding of FSH to granulosa cells was highest (38.1 ± 1.2 cpm x $10^3$/mg) in smallest follicles (< 2 mm) examined and decreased as follicle diameter increased to 22.3 cpm x $10^3$/mg for larger follicles (4-6 mm). Specific binding of $^{125}$I-FSH to theca was limited in comparison to binding to granulosa cells, and increased slightly with follicle diameter. In contrast, specific binding of $^{125}$I-hCG to theca was highest in small (< 2 mm) follicles (84.3 ± 17.9 cpm x $10^3$/mg) and significantly lower in large (4-6 mm) follicles (60.1 ± 15.7 cpm x $10^3$/mg) ($p < 0.05$). There was some binding (5-10 x $10^3$ cpm/mg) of $^{125}$I-hCG to granulosa cells of small follicles (< 2 mm) and a marked increase in binding to granulosa cells of follicles > 4 mm (102 x $10^3$ cpm/mg) (Figure 1).

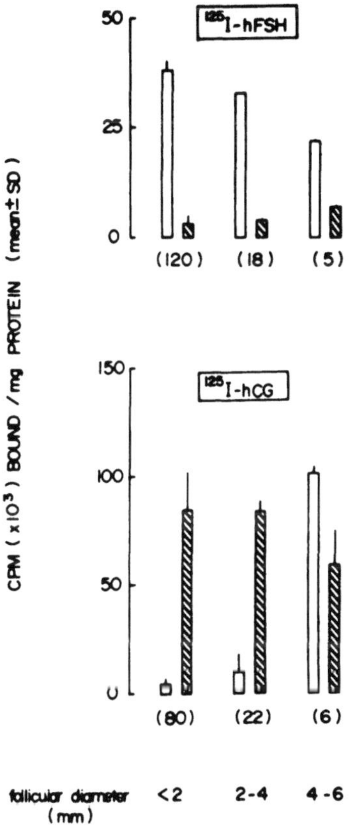


FIGURE 1: Mean specific binding of $^{125}$I-hFSH and $^{125}$I-hCG to granulosa ☐ and theca ▧ of ovine follicles *in vitro* as a function of follicular diameter.

*Binding as a function of atresia:* A significant decrease in mean specific binding of $^{125}$I-hFSH and $^{125}$I-hCG to granulosa cells was associated with increased degrees of atresia within any size group. Analysis of variance revealed that binding of $^{125}$I-FSH to granulosa varied as a function of atresia rather than follicular diameter ($p < 0.05$). In contrast, binding of $^{125}$I-hCG to granulosa varied as a function of follicular diameter rather than atresia ($p < 0.01$) (Figure 2).

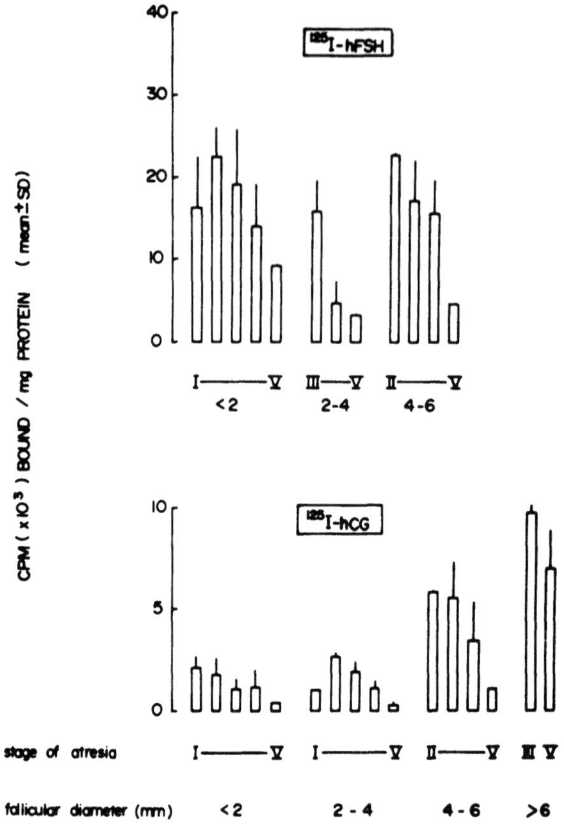


FIGURE 2: Mean specific binding of $^{125}$I-hFSH and $^{125}$I-hCG to granulosa cells of ovine follicles *in vitro* as a function of atresia.

## DISCUSSION

These data provide evidence for changes in specific binding of FSH and LH (hCG) to granulosa cells of ovine follicles as a function of follicular atresia and diameter. Similar changes in the binding of FSH and LH to granulosa cells of the rat have been observed (1). The detection of maximal specific binding of $^{125}$I-FSH to granulosa cells of the smallest follicles examined, infers that granulosa cells of ovine follicles acquire receptors at an even earlier growth stage (< 1 mm in the preantral stage) than examined in this instance. Conversely, significant numbers of LH receptors are not observed on granulosa cells until follicles reach 4 mm diameter. The role of oestrogen in the induction of these LH receptors on granulosa cells is presently under examination. In contrast to the granulosa cell, binding of gonadotrophins to the theca did not vary significantly; remaining relatively high for LH (hCG) and low for FSH.

High levels of LH receptor in thecal tissue and FSH receptor on granulosa are in accordance with the two-cell hypothesis of follicular steroidogenesis (14,15). This hypothesis states that thecal cells provide androgen precursor which is subsequently aromatised to oestrogen by granulosa cells under the stimulatory influence of FSH. The presence of LH receptors on thecal cells and FSH receptors on granulosa cells of small follicles suggests they have the capacity to produce oestrogen before LH receptors are present on the granulosa. Alternatively, follicular oestrogen may be of thecal origin (15) since in the sheep, small episodic releases of LH are accompanied by episodic secretion of androstenedione and oestradiol-17β (16), and theca cells have an apparent aromatase activity (17,18). The demonstration of LH receptors on thecal cells strengthens this argument.

The relatively long incubation time (27-36 h) needed to equilibrate the $^{125}I$-labelled gonadotrophin between incubation medium and follicular fluid is most probably a consequence of passive diffusion of gonadotrophin across the theca externa and interna, the basement membrane and finally the granulosa. This is in marked contrast to the situation *in vivo* where gonadotrophins are presented to the follicle via the vasculature which penetrates the theca interna to the basement membrane, thus eliminating the necessity to cross the entire thecal structure. No data in respect of the time required for endogenous gonadotrophin to penetrate the follicle is presently available, however the finding that levels of FSH in follicular fluid approximate but never exceed those in plasma (Findlay, Carson, McNatty, Mueller and Cumming, unpublished observations) suggests that the same diffusional process is operative *in vivo*.

Detailed morphological studies (3,4,7) revealed that degenerative changes associated with early atresia of ovine follicles is limited to the membrana granulosa directly adjacent to the antral cavity, whereas widespread degeneration of granulosa cells together with disappearance of the cumulus-oocyte unit characterises secondary atresia, and decreased thecal integrity are observed in tertiary atresia. These cellular changes are accompanied by a change in blood flow within the thecal microcirculation (7) and a reduction in the steroidogenic capacity of the follicle, particularly its ability to synthesize oestrogen (8). Moor et al. (8) postulated that limited oestrogen synthesis is a predisposing factor in atresia. Our studies have demonstrated a decrease in FSH binding to granulosa from atretic follicles. As FSH is required for oestrogen synthesis by granulosa (19) the loss of FSH binding in atretic follicles might explain the loss of aromatising capacity of these follicles (8). This is not to suggest that loss of binding sites is the cause of atresia. Indeed many of the morphological signs of atresia are apparent before any significant changes in binding are observed. The decrease in FSH binding associated with atresia may also explain the decrease in FSH binding to large follicles, since a greater proportion of large follicles are atretic (classes III-V) (20).

Currently, the correlations of the biological response of follicles and gonadotrophin binding capacity of granulosa cell suspensions to the stage of follicular atresia are being conducted.

## REFERENCES

(1) J.S. Richards and A.R. Midgley Jnr. Biol. Reprod. 14(1976)82.
(2) H.J. Rajaniemi, A.R. Midgley Jnr., J.A. Duncan and L.E. Reichert Jnr. Endocrinology 101(1977)898.
(3) A. Brand and W.H.R. de Jong. J. Reprod. Fert. 33(1973)431.
(4) K.E. Turnbull, A.W.H. Braden and P.E. Mattner. Aust. J. Biol. Sci. 30(1977)229.
(5) J.J. Peluso, R.W. Steger and E.S.E. Hafez. J. Reprod. Fert. 49(1977)215.
(6) R.M. Moor, M.F. Hay, J.E.A. McIntosh and B.V. Caldwell. J. Endocrinol. 58(1973)599.
(7) M.F. Hay, D.G. Cran and R.M. Moor. Cell. Tiss. Res. 169(1976) 575.
(8) R.M. Moor, M.F. Hay, H.M. Dott and D.G. Cran. J. Endocrinol. (77(1978) in press.
(9) Y. Miyachi, J.L. Vaitukaitis, E. Nieschlag and M.B. Lipsett. J. Clin. Endoc. Metab. 34(1972)23.
(10) S. Kammerman and R.E. Canfield. Endocrinology 90(1972)384.
(11) K.W. Cheng. J. Clin. Endoc. Metab. 41(1975)58.
(12) C.Y. Lee and R.J. Ryan. Endocrinology 89(1971)1515.
(13) O.H. Lowry, N.J. Rosebrough, A.L. Farr and R.J. Randall. J. Biol. Chem. 193(1951)265.
(14) B. Falck. Acta. Physiol. Scand. (Suppl.) 47(1959)163.
(15) D.T. Baird in: Endocrinology, Vol. 1, ed. V.H.T. James (Excerpta, Amsterdam, 1977) p. 330.
(16) D.T. Baird. J. Reprod. Fert. 50(1977)183.
(17) M.F. Hay and R.M. Moor. J. Reprod. Fert. 43(1975)313.
(18) C.P. Channing and S.P. Coudert. Endocrinology 98(1976)590.
(19) J.H. Dorrington, Y.S. Moon and D.T. Armstrong. Endocrinology (97(1975)1328.
(20) R.S. Carson, J.K. Findlay, H.G. Burger and A.O. Trounson. Biol. Reproduction - submitted for publication (1978).

Effect of Oestradiol-17β and Gonadotrophins on Prostaglandin production by Pre-Ovulatory Pig Follicles Superfused *In Vitro*

Kay S. Dodson and John Watson

Department of Biochemistry

University of Strathclyde, Glasgow G4 0NR

Abstract

Pre-ovulatory pig follicles were superfused *in vitro* with oestradiol, LH or FSH in the presence and absence of indomethacin. Prostaglandin $E_2$ and $F_{2\alpha}$ and oestradiol were measured in the same samples of superfusate. LH and FSH caused an increase in levels of both prostaglandins (approximately 2-fold) but the most marked effect was of oestradiol in $PGE_2$ where a 4-fold increase in levels above controls was observed. In all cases indomethacin inhibited the effects. Both gonadotrophins caused an increase in oestradiol production although the response to LH was much slower than to FSH. The results show a possible relationship between FSH, oestradiol and PG synthesis in the pre-ovulatory pig follicle.

Introduction

There is now considerable evidence that prostaglandins are involved in the series of events resulting in ovulation. The concentration of PGE and PGF in rabbit graafian follicles (1), $PGF_{2\alpha}$ in rat ovaries (2) and $PGF_{2\alpha}$ in porcine follicular fluid (3) increases towards the time of ovulation and in rabbits follicular granulosa cells have been shown to be the site of production of $PGF_{2\alpha}$ (4). Also, doses of indomethacin which block ovulation inhibit the pre-ovulatory rise in follicular PG levels in rats (2) and rabbits (5). Gonadotrophins appear to stimulate follicular PG synthesis and c AMP has been shown to be one possible mediator of this action (17) and protein synthesis appears also to be involved (16).

LH stimulates PG synthesis in rabbit pre-ovulatory follicles studied in vivo (5) whilst LH/FSH treatment of cultured granulosa cells from pre-ovulatory follicles has no effect on PG levels (4). One difference between the two systems is the presence of theca cells in the former, and an explanation for these observations might be that in the rabbit a steroid, possible oestradiol, could be produced by the theca cells in response to gonadotrophin and lead to the secretion of prostaglandins prior to ovulation (4). The present study was designed to investigate possible pre-ovulatory interrelationships between gonadotrophin, oestradiol and prostaglandin production in the pig follicle, where we have previously shown that in theca cells LH causes a rapid transitory rise in oestradiol secretion (6), using an in vitro superfusion technique. This technique allows incubation of follicular tissue with continuous exposure to fresh medium, controlled addition to the tissue of substances under investigation and automatic collection of medium after exposure to the tissue for analysis. It is therefore a useful method of studying these interrelationships.

Materials and Methods.

Materials: All chemicals were of A R grade and solvents were further purified by distillation in an all glass system. Tritiated steroid and prostaglandins were obtained from the Radiochemical Centre, Amersham, and had the following specific activities: [2, 4, 6, 7 - $^3$H] oestra - 1, 3, 5 (10) -triene - 2, 17β -diol (oestradiol), 85 Ci/m mol. [5, 6, 8, 11, 12, 14, 15 (n) - $^3$H] prostaglandin $F_{2\alpha}$, 130 Ci/m mol. and [5, 6, 11, 14, 15, (n) - $^3$H] prostaglandin $B_2$, 130 Ci/m mol. Authentic oestradiol was obtained from the Sigma Chemical Company. Bovine LH (NIH-LH-B9) and bovine FSH (NIH-FSH-B1) were used. Medium 199 (type 45) was obtained from Wellcome Reagents Ltd., Beckenham, Kent. Indomethacin was obtained from Sigma Chemical Company and a concentrated solution (10 mg/ml) was prepared in ethanol, a small volume being dissolved in medium when required. Scintillator for radioimmunoassay consisted of PPO (2 g) and POPOP (0.1 g Koch-Light Laboratories Ltd.) dissolved in toluene (800 ml) and Triton X-100 (200 ml, Hopkins and Williams Ltd.)

Tissue: Intact ovaries, obtained from mature sows immediately after slaughter in the local abattoir, were chilled and transported to the laboratory. Large pre-ovulatory follicles

0.8 cm to 1.0 cm in diameter were removed from the ovaries and freed of connective tissue. The follicles were cut in half and the follicular fluid allowed to drain out. Approximately 200 mg of tissue was used in each superfusion chamber.

Superfusion: The superfusion system (7) was set up in conjunction with a fraction collector and incorporated all the details previously described (8). Ten chambers were used in each experiment; two blank (medium only with no tissue present), two controls (tissue with medium only and no additional substances) and two for each substance under investigation. The system was run for 90 min with tissue present before collection began. Superfusate was then collected initially at 20 min intervals and later at 45 min intervals for up to 18 h. For the first 2 h Medium 199 was passed through the system. Addition of substances under investigation followed for 3 h and then superfusion continued with Medium 199 alone. LH, FSH or oestradiol were added at a concentration of 10 μg/ml Medium 199 and indomethacin at a concentration of 1 μg/ml. A flow rate of 20 ml/h was used in all experiments and for all chambers. All samples were stored at -20°C to await extraction and assay.

Analysis of Samples: In the case of LH and FSH treatment, samples were assayed for oestradiol, $PGF_{2\alpha}$ and $PGE_2$ and in the case of oestradiol treatment, samples were assayed only for the two prostaglandins. All analyses were by radioimmunoassay; the methods used having been fully described previously (9, 10). After counting the radioactivity measurements were processed automatically by computer which plotted a standard curve for each assay and gave values for each sample (11). Blank values, quality control values and inter- and intra-assay coefficients of variation have been recorded previously for each method (9, 10).

The final results were obtained by subtracing values for blank chambers from control and experimental chambers and results calculated on the basis of tissue weight.

Results

Representative results from at least three experiments performed for each test substance are presented in Fig. 1. and 2. The data in these Figs. was obtained in the same experiment, oestradiol and PG levels being measured in the same samples in order to correlate the results. Fig. 1a and b shows the effect on prostaglandin levels of superfusing pig follicles with LH and LH plus indomethacin for 3 h. The control tissue produces $PGE_2$ (fig. 1a) reaching a maximum after 4 h of superfusion. LH causes a rapid increase in $PGE_2$ production to twice control levels, which is maintained for 4 h after LH treatment ceases. The presence of indomethacin inhibits this effect and depresses $PGE_2$ levels to below control values. $PGF_{2\alpha}$ production (Fig. 1b) is also increased by LH treatment although a longer time is taken to achieve maximum production. The effect is inhibited by indomethacin.

Treatment with FSH (Fig. 1c and d) causes a rapid increase in levels of both PGs to approximately 3 times control levels. In both cases maximum production is reached after 2 h of FSH treatment, a drop is then observed followed by a secondary rise reaching maximum levels 4 h after FSH treatment has ceased. Indomethacin inhibits the effects.

Fig. 2a and b show the effects of oestradiol on PG levels. $PGE_2$ levels (Fig. 2a) increase rapidly and reach values of 4 times control levels which are sustained for 4 h after treatment with oestradiol stops. $PGF_{2\alpha}$, on the other hand, (Fig. 2b) takes longer to respond to oestradiol treatment and levels reach only twice control values.

The effect of gonadotrophin on oestradiol is shown in Fig. 2c and d. FSH causes a rapid increase in oestradiol reaching 2 to 3 times control levels during the first hour of FSH treatment (Fig. 2c). LH elicits a much slower response, (Fig. 2d), oestradiol levels rising after an hour of treatment and reaching less than twice control levels by the time treatment stops. The initial high level of oestradiol seen in both cases at the beginning of superfusion is presumably endogenous steroid still present in the tissue in spite of 90 min. washing.

Discussion.

The results demonstrate the ability of pig pre-ovulatory follicular tissue to produce increased levels of $PGE_2$ and $F_{2\alpha}$ in response to LH, FSH and oestradiol <u>in vitro</u>. Because of the large amount of tissue required to carry out these experiments, ovaries from several animals were used which probably accounts for the variation in control levels of $PGE_2$ found (compare $PGE_2$ control levels in Fig. 1a with Fig. 2a). Control levels of $PGF_{2\alpha}$ however tend to be more constant.

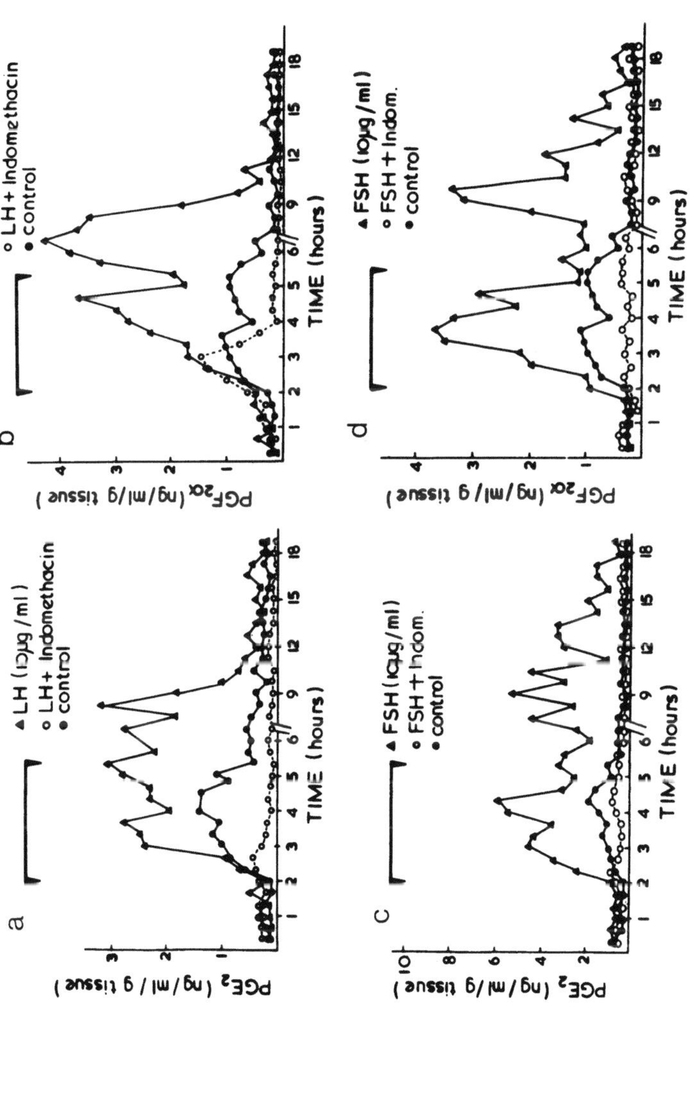


Fig. 1. Prostaglandin secretion by pre-ovulatory pig follicles during superfusion _in vitro_. The horizontal bar represents the time when test substances were present in the superfusing medium.

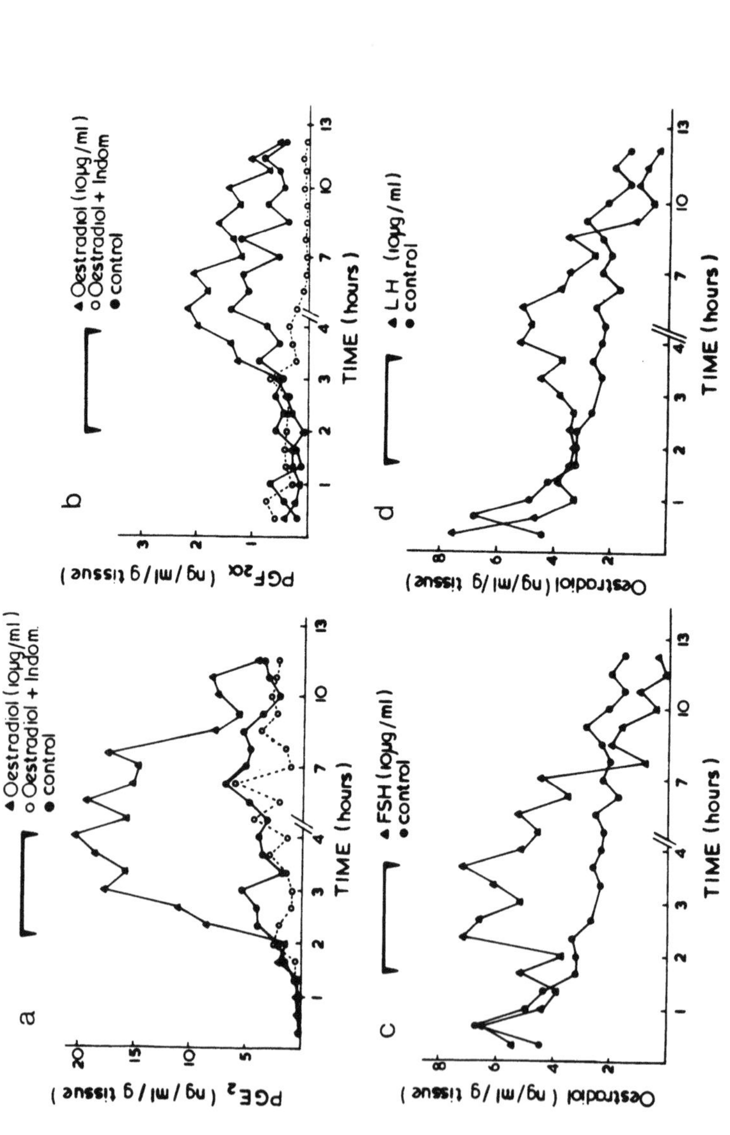


Fig. 2. Prostaglandin and oestradiol secretion by pre-ovulatory pig follicles during superfusion in vitro. The horizontal bar represents the time when test substances were present in the superfusing medium.

All the pig material was obtained from the abattoir and was therefore not accurately dated, although there is considerable literature on the size of follicles in relation to the time of the cycle (12, 13) on which the classification used here was based.

LH, FSH and oestradiol may stimulate <u>de novo</u> synthesis of both prostaglandins as, in all cases, the effects are inhibited by adding indomethacin along with the test substances. The maintenance of maximum levels of PG production up to 4 h after treatment with gonadotrophins or oestradiol stops is presumably due to gonadotrophin and oestradiol binding to receptors on the follicular cells (14, 13), thus maintaining their physiological effect during this time.

The effect of LH on oestradiol in the present study differs from the effect we reported in isolated theca cells studied by the same technique (7). In isolated cells LH caused a rapid stimulation of oestradiol secretion followed by inhibition after 2 h of superfusion. In the follicle <u>in vitro</u> it appears that oestradiol responds much more slowly to LH. The effect of FSH on oestradiol is rapid and more marked. It is well established in rats FSH stimulates the aromatase enzyme system and hence oestradiol synthesis (15) and a similar mechanism may be active in the pig.

In most cases $PGF_{2\alpha}$ responds more slowly to gonadotrophins and oestradiol than $PGE_2$ where levels increase rapidly after treatment begins. The most marked effect observed is that of oestradiol on $PGE_2$ levels. The results show that under FSH stimulation the tissue has the ability to respond rapidly by producing oestradiol. Hence treatment with FSH could bring about increases in PG levels by producing oestradiol as an intermediate. PG synthesis is stimulated by LH more rapidly than oestradiol synthesis and on the basis of this <u>in vitro</u> study it seems less likely that oestradiol could act as a mediator of PG production by LH. This observation is supported by the demonstration in the rat <u>in vivo</u> that inhibition of steroid biosynthesis with aminoglutethimide phosphate does not inhibit LH-induced PG synthesis (18), suggesting that a steroid is not an obligatory mediator of this action.

The results of FSH stimulation of PGs were unexpected as it has previously been shown in rabbits that FSH is not effective in increasing PG synthesis (17). In the pig a separate system of

PG stimulation might be envisaged, induced by FSH and possibly mediated via oestradiol and it is tentatively suggested that one of the roles of FSH prior to ovulation could be this stimulation of PG synthesis.

Acknowledgements

We thank Mrs. Joanne Raeburn and Miss Ruth MacCormack for extensive technical assistance; Dr. J.E. Pike and the Upjohn Company for authentic prostaglandins; Professor E.W. Horton, University of Edinburgh for antibodies to prostaglandins $F_{2\alpha}$ and $B_2$; Dr. B.V. Caldwell, Yale University, U.S.A. for the oestradiol-17β antibody and the National Institute of Health, Bethesda, Maryland, U.S.A. for the LH and FSH.

The work was supported by a grant from the World Health Organisation (Project No. 75181) for which we are very grateful.

References

1. Le Maire, W.J., N.S.T. Yang, H.R. Behrman and J.M. Marsh. Prostaglandins 3 : 367. 1973.
2. Armstrong, D.T. and J. Zamecnik. Mol. Cell Endocr. 2 : 125. 1975.
3. Ainsworth, L., R.D. Baker and D.T. Armstrong. Prostaglandins 9 : 915. 1975.
4. Challis, J.R.G., G.F. Erickson and K.J. Ryan. Prostaglandins 7 : 183. 1974.
5. Yang, N.S.T., J.M. Marsh and W.J. Le Maire. Prostaglandins 4 : 395. 1973.
6. Watson, J., and K.S. Dodson. IRCS Medical Science. 4 : 12. 1976.
7. Rodway, R.G., K.S. Dodson and J.Watson. Europ. J. Obstet. Gynec. Reprod. Biol. 6 : 201. 1976.
8. Watson, J. and F.M. Maule Walker. J. Reprod. Fert. 51: 393 1977.
9. McDougall, A.N., F.M. Maule Walker and J. Watson, Br. J. Pharmac. 60 : 425. 1977.
10. Patek, C.E. and J.Watson. Prostaglandins 12 : 97. 1976.
11. Seaton, B., J. Lusty and J. Watson. J. Steroid Biochem. 7 : 511. 1976.
12. Channing, C.P. and F. Ledwitz-Rigly in Methods in Enzymology Vol. XXXIX. Hormone Action Part D (J.G. Hardman and B.W. O'Malley, editors) Academic Press, London, 1975. p. 186.
13. Channing, C.P. and S. Kammerman. Endocrinology 92 :

531. 1973.

14. King, R.J.B. and W.I.P Mainwaring, editors, Steroid-Cell Interactions, Butterworths, London, 1974. p.247

15. Moon, Y.H., J.H. Dorrington and D.T. Armstrong. Endocrinology 97 : 244. 1975.

16. Clark, M.R., Marsh, J.M. and Le Maire, W.J. Prostaglandins 12 : 209. 1976.

17. Marsh, J.M. Yang, N.S.T. and Le Maire, W.J. Prostaglandins 7 : 269. 1974.

18. Armstrong, D.T., Dorrington, J.H. and Robinson, J., Can. J. Biochem. 54 : 796. 1976.

# LH-STIMULATED cAMP FORMATION IN RAT GRANULOSA CELLS DURING FOLLICULAR MATURATION - A NON-REFRACTORY RESPONSE

L. Hamberger, L. Nilsson, K. Nordenström and A. Sjögren
Dept. of Physiology, University of Göteborg, Sweden

In early experiments on isolated granulosa cells from porcine ovaries Kolena and Channing (1,2) reported that the in vitro administration of either FSH or LH caused an acute stimulation of cAMP formation by these cells. The same group (3) also found that there was an induction of LH receptors on the surface of granulosa cells that parallelled with follicular maturation. Both FSH and estradiol are believed to be essential for the development of LH receptors in rat granulosa cells (4, 5, 6). In recent studies on granulosa cells from immature rat ovaries (7, 8) and on granulosa cells from small follicles isolated from sheep ovaries (9), LH was found to have a lack of effect in vitro on cyclic AMP formation. With increasing maturation of follicles from which the granulosa cells were derived, LH was not, in the case of the rat, able to stimulate cyclic AMP formation until the preovulatory stage. Exposure to LH in vitro of intact preovulatory follicles from this species caused acute stimulation of cAMP formation (10) but in addition refractoriness to a new stimulation by the same hormone is induced for several hours (11). A similar mechanism seems to be active in vivo with a decreased sensitivity to LH after the endogenous LH-FSH surge (12, 13). In general, the mechanism(s) for hormonal refractoriness is poorly understood and has in the above-mentioned context not been investigated in detail (14). The present study was intended to explore if a similar refractoriness to LH is present in granulosa cells separated and isolated from preovulatory rat follicles obtained from a PMSG rat model.

## Material and Methods

Animals: Female Sprague-Dawley rats were obtained from Anticimex Ltd., Sweden, and kept in rooms with constant temperature (24-26 °C) and relative humidity (50-55 %). They were exposed to 14 h of controlled illumination starting at 6 a m. The rats were fed a standardized pellet diet and water ad lib. At 28 days of age the rats were injected subcutaneously with a single dose (8 I. U.) of PMSG. This is known to induce follicular growth and an endogenous LH-FSH peak in the afternoon (4-7 p m) of day 30. Subsequently 8-12 follicles rupture with the release of ova in the early morning (2-4 a m) of day 31. For details concerning the PMSG model see Herlitz et al. (15). Rats were killed either before or after the endogenous LH-FSH peak by cervical fracture and the ovaries immediately removed. In one series of experiments the rats were injected intra-peritoneally with FSH or LH (10 μg/rat) or saline on the morning of day 30 before the endogenous peak, and the ovaries removed 2 h later.

Isolation procedure: After removal, the ovaries were placed in ice-cold Krebs bicarbonate buffer (composition see below). Eight to twelve preovulatory follicles were isolated by careful free-hand dissection under a stereo-microscope. The follicles were transferred to hollow glass cups containing plain buffer and the follicle wall opened by stainless steel scalpels and needles. Using gentle pressure and a scraping motion, most of the granulosa cells were released from the theca capsule. Granulosa cells were collected with a Pasteur pipette and placed in glass tubes. These tubes were centrifuged in a swing-out rotor for 20 sec that gives a maximal speed corresponding to 900 x g for 5 sec. After centrifugation, the supernatant was decanted and the loose cell pellet resuspended in fresh chilled buffer containing the appropriate test substance (LH or FSH, usually in a final concentration of 1 μg/ml). Each granulosa cell sample, collected from 8 preovulatory follicles, was incubated under continuous shaking at 37° C in a 10 ml glass tube containing 0.5 ml medium and a gas phase of 95 % $O_2$-5 % $CO_2$. After incubation, the glass tubes were immediately placed on ice, and 0.5 ml of 10 % trichloro-acetic acid was added. Both the medium and the cells were homogenized in a glass homogenizer. Aliquots were taken for protein determination according to Lowry et al. (16), using bovine serum albumin as the standard. The content of cyclic adenosine 3′,5′-monophosphate (cAMP) in tissue plus medium was determined in duplicate according to the protein binding technique described by Gilman (17).

Incubation medium: A modified Krebs Ringer bicarbonate buffer ($CaCl_2$ 1.25 mM), pH 7.4 was used for the preparative work, for incubation it was fortified with 5.5 mM D-glucose.

Hormones: Bovine LH (NIH-LH-B9) and ovine FSH (NIH-FSH-S11) were donated by the Hormone Distribution Office of the National Institutes of Health, Bethesda, Md., USA.

Statistical procedure: Results have been expressed as the mean ± SEM and the mean values were compared using either a Student's t-test (2 groups) or an analysis of variance (with one criterion of classification) followed by the Student-Newman-Keul's test, when more than two groups were compared. A p-value of less than 0.05 was considered significant.

## RESULTS

### Gonadotrophin effects before the endogenous LH-FSH peak

Addition of LH *in vitro* (1 μg/ml) to the incubation medium caused a significant increase in granulosa cell cAMP after 15 min with a maximal stimulatory effect after 60 min and a decline after 120 min of incubation (Fig. 1). An even

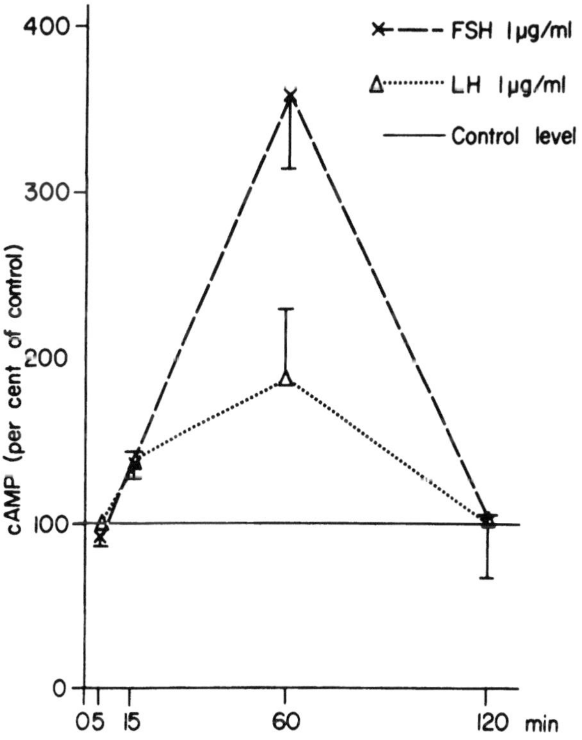


Fig. 1 Time-response curves illustrating the effects caused by *in vitro* addition of LH and FSH on cAMP formation in isolated granulosa cells from preovulatory follicles.

more pronounced stimulation was elicited by the addition of FSH (1 μg/ml) with a similar time-response curve as for LH (Fig. 1). As maximal stimulation occurred then, a one-hour incubation was chosen in the next series of experiments, in which the rats were pre-treated with a single intra-peritoneal injection of either LH or FSH (10 μg/rat) or saline 2 h prior to their sacrifice and the isolation of granulosa cells. This pre-treatment caused a potentiation of the LH (1 μg/ml) effect in vitro that was most pronounced when the rats were injected with LH (Fig. 2). No such potentiation was found when the cells were incubated with FSH (1 μg/ml) after pre-treatment with LH or FSH (Fig. 2). The injection of LH or FSH 2 h before incubation without hormones in the incubation medium did not give any significant stimulation of granulosa cell cAMP formation. One possible explanation of the lack of effect of LH and FSH in vivo is that the granulosa cells, at the time of isolation, already have had their maximal response the gonadotrophins injected; i.e., within 2 h similar to the granulosa cells in vitro (Fig. 1). However, this will be the subject of further investigations in our laboratory.

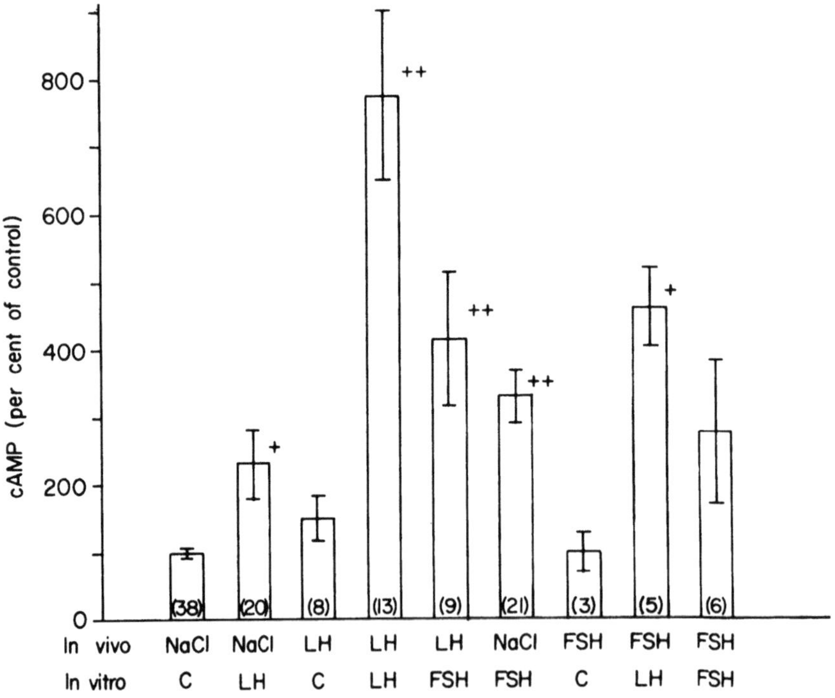


Fig. 2 Effects of gonadotrophins in vitro on cAMP formation in isolated granulosa cells after in vivo injection of LH, FSH or saline.

Gonadotrophin effects after the endogenous LH-FSH peak

As can be seen from Fig. 3, granulosa cells isolated and incubated in the evening (8-10 p m) after the endogenous LH-FSH peak showed a statistically significant increase in their cAMP formation as compared with controls isolated before the gonadotrophin surge (Fig. 3, group H-A). The addition of LH *in vitro* caused a marked increase in granulosa cell cAMP (Fig. 3, group I, K) of the same magnitude as that elicited by LH *in vivo* + LH *in vitro* before the endogenous gonadotrophin peak (group F). FSH *in vitro* in the evening (8-11 p m) was much less potent than LH.

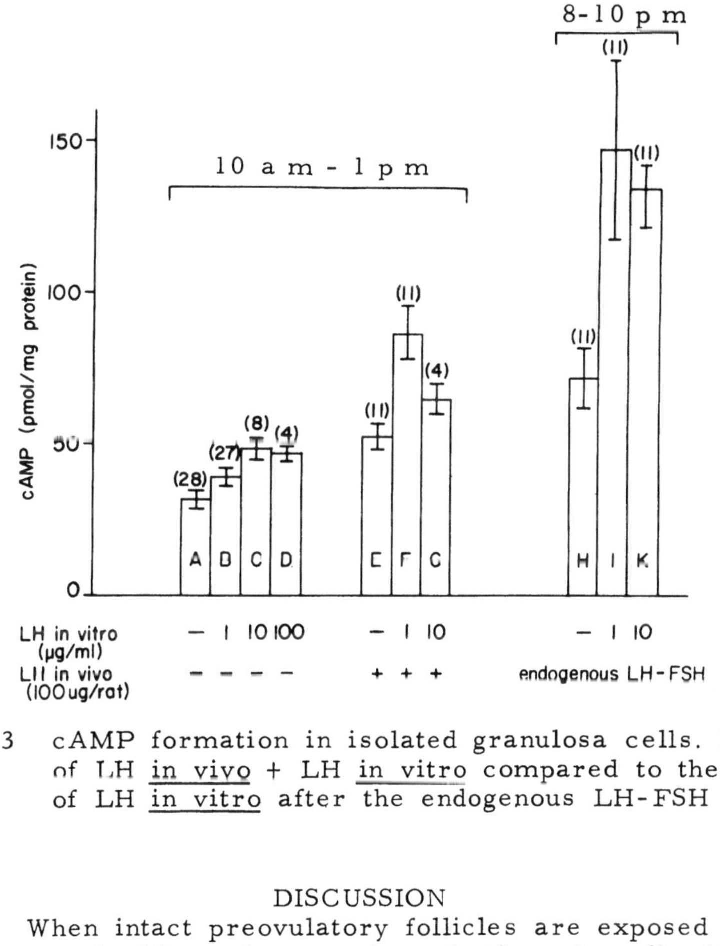


Fig. 3 cAMP formation in isolated granulosa cells. Effect of LH *in vivo* + LH *in vitro* compared to the effect of LH *in vitro* after the endogenous LH-FSH peak.

## DISCUSSION

When intact preovulatory follicles are exposed to LH they respond with an increased production of cyclic AMP and this stimulation is accompanied by a desensitization of the cyclic AMP system to further action by the same hormone. This type of response can be seen when LH is added to follicles *in vitro* or when it reaches its target *in vivo* by injection or by the impact of the preovulatory gonadotrophin surge (18, 11, 12, 13, 19). The mechanism of desensitization is not yet completely understood but in the physiological situation may involve actions directly on the adenylate cyclase enzyme and its coupling to gonadotrophin receptor (12, 20), and/or on receptor occupancy and turnover (21). Even less is understood about the physiological significance of the desensitization phenomenon, but theoretically it may cause a drop in follicular cyclic AMP levels just before ovulation; a feature which might be involved in the maturation process of the oocyte. The rise in the intrafollicular cAMP level following the gonadotrophin surge is essential for the numerous effects of LH on the follicle, including the resumption of oocyte meiosis. Since cAMP in itself inhibits oocyte maturation (22) it is of the utmost importance that the intrafollicular cAMP level then declines or at least does not reach higher levels that allow maturation of the oocyte, (23). Nilsson et al. (13) were not able to demonstrate any decrease in follicular cAMP prior to ovulation in the rat. If desensitization and cAMP decrease were important for luteinization, these events would be expected to be demonstrable in follicular granulosa cells prior to ovulation, and the present study was undertaken to investigate the sensitivity of the cAMP system of such cells in isolation.

As can be seen from our results rat granulosa cells, isolated before the endogenous LH-FSH surge, increased their production of cAMP in response to both LH and FSH in vitro, the latter being the more effective stimulator, thereby confirming the results of earlier studies (7, 8). When isolated in the evening after the gonadotrophin peak the granulosa cells exhibited a greater response to LH than to FSH. This is not indicative of any desensitization to the action of LH but is in accordance with the development of gonadotrophin binding, reported by Channing and Kammerman (3). Furthermore, if the granulosa cells were removed (on the day before ovulation) before the LH-FSH peak but after the injection of a high dose of gonadotrophin no decrease in sensitivity to LH in vitro was seen. On the contrary, in vivo exposure to LH clearly facilitated the response to a subsequent in vitro exposure to the same hormone. Since the gonadotrophins used in this study cannot be considered to be completely pure, it is not possible to differentiate between the actions of LH and FSH. However, the most powerful potentiation was seen when LH in vivo was followed by LH in vitro. Also, the preovulatory gonadotrophin peak, which in the applied PMSG model is a combined LH-FSH surge, elicited the same response. The cyclic AMP formation in rat granulosa cells is increased by the preovulatory gonadotrophin surge analagous to the situation in the intact follicles but, in contrast to follicular cyclic AMP formation, it can be further raised by the addition of new hormone. This discrepancy must mean either that the desensitization mechanism of the follicle resides exclusively in the theca cells or that the experimental situation fails to reveal its existence in the granulosa cells. Desensitization in the granulosa cells might be part of a biphasic response with a different time course than in the follicles. Or the isolation procedure may create an artificial response either by removing an inhibitory substance from the follicular fluid, or by breaking the contact between granulosa and theca cells. Another possibility is that the isolated cells and the intact follicles have differences in receptor kinetics that interfers with the results, but this appears less likely after considering the results of Nimrod et al. (6) who observed the same discrepancy. With these reservations, it seems probable that the preovulatory LH-FSH peak is one of the factors that governs the transformation of the primarily FSH-responsive granulosa cell into a primarily LH-responsive luteal cell. It remains to be clarified whether this effect is mediated by changes in the coupling of adenylate cyclase to receptors, by changes in receptor turnover or by other, hitherto unknown mechanisms.

## REFERENCES

(1) J. Kolena and C. P. Channing. Biochim. Biophys. Acta 252 (1971) 601.

(2) J. Kolena and C. P. Channing. Endocrinology 90(1972)1543.

(3) C. P. Channing and S. Kammerman. Endocrinology 92 (1973)531.

(4) A. J. Zelesnik, A. R. Jr. Midgley and L. E. Jr. Reichert. Endocrinology 95(1974)818.

(5) J. S. Richards. Endocrinology 97(1975)1174.

(6) A. Nimrod, E. Bedrak and S. A. Lamprecht. Biochim. Biophys. Res. Commun. 78(1977) 977.

(7) A. K. Goff and D. T. Armstrong. Endocrinology 101(1977) 1461.

(8) L. Hamberger, K. Nordenström, S. Rosberg and A. Sjögren. Acta Endocrinologica (1978) in press.

(9) T. J. Weiss and D. T. Armstrong, ASRB meeting, Australia (1977).

(10) L. Nilsson, S. Rosberg and K. Ahrén. Acta Endocrinologica (Kbh.) 77(1974) 559.

(11) S. A. Lamprecht, U. Zor, A. Tsafriri and H. R. Lindner. J. Endocrinol. 57(1973) 217.

(12) M. Hunzicker-Dunn and L. Birnbaumer. Endocrinology 99(1976) 211.

(13) L. Nilsson, T. Hillensjö and C. Ekholm. Acta Endocrinologica (Kbh.) 86 (1977a) 384.

(14) S. A. Lamprecht, U. Zor, Y. Salomon, Y. Koch, K. Ahrén and H. R. Lindner. J. Cyclic Nucleotide Research 3(1977) 69.

(15) H. Herlitz, Y. Koch, M. I. Khan and K. Ahrén. Eur. J. Obstet. Gyn. Reprod. Biol. 6(1976) 175.

(16) O. H. Lowry, N. J. Rosebrough, A. L. Farr and R. J. Randall. J. Biol. Chem. 193(1951) 265.

(17) A. G. Gilman. Proc. Natl. Acad. Sci. (USA) 67(1970) 305.

(18) J. M. Marsh, T. M. Mills and W. J. Lemaire. Biochim. Biophys Acta 304 (1973) 197.

(19) L. Nilsson, L. Hamberger, T. Hillensjö and C. Ekholm. Acta Endocrinologica (Kbh.) 85(1977b) 39.

(20) U. Zor, S. A. Lamprecht, Z. Misulovin, Y. Koch and H. R. Lindner. Biochim. Biophys. Acta 428(1976) 761.

(21) M. Conti, J. P. Harwood, A. J. W. Hsueh, M. L. Dufau and K. K. J. Catt. J. Biol. Chem. 251(1976) 7729.

(22) C. Magnusson and T. Hillensjö. J. exp. Zool. 201(1977) 139.

(23) T. Hillensjö, C. Ekholm and K. Ahrén. Acta Endocrinologica 87 (1978) 377.

# EFFECT OF THE ADMINISTRATION OF AN OVULATORY DOSE OF HCG ON THE SUBCELLULAR DISTRIBUTION OF FOLLICULAR PROTEIN KINASE

Mary Hunzicker-Dunn, Ming-ann Wang, and Richard Jungmann

Department of Biochemistry, Northwestern University

Medical School, Chicago, Illinois 60611

## INTRODUCTION

Most protein hormones initate a biological response in their target cells by initially binding to a membrane-localized receptor, thereby activating adenylyl cyclase and raising intracellular levels of cAMP. Currently, the only known manner in which cAMP can mediate cellular function is through cAMP-dependent protein kinases (PK). cAMP has been shown to activate cAMP-dependent PK by binding to the regulatory subunit of the inactive holoenzyme, thereby releasing the active, cAMP-insensitive catalytic subunit which can catalyze the transfer of phosphate form ATP to cellular protein (1). Presumably, the phosphorylation of specific cellular protein(s) regulates cellular activity in such a way that the known hormonal response ensues.

In estrous rabbits, the administration of an ovulatory dose of hCG promotes a rapid increase in the intracellular cAMP levels in ovarian follicles (2) via activation of the follicular adenylyl cyclase system (3). The effects of LH which are thought to be mediated by cAMP in follicles have been recently reviewed (4-6). However, little is known concerning the mechanism by which cAMP mediates the actions of LH in follicles. One study in preovulatory rat follicles demonstrated that this tissue contains PK activity whose phosphorylative capacity was enhanced by cAMP (7). In view of the current obligatory role of cAMP-dependent PK in cAMP-mediated hormone action, we have investigated the effect of the administration of an ovulatory dose of hCG to estrous rabbits on the subcellular distribution, activities, and cAMP-dependency of PK, and cAMP binding activities in preovulatory and postovulatory follicles.

RESULTS AND CONCLUSIONS

We initially determined the subcellular distribution and cAMP-dependency of PK in popped, Graafian follicles from untreated estrous rabbits. Fifteen percent of the total cellular PK activity was compartmentalized in the nuclear, mitochondrial and microsomal fractions (Fig. 1). About 50% of this particulate PK activity was not affected by the heat-stable PK inhibitor, which forms a catalytically inactive complex with the catalytic subunit of cAMP-dependent PK (8), and was thus cAMP-independent (Fig. 2). The vast majority of follicular PK was identified in the 105,000 x g supernatant (cytosol) as cAMP-dependent (Figs. 1,2). Previous studies of PK in bovine CL have also demonstrated a similar pattern of subcellular distribution, with the great majority of PK localized in the cytosol fraction (9).

The administration of an ovulatory dose of hCG (100 IU, iv) resulted in marked changes in the specific activities of particulate PK. Nuclear PK was reduced 50% ($P<0.005$) within 1 h of hCG injection and remained low for the next 4 h ($P<0.001$) (Fig. 3). Nuclear PK activities then rose after ovulation to preinjection levels. The microsomal PK activity also declined, reaching significance ($P<0.005$) by 5 h post hCG (Fig. 4). By 48 h, microsomal PK activities had also risen to preinjection levels. No significant changes in the mitochondrial or cytosol PK specific activities were observed by 48 h following hCG injection (Figs. 5,6).

The hCG-induced changes in PK activities affected cAMP-dependent and cAMP-independent PK to a similar degree, as revealed by studies with the heat-stable PK inhibitor, and no changes in the relative distribution of cAMP-dependent vs. cAMP-independent PK, as reported in Fig. 2, were observed following the administration of hCG (not shown). PK in the particulate and soluble fractions was not markedly stimulated by cAMP (Figs. 3-6). However, the presence in these fractions of the dissociated cAMP-insensitive catalytic subunit of PK was confirmed by the PK inhibitor studies (Fig. 2).

Studies on the subcellular distribution of cAMP-binding activities in Graafian follicles obtained from untreated estrous rabbits revealed that approximately 90% of the cAMP-binding activity was localized in the cytosol, and the remaining 10% was distributed among the three particulate fractions (Fig. 7). The administration of an ovulatory dose of hCG did not promote any detectable alterations in the activities or distribution of cAMP-binding activities at 2 h post hCG (Fig. 7).

These results indicate that the particulate fractions of mature, preovulatory and postovulatory follicles contain compartmentalized regulatory and catalytic subunits of the cAMP-dependent PK as well as cAMP-independent PK. Even though cAMP-dependent PK was generally dissociated at the time of the PK assays, as indicated by the small

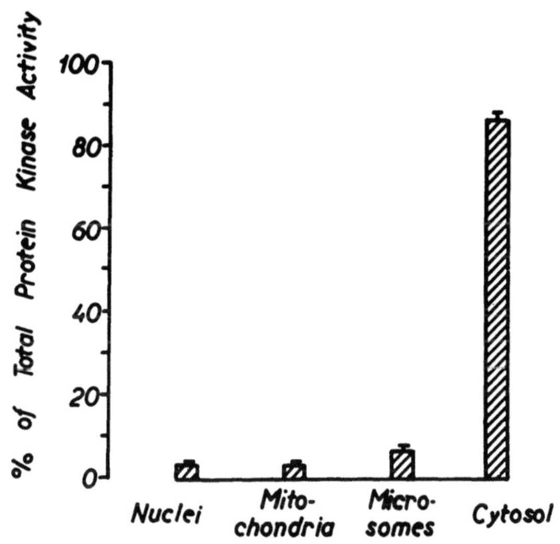


Fig. 1. Distribution of PK in subcellular fractions of follicles obtained from estrous rabbits. Ovaries of rabbits, sacrificed by cervical dislocation, were removed and cooled to 0 C in buffer consisting of 10 mM TRIS-HCl, 3 mM $MgCl_2$, 0.32 M sucrose, pH 7.0. Graafian follicles were dissected, popped, and homogenized in 10 volumes of the same buffer. The homogenate was centrifuged at 1000 x g for 10 min. Then the 1000 x g pellet was suspended by homogenization in 1 ml buffer consisting of 10 mM TRIS-HCl, 0.15 M NaCl, 3 mM $MgCl_2$, 0.25 M sucrose, pH 7.0, mixed with 4 ml 2 M sucrose, and centrifuged at 105,000 x g for 60 min. The resulting pellet was resuspended in 10 mM TRIS-HCl, 0.15 M NaCl, 3 mM $MgCl_2$, 0.1% Triton X-100, pH 7.0, and centrifuged at 10,000 x g for 10 min. This final nuclear pellet was resuspended by homogenization in 10 mM TRIS-HCl, 3 mM $MgCl_2$, 0.1% Triton X-100, pH 7.0, and assayed for PK activity 1 h after the introduction of Triton X-100 to the final nuclear pellet. The 1000 x g supernatant fraction was centrifuged at 10,000 x g for 30 min. The resulting pellet was resuspended in homogenizing buffer and recentrifuged. The final mitochondrial pellet was resuspended by homogenization in 10 mM TRIS-HCl, pH 7.0, 3 mM $MgCl_2$, and assayed. The 10,000 x g supernatant fraction was centrifuged at 105,000 x g for 60 min. The resulting microsomal pellet was rinsed with the original homogenization buffer and resuspended by homogenization in 10 mM TRIS-HCl, pH 7.0, 3 mM $MgCl_2$, 0.5 % Triton X-100, and assayed 1 h after the introduction of Triton X-100. The 105,000 x g supernatant, designated cytosol, was assayed. PK assays were for 10 min at 37 C in 0.225 ml of a medium containing 44 μM [$\gamma$-$^{32}$P]ATP, 100 μg protamine, 22 mM $\alpha$-glycerol phosphate, 4.4 mM NaF, 4.4 mM magnesium acetate, 0.89 mM theophylline, 0.44 mM dithiothreitol, and 10-40 μg crude enzyme protein in the absence and presence of $4.5 \times 10^{-8}$M cAMP (a maximally stimulating concentration). The reaction was terminated with 1 ml

15% trichloroacetic acid containing 1% sodium dodecyl sulfate. [$^{32}$P] Protamine was isolated on Millipore filters. Results are expressed as mean percentage of duplicate determinations of total PK activity ± SEM of 7 separate experiments each of which required a minimum of 8 rabbits.

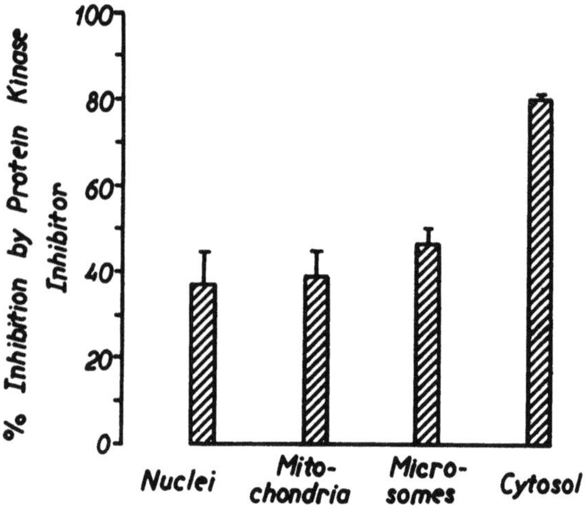


Fig. 2. Effect of heat-stable inhibitor on PK activity in subcellular fractions of estrous rabbit follicles. Saturating concentrations of the heat-stable inhibitor, isolated from rabbit skeletal muscle (10,11) were added to PK assays performed in the absence and presence of cAMP. For rest of conditions, see legend to Fig. 1.

degree of activation by cAMP, the state of activation of these PK *in vivo* cannot be discerned from these experiments in view of the possible effects of our subcellular fractionation conditions upon the isoenzymes of PK (low ionic strength and a time lapse of approximately 3 h between animal sacrifice and PK assays). We do not know which isoenzyme(s) of cAMP-dependent PK is present in the particulate fractions of ovarian follicles but have determined that the cytosol fractions of estrous follicles contain only the type II isoenzyme of cAMP-dependent PK (12-15). Experiments are currently in progress to ascertain the isoenzyme species of the cAMP-dependent PK present in the particulate fractions of ovarian follicles.

Our observation that the decrease in nuclear and, to a lesser extent, in microsomal PK activities is not accompanied by decreases in cAMP binding activities indicates that hCG promotes a specific release of the catalytic subunit of PK from these fractions. The fate of the catalytic subunits is not known. We did not observe any

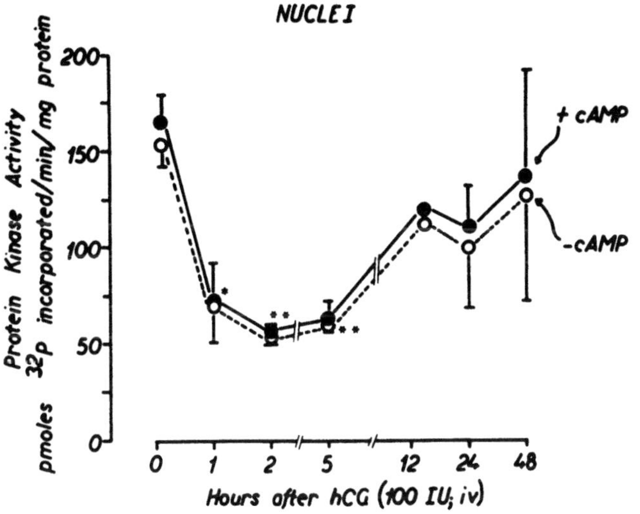


Fig. 3. Effect of an ovulatory dose of hCG on PK activity in the nuclear fraction of estrous rabbit follicles. Rabbits were injected iv with hCG (100 IU/3.5-4.5 kg body weight) and sacrificed at the indicated times. *, P<0.005; **, P<0.001 compared to untreated rabbits. Results are expressed as pmoles $^{32}$P incorporated/min/mg protein ± SEM of 8, 3, 2, 2, 1, 5 and 3 experiments at times 0, 1, 2, 5, 12, 24, and 48 h post hCG, respectively. For rest of conditions, see legend to Fig. 1.

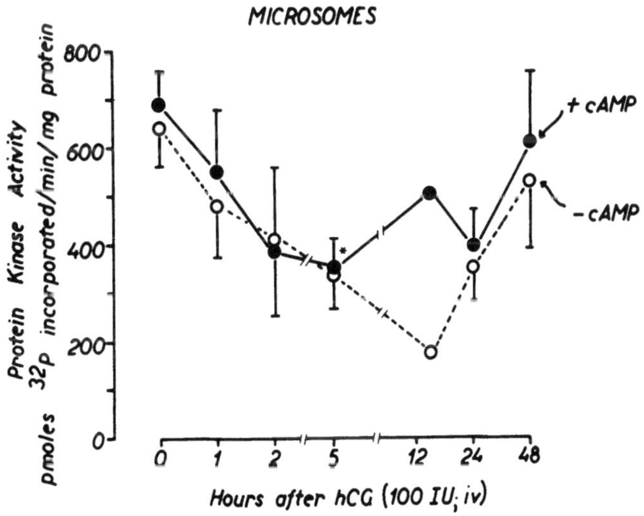


Fig. 4. Effect of an ovulatory dose of hCG on PK activity in microsomal fractions of estrous rabbit follicles. For details, see legends to Figs. 1 and 3.

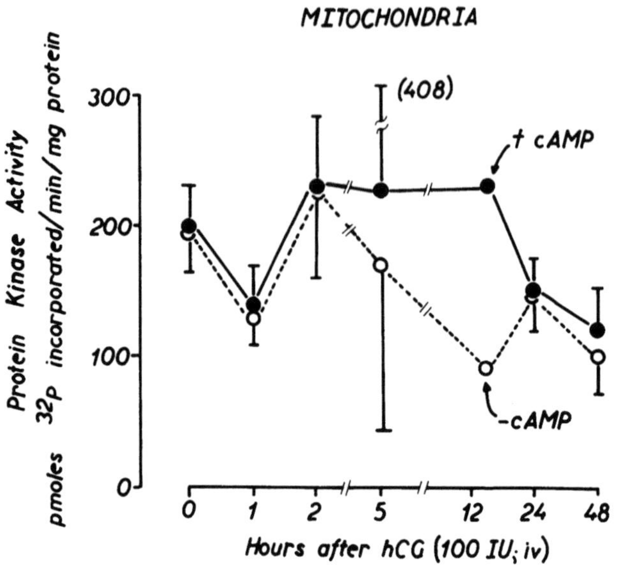


Fig. 5. Effect of an ovulatory dose of hCG on PK activity in mitochondrial fractions of estrous rabbit follicles. For details, see legends to Figs. 1 and 3.

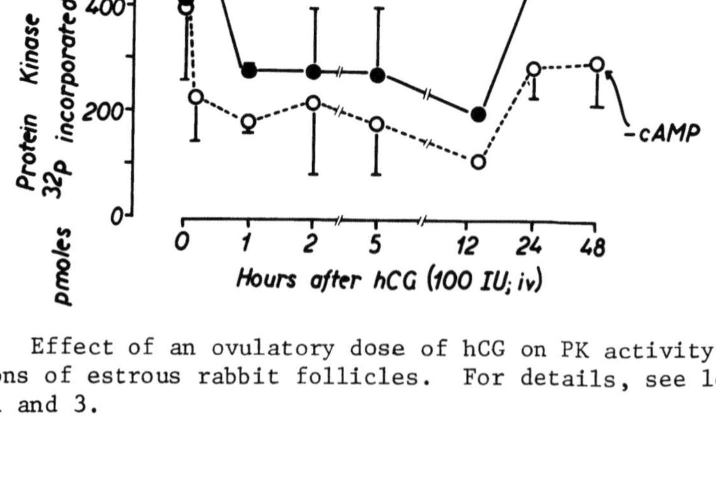


Fig. 6. Effect of an ovulatory dose of hCG on PK activity in cytosol fractions of estrous rabbit follicles. For details, see legends to Figs. 1 and 3.

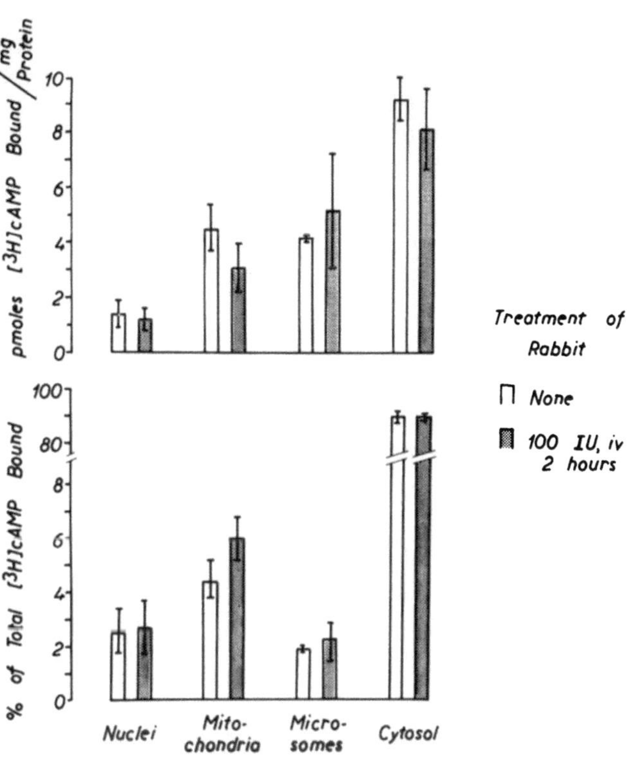


Fig. 7. Effect of an ovulatory dose of hCG on cAMP-binding activity and on the distribution of cAMP-binding activity in subcellular fractions of estrous rabbit follicles. Rabbits were sacrificed before or 2 h after the injection of hCG. cAMP binding assays were for 1 h at 0 C in 0.3 ml of a medium containing 50 mM TRIS HCl, pH 7.4, 8.3 mM $MgCl_2$, 5.8 mM theophylline, 83 nM [$^3$H]cAMP, and 10-50 μg ovarian protein extract. [$^3$H]Proteins were isolated on Millipore filters.

specific increases in PK activity in any of the other fractions; however, movement of the catalytic subunits into the soluble fraction, at least, probably would not have been detected due to the already large percentage of PK in this fraction and the relatively small contribution of the particulate enzymes. Recent studies by Corbin et al. (16) on heart tissue of various species have also shown that cAMP promotes a release of the catalytic subunit of PK from particulate fractions upon formation of the cAMP-regulatory subunit complex in the particulate fractions. Moreover, these authors demonstrated a concomitant increase in cytosol PK activity. However, in these tissues, a much larger percentage (30-50%) of PK was initially associated with the particulate fractions.

In conclusion, the activity and intracellular distribution of cAMP-dependent PK in ovarian follicles are subject to regulation by hCG. That the apparent reduction in the catalytic activity of PK in nuclear and microsomal fractions occurs in a time-dependent manner after hCG administration under identical tissue manipulations strongly suggest that these changes are of physiological significance. However, the nature of the involvement of these alterations of PK in preovulatory and/or postovulatory follicular events is yet to be identified.

## ACKNOWLEDGEMENTS

This work was supported by the Research and Educational Fund, Northwestern Memorial Hospital, and by Grant HD 111356 from the National Institutes of Health.

## REFERENCES

1. T. A. Langan, in: Advances in Cyclic Nucleotide Research, Vol. 3, ed. P. Greengard and G.A. Robison (Raven Press, New York, 1973) p. 99.
2. J.M. Marsh, T.M. Mills, and W.J. LeMaire. Biochem.Biophys. Acta 304(1973)197.
3. M. Hunzicker-Dunn and L. Birnbaumer. Endocrinology 99(1976)185.
4. J.M. Marsh, in: Advances in Cyclic Nucleotide Research, Vol. 6, ed. P. Greengard and G.A. Robison (Raven Press, New York, 1975) p. 137.
5. M.R. Clark, J.M. Marsh, and W.J. LeMaire. Endocrinology 102 (1978)39.
6. M. Hunzicker-Dunn, R.A. Jungmann, and L. Birnbaumer, in: Ovarian Follicular Development and Function, ed. A.R. Midgley, Jr. and W.A. Sadler (Raven Press, New York, in press).
7. S.A. Lamprecht, U. Zor, A. Tsafriri, and H.R. Lindner. J. Endocrinol. 57(1973)217.
8. C.D. Ashby and D.A. Walsh. J.Biol.Chem. 247(1972)6637.
9. S. Goldstein and J.M. Marsh, in: Protein Phosphylation in Control Mechanisms, ed. F. Huijing and E.Y.C. Lee (Academic Press, New York, 1973) p. 123.
10. M. Hunzicker-Dunn and R.A. Jungmann. Endocrinology (1978) in press.
11. D.A. Walsh, C.D. Ashby, C. Gonzalez, D. Calkins, E.H. Fischer and E.G. Krebs. J.Biol.Chem. 246(1971)1977.
12. M. Hunzicker-Dunn and R.A. Jungmann. Endocrinology (1978) in press.
13. M. Hunzicker-Dunn and R.A. Jungmann. Endocrinology (1978) in press.
14. J.D. Corbin, S.L. Keely, and C.R. Park. J.Biol.Chem. 250 (1975)218.

15. J.D. Corbin and S.L. Keely, J.Biol.Chem. 252(1977)910.
16. J.D. Corbin, P.H. Sugden, L.M. Lincoln, and S.L. Keely. J.Biol.Chem. 252(1977)3854.

# STRUCTURAL ANALYSIS OF GRANULOSA CELLS FROM HUMAN OVARIES IN CORRELATION WITH FUNCTION

W. Mestwerdt*, O. Müller** and H. Brandau*

Departments of Gynecology and Anatomy, University of Würzburg/Germany* and University of Berne/Switzerland**

The searching histological studies of Stieve on human ovaries permitted, for the first time, a classification of follicles into 6 stages (see Mestwerdt et al. 1977). These purely light-microscopic investigations afforded, however, only partial insight into the specific cellular activities of different tissue layers in the individual follicle stages.

Electron microscopy, by its outstanding demonstration of characteristic cytological structures both in detail and in terms of morphological organization, lends itself well to function-related morphological studies. The use of morphometric methods further permits the quantitative detection of submicroscopic structures and thus their improved functional interpretation. With this in mind, we conducted an electron microscopic, morphometric study on granulosa cells harvested from human ovaries during the various follicular stages. Our findings on follicular granulosa cells were compared with those obtained from granulosa cells of freshly ruptured follicles and young corpora lutea, respectively.

## Material and methods

1. The material under study was obtained from 50 patients who underwent surgery for a variety of gynecologic diseases. The age of the menstruating patients was between 21 and 50 years.
2. In 13 of the 50 patients, the follicles removed between day 10 and 15 of the menstrual cycle were found to be either developing, mature or on the point of rupture (preovulatory follicles). All follicles were $>1$ cm in diameter. We also examined 4 freshly ruptured follicles and young corpora lutea, respectively, taken between day 17 and 22 of the cycle. To determine the time of follicular rupture as nearly accurate as possible,

the luteinizing hormone (LH) levels in serum were measured in all 13 patients at intervals of 4 to 6 hours, starting at least 24 hours before the day of follicular removal. Measurements were done by radioimmunologic methods (Mestwerdt et al. 1977).

3. The tissue was processed by standard electronmicroscopic methods (Mestwerdt et al. 1977).
4. Full details of the methods used for evaluating electron microcopic, morphometric studies, including preparation of micrographs and calculation of absolute values, have been reported recently (Mestwerdt and Müller 1978).

## Results and Discussion

### A. Primary, secondary, and resting tertiary follicles (group I)

In electron micrographs, the granulosa of primary, secondary, and resting tertiary follicles ($<$1 cm in diameter) exhibited characteristic submicroscopic features involving the cytoplasm and the nucleus (Mestwerdt et al. 1977). A striking morphometric finding in the granulosa cells of this group of follicles was the small mean volume of cells and nuclei. The cytoplasmic structures, when viewed in relation to cell volume, showed abundant endoplasmic reticulum, the rough reticulum being twice that of the smooth reticulum. Another prominent feature was the small lipid volume in the cytoplasm of these cells (see Fig. 1 and Table).

In terms of specific functional properties, the abundance of rough endoplasmic reticulum would seem to indicate that these cells are heavily engaged in protein synthesis. The production of follicular fluid, supply of nutriment to the ovum and, more recently, the formation of a factor inhibiting the effect of LH have been discussed (Ledwitz-Rigby et al. 1977, Mestwerdt et al. 1977).

In addition, very thin filaments were seen in the cytoplasm of granulosa cells harvested from primary and secondary follicles. As these filaments were primarily found in granulosa cells in the immediate vicinity of the zona pellucida, it is reasonable to conclude that they are a likely factor in the formation of the zona pellucida (Hertig and Adams 1976, Martinek and Krausova 1972, Mestwerdt et al. 1977).

It is significant that in these stages of follicular development the granulosa rarely exhibits submicroscopic features characteristic of luteinization and steroid biosynthetic functions, respectively.

### B. Developing, mature and ready-to-rupture follicles (preovulatory follicles), group II

When correlating 13 preovulatory follicles and 4 freshly ruptured follicles with the serum LH levels measured over a short lenght of time, 6 follicles - group IIa - were associated with low LH levels (LH$<$5 ng/ml serum). In the IIb group, 7 follicles were found to correlate with rising LH levels or LH peaking (LH

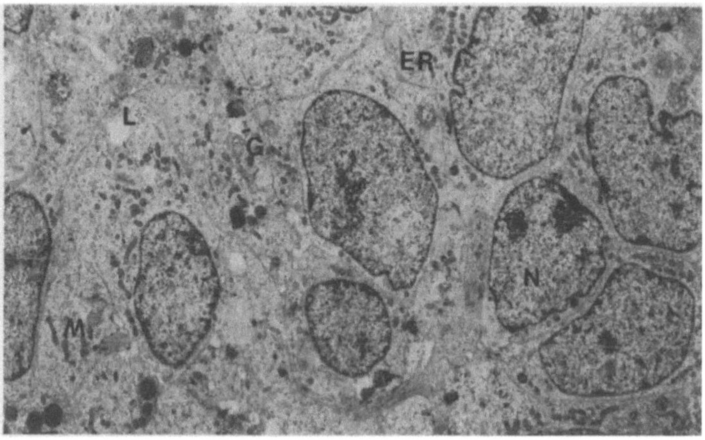


Fig. 1: Representative electronmicrograph of granulosa cells harvested in group I. N = nucleus, M = mitochondria, ER = endoplasmic reticulum, Li = lipid, G = Golgi complex. Magnification 4,060 x.

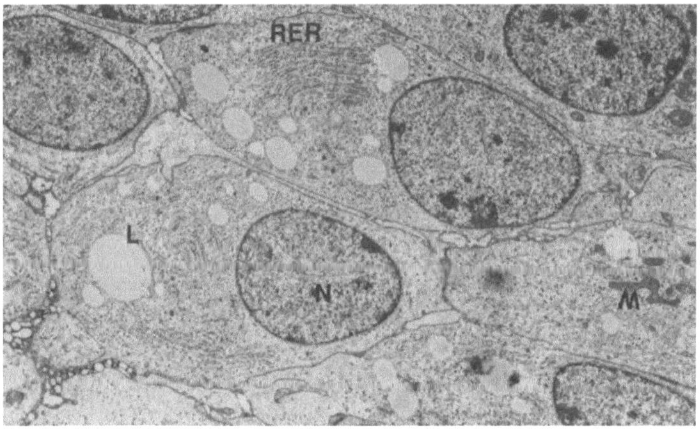


Fig. 2: Representative electron micrograph of granulosa cells harvested in group IIa. RER = rough endoplasmic reticulum. Magnification 4,060 x.

$>$25 ng/ml serum).

IIa: The results of our submicroscopic-morphometric studies on granulosa cells from preovulatory follicles (LH $<$5 ng/ml serum) were comparable to those obtained in resting tertiary follicles. Salient findings were the presence of large lipid droplets and the predominance of rough endoplasmic reticulum over the smooth reticulum (see Fig. 2 and Table).

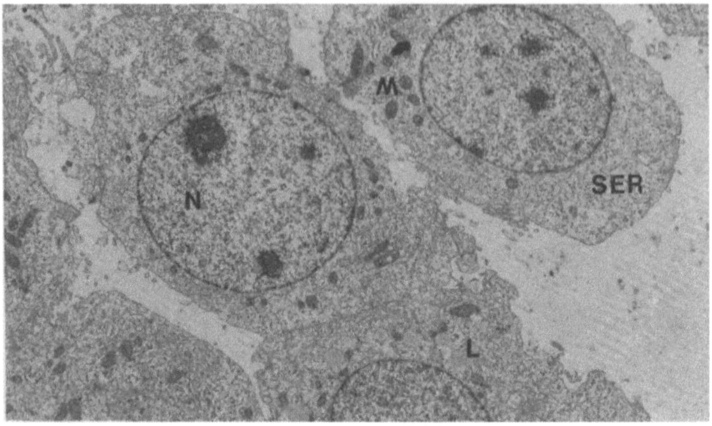


Fig. 3: Representative electron micrograph of granulosa cells harvested in group IIb. SER = smooth endoplasmic reticulum. Magnification 4,060 x.

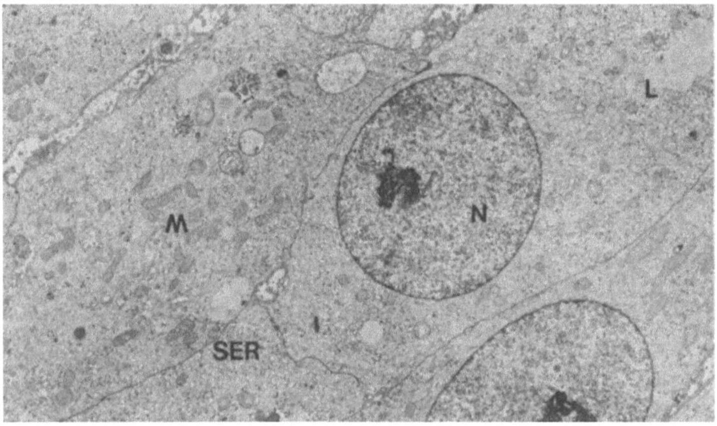


Fig. 4: Representative electron micrograph of granulosa cells harvested in group III. Magnification 4,060 x.

IIb: Our studies on granulosa cells from preovulatory follicles (serum LH levels rising or reaching a peak) revealed changes in fine structure suggesting transformation of granulosa cells active in protein synthesis to cells participating in steroid biosynthesis. The most prominent submicroscopic finding was the predominance of the smooth endoplasmic reticulum over the rough reticulum in the cytoplasm of granulosa cells. The lipid volume showed a threefold increase but the individual lipid droplet decreased slightly in diameter, a feature pointing to a progressive dispersion of lipid in small droplets (see Fig. 3 and Table).

Table

Structural analysis of granulosa cells in correlation with function
(electron microscopic, morphometric findings)

| | group I<br>primary, secondary resting tertiary follicles<br>Ø < 1 cm | group II<br>developing, mature, ready to rupture follicles (preovulatory follicles)<br>Ø > 1 cm | | group III<br>freshly ruptured follicles (young corpora lutea)<br>Ø > 1 cm |
|---|---|---|---|---|
| | | IIa | IIb | |
| cell volume | ≦ 750 μ³ | 750 μ³ | 1.300 μ³ | 5.500 μ³ |
| nuclear volume | ≦ 150 μ³ | 150 μ³ | 250 μ³ | 450 μ³ |
| mitochondrial volume | ≦ 35 μ³ | 35 μ³ | 70 μ³ | 360 μ³ |
| lipid volume | ≦ 20 μ³ | 20 μ³ | 70 μ³ | 220 μ³ |
| lipid (%)/cell | ≦ 2.31 % | 2.31 % | 4.76 % | 4.33 % |
| mean Ø of lipid droplet | ≧ 0.90 mμ | 0.90 mμ | 0.86 mμ | 0.67 mμ |
| surface density of: | | | | |
| rough endopl. retic. | ≦ 1.660 μ² | 1.660 μ² | 2.460 μ² | 7.900 μ² |
| smooth endopl. retic. | ≦ 900 μ² | 900 μ² | 3.130 μ² | 19.200 μ² |
| Golgi complex | ≦ 570 μ² | 570 μ² | 1.580 μ² | 5.980 μ² |
| cell membrane | ≦ 930 μ² | 930 μ² | 2.170 μ² | 5.500 μ² |

C. Freshly ruptured follicles and young corpora lutea, respectively (group III)

In the patients with freshly ruptured follicles and young corpora lutea, respectively, the serum LH levels were found to have decreased ( $<$ 25 ng and 10 ng/ml serum).

It became apparent from our electronmicroscopic-morphometric studies that in this group the process involving conversion of granulosa cells engaged in protein synthesis to cells active in steroid biosynthesis (luteinization) had largely gone to completion. The cellular and nuclear volumes showed fourfold and twofold increases. The surface density of the smooth endoplasmic reticulum increased even sixfold over that of granulosa cells from preovulatory follicles during LH stimulation. It was further conspicuous that the lipid were increased in volume, even though the percentage lipid content per cell was virtually unchanged when compared to that of granulosa cells from preovulatory follicles. Lipid droplets also continued to decrease in diameter (see Fig. 4 and Table). It is exactly this reduction in size that is discussed in association with an increasing mobilization of the lipid pools; these are a probable requirement for the synthesis of the greatly increased membrane system of the endoplasmic reticulum and for progesterone synthesis. We have found an increase in serum progesterone to occur in correlation with the morphological parameter of the granulosa from preovulatory follicles (Mestwerdt et al. 1977). Our submicroscopic findings above the luteinization-process of the granulosa of preovulatory follicles, which emerged under physiologic conditions were found to be comparable in principle to those obtained under experimental conditions by numerous authors (Blanchette 1966, Channing and Crips 1972, Channing and Tsafriri 1977).

With electron microscopic, morphometric methods we exactly quantified the morphologic substrate of the luteinization-process for the first time. It could be demonstrated, that the morphologic feature of the granulosa cell changed fundamentally within a short time both qualitatively and quantitatively respect.

## References

(1) E.J. Blanchette. J.Cell Biol. 31 (1966) 501.
(2) C.P. Channing and T.M. Crisp. Gen. Comp. Endocr. 3 (1972) 617.
(3) C.P. Channing and A. Tsafriri. Metabolism 26 (1977) 413.
(4) A. Hertig and E.C. Adams. J. Cell. Biol. 34 (1967) 647.
(5) F.L. Ledwitz-Rigby, K.L. Gay, J. Young, M. Stetson and C.P. Channing. J. Endocrinol 1977 (in press)
(6) J. Martinek and H. Krausova. Fol. Morph. 20 (1972) 73.
(7) W. Mestwerdt, O. Müller and H. Brandau. Arch. Gynäk. 222 (1977) 45 and 115.
(8) W. Mestwerdt and O. Müller. Arch. Gynäk. 225 (1978) 51.

# REGULATION OF LH(hCG)-RECEPTOR AND ADENYLATE CYCLASE IN LUTEINIZING GRANULOSA CELLS

Hannu J. Rajaniemi and Kalle Jääskeläinen

Departments of Anatomy and Biochemistry, University of Kuopio, 70101 Kuopio 10, Finland

The exposure of luteal and testicular tissue in vivo to hCG has been shown to induce a rapid loss of sensitivity of adenylate cyclase (AC) to subsequent LH- or hCG-stimulations. The explanation for this phenomenon appears to be the loss of available receptor sites (1-4). However, the molecular mechanisms of the receptor loss are completely unknown.

Since a similar high and transient gonadotropin stimulation occurs physiologically in granulosa cells prior to ovulation, we decided to investigate the effect of hCG on gonadotropin sensitive AC and LH(hCG)-receptor sites in granulosa cells. To elucidate the molecular mechanisms involved in the regulation of LH(hCG)-receptor sites we have also analysed the catabolic fate of hCG in granulosa cells following the initial *in vivo* uptake.

## MATERIALS AND METHODS

Immature 26 day old rats (Long-Evans) were injected s.c. at 0900 h with 15 IU of PMSG (Organon, Oss, Holland) to initiate and synchronize the follicular development. Two days later at 1300 h the animals received a single tail vein injection of 1, 5 or 100 μg of hCG (10500 IU/mg, Organon, Oss, Holland) in 0.2 ml of 0.01 M phosphate buffered 0.14 M NaCl (pH 7.0). Each dose of hCG was mixed with $10^6$ c.p.m. of $^{125}$I-labelled hCG prior to the injection. The hormone was radioiodinated by the chloramine-T-method (5). The specific activity of the preparations averaged 25 μCi/μg. Granulosa cells were isolated in 0.01 M Tris-HCl (pH 7.5) containing 1 mM EDTA and 0.79 M sucrose 15 and 30 min and 1, 2, 6 and 24 h after the hCG-injections. The cells were washed once and homogenized in the same medium to give a protein concentration of about 5 mg/ml. Changes in basal, LH-, FSH- and fluoride sensitive AC activity were measured by quantitating the formation of $^{32}P$ cyclic AMP from α-labelled $^{32}P$-ATP (6). The hormone stimulations were carried out including 1 μg of ovine LH (NIH-LH-S16), hCG (10500 IU/mg), ovine FSH (NIH-FSH-S9) or 10 mM NaF in the incubations. The incubation time was 20 min at 30°C, when the reaction was still linear.

The occupied receptor sites were quantitated by means of the *in vivo* bound labelled hCG and the free receptor sites by measuring the remaining capacity of the homogenate to bind labelled hCG.

## Degradation of hCG in Granulosa Cells *in Vivo*

PMSG-primed immature rats were injected with 5 μg of hCG mixed with $10^6$ c.p.m. of labelled hCG. The animals were killed 6 h after the hormone injection and granulosa cells were isolated and washed in PBS. Subsequently, the cells were homogenized in 0.5 % Triton X-100 in PBS. Undissolved particles were sedimented by centrifuging at 106.000 x g for 2 h at 4°C in MSE-ultracentrifuge. Approximately 90 % of the bound radioactivity was obtained in the supernatant. An aliquot of the supernatant was subjected to a gel filtration on a Sepharose 6B column (2.6 x 63 cm) equilibrated and eluted with 0.5 % Triton X-100 in PBS at 4°C. The column was calibrated using blue dextran, labelled hCG, hCG-α- (UCB, Bruxelles, Belgium) and -β-subunit (NIAMDD, NIH, Bethesda, U.S.A.) and radioiodine as standards. The radioactive peaks were identified by means of partition coefficients, biological and immunological activity as described elsewhere (7). The low molecular weight radioactivity was analysed by fractionation on a Copper-Sephadex G-25 column as described by Fazakerley and Best (8).

## Degradation of Labelled hCG in Granulosa Cell Homogenate *in Vitro*

Granulosa cells were isolated from ten PMSG-primed immature rats washed in PBS and homogenized in 1 ml of 0.01 M phosphate buffer (pH 6.5) containing 0.05 % Triton X-100. Labelled hCG ($10^5$ c.p.m.) was added to the homogenate and the mixture was incubated for 2 h at 37°C. The incubation was terminated by the addition of 2 ml of ice-cold 0.5 % Triton X-100 in PBS. The particles were re-homogenized and the soluble fraction was separated by centrifuging at 106.000 x g for 2 h at 4°C in MSE-ultracentrifuge and fractionated on a Sepharose 6B column as described above.

## RESULTS AND DISCUSSION

The administration of 5 µg of hCG to PMSG-primed immature rats results in a 6-fold increase in basal AC activity within 6 h after the hormone injection, associated with a marked decrease in LH-sensitivity of the enzyme as early as 1 h from the injection (Fig. 1). The FSH-sensitivity of AC decreased more slowly, reaching a low level by 6 h. In contrast, the fluoride sensitivity of AC only fell by about 15 % within the first hour and remained thereafter at the reduced level for the subsequent 23 h. To determine whether the loss of LH-sensitivity of AC is due to corresponding changes in LH(hCG)-receptor sites the remaining capacity of the granulosa cell homogenate to bind labelled hCG was analysed after different doses of hCG. The in vitro binding of labelled hCG decreased rapidly after hCG-injections (Fig. 2). The administration of 100 µg of hCG induced a complete loss of hCG-binding within 15 min, 5 µg of hCG within 6 h and 1 µg of hCG within 24 h. It is interesting to note that although 68 % of the receptors were unoccupied, at 15 min after the injection of 5 µg of hCG there was a pronounced decrease in LH-stimulatable AC activity. This disparity suggests that besides the receptor loss there may be other mechanisms involved in the desensitization of AC to luteinizing hormones in the granulosa cells.

These results suggest that the desensitization of AC to LH-stimulation in granulosa cells after hCG-injections is associated with the loss of available LH(hCG)-receptor sites. However, other mechanisms may also be involved in the initial desensitization. One possible explanation for the loss of LH(hCG)-receptors would be the occupancy of the receptor sites by the injected hCG (9). We tested

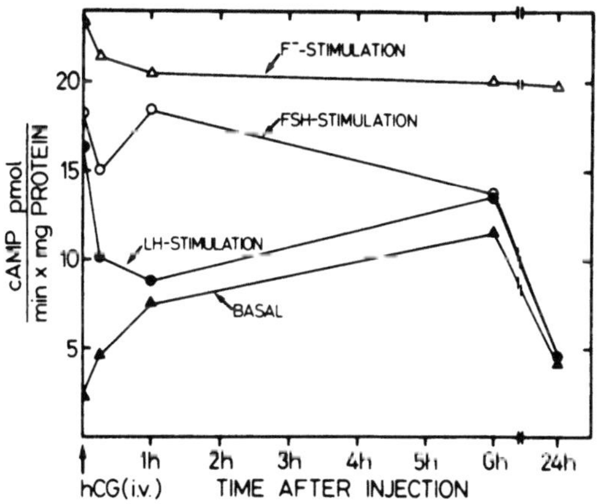


Fig. 1. Time course of changes in basal, LH-, FSH- and fluoride sensitivity of AC in granulosa cells following the injection of 5 µg of hCG.

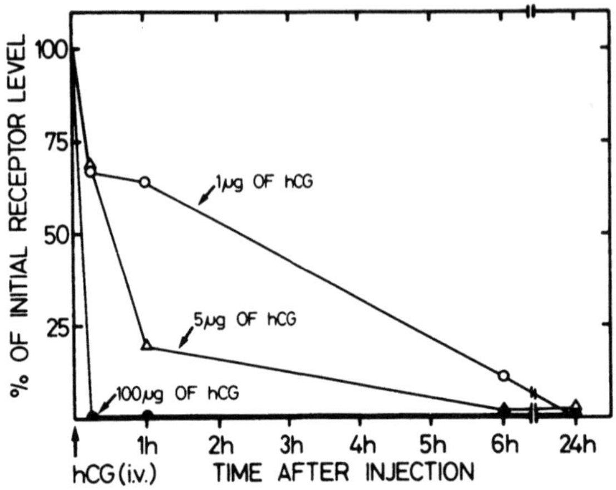


Fig. 2. Time course of changes in hCG-binding capacity of granulosa cells following the injection of 1, 5 or 100 µg of hCG.

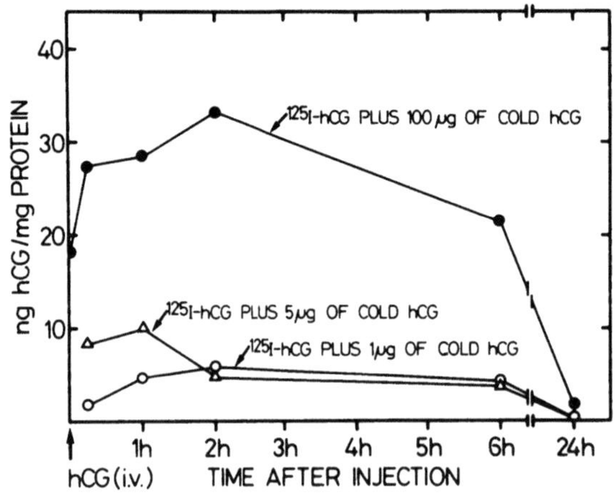


Fig. 3. Concentration of hCG in the granulosa cells at different times after the injection of 1, 5 or 100 µg of hCG mixed with $10^6$ c.p.m. of labelled hCG.

this possibility by measurement of the *in vivo* occupancy of the receptors after different doses of hCG. The bound hormone was quantitated by means of the labelled hCG mixed with the "cold" hormone prior to the injection. The occupancy increased to a maximum within 1-2 h after each dose of hCG, began to decrease thereafter and reached a low level by 24 h (Fig. 3). There was no corresponding increase in free receptor sites associated with the disappearance of the *in vivo* occupancy. These findings suggest that the initial decrease in LH-sensitivity of AC is associated with the occupancy of the receptors. However, the prolonged loss cannot be attributed to the occupancy but rather to a destruction of both the receptors once occupied by the hormone as well as the free receptor sites. Thus, the receptor occupancy possibly inititates a process which leads to an inactivation not only of the occupied sites but also of free receptor sites.

The biochemical mechanisms involved in the inactivation of the receptor sites are not known. Some reports about a possible internalization and subsequent degradation of the hormone have been published (10). We studied also in more detail the in vivo degradation of labelled hCG after the initial uptake. The gel filtration of the Triton X-100 extracts prepared from the granulosa cells 6 h after the injection of $10^6$ c.p.m. of labelled hCG alone or mixed with 1 or 5 μg of "cold" hCG displayed three radioactive peaks (Fig. 4). The first peak was identified as a hormone-receptor complex. The radioactive material in the second peak was eluted at the same volume with intact labelled hCG and it bound to LH(hCG)-receptor of pseudopregnant rat ovaries and to anti-hCG and anti-α-subunit antibodies suggesting that it contained free labelled hCG. The elution volume of the third peak was same as labelled α-subunit and the material bound to anti-α-subunit antibodies but not to LH(hCG)-receptor. This suggests strongly that it represents labelled α-subunit released from the hormone. To determine whether the labelled hormone is catabolized in vitro by granulosa cells, labelled hCG was incubated in granulosa cell homogenate for 2 h at 37°C. After the incubation the radioactivity was distributed to three

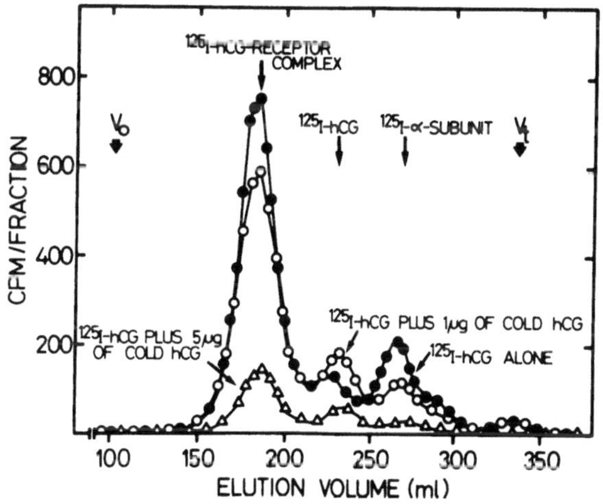


Fig. 4. Gel filtration of Triton X-100 extract of granulosa cells on a Sepharose 6B column. The animals received 6 h before the killing $10^6$ c.p.m. of labelled hCG alone or mixed with 1 or 5 μg of "cold" hCG.

radioactive peaks in gel filtration (Fig. 5). The first peak was identified as labelled hCG and the second one as α-subunit of the hormone. The third peak was further analysed on a Copper-Sephadex G-25 column and it was eluted at the same volume with radioiodonated tyrosine suggesting that a part of the hormone was hydrolysed to amino acids during the incubation.

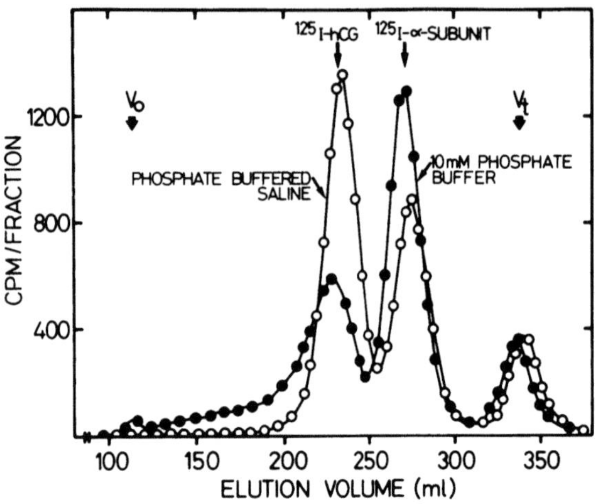


Fig. 5. Gel filtration of Triton X-100 extract of granulosa cell homogenate after *in vitro* incubation with labelled hCG in the presence and absence of 0.14 M NaCl.

Our results suggest that the LH(hCG)-receptors in granulosa cells are regulated negatively by hCG. The regulation appears to involve both occupied and free receptor sites. The discrepancy between the rate of disappearance of LH(hCG)-receptor and LH-sensitivity of AC suggests that besides the loss of LH(hCG)-receptor other mechanisms may also be involved in the desensitization of AC to LH stimulation. The extensive degradation of hCG to subunits and amino acids by the granulosa cells *in vivo* and *in vitro* may indicate that the receptor regulation involves an internalization and catabolism of the hormone or hormone receptor complex.


ACKNOWLEDGMENTS

The authors wish to thank Dr. A. Rees Midgley, Jr., Univ. of Michigan for the gift of purified hCG for radioiodinations, NIAMDD, NIH, for hormone preparations and antiserums for hCG-subunits. This investigation was supported by a grant from The National Research Council for Medical Sciences, Finland.

## REFERENCES


(1) R.M. Sharpe. Nature 264(1976)644.
(2) A.J.W. Hsueh, M.L. Dufau and K.J. Catt. Biochem.biophys.Res. Commun. 72(1976)1145.
(3) M. Conti, J.P. Harwood, M.L. Dufau and K.J. Catt. Molec. Pharmacol. 13(1977)1024.
(4) M.C. Rao, J.S. Richards, A.R. Midgley, Jr. and L.E. Reichert, Jr. Endocrinology 101(1977)512.
(5) F.L. Leidenberger and L.E. Reichert, Jr. Endocrinology 91 (1972)135.
(6) C. Nakai and G. Brooker. Biochem.biophys.Acta (Amst.) 391 (1975)222.
(7) S. Markkanen, K. Töllikkö, T. Vanha-Perttula and H. Rajaniemi. Submitted for publication (1978).
(8) S. Fazakerley and D.R. Best. Anal.Biochem. 12(1965)290.
(9) M. Conti, J.P. Harwood, A.J.W. Hsueh, M.L. Dufau and K.J. Catt. J.Biol.Chem. 251(1976)7729.
(10) M. Ascoli and D. Puett. FEBS Letters 75(1977)77.

# CHANGES IN RESPONSIVENESS OF PORCINE GRANULOSA CELLS TO PROSTAGLANDINS AND LUTEINIZING HORMONE IN TERMS OF CYCLIC-AMP ACCUMULATION DURING FOLLICULAR MATURATION

Alison Schwartz-Kripner and Cornelia P. Channing

Dept. of Physiology, University of Maryland
School of Medicine
660 West Redwood Street, Baltimore, Md. 21201

## INTRODUCTION

Kolena and Channing (1) have shown previously that: 1.) luteinizing hormone (LH) and prostaglandin $E_2$ ($PgE_2$) can stimulate cyclic-AMP accumulation by granulosa cells from medium-sized porcine follicles and 2.) the effects of these substances when mixed together were more than additive. Such a finding suggests but does not prove they acted at different receptor sites. Channing (2) and Lindsey and Channing (3) have found that granulosa cells obtained from small (1-2 mm) porcine follicles responded minimally to LH while cells harvested from medium and large follicles were more responsive to LH. Cells from large follicles produced 6-15 fold more cyclic-AMP in comparison to controls when stimulated by LH, while cells from small follicles exhibited not more than a three-fold increase in cyclic-AMP accumulation in response to LH. If it is true that prostaglandins of the E and F series have separate receptor and/or adenyl cyclase systems compared to LH, then it would be expected that changes in responsiveness of porcine granulosa cells obtained from different sized follicles to prostaglandins would be different compared to LH.

To test this hypothesis, the ability of $PgF_{2\alpha}$, $PgE_2$ and LH to stimulate cyclic-AMP accumulation in granulosa cells obtained from small (1-2 mm), medium (3-5 mm) and large (6-12 mm) porcine follicles was examined.

## MATERIALS AND METHODS

Granulosa cells were harvested by aspiration using ovaries obtained from a local slaughterhouse as outlined in detail previously (4). The cells were centrifuged at 500 g to separate them from follicular fluid and washed three times in Eagle's medium. The final washed granulosa cells were counted in a hemocytometer in 0.06% trypan blue in Hank's solution and the live cells (cells which excluded the dye) and the dead cells (cells which had a distinct nucleus and cytoplasm but took up the dye) enumerated. The cells were resuspended in Eagle's medium plus 25 mM Hepes buffer and 1% bovine serum albumin at a concentration of 1 to 10 x $10^7$ cells/ml/medium. The 1% BSA (Metrix Corp.) contained nondetectable amounts of LH activity in a radioreceptor assay (5); <1 ng/ml in terms of NIH ovine LH. The cells were distributed in 1 ml of medium into Packard glass scintillation vials and incubated for 30 minutes with either nothing, PG $E_2$, Pg $F_{2\alpha}$ or ovine LH in 5% $CO_2$ in air at $37^o$ C with shaking for 30 minutes. At the end of the incubation the vials were placed on ice. The cells were separated from the medium by centrifugation at 1000 g for 5 minutes. The cells were mixed with 0.5 ml boiling 2 mM Na Acetate followed by sonication and centrifugation at 2000 g for 15 minutes. The soluble supernatant was assayed for cyclic-AMP by a modification of the Gilman assay (1,4,6). Each sample was assayed in duplicate. Each dose of prostaglandin or LH was added to 3-4 replicate vials of cells in each incubation using a given batch of cells. Three batches of cells were examined.

Ovine LH (NIH LH S 19) was donated by the Hormone Distribution Offices of the National Institute of Arthritis and Metabolic Diseases, NIH. Prostaglandins, Pg $E_2$, and the THAM salt of PG $F_{2\alpha}$ were generously donated by Dr. J.E. Pike of the Upjohn Co., Kalamazoo, Michigan. The LH and Pg were dissolved in water or absolute ethanol immediately prior to use. The Pg $E_2$ was added in a volume of 10 µl absolute ethanol per 1.0 ml medium. This amount of ethanol had no effect on cyclic-AMP accumulation. The Pg $F_{2\alpha}$ THAM salt was added in 20 µl of distilled water. Statistical differences were analyzed by a Students t test.

## RESULTS

The effect of various doses of LH and prostaglandins upon cyclic-AMP accumulation using two separate batches of granulosa cells is shown in Table 1 and 2.

TABLE 1. Relative responsiveness of granulosa cells from small, medium and large porcine follicles to oLH, $P_gE_2$ and $P_gF_{2\alpha}$ in terms of % control of cyclic 3',5'-AMP production. (Experiment 1)

| Addition to Cells | Cyclic AMP Accumulation, Follicular source of cells: SMALL | | MEDIUM | | LARGE | |
|---|---|---|---|---|---|---|
| | pMoles cAMP/ 5 x $10^7$ cells | % Control | pMoles cAMP/ 5 x $10^7$ cells | % Control | pMoles cAMP/ 5 x $10^7$ cells | % Control |
| Control | 9.5 ±0.74 | ------- | 3.2 ±0.33 | ------- | 102.2± 9.3 | ------- |
| 0.1 LH | 11.98 ±0.49 | 125±5[c] | 5.89±0.91 | 185±28[c] | 1038 ±91.4 | 1114±98[a] |
| 1.0 LH | 35.7 ±3.22 | 373±34[a] | 13.64±0.88 | 429±28[a] | 1555 ±72.9 | 1670±78[a] |
| 10 $P_gE_2$ | 19.79±0.8 | 207±8.4[a] | 3.62±0.16 | 111±6.1[d] | 133.21±6.17 | 143±6.6[b] |
| 5.0 $P_gE_2$ | 18.73±1.6 | 196±17[b] | 4.36±1.15 | 137±36[d] | 118.6 ±8.17 | 127±8.6[c] |
| 10.0 $P_gE_2$ | 31.66±3.7 | 331±38[a] | 3.15±0.11 | 99±3.5[d] | 119.3 ±7.2 | 128±7.6[c] |
| 100 $P_gE_2$ | 60.31±0.41 | 631±4.2[b] | 15.02±1.72 | 472±54[d] | 138.9 ±4.2 | 149±4.7[b] |
| 5.0 $P_gF_{2\alpha}$ | 9.13±0.72 | 95±7.6[d] | 3.69±0.24 | 116±7.5[d] | 106.9 ±5.4 | 115±5.7[d] |
| 10.0 $P_gF_{2\alpha}$ | 8.17±0.2 | 85±2.2[d] | 3.58±0.31 | 115±7.5[d] | 107.02±9.1 | 115±9.7[d] |
| 100 $P_gF_{2\alpha}$ | 12.91±1.56 | 135±16[d] | 2.95±0.13 | 92±4.2[d] | 106.8 ±9.1 | 115±9.7[d] |

All doses expressed refer to μg/ml quantities.
The granulosa cells were incubated for 30 min in Eagle's medium plus 1% BSA at 37° C.
Values shown are the means of 4 replicate incubations.

| p vs. control | |
|---|---|
| a | p = <0.001 |
| b | p = <0.01 |
| c | p = <0.05 |
| d | p = N.S. |

TABLE 2. Relative Responsiveness of granulosa cells from small, medium and large porcine follicles to oLH, $P_gE_2$ and $P_gF_{2\alpha}$ in terms of stimulation of cyclic AMP accumulation. (Experiment 2)

| Addition to Cells | Cyclic AMP Accumulation — Follicular source of cells: SMALL pMoles cAMP/ $5 \times 10^7$ cells | SMALL % Control | MEDIUM pMoles cAMP/ $5 \times 10^7$ cells | MEDIUM % Control | LARGE pMoles cAMP/ $5 \times 10^7$ cells | LARGE % Control |
|---|---|---|---|---|---|---|
| Control | $16.1 \pm 0.6$ | ----- | $16.9 \pm 1.5$ | ----- | $144.6 \pm 8.9$ | ----- |
| 0.01 oLH | $11.7 \pm 0.9^c$ | 73 | $17.1 \pm 1.5^d$ | 101 | $120.7 \pm 6.1^d$ | 83 |
| 0.1 oLH | $23.2 \pm 3.1^c$ | 144 | $30.4 \pm 1.2^c$ | 180 | $292 \pm 37.7^a$ | 202 |
| 1.0 oLH | $22.5 \pm 0.9^c$ | 139 | $50.9 \pm 2.4^a$ | 302 | $813 \pm 252$ | 563 |
| 1.0 $P_gE_2$ | $26.3 \pm 1.4^c$ | 163 | $23.6 \pm 1.1^c$ | 140 | $148 \pm 4.1^d$ | 103 |
| 5.0 $P_gE_2$ | $23.1 \pm 1.5^c$ | 142 | $22.9 \pm 0.9^c$ | 136 | $177 \pm 17^d$ | 122 |
| 10.0 $P_gE_2$ | $20.4 \pm 1.2^c$ | 126 | $22.9 \pm 1.4^c$ | 136 | $176 \pm 11^d$ | 122 |
| 100 $P_gE_2$ | $47.9 \pm 7.6^a$ | 297 | $85.1 \pm 5.6^a$ | 504 | $214 \pm 28^d$ | 148 |
| 5 $P_gF_{2\alpha}$ | $10.5 \pm 0.7^a$ | 65 | $16.3 \pm 2.3^d$ | 100 | $125 \pm 6.5^d$ | 87 |
| 10 $P_gF_{2\alpha}$ | $10.3 \pm 1.3^c$ | 64 | $15.6 \pm 0.5^d$ | 93 | $125 \pm 7.5^d$ | 86 |
| 100 $P_gF_{2\alpha}$ | $11.0 \pm 1.1^c$ | 68 | $21.1 \pm 1.2^c$ | 125 | $119 \pm 6.4^d$ | 83 |

All doses expressed refer to µg/ml quantities.
The granulosa cells were incubated for 30 min in Eagle's medium plus 1% BSA at 37° C.
Values shown are the means of 4 replicate incubations.

p vs. control

a p = <0.001
b p = <0.01
c p = <0.05
d p = N.S.

A summary of these data plus data from a third experiment is presented in Table 3.

TABLE 3. Summary of effect of LH, $P_gE_2$ and $P_gF_{2\alpha}$ upon cyclic-AMP accumulation.

| | Cyclic AMP accumulation (Percent of Control) | | | | | |
|---|---|---|---|---|---|---|
| Addition of Cells | Follicular Source of cells: SMALL | MEDIUM | | | LARGE | |
| | $\bar{X}$ SE (n) | $\bar{X}$ SE (n) | P vs small | P vs large | $\bar{X}$ SE (n) | P vs small |
| 1.0 μg LH | 243$\pm$43(9) | 356$\pm$26(9) | 0.04 | <0.001 | 1054$\pm$197(9) | <0.001 |
| 10 μg $P_gE_2$ | 210$\pm$30(12) | 154$\pm$20(12) | NS | NS | 119$\pm$5 (12) | <0.01 |
| 100 μg $P_gE_2$ | 367$\pm$63(12) | 489$\pm$29(9) | NS | <0.001 | 128$\pm$13(12) | 0.003 |
| 10 μg $P_gF_{2\alpha}$ | 73$\pm$6 (9) | 101$\pm$6 (9) | 0.003 | NS | 91$\pm$7 (12) | NS |
| 100 μg $P_gF_{2\alpha}$ | 101$\pm$15(8) | 110$\pm$70(9) | NS | NS | 97.7 (9) | NS |

The data shown are the composite of 3 batches of granulosa cells.
Each batch of cells was incubated for 30 minutes in Eagle's medium plus 1% BSA at 37° C.

As the follicle matured there was an increase in the ability of LH to stimulate cyclic-AMP which was significantly greater in large compared to medium and small follicles. In the case of cells from large follicles, 1.0 μg/ml LH led to a 10-fold stimulation in cyclic-AMP accumulation which is to be contrasted to a 2.4 and 3.5-fold stimulation in the case of cells obtained from small and medium ($p<0.001$) follicles. A dose of 1.0 μg/ml LH was also effective in stimulating cyclic-AMP accumulation but to a lesser degree. Granulosa cells obtained from large follicles were less responsive to $P_gE_2$ (10 or 100 μg) compared to cells obtained from small and medium follicles ($p < 0.001$) (Table 3). Cells obtained from large follicles 100 μg $P_gE_2$ produced an average of 1.28-fold stimulation whereas addition of a similar dose produced 3.67 and 4.88-fold stimulation in cells from small and medium follicles ($p < 0.01$ large vs. small or medium follicles) in each experiment. Neither 1, 10 nor 100 μg $P_gF_{2\alpha}$ significantly altered cyclic-AMP levels in cells from small, medium or large follicles.

## DISCUSSION

These findings demonstrate that as the porcine follicle matures there is a decrease in granulosa cell responsiveness to prostaglandin $E_2$ concurrent with an increase in LH responsiveness in terms of percent stimulation of cyclic-AMP accumulation. This change was most marked in the case of cells harvested from small and medium compared to those harvested from large follicles. This finding, taken in conjunction with earlier observations of Kolena & Channing (1) of an additive effect of $P_gE_2$ and LH in granulosa cells from medium-sized follicles is best explained by the existence of two separate receptors: one for LH and another for prostaglandin $E_2$. Whether there is a separate adenyl cyclase responding to each stimulatory agent cannot be answered by these findings. The findings of an increase in responsiveness of granulosa cells to LH as the follicle matures is similar to our preliminary observations (7,3) and those of Lee (8). It is noteworthy that responsiveness of granulosa cells to $P_gE_2$ somewhat resembles that of FSH; the responsiveness of granulosa cells to FSH was greater in the case of granulosa cells obtained from small and medium-sized follicles. A more than additive response in terms of cyclic-AMP accumulation was produced and would indicate that the two agents may not act upon the same receptor. The relationship between FSH and prostaglandin $E_2$ in activation of adenyl cyclase merits further investigation.

In terms of actual pMoles of cyclic-AMP rather than percent of control values, the changes of responsiveness of granulosa cells to $P_gE_2$ as a function of follicular maturity were small. In the case of LH, the increase in responsiveness with follicular maturity was many fold greater, supporting the concept of a basic difference in the action of the two agents. Differences in phosphodiesterase levels were probably not the causal agent in the changes of LH vs. prostaglandin responsiveness since addition of a phosphodiesterase inhibitor, theophylline, only exerted a 10-15% increase in the effect of LH and prostaglandin (1, unpublished observations).

Lack of response of granulosa cell cyclic-AMP levels to $P_gF_{2\alpha}$ could indicate a lack of cyclic-AMP mediation in its effects upon the granulosa cell.

---

The support of grant HD 08834 from the National Institute of Child Health and Human Development is gratefully acknowledged. We thank Sandy Fowler for her expert technical assistance.

## REFERENCES

(1) J. Kolena and C.P. Channing. Endocrinology 90(1972)1543.

(2) C.P. Channing. in: Internat'l Congress Series No. 273, (Excerpta Medica Amsterdam, 1972) p. 914.

(3) A.M. Lindsey and C.P. Channing. Abstr. in Program of 9th Ann. Meeting of Soc. for Study of Reproduction (1976).

(4) C.P. Channing and F. Ledwitz-Rigby, in: Methods in Enzymology (Academic Press, New York, 1975)183.

(5) C.P. Channing, in: Gonadotropins and Gonadal Function (Academic Press, New York, 1974)185.

(6) I. Mashiter, G.D. Mashiter, R.L. Hanger and J.B. Field, Endocrinology 92(1973)541.

(7) C.P. Channing, in: International Congress Series No. 273, Endocrinology, Proc. of 4th International Congress of Endocrinology (Excerpta Medica, Amsterdam, 1972)914.

(8) C.Y. Lee. Endocrinology 99(1976)42.

# SOME MORPHOLOGICAL AND FUNCTIONAL CHARACTERISTICS OF CELLS OF THE PORCINE THECA INTERNA IN TISSUE CULTURE

S. Stoklosowa, J. Bahr[1], and E. Gregoraszczuk

Department of Animal Physiology, Institute of Zoology

Jagiellonian University, Krakow, Poland

[1]Department of Animal Science, Animal Genetics Laboratory

University of Illinois, Urbana, Illinois 61801

## ABSTRACT

Theca interna cells were collected from large porcine follicles in the follicular phase of the cycle. The theca interna layer was manually separated from the theca externa, washed thoroughly, occasionally treated with hyaluronidase solution and trypsinized. The resulting cell suspension consisted of approximately 70% theca interna cells, 8% granulosa cells, and 23% nonsteroidogenic cells. In one experiment, cells collected from the granulosa layer and the theca interna of the same follicles were compared. Specific cell types were identified by: 1) staining with Oil Red O (ORO) to visualize lipid droplets in the cytoplasm, 2) measuring activity of the $\Delta^5$, 3β-hydroxysteroid dehydrogenase ($\Delta^5$,3β-OH SH) enzyme, and 3) quantitating estrogen production by the cultured cells. Nonsteroidogenic cells do not stain by any of the applied techniques.

Theca interna cells can be distinguished from granulosa cells in suspension prior to culture by the following criteria:

1) more intense staining with ORO,
2) stronger $\Delta^5$,3β OH SH activity,
3) larger cell size.

Theca interna cells in culture can be differentiated from granulosa cells by the following parameters:

1) more elongated shape,
2) slower and more chaotic growth,

3) weaker enzyme activity,
4) increased estrogen production.

## Introduction

The techniques for the collection of granulosa cells and their subsequent culturing are well established (1-7). In contrast the isolation of theca cells has met with many difficulties. In the past explants of theca interna tissue have been used in either short term incubation (8) or in organ culture (4, 9, 10). This communication describes a method used to isolate theca interna cells from porcine follicles and the development of primary cultures.

## Materials and Methods

Ovaries were obtained from pigs killed at a slaughter house and were immediately placed in cold sterile saline containing 240 U penicillin, 0.2 mg streptomycin, and 20 IU mucostatin/ml. Ovaries with large follicles (1.0 - 1.2 cm in diameter) were selected for use. To isolate the theca interna tissue, follicles were slit and granulosa cells removed with a platinum loop. Then the follicles were excised, placed in a Petri dish and cut open exposing the interior of the follicle. The remaining granulosa cells were scraped off using a blunt spatula and then flushed with a vigorous stream of phosphate buffered saline (PBS). To remove the remaining granulosa cells, the tissue was rinsed in a solution containing 10 IU/ml of hyaluronidase. With the aid of a dissecting microscope, the theca interna was manually separated from the theca externa. The theca interna tissue was then washed in Medium 199 to remove any hyaluronidase contamination. The tissue was then transferred to fresh Medium 199 and energetically pipetted many times with a capillary pipette having a large orifice. This procedure was repeated until no granulosa cells were visible in the medium when checked under a microscope.

To prepare the cell suspension, the theca interna tissue was minced with a scissors and placed in a 0.25% trypsin solution for 10 min at 37°C followed by a washing and centrifugation. After this procedure was repeated 3 times, the theca interna tissue was dispersed. Cells were tested for viability with trypan blue stain. The uptake of Oil Red O (ORO) (Casselman, 1959) and the presence of $\Delta^5$,3β-hydroxysteroid dehydrogenase (3β-OH SD) (Fischer and Kahn, 1972) were markers used to differentiate theca interna cells from granulosa cells and nonsteroidogenic cells in the cell suspension. In the resulting cell suspension, there were approximately 70% theca interna cells, 8% granulosa cells, and 23% nonsteroidogenic cells.

Cultures of cells obtained from theca interna layers and granulosa cells were established using Medium 199 supplemented with 15% calf serum. Though no quantitative measurements of LH and FSH in the serum were made, the levels of gonadotropins were apparently of sufficiently low levels because reduction of the serum in the Medium 199 from 15% to 7% made no difference in cell growth or morphology. According to radioimmunoassay measurements, progestins were nondetectable while 6-8 picograms/ml of estrogens were measured. Cells were grown in monolayers in Leighton tubes at 37$^{o}$C. Medium was changed every second or third day and saved for estrogen determination.

## Results and Discussion

The presence of distinctive morphological and histochemical characteristics permits one to distinguish cells from theca interna tissue and granulosa cells both while preparing cell cultures and after cultures have been established. Prior to culturing, cells of theca interna tissue stain more intensely with ORO while granulosa cells take up less ORO and these granules appear closer to the surface. Another means of distinguishing these two cell types is by size. Granulosa cells are smaller, being approximately 10 to 14 μ in diameter whereas cells from theca interna tissue range in diameter from 17 to 23 μ. In culture the growth of the cells from the theca interna layer is more chaotic and cells have elongate cell bodies whereas granulosa cells grow in compact colonies and have a luteal morphology. While prior to culture the cells from the theca interna layer demonstrate a stronger 3β-OH SD activity, after culturing granulosa cells exhibit a more intense 3β-OH SD activity which may be attributed to the fact that granulosa cells luteinize spontaneously in culture (11).

Measurements of estrogen secreted into the medium indicate that theca interna cells secrete significantly more estrogen than granulosa cells. On day 2 of culture, theca interna cells secreted approximately 7 to 10 times more estrogen than did the granulosa cells. By day 4 of culture, estrogen content of the medium of the theca interna cells had decreased significantly. This decrease is partially caused by the fact that many thecal cells have not attached to the glass and are lost during the medium change. The growth of thecal cells in culture is satisfactory but certainly not as rich as that of the granulosa cells. Cultures have been maintained for 15 to 20 days at which time both granulosa cells and thecal cells become fibroblastic. This phenomenon is probably connected with the aging and degeneration of cultured cells.

In summary, we have reported a method for the preparation of cells from porcine theca interna tissue for culture. In addition, we have suggested specific criteria which can be used to distinguish theca interna cells from granulosa cells prior to and after the establishment of cultures.

(Supported by NIH PL 480 Research Agreement No. 05-001-0, a-nex 05-040-N.)

## REFERENCES

(1) L. Bjersing. Acta Pathol. Microbiol. Scand. 55(1962)127.

(2) C. P. Channing. Nature. 210(1966)1266.

(3) C. P. Channing. J. Endocr. 43(1969)403.

(4) C. P. Channing. J. Endocr. 45(1969)297.

(5) C. P. Channing. Endocrinology. 87(1970)49.

(6) T. V. Fischer and R. H. Kahn. In Vitro. 7(1972)201.

(7) S. G. Korenman, P. M. Lolc, I. Beceiro, B. M. Sherman, and D. K. Granner. Endocrinology. 93(1973)1423.

(8) E. Lacroix, W. Eechaute, and I. Leusen. Steroids. 23(1974)337.

(9) M. Ryle, J. Kent, and J. Boggis. Ann. Biol. Anim. Bioch. Biophys. 13(1973)223.

(10) P. Rondell. Biol. Reprod. 10(1974)199.

(11) C. P. Channing. Endocrinology. 87(1970)156.

# THE CONTROL OF OVULATION IN THE COW BY PROSTAGLANDIN INHIBITOR

M. Shemesh

Dept. of Hormone Research, Kimron Veterinary Institute, Bet-Dagan, Israel

## INTRODUCTION

Recently, it has been shown that follicular fluid from small porcine follicles diminished accumulation of cyclic AMP in response to LH and decreased secretion of progesterone by cultured granulosa cells obtained from large follicles (1). Prostaglandins (PG) have been implicated in the functioning of the gonads mainly in the induction of ovulation-related events. Several investigations have presented evidence demonstrating that prostaglandins are involved in the ovulatory action of luteinizing hormone (LH). Indomethacin, a potent inhibitor of prostaglandin synthesis (2) prevents ovulation in the rat (3-5) and in the rabbit (6, 7). Several authors have also reported that in the rat the ovarian content of PGE (8, 9) and PGF (10) rises shortly after the preovulatory gonadotropin surge. The present study was designed to investigate the possible existence of a factor which might participate in the control of ovulation in non-ovulating bovine follicles by inhibiting PG synthesis by the cultured granulosa cells. The inhibitory effect of follicular fluid on progesterone secretion was also studied.

## MATERIALS AND METHODS

A pool of follicular fluid was collected from cows 20 min after slaughter by aspiration from preovulatory and midcycle follicles. Follicular fluid and control samples of calf serum were mixed and stirred with charcoal (50 mg/ml) for 1 h at room temperature and subsequently centrifuged three times at 20,000 g for 30 min. Progesterone, prostaglandin and estradiol concentrations in the follicular fluid as measured by radioimmunoassay (11) were undetectable (<25 pg/ml). Four to six trials, each with 10 replicate samples, were carried out for each parameter studied.

## RESULTS

To examine the effects of midcycle, preovulatory follicular fluids and calf serum on prostaglandin F and progesterone secretion, granulosa cells from large follicles were incubated for 24 h in Medium 199, supplemented with charcoal-extracted follicular fluid or serum. Both progesterone and prostaglandin F were measured by radioimmunoassay. The results of this experiment are shown in Table 1. Addition of 5%, 10% and 20% of charcoal-extracted midcycle follicular fluid to the culture medium caused a two-fold decrease in both progesterone and prostaglandin F accumulation by the cultured cells ($p<0.01$). However, when preovulatory (on day of oestrus) follicular fluid was used, the inhibitory effect on progesterone and prostaglandin F secretion was not apparent.

TABLE 1

Effect of 5% to 20% bovine follicular fluid (BFF) upon progesterone and prostaglandin F secretion by cultured granulosa cells derived from large follicles during 24 h incubation (Mean $\pm$ S.E.M.; n=40 - 60)

| Compound | pg/2 x $10^5$ cell/24 h | | | | |
|---|---|---|---|---|---|
| | | midcycle | | | preovulatory |
| | Control | 5% BFF | 10% BFF | 20% BFF | 10% BFF |
| Progesterone | $1688\pm83^{a}$ | $860\pm43^{b}$ | $1098\pm80^{b}$ | $862\pm86^{b}$ | $1720\pm88^{a}$ |
| Prostaglandin F | $553\pm41^{a'}$ | $191\pm95^{b'}$ | $233\pm30^{b'}$ | $254\pm16^{b'}$ | $575X48^{a'}$ |

Non-matching superscripts (a,b and a',b') are significantly different ($p<0.05$, from Shemesh, 1978, submitted).

The possibility that LH might overcome the inhibitory effect of midcycle follicular fluid was also examined. As shown in Table 2, addition of 1 μg bovine LH/ml to the culture medium caused a significant ($p<0.01$) increase in both progesterone and prostaglandin F secretion by the cultured granulosa cells. However, in the presence of 20% follicular fluid the inhibitory effect of the follicular fluid as well as the stimulatory effect of LH on progesterone and prostaglandin F secretion were not apparent.

TABLE 2

The effect of midcycle bovine follicular fluid (BFF) combined with LH (1 μg/ml) upon progesterone secretion by cultured granulosa cells derived from large follicles (diameter of 2 cm) during 24 h incubation (Mean S.E.M.; n=40 - 60)

| Compound | pg/2 x $10^5$ cell/24 h | | | |
|---|---|---|---|---|
| | Control | LH | 20% BFF | LH + 20% BFF |
| Progesterone | $1517 \pm 48^{a}$ | $1780 \pm 129^{b}$ | $813 \pm 62^{c}$ | $1160 \pm 199^{d}$ |
| Prostaglandin F | $494 \pm 28^{a'}$ | $706 \pm 56^{b'}$ | $194 \pm 38^{c'}$ | $402 \pm 72^{d'}$ |

Non-matching superscripts (a,b,c,d,a',b',c',d') are significantly different ($p<0.05$) (From Shemesh, 1978, submitted).

These findings suggest the presence of a prostaglandin synthetase inhibitor as well as a luteinization inhibitor in the bovine follicular fluid from midcycle follicles which disappear in the preovulatory follicle. Since LH overcomes the inhibitory effects of follicular fluid it is also possible that preovulatory follicles contain high levels of LH which overcome these inhibitory effects. The high level of LH in large follicles compared with undetectable levels in small follicles (1) support this concept. Recently (12) it has been reported that LH-sensitive adenylate cyclase was inhibited by bovine follicular fluid. It is not clear whether the follicular inhibitor acts directly on adenylate cyclase which suppresses cyclic AMP and thus PGF and P, or whether two different inhibitors are involved. Since the inhibitory effect of follicular fluid on prostaglandin F secretion is also apparent in other tissues such as cultured rat hemipituitaries (Shemesh and Ailenerg, unpublished data), the possibility exists that midcycle follicular fluid contains a specific inhibitor of prostaglandin synthetase. This activity declines with maturation and failure of the follicle to alter this inhibitor might result in a cystic or atretic follicle. Support for this concept emerges from our present studies which indicate that only ovulating follicles lack the inhibitory activities, whereas activity still exists in the other follicles on the day of ovulation.

## REFERENCES

1. Ledwitz-Rigby, F., Rigby, B.W., Gay, V.L., Stetson, Youngs, Y. and Channing, C.P. J. Endocr. 74(1977) 175.

2. Vane, J.R. New Biology 43 (1971) 231.

3. Armstrong, D.T. and Grinwich, D.L. Prostaglandins 1 (1972) 21.

4. Tsafriri, A., Lindner, H.R., Zor, U. and Lamprecht, S.A. Prostaglandins 2 (1972) 1.

5. Orczyk, G.P. and Behrman, R. Prostaglandins 1 (1972) 3.

6. Grinwich, D.L., Kennedy, P.G. and Armstrong, D.T. Prostaglandins 1 (1972) 89.

7. O'Grady, J.P., Caldwell, B.V., Auletta, F.J. and Speroff, L. Prostaglandins 1 (1972) 97.

8. Lindner, H.R., Tsafriri, A., Lieberman, E., Zor, U., Koch, Y., Bauminger, S. and Barnea, A. Prog. Horm. Res. 30 (1974) 79.

9. Bauminger, S., Lieberman, E. and Lindner, H.R. Prostaglandins 9 (1975) 753.

10. Armstrong, D.T. Recent Prog. Horm. Res. 30 (1974) 131.

11. Shemesh, M. and Hansel, W. Proc. Soc. Exp. Biol. Med. 148 (1975) 243.

12. Amsterdam, A., Shemesh, M., Riesel, Ramit, Amir, Yehudit and Salomon, Y. Abst. in Isr. J. Med. Sci. 13 (1977) 968.

SESSION NUMBER: III

TITLE: STEROIDOGENESIS

PRINCIPAL SPEAKER: David Schomberg

PRESENTED ABSTRACTS:

3.) K. Shiota and W.G. Wiest. On the Mechanism of Prolactin Stimulation of Steroidogenesis.

4.) N. Shahabi, J. Bahr, O.K. Dial and S. Glenn. Steroid Secretion by Perifused Ovarian Follicles of Immature, PMSG-Treated Immature, and Mature Rabbits.

5.) L.D. Anderson, F.W. Schaerf and C.P. Channing. Effects of Follicular Development on the Ability of Cultured Porcine Granulosa Cells to Convert Androgens to Estrogens.

6.) E. Su-Rong Huang and A.V. Nalbandov. Testosterone Synthesis by Chicken Follicular Cells.

7.) J.E. Fortune and W. Hansel. Modulation of Thecal Progesterone Secretion by Estradiol-17β.

CHAIRPERSON: Roy Greep

# STEROIDAL MODULATION OF STEROID SECRETION *IN VITRO*: AN EXPERIMENTAL APPROACH TO INTRA-FOLLICULAR REGULATORY MECHANISMS

David W. Schomberg

Departments of Obstetrics and Gynecology and Physiology

Duke University Medical Center, Durham, N.C. 27710

## INTRODUCTION

The secretion of steroid hormones by the pre-ovulatory ovarian follicle is a complex developmental process which integrates several modalities of endocrine regulation. In addition to the well-recognized participation of two cell types in steroid secretion (1,2), both of which are gonadotropin-dependent (3), it is now becoming apparent that follicular steroidogenesis is also modulated by the steroid hormones themselves. Based primarily upon studies *in vitro*, two types of modulation have been demonstrated to date: 1) a substrate model in which the conversion of androgen to estrogen is a rate-limiting step dependent upon follicle stimulating hormone (FSH), and 2) a regulatory model in which direct stimulatory or inhibitory steroidal action alters steroid secretion via mechanisims not yet understood. In this chapter a brief review of the studies which formed these lines of research will be presented followed by a discussion of on-going work in our laboratory which addresses certain aspects of both models. The foci of these studies involve: 1) developmental considerations in granulosa cell estrogen biosynthesis, 2) examination of the effects of estradiol ($E_2$) and dihydrostestosterone (DHT) upon progesterone secretion *in vitro* by highly differentiated porcine granulosa cells from large follicles, 3) characterization of the effect of intra-ovarian implants of anti-androgen (AA) upon subsequent granulosa cell progesterone secretion *in vitro*, and 4) preliminary investigation of the possible mechanism(s) of DHT-stimulated progesterone production *in vitro*.

## CELLULAR AND ENZYMATIC BASIS OF FOLLICULAR STEROIDOGENESIS

### Studies *In Vivo* and *In Vitro*

As is well known to students of ovarian physiology, incubation studies of the type which established that progestin and androgen were the principal products of the luteal and interstitial compartments, respectively (4), were not applicable to the follicular compartment because of the difficulty in separating thecal and granulosa elements. The cellular basis of follicular estrogen production in particular has thus enjoyed an interesting recent history beginning with the experiments *in vivo* of Falck in 1959 (5) wherein evidence for estrogen production was obtained only with combinations of rat granulosa and thecal tissue. Isotopic studies performed in several different additional species, both *in vivo* and *in vitro* during the next decade, however, indicated that the thecal cells possessed all of the requisite enzyme assemblies necessary to produce estrogens from acetate (6, 7). The granulosa cells on the other hand, were shown to be extremely deficient in converting $C_{21}$ to $C_{19}$ steroids; negligible or relatively low quantities of androgen and thus estrogens were therefore detected in these studies.

An important addition to our understanding of the role of the granulosa cell in estrogen production was provided by Moon et al (8) and Dorrington et al (9) who showed that the conversion of androgen to estrogen is under specific regulation by FSH. Further work by this laboratory demonstrated that the thecal compartment is a significant source of androgen (10, 11). Either androgen or FSH can be rate-limiting (for earlier references, see Ref. 2). In addition, Erickson and Hsueh have recently shown that FSH *in vivo* and *in vitro* stimulates the appearance of aromatase activity in granulosa cells of the primary follicles of the immature, hypophysectomized, estrogen-treated rat (12).

Since isolated granulosa cells in the presence of FSH and adequate substrate produce substantially greater quantities of estrogen than was heretofore appreciated, the concept of the theca as the principal estrogen-secreting compartment of the ovary has now been called into question.

### Results *In Vitro* vs. Results *In Vivo*

Table 1 lists the most-quoted studies *in vivo*, including that of Falck (5), which were designed to determine the predominant source of follicular estrogen. When estrogen production was detected, the thecal contribution was substantially greater. (For a critique of the assumptions and procedures used in these studies, see Ref. 2).

Recent studies *in vitro* utilizing radioimmunoassay of the products formed by culture of isolated rat (15), sheep (16), and human (17) follicles all indicate that significant amounts of estrogen are

Table 1. Follicular estrogen production *in vivo*: Theca vs. Granulosa

| Study (Ref.) | Preparation | Result |
|---|---|---|
| Falck, 1959 (5) | Tissue transplants/ Rat | only cell combinations + |
| Younglai & Short, 1970 (13) | Isotopic/Mare | T > G |
| Coudert & Channing, 1976 (14) | Surgical/Monkey | T >> G |

Table 2. Aromatizing activity *in vitro*: Theca vs. Granulosa

| Study (Ref.) | Preparation | Result |
|---|---|---|
| Ryan & Short, 1965 (18) | Isotopic/incubation/mare/ slaughterhouse | T < G |
| Lacroix et al, 1974 (19) | Isotopic/incubation/cow/ slaughterhouse | T < G |
| Makris & Ryan, 1977 (20) | Culture/normal hamster | T < G |
| Moor, 1977 (21) | Culture/PMSG-Rx sheep | only T + G |
| Fortune & Armstrong, 1978 (22) | Culture/PMSG-Rx rat | T << G |

produced by these preparations. In contrast to the indications of the studies *in vivo*, however, when estrogen production by isolated granulosa cells vs. thecal preparations was examined *in vitro* under a variety of conditions and the results expressed on a per follicle basis, the vast majority of studies indicated that aromatase activity of the theca was less than that of the granulosa (Table 2).

At present there appears to be no straightforward explanation for this dichotomy. Species differences are one possibility although the results of Table 2 indicate a rather uniform response across several species. Our studies (Figure 1) with isolated porcine granulosa cells *in vitro* are an exception to those listed in Table 2 in that they provide indirect evidence suggesting the theca is a significant source of estrogen, at least during later stages of follicle development (see Ref. 23 for characteristics of the estrogen antibody utilized. Samples of culture media were assayed directly).

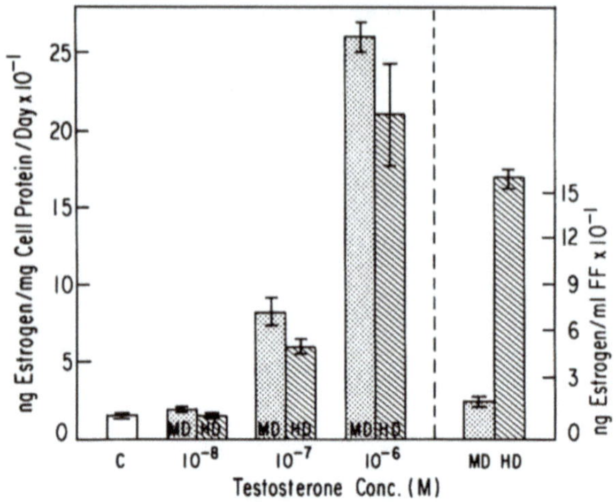


Figure 1. Estrogen production from testosterone (T) by moderately differentiated (MD) and highly differentiated (HD) porcine granulosa cells from 3-5 and 7-9 mm follicles, respectively (left panel), and follicular fluid (FF) estrogen concentrations from the same follicles (right panel). Although FF estrogen levels increased dramatically during this phase of development, granulosa cell aromatising capacity (MD vs. HD) over a wide range of T remained unchanged.

These results indicate that porcine granulosa cells possess an aromatase which is very responsive over a wide range in substrate concentration. Activity did not increase, however, during the period of follicular maturation which occurs between luteal regression and the onset of the LH surge in this species. The sevenfold increase in follicular fluid estrogen concentration which occurred simultaneously cannot be ascribed to simple accumulation of granulosa-derived estrogen since follicular fluid volume also increases about four-fold during this period (24). The increased estrogen concentration is thus most easily explained by postulating a substantial contribution from theca-derived estrogen. These findings and those of others presented at this workshop point to another consideration which may help clarify the apparent *in vitro*/*in vivo* controversy, namely, differential developmental changes in thecal and granulosa estrogen biosynthesis during follicular maturation. The culmination of developmental changes, as far as the pre-ovulatory phase is concerned, is brought about by the ovulatory surge of luteinizing hormone (LH). The results of several studies show that ovulatory levels of endogenous or exogenous gonadotropin result in decreased estrogen production (2, 15, 16, 25, 26). This is apparently the result of a dual mechanism in which thecal production of androgen is markedly reduced following an initial stimulation (25), while granulosa cell aromatase capability *per se* decreases (26) as progesterone production is simultaneously accelerated (15, 25).

Another consideration which is relevant to the studies *in vitro* is that of the technical aspects of thecal-layer culture. In order to make direct quantitative comparisons of steroidogenesis *in vitro* by thecal and granulosa cell types, it may be necessary to establish their metabolic and/or endocrine requirements in greater detail. For example, even though monolayer culture of granulosa cells is technically very simple and the procedure has gained widespread acceptance, we have not yet characterized sufficiently well those factors *in vitro* regulating gonadotropin receptor populations and other cell functions which optimize steroidogenesis (Ref. 27, and next section). Also, current work in our laboratory which is still in a preliminary stage (Haney and Schomberg, unpublished observations) indicates that progesterone and estrogen production by isolated thecal tissue from hemi-sected 7-8 mm porcine follicles are significantly altered by culture conditions. Production of both steroids is significantly reduced by culturing the tissue under explant culture conditions versus organ culture conditions. (The media utilized, gas phase, and other conditions were as previously described for granulosa cell culture, Ref. 28).

## STEROIDAL MODULATION OF STEROID SECRETION: AN ASPECT OF ANDROGEN ACTION DURING FOLLICLE DEVELOPMENT

The results of several recent studies demonstrating androgen-stimulated progesterone production by isolated follicles or granulosa cells *in vitro* (Table 3) are of considerable interest because of the contrast between these findings and the long history associating several aspects of ovarian dysfunction and/or anti-mitotic sequelae with androgen action, e.g., polycystic ovarian disease, death of the ovum and granulosa cells, and follicular atresia (30-32). At this point in time it is difficult to integrate these positive and negative roles of androgen into any coherent schemata other than possibly a developmental one (see Ref. 29 for example). Since it is theoretically possible that the steroidogenic and the anti-mitotic effects of androgen reflect very disparate end points involving different cellular mechanisms, it may be valid from an experimental point of view to approach these aspects separately, at least in the initial stages. It is worth pointing out in this context that although direct demonstrations of androgen-stimulated (non-aromatizable and aromatizable androgen) progesterone synthesis in isolated granulosa cell systems have been obtained in the presence or absence of gonadotropin, no concomitant gross effects upon cell morphology or cell viability were noted in any of the studies (Table 3). Based upon this observation, an hypothesis which perhaps constitutes a "simplest-case" alternative among several possibilities comes to mind, namely, that the positive effects of androgen are direct and may also be androgen-receptor mediated while the negative effects are indirect or passive and will involve more complicated mechanistic models. (A "passive" model would not preclude androgen receptor-mediated action. An action which negated the positive effects of estrogen by reducing estrogen receptor function (38) could produce an anti-mitotic effect without the involvement of a specific androgen-dependent growth-inhibiting protein for example). The approach taken in our laboratory to date has been to investigate the more direct aspect of androgen-stimulated steroidogenesis as outlined in the introduction.

Table 3. Androgen-stimulated ovarian progesterone synthesis: studies *in vitro*

| Study (Ref.) | Preparation | Compounds |
|---|---|---|
| Moor, et al, 1975 (33) | ovine follicles/organ culture | AD, T |
| Schomberg, et al, 1976 (28) | MD g. cells/porcine/ monolayer | DHT |
| Nimrod & Lindner, 1976 (34) | hypox-DES-Rx rat/ pre-antral g. cells/ monolayer | AD, T, DHT FSH |
| Armstrong & Dorrington, 1976 (35) | hypox-$E_2$-Rx rat/ pre-antral g. cells/ monolayer | T, DHT, FSH |
| Lucky, et al, 1977 (36) | hypox-DES-Rx rat/ pre-antral g. cells/ monolayer | T, DHT |
| Shemesh & Ailenberg, 1977 (37) | bovine follicles/ organ culture | AD |

## Studies *In Vitro*: Effects of DHT and Estradiol Upon Progesterone Synthesis by Granulosa Cells from Large Porcine Follicles

Since many of the studies listed in Table 3 above utilized isolated granulosa cells from pre-antral or moderately differentiated antral follicles, the physiologic significance of androgen-stimulated progesterone secretion in terms of continued follicle development may be open to question since the majority of such follicles are destined to become atretic. This caveat also applies to those studies *in vitro* showing an inhibitory effect of estradiol (E) upon granulosa cell progesterone synthesis (28, 39). Studies were undertaken, therefore, to determine whether the modulatory effects of DHT and E upon progesterone secretion by highly differentiated (HD) granulosa cells from 8-10 mm porcine follicles were similar to those described for cells of the immature follicles (28). The androgen and E concentrations tested are comparable to those of porcine follicular fluid at later stages of follicle development. The results show that the progesterone secretory responses are indeed similar, thus demonstrating a role for these steroids in the modulation of progesterone secretion throughout follicle development

(Figure 2). Again, no gross morphologic or lytic cellular changes were observed in either the DHT- or E-treated groups.

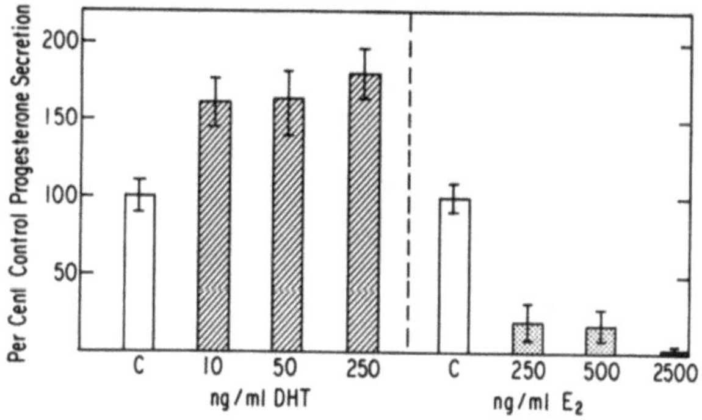


Figure 2. Progesterone secretion by granulosa cell monolayers prepared from 10 mm pre-ovulatory porcine follicles: stimulation by dihydrotestosterone (DHT) and inhibition by estradiol ($E_2$).

## Studies *In Vivo* and *In Vitro*: Effects of Intra-Ovarian Implants of Antiandrogen Upon Progesterone Secretion by Granulosa Cell Cultures

Although the studies *in vitro* presented above suggest a biochemical role for androgen in granulosa cell steroidogenesis, they do not constitute direct evidence for the action of intra-ovarian androgen *in vivo*. Studies were therefore designed to determine whether a requirement for local androgen in granulosa cell steroidogenesis could be demonstrated by utilizing intra-ovarian Silastic implants containing the antiandrogens Flutamide or its active hydroxylated metabolite (40). The results showed that the presence of antiandrogen during a 4-day period of follicular development (days 15-19 of the porcine estrous cycle) reduced the level of progesterone secretion in porcine granulosa cell cultures by approximately 50% (Figure 3). These results thus complement the studies *in vitro* suggesting a role for androgen in granulosa cell steroidogenesis during follicular development.

As discussed previously (40), results of this nature imply that pre-ovulatory antiandrogen treatment may significantly alter subsequent ovarian functions related to androgen-dependent granulosa cell development. Non-steroidogenic processes such as ovulation have also been affected by antiandrogen treatment (anti-testosterone serum) as shown by Mori, et al, (41). Identification of the locus of follicular action was not the objective of this study (41), but it is possible that granulosa cells could be involved since they possess androgen receptors (42) and secrete plasminogen activator (43). Observations of this type also seem to contradict the earlier studies which demonstrated inhibition of follicle development by androgens (32, 44). However, since markedly different follicular stages were being studied, the possibility that the responses observed were a function of different developmental states is again emphasized.

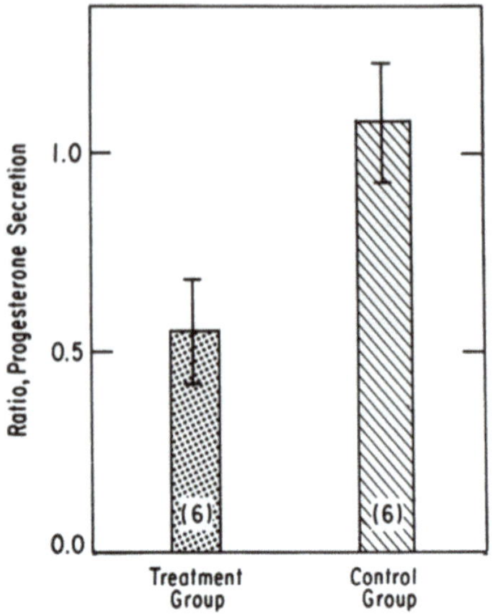


Figure 3. Effect of intraovarian antiandrogen (AA) during a limited period of follicle development upon subsequent progesterone secretion by porcine granulosa cell cultures. A significant reduction in the ratio of progesterone secretion ($p < 0.025$) was noted for cells obtained from animals in which each ovary received an active or an inactive AA-containing implant (Treatment group) as compared to those in which both ovaries received an inactive AA-containing implant (Control group). The number of animals in each group is indicated within the parenthesis; an average of 4 pairs of individual cultures was obtained from each animal. Brackets represent ± SEM. (Taken from Schomberg et al, Ref. 39).

## Cellular Mechanisms Involved in Steroidal Modulation of Steroid Secretion: Preliminary Studies

The preceding studies demonstrating DHT stimulation of progesterone secretion throughout follicle maturation, and sub-normal production *in vitro* following reduced functional levels of local ovarian androgen support the concept of a physiological role(s) for androgen in normal follicular steroidogenesis. The cellular mechanisms involved are unknown. Since an interaction of steroid and polypeptide hormones is involved in the induction of the granulosa cell gonadotropin receptor population (45), it would seem logical to examine the possibility that androgen-modulated steroidogenesis might in turn involve modulation of the LH/hCG receptor population. Results of preliminary studies suggest that the combination of hCG (kindly provided by Dr. D. Puett, Vanderbilt University; 13,588 IU/mg in the rat leydig cell system) and DHT stimulates progesterone synthesis *in vitro* to a greater degree than either compound alone (Figure 4). The effect may be more pronounced during the second 48 hour culture period. Progesterone production in these studies was quantitated as previously described (28) except that sheep anti-11α-hydroxyprogesterone serum generated in this laboratory was utilized (SF-36, #13). Full details of its characteristics will be published elsewhere.

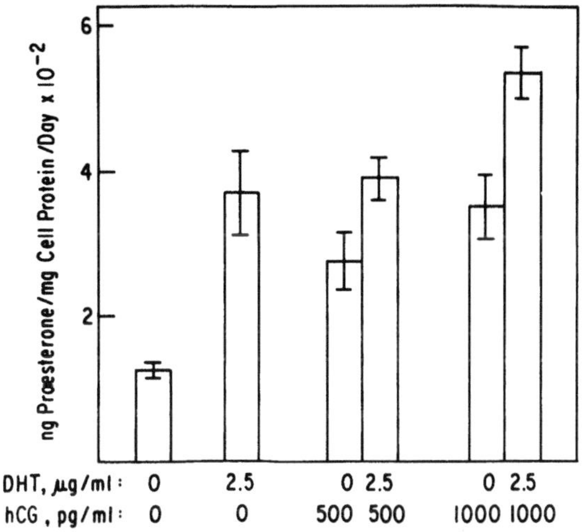


Figure 4. Interaction of DHT and hCG in the stimulation of progesterone synthesis by moderately differentiated granulosa cells from days 2-4 of monolayer culture. Cells were maintained in Ham's $F_{12}$ nutrient media: fetal calf serum (90: 10). Each bar graph and associated bracket represent the mean and SEM of 4 cultures.

The facilitory effect of DHT upon hCG action is not due to increased numbers of LH/hCG receptor binding sites (unpublished observations); i.e., DHT alone does not prevent the net loss of LH/hCG receptors observed in this system (27). The question of whether the combination of DHT and hCG increases LH/hCG receptor binding relative to the appropriate controls has not been examined.

Progress has been made towards the goal of achieving a steady-state level of LH/hCG receptor binding with time in culture and thus eliminating the significant net loss of receptor sites as mentioned above. Figure 5 shows the loss of binding activity typically seen over a 4-day culture period and a significant reduction in net loss by the addition of 25mU/ml insulin (pork insulin, 25 U/mg, Eli Lilly ID G04-94-193), 8 ng/ml thyroxine, and 10 ng/ml hFSH (LER-1577) to the basal medium (90% Ham's $F_{12}$ nutrient media: 10% fetal calf serum). The binding studies were carried out as described by Stouffer, et al, (27). The net receptor gain due to I-T-FSH treatment is actually less than that indicated in Figure 5, however, because the estimate of initial cellular protein includes a contribution from red blood cells and non-viable granulosa cells. The differential between initial binding and that observed after a given culture period when only viable cells remain, is thus somewhat artifically reduced. In order to obtain an accurate assessment of initial binding in these type studies, methodology will have to be developed which will separate the contaminating red blood cells and non-viable granulosa cells from the viable ones. This consideration does not affect the comparison of binding between control I-T-FSH-containing media and that containing steroids. As indicated in Figure 5, neither DHT nor $E_2$ altered binding, providing another indication that the stimulatory and inhibitory effects, respectively, of these compounds upon progesterone secretion are not mediated through changes in LH/hCG receptor number *per se*.

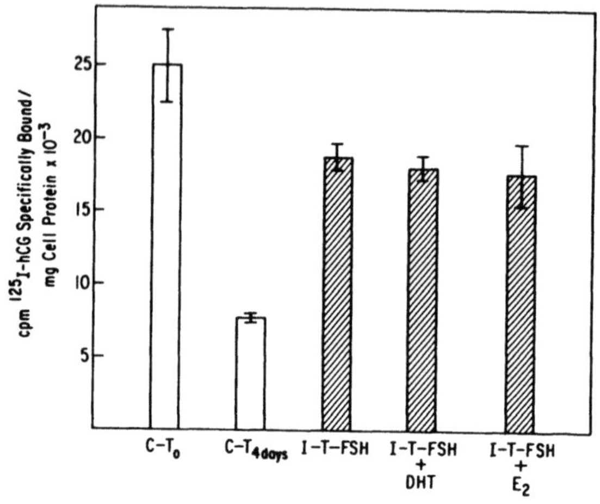


Figure 5. LH/hCG receptor binding during monolayer culture of moderately differentiated porcine granulosa cells. All data with the exception of initial binding at the time of cell harvest ($C-T_O$) were obtained after 4 days. Net loss of receptor ($C-T_O$ vs. $C-T_{4\ days}$) was significantly reduced by a combination of insulin (I), thyroxine (T) and FSH. See text for additional details. This effect was not altered by the addition of either dihydrotestosterone (DHT, 2.5 ng/ml) or estradiol ($E_2$, 25 ng/ml). Each bar graph and associated bracket represent the mean and SEM of 3 cultures.

The mechanism underlying the synergism between androgen and FSH in stimulating progestin synthesis by rat pre-antral granulosa cells has been extensively investigated by Nimrod (46, 47). Similar to the results above, no evidence was found for androgen (androstenedione) involvement in stimulating FSH receptor numbers *per se*. Furthermore, no indication of an enhancement of the activation of the coupling of FSH receptor to adenylate cyclase was obtained, nor was a significant change in cyclic AMP degredation by cyclic 3', 5'-nucleotide phosphodiesterase observed. These studies also ruled out the possibility that the effect of androgen involved significant competitive inhibition for 5α-reductase, thus resulting in an indirect accumulation of progesterone and 20α-hydroxy-progesterone rather than enhanced production. Importantly, androstenedione did exert a synergistic effect on the stimulation of progestin by dibutyryl cyclic AMP (DBC) which suggests that the locus of action is at a step distal to cyclic AMP production.

The converse situation, i.e., inhibition of progestin production by estradiol-17β ($E_2\beta$), also appears to involve more distal steps. Williams et al (48) found that $E_2\beta$ inhibition of LH-stimulated progesterone synthesis by dispersed bovine luteal cells did not involve an impairment of cellular cyclic AMP generation. In addition, the stimulatory effect of exogenous DBC upon progesterone synthesis was blocked by $E_2\beta$.

The above results point toward the possibility of some type of interaction between steroid hormone and cyclic AMP action in the modulation of steroid secretion. Greengard has postulated that the locus of such interactions may prove to be a unique cyclic AMP-dependent protein kinase (49). Fuller et al have shown that the total amount of type I cytoplasmic cyclic AMP-dependent protein kinase, especially of the ventral prostate and levator ani, is controlled to a considerable extent by DHT (50). These studies suggest additional possibilities for further investigation of the molecular mechanisms involved in steroidal modulation of steroid secretion.

## SUMMARY

Studies *in vitro* and *in vivo* indicate that steroid secretion by granulosa cells of the developing follicle is modulated in part by steroid hormones themselves. Androgens stimulate progestin synthesis *in vitro* by granulosa cells of all developmental stages independently of their role as an estrogen precursor; generally, the action of estradiol results in suppression of progesterone secretion. The nature of the relationship between steroid secretion by granulosa cells and the physiological processes of follicular development or atresia remains to be investigated in greater detail. The cellular mechanism(s) of steroidal modulation of steroid secretion is unknown but apparently does not involve changes in the number of granulosa cell LH receptors *per se*.

## ACKNOWLEDGEMENTS

The collaboration of L. Tyrey, Ph.D., A. H. Haney, M.D., M. A. Brown, R. Varney, and B. Gunn in several aspects of these studies is most gratefully acknowledged. I thank Drs. D. Puett and L. E. Reichert, Jr., for the gifts of purified hCG and hFSH, respectively, and the Eli Lilly Company for their generous gift of insulin. I also thank D. Sanders for typing the manuscript. Supported in part by grant HD 11827 from the NICHHD, USPHS.

## REFERENCES


(1) R.V. Short, J. Endocr. 24(1962) 59.

(2) D.T. Armstrong and J.H. Dorrington, in: Regulatory Mechanisms Affecting Gonadal Hormone Action, Advances in Six Hormone Research, Vol. 3, eds. J.A. Thomas and R.L. Singhal (University Park Press, Baltimore, 1977) p. 217.

(3) G.S. Greenwald, in: Handbook of Physiology, Section 7, Endocrinology, Vol. IV, ed. R.O. Greep (American Physiological Society, Washington, D.C., 1974) p. 193.

(4) K. Savard, J.M. Marsh, and B.F. Rice, Rec. Progr. Hormone Res. 21(1965)285.

(5) B. Falck, Acta Physiol. Scand. (Suppl. 163) 47(1959)1.

(6) K.J. Ryan and K. Petro. J. Clin. Endo. Metab. 28(1968)355.

(7) L. Bjersing and H. Carstensen. Biochim. Biophys. Acta 86(1964)639.

(8) Y.S. Moon, J.H. Dorrington, and D.T. Armstrong. Endocrinology 97(1975)244.

(9) J.H. Dorrington, Y.S. Moon, and D.T. Armstrong. Endocrinology 97(1975)1328.

(10) J.E. Fortune and D.T. Armstrong. Endocrinology. 100(1977)1341.

(11) J.E. Fortune and D.T. Armstrong, in: Proceedings of the NIH Workshop on Ovarian Follicular Development and Function, Santa Cruz, Calif., 1977 (In press).

(12) G.F. Erickson and A.J.W. Hsueh. Endocrinology 102(1978)1275.

(13) E.V. Younglai and R.V. Short. J. Endocrinol. 47(1970)321.

(14) C.P. Channing and S.P. Coudert. Endocrinology 98(1976)590.

(15) H.R. Lindner, A. Tsafriri, M.E. Lieberman, U. Zor, Y. Koch, S. Bauminger, and A. Bornea. Rec. Progr. Hormone Res. 30(1974)79.

(16) R.M. Moor. J. Endocr. 61(1974)455.

(17) J.S.G. Biggs, T.G. Baker, and D. Hutchon. Brit. J. Obstet. Gynec. 85(1978)211.

(18) K.J. Ryan and R.V. Short. Endocrinology 76(1965)108.

(19) E.W. Lacroix, W. Eeuchaute, and I. Leusen. Steroids 23(1974) 337.

(20) A. Makris and K.J. Ryan. Steroids 29(1977)65.

(21) R.M. Moor. J. Endocr. 73(1977)143.

(22) J.E. Fortune and D.T. Armstrong. Endocrinology 102(1978)227.

(23) S. Aksel, D.W. Schomberg, L. Tyrey, and C.B. Hammond. Am. J. Obstet. Gynecol. 126(1976)165.

(24) S.S.C. Chang, J.D. Jones, R.D. Ellefson, and R.J. Ryan. Biol. Reprod. 15(1976)321.

(25) C.R. Parker, Jr., A. Costoff, T.G. Muldoon, and V.B. Mahesh. Endocrinology 98(1976)129.

(26) Y. Katz and D.T. Armstrong. Endocrinology 99(1976)1442.

(27) R.L. Stouffer, L. Tyrey, and D.W. Schomberg. Endocrinology 99(1976)516.

(28) D.W. Schomberg, R.L. Stouffer, and L. Tyrey. Biochem. Biophys. Res. Commun. 68(1976)77.

(29) H.R. Lindner, A. Amsterdam, Y. Salomon, A. Tsafriri, A. Nimrod, S.A. Lamprecht, U. Zor, and Y. Koch. J. Reprod. Fert. 51(1977)215.

(30) I.F. Stein and M.L. Leventhal. Am. J. Obstet. Gynec. 29(1935) 181.

(31) J.H. Gaarenstroom and S.E. De Jongh. A Contribution to the Knowledge of the Influences of Gonadotropic and Sex Hormones on the Gonads of Rats (Elsevier, New York, 1946) p. 87.

(32) R.W. Payne and R.H. Runsen. Endocrinology 62(1958)313.

(33) R.M. Moor, M.F. May, and R.F. Seamark. J. Reprod. Fert. 45(1975)595.

(34) A. Nimrod and H.R. Lindner. Mol. Cell. Endocrinol. 5(1976)315.

(35) D.T. Armstrong and J.H. Dorrington. Endocrinology 99(1976) 1411.

(36) A.W. Lucky, J.R. Schreiber, S.G. Hillier, J.D. Schulman, and G.T. Ross. Endocrinology 100(1977)128.

(37) M. Shemesh and M. Ailenberg. Biol. Reprod. 17(1977)499.

(38) S. Saiduddin and H.P. Zassenhaus. Endocrinology 102(1978) 1069.

(39) S.G. Hillier, R.A. Knazek, and G.T. Ross. Endocrinology 100 (1977)1539.

(40) D.W. Schomberg, R.F. Williams, L. Tyrey, and L.C. Ulberg. Endocrinology 102(1978)984.

(41) T. Mori, A. Suzuki, T. Nishimura, and A. Kombegawa. Endocrinology 101(1977)623.

(42) J.R. Schreiber and G.T. Ross. Endocrinology 99(1976)590.

(43) S. Strickland and W.H. Beers. J. Biol. Chem. 251(1976)5694.

(44) J.P. Louvet, S.M. Harman, J.R. Schreiber, and G.T. Ross. Endocrinology 97(1975)366.

(45) J.S. Richards and A.R. Midgley, Jr. Biol. Reprod. 14(1976) 82.

(46) A. Nimrod. Mol. Cell. Endocrinol. 8(1977a)189.

(47) A. Nimrod. Mol. Cell. Endocrinol. 8(1977b)201.

(48) M.T. Williams, M.R. Clark, and J.M. Marsh. Proc. 59th Ann. Meeting Endocrine Soc. Abstr. 231(1977)172.

(49) P. Greengard. Science 199(1978)46.

(50) J.M. David, C.V. Byers, and D.H. Russell. Proc. Natl. Acad. Sci. USA 75(1978)223.

# ON THE MECHANISM OF PROLACTIN STIMULATION OF STEROIDOGENESIS

Kunio Shiota and Walter G. Wiest

Department of Obstetrics and Gynecology
Washington University School of Medicine
St. Louis, Missouri 63110

## INTRODUCTION

Rescue of the rat corpus luteum was shown by Neill and his co-workers (1) to be prolactin (PRL) dependent. Morishige and Rothchild (2) were able to demonstrate effectively that the maintenance of pregnancy in the rat is virtually under PRL control for the first 6 days. In both instances (1,2), the PRL effect was mediated through increased secretion of luteal progesterone. Crisp has demonstrated more recently (3) that cultured rat granulosa cells, when maintained with PRL, secreted more progesterone than controls maintained in hormone-free medium. This observation was interpreted as a demonstration of the luteotropic action of PRL. In our attempts to understand the stimulatory action of PRL on luteal steroidogenesis, we have employed the technique of perifusion (continuous flow *in vitro* incubation) (4) together with the use of enzyme dispersed rat luteal cells taken at specific times in early pregnancy and pseudopregnancy (5-7). Our results indicate that during early pregnancy, progesterone secretion by dispersed luteal cells is stimulated most effectively by appropriate combinations of both LH and PRL (5-7). PRL alone maintained but did not stimulate progesterone secretion (5). The purpose of this presentation is to report recent findings related to PRL regulation of luteal steroidogenesis.

## Source of Dispersed Luteal Cells

Two month old, female rats purchased from Holtzman Co., Madison were maintained in an air conditioned room under 14 hours light

(5:00 am - 7:00 pm) 10 hours dark and were given Purina Rat Chow and water without restriction. Pseudopregnancy was induced by cervical stimulation with a glass rod either on the evening of proestrus or the morning of estrus. Day-1 of pseudopregnancy coincided with the first day of diestrus after cervical stimulation. Rats were killed by decapitation under light ether anesthesia between 9:00-9:30 am on the 4th day of pseudopregnancy. Ovaries from 4 rats were collected immediately. Newly formed corpora lutea (40-50) were removed and transferred to Dulbecco's modified Eagle medium containing 0.3% Penicillin G saturated with 5% $CO_2$ 95% $O_2$ gas mixture. Dispersion of luteal cells was done at room temperature by the method of Wu et al. (5).

## Perifusion Procedure

Dispersed cells were perifused according to the method previously described (5) at a flow rate of 3 ml/10 minutes. Effluent samples were collected every 15 minutes. During the first 15 minutes only control medium (Dulbecco's modified Eagle medium containing 0.3% Penicillin G and 0.25% bovine serum albumin) was used in order to establish the basal level of steroid secretion. Following collection of the first fraction, the perifusion medium was usually changed by the addition of gonadotropins or other additives detailed for each experiment in the following section.

Quantities of progesterone and 20α-dihydroprogesterone secreted during perifusion are expressed in terms of the "accumulation ratio" derived from the concentration of steroid present in the first (control) fraction reduced to unity and the concentration of steroid in each succeeding fraction normalized to the first. The accumulation ratio comprised the sum of successive fractions. Thus, an accumulation ratio of 60 established that 60 times the steroid concentration contained in the first (control) fraction had been secreted within the perifusion period.

## Static Incubation Procedure

Dispersed luteal cells (90,000 to 420,000 cells) equivalent to 2.8 - 13.0 μg protein were incubated for up to 60 minutes at 37C in 4.0 ml of Dulbecco's modified Eagle medium containing 0.3% Penicillin G and 0.25% bovine serum albumin saturated with 5% $CO_2$ - 95% $O_2$. Additions of components to this medium were made as indicated in the experiments described. The incubation was initiated by the addition of the cell suspension, and aliquots were removed from the incubation medium for analysis as required.

### Analyses

Progesterone and 20α-dihydroprogesterone were analyzed in effluent media from the perifusion experiments and in aliquots from the static incubations by radioimmunoassay methods previously described (4).

No attempt was made to quantify c-AMP in perifused cells or in the effluent perifusion medium. However, c-AMP was determined in cells and suspending medium obtained from static incubation of dispersed luteal cells using the c-AMP ($^{125}I$) RIA kit supplied by New England Nuclear. Sensitivity of the assay was increased by acetylation of reference standards and unknown samples. Cyclic AMP was quantified in total incubation medium (suspending buffer plus dispersed cells), in cells rapidly separated from suspending medium by centrifugation, and in the incubation medium freed of cells.

## RESULTS AND DISCUSSION

### LH Stimulation of Luteal Cell Steroidogenesis

The LH dose-response previously obtained (5) with dispersed luteal cells from day-2 pregnant rats resembles the more recent observations made with cells from day-4 pseudopregnant rat shown in Fig. 1. Because of the possible involvement of c-AMP as an intermediate in LH stimulation, the response of dispersed cells to increasing levels of perifused dibutyryl cyclic AMP (dbc-AMP) was also measured (see Fig. 2). Perifusion of 0.1-0.3 mM isobutyl methylxanthine (MIX), an inhibitor of the phosphodiesterase responsible for reducing intracellular c-AMP levels, increased the effectiveness of 20 ng/ml LH; however, too high doses of MIX were less effective due probably to an inhibitory action of MIX on protein synthesis (8,9). Thus, steroidogenesis by perifused luteal cells was subject to LH stimulation by a mechanism probably involving c-AMP acting as a second messenger and making this system comparable to others previously studied (10-20).

The direct correlation between LH-stimulated steroidogenesis and levels of endogenous c-AMP was examined during static incubation of day-4 dispersed luteal cells. The effect of added LH on both steroidogenesis and endogenous c-AMP is shown in Fig. 3. In the presence of 0.1 mM MIX, maximum synthesis of progesterone and 20α-dihydroprogesterone was obtained with LH added at a concentration of 10 ng/ml. At that level of LH, only marginal increases in intra-

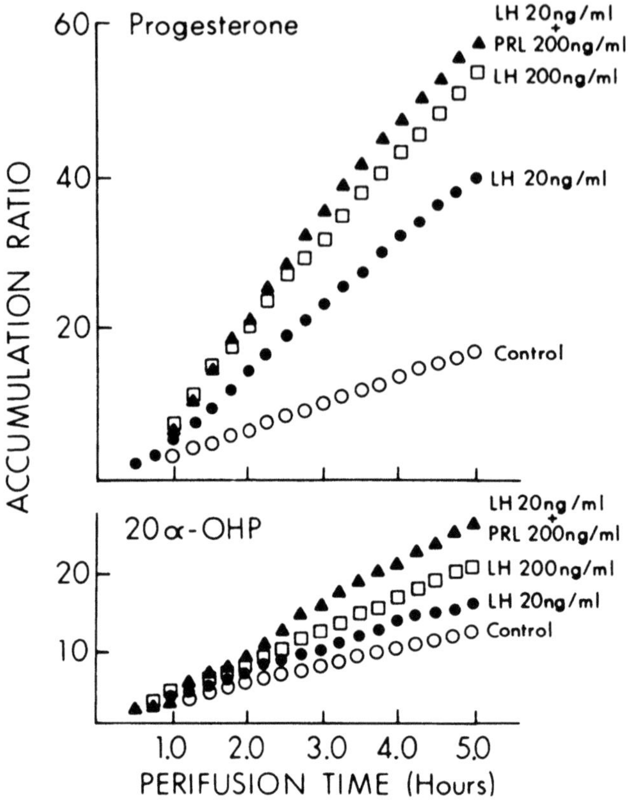


Fig. 1. LH-responsiveness and the synergism of PRL on the stimulation of steroidogenesis by dispersed day-4 luteal cells during perifusion. Progesterone and 20α-dihydroprogesterone (20α-OHP) secreted under control conditions were increased by the addition of LH (20 and 20 ng/ml) and by the combination of LH (20 ng/ml) and PRL (200 ng/ml).

cellular and total c-AMP were detected; larger doses of LH increased both intracellular and total c-AMP by several fold without any further increase in steroidogenesis. Thus, the disproportionality between LH stimulation and the level of endogenous c-AMP observed in other endocrine systems (21-24) is also apparent in these findings.

## PRL Stimulation of Luteal Cell Steroidogenesis

While the effects of LH on dispersed luteal cells just detailed do not bear directly on the question of the PRL regulation of steroidogenesis, they do provide a base for considering the synergistic relationship of LH and PRL which we have previously reported (5) and will extend in this report.

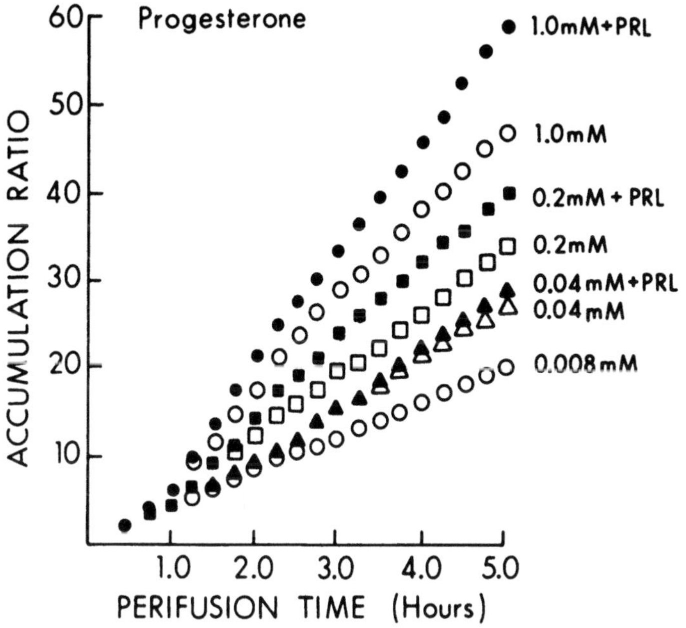


Fig. 2. dbc-AMP-responsiveness and the synergism of PRL (200 ng/ml) on the stimulation of progesterone secretion by dispersed day-4 luteal cells during perifusion at the indicated dbc-AMP concentrations.

Synergism was observed when PRL was added to suboptimal levels of LH during perifusion as shown in Fig. 1. The addition of PRL produced a stimulatory effect on progesterone secretion greater than that achieved by a 10-fold increment in LH alone. With the probable intermediary role of c-AMP in mind, it was considered appropriate to test the possible effect of added PRL on the action of perifused dbc-AMP. As shown in Fig. 2 synergism was again observed in the steroidogenic response.

Stimulation of steroidogenesis by combinations of PRL with LH and dbc-AMP contrast with those seen when PRL alone was added to the perifusion medium (see Fig. 4). We had noted previously (5) that the initial rate of steroidogenesis by perifused cells was effectively maintained up to 5 hours by PRL, but stimulation of steroidogenesis was not observed. Thus, it appeared that concentrations of PRL which had been stimulatory when used together with LH and dbc-AMP had no effect on the rate of steroidogenesis when added alone to the perifusion medium. Such ineffectiveness of PRL administered _in vitro_ has been previously noted (25).

A luteotropic action of PRL given in vivo seemed to be responsible for the ability of luteal cells to synthesize progesterone _in vitro_ in response to LH stimulation (26). The capacity of PRL

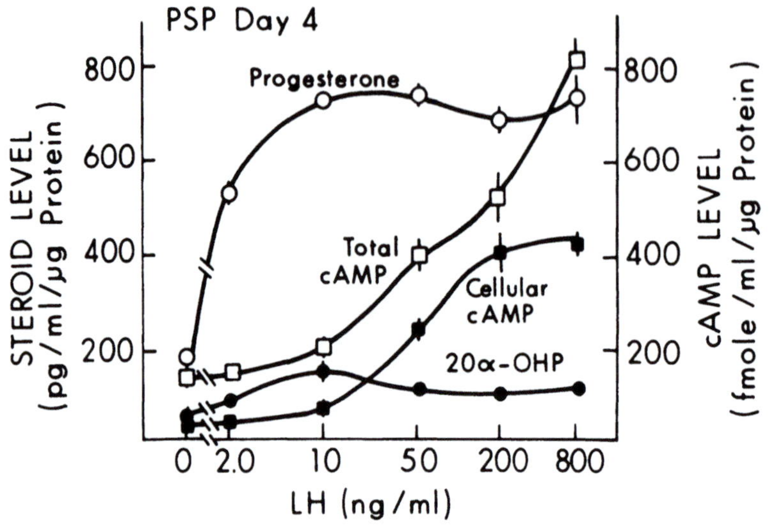


Fig. 3. LH stimulation of steroidogenesis and c-AMP formation during static incubation of dispersed day-4 luteal cells (5.33 µg protein equivalent) in 4.0 ml of Dulbecco's modified Eagle medium containing 0.3% Penicillin G, 0.25% bovine serum albumin, and 0.1 mM MIX. Incubation was carried out at 37C for 1 hour under an atmosphere of 5% $CO_2$ - 95% $O_2$. Each point represents the mean derived from 3 separate incubations. Vertical lines through the points indicate standard errors of the means.

to increase the number of LH receptors has been noted (27), and it has been suggested that the mechanism of PRL-LH interaction resides in the ability of PRL to maintain the function of luteal cell LH receptors (28). The "protective" action of PRL on dispersed luteal cells which resulted in the prolonged retention of viability and functional capacity (5) might operate through such a mechanism. However, Crisp's demonstration (3) suggested a direct luteotropic action of PRL. Furthermore, Lahav et al. (29) have observed that addition of PRL to cultured rat luteal cells suppressed 20α-hydroxysteroid dehydrogenase activity. Since this enzyme catalyzes the reduction of progesterone to 20α-dihydroprogesterone, its inhibition could result in higher concentrations of progesterone in the culture medium. These latter reports (3,29) suggests a direct action of PRL related to steroidogenesis by luteal cells which might not be mediated through LH or LH receptors.

Clarification of the mechanism of the PRL augmentation of the LH-stimulation of steroidogenesis was sought by analyzing both

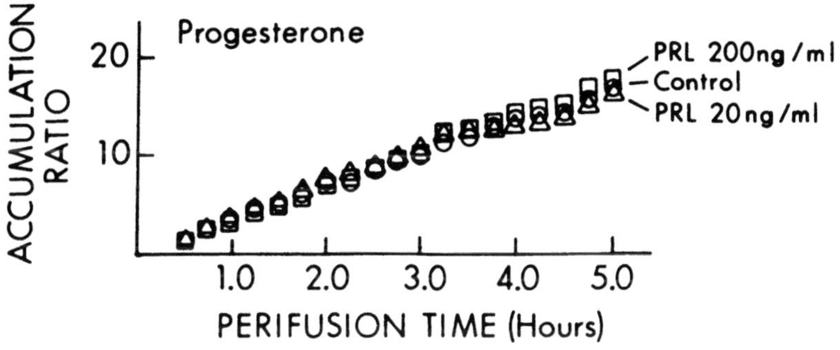


Fig. 4. Refractoriness of dispersed day-4 luteal cells to PRL perifused in the concentrations indicated.

steroid synthesis and the accumulation of c-AMP during static incubation. As shown in Figure 5, addition of PRL alone to incubating luteal cells failed to increase either endogenous c-AMP levels or steroid synthesis. Although stimulation of steroidogenesis and c-AMP formation by LH alone was again noted (Fig. 5), combination of PRL and LH did not increase c-AMP production above that given by LH alone. Unfortunately, the insignificant increase in progesterone synthesis obtained by the addition of PRL to the LH-medium did not duplicate results obtained during perifusion (cf. Figs. 1 and 5). This lack of agreement, which probably resulted from using too high levels of LH in the static incubation, weakens the significance of the observation. However, it does appear that the acute action of PRL on dispersed luteal cells did not result in stimulation of adenylyl cyclase activity. A similar observation has been made by others (30-32). Repeated injections of PRL to cycling rats has been shown to elevate adenylyl cyclase activity in ovarian tissue (33), but this observation did not exclude the possible contribution by endogenous LH to cyclase activation.

Thus, in view of the fact that the maximum rate of steroidogenesis proceeds at relatively low levels of intracellular c-AMP, and since we have failed to stimulate c-AMP formation by PRL added _in vitro_ and have shown that PRL augments the steroidogenic action of dbc-AMP, we conclude that PRL augmentation probably does not involve stimulation of adenylyl cyclase activity. Instead, it seems more probable to us that PRL acts at a locus within the luteal cell in a way that increases the effectiveness of the c-AMP compartment involved in steroidogenesis.

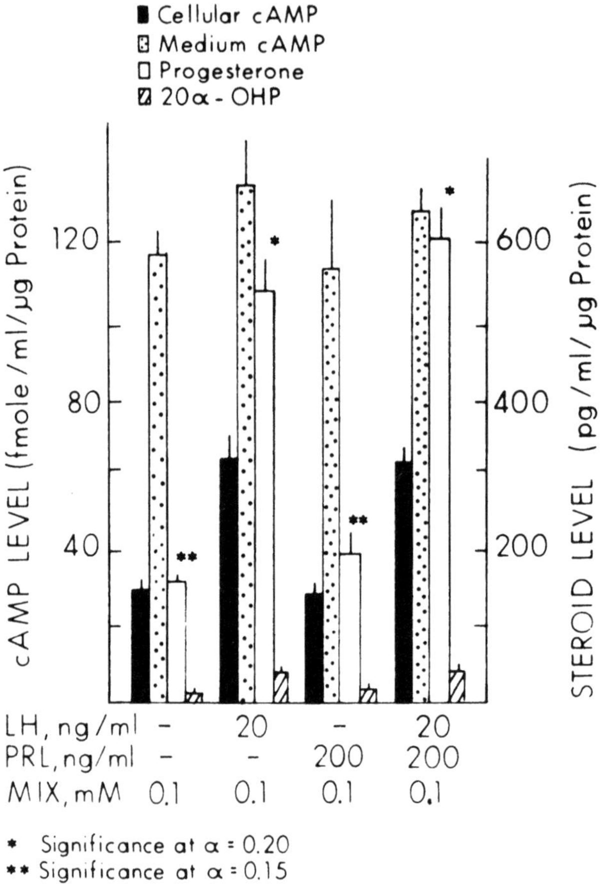


Fig. 5. Effects of combined LH and PRL on steroid and c-AMP formation. Static incubation of dispersed day-4 luteal cells (5.13 μg protein equivalent) in 4.0 ml of Dulbecco's modified Eagles medium containing 0.3% Penicillin G., 0.25% bovine serum albumin and the additions indicated above. Incubation was performed at 37C for 1 hour under an atmosphere of 5% $CO_2$ - 95% $O_2$. Each point represents the mean derived from 3 separate incubations. Vertical lines through the points indicate standard errors of the means.

## ACKNOWLEDGEMENTS

LH (NIH-LH-B-9 bovine) and prolactin (NIH-P-bovine) were obtained through the courtesy of the Hormone Distribution Officer, NIAMDD, NIH.

The authors are indebted to Mrs. Janey Jackson for typing the manuscript.

This work is supported in part by NICHD Specialized Population Research Center Grant P50-HD 08235.

## REFERENCES

(1) M.S. Smith, M.E. Freeman and J.D. Neill. Endocrinology 96 (1975)219.
(2) W.K. Morishige and I. Rothchild. Endocrinology 95(1974)260.
(3) T.M. Crisp. Endocrinology 101(1977)1286.
(4) I. Hashimoto, T. Asano and W.G. Wiest. Endocrinology 96 (1975)421.
(5) D.H. Wu, W.G. Wiest and A.C. Enders. Endocrinology 98(1976) 1378.
(6) H. Naito and W.G. Wiest. Program 59th Annual Meeting, The Endocrine Society, June 8-10, 1977, Abst. No. 579.
(7) D.H. Wu and W.G. Wiest. Endocrinology, in press.
(8) C. Mendelson, M. Dufau and K. Catt. J. Biol. Chem. 250(1975) 8818.
(9) A. Peytreman, W.E. Nicholson, G.W. Liddle, J.G. Hardman and E.W. Sutherland. Endocrinology 92(1973)525.
(10) J.M. Marsh, R.W. Butcher, K. Savard and E.W. Sutherland. J. Biol. Chem. 241(1966)5436.
(11) F.A. Juel Jr, J.L. Humes, J. Tarnoff, V.J. Cirillo and E.J. Ham. Science 169(1973)883.
(12) J.H. Dorrington and B. Baggett. Endocrinology 83(1969)989.
(13) D.A. Stanfield and D.J. Franks. Biochem. Biophys. Acta 242 (1971)606.
(14) Y.A. Fontaine, E. Fontaine-Bertrand, N. Delerve-LeBelle and C. Salmon. J. Physiol. (Paris) 63(1971)49A.
(15) J.M. Marsh and K. Savard. Steroids 8(1966)133.
(16) J.H. Dorrington and R. Kilpatrick. Biochem. J. 104(1967)725.
(17) J. Kolena and C.P. Channing. Endocrinology 90(1972)1543.
(18) C.P. Channing and S. Kammerman. Biol. Reprod. 10(1974)179.
(19) A. Tsafrifri, H.R. Lindner, U. Zor and S.A. Lamprecht. J. Reprod. Fertil. 31(1972)39.
(20) J.B. Miller and P.L Keyes. Endocrinology 95(1974)253.
(21) D.L. Grinwich, E.A. Ham, M. Hichens and H.R. Behrman. Endocrinology 98(1976)146.

(22) W.R. Moyle and J. Ramachandran. Endocrinology 93(1973)127.
(23) R.J. Beall and G. Sayers. Arch. Biochem. Biophys. 148(1972)70.
(24) K.J. Catt and M.L. Dufau. Nature New Biol. 244(1973)219.
(25) W.Y. Huang and W.H. Pearlman. J. Biol. Chem. 237(1962)1060.
(26) D.W. Armstrong, L.S. Miller and K.A. Knudsen. Endocrinology 85(1969)393.
(27) J.S. Richards and J.J. Williams. Endocrinology 99(1976)1571.
(28) D.L. Grinwich, M. Hichens and H.R. Behrman. Biol. Reprod. 14(1976)212.
(29) M. Lahav, S.A. Lamprecht, A. Amsterdam and H.R. Lindner. Mol. Cell. Endocrin. 6(1977)293.
(30) N.R. Mason, R.J. Schaffer and R.E. Toomey. Endocrinology 93 (1973)34.
(31) J.M. Marsh. J. Biol. Chem. 245(1970)1596.
(32) H. Herlitz, L. Hamberger, S. Rosberg and K. Ahren. Acta Endocrinologica 77(1974)737.
(33) M. Hunzicker-Dunn and L. Birnbaumer. Endocrinology 99(1976) 198.

DISCUSSION - CHAIRPERSON - R.O. GREEP

G. GIBORI: LH in a nonacute situation, and in an in vivo system stimulates progesterone synthesis, not directly but by providing the corpora lutea with estrogen, and in the final analysis, it's estrogen that is the luteotropic hormone. Prolactin could maintain LH and estrogen receptors and could allow the final luteotropic hormone, estrogen, to stimulate progesterone synthesis.

W.G. WEIST: I can't refute that. We have not done perfusion studies with estrogen. We intend to do so.

R. GREEP: Perhaps those who are burning to ask questions can hold them until the general discussion.

L. BIRNBAUMER: I would like just to make a very small comment and that is, that we have added prolactin to adenyl cyclase assays and have not been able to demonstrate any effect.

STEROID SECRETION BY PERIFUSED OVARIAN FOLLICLES OF IMMATURE, PMSG-TREATED IMMATURE, AND MATURE RABBITS

N. Shahabi[1], J. Bahr[2], O. K. Dial[2], and S. Glenn[3]

[1]Department of Animal Science, Pahlavi University
Shiraz, Iran

[2]Departments of Animal Science Physiology
University of Illinois, Urbana, Illinois 61801

[3]Department of Biochemistry, University of Texas
Houston, Texas 77030

ABSTRACT

The follicle changes morphologically as it matures. To determine if this anatomical change is coincident with changes in steroidogenesis, follicles were removed from immature (Im) (2.0 kg), immature-PMSG (Im-P) (2.0 kg) and mature (M) (4.0 kg) rabbits and perifused constantly with Medium 199-0.1% BSA. Medium was assayed for estradiol-17-β ($E_2$), testosterone (T), and progesterone ($P_4$). Follicular sizes were 1.0 mm (Im) and 1.5-2.0 mm (Im-P and M). After the establishment of a baseline, 100 ng of LH (NIH-S20) and 50 ng of FSH (50.3 x NIH-FSH-S1) were injected into the chambers. To eliminate the effect of follicular size on steroid secretion, the data were evaluated by considering the ratios of $E_2$:T:$P_4$ with $E_2$=1.0. Independent "t" test and Behrens-Fisher tests were used to test for significance. Steroid secretion by Im-P follicles did not increase following LH:FSH but rather decreased gruduully with time. The $E_2$:T:$P_4$ ratios before and 1 h after LH:FSH were 1.0:0.38:0.25 and 1.0:0.38:0.37 for Im-P follicles. Before gonadotropin stimulation there was no difference in the ratios $E_2$:T (1.0:0.94 vs 1.0:).52) and $E_2$:P (1.0:0.26 vs 1.0:0.47) for Im and M follicles respectively. One h after LH:FSH, the $E_2$:T and $E_2$:$P_4$ for Im vs M follicles were 1.0:9.8 vs 1.0:3.8 and 1.0:0.41 vs 1.0:0.88 respectively ($P < 0.05$). There was significantly less T and more $P_4$ relative to $E_2$ secreted by M than Im follicles. These data suggest that as a follicle matures, the steroid profile changes.

## Introduction

As the ovary matures it changes morphologically and biochemically. Follicles increase in size with the formation of an antrum and distinct granulosa and theca cellular components. During this follicular maturation, which is induced by FSH, follicles become responsive to LH (1,2,3) which is mediated by the adenylyl cyclase system (4). During follicular development, there is a gradual increase in adenylyl cyclase activity since the levels of this enzyme are significantly less in follicles removed from immature rabbits and rats (5, 6). However, when follicular maturation in immature animals is hastened by PMSG-treatment there is a significant increase in adenylyl cyclase activities (6).

The aim of this study was to investigate if the steroid profile of the follicle changes during the maturational process. The endpoint used was steroid synthesis by perifused isolated ovarian follicles removed from immature, mature, and PMSG-treated immature rabbits.

## Materials and Methods

<u>Animals</u>. Three groups (6-8/group) of New Zealand White Rabbits, immature (Im) (2-0 kg), immature-PMSG (Im-P), (2.0 kg) and mature (M) (4.0 kg), which had been housed separately for at least 3 weeks, were used. The Im-P were injected sc 56 h earlier with 150 U of PMSG to promote follicular maturation. Rabbits were killed by cervical dislocation and the ovaries were immediately removed and put in ice-cold phosphate buffered saline (PBS). Follicles were removed by the use of tissue forceps under a stereoscope. At all times during the preparation of the follicles, they were maintained in cold PBS. Follicles from Im rabbits were approximately 1 mm in diameter whereas those from Im-P and M rabbits were approximately 1.0 2 mm in diameter.

Four follicles, each from a different animal within the treatment group, were placed in a chamber and perifused with Medium 199 containing 0.1% BSA (TCM). The TCM was maintained at a pH of 7.2-7.4 by aerating it with 95%-5%, $O_2$-$CO_2$ before it entered the individual chambers. Chambers were maintained at 37°C by placing them in a water bath in which water was heated and constantly circulated by a circulating heater pump (Haake E52). After flushing the chambers for 30 min at the rate of 1 ml/20 min, a 20 min collection was taken to establish basal secretion. Then LH (100 ng, NIH-S-20;NIAMMD) and FSH (50 ng, 50.3 x NIH-SI-FSH: Papkoff) were injected into each chamber. Succeeding samples were collected at 20 min intervals for 3 h.

Steroid measurements. Samples were assayed for progesterone ($P_4$), testosterone (T), and estradiol-17β ($E_2$) without extraction. The specificity of the antibodies eliminated the need for chromatography. The volume of TCM in each assay tube was maintained at a constant volume by adding freshly prepared TCM to every tube in the assay.

Statistics. To determine differences within a group, the Student "t" test was used whereas differences between independent groups were tested using the independent "t" test. Changes in the ratio of steroids, secreted by the same group with time were determined first by converting hormone values to natural logs, second by finding the difference between the ratio and then testing for significance by the Student "t" test. To determine differences among different groups, the independent "t" test or Behrens-Fisher test was used. The latter test is appropriate because variances are not pooled but considered separately.

## Results

Steroid secretion by perifused follicles of Im, Im-P, and M rabbits are presented in Fig. 1. Unstimulated follicles (basal secretion) of Im rabbits secreted low quantities of steroids, as compared to Im-P and M follicles, with approximately equal amounts of $E_2$ and T. The steroid profiles for Im-P and M were similar with $E_2$ being the predominant steroid secreted. Following stimulation

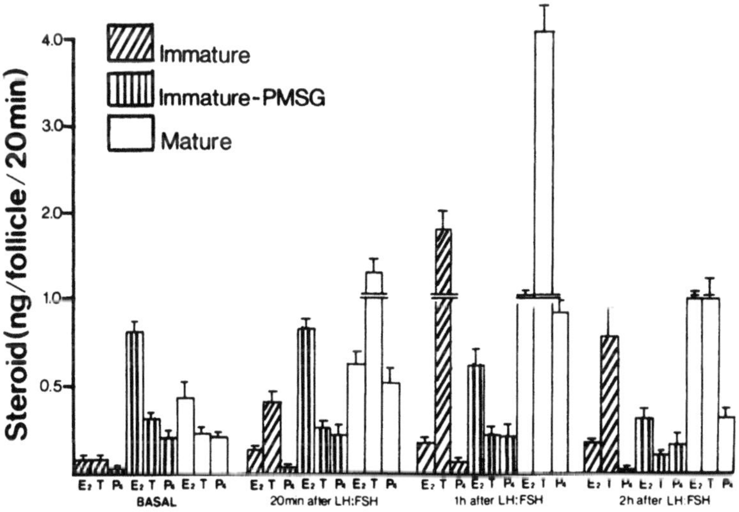


Fig. 1. Steroid secretion by perifused follicles of Im, Im-P, and M rabbits before (basal) and at various times after gonadotropin challenge.

by gonadotropins, steroid secretion by Im and M increased dramatically with significant increases ($P < 0.05$) in $E_2$ and T by Im follicles and in $E_2$, T, and $P_4$ by M follicles 20 min after LH:FSH, reaching maximum levels at 1 h. In contrast, Im-P did not respond to gonadotropin challenge; rather a gradual decrease in steroid secretion occurred with time.

Because of differences in follicular size among the three groups, it was impossible to perform a comparative analysis of steroid secretion by Im, Im-P, and M follicles. Therefore to circumvent this variable, changes in the ratio of $E_2$:T:$P_4$ were considered with the arbitrary choice of $E_2$=1 at each sampling time (Fig. 2). The $E_2$:T

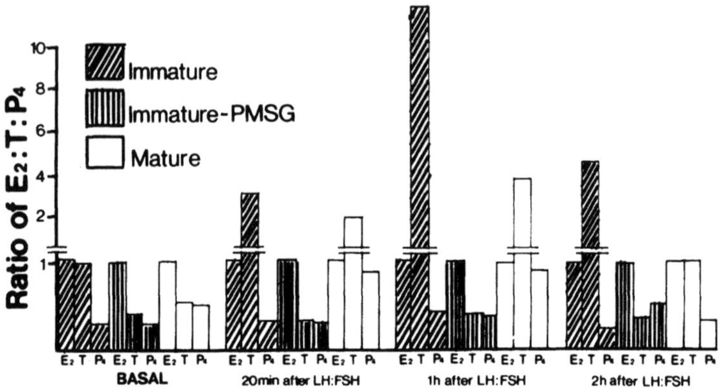


Fig. 2. The ratio of $E_2$:T:$P_4$ for Im, Im-P and M rabbits before (basal) and at various times after gonadotropin challenge. At each specific time, $E_2$=1.

and $E_2$:P ratios for Im and M were similar for unstimulated follicles but were significantly different ($P < 0.01$) for 1 h after LH:FSH. The Im follicles secreted approximately 3-fold more T relative to $E_2$ than did the M follicles while the latter group secreted about 2-fold more $P_4$ relative to $E_2$ than did the former group. A comparison of basal steroid secretion by Im-P follicles with that of Im and M indicates that $E_2$:P for Im-P vs M was different ($P < 0.01$) with M follicles secreting more $P_4$ relative to $E_2$ than Im-P follicles. Because the Im-P did not respond to gonadotropin stimulation, no further comparisons were made.

## Discussion

The results of this experiment show that follicles of Im rabbits have a significantly different steroidogenic capacity than those follicles of M and Im-P rabbits. While unstimulated Im follicles secrete approximately equal, albeit low, amounts of $E_2$ and

and T, unstimulated M follicles secrete primarily $E_2$. This same response is observed when the maturation of Im follicles is accelerated by PMSG. After challenge with gonadotropin, there is a significant change in the ratio of steroids secreted by Im and M follicles. Although T is the predominant steroid secreted by both Im and M follicles, Im follicles secrete significantly more T than do M follicles. On the other hand, M follicles secrete more $P_4$ relative to $E_2$ than do the Im follicles. At the present time there is no acceptable explanation for the abundant secretion of T in the rabbit despite the fact that it has been reported frequently (7-10) whereas in the cycling rat, it has been clearly demonstrated that T has a definite role in the feedback regulation of FSH (11-13).

The significant change in the ratio of $E_2$:T:$P_4$ as the ovary matures may reflect morphological changes in the follicle. Greenwald (14) has demonstrated by histological studies that as follicles mature there is an increase in the number of granulosa layers. Since granulosa cells convert T to $E_2$, an increase in granulosa cells may result in a greater conversion of T to $E_2$, thus explaining the shift in the ratios of $E_2$:T and $E_2$:$P_4$ as the follicle matures (15).

In this experiment follicles removed from Im-P rabbits were unresponsive to gonadotropin stimulation as monitored by steroid secretion. At this time we are unable to explain this lack of response because rabbits treated with the same dose of PMSG do ovulate in response to LH (personal observation). Immature rats, treated with PMSG 48h earlier, are responsive to LH as measured by stimulated adenylyl cyclase activity which would suggest, if one can extrapolate from the rat to the rabbit, that the lack of steroidogenic response is apparently not the result of desensitized follicles or a lack of gonadotropin receptors (6).

In conclusion we have demonstrated that the steroidogenic profile of a follicle changes during the maturational process. There is less T and more $P_4$ relative to $E_2$ secreted by follicles of M and Im-P rabbits as compared to Im rabbits.

(Supported by NSF BMS75-18994.)

REFERENCES

(1) M. Hunzicker-Dunn and L. Birnbaumer. Endocrinology. 99(1976) 198.

(2) Y. Salomon, A. Yanovsky, Y. Mintz, Y. Amir and H. R. Lindner. J. Cyclic Nucleotide Res. (1978) In press.

(3) J. S. Richards and A. R. Midgley. Biol. Reprod. 14(1976)82.

(4) H. R. Lindner, A. Tsafriri, M. E. Lieberman, U. Zor, Y. Koch, S. Bauminger and A. Barnea. Recent Prog. Horm. Res. 30(1974)79.

(5) M. Hunzicker-Dunn and L. Birnbaumer. Ovarian Follicular Development and Function (Raven Press, New York, 1978). in press.

(6) Y. Salomon. FEBS. 45(1978)299.

(7) E. V. YoungLai. J. Reprod. Fert. 40(1974)95.

(8) J. Hilliard, R. J. Scaramuzzi, C. N. Pang, R. Penardi and C. H. Sawyer. Endocrinology. 94(1974)267.

(9) J. Bahr. Biol. of Reprod. 18(1978)193.

(10) T. M. Mills and K. Savard. Endocrinology. 92(1973)788.

(11) C. Beyer, R. B. Jaffe, and V. L. Gay. Endocrinolgoy. 91(1972) 1372.

(12) C. Beyer, M. L. Cruz, V. L. Gay, and R. B. Jaffe. Endocrinology. 95(1974)772.

(13) V. L. Gay and R. L. Tomacari. Science. 184(1974)78.

(14) G. S. Greenwald. Anat. Rec. 178(1974)139.

(15) J. E. Fortune and D. T. Armstrong. Endocrinology. 102(1978) 227.

## DISCUSSION - CHAIRPERSON - R.O. GREEP

N.B. SCHWARTZ: Last night, there was a controversy between Drs. Goodman and Richards which was unresolved but may be cleared up now. The issue was, why is there a difference between post-partum and cyclic ovulation in the rat. It seems to me that from the data you have shown in the rabbit you could predict that as the progesterone levels drop as happens toward the end of pregnancy that the follicles in the ovary should have a different responsiveness to LH and FSH, and results in an altered estrogen profile.

J. BAHR: Surely to complete this study we are measuring LH in these animals.

R.O. GREEP: A final comment, Dr. Weiss.

J. Weiss: Perifusion or continuous flow incubation, obviously, is coming into vogue in studies of the ovary. It may be appropriate and helpful for me to mention some of the things that we learned over the past few years, having worked with this system in the pituitary. It is made to sound very easy by both of the previous speakers to carry out these experiments but they are not as simple as they sound. There are a number of things that have to be considered in order to get the kind of highly reproducible results that we have been able to obtain with the anterior pituitary. The size of the chamber is of importance since it determines the turnover time which in turn determines the half-life of the infused substances. I'm a little worried by the use of syringes unless the size is controlled very carefully. A second important variable is the flow rate, 3 ml/h is rather slow. I would anticipate a significant degradation of the infused peptide hormones with a flow rate of 3 ml/h at 37°C. I would suggest that it is worthwhile to test the survival of the particular agents that one introduces and the survival of the product, that one is trying to isolate, under the conditions of the experiment. Now, an even more serious problem arises when one is trying to compare different tissue compartments, for example follicles derived from different sources. In a completely submerged state, it is only approximately 200μ or so of the peripheral cells that are really surviving optimally. Oxygen transfer through the fluid to the tissue is poor. That is, the gas penetration from liquid to solid is poor. Furthermore the penetration of the agents that one introduces might also differ, depending on the shape, the form, the size, the kind of tissue one is dealing with. Consequently, direct comparisons, between different subsets..different ovarian compartments, for example, should be made with care, taking into consideration that the penetration of your agents and the survival of your cells might differ. It is useful in fact to check the morphology of the tissue at the end of the experiment. These are just some general principles that may be helpful to those embarking in this field.

J. BAHR: Thank you, Judy. You've done a very fine job with the use of the pituitary tissue in perifusion systems. I will agree that after we have done them most things look simple--but I must admit that we spent 9 months or more studying the system and validating it! However, one comment, I'll agree that if one puts peptide hormones into the medium and pumps them at a very slow rate, degradation of the hormone may occur. However, in the studies I have reported here we inject the LH directly into the chambers.

EFFECTS OF FOLLICULAR DEVELOPMENT ON THE ABILITY OF CULTURED PORCINE GRANULOSA CELLS TO CONVERT ANDROGENS TO ESTROGENS

Larry D. Anderson, Frederick W. Schaerf and Cornelia P. Channing
Department of Physiology, University of Maryland School of Medicine
660 West Redwood Street, Baltimore, Maryland 21201

## INTRODUCTION

It has been shown by Channing and her colleagues that as the antral porcine follicle grows from a 2 mm follicle to a 6-12 mm follicle, the granulosa cells exhibit up to a 100 fold increase in LH/hCG receptors (1), an increase in ability to secrete progesterone in culture (2), as well as an increase in ability to respond to LH in terms of cyclic AMP accumulation (3,4). However, changes in the ability to secrete estrogen have not been measured. It has been shown that granulosa cells of many species can secrete estrogen provided they are given androgen substrate (sow (5), mare (6), cow (7), rabbit (8), sheep (9), and rat (10).

The purpose of the present study was to examine changes in the ability of porcine granulosa cells to convert androgens to estrogens as a function of the stage of follicular maturation. In addition, influences of gonadotrophins and follicular fluid upon the ability of the granulosa cells to convert androgens to estrogens and secrete progesterone was examined. Since it has been shown by Eiler and Nalbandov that the estrogen content of porcine follicular fluid rises as follicular development proceeds (11), there is a need to examine whether the ability of the granulosa cells to convert androgens to estrogens proceeds in a similar fashion.

## MATERIALS AND METHODS

Porcine ovaries were obtained within fifteen minutes of slaughter at a local meat packing plant and transported to the laboratory for cell harvesting. Granulosa cells were harvested from small 1-2mm, medium 3-5mm, and large 6-12mm antral follicles.

Granulosa cells were separated from follicular fluid by centrifugation at 600 x g for five minutes and were washed three times in tissue culture medium 199. The cells were counted using a hemocytometer and viable cells were indicated by their ability to exclude trypan blue. Granulosa cells were incubated at $37^{\circ}C$ in Falcon multiwell test plates at a concentration of $1 \times 10^6$ cells per well with four wells per variable. Tissue culture medium 199 containing 10% pig serum and antibiotics were added. Purified human FSH (LER 1951, generously donated by Dr. Leo Reichert) and ovine LH (NIH S-19) were added to the culture media at a concentration of 1.0, 0.1 and 0.01 μg/ml. In addition, testosterone (T), androstenedione (4 Androsten 3,17 Dione) (A) and dihydrotestosterone (DHT) (Sigma Chemical Co., St. Louis, Mo.) were added at a concentration of 1.4 and 0.14 μg/ml with and without gonadotrophins. Porcine follicular fluid obtained from small antral follicles and charcoal-treated to remove steroids was added with media at a concentration of 50%. Follicular fluid was assayed for steroids after charcoal treatment, and contained neglible amounts of progesterone (<100 pg/ml) and total estrogens (<25 pg/ml). Cultures were terminated after either two or four days and the cells were stained with hemotoxylin and oil red O for later cell counting. Media was analyzed by radioimmunoassay using antisera for estrogens and a highly specific antiserum generated against the 11 OH progesterone BSA conjugate (kindly donated by Dr. Gordon Niswender) for the progesterone determinations. Estrogen radioimmunoassay was carried out according to a modification of the method of Hotchkiss et al (12) using an antisera generously donated by Dr. Burton Caldwell. The antisera was generated in sheep against Estradiol-hemisuccinyl BSA and cross-reacted equally against estrone and estradiol as shown by us previously (13). Both steroids were assayed in diluted, unextracted aliquots of the media as previously described (14). The sensitivity of the radioimmunoassay was 10 picograms per culture for total estrogens and 100 pg progesterone per culture. Medium blanks were consistently lower than these values. Statistical analyses were performed using the Student's T Test. The data was expressed as nanograms of steroid per day per one million cells, and represents the mean plus or minus the standard error of four replicates per variable.

## RESULTS

Granulosa cells from all sized follicles secreted low levels of estrogen in both control media (C) and media supplemented with FSH alone (Figure 1). However, the addition of testosterone (T) or androstenedione (A) to the culture media resulted in a significant ($p<0.01$) 8 to 420 fold increase in estrogen secretion. The magnitude of estrogen secreted by granulosa cells increased as the size of the follicle increased. The lowest conversion of androgen was seen in cells from small follicles, while the highest was seen in cells from medium and large follicles. Dihydrotestos-

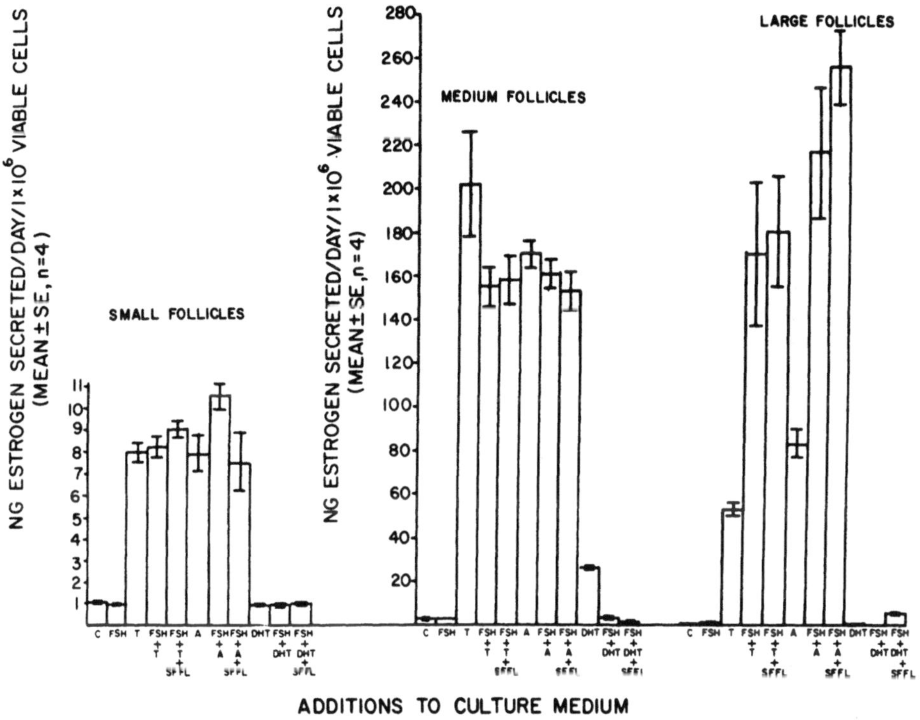


FIGURE 1. Capacity of granulosa cells from various sized follicles to secrete estrogens. Nanograms of estrogen secreted per day per million cells is plotted on the ordinate while the various additions to the culture media are shown on the absicca. Data are expressed as the mean $\pm$ SE of four replicates per variable. The ordinate for granulosa cell from small follicles is ten fold smaller than that for granulosa cells from medium and large follicles.

terone (DHT), either alone or in the presence of FSH had no effect on estrogen secretion. The addition of FSH with either testosterone or androstenedione had no effect on the secretion of estrogen by granulosa cells from small and medium follicles. However, FSH increased estrogen secretion in the presence of both testosterone and androstenedione in granulosa cells from large follicles.

The effects of various doses of FSH, LH and testosterone on both progesterone and estrogen secretion by granulosa cells from medium follicles after two days of culture is shown in figure 2. Both FSH and LH stimulated progesterone secretion, while the presence of testosterone alone or in combination with FSH or LH had no effect on progesterone secretion above that found for controls. In contrast to the progesterone data, the addition of either FSH or LH was unable to stimulate estrogen secretion at any concentration. As previously shown in this study, the addition of testosterone, either alone or in the presence of FSH, dramatically increased estrogen secretion. However, the addition of LH with testosterone at two concentrations, 1.4 and 0.14 μg per milliliter, significantly ($p<0.01$) decreased the secretion of estrogen, as compared to the amount of estrogen secreted when testosterone was added alone.

After four days of culture, granulosa cells from medium follicles were unable to secrete estrogen in detectable amounts, irrespective of whether FSH or testosterone was added either alone or in combination. Progesterone secretion in these cells remained elevated however, (Figure 3). The presence of FSH alone had a slight stimulatory effect on progesterone secretion, whereas the addition of either testosterone or androstenedione alone had no significant effect. When FSH was added in combination with testosterone, androstenedione or dihydrotestosterone, progesterone secretion was augmented two to three fold above FSH alone. The addition of 50% follicular fluid from small follicles to cultures containing FSH and androgens resulted in a dramatic stimulation ($p<0.01$) of progesterone secretion above that found for FSH and androgens alone, and FSH by itself. Charcoal-treated pig serum used as a control for follicular fluid had no such stimulatory effect.

## DISCUSSION

These studies show that as the porcine follicle matures there is an increased ability of granulosa cells to convert androgens to estrogen. This increase was greatest between granulosa cells obtained from medium versus small follicles. The ability to secrete progesterone also increased during follicular maturation but the increase in ability to secrete progesterone was greatest, however, in cells from large versus medium follicles. Armstrong and his colleagues observed that granulosa cells from large pre-

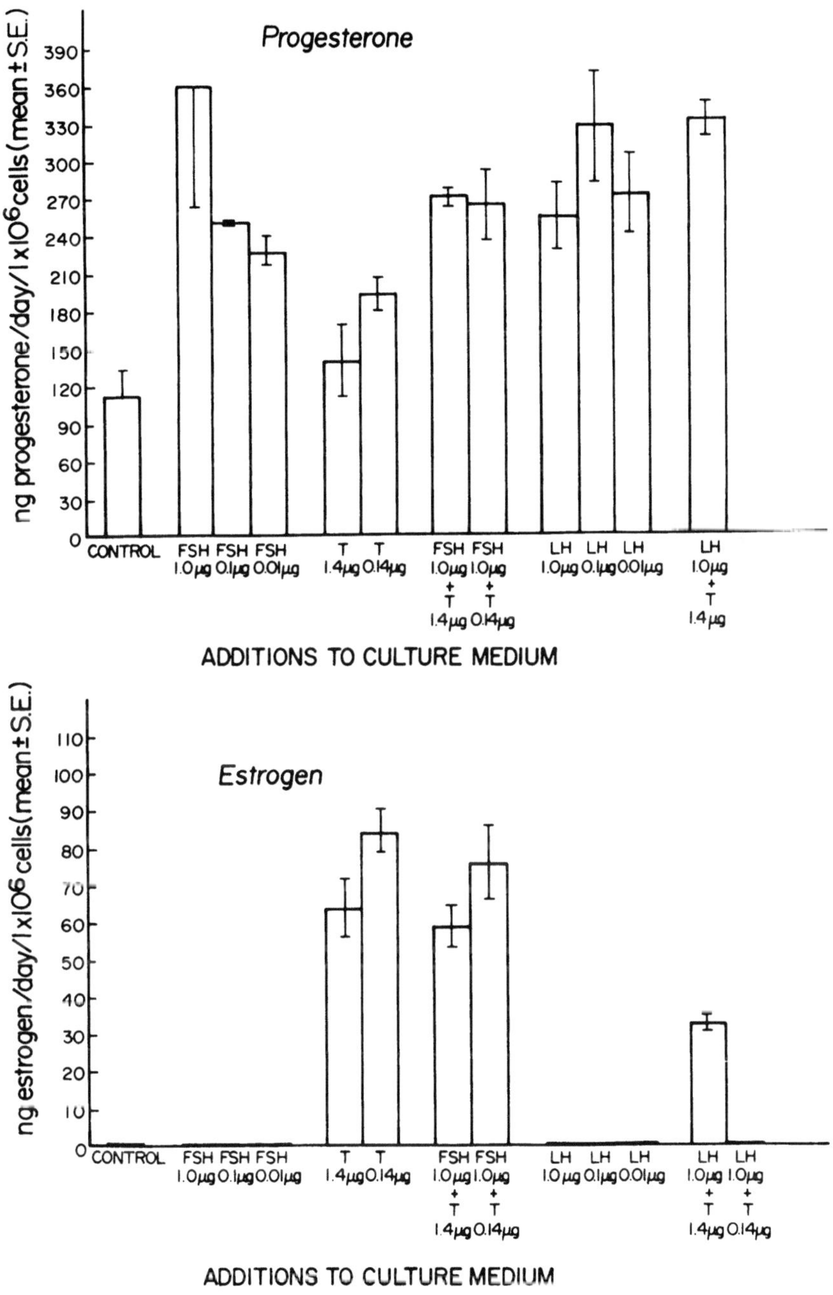


FIGURE 2. Capacity of Porcine Granulosa cells from medium follicles to secrete progesterone and estrogen after two days of culture. The upper panel depicts progesterone secretion while the lower panel depicts estrogen secretion. Data are expressed as the mean ± SE of four replicates per variable.

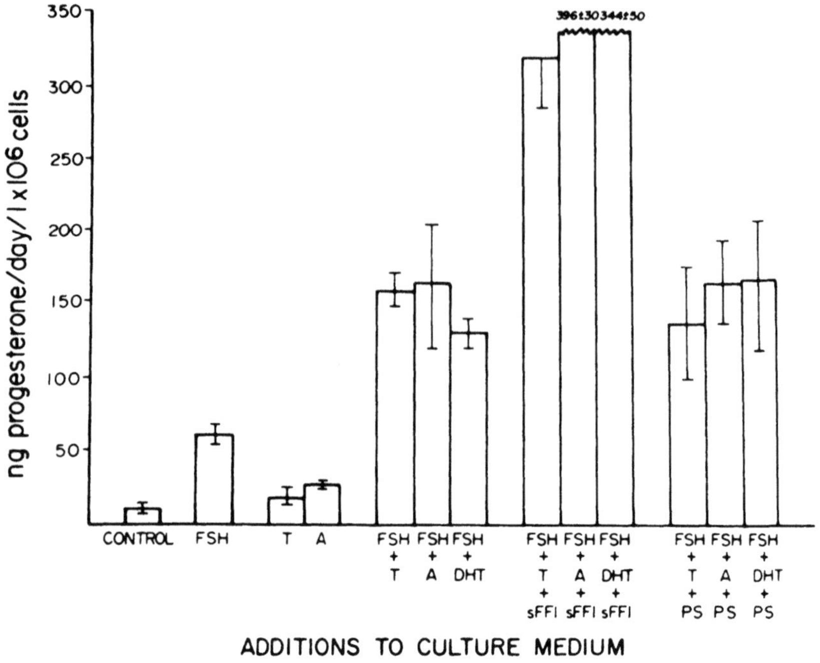


FIGURE 3. Capacity of porcine granulosa cells from medium follicles to secrete progesterone after four days of culture. Data are expressed as the mean ± SE of four replicates per variable. Estrogen secretion after four days was nondetectable and is not shown.

ovulatory follicles of the rat and pig could convert androgens to estrogens but did not examine granulosa cells from other stages of follicular development (10,15). It is possible that the increase in follicular fluid estrogen in medium and large follicles (11) is therefore in part due to a contribution of granulosa cells. Thecal estrogen may also be a source of this follicular fluid estrogen and is the subject of future investigation.

The fact that only granulosa cells obtained from large follicles responded to FSH with an increase in conversion of androgen to estrogen is not readily explained by changes in FSH receptors which are lower in number in granulosa cells from large versus small follicles (16). It is possible that in the present study, the percent of FSH receptors coupled to a response which results in stimulation of aromatase activity is greater in cells from large compared to small follicles. As well, the possibility exists that some step other than the binding of FSH to its receptor, such as the aromatase enzyme itself, is not fully compotent to respond until the granulosa cells are in the more mature state such as that found in large follicles.

Granulosa cells from medium follicles were unable to secrete estrogens between 2 and 4 days of culture while progesterone secretion remained elevated, suggesting that the aromatase enzyme system is short-lived, or that the proper maintainance conditions were absent in the culture media. In contrast, progesterone secretion remained elevated throughout the 4 day culture period. The addition of androgens in the presence of FSH augmented progesterone secretion only after two days of culture, and not prior to this time period. Similar observations of androgen augmentation of progesterone secretion in granulosa cells from antral and preantral follicles have been reported (17,18).

In a similar fashion, the addition of charcoal-treated follicular fluid from small antral follicles, resulted in a further increase in progesterone secretion in the presence of FSH and androgens. These results are in contrast to previous reports of Ledwitz-Rigby and Rigby who observed that follicular fluid from small follicles inhibits progesterone secretion by granulosa cells from medium and large porcine follicles (19). These investigators added 50% small follicular fluid alone to the cultures, where in the present studies small follicular fluid was mixed with androgens and FSH. Since dissimilar culture conditions were employed, this may partially explain these disparate results.

Several reports have suggested that LH given *in vivo* and *in vitro*, can decrease the concentration of estrogens in ovary and blood (20,21). Our results show more specifically that LH can inhibit estrogen secretion by granulosa cells *in vitro* when androgens and LH are added to the culture media.

## SUMMARY

Porcine granulosa cells from different stages of follicular development were examined for their ability to convert androgens to estrogens and their ability to secrete progesterone. Granulosa cells from all stages of follicular development can convert androgens to estrogens, and the addition of exogenous androgen is required for estrogen secretion. Granulosa cells obtained from medium and large follicles have a greater capacity to convert androgens to estrogens than do granulosa cells obtained from small follicles. The addition of FSH brought about an increase in estrogen secretion in the presence of androgen only in granulosa cells from large follicles. LH attenuated estrogen secretion in granulosa cells from medium follicles treated with testosterone. Granulosa cells from medium follicles were unable to secrete estrogen from days two to four irrespective of treatment.

Androgens augment FSH stimulated progesterone secretion in granulosa cells from medium follicles from days two to four, and the addition of follicular fluid from small follicles stimulates progesterone secretion in the presence of FSH and androgens.


## Acknowledgements

These studies were supported in part by NIH Training Grant HD 00435 and NIH Grant HD 08834 to C.P.C. The typing of the manuscript by Ms. Gail Williams is gratefully acknowledged.


## References


(1) Channing, C.P. and S. Kammerman, Endocrinol. 92(1973)531.

(2) Channing, C.P. Endocrinol. 87(1970)156.

(3) Channing, C.P. in: Proceedings of the IV International Congress of Endocrinology held at Washington, D.C., June 19-24, 1974, ed. R.O. Scow, Excerpta Medica, Amsterdam, pp. 914-919.

(4) Lindsey, A.M. and C.P. Channing, Biol. of Reprod. (in press) 1978.

(5) Bjersing, L. and H. Carstensen, J. Reprod. Fertil. 14(1967)101.

(6) Channing, C.P., J. Endocrinol. 43(1969)403.

(7) Lacroix, E., W. Eechaute and I. Leusen, Steroids 23(1974)337.

(8) Erickson, G.F. and K.J. Ryan, Endocrinol. 97(1975)108.

(9) Moor, R.M., J. Endocrinol. 73(1977)143.

(10) Fortune, J.E. and D.T. Armstrong, Endocrinol. 102(1978)227.

(11) Eiler, H. and A.V. Nalbandov, Endocrinol. 100(1977)331.

(12) Hotchkiss, J., L.E. Atkinson and E. Knobil, Endocrinol. 89(1971)177.

(13) Channing, C.P. and S.P. Coudert, Endocrinol. 98(1976)590.

(14) Channing, C.P., V. Tsai and D. Sachs, Biol. of Reprod. 15(1976)235.

(15) Tsang, B.K., Y.S. Moon and D.T. Armstrong, Fed. Proceedings, Vol. 37, p. 723 (Abstract #2680) 1978.

(16) Nakano, R., T. Akahori, K. Katayama and S. Tojo, J. Reprod. Fert. 51(1977)23.

(17) Schomberg, D.W., R.L. Stouffer and L. Tyrey, Biochem. Biophys. Res. Commun. 68(1976)77.

(18) Lucky, A.W., J.R. Schreiber, S.G. Hillier, J.D. Schulman and G.T. Ross, Endocrinol. 100(1977)128.

(19) Ledwitz-Rigby, F., B.W. Rigby, V.L. Gay, M. Stetson, J. Young and C.P. Channing, J. Endocrinol. 74(1977)175.

(20) Katz, Y and D.T. Armstrong, Endocrinol. 99(1976)1442.

(21) Moor, J. Endocrinol. 61(1974)455.

# TESTOSTERONE SYNTHESIS BY CHICKEN FOLLICULAR CELLS

Eileen Su-Rong Huang and A. V. Nalbandov

Animal Genetics Laboratory, University of Illinois

Urbana, Illinois 61801

## ABSTRACT

The roles of theca and of granulosa cells in the synthesis of testosterone (T) in ovarian follicle of hens have been studied. Theca cells alone synthesized no progesterone (P), and no or only negligible amounts of T. Granulosa cells, in contrast, without stimulation produced large amounts of P and small amounts of T. When incubated with ovine gonadotrophins, theca cells showed greater response than granulosa cells in synthesizing more T (2- to 2.5-fold increase). Combination of the two follicular cell types decreased synthesis of P, but increased that of T (9- to 17-fold). When 0.1 μg of ovine luteinizing hormone (oLH) was added to the combined cells, T concentrations increased most dramatically, a 83-fold increase was seen after 12 h of incubation. Theca cells incubated with exogenous P also produced large amounts of T. These results indicated that in hens, theca cells may be the major site of follicular T synthesis, however, theca cells alone cannot synthesize T unless granulosa cells or P were added. It is suggested that granulosa cells may use P as a precursor, and that P is converted to T in the theca cells.

## INTRODUCTION

Thecal tissue isolated from hamster, rabbit or rat ovarian follicles spontaneously secreted androgen *in vitro*, whereas granulosa cells collected from the same follicles produced little or negligible amounts of androgen (1, 2, 3). Although the wall of the three largest preovulatory follicles of hens contains T (4), the cellular source for synthesis of this follicular steroid has never been studied *in vitro*.

Obtaining pure mammalian thecal tissues without contaminating granulosa cells has been difficult. Furthermore, in rats, more T synthesized by crude theca preparations than by more clean theca tissues (5). The question then raised is what is the contribution of contaminated granulosa cells to the synthesis of T by theca cells. Obviously, this question can only be answered by studying pure theca preparations. Ovarian follicles of hens are especially suitable for studies of this kind. Mature chicken follicles bulge and ultimately hang down individually from the ovary. No interstitial tissue is present. In addition, these follicles do not have an antrum as do mammals, and the ovum (yolk) is surrounded by a single layer of granulosa cells. The granulosa cells, at one end, form cytoplasmic processes which penetrate deeply into the zona radiata where they interweave with cytoplasmic processes from the ovum. At the other end, they are separated from the thecal layer by a thick basement membrane (6). These unique anatomical and histological features facilitated the isolation of clean thecal tissues and made it possible (7).

This study was done to investigate the cellular site of T synthesis in chicken follicles, and to examine the role of granulosa cells in T synthesis.

## MATERIALS AND METHODS

Mature, regularly laying hens, exposed to 14 h light (0700 to 2100 h), were used. The time of oviposition of all experimental birds was determined daily in order to predict the time of ovulation. The 3 largest preovulatory follicles were collected at 15 to 16 h before expected ovulation, a time when endogenous LH was low. In order to separate the granulosa layer from the theca layer, a slit was made with a surgical blade on these follicles. The yolk and the enclosing granulosa layer emerged leaving the clean theca tissue behind. The granulosa layer was identified as the topmost layer covering the yolk and was easily pulled away as a complete sheet from the yolk. Theca tissues were minced, pooled and incubated with medium 199 (Grand Island Biological Co.) containing collagenase (500 U/ml, Sigma, type I) for 2 h, while pooled granulosa layers were incubated with the same medium for only 10 min. After incubation, follicular cells were centrifuged and washed 4 times with incubation medium containing 0.4% bovine serum albumin, penicillin G (50 U/ml, Sigma) and streptomycin sulfate (50 U/ml, Sigma). A viability count of cells was done with a hemacytometer using the vital staining procedure. The cell suspension was then diluted with incubation medium to a final concentration of $1 \times 10^6$ cells/ml. When the cell suspensions were prepared to study the effect of gonadotrophins, a decreasing concentration of trypsin inhibitor (Type II-L, Lima bean, Sigma) was added to the 4 washings (0.02%, 0.002%, 0.001%, 0.001%), and cells were diluted with medium containing 0.001% trypsin inhibitor. One-ml aliquot of granulosa or theca cells were incubated at 40°C with or without 10 μg of ovine oFSH (oFSH, NIH-FSH-S8), or 1 μg of oLH (NIH-LH-S20); or both cell types were incubated together with or without 0.1 μg of oLH; or theca cells with 0.1 μg of P (Sigma). The total content of P and T in the incubated samples were measured by radioimmunoassay.

## RESULTS AND DISCUSSION

Our preliminary data showed that in spite of differences in follicle sizes, the synthesis patterns of T of theca or granulosa cells were similar. During a 3 h incubation period, granulosa cells spontaneously synthesized small amounts of T, while theca cells synthesized no or only negligible amounts of T (Table 1, control group). In all cases, the concentration of T before incubation was between 0.030 to 0.045 ng/1 x $10^6$ cells.

Table 1. Effect of oFSH (10 μg) or oLH (1 μg) on synthesis of testosterone by theca (Th) or granulosa (Gr) cells from hen follicles.

| Cell type | Follicle[a] | Treatment: Control | FSH | LH |
|---|---|---|---|---|
| Th | $F_1$ | 0.031±0.004[b] | 0.043±0.002 | 0.060±0.006* |
| | $F_2$ | 0.053±0.005 | 0.136±0.002* | 0.136±0.001* |
| | $F_3$ | 0.048±0.002 | 0.094±0.003* | 0.116±0.006* |
| Gr | $F_1$ | 0.045±0.004 | 0.061±0.008* | 0.117±0.003* |
| | $F_2$ | 0.140±0.010 | 0.141±0.012 | 0.185±0.015 |
| | $F_3$ | 0.133±0.007 | 0.170±0.015* | 0.164±0.004 |

Theca cells were incubated for 3 h, and granulosa cells for 2 h.
[a]$F_1$ = the largest follicle; $F_2$ = the second largest follicle; $F_3$ = the third largest follicle.
[b]Results were expressed in ng/1 x $10^6$ cells SEM, n = 3.
*Significantly different from the corresponding control group.

It was noted that when cells were incubated without any ovine gonadotrophins, the T concentrations were higher in granulosa cells of the more immature follicles, the second ($F_2$) and third largest ($F_3$) follicles, those not destined to ovulate next. Interestingly, theca cells were able to synthesize T after incubated with exogenous oFSH or oLH (Table 1). In general, T concentrations increased 2- to 2.5-fold in theca cells treated with ovine gonadotrophins. In granulosa cells, only the largest follicle ($F_1$), which would ovulate 15 to 16 h later, showed prominent response to the treatment of gonadotrophins (Table 1).

Theca cells do not synthesize any P after 3 h of incubation (Table 2). The P concentrations in the control group were no higher than those before the incubation started. Neither did these theca cells respond to gonadotrophins (Table 2). In contrast, large amounts of P were synthesized by granulosa cells even without exogenous oFSH or oLH (Table 2). Incidentally, in all follicle sizes P concentration in granulosa cells before incubation was below 4 ng/1 x $10^6$ cells. Both oFSH and oLH had significant stimulatory effect on P synthesis by granulosa cells (Table 2).

Although the data presented above indicate that both follicular cells types can synthesize T, theca cells are more responsive to gonadotrophic stimulation. Yet, the incremental changes of T concentration in theca cells are no more than 2.5 fold. A much higher rate of T synthesis was seen when these two follicular cell types were incubated together (Fig. 1). In this experiment, follicular cells collected from $F_3$ were studied. The T concentration increased 9-fold or 17-fold after 3 h or 6 h of incubation respectively. The T synthesis by the combined cells was significantly enhanced by 0.1 μg of oLH ($P < 0.001$). When the combined cells were incubated with oLH for 12 h, T concentration increased most dramatically (83-fold increase).

The combination of two follicular cell types, however, has a negative effect on P synthesis (Fig. 2). When cells were incubated alone, large amounts of P were produced by granulosa cells, while

Table 2. Effect of oFSH (10 μg) or oLH (1 μg) on synthesis of progesterone by theca (Th) or granulosa (Gr) cells from hen follicles.

| Cell type | Follicle[a] | Treatment | | |
|---|---|---|---|---|
| | | Control | FSH | LH |
| Th | $F_1$ | 0.27±0.04 | 0.25±0.02 | 0.33±0.01 |
| | $F_2$ | 0.18±0.07 | 0.38±0.04 | 0.18±0.02 |
| | $F_3$ | 0.10±0.00 | 0.12±0.02 | 0.12±0.01 |
| Gr | $F_1$ | 60.9±1.9 | 155.3±6.3* | 544.8±14.0* |
| | $F_2$ | 89.0±1.5 | 155.8±7.0* | 221.0± 0.7* |
| | $F_3$ | 37.9±1.0 | 72.6±1.1* | 118.7± 7.5* |

For explanation of footnotes, see Table 1.

no P synthesis could be detected in theca cells. After combining and incubating the two cell types together, P concentration decreased strikingly. Ovine LH was able to enhance more P synthesis of the combined theca and granulosa cells, but the effect of oLH was restricted to the first 3 h of incubation.

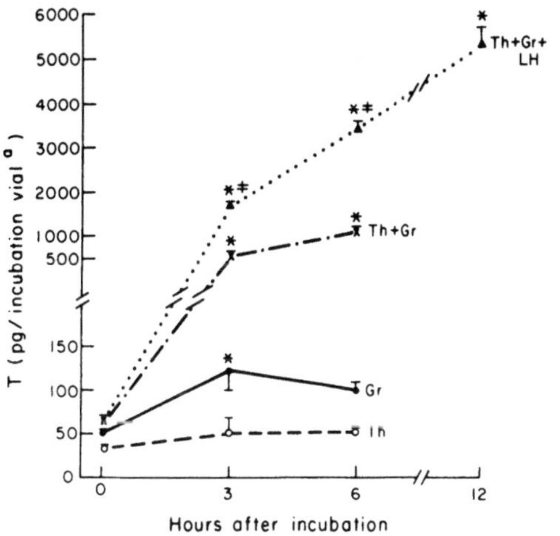


Fig. 1. Effect of combination of the two follicular cell types on testosterone synthesis. Th = Theca cells; Gr = Granulosa cells. a = 1 x $10^6$ cells/vial when Th or Gr was incubated alone; and 1 x $10^6$ Th plus 1 x $10^6$ Gr cells/vial when cells were incubated together. * = Significantly different from previous point. ǂ = Significantly different from corresponding point of Th + Gr. Results were Mean ±SEM, n = 3.

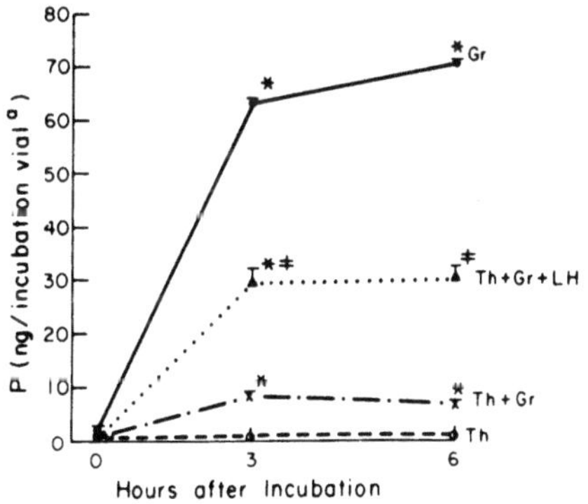


Fig. 2. Effect of combination of the two follicular cell types on progesterone synthesis. For explanation of symbles, see Fig. 1.

This is the first evidence showing that both theca and granulosa cells are required for synthesizing large amounts of T. The finding that the concentration of P declined while that of T increased in incubations in which the two cell types are combined imply the possibility of P being converted to androgens by these cells, thus causing P concentration to drop. However, by which cell type? Since granulosa cells alone synthesized large amounts of P, but only little T, and since theca cells alone could not synthesize P, but were able to synthesize T, the most likely candidate for converting P to T are theca cells. This possibility was supported by experiment where large amounts of T were produced by theca cells when incubated with exogenous P (Fig. 3). These theca cells were collected from $F_2$ and $F_3$ follicles. Ovine LH increased synthesis of both P and T in the combined cells. Since P can be converted to T by theca cells, it is possible that at least part of the stimulatory effect of oLH on synthesis of T is accomplished by increasing concentration of P.

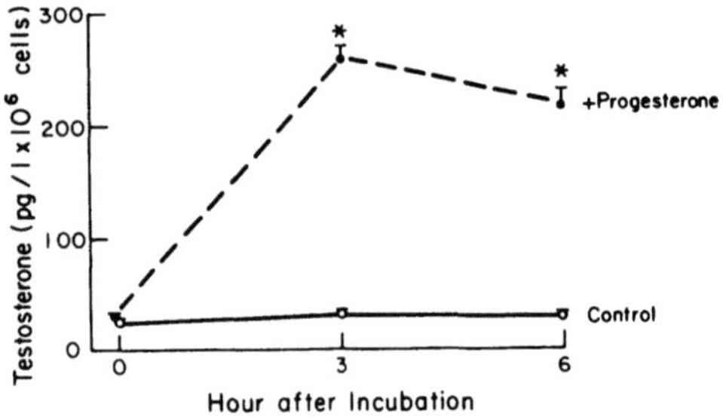


Fig. 3. Effect of exogenous progesterone (0.1 μg) on synthesis of testosterone by theca cells. Each point represents mean SEM, n = 5. o = Control group; • = Progesterone treated group. * = Significantly different from the corresponding point in the control group ($P < 0.001$).

REFERENCES

(1) A. Makris and K. J. Ryan. Endocrinology 96(1975)694.

(2) G. F. Erickson and K. J. Ryan. Endocrinology 99(1976)452.

(3) J. E. Fortune and D. T. Armstrong. Endocrinology 100(1977)1341.

(4) N. A. Shahabi, H. W. Norton and A. V. Nalbandov. Endocrinology 96(1975)962.

(5) J. E. Fortune and D. T. Armstrong. Endocrinology 102(1978)227.

(6) G. M. Syburn and A. H. Baillie. in: Physiology of the Domestic Fowl, eds. C. Horton-Smith and E. C. Amoroso (Oliver and Boyd, Edinburgh and London, 1966) P. 30.

(7) E. S. Huang. In Vitro Steroidogenesis by Ovarian Follicular Cells of Hens. Ph.D. thesis. (University of Illinois, Urbana, Illinois, 1978).

# MODULATION OF THECAL PROGESTERONE SECRETION BY ESTRADIOL-17β

J.E. Fortune and W. Hansel

Section of Physiology and Department of Physical Biology

Cornell University, Ithaca, New York

## ABSTRACT

The effects of estradiol-17β on progesterone secretion by isolated bovine theca interna were studied. Preparations of theca interna were obtained from preovulatory follicles of Holstein heifers ($n = 5$) prior to the endogenous LH surge and cultured for 3 days in a defined medium containing various concentrations of estradiol (0, 0.001, 0.01, 0.1, 1, or 10 μg/ml). Secretion of progesterone by control theca was low during the first 24 h of culture and increased steadily during the subsequent 48 h. The lowest concentration of estradiol (0.001 μg/ml) significantly stimulated thecal progesterone production, while higher concentrations (0.1, 1, and 10 μg/ml) significantly inhibited secretion of progesterone. Estradiol content of the follicular fluid of these preovulatory follicles and several others ($n = 13$) averaged 1.5 μg/ml ± 0.1 (SEM). It is concluded that bovine theca interna is capable of secreting progesterone, that high intrafollicular concentrations of estradiol may inhibit thecal progesterone production *in vivo* prior to the LH surge, and that estradiol concentration may be be an important modulating factor in the control of follicular progesterone secretion.

## INTRODUCTION

Mammalian ovaries secrete progesterone primarily during the luteal phase of the estrous cycle; plasma progesterone concentrations are low during the follicular phase (1,2). In normal cycles preovulatory follicles, which secrete primarily estrogen and androgen prior to the LH surge (see 3 for review), differentiate after ovulation into corpora lutea that secrete large quantities of progesterone. It is known that both luteal cells and luteinized granulosa cells of many species can secrete progesterone and that progesterone production is enhanced by LH (4,5). Although there is evidence for large mammals that theca interna, as well as granulosa cells, become incorporated into the corpus luteum (6,7), it is not known if theca-lutein cells contribute to luteal progesterone secretion. Estradiol has been reported to both increase (8) and decrease (9,10) progesterone production by cultured granulosa cells. The experiments reported here were performed to determine if isolated bovine theca interna can secrete progesterone and to study the effects of different concentrations of estradiol on thecal progesterone secretion.

## MATERIALS AND METHODS

Holstein heifers with regular estrous cycles were ovariectomized on Day 19 or 20 of the estrous cycle (Day 0 = estrus; average bovine cycle length = 21 days). Blood samples were collected by jugular puncture daily beginning with Day 16 and every 6 h for 24 h prior to surgery. Animals were observed twice daily for estrous behavior. Plasma was assayed for progesterone by radioimmunoassay (RIA) on Day 18 and/or 19. Ovariectomies were performed only when plasma progesterone had fallen to preovulatory levels (~1 ng/ml) but estrous behavior had not been observed. Subsequent measurement of plasma LH by solid phase radioimmunoassay (11) confirmed that ovaries were removed prior to the LH surge.

The preovulatory follicle (i.e. the largest follicle, 12) was dissected from each pair of ovaries. After aspiration of the follicular fluid, the follicle was cut in half and the theca interna and adhering granulosa cells were separated from the surrounding theca externa and stroma by dissection (13). The theca was scraped gently but firmly with a glass needle to remove granulosa cells. Histological examination of scraped theca showed that this method yielded theca free of contaminating granulosa cells. The theca interna was cut into pieces and the pieces distributed at random among treatment dishes and cultured in a completely defined medium (modified Eagle's Minimum Essential Medium, 14) in the presence or absence of highly purified bovine LH (LH-HP, lot #G3-235A, 10 ng/ml) in one of the following concentrations of estradiol: 0, 0.001, 0.01, 0.1, 1, or 10 μg/ml. Medium containing estradiol was prepared by adding aliquots of stock solutions of estradiol in ethanol to culture medium (0.1 ml/30 ml medium); the same concentration of ethanol was added to control medium. Thecal preparations were cultured in a humidified incubator at 36°C in an atmosphere of 95% air and 5% $CO_2$. Medium was collected and replaced at 3, 6, 12, 24, 48, and 72 h of culture. Treatments were applied in duplicate to theca from each heifer.

Petroleum ether extracts of plasma amd unextracted aliquots of culture media were measured for progesterone by RIA as described by Fortune and Armstrong (15). The antiserum used did not exhibit detectable cross reaction (<0.01%) with estradiol-17β. Aliquots of follicular fluid were extracted with 10 volumes of diethyl ether and measured for estradiol as previously described (15). Protein was measured by the method of Lowry et al. (16).

Analysis of variance and Scheffé's method for multiple comparisons were used for statistical analyses of data. Data were log transformed before analysis when heterogeneity of variance was present.

RESULTS

Progesterone secretion by control thecal preparations was low during the first day of culture and much higher during the second and third days (Table 1). The lowest concentration of estradiol (0.001 μg/ml) significantly increased ($P < 0.01$) the rate of progesterone secretion as compared with the control, while higher doses of 0.1, 1, and 10 μg/ml significantly inhibited ($P < 0.05$, 0.001, 0.001, respectively) thecal progesterone secretion. Estradiol at 0.01 μg/ml had no significant effect ($P > 0.10$) on progesterone

Table 1. Progesterone Secretion (pg/μg protein) by Bovine Theca Interna Cultured in Various Concentrations of Estradiol-17β (n = 10 thecal preparations)

| Estradiol (μg/ml) | 0-3 h | 3-6 h | 6-12 h | 12-24 h | 24-48 h | 48-72 h |
|---|---|---|---|---|---|---|
| 0 | 5.7<br>±1.0 | 3.5<br>±0.5 | 4.9<br>±1.3 | 13.1<br>±2.8 | 82.1<br>±19.7 | 212.6<br>±58.2 |
| 0.001 | 16.1<br>±5.9 | 8.5<br>±2.2 | 14.4<br>±2.1 | 41.6<br>±6.3 | 194.1<br>±35.6 | 277.6<br>±69.8 |
| 0.01 | 6.4<br>±1.7 | 2.0<br>±0.5 | 4.3<br>±1.1 | 18.4<br>±3.6 | 138.1<br>±31.4 | 284.6<br>±66.6 |
| 0.1 | 5.4<br>±0.9 | 2.4<br>±0.6 | 2.8<br>±0.8 | 8.6<br>±2.4 | 61.8<br>±16.4 | 91.0<br>±26.7 |
| 1.0 | 5.7<br>±2.0 | 3.1<br>±1.1 | 3.4<br>±1.2 | 6.8<br>±2.8 | 6.9<br>±2.7 | 14.2<br>±6.0 |
| 10.0 | 0.9<br>±0.5 | 0.5<br>±0.3 | 0.4<br>±0.2 | 0.1<br>±0.1 | 0.3<br>±0.2 | 3.5<br>±3.2 |

production. Although LH slightly increased progesterone secretion by thecal preparations ($P < 0.10$), the effects of the various concentrations of estradiol were similar in the presence or absence of LH (data not shown).

In two of the five experiments the initial 72 h of culture in various doses of estradiol was followed by an additional 3 days of culture in the absence of estradiol. Under these conditions thecal preparations that had been exposed initially to 0, 0.001, 0.01, 0.1, 1, and 10 μg/ml estradiol secreted an average of 163 ± 35, 349 ± 86, 184 ± 42, 242 ± 74, 347 ± 93, and 183 ± 61 pg progesterone/μg protein/day ($n = 4$), respectively.

Estradiol was measured in the follicular fluid of 13 bovine preovulatory follicles, including those used in this study. Estradiol concentration in the follicular fluid averaged 1.5 μg/ml ± 0.1 (SEM).

## DISCUSSION

These results demonstrate for the first time that bovine theca interna, free of granulosa cells, can secrete progesterone. There is histological evidence that theca cells contribute structurally to the bovine corpus luteum (6,7). The present results provide evidence that theca cells are capable of contributing to follicular and luteal progesterone secretion. The data show that estradiol can modulate thecal progesterone secretion in either a stimulatory or inhibitory fashion, depending on dosage.

When theca that had been exposed to various levels of estradiol for 3 days was cultured for an additional 3 days in the absence of estradiol, the inhibitory effect of the higher doses of estradiol was reversed. Preliminary electron microscopic examination of theca cultured for 3 days has shown that although localized spots of degeneration are present within theca cells, the tissue is generally healthy and free of contaminating granulosa cells. Theca cultured in 10 μg/ml estradiol exhibited morphological characteristics similar to those of theca cultured without estradiol. The reversibility of the inhibitory effects of estradiol and the normal morphology of theca exposed to high concentrations of estradiol suggest that the inhibition of progesterone secretion does not result from toxic effects on the tissue.

Although the doses of estradiol that inhibit progesterone secretion seem high and "non-physiological" compared with blood levels, estradiol concentration in the follicular fluid of 13 bovine preovulatory follicles averaged 1.5 μg/ml. Other investigators have also found high concentrations of estradiol in follicular fluid from monkey, pig, and human preovulatory follicles (17-20). Although it is difficult to determine steroid concentrations in the immediate environment of the theca cells in vivo, the presence of high concentrations of estradiol in the follicular fluid prior to the endogenous LH surge suggests that the theca may be exposed to concentrations much higher than plasma levels of estradiol and that high levels of estradiol within the follicle may inhibit progesterone secretion prior to the LH surge. There is evidence that estradiol concentration in the follicular fluid of pig, monkey, and cow preovulatory follicles decreases considerably after the LH surge (17; 18; Fortune and Hansel, unpublished observations). We propose that intrafollicular estradiol concentration may be an important factor in the regulation of follicular progesterone secretion.

## ACKNOWLEDGMENTS

We wish to thank S. Vincent, P. Tsang, and R. Saatman for technical assistance and Y. Eisner, R. Milvae, and A.J. Sherman for help with blood collection and surgery. The highly purified bovine LH was generously supplied by Dr. Harold Papkoff (Univ. of Calif., San Francisco). We are indebted to Dr. John Eppig (The Jackson Laboratory, Bar Harbor, Maine) for electron microscopy of thecal tissue.

## REFERENCES

(1) W. Hansel, P.W. Concannon, and J.H. Lukaszewska. Biol. Reprod. 8(1973)222.
(2) M.S. Smith, M.E. Freeman, and J.D. Neill. Endocrinology 96(1975)219.
(3) W. Hansel and J.E. Fortune. Control of Ovulation, eds. D.B. Crighton, G.R. Foxcroft, N.B. Haynes, and G.E. Lamming (Butterworths, Woburn, Mass., 1978) p. 237.
(4) W. Hansel. Karolinska Symposia on Research Methods in Reproductive Endocrinology, 3rd Symposium, ed. E. Diczfalusy (The Reproductive Endocrinology Unit, Stockholm, Sweden, 1971) p. 295.
(5) C.P. Channing, Rec. Prog. Horm. Res. 26(1970)589.
(6) L. Donaldson and W. Hansel. J. Dairy Sci. 48(1965)905.
(7) B.L. Lobel and E. Levy. Acta Endocrinol. [Suppl] (KHB) 132, 59(1968)1.
(8) J. Bernard. J. Reprod. Fertil. 43(1975)453.
(9) K.H. Thanki and C.P. Channing. Endocrine Res. Commun. 3(1976)319.
(10) D.W. Schomberg, R.L. Stouffer, and L. Tyrey. Biochem. Biophys. Res. Commun. 68(1976)77.
(11) W. C. Hobson and W. Hansel. Endocrinology 91(1972)185.
(12) J. Dufour, H.L. Whitmore, O.J. Ginther, and L.E. Casida. J. Anim. Sci. 34(1972)85.

(13) S. Stoklosowa. Tissue Culture Association Manual. 1(1975)205.
(14) J.H. Dorrington, Y.S. Moon, and D.T. Armstrong. Endocrinology 97(1975)1328.
(15) J.E. Fortune and D.T. Armstrong. Endocrinology 102(1978)227.
(16) O.H. Lowry, N.J. Rosebrough, A.L. Farr, and R.J. Randall. J. Biol. Chem. 193(1951)265.
(17) C.P. Channing and S.P. Coudert. Endocrinology 98(1976)590.
(18) H. Eiler and A.V. Nalbandov. Endocrinology 100(1977)331.
(19) R.E. Fowler, S.T.H. Chan, D.E. Walters, R.G. Edwards, and P.C. Steptoe. J. Endocrinol. 72(1977)259.
(20) K.P. McNatty, D.T. Baird, A. Bolton, P. Chambers, C.S. Corker, and H. McLean. J. Endocrinol. 71(1976)77.

## GENERAL DISCUSSION - STEROIDOGENESIS
## CHAIRPERSON - R. O. Greep

J. K. FINDLAY: I think it pertinent to mention in the light of the discussion so far, that in the normal animal, at least two-thirds of the follicles present in the ovary are in various states of atresia. I should like to present some data from a graduate student, Ron Carson, who has investigated the binding of gonadotropins to ovine ovarian follicles of various sizes and states of atresia (Carson et al., this workshop).

Follicles were dissected from the ovary and then classified according to size and state of atresia, and then incubated as whole follicles in the presence of labelled gonadotropin, with and without excess gonadotropin to give specific binding. After incubation the various follicular components were separated and counted for $\gamma$-activity.

There was a decrease in binding of FSH to granulosa from the larger follicles and there was very little binding of FSH to theca. There was considerable binding of LH to theca from follicles of 2mm diameter, but very little binding to granulosa, and as larger follicles were examined, the granulosa component of hCG binding increased as we would expect. We were able to classify the follicles according to states of atresia, State I being essentially non-atretic, State V being atretic. This was done by examining the dissected follicle under a stereomicroscope by transmitted light and looking at the continuity of the granulosa, vascularity of the theca, atretic bodies etc. (Moor et al., J. Endocr. 77:309, 1978). As increasing degrees of atresia were examined, the FSH binding decreased. An analysis of variance revealed that the amount of FSH binding was significantly related to the state of atresia; with hCG there is some decrease in binding associated with the state of atresia, but the hCG binding is more related to the size of the follicle.

Our data is pertinent in view of the recent publication of Moor et al. which demonstrates that atretic follicles lose their capacity to aromatize androgens to estrogens. It also demonstrates that stage of atresia should be kept in mind when studies of gonadotropin binding are being carried out on follicular components of ovaries.

C.P. CHANNING: This brings to mind that in order to examine atresia in our studies, we require better criteria to define this state. In our monkey studies we showed that in vivo when the granulosa cells are rinsed out of the preovulation follicle, the theca layer still secrete estrogen into the ovarian vein. (Endocrinology 98: 1568, 1976) The possibility that some granulosa cells are left behind in the follicle could present a problem, so we did further studies with Dr. Thomas Crisp and found that at the EM level, the theca of the rinsed follicle was naked and had no granulosa cells left on it. Subsequently we took the same follicle cell types, theca and granulosa, and cultured them in vitro and found that theca made most of the estrogen whereas the granulosa described neglible amounts of estrogen in the absence of androgen substrate. At the EM level, Dr. Crisp found the theca had highly developed endoplasmic reticulum, and the mitochrondria christae were also highly developed, whereas in the granulosa cell, they were not.

Studies in the human were done with Dr. Ann Wentz, and as in the monkey the theca alone seems to make estrogen. It depends on how mature the follicle is as to how much estrogen can be made. If you mix the granulosa together with the theca you can increase estrogen production. In a similar study in the Rhesus monkey we have expressed estrogen secretion per follicle. It's difficult to express estrogen secretion on a per cell basis in thecal culture since in the theca there is connective tissue and blood vessels plus steroidogenically active cells. One way to get around this problem is to express estrogen secretion per follicle which is about the only way one can express this. (Channing et al. in Endocrinology of the ovary (R. Scholler ed) Proceedings of a conference held in Fresnes, France 1976, pp. 71-86. Sepe Editions, Paris 5.

In the intact animal, the theca still makes the estrogen for secretion into the ovarian vein and the granulosa doesn't make very much. The granulosa may aromatize some androgen supplied by the theca with this estrogen remaining within the follicle. Estrogen in follicular fluid may keep the granulosa from secreting too much progesterone as Dr. Schomberg has shown, and as we have just learned from Dr. Fortune.

E. SU-RONG HUANG: I would like to show you some data that demonstrate a phenomenon similar to the one that Dr. Channing just showed you. In a chicken follicle system, we found that only the theca cells, but not the granulosa cells, synthesize estrogen. In the first experiment theca or granulosa cells were incubated either alone or together for 6 hours, and estrogen synthesized by these cells was measured. The results show that neither the granulosa nor the theca cells alone synthesized estrogen. However, a combination of the two cell types produced large amounts of estrogen.

Next we incubated theca cells with increasing concentrations of exogenous testosterone and measured estrogen produced. The results show clearly that theca cells from small, immature follicles can synthesize estrogen. A dose-response relationship was noted between estrogen production and the amount of testosterone added. Estrogen production was not, however, observed either in the more advanced preovulatory follicles or in the post-ovulatory follicles.

A similar study was performed with granulosa cells collected from the same follicles. We found no estrogen synthesis by any of the granulosa cells from all follicles studied.

F. LEDWITZ-RIGBY: I have two separate comments on two separate papers. The first relates to Dr. Weist's presentation. I have looked at progesterone secretion by porcine granulosa cells in vitro in response to prolactin and LH. Prolactin by itself in concentrations up to 15 ng/ml stimulated progesterone secretion. When prolactin was added in combination with 100 nanograms of LH, the amount of progesterone secreted decreased as the concentration of prolactin increased. So as opposed to its action in rat cells prolactin appears to suppress LH stimulation of porcine cells. The other comment, is a question for Dr. Anderson. I was surprised that he saw stimulation of progesterone secretion by fluid from small follicles using granulosa cells from medium follicles in the presence of androgens and FSH. I have seen inhibition by the fluid from small follicles without the other hormones present. I'm wondering if Dr. Anderson has looked at progesterone secretion by cells from medium follicles in the presence of fluid from small follicles alone?

L. ANDERSON: We have not examined the effect of addition of small follicular fluid alone on progesterone secretion by granulosa cells from medium porcine follicles.

J.E. FORTUNE: In regard to the data presented by Dr. Channing, I would like to report that unlike the monkey and human theca interna from bovine preovulatory follicles appears incapable of secreting estradiol, while granulosa cells do secrete estradiol if they are provided with an androgen precursor. The evidence to date indicates that granulosa cells cannot synthesize androgens from cholesterol or progestins, (D.T. Armstrong and J.H. Dorrington. in Regulatory Mechanisms Affecting Gonadal Hormone Action, Volume 3, eds. J.A. Thomas and R.L. Senghal (University Park Press, Baltimore, 1977) p. 217.) (W. Hansel and J.E. Fortune. in Control of Ovulation, eds. D.B. Crighton, G.R. Foxcroft, N.B. Haynes, and G.E. Lamming (Butterworths, Woburn, Mass., 1978) p. 237).

C.P. CHANNING: Monkey granulosa cells also aromatized testosterone. A possible working hypothesis is that the theca can serve as a principle source of estrogen for ovarian venous estrogen, whereas perhaps the granulosa cell can interact somewhat to contribute to intrafollicular estrogen. Perhaps we should say that estrogen may end in the follicular fluid or the ovarian venous effluent. Both of these compartments should be considered.

G. NISWENDER: I would like to make one comment concerning the excellent presentation of Dr. Fortune. Jim Caffrey, a post-doctoral Fellow in our lab has currently a publication on 3-beta-hydroxysteroid-dehydrogenase isomerase enzyme complex (in press). He has shown that among other things, this enzyme system in the sheep is strongly inhibited by either estradiol or testosterone directly. This finding may explain Dr. Fortune's data, that is, if the sheep and the cow are similar.

K.J. RYAN: With respect to the comments about granulosa and theca, I think we ought to recognize here that we're talking about different species and different experimental conditions. I suspect that those investigators working with cows and monkeys are both correct in their observations. They're just not comparing the same things. The observations that Channing has made on the monkey and the human have been made about ten years ago. People have been neglecting the fact that they are dealing with different species, and they should pay more attention to it.

R. GREEP: That certainly is true this morning. Species seems to be the determining factor as to what the theca and the granulosa do.

G. ROSS: I would like the Chairman of this session to speak to the issue of the interaction of FSH and LH in stimulating the production of estrogen. To my knowledge, there is no mammalian system in which those two hormones do not synergize. With the exception of Jack Findlay's slide, I've never seen a demonstration of very much FSH binding by thecal cells. From a receptor point of view, then, the only cell that has the capacity to react with both hormones would appear to be the granulosa cell.

R. GREEP: In as much as LH and FSH are present in the blood stream at all times during the ovarian cycle it is unlikely that under normal circumstances one hormone ever acts entirely alone. The separate actions of FSH and LH have been studied but under artificial circumstances as in the post hypophysectomized immature female rat. There LH alone does not elicit the secretion of estrogen but when administered immediately following or in conjunction with FSH treatment then estrogen is secreted in abundance. On the basis of this and related cytochemical evidence I concluded many years ago that while the theca was primarily responsible for the secretion of estrogen it must be in some way sensitized by the prior or simultaneous action of FSH. Later, Falck demonstrated by means of separate ocular transplant of theca and granulosa that neither alone was capable of producing an estrogenic response in an adjacent transplant of a piece of the vagina. When theca and granulosa were both transplanted to the same eye chamber estrogenic responses were obtained. These results suggested very strongly that the theca and granulosa must act in concert to elicit ovarian steroidogenesis but they did not reveal which tissue actually produced the hormone. These studies likewise did not cast any light on the specificity of action of the gonadotropins. They did however by implication strengthen the view that FSH and LH must act in unison to achieve an output of estrogen. No conclusion could be reached from any of the foregoing studies as to whether FSH and LH act synergistically to effect estrogen secretion. Still the presumption is that they do.

Although it is of interest to try to learn what specific responses FSH and LH elicit when acting alone such information may well be academic. There is overwhelming physiological evidence that they act in concert and often synergistically but the specificity of their actions remains a problem. Binding studies, which may shed much light on the matter, appear to show that these two gonad stimulating hormones have a considerable degree of specificity of action. About all one may safely conclude from the evidence to date is that there are certain ovarian responses that are due predominantly to LH and others that fall predominantly under the influence of FSH. We have much yet to learn as to how the gonadotropins share the task of regulating ovarian morphologic and secretory responses.

It has been the central objective of this particular session to gain enlightenment as to what the steroidogenic properties of the theca and the granulosa are under conditions of effective gonadotropic stimulation. What has become most evident is that variation in response among different species is so great that no general conclusions can as yet be formulated. That in itself is valuable information. The aim must now be to fully understand the species under study be it rat, rabbit, guinea pig, sheep, monkey or human.

A. SHAIKH: Today we talked about steroid regulation, steroid secretion and gonadotropic regulation of steroid secretion in all in vitro systems. I wish to speak of some in vivo studies. I would like to show how the ovaries function during the menstrual cycle of the baboon. This gives us some clues as to the local effects of the steroids produced by the ovaries.

FIGURE D shows the estrogen and progesterone secretion in the utero-ovarian vein plasma from both sides and peripheral plasma during the menstrual cycle of the baboon. The line divides the luteal phase of the previous cycle from the follicular phase of the subsequent cycle. The top panel shows luteal phase steroids in the utero-ovarian vein plasma of the nonovulating ovary. This ovary contained the follicle which ovulated in the subsequent cycle. The middle panel shows the luteal phase steroids in the utero-ovarian vein plasma of the ovulating ovary. In the subsequent follicular phase this side of the ovary was nonovulatory. The bottom panel shows steroids in the peripheral plasma. All the steroid levels are plotted on a log scale. The solid lines show estrogen values and the broken lines progesterone. The interesting observation here is that during the luteal phase progesterone in

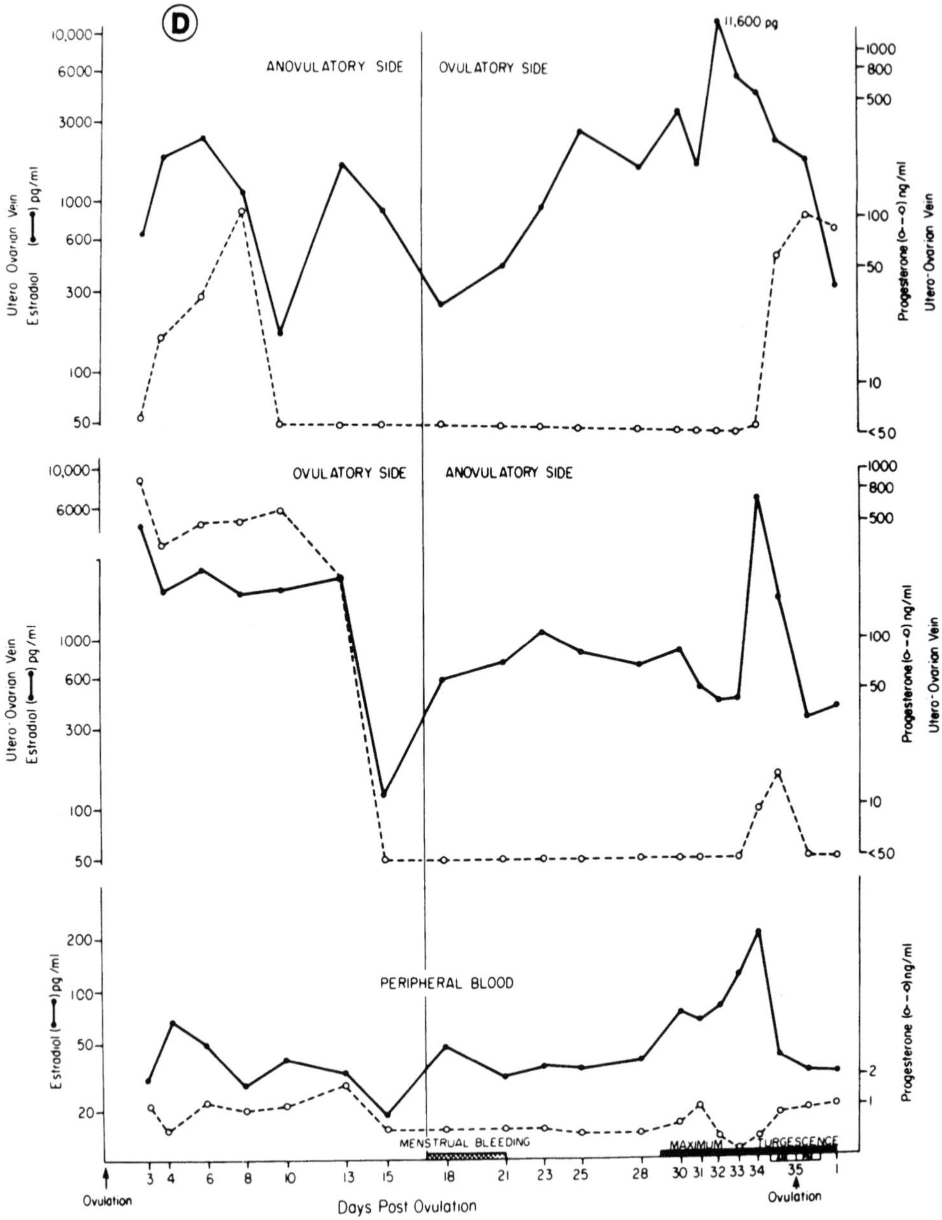


the anovulatory side utero-ovarian vein plasma is also very high, which drops to undetectable levels about 5 days before the progesterone levels drop in the vein on the ovulatory side. The question therefore arises: does this earlier drop in progesterone on the nonovulated side, which is around Day 10 of the cycle, give that ovary a head start in follicular development? The solid line on the same side shows increases in estrogen secretion, right after the drop in progesterone. Does this mean that at this stage the follicles in this ovary have already started developing? Where does this estrogen come from? Does the higher concentration of progesterone in the utero-ovarian vein from the ovulating ovary indicate local suppression of follicular development? Immediately after the drop in progesterone to undetectable levels, the ovarian production of estrogens on the ovulated side increases. Now, of course, there is follicular development in both the anovulatory and ovulatory ovary, but the anovulatory ovary of the previous cycle then goes on to ovulate with the preovulatory levels of estrogen reaching about 11,000 pg/ml.

R. GREEP: Please be brief and make your point.

A. SHAIKH: This shows that the local concentrations do play an important role in follicular development. Thus while peripheral levels play an important role in regulating feedback mechanisms by acting on organs away from the ovary, local concentrations regulate steroid secretion and follicular development.

S. DAY: We have recently been studying LH stimulability of the cyclase system after successively later administration of prolactin during the cycle in the rat. We have been able to show that if prolactin administration is started before 12 noon on diestrus 2 then LH stimulability of the cyclase system increases. When LH stimulability of cyclase activity is increased by prolactin, then we see increased progesterone levels in those animals on proestrus. If cyclase activity is not stimulated in response to prolactin administration, then the progesterone levels are low, as you would expect to see on proestrus. This is just supporting the idea of synergism between prolactin and LH. Perhaps, prolactin is maintaining or enhancing LH stimulability through maintaining or inducing cyclase system components. LH can then exert its effect on progesterone synthesis through this cyclase system.

R. GREEP: Do any of the speakers want to respond to that?

W. WIEST: You reason that your observations in the cycling animal also pertain to the psuedopregnant corpus luteum, I presume. Do you have any evidence of that being the case? Our failure to find an effect on intracellular cyclic-AMP concentration by prolactin alone persuaded us against the point of view that prolactin acted directly on adenyl cyclase.

I suppose there may be a more complex rationalization to the effect of prolactin than that, and we're looking for it.

M. RAJ: My question is to Dr. Weist. When you incubate luteal cells obtained from corpora lutea prior to Day 6, do you observe any stimulation of progesterone with prolactin alone?

W. WEIST: No. We have used day 2 and day 4 cells.

S.K. BATTA: In regard to the controversy concerning estrogen production by the granulosa cells, I have cultured human granulosa cells in collaboration with Dr. Wentz, and with Dr. Channing, and found that the granulosa cells when cultured alone do produce some estrogen and progesterone. But, when we add testosterone to the cultures of granulosa cells, there is almost a fourfold increase in estrogen production. Progesterone production on the contrary goes down.

R. GREEP: With that, I think we must conclude the session.

DISCUSSION - CHAIRPERSON - R.O. GREEP

G. ROSS: Over the two years that we have been interested in what androgens do in stimulating follicle growth and atresia, we have been troubled by the observation that the TFM mouse which lacks the classical androgen receptor present in other androgen target tissues seems to reproduce quite well. In order to check the validity of that concept for the rat, Zeleznik and Hillier have treated rats with the metabolically active analogue of flutamide and demonstrated that follicle growth, antrum formation, ovulation, and corpus luteum formation proceed very well. This argues then that an androgen receptor, seems not to be an obligatory requirement for follicular maturation, ovulation and corpus luteum formation.

D. SCHOMBERG: I would agree with everything you said, Griff. As I understand it, the TFM mouse or rat isn't totally devoid of androgen receptor, but has approximately 15% of that seen in the normal condition. The only thoughts I have vis-a-vis reproduction in these TFM animals is that perhaps they manage with this 15% in conjunction with the high concentration of local testosterone in the ovary.

S.K. BATTA: Dr. Schomberg, I am always afraid in vitro culture studies do not always show what's happening in vivo. First, we do not observe the total number of cells in a small compartment with a volume of 50 microliters to 1 milliliter, and contains a large number of cells which are not luteinized and are surrounded by follicular fluid. We can not accurately compare the total number of cells present in a follicle to the number of cells used in culture for steroidogenesis. If the total number of granulosa cells present in a follicle are studied for steroidogenesis in a single compartment it might reveal that a contact inhibition, or that the cells are in an antimitotic stage, and produce less steroids as compared to luteinized cells. Secondly, among cells cultured in vitro, this is not a normal state of affairs. In cultured cells we lose the follicular fluid and the luteinization inhibition factor. Finally the granulosa cells in culture may not be similar to the granulosa cells in follicles histologically or enzymically.

D. SCHOMBERG: I can't disagree. The approach we have taken in our work is to do some in vivo and in vitro comparisons. As to the comment about antimitotic activity and contact inhibition, I've never observed contact inhibition in the granulosa cell culture system.

DISCUSSION - CHAIRPERSON - R.O. GREEP

G. GIBORI: I wanted to point out the differences between follicles and corpus luteum. While follicles need gonadotropin to convert testosterone to estrogen, the corpus luteum will aromatize androgen to estrogen with no gonadotropin at all, and also, while in the follicles androgen stimulates progesterone synthesis. In the corpora lutea, only estrogen will stimulate progesterone synthesis.

D. SCHOMBERG: In a developmental context, when we add FSH to moderately and highly differentiated granulosa cells to examine the conversion of testosterone to estrogen, we haven't noted much stimulation of aromatase activity. I have the opinion that FSH action is perhaps most effective at the peri-antral stage. We also know that FSH receptors decrease as the granulosa cell matures. So in the developmental context, the granulosa cell seems to be dissociating itself from FSH control.

G. GIBORI: But what about LH?

D. SCHOMBERG: LH, via the theca, is providing more androgen substrate for the granulosa cells. Maybe a similiar situation exists in the corpus luteum also.

G. GIBORI: Isn't it possible that LH is stimulating the conversion of androgen to estrogen in systems other than follicles?

D. SCHOMBERG: Yes. It is entirely possible that qualitatively different responses to gonadotropic or steroidal stimuli will be observed with non-luteinized granulosa and thecal cells on one hand and granulosa-lutein and thecal-lutein cells on the other.

STEROID SECRETION BY OVARIAN FOLLICLES OF RABBITS IN DIFFERENT REPRODUCTIVE STATES

J. Bahr, N. Shahabi[1], R. Gardner, and L. Critchlow

Department of Animal Science, University of Illinois

Urbana, Illinois 61801

[1]Pahlavi University, Shiraz, Iran

ABSTRACT

Adenylyl cyclase (AC) activity of follicles of estrous rabbits (C) is greater than that of follicles of pseudopregnant (Ps) and pregnant (P) rabbits. To determine if there is a concomitant change in steroidogenesis and if the reproductive condition of the animal alters follicular steroid secretion, we removed follicles from C, Ps (7 or 15 days), or P (7, 15 or 28 days) rabbits and placed them in perifusion chambers. After the establishment of a baseline, LH (100 ng, NIH-S20; NIAMMD) and FSH (50 ng, 50.4 x NIH-FSH-SI; Papkoff) were injected into the chambers. The medium was assayed for progesterone ($P_4$), testosterone (T), and estradiol-17β ($E_2$). Steroid concentrations are expressed as pg/follicle/20 min. In the unstimulated follicles, T secretion (225) was similar for all groups except less for 15Ps and 28P (120;$P < 0.025$); $E_2$ secretion by C (442) was greater ($P < 0.001$) than for Ps and P groups (164) which were all similar. After LH:FSH, T (4790) and $E_2$ (910) secretion by C was elevated ($P < 0.025$) over Ps and P (T-2690; $E_2$-319). There were no differences among Ps and P groups except that 28P secreted significantly less T than 15P ($P < 0.05$). $P_4$ secretion by unstimulated follicles were similar for C, Ps, and P. After LH: FSH, differences were observed. With increasing length of pseudopregnancy or pregnancy, follicles secreted significantly less $P_4$ (C-1046, 7Ps-755, 15Ps-448, 7P-885, 15P-561, 28P-360). C, 7Ps and 7P were different ($P < 0.01$) from 15Ps and 15P and 15P was different ($P < 0.01$) from 28P. These results indicate that follicles removed from rabbits in various reproductive states have different steroid profiles. Possibly high intragonadal $P_4$ and circulating gonadotropins have a modulating effect on follicular steroidogenesis in Ps and P rabbits since AC activities are similar.

## Introduction

Ovaries of pseudopregnant and pregnant rabbits contain not only numerous corpora lutea but also many Graafian follicles. Secretion of estradiol by these follicles is necessary for the maintenance of the corpora lutea (1). These follicles are responsive to LH in that they can be ovulated by exogenous LHRH or LH. Basal and stimulated activities of adenylyl cyclase have been measured in ovarian follicles of estrous, pseudopregnant, and pregnant rabbits (2). Follicles removed from estrous rabbits have higher basal and stimulated levels of adenylyl cyclase than do follicles removed from pseudopregnant and pregnant rabbits. However, the relative increase in adenylyl cyclase activities following LH stimulation is similar for estrous and pseudopregnant and pregnant rabbits. Furthermore basal and stimulated activities of adenylyl cyclase are similar for pseudopregnant and pregnant rabbits and are not affected by the day of pseudopregnancy or pregnancy.

The mediatory role of cAMP in the action of LH in the ovary, specifically that of steroidogenesis, has been well documented (3-6). Following exposure to LH there is a rapid increase in cAMP (7-11) with a concomitant increase in steroid secretion by the ovarian follicles (6, 12-14). After this initial challenge by LH, follicles become desensitized to further gonadotropin stimulation (8, 11) and steroid secretion becomes minimal (12, 15 and 16). Against the backdrop of this information we designed an experiment to determine if basal and stimulated steroid secretion by follicles of estrous, pseudopregnant, and pregnant rabbits correlated with the adenylyl cyclase activities of these follicles and if the reproductive condition of the animal altered follicular steroid secretion.

## Materials and Methods

Mature female rabbits of mixed breeds which had littered at least once were divided into six groups with 6 to 8 animals in each group: control (C), 7 day pseudopregnant (7Ps), 15 day pseudopregnant (15Ps), 7 day pregnant (7P), 15 day pregnant (15P), and 28 day pregnant (28P). All rabbits had been isolated in individual cages for at least 21 days prior to their use. Rabbits were made pseudopregnant by injecting them IM with 50 IU of hCG whereas rabbits were made pregnant by mating.

To obtain ovarian follicles, rabbits were killed by cervical dislocation and ovaries immediately removed and placed in cold phosphate buffered saline (PBS) ($4^{o}C$). Follicles were dissected from the ovary and cleared of interstitial tissue using tissue forceps and with the aid of a stereoscope. At all times during the preparation, follicles were maintained in cold PBS.

To determine steroid secretion by these follicles before and after stimulation by gonadotropins, 4 follicles were placed in a chamber and perifused with Medium 199 containing 0.1% BSA (TCM). The TCM was maintained at a pH of 7.2 to 7.4 by aerating it with 95%-5%, $O_2$-$CO_2$ before it entered the individual chambers. Chambers were maintained at 37°C by placing them in a water bath in which water was heated and constantly circulated by a circulating heater pump (Haake E52). The TCM flowed through the chambers at a rate of 1 ml/20 min. After flushing the chambers for 30 min, two baseline samples were collected after which 100 ng of LH (1.03 x NIH-SI; NIAMDD) and 50 ng of FSH (50.3 x NIH-SI; Dr. H. Papkoff), freshly prepared in PBS containing 0.1% gelatin, were injected into the chambers in a 50 µl volume. Samples were collected at 20 min intervals during the subsequent 2 h period. The TCM was assayed for progesterone ($P_4$), testosterone (T), and estradiol-17β ($E_2$) without extraction. Hormone concentration was determined by RIAPAC (S. Glenn). Data was analyzed by the independent "t" test.

## Results

Steroid secretion by follicles removed from C, Ps, and P rabbits show different profiles for $P_4$, T, and $E_2$ (Table 1). Basal secretion of $P_4$ is similar for all groups. Following gonadotropin stimulation significant differences in $P_4$ secretion are observed. Follicles removed from C rabbits secreted the largest amount of $P_4$ whereas 28P follicles secreted the lowest amounts of $P_4$. Follicles removed from 7Ps and 7P secreted less $P_4$ than C follicles, though not significantly different ($P>0.05$), but secreted significantly more $P_4$ ($P < 0.01$) than 15Ps, 15P, and 28P follicles. Furthermore, $P_4$ secretion by 28P follicles was significantly less ($P < 0.01$) than that observed for 15P.

In contrast $E_2$ secretion by C, Ps, and P follicles is distinctly different from the profile observed for $P_4$. $E_2$ output, by the C follicles both before and after gonadotropin challenge, was significantly greater ($P < 0.001$) than that of the Ps and P follicles which all secreted similar amounts. On the other hand, follicular secretion of T produced still another profile, an intermediate between that for $P_4$ and $E_2$. Basal T secretion was similar for all groups, except significantly less ($P < 0.025$) for 15Ps and 28P. However after LH:FSH, C follicles secreted significantly more T ($P < 0.025$) than did Ps and P follicles. Secretion of T by 28P was still significantly less ($P < 0.05$) than that observed for the other Ps and P groups.

## Discussion

The aim of this experiment was to determine whether steroid secretion by follicles isolated from estrous, pseudopregnant and

TABLE 1. STEROID SECRETION BY PERIFUSED FOLLICLES REMOVED FROM ESTROUS, PSEUDOPREGNANT, AND PREGNANT RABBITS BEFORE AND AFTER GONADOTROPIN STIMULATION

| Reproductive state of the animal* | Progesterone pg/follicle/20 min | | Testosterone ng/follicle/20 min | | Estradiol-17-β pg/follicle/20 min | |
|---|---|---|---|---|---|---|
| | Basal | Stimulated | Basal | Stimulated | Basal | Stimulated |
| Estrous | 206.29 ± 35.70†$^{e}$ | 1046.53 ± 162.63$^{e}$ | 0.23 ± 0.03$^{e}$ | 4.79 ± 0.55$^{e}$ | 442.15 ± 76.29$^{e}$ | 910.05 ± 105.11$^{e}$ |
| 7 day pseudopregnant | 240.88 ± 36.29$^{e}$ | 755.50 ± 102.30$^{e}$ | 0.18 ± 0.02$^{e}$ | 2.88 ± 0.40$^{f}$ | 177.69 ± 26.35$^{f}$ | 346.45 ± 31.99$^{f}$ |
| 15 day pseudopregnant | 154.88 ± 34.94$^{e}$ | 448.38 ± 70.63$^{fg}$ | 0.13 ± 0.02$^{f}$ | 2.45 ± 0.32$^{f}$ | 178.80 ± 23.31$^{f}$ | 318.00 ± 50.49$^{f}$ |
| 7 day pregnant | 184.77 ± 30.48$^{e}$ | 885.85 ± 140.96$^{e}$ | 0.22 ± 0.03$^{e}$ | 2.89 ± 0.25$^{f}$ | 166.35 ± 22.11$^{f}$ | 265.82 ± 38.41$^{f}$ |
| 15 day pregnant | 264.47 ± 29.69$^{e}$ | 561.52 ± 44.24$^{f}$ | 0.25 ± 0.06$^{e}$ | 3.00 ± 0.34$^{f}$ | 187.33 ± 27.82$^{f}$ | 416.93 ± 55.26$^{f}$ |
| 28 day pregnant | 166.50 ± 35.95$^{e}$ | 360.22 ± 66.77$^{g}$ | 0.11 ± 0.01$^{f}$ | 2.15 ± 0.19$^{g}$ | 113.15 ± 32.05$^{f}$ | 252.66 ± 66.70$^{f}$ |

* 6-8 animals/group

† Means ± SEM

e,f,g Differences between values with the same superscript(s) in the same column are nonsignificant ($P > 0.05$).
Differences between values with different superscript(s) in the same column are significant ($P < 0.05$).

pregnant rabbits reflects adenylyl cyclase activities. These specific steroids, $P_4$, T, and $E_2$, were assayed because they have biological significance in the nonpregnant and pregnant animal and are located at key sites in the biosynthetic pathway. Because of the presence of distinct steroid secreting compartments in the total ovary, i.e. follicles, CL, interstitial tissue, it was necessary to isolate the ovarian follicles to study their responses in an in vitro situation. The use of a perifusion system, in which the medium is constantly flowing through the chambers allows one to study temporal changes in steroidogenesis without concern that there is an accumulation of metabolites in the medium.

Results obtained in this experiment suggest that there appears to be a correlation between follicular adenylyl cyclase activities and the steroidogenic capacity of the follicle to secrete T after gonadotropin stimulation and $E_2$ both before and after gonadotropin stimulation. While C follicles have higher basal and stimulated adenylyl cyclase activities than Ps and P follicles, likewise C follicles secrete more T and $E_2$. From our results we cannot identify whether this difference is attributable to an elevated adenylyl cyclase activity, which is an indirect measurement of receptor activity, and or is the result of differences in the levels of steroid synthetic enzymes and the presence of CL.

To further complicate the analysis of these results, the $P_4$ secretion profile is distinctly different from that of $E_2$. While basal levels of $P_4$ are similar for C, Ps, and P, the pattern changes following gonadotropin stimulation. Follicular secretion of $P_4$ do not reflect adenylyl cyclase activities (2), rather there is a progressive decrease in responsiveness to LH:FSH with increasing length of pseudopregnancy or pregnancy. In other words, the longer the period of time the follicles are located in close proximity to CL, the more depressed is their $P_4$ secretion following LH:FSH. Though we do not have any evidence at the present time to support this conclusion, these results may imply that CL exert an inhibitory influence on follicular $P_4$ secretion and in some cases T secretion (Table 1) through the elevation of the intragonadal levels of $P_4$. These significant differences in steroid secretion may suggest that the steroid synthetic capacity of a follicle is regulated not only by circulating gonadotropins but also by the intragonadal levels of steroids. From other systems, there is evidence that estrogen, synthesized by follicles in response to FSH, plays a regulatory role in follicular maturation through the induction of LH receptors (17). In in vitro systems, it has been demonstrated that addition of T to the medium increases $P_4$ secretion by granulosa cells (18, 19). Although these reported results demonstrate a positive effect of steroids on ovarian function, it is also plausible that steroids may regulate their own synthesis through a local negative feedback system. These results, though preliminary, suggest that the presence of high intragonadal $P_4$ levels appear to depress follicular steroid secretion.

## References

(1) P. L. Keyes and A. V. Nalbandov. Endocrinology. 82(1968)799.

(2) M. Hunzicker-Dunn and L. Birnbaumer. Ovarian Follicurlar Development and Function. (Raven Press, New York, 1978) in press.

(3) J. M. Marsh and K. Savard. Steroids. 8(1966)133.

(4) J. M. Marsh, R. W. Butcher, K. Savard and E. W. Sutherland. J. Biol. Chem. 241(1966)596.

(5) J. F. Dorrington and B. Baggett. Endocrinology. 84(1969)989.

(6) T. Mills. Endocrinology 96(1975)440.

(7) A. Tsafriri, H. R. Lindner, U. Zor, and S. A. Lamprecht. J. Reprod. Fert. 31(1972)39.

(8) S. A. Lamprecht, U. Zor, A. Tsafrifi, and H. R. Lindner. J. Endocrinol. 57(1973)217.

(9) H. R. Lindner, A. Tsafriri, M. W. Lieberman, U. Zor, Y. Koch, S. Bauminger and A. Barnea. Rec. Prog. Horm. Res. 30(1974)79.

(10) K. Ahrén, H. Herlitz, L. Nilsson, T. Perklev, S. Rosberg, and G. Selstom. Gonadotropins and Gonadal Function (Academic Press, New York, 1974) p. 364.

(11) M. Hunzicker-Dunn, and L. Birnbaumer. Endocrinology 99(1976)185.

(12) J. Hilliard and L. W. Eaton, Jr. Endocrinology 89(1971)522.

(13) E. V. YoungLai. J. Reprod. 18(1978)193.

(14) J. Bahr. Biol. Reprod. 18(1978)193.

(15) M. E. Lieberman, A. Barnea, S. Bauminger, A. Tsafriri, W. P. Collins, and H. R. Lindner. Endocrinology 96(1975)1533.

(16) M. S. Smith, M. E. Freeman and J. D. Neill. Endocrinology 96(1975)219.

(17) J. S. Richards and A. R. Midgley. Biol. Reprod. 14(1976)82.

(18) A. W. Lucky, J. R. Schreiber, S. G. Hillier, J. S. Schulman, and G. T. Ross. Endocrinolgoy 100(1977)128.

(19) D. W. Schomberg, R. L. Stouffer and S. Tyrey. Biochem. Biophys. Res. Commun. 68(1976)77.

# HORMONE REQUIREMENTS FOR LONG-TERM MAINTENANCE OF RAT GRANULOSA CELL CULTURES

Grace M. Centola

Department of Anatomy University of Maryland

Dental School, Baltimore, Maryland 21201

## INTRODUCTION

Since the pioneering work of Astwood (1), prolactin has been considered the major luteotrophic agent in the rat. Maintenance of rat luteal function *in vitro* beyond 6-8 days has been difficult (2,3). The relationship between luteal function and the role played by the other pituitary gonadotrophins, steroids and placental hormones is not entirely understood. This is further complicated by the apparent fluctuations in the levels of the luteotrophic hormones required for *in vivo* maintenance of luteal function (6, 7, 10, 11). The present study was undertaken to determine the hormone requirements for luteinization and long-term maintenance of rat granulosa cell cultures. An attempt was made to reconcile the apparent conflicting results from *in vivo* and *in vitro* maintenance of luteal function.

## MATERIALS AND METHODS

Twenty-eight day old, immature female rats (Lewis/Fmai inbred strain) were given a single subcutaneous (sc) injection of 5 IU Pregnant Mares' Serum Gonadotrophin (PMSG) (NIAMDD-PMSG-2, 2200 IU/mg) at 1200 h in order to induce the first estrous cycle and to initiate development of ovarian follicles (2). Fifty hours later, all rats received a single sc injection of 100μg bovine luteinizing hormone (bLH) (NIH-LH-B9) in order to mimic the proestrous gonadotrophin surge and to insure development of follicles for harvesting granulosa cells (2). The animals were sacrificed and the granulosa cells harvested and cultured according to the technique of Crisp and Denys (2) and Crisp (3).

In order to determine the hormone requirements for maintenance of progesterone secretion by rat granulosa cells in vitro beyond 6 days, the following hormones and concentrations were used: 1.0μg rPRL/ml (NIAMD-Rat-Prolactin-B-1) (2,3); 0.2mM dbcAMP(4); 0.02mM dbcGMP (5); 0.1μg bLH/ml (NIH-LH-B9) (3); 0.1μg hCG/ml (hCG CR 115:50 MCG lyophilized); 5 ng Estradiol-17β/ml (3); 1.0μg hPL/ml (2,3). In order to determine if luteinized rat granulosa cells in vitro might respond to testosterone by an increase in progesterone secretion as shown by others for preantral granulosa cell cultures (8,20), rat granulosa cells were cultured in the presence of 1.0μg rPRL/ml plus $10^{-4}$M testosterone proprionate.

Progesterone secretion was measured in spent culture medium by specific radioimmunoassay using a modification (3) of the technique of Furr (9). All data were expressed as the mean ± the standard error in at least 4 replicate cultures per group. Significant differences between the means were obtained using the students' "t" test.

## RESULTS

The effects of purified hormones on progesterone secretion by rat granulosa cell cultures are shown in figures 1-6. During the culture period, rPRL significantly ($p<0.001$) increased progesterone secretion over that of controls. Secretion rose to a peak after 4 days of culture and gradually decreased throughout the remainder of culture.

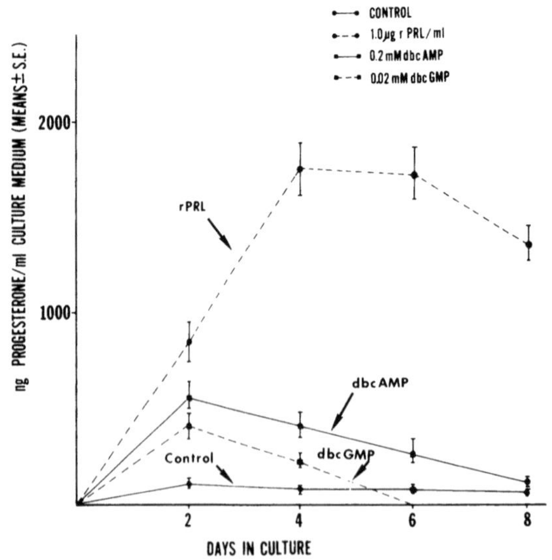


Figure 1. Effect of dibutyryl cyclic AMP and dibutyryl cyclic GMP on progesterone secretion by rat granulosa cell cultures.

Although dbcAMP or dbcGMP added separately to the culture medium stimulated progesterone secretion slightly but significantly ($p < 0.001$) above that of controls for at least 4 days, they failed to mimic the secretory pattern afforded by rPRL (figure 1). Furthermore, dbcAMP was more effective than dbcGMP. A combination of rPRL plus both dbcAMP and dbcGMP extended elevated levels of progesterone by two days as compared to cultures maintained on rPRL alone (figure 2). Progesterone secretion in cultures maintained on rPRL alone for days 0-4, and rPRL + dbcAMP for days 4-12 was low for the first 4 days of culture. Thereafter, secretion increased to a peak on day 6 and remained significantly ($p < 0.001$) greater than that in cultures maintained solely with rPRL (figure 3).

In another experiment, granulosa cells were grown in the presence of rPRL for days 0-4, followed by rPRL + bLH for days 4-12 (figure 3). Progesterone secretion in this group peaked on day 6, and remained significantly ($p < 0.001$) greater than that in the cultures maintained on rPRL alone. The secretory pattern of these cultures very closely resembled that of cultures maintained with rPRL + dbcAMP (figure 3).

The effects of hCG, $E_2$, and hPL are shown in figures 4 and 5. Progesterone secretion in cultures maintained on rPRL for days 0-4, and rPRL + hCG + $E_2$ + hPL (days 4-12) very closely resembled the

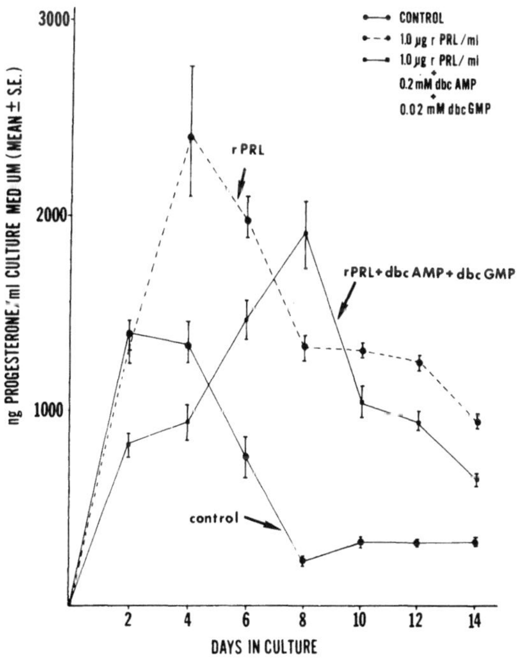


Figure 2. Effect of a combination of rPRL, dbcAMP and dbcGMP on progesterone secretion by rat granulosa cell cultures.

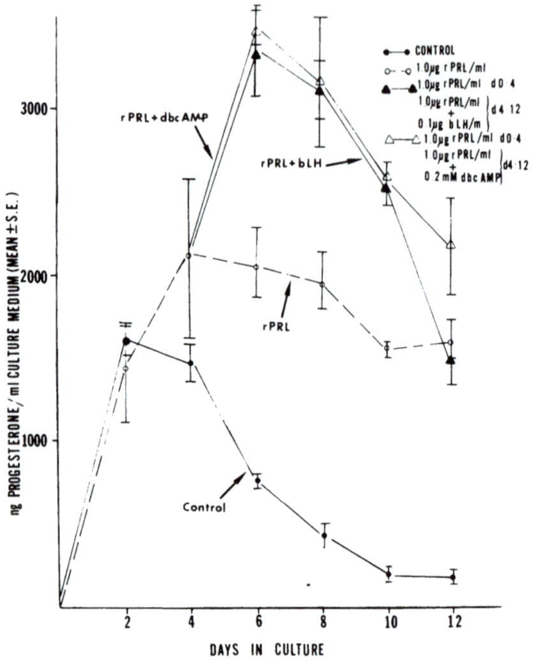


Figure 3. Effect of a combination of rPRL + bLH and rPRL + dbcAMP on progesterone secretion by rat granulosa cell cultures.

pattern of secretion in cultures maintained with rPRL alone for the entire culture period (figure 4). When rPRL was replaced by hCG and $E_2$ on days 6-12 and hPL and $E_2$ on days 12-18, progesterone secretion decreased immediately to low levels which were significantly ($p < 0.001$) different than that for cultures maintained on rPRL alone (figure 5).

The effects of rPRL + testosterone proprionate are shown in figure 6. Progesterone levels in this group peaked on day 6 and remained significantly ($p < 0.001$) greater than that in the control and rPRL groups for the entire 14 day culture period.

## DISCUSSION

This study along with previous *in vitro* studies (2, 3) confirms various *in vivo* studies (6, 7, 10) that prolactin is the primary luteotrophic stimulus for early maintenance of luteal function in the rat. Based upon the present study, luteal function beyond 6-8 days *in vitro* is a reflection of the hormonal content of the culture medium.

When dbcAMP was added to rat granulosa cell cultures, secretion was not maintained nor was the effect of rPRL mimicked, confirming

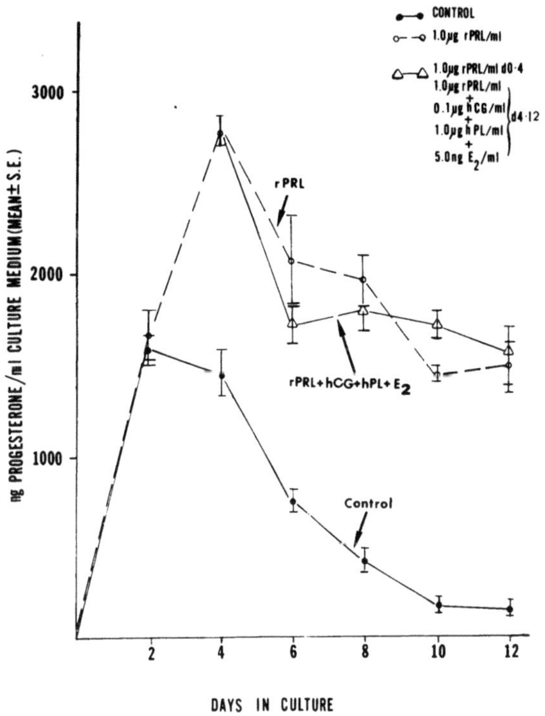


Figure 4. Effect of a combination of rPRL + hCG + hPL + $E_2$ on progesterone secreion in rat granulosa cell cultures.

the work of Crisp (3). However, the present study suggests that cGMP alone is not the intracellular mediator for prolactin as had been previously suggested (3). Perhaps a combination of cyclic nucleotides or other intermediates is required during some period of luteal function. Furthermore, other concentrations of cyclic nucleotides may be optimal for in vitro luteal maintenance. Since the effects of rPRL + LH and rPRL + dbcAMP on progesterone secretion by rat granulosa cells in culture were similar, it is suggested that cAMP might be the intracellular mediator for LH in rat granulosa cells in vitro as has been shown in granulosa cells of several other species (12, 13). Perhaps LH is a necessary component of the luteotrophic complex for maintenance of luteal function for a short time from day 6-8 or 10 of culture as demonstrated by Morishige and Rothchild (7) for days 8-12 of pregnancy. It may be postulated that while prolactin is the major luteotrophin for early (up to day 6) in vitro maintenance of luteal function, LH may be important during a critical period from days 6-10 of culture. Since it was suggested that hCG (or LH), rPL and estradiol may play some role in maintenance of luteal function without PRL during late pregnancy (6, 14, 15) the effect of sequential addition of these hormones on in vitro luteal function was examined. These combinations were not effective in maintaining progesterone secretion. Removal of prolactin on day 6 and the subsequent reduction in progesterone secretion suggests that reduced concentrations of this gonadotrophin result in a premature regression of the luteal cells. Alternately, the hCG or hPL may not be biologically active in the rat due to species differences. However addition of hPL to rat granulosa cells early in culture clearly enhances progesterone secretion (3, 4). The rat placenta may also secrete an unknown factor which may be involved in regulation of luteal function (15), and which is missing in our culturing system. Perhaps there is a specific sequence of hormonal interactions necessary for maintenance of *in vitro* progesterone secretion, which may involve induction or unmasking of receptors during *in vivo* or *in vitro* maturation (17, 18, 19). Furthermore, the influence, if any, of the thecal compartment in regulation of luteal function is unknown since this ovarian element is missing in our system. This may explain our inability to maintain progesterone secretion in rat granulosa cell cultures beyond 6-8 days. A most exciting observation in the present study was the finding of enhanced progesterone secretion and extended maintenance of rat granulosa cell function when testosterone was added with rat prolactin. These results along with others suggest that luteinized rat granulosa cells may be capable of responding to androgen

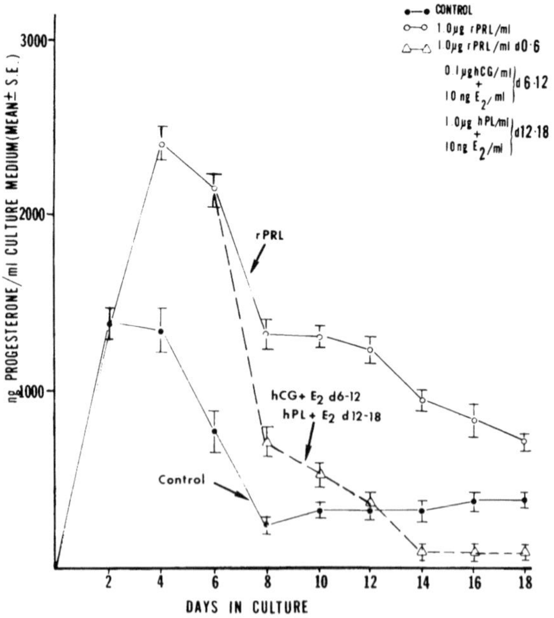


Figure 5. Effect of sequential addition of hCG + $E_2$ and hPL + $E_2$ on progesterone secretion in rat granulosa cell cultures.

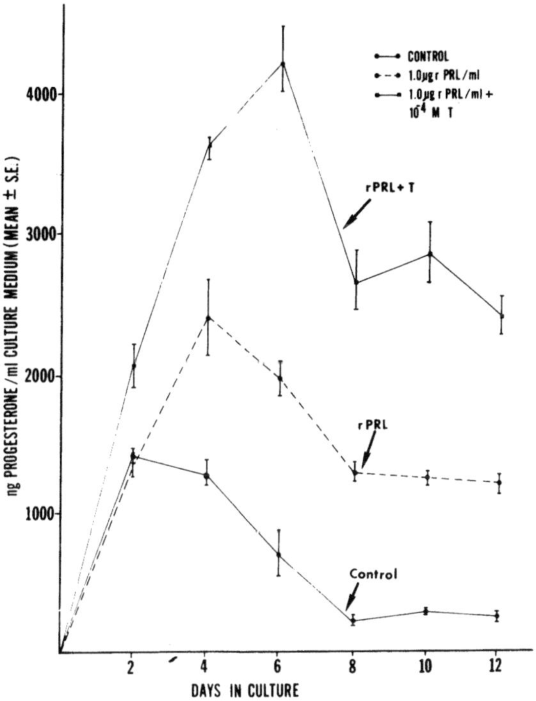


Figure 6. Effect of rPRL + testosterone proprionate on progesterone secretion in rat granulosa cell cultures.

stimulation following luteinization in culture (8, 20). However, Gibori et al. (15) have suggested that testosterone is converted to estradiol which is luteotrophic. Furthermore, Bernard (21) has shown that estradiol stimulates progesterone secretion by cultures of rat preovulatory granulosa cells. The effects of testosterone, however, on progesterone secretion are not believed to be due to its aromatization to estradiol, since dihydro-derivatives of testosterone, which cannot be converted to estradiol, enhance progesterone secretion by rat granulosa cells in culture (8, 20). Further studies, utilizing the addition of testosterone in varying concentrations, as well as dihydrotestosterone and other steroids, with and without PRL are needed to clarify the luteotrophic role of testosterone in the rat.

ACKNOWLEDGEMENT

The author gratefully acknowledges the assistance of Dr. Thomas M. Crisp throughout the course of this study. The author also thanks

Dr. Larry Anderson for help in preparing this manuscript. Special thanks are given to Mrs. Mary Walls for photographic work.

REFERENCES

(1) E.B. Astwood. Endocrinology 28(1941)309.

(2) T.M. Crisp and F.R. Denys, in: Electron Microscopic Concepts of Secretion: Ultrasctructure of Endocrine and Reproductive Organs, ed. M. Hess (John Wiley and Sons, Inc., New York, 1975) pp. 3-33.

(3) T.M. Crisp. Endocrinology 101(1977)1286.

(4) C.P. Channing. Endocrinology 94(1974)1215.

(5) N.D. Goldberg, R.F. O'Dea and M.K. Haddox. Adv. in Nucleotide Res. 3(1973)155.

(6) M. Takayama and G.S. Greenwald. J. Endoc. 56(1973)421.

(7) W.K. Morishige and I. Rothchild. Endocrinology 95(1974)260.

(8) A.W. Lucky, J.R. Schreiber, S.G. Hillier, J.D. Schulman and G.T. Ross. Endocrinology 100(1977)128.

(9) B.J.A. Furr. Acta Endocrinol. Kbh 72(1973)89.

(10) I. Rothchild, G.J. Pepe, and W.K. Morishige. Endocrinology 95(1974)280.

(11) H.G. Madhwa Raj and N.R. Moudgal. Endocrinology 86(1970)874.

(12) C.P. Channing. Endocrinology 87(1970)49.

(13) C.P. Channing. Rec. Prog. Horm. Res. 26(1970)589.

(14) J.S. Richards. Endocrinology 96(1975)227.

(15) G.E. Gibori, E. Antczak and I. Rothchild. Endocrinology 100 (1977)1483.

(16) W.K. Morishige, G.J. Pepe and I. Rothchild. Endocrinology 92(1973)1527.

(17) C.P. Channing and S. Kammerman. Endocrinology 92(1973)531.

(18) C.P. Channing and S. Kammerman. Biol. Reprod. 10(1974)179.

(19) R.J. Ryan and C.Y. Lee. Biol. Reprod. 14(1976)16.

(20) D.T. Armstrong and J.H. Dorrington. Endocrinology 99(1976)1411.

(21) J. Bernard. J. Reprod. Fertil. 43(1975)453.

LONG TERM PERIFUSION OF RABBIT OVARIAN FOLLICLES

C. D'Amato, J. Bahr, B. Stockert, F. Calvo and D. Kesler

Departments of Animal Science and Physiology, Animal

Genetics Laboratory, University of Illinois, Urbana, IL

ABSTRACT

Preovulatory follicles were dissected from ovaries of estrous rabbits. Four follicles were placed in each chamber of a perifusion system. Medium 199, containing 10% rabbit serum and antibiotics, eluted through the chambers at a rate of 1 ml/10 min. Every 6 h for 36 h, 1 µg of LH and 50 ng of FSH were injected into each chamber. $P_4$, T and $E_2$ were measured in the medium. $P_4$ exhibited a peak after each of the injections at 6, 12, 18, 24 and 30 h. $E_2$ and T gave major peaks after the injection at 0 time, with minor peaks after the injection at 6 h; thereafter T and $E_2$ secretion gradually decreased to levels below those of baseline.

## Introduction

The adenylyl cyclase system of preovulatory follicles loses its capacity to be stimulated by a second challenge of LH when a large dose of gonadotropin had been previously administered (1,2). Hunzicker-Dunn and Birnbaumer (3) reported that when estrous follicles receive an ovulatory dose of hCG this desensitization phenomenon is complete within 10-11 hrs following hCG and persists for 48 hrs after the hCG. Within 72 hrs the CL which is formed develops an adenylyl cyclase system that is sensitive to gonadotropins. This particular study was conducted to investigate the desensitization phenomenon of preovulatory follicles using steroid secretion as an end point during the gradual transition of a preovulatory follicle to a CL.

## Materials and Methods

Preovulatory follicles were dissected from the ovaries of estrous rabbits. Four follicles were placed in each of eight chambers of a perifusion system. Medium 199, buffered to pH 7.2-7.4 and containing 10% rabbit serum and antibiotics, eluted through the chambers at a rate of 1 ml/10 min. One μg of LH (NIH-LH-S-20; NIAMDD) and 50 μg of FSH (50.3 x NIH-FSH-S-1; Dr. H. Papkoff) were injected into each chamber every 6 h for a total of 36 h. Preliminary dose response studies indicated that those doses of LH and FSH were maximum doses. Two ml of medium were collected and assayed for progesterone ($P_4$), testosterone (T), and estradiol-17-β ($E_2$), before the first injection (basal) 20 min prior to and at 30 min, 2 h and 3 h after each injection of gonadotropins. Data were subjected to analysis of variance and Kramer's Multiple range test (4). To simplify presentation of the results, the use of significance implies that steroid concentrations were different at the $P < 0.05$ level.

## Results

Preovulatory rabbit follicles placed in a perifusion system were repeatedly challenged every 6 h for 36 h with a maximum dose of LH and FSH (Figure 1). While the steroidogenic profiles for T and $E_2$

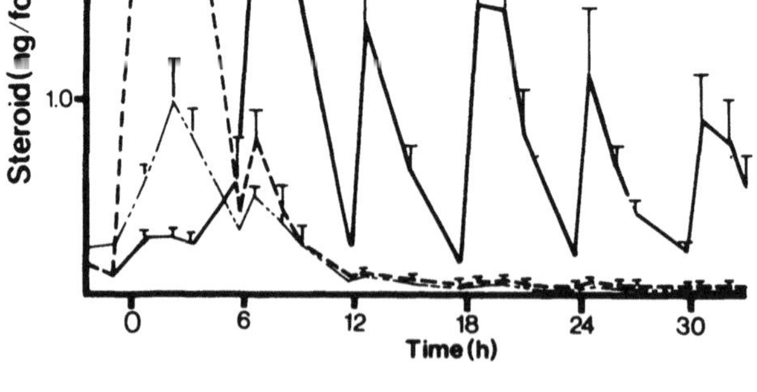


Fig. 1. Steroidogenic profile from perifused preovulatory ovarian follicles of estrous rabbits which were stimulated with LH:FSH every 6 h for 36 h.

were similar, those for $P_4$ were different. Following the first injection of LH:FSH, follicular secretion of all three steroids increased, with T and $E_2$ reaching significant levels 2 h after exposure to LH:FSH. However the increase in $P_4$ output was not significant when compared to the baseline value. Subsequent to the second exposure to gonadotropins, there was a significant alteration in follicular steroid secretion. This time the concentration of $P_4$ in the perifusion medium was significantly greater when compared to baseline and to the peak observed after the first injection of LH:FSH. There were also concomitant decreases in secretion of T and $E_2$ which were significantly less that those observed following the first challenge with LH:FSH. Thereafter a gradual decrease in T and $E_2$ secretion was observed with levels decreasing below those of baseline by 10 h following the first exposure to gonadotropins. In contrast, $P_4$ increased significantly following each challenge of gonadotropins except following the first injection of LH:FSH. Histological examination of stimulated follicles demonstrated that lutenization has been initiated but that ovulation had not occurred whereas unstimulated follicles showed no luteinization.

## Discussion

The use of the perifusion system, as modified in our laboratory, is a new method for the study of hormone regulation of steroidogenesis in isolated intact rabbit follicles. This dynamic system allows the investigator to study the changes which occur in steroid secretion upon repeated stimulation with gonadotropins without interference from metabolites which may have an inhibition on steroid synthesis and secretion.

There are a number of reports in the literature employing both _in vivo_ and _in vitro_ systems which state that there is complete desensitization of the adenylyl cyclase system in preovulatory follicles in response to an ovulatory dose of LH (1-3,5). Our studies, however, seem to indicate that this refractoriness is not complete but instead is differential with respect to cell type as monitored by steroid secretion from preovulatory follicles which are repeatedly challenged with an ovulatory dose of gonadotropins. As seen in Figure 1, initial gonadotropin stimulation results in a statistically significant release of T and $E_2$ but not $P_4$ when compared to baseline values. Subsequent stimulations with LH:FSH results in significant but decreasing amounts of $P_4$ released from the follicles whereas no significant response was seen for T and $E_2$ for the duration of the experiment.

Younglai (6) has demonstrated that when rabbit preovulatory follicles are exposed to 5 μg of LH for 15 min in a static incubation system, there is an increase in $P_4$, T, and estrogen in the medium. Progesterone levels remained elevated during the 3 h sampling time after LH stimulation while those of T and estrogen peaked once and then decreased. Mills _et al_.(7), have shown that isolated rabbit preovulatory follicles are capable of *de novo* synthesis of estrone and estradiol-17-β from acetate [1-14C] upon *in vitro* stimulation with LH. Long term studies in which *in vivo* and *in vitro* systems have been used to study the pattern of steroid release from rat (8), and sheep follicles (9) have shown that gonadotropin challenge initially causes a transitory stimulation of estrogen secretion followed by inhibition. Younglai (6), has postulated that high levels of LH could be inhibitory to the aromatizing system. The lack of $E_2$ and T response to repeated stimulation of LH which is observed in our perifusion system may be due to an inhibitory effect on the aromatizing enzymes or to a shifting of the enzymatic pathways toward production of $P_4$.

An ovulatory dose of LH has been shown to increase the levels of $P_4$ secreted by preovulatory follicles of sheep (9), rats (8), and rabbits (10,11). Although initial stimulation with LH:FSH did not result in a significant $P_4$ release, subsequent injections of gonadotropins into the chambers caused significant increases in $P_4$ secretion concomitant with the initiation of luteinization of the granulosa cells as shown histologically. This repeated response of $P_4$ to multiple injection of LH:FSH may be due to the fact that the integrity of the follicles was preserved, in other words, both theca and granulosa cell layers were present. This is in contrast to published reports of complete desensitization of the adenylyl cyclase system in which popped follicles were utilized resulting possibly in a significant loss of the granulosa cells (3). It is possible that this observed desensitization occurs primarily in the theca layer with a concurrent partial refractoriness of the granulosa cells which at this time are undergoing both biochemical and morphological luteinization.

(Supported in part by NSF-BMS75-18994 and PHS-IT 32 GMO 7143-03).

REFERENCES

(1) J. M. Marsh, T. M. Mills, and W. J. LeMaire. Biochim. Biophys. Acta. 304(1973)197.

(2) S. A. Lamprecht, U. Zor, A. Tsafriri, and H. R. Lindner. J. Endocrinol. 57(1973)217.

(3) M. Hunzicker-Dunn, and L. Birnbaumer. Endocrinology. 99(1976) 185.

(4) C. Y. Kramer. Biometrics. 12(1956)307.

(5) M. Hunzicker-Dunn and L. Birnbaumer. Endocrinology. 99(1976) 198.

(6) E. V. Younglai. J. Reprod. Fert. 40(1974)95.

(7) T. M. Mills, P. J. Davies, and K. Savard. Endocrinology 88 (1971)857.

(8) T. H. Hillensjö, S. Bauminger, and K. Ahrén. Endocrinology. 99(1976)996.

(9) R. M. Moor. J. Endocrinol. 61(1974)455.

(10) T. M. Mills and K. Savard. Endocrinology. 92(1973)788.

(11) J. Bahr. Biol. of Reprod. 18(1978)193.

# ESTROGEN INHIBITS PORCINE THECAL ANDROGEN PRODUCTION *IN VITRO*

P. C. K. Leung, B. K. Tsang and D. T. Armstrong

Departments of Physiology and of Obstetrics & Gynaecology

University of Western Ontario, London, Canada N6A 5A5

## INTRODUCTION

The importance of the pituitary gonadotropins in the regulation of ovarian steroidogenesis has long been recognized (1,2). Recently, we have reported an inhibitory action of estradiol-17β on luteinizing hormone (LH)-induced ovarian androgen levels *in vivo*, and suggested that this may be the result of an intra-ovarian feedback regulation by estrogen of its own production (3). The present communication extends this proposal to porcine thecal androgen production *in vitro*.

## MATERIALS AND METHODS

Medium-sized follicles (with diameters ranging between 3.5 and 5 mm) were removed from ovaries obtained from individual gilts (4-6 months of age) at a nearby slaughterhouse. They were cleaned of connective tissues, cut into small pieces and stirred gently for 5 min on a magnetic stirrer in Hanks' Balanced Salt Solution (modified, Flow Laboratories) without magnesium and calcium and supplemented with HEPES buffer (20 mM). The released granulosa cells were removed with a Pasteur pipet. The stirring process was repeated twice, and the follicular wall preparation (approximately 300 μg protein per dish) following granulosa cell removal was incubated at 37°C under an atmosphere of 5% $CO_2$ in air. The incubation medium (0.3 ml per dish) was Eagle's Minimum Essential Medium (MEM) containing penicillin (50 units/ml), streptomycin (50 μg/ml), fungizone (0.62 μg/ml), glutamine (2 mM) and non-essential amino acids. Purified LH (LH-HP, 1 μg/ml, provided by Dr. Harold Papkoff, University of California, San Francisco, Calif.)

was added in 3 µl saline, and estradiol-17β ($E_2$) of various concentrations was added in 3 µl absolute ethanol. The incubation was stopped 4 h later by adding 1 ml ice-cold absolute ethanol for extraction of steroids. Androgens (testosterone + dihydrotestosterone) and progesterone were measured without further purification by radioimmunoassays (3).

## RESULTS AND DISCUSSION

As shown in Figure 1, the presence of LH in the incubation medium significantly ($P < 0.01$) stimulated thecal androgen production as compared to non-LH-treated controls (4.98 ± 1.04 *vs.* 1.79 ± 0.46 ng/mg protein). Addition of increasing concentrations of $E_2$ to the incubation medium resulted in a highly significant ($P < 0.01$) inhibition of the LH-induced androgen production by the thecal cells; this inhibition did not deviate from a linear relationship. Progesterone production by porcine thecal cells in the same experiment is also shown in Figure 1. Presence of LH in the incubation medium significantly stimulated progesterone pro-

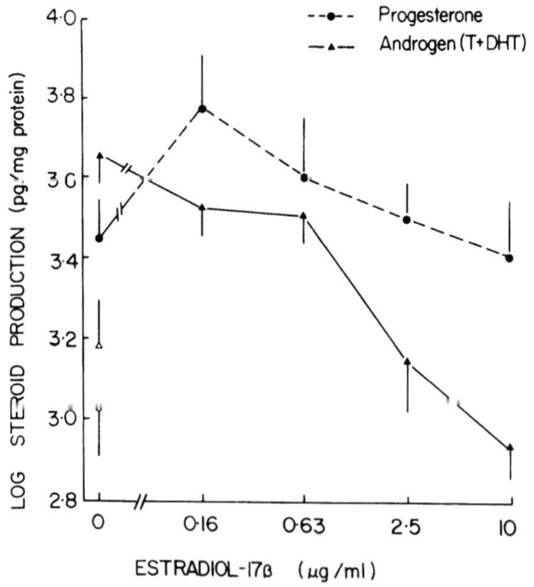


FIGURE 1. Effect of different concentrations of $E_2$ on LH-stimulated steroid production by porcine theca *in vitro*. Each point represents the mean ± S.E.M. for 6 incubations. Progesterone and androgen levels in non-LH-treated incubations are indicated by (○) and (△), respectively.

duction as compared to non-LH-treated controls (3.26 ± 0.64 *vs.* 1.32 ± 0.49 ng/mg protein). Addition of different concentrations of $E_2$ did not affect (ANOVA, $P > 0.05$) the ability of LH to stimulate progesterone production by the thecal cells.

In the rat ovarian follicle, there is evidence that LH acts upon the thecal cells to stimulate the production of androgens; the latter then cross to the granulosa cells where, under the influence of follicle-stimulating hormone (FSH), they are aromatized to estrogens (4-6). It appears that the same mechanism may occur in the porcine follicle. Granulosa cells from medium-sized porcine follicles secrete large quantities of $E_2$ when cultured in the presence of aromatizable androgens, whereas thecal preparations from these same follicles do not (manuscript in preparation, Tsang and Armstrong). Thus, the present demonstration that $E_2$ directly inhibits thecal androgen production suggests that not only does the theca interact with the granulosa through the production of androgens, but also the granulosa interacts with the theca through the production of estrogen. The mechanism(s) by which estrogen acts is not known. Since progesterone production is not inhibited by $E_2$ (Figure 1), it appears that the inhibition of the thecal androgen response to LH may be exerted, at least partly, at an enzymatic step(s) in the biosynthetic pathway between androgens and their $C_{21}$-precursors.

## REFERENCES

(1) P.E. Smith. Am. J. Anat. 45(1930)205.

(2) R.O. Greep, H.B. Van Dyke and B.F. Chow. Endocrinology 30 (1942)635.

(3) P.C.K. Leung, A.K. Goff, T.G. Kennedy and D.T. Armstrong. Biol. Reprod. (1978) in press.

(4) D.T. Armstrong and J.H. Dorrington, in: Regulatory Mechanisms Affecting Gonadal Hormone Action, Vol. 3: Advances in Sex Hormone Research, eds. J.A. Thomas and R.L. Singhal (University Park Press, Baltimore, 1977) p. 217.

(5) J.E. Fortune and D.T. Armstrong. Endocrinology 100(1977)1341.

(6). J.E. Fortune and D.T. Armstrong. Endocrinology 102(1978)227.

# ANTRAL FOLLICULAR AND INTERSTITIAL CONTRIBUTIONS TO PROGESTERONE SECRETION IN THE PROESTROUS HAMSTER

Srinivas K. Saidapur and Gilbert S. Greenwald

Department of Physiology, University of Kansas Medical Center
Kansas City, Kansas 66103

In the proestrous hamster the LH surge is usually initiated by 1500 h, preceding by one hour the increase in serum FSH (1, 2). Concomitant with the onset of the LH surge, an equally prompt and sustained rise in serum progesterone is observed (3). Previous experiments indicate that there is a causal relationship between LH and progesterone secretion (4).

Several lines of evidence indicate that the interstitium may be one of the primary sources for progesterone production in the hamster. Firstly, after unilateral irradiation of an exteriorized ovary (to eliminate antral follicles or corpora lutea), the concentration of progesterone during the LH surge is identical between the irradiated and normal ovary (4). Moreover, in hamsters hypophysectomized for one week the ovary contains only preantral follicles with at most 5-6 layers of granulosa cells and interstitial cells. Following a single injection of LH to these long term hypophysectomized hamsters ovarian levels of progesterone rise within 3 h (4) whereas there is no increase in serum estradiol (5).

In the intact proestrous hamster, the corpora lutea (CL) have regressed both morphologically (6) and biochemically (7). Similarly, CL removed from the ovaries of proestrous hamsters are unable, in vitro, to synthesize appreciable amounts of progesterone, estradiol, or testosterone (8). Moreover, histochemically the preantral follicles throughout the cycle do not contain lipids and enzymes related to steroidogenesis; in contrast both the antral follicles (AF) and interstitial gland cells are active steroidogenic tissues.

In the present study, we present quantitative information comparing the in vitro production rates of progesterone by isolated antral follicles (AF) and the nonantral follicular portion of the ovary (NAO). The latter component consists of preantral follicles, regressing CL and interstitial cells. Ovaries were obtained from proestrous hamsters killed before (1145 h) during (1445 h, 60-90 min after the onset) and after (1745 h) the LH surge is initiated. Details on material and methods are described elsewhere (8, 9, 10).

Table 1. Production rate of progesterone by the hamster ovary on the afternoon of proestrus

| Incubation * started at | Production rate° of progesterone (ng/mg/h ± SEM) | |
|---|---|---|
| | Non-antral ovary (NAO) | Antral follicles (AF) |
| 1200 h | 1.0 ± 0.3 (6) | 3.1 ± 1.6 (4) |
| 1500 h | 21.1 ± 6.2 (5) | 20.9 ± 1.4 (5) |
| 1800 h | 38.9 ± 5.4 (6) | 37.1 ± 5.0 (6) |

* The ovaries were removed at the specified times and incubated for 2 h.

° Production rate = $\frac{(A + B) - C}{2}$ where A = steroid content/mg tissue after incubation, B = steroid content of the medium/mg tissue incubated, C = mean steroid content/mg tissue before incubation; 2 = time of incubations in hours.

Before the LH surge the production rate of P is minimal in both AF and NAO (Table 1). However, an approximate 20 fold increase in P occurs at the beginning of the LH surge and there is still a further increase in production rate of P by 1800 h. At all time intervals the production rate of P is similar between the AF and NAO compartments. What has to be taken into consideration is that the total weight of AF is about 2 mg per ovary whereas the NAO weight is about 10 mg per ovary. Presumably the five fold difference in weight between the NAO and AF compartments reflects the difference in their steroidogenic capability to secrete progesterone. Thus the bulk (approximately 75%) of the progesterone secreted during the proestrous surge of the hamster apparently represents interstitial production of the hormone.

## Acknowledgments

The research was supported by grants from NIH (HD 00596) and the Ford Foundation. S.K.S. (present address: Department of Zoology, Karnatak University, Dharwar-580003, India) was supported as a Ford Foundation postdoctoral fellow in reproductive biology.

## References

1. Bast, J. D. and G. S. Greenwald. Endocrinology 94(1974) 1295-1299.
2. Siegel, H. I., J. D. Bast and G. S. Greenwald. Endocrinology 98(1976)48-55.
3. Saidapur, S. and G. S. Greenwald. Biol. Reprod. 18(1978a) 401-408.
4. Norman, R. and G. S. Greenwald. Endocrinology 89(1971) 598-605.
5. Baranczuk, R. and G. S. Greenwald. Endocrinology 92(1973) 85-812.
6. Saidapur, S. and G. S. Greenwald. Am. J. Anat. 151(1978b) 71-86.
7. Chatterjee, S. and G. S. Greenwald. J. Endocr. 68(1976) 251-256.
8. Terranova, P. F., Connor, J. S. and G. S. Greenwald. Biol. Reprod. In press, 1978.
9. Saidapur, S. K. and G. S. Greenwald. Biol. Reprod., In press, 1978c.
10. Terranova, P. F. and G. S. Greenwald. Biol. Reprod. 18 (1978) 170-175, 1978.

SESSION NUMBER: IV

TITLE: OOCYTE MATURATION

PRINCIPAL SPEAKER: Paul Wasserman

INVITED DISCUSSANT: Alexander Tsafriri

PRESENTED ABSTRACTS:

8.) T. Hillensjö, A. Schwartz-Kripner, S.H. Pomerantz and C.P. Channing. Action of Porcine Oocyte Maturation Inhibitor In Vitro: Possible Role of the Cumulus Cells.

9.) C. Readhead, M.H. Kaufman, A.W. Schuetz and G.E. Abraham. Relationship between Steroidogenesis and Oocyte Maturation in Rat Graafian Follicles Cultured In Vitro.

CHAIRPERSON: Allen Schuetz

MEIOTIC MATURATION OF MOUSE OOCYTES *IN VITRO*

P.M. Wassarman, R.M. Schultz, G.E. Letourneau,
M.J. LaMarca, W.J. Josefowicz, and J.D. Bleil
Dept. Biological Chemistry and Lab Human Reproduction
and Reproductive Biology, Harvard Medical School
45 Shattuck St., Boston, Massachusetts 02115, U.S.A.

"Meiotic maturation" of mouse oocytes takes place during the final phase of oogenesis and refers specifically to the process of nuclear progression from the diplotene ("dictyate") stage of the first meiotic prophase to metaphase II. Meiotic maturation is characterized by dissolution of the nuclear (germinal vesicle) membrane, condensation of diffuse chromosomes into distinct bivalents, alignment of chromosomes on the metaphase I spindle, separation of homologous chromosomes and emission of the first polar body, and arrest of nuclear progression at metaphase II (Fig. 1). In the sexually mature female mouse, fully-grown oocytes in Graafian follicles resume meiosis and undergo meiotic maturation just prior to ovulation. The resumption of meiosis can be mediated by a hormonal stimulus *in vivo* or simply by the release of oocytes from their ovarian follicles into a suitable culture medium *in vitro* (1,2).

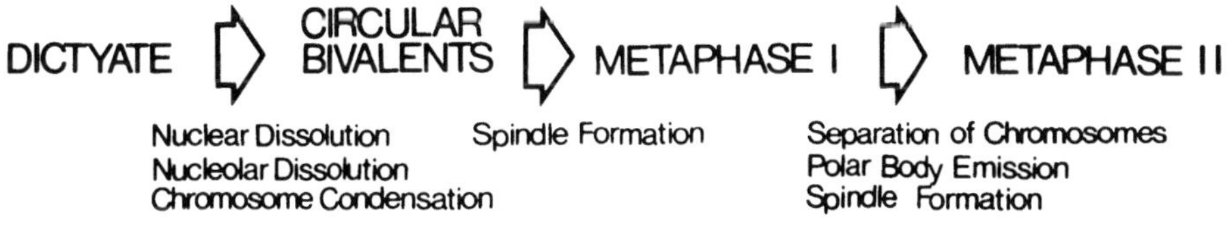


Fig. 1. Meiotic maturation

The process of meiotic maturation is of fundamental importance in mammalian development, since it is only after nuclear progression to metaphase II that oocytes are competent to be fertilized.

Furthermore, this relatively brief period of oogenesis includes not only meiotic events, but also an extensive reprogramming of oocyte metabolism, occasionally termed "cytoplasmic maturation" (2,3). While it is clear that cytoplasmic maturation of the oocyte is an essential step on the path to successful early mammalian development, the exact nature of this process is not well understood.

In our own laboratory, we have been investigating several aspects of spontaneous meiotic maturation of mouse oocytes in vitro. In particular, we have focused on the following questions: What is the biochemical basis of the acquisition of meiotic competence during oocyte growth? What are the biochemical consequences of the morphological events that characterize meiotic maturation of the oocyte? What biochemical changes constitute cytoplasmic maturation of the oocyte and how are these changes induced during meiotic maturation? In what manner and to what degree do the biochemical events associated with cytoplasmic maturation influence early embryonic development of the mouse? Here we briefly review some of the progress made in our attempt to obtain answers to these questions.

Meiotic Maturation of Mouse Oocytes In Vitro. We obtain fully-grown oocytes from adult (8-12 weeks-of-age), randomly bred female Swiss albino mice (CD-1, Charles River Labs) by puncturing excised ovaries with fine steel needles under a dissecting microscope.

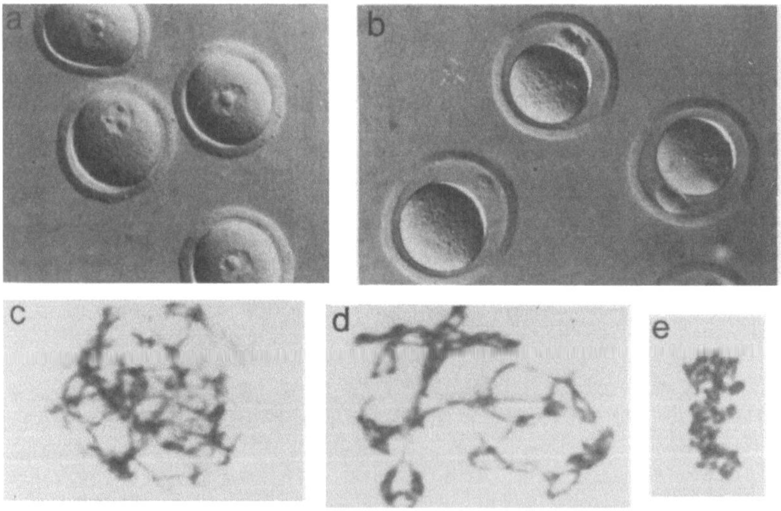


Fig. 2 Photomicrographs of mouse oocytes during culture in vitro (A,B) and of Giemsa-stained chromosome spreads from mouse oocytes during condensation of dictyate chromatin into compact bivalents in vitro (C-E). Shown are oocytes in dictyate (A) and metaphase II (B).

Oocytes containing an intact germinal vesicle (GV) and free of cumulus cells are harvested using a mouth-operated micropipette and washed in culture medium containing 100 µg/ml of dibutyryl cyclic AMP (dbcAMP) (4,5). Cell culture is carried out in either plastic Falcon dishes or embryological watchglasses in 50-250 µl of medium under paraffin oil at 37°C in a humidified atmosphere of 5% $CO_2$ in air.

When isolated oocytes are transferred to medium not containing dbcAMP spontaneous meiotic maturation takes place. The time sequence of meiotic maturation *in vitro* can be approximated as follows using Nomarski differential-interference microscopy: GV breakdown (GVBD) takes place with 1-5 hr, metaphase I is reached in 5-10 hr, and metaphase II is reached in 10-15 hr; some of these events are shown in Fig. 2. Under the experimental conditions used in our laboratory, approximately 80% of the oocytes placed in culture undergo GVBD within 3 hr and, of these, about 70% subsequently emit first polar bodies.

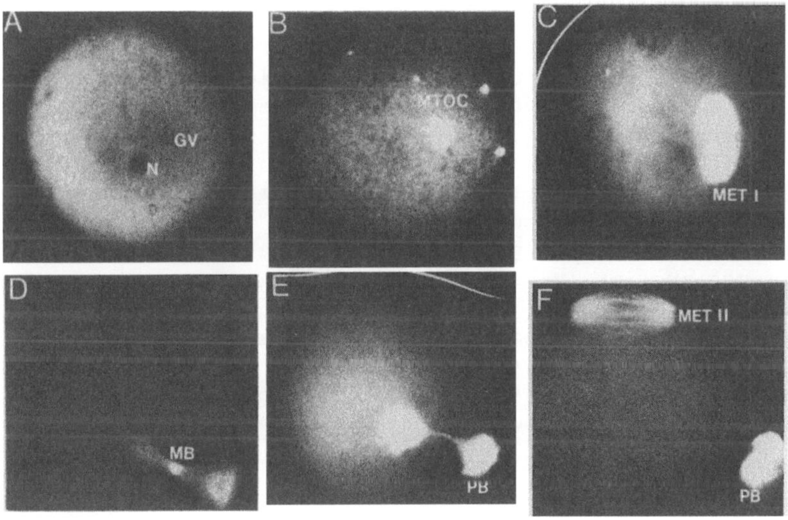


Fig. 3 Immunofluorescent staining of mouse oocytes during meiotic maturation *in vitro*.

In addition to Nomarski microscopy, we have used immunofluorescent anti-tubulin staining to follow nuclear progression from dictyate to metaphase II during meiotic maturation of mouse oocytes *in vitro* (6). Antibody directed against tubulin isolated from sea urchin eggs decorates the metaphase I and metaphase II spindles, as well as the cytoplasmic bridge, midbody, and polar body of the maturing mouse oocyte (Fig. 3). The time sequence for meiotic maturation deduced from anti-tubulin staining of mouse oocytes is in good agreement with that proposed on the basis of light microscopy. Furthermore, this technique has allowed us to follow the formation of "microtubule organizing centers" (MTOC) in the oocyte's cytoplasm just prior to GVBD and to examine their involvement in spindle assembly.

Inhibition of Meiotic Maturation of Mouse Oocytes at Specific Stages of Nuclear Progression. Meiotic maturation can be reversibly blocked at discrete stages prior to metaphase II when mouse oocytes are cultured in the presence of various drugs (7). GVBD, the first morphological feature easily recognizable during meiotic maturation, does not take place in the presence of dbcAMP; whereas, chromosome condensation is initiated under these conditions, but aborts at a stage far short of compact bivalents. Similarly, chloroquine, a well characterized inhibitor of lysosomal enzymes, also prevents GVBD. The inhibitory effects of dbcAMP and chloroquine suggest that GVBD may occur via a cyclic AMP controlled protease activation mechanism, analogous to that proposed for a variety of polypeptide hormone mediated biological phenomena. GVBD and chromosome condensation take place in an apparently normal manner in the presence of puromycin, Colcemid, or cytochalasin B. However, nuclear progression is blocked at the circular bivalent stage when oocytes are cultured continuously in the presence of either puromycin or Colcemid, while oocytes cultured in the presence of cytochalasin B proceed to metaphase I and arrest. The effect of puromycin and Colcemid on meiotic maturation indicates that GVBD and chromosome condensation are not dependent upon either concomitant protein synthesis or microtubule assembly. Failure of homologous chromosomes to separate after alignment on the metaphase I spindle in the presence of cytochalasin B suggests that microfilaments are involved in this phase of nuclear progression. Cytokinesis, in the form of polar body emission, is blocked when any of the earlier events of meiotic maturation fail to take place (Table 1). The inhibitory effects of these drugs on meiotic events are reversible to varying degrees depending upon the duration of exposure to the particular drug.

TABLE I INHIBITION OF MEIOTIC MATURATION *IN VITRO*

| DRUG | DOSE (μg/ml) | STAGE OF MEIOTIC ARREST |
|---|---|---|
| NONE | - | METAPHASE II |
| DIBUTYRYL CYCLIC AMP | 100 | DICTYATE (GV) |
| CHLOROQUINE | 5 | DICTYATE (GV) |
| PUROMYCIN | 10 | CIRCULAR BIVALENTS (GVBD) |
| COLCEMID | 10 | CIRCULAR BIVALENTS (GVBD) |
| CYTOCHALASIN B | 5 | METAPHASE I |

Relationship Between Oocyte Growth and the Acquisition of Meiotic Competence. We use the ovaries of juvenile mice, 5-21 days-of-age, to obtain growing oocytes ranging in diameter from about 40 to 80 μm (8,9). The procedure employed to isolate these oocytes consists of two steps - a short incubation of excised ovaries in phosphate-buffered saline containing either trypsin or a mixture of hyaluronidase, collagenase, and lysozyme, followed by physical disruption of the ovaries using fine steel needles and collection of the oocytes with a mouth-operated micropipette. The diameters of the oocytes harvested have been determined using hundreds of animals and we have found that there is a linear relationship between the age of the donor animal and the average diameter of the oocytes recovered. Between 9 and 21 days-of-age the average oocyte diameter increases from less than 50 μm to more than 80 μm, corresponding to about a four-fold increase in oocyte volume (Fig. 4). These oocytes are viable, based upon a variety of morphological (10,11) and biochemical (9,12,13) criteria, and can be cultured successfully for at least 24 hr in a chemically defined culture medium.

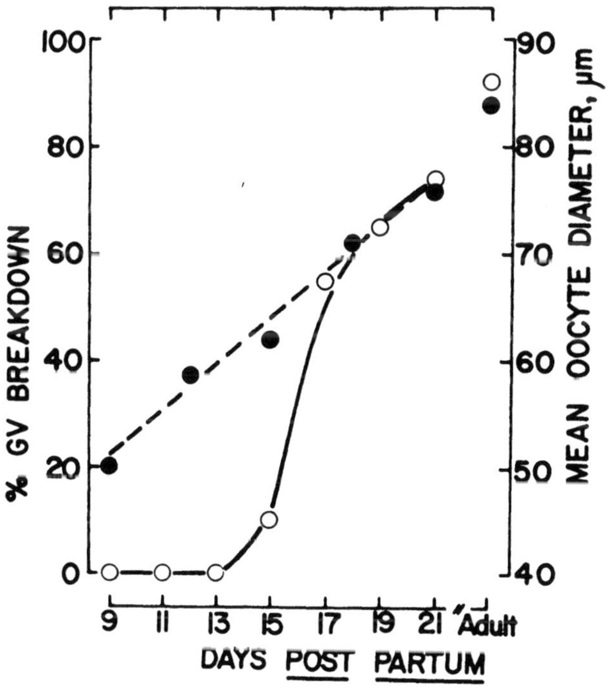


Fig. 4 Relationship between oocyte size, meiotic competence, and the age of the donor mice. Mean diameters (● - ●) and percentage of GVBD (O - O) are plotted against days post partum.

Oocytes recovered from juvenile mice 9, 11, and 13 days-of-age fail to resume meiosis when cultured overnight _in vitro_, as evidenced by the retention of intact GVs. On the other hand, oocytes recovered from mice 15 day-of-age or older resume meiosis at a frequency which increases with the age of the mice (Fig. 4). Within individual litters, the mean diameter of oocytes which fail to mature ("incompetent oocytes") is significantly less than that of the oocytes that mature ("competent oocytes"). These results correlate with the increase in mean diameter of oocytes recovered over the same age range. Of the oocytes which resume meiosis _in vitro_, those from younger animals exhibit a high frequency of incomplete meiotic maturation with arrest after completion of GVBD. Accordingly, the ratio of metaphase II to GVBD arrest changes markedly with the age of the mice (with oocyte diameter), increasing from 0.16 for the 15 day animal to 9.0 for the adult. These results suggest that the ability to resume meiosis ("meiotic competence") is acquired at a specific stage of oocyte growth in the juvenile mouse and that the ability to complete meiotic maturation is acquired subsequently (8).

_Morphological and Biochemical Changes During Oocyte Growth._ Growth of the mouse oocyte involves not just tremendous enlargement of the cell, but also extensive alterations in its overall metabolism as reflected in the ultrastructure of the oocyte at various stages of growth (10). Electron microscopic examination of fixed and sectioned _isolated oocytes_ reveals progressive ultrastructural changes in the nucleoli, ribosomes, mitochondria, endoplasmic reticulum, Golgi complex, and other organelles and inclusions during oocyte growth (Fig. 5). Early oocyte growth (3-5 days-of-age, diameter 30-40 µm, meiotically incompetent) is characterized by a thin, diffuse zona pellucida, a few relatively short microvilli, an extensive network of smooth endoplasmic reticulum, elongated and dumbbell-shaped mitochondria which are intimately associated with the endoplasmic reticulum and have cristae in an "orthodox" configuration, numerous clusters of free ribosomes, a Golgi apparatus consisting solely of flattened stacks of lamellae arranged in a parallel manner, a few small multivesicular bodies, and a rapidly enlarging fibrillo-granular nucelolus. During the middle stages of oocyte growth (8-14 days-of-age, diameter 45-60 µm, meiotically incompetent) the zona pellucida thickens and becomes more dense, more microvilli are present per unit of surface area, the smooth endoplasmic reticulum becomes vesicular and is less abundant than before, the mitochondria are round or oval, are smaller but more abundant than before and possess columnar-shaped cristae, ordered lattices begin to appear while free ribosomes decrease in number, the Golgi apparatus consists of parallel lamellae, vacuoles, and granules, and the nucleolus becomes larger, fibrillar, and dense. During the late stages of oocyte growth (beyond 14 days-of-age, diameter 60-80 µm, meiotically competent) the microvilli are longer than before, the zona pellucida is thicker and denser, lipid droplets appear, sometimes in association with the Golgi complex which is now highly

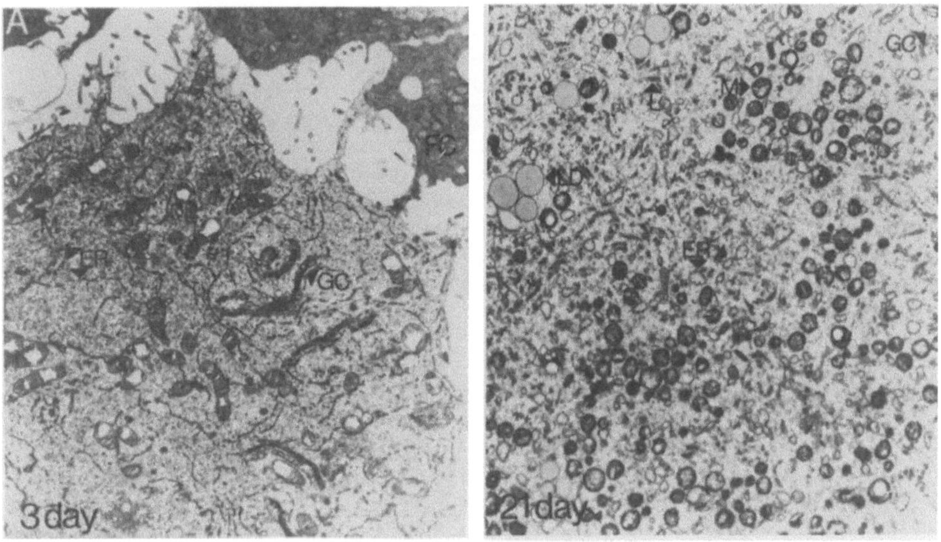


Fig. 5 A comparison of the ultrastructure of oocytes isolated from mice 3 and 21 days-of-age. ER, endoplasmic reticulum; M, mitochondria; GC, Golgi complex; ZP, zona pellucida; L, lattices; LD, lipid droplets; FC, follicle cells.

vacuolated and granular, the mitochondria are round or oval and possess arched or concentrically arranged cristae, small vesicles of smooth endoplasmic reticulum are found free in the cytoplasm and in association with mitochondria, ordered lattices are abundant, large multivesicular bodies are plentiful, and the nucleolus is very dense. Generally, these ultrastructural changes seem to reflect a high degree of metabolic activity which is specifically associated with rapid growth of the oocyte and probably occurs as a consequence of the activation of a developmental program which prepares the oocyte for the final stages of meiosis and early embryogenesis.

Measurements made with oocytes isolated from juvenile mice indicate that the protein content of growing mouse oocytes is related linearly to oocyte volume, with each doubling of volume accompanied by a doubling of protein (9,14). The protein content of the mouse oocyte increases throughout growth from less than 1 ng to about 25 ng (exclusive of zona pellucida), such that the protein concentration of the fully-grown oocyte is close to 10%.

To begin to determine whether the increase in protein content during oocyte growth is accompanied by the stage-specific synthesis of different classes of proteins or of the same set of proteins at all stages, we have examined electrophoretically the classes of proteins synthesized in mouse oocytes at several stages of growth (9). Such an analysis is of particular interest since the mouse oocyte undergoes significant ultrastructural changes during growth and since the oocyte becomes competent to resume meiosis at a specific stage of growth (see above). Both 1-dimensional and 2-dimensional electrophoresis of [$^{35}$S]methionine-labeled oocyte proteins has been performed at various stages of oocyte growth and examination of the fluorograms reveals two major kinds of changes in protein synthesis: (i) the appearance or disappearance of particular proteins and (ii) an increase or decrease in the relative rates of synthesis of particular proteins. However, the overall qualitative pattern of protein synthesis remains remarkably constant throughout oocyte growth (Fig. 6).

<u>Qualitative Changes in Protein Synthesis During Meiotic Maturation of Mouse Oocytes</u>. Mouse oocytes cultured in the presence of dbcAMP do not undergo GVBD or the subsequent morphological events associated with meiotic maturation. Therefore, to determine whether changes in the pattern of protein synthesis occur during meiotic maturation, mouse oocytes were cultured overnight in medium contain-

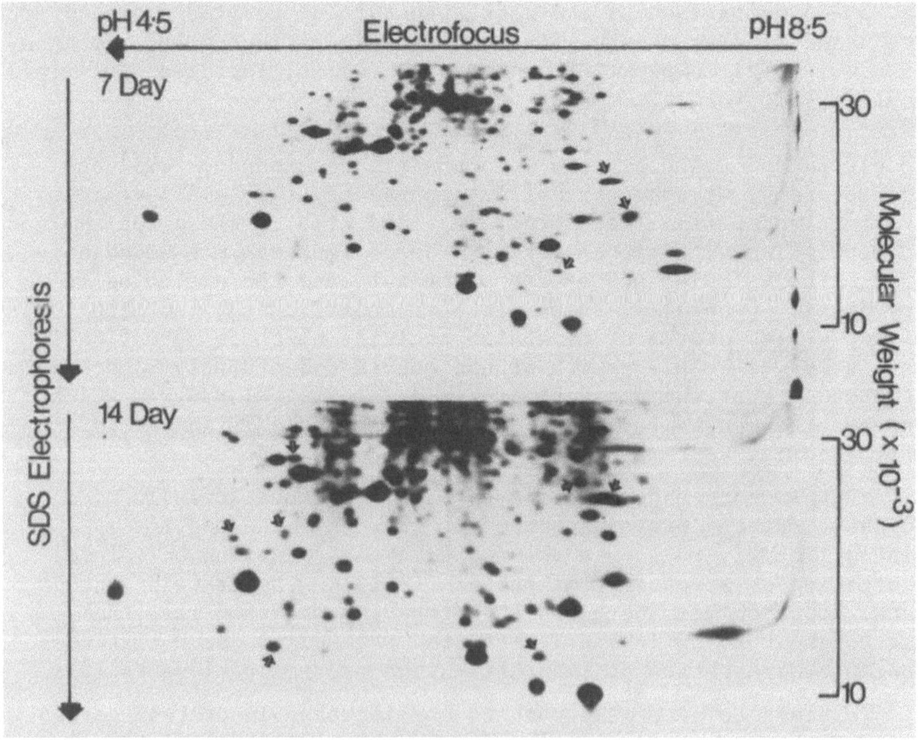


Fig. 6 High resolution two-dimensional gel analysis of proteins synthesized in oocytes from mice 7 and 14 days-of-age.

ing [$^{35}$S]methionine in the presence or absence of dbcAMP (15). A comparison of fluorograms prepared from oocytes which underwent spontaneous meiotic maturation with fluorograms prepared from oocytes which were arrested in dictyate by dbcAMP reveals a large number of differences (Fig. 7). It should be noted that the addition of dbcAMP to the culture medium after GVBD has occurred does not prevent the changes in protein synthesis associated with meiotic maturation from taking place. Furthermore, the pattern of [$^{35}$S]-methionine-labeled proteins obtained with oocytes cultured continuously in the presence of dbcAMP is virtually indistinguishable from that obtained with oocytes which fail to undergo spontaneous meiotic maturation.

In an attempt to identify changes in the pattern of protein synthesis which correlate with particular stages of nuclear progression, mouse oocytes were exposed to [$^{35}$S]methionine at 5 hr intervals. This protocol permitted an analysis of those proteins synthesized during the period of GVBD (0-5 hr), of metaphase I spindle formation (5-10 hr), and of polar body emission (10-15 hr). Although minor changes in protein synthesis were detected during the first 5 hr of culture, the major changes associated with meiotic maturation take place during the 5-10 hr period following GVBD and are then accentuated during the 10-15 hr period (15).

Role of Protein Synthesis During Meiotic Maturation of Mouse Oocytes. The results of experiments described above suggest that, unlike amphibian and echinoderm oocytes, fully-grown mouse oocytes are poised to resume meiosis such that concomitant protein synthesis is not necessary for GVBD and chromosome condensation to take place *in vitro*. On the other hand, we have found that protein synthesis is necessary for a brief period just following GVBD in order that meiotic maturation proceed beyond the circular bivalent stage; this evidence is summarized in Table 2.

In addition to dbcAMP, Colcemid and cytochalasin B have also been used to determine whether the changes in protein synthesis which accompany meiotic maturation are linked to particular morphological events, such as spindle formation or polar body emission. The results shown in Fig. 8 indicate that these changes in protein synthesis do take place when oocytes are cultured continuously in the presence of either Colcemid or cytochalasin B. Therefore, while changes in the pattern of protein synthesis must take place for nuclear progression to proceed beyond GVBD, these changes are not dependent on morphological events subsequent to GVBD.

Of considerable interest is the observation that virtually all of the changes in protein synthesis which occur during meiotic maturation persist during early embryogenesis. Four different proteins are shown in Fig. 9 whose synthesis is "turned-on" during meiotic maturation and continues in the 1-cell embryo.

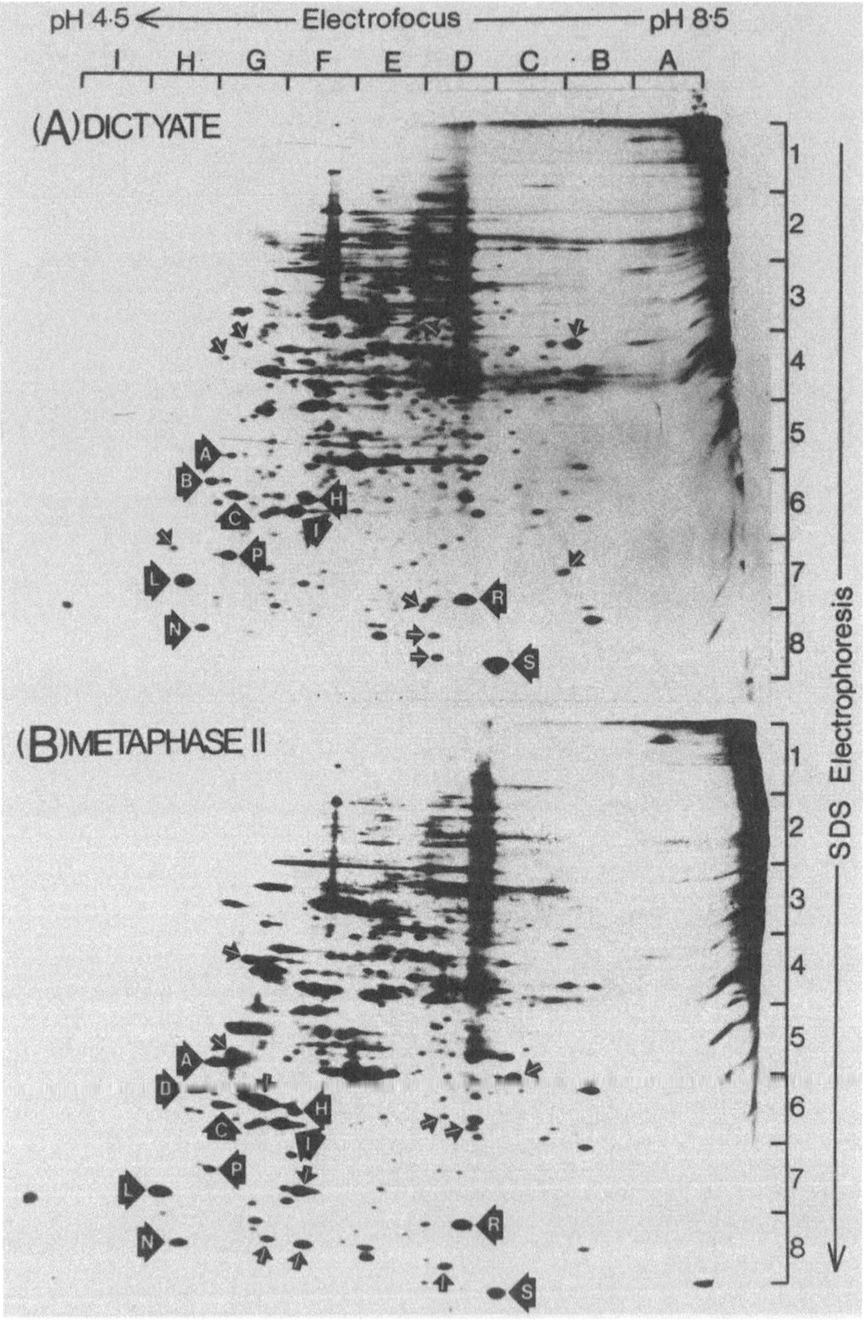


Fig. 7 High resolution two-dimensional gel analysis of proteins synthesized in fully-grown mouse oocytes at dictyate of first meiotic prophase (A) and during meiotic maturation (B) _in vitro_.

TABLE II EFFECT OF PUROMYCIN ON MEIOTIC MATURATION *IN VITRO*

| OOCYTE CULTURE CONDITIONS | % GVBD | % POLAR BODY EMISSION |
|---|---|---|
| MEDIUM (16 HR) | > 80 | > 70 |
| PUROMYCIN (16 HR) | > 80 | < 5 |
| MEDIUM (5 HR) → PUROMYCIN (11 HR) | > 80 | < 5 |
| MEDIUM (8 HR) → PUROMYCIN ( 8 HR) | > 80 | > 70 |
| $Bt_2cAMP$ (8 HR) → MEDIUM (16 HR) | > 80 | > 70 |
| $Bt_2cAMP$ (8 HR) → PUROMYCIN (16 HR) | > 80 | < 5 |

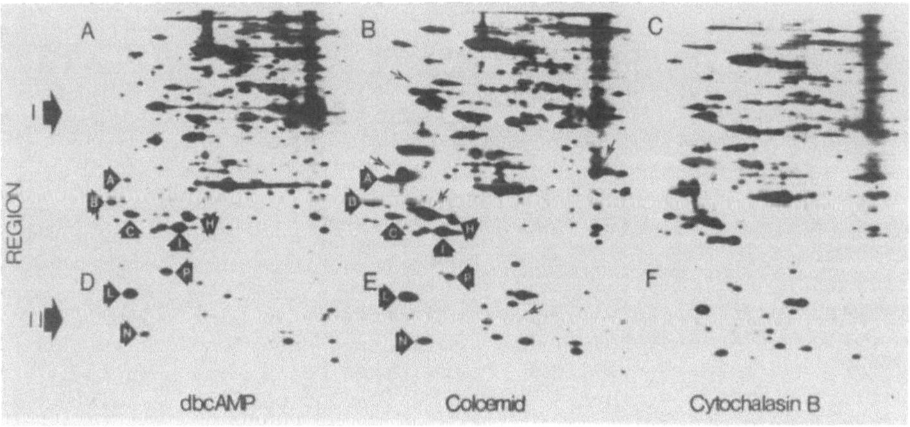


Fig. 8 High resolution two-dimensional gel analysis of proteins synthesized in fully-grown mouse oocytes arrested at various stages of meiosis.

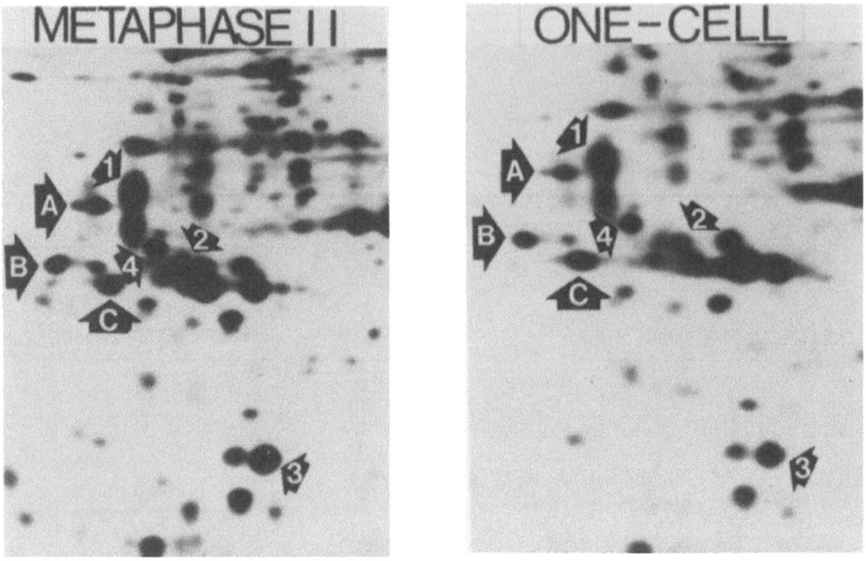


Fig. 9 High resolution two-dimensional gel analysis of proteins synthesized in unfertilized mouse eggs and in one-cell mouse embryos.

Control of Protein Synthesis During Meiotic Maturation of Mouse Oocytes. Nucleate and anucleate fragments of mouse oocytes have been isolated following treatment of fully-grown oocytes with cytochalasin B (16,17). The nucleate oocyte fragments resume meiosis *in vitro*, progressing from dictyate to metaphase II in an apparently normal manner (Fig. 10), and exhibit all of the changes in protein synthesis normally associated with meiotic maturation of mouse oocytes (18). Similarly, many of these changes in protein synthesis also occur in anucleate oocyte fragments in the absence of nuclear progression (Fig. 11). These results strongly suggest that the reprogramming of protein synthesis during meiotic maturation of mouse oocytes is directed by RNA templates already present in the oocyte's cytoplasm. Results of other experiments show that the pattern of protein synthesis seen with denuded oocytes is identical to that seen with oocytes cultured with surrounding cumulus cells and that changes in protein synthesis observed during meiotic maturation cannot be attributed to post-translational modification of proteins synthesized prior to GVBD.

Absolute Rates of Protein Synthesis During Meiotic Maturation. Measurements of the rates of incorporation of [$^{35}$S]methionine into protein and the specific activities of endogenous free-methionine pools have been used to calculate the absolute rates of protein synthesis in mouse oocytes during spontaneous meiotic maturation *in vitro* (19). [$^{3}$H]Fluorodinitrobenzene (FDNB) was used to determine the specific activity of the oocyte's free-methionine pool.

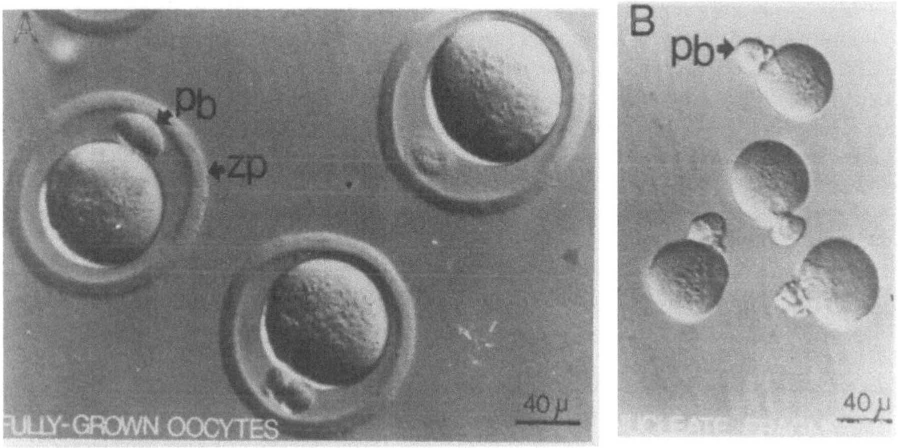


Fig. 10 Photomicrographs of fully grown mouse oocytes (A) and nucleate mouse oocyte fragments (B) which have undergone meiotic maturation _in vitro_. PB, polar body; ZP, zona pellucida.

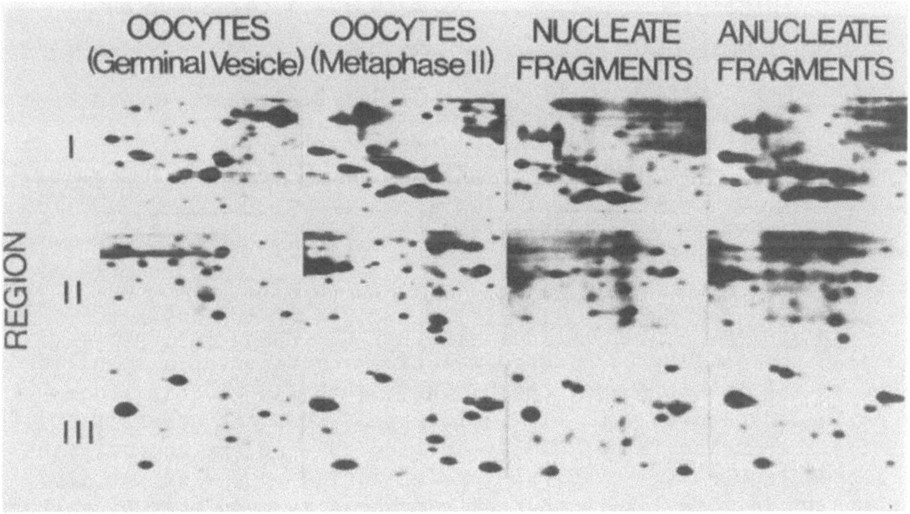


Fig. 11 High resolution two-dimensional gel analysis of proteins synthesized in fully grown mouse oocytes or in fragments of fully-grown mouse oocytes.

It was found that the absolute rate of protein synthesis decreased from 43 to 31 pg/hr/ oocyte during meiotic progression from dictyate to metaphase II, while the size of the intracellular free-methionine pool decreased from 62 to 35 fmol/oocyte during the same period (Table 3). Comparable measurements made on ovulated mouse oocytes which had undergone meiotic maturation in vivo, strongly suggest that the decrease in the absolute rate of protein synthesis observed during meiotic maturation in vitro is physiologically significant. An alternative method that depends upon differential expansion of the oocyte's endogenous methionine pool was also used to determine absolute rates of protein synthesis. The results of these experiments are in excellent agreement with those obtained using [$^3$H]FDNB, indicating that the oocyte's free-methionine pool is not compartmentalized.

TABLE 3. ABSOLUTE RATES OF PROTEIN WYNTHESIS DURING MEIOTIC MATURATION

| STAGE OF MEIOSIS | ABSOLUTE RATE (AVE. ± SD) (pg protein/hr/oocyte) |
|---|---|
| DICTYATE | 42.9 ± 4.4 (41.8 ± 5.5)[a] |
| GVBD - METAPHASE I | 36.3 ± 1.4 |
| METAPHASE I - METAPHASE II | 31.1 ± 3.3 |
| METAPHASE II (OVULATED) | 33.0 ± 2.2 |

[a] This value determined using a "kinetic" method (19).

Discussion. While it is clear that the mammalian oocyte acquires the ability to undergo meiotic maturation at a specific stage in its growth phase, the biochemical basis of this process is not understood as yet. It would appear that the acquisition of meiotic competence is not simply attributable to the accumulation of macromolecules to "critical levels" during oocyte growth (18). Rather, it is more likely that this process is regulated by certain gene products which are synthesized at a specific stage of oocyte growth. It is also clear that, although the mammalian oocyte is meiotically competent even before it achieves full growth, it does not undergo meiotic maturation until follicular development is completed. Here again, the mechanism for maintaining the oocyte in dictyate of the first meiotic prophase is not known. However, it is tempting to suggest that the fact that dbcAMP reversibly inhibits the onset of nuclear progression in isolated mouse oocytes may be related to the proposed regulatory role for cAMP during the mitotic cell cycle (20,21; see ref. 9 for discussion of this point). Since the oocyte and its surrounding cumulus cells are coupled via junctions (22,23), which may permit "biochemical communication" between the two cell types (23,24), the follicle cells may maintain intracellular cAMP levels in the oocyte which prevent the resumption of meiosis. Disruption or alteration of these channels of communication at the time of ovulation (or at the time of isolation of oocytes) could result in a lowering of cAMP levels in the oocyte and the onset of meiotic maturation.

Unlike amphibian and echinoderm oocytes, fully-grown mouse oocytes are poised to resume meiosis, so that no new proteins need be synthesized for GVBD and chromosome condensation to take place in vitro. However, for nuclear progression to proceed beyond the circular bivalent stage of meiosis, protein synthesis is required for a brief period following GVBD. This is of particular interest since the overall pattern of protein synthesis in the mouse oocyte changes dramatically at the time of GVBD, suggesting that certain of these changes may be involved in subsequent events, such as spindle assembly. This reprogramming of protein synthesis is apparently not dependent on new transcription of the nuclear genome, but rather on informational RNA already present in the oocyte cytoplasm. It is not clear whether certain of the proteins which are

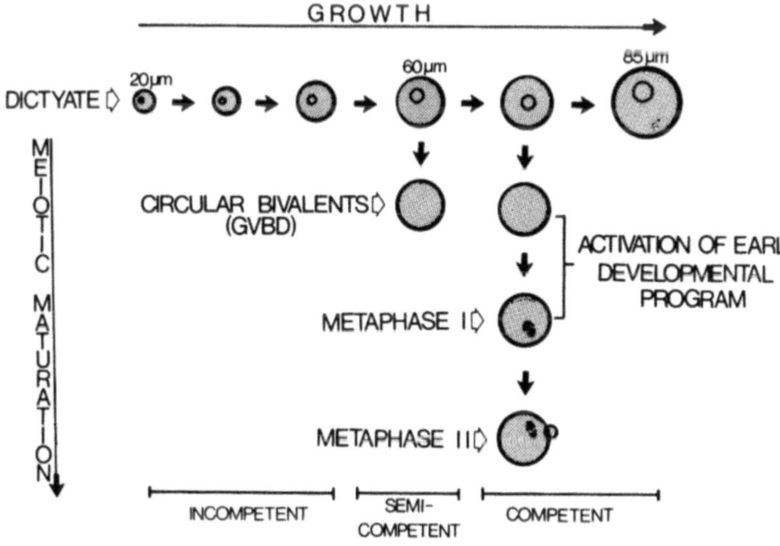


Fig. 12

synthesized following GVBD represent translation products of hitherto untranslated, stored, informational RNA. In any case, many of the proteins whose synthesis is initiated or enhanced during meiotic maturation of the oocyte continue to be synthesized in the early mouse embryo. This observation is consistent with the idea that the program for early mammalian development is actually "activated" during the final stages of oogenesis.

The relationships between oocyte growth, meiotic maturation, and early embryogenesis are now beginning to be understood at the molecular level (Fig. 12). Since "embryogenesis begins during oogenesis" this understanding is necessary if we hope to solve health related problems in the areas of human reproduction and human development. For without a thorough understanding of the molecular basis of oogenesis and embryogenesis in the mammal, it is unlikely that an effective technology designed to deal with abnormalities in human reproduction and development will be forthcoming.

Acknowledgement. The research carried out in Dr. Wassarman's laboratory was supported by grants from The National Institute of Child Health and Human Development, The National Science Foundation, and The Rockefeller Foundation. R.M.S. is a postdoctoral fellow of The Rockefeller Foundation, M.J.L. is on sabbatical leave from the Biology Department, Lawrence University, Appleton, Wisconsin, and J.D.B. is a predoctoral trainee supported by a National Research Service Award in Cell and Developmental Biology.

Summary. The capacity to undergo spontaneous meiotic maturation *in vitro* ("meiotic competence") is acquired at a specific stage of oocyte growth in the juvenile mouse. Growth involves not just tremendous enlargement of the oocyte, but also extensive alterations in the ultrastructure of various organelles and inclusions. These ultrastructural changes reflect alterations in the overall metabolism of the oocyte, including a reprogramming of the pattern of protein synthesis. It would appear that the acquisition of meiotic competence during growth of the mouse oocyte is attributable to changes in the quality, rather than the quantity, of cytoplasm. Meiotic maturation *in vitro* is characterized by many qualitative changes in protein synthesis, especially during the period following germinal vesicle breakdown. Concomitant protein synthesis is not required for germinal vesicle breakdown to take place. Mixing of the oocyte's nucleoplasm and cytoplasm must occur for those changes in protein synthesis which characterize meiotic maturation to take place and these changes are necessary for nuclear progression beyond the circular bivalent stage. The reprogramming of protein synthesis during meiotic maturation does not depend upon concomitant transcription of the nuclear genome and the changes observed are not simply due to post-translational modification of proteins. Many of the newly synthesized proteins that first appear during meiotic maturation of the oocyte continue to be synthesized in the 1-cell embryo. The absolute rate of protein synthesis decreases from 43 to 31 pg/hr/oocyte during meiotic progression from dictyate to metaphase II, while the size of the intracellular free-methionine pool decreases from 62 to 35 fmol/oocyte during the same period. The oocyte's free-methionine pool is not compartmentalized. Differences noted between meiotic maturation of mammalian and non-mammalian oocytes underscore the potential problems which may be encountered in extrapolating the results of experiments on lower species to the mammal.

## References

(1) T.G. Baker, in: Reproduction in Mammals, Vol. 1, ed. C.R. Austin and R.V. Short (Cambridge University Press, 1972) p. 14.
(2) A. Tsafriri, in: Evolution of the Vertebrate Ovary, ed. E. Jones (Plenum Publishing Corp., in press).
(3) L.D. Smith, in: The Biochemistry of Animal Development, Vol. 3, ed. R. Weber (Academic Press, N.Y., 1975) p. 1.
(4) S. Stern and P.M. Wassarman, J. Cell Biol. 59 (1973) 335a.
(5) P.M. Wassarman and G.E. Letourneau, J. Cell Sci. 20 (1976) 549.
(6) P.M. Wassarman and K. Fujiwara, J. Cell Sci. 29 (1978) 171.
(7) P.M. Wassarman, W.J. Josefowicz, and G.E. Letourneau, J. Cell Sci. 22 (1976) 531.
(8) R.A. Sorensen and P.M. Wassarman, Devel. Biol. 50 (1976) 531.
(9) R.M. Schultz and P.M. Wassarman, J. Cell Sci. 24 (1977) 167.
(10) P.M. Wassarman and W.J. Josefowicz, J. Morphol. 156 (1978) 209.
(11) J.J. Eppig, Devel. Biol. 60 (1977) 371.
(12) F. Mangia and C.J. Epstein, Devel. Biol. 45 (1975) 211.
(13) J.J. Eppig, J. Exptl. Zool. 198 (1976) 375.
(14) R.M. Schultz and P.M. Wassarman, Analyt. Biochem. 77 (1977) 25.
(15) R.M. Schultz and P.M. Wassarman, Proc. Natl. Acad. Sci., USA 74 (1977) 538.
(16) P.M. Wassarman, D.F. Albertini, W.J. Josefowicz, and G.E. Letourneau, J. Cell Sci. 21 (1976) 523.
(17) P.M. Wassarman, T.E. Ukena, W.J. Josefowicz, G.E. Letourneau, and M.J. Karnovsky, J. Cell Sci. 26 (1977) 323.
(18) R.M. Schultz, G.E. Letourneau, and P.M. Wassarman, J. Cell Sci. 30 (1978) 251.
(19) R.M. Schultz, M.J. LaMarca, and P.M. Wassarman, Proc. Natl. Acad. Sci., USA 75 (1978) in press.
(20) M.W. Bitensky and R.E. Gorman, Prog. Biophys. Molec. Biol. 26 (1973) 409.
(21) M.M. Burger, B.M. Bombik, B.M. Breckenridge, and J.R. Sheppard, Nature (New Biol.) 239 (1972) 161.
(22) D.F. Albertini and E. Anderson, J. Cell Biol. 63 (1974) 234.
(23) N.B. Gilula, in: International Cell Biology, ed. B.R. Brinkley and K.R. Porter (Rockefeller University Press, 1977) p. 61.
(24) P.M. Wassarman and G.E. Letourneau, Nature 361 (1976) 73.

DISCUSSION - CHAIRPERSON - A.W. SCHUETZ

A.R. MIDGELY: Dr. Wasserman, is it possible that the changes you have observed in the oocytes actually are the result of maturational changes occurring in the granulosa cells, which then influence the oocytes?

P. WASSERMAN: Obviously, one should take into consideration the possible influence of surrounding cells on the oocyte itself. I think that John Eppig's work showing that one can obtain growth of isolated mouse oocytes in vitro suggests that hormones may not be directly involved in growth of the oocyte. I think there is a variety of other evidence to support this suggestion, but certainly, at this time it would be premature to state the follicle cells are not at all involved in regulating the progress of the oocyte including the acquisition of meiotic competence.

# MAMMALIAN OOCYTE MATURATION: MODEL SYSTEMS AND THEIR PHYSIOLOGICAL RELEVANCE

A. Tsafriri

Department of Hormone Research, The Weizmann Institute of Science, Rehovot, Israel

## INTRODUCTION

The study of oocyte maturation *in vitro* illustrates a dilemma often facing the physiologist: the simplest system, consisting of cultured isolated oocytes, does not appear to match the physiological events *in vivo* in that the oocytes mature spontaneously in hormone-free media (1,2,3). On the other hand, the oocyte explanted within the follicle while maintaining its normal association with the cumulus oophorus matures in culture only following a hormonal stimulus (4,5), closely resembling the requirements prevailing *in vivo*. However, being a multicompartmental model it does not allow to draw unequivocal conclusions about the primary site(s) of gonadotropin action in inducing oocyte maturation and the location of the primary response to this stimulus. A third approach to the study of oocyte maturation *in vitro*, namely the culture of oocyte together with follicular constituents (6,7) was adopted in order to test the role of follicle cells in the control of oocyte maturation. In this system, the dependence of maturation on hormonal stimulation is restored, and this approach affords an opportunity to analyze events in the oocyte and follicular compartment separately and to define the primary site of hormone action. Some of the results obtained by each of these approaches will be reviewed with special reference to their physiological relevance.

### Time Course of Germinal Vesicle Breakdown (GVB)

Germinal vesicle breakdown of rat oocytes occurs *in vivo* 2.5 - 3.5 h following an endogenous or exogenous gonadotropic stimulus (8,9,10). Whereas GVB is observed in isolated oocytes 1.5 h after explantation, 2.75 h are required for follicle-enclosed oocytes to reach the same degree of maturation (Fig. 1). Thus, GVB breakdown is achieved in spontaneously maturing oocytes significantly earlier than the corresponding change in hormone-induced maturation of oocytes *in vivo* or *in vitro*. This discrepancy suggests that in the isolated oocyte some regulatory step is by-passed which is essential for the triggering of meiosis under physiological conditions.

## Is Cyclic AMP Involved in the Control of Meiosis?

The involvement of cyclic AMP (cAMP) in the control of the resumption of maturation is suggested on the basis of four lines of evidence: (a) LH, the physiologic trigger of maturation induces an immediate rise in follicular cAMP level (3,11); (b) chemically unrelated substances, such as hCG, FSH, $PGE_2$ (4,12,13) and cholera en-

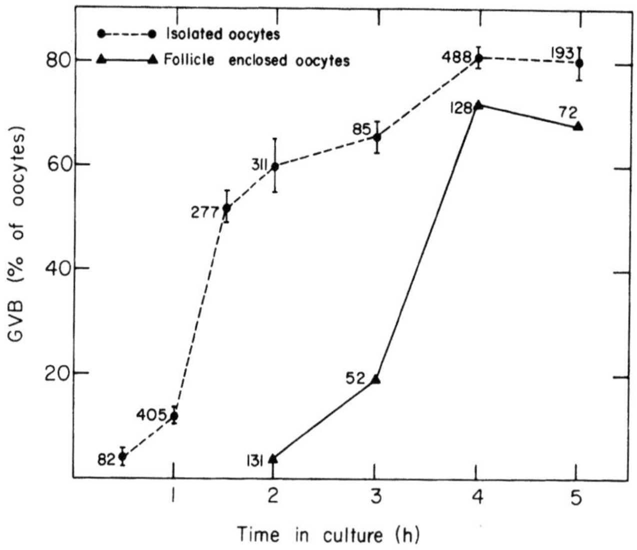


Fig. 1. Time course of germinal vesicle breakdown of isolated and follicle-enclosed oocytes. Isolated oocytes were obtained 48 h after PMSG (15 i.u./rat) administration to 26-day-old rats and cultured in hormone-free medium for the period indicated. Follicles were excised from either mature rats on the morning of the day of proestrus or from immature rats treated with PMSG 48-52 h earlier. To the cultured follicle-enclosed oocytes LH (5 µg/ml) was added. There was no difference in the time course of GVB, in oocytes obtained from mature or immature animals, hence the data were pooled. The number of oocytes examined at each time-interval is indicated (from Bar-Ami and Tsafriri, in preparation).

terotoxin (14) which augment follicular cAMP accumulation, also induce GVB; (c) injection of dibutyryl cAMP (dbcAMP) into the follicular antrum (4) or addition of 8-bromo-cAMP to the culture medium (13) brought about GVB by itself; (d) addition of an inhibitor of phosphodiesterase, ICI 63.197, induces GVB (12,13). However, two sets of observations appeared to contradict this hypothesis: (a) addition of cAMP or dbcAMP and of some inhibitors of phosphodiesterase, such as theophylline or IBMX, to cultured follicle-enclosed oocytes prevented the LH-induced maturation (4,12,13); (b) addition of phosphodiesterase inhibitors or dbcAMP prevents the spontaneous maturation of isolated oocytes (15,16). Several explanations may be put forward to reconcile these seemingly discrepant findings. The contrasting effects of dbcAMP and of phosphodiesterase inhibitor on isolated and follicle-enclosed oocytes may be interpreted as resulting from compartmentalization of cAMP in the follicle: it is possible that LH induces a rise in cAMP only in granulosa cells, whereas phosphodiesterase inhibitors cause cAMP accumulation indiscriminately in all compartments of the follicle, including the oocyte itself. The increased concentration of the nucleotide in germ cell might be inhibitory to the resumption of meiosis. The inhibition of spontaneous maturation of isolated oocytes is apparently related to the high intracellular concentration of the nucleotide (17). Another interpretation is suggested by the results of an elegant experiment of Hillensjö *et al.* (13): when dbcAMP was present in the medium for only the first 2 h of incubation and the follicles were washed and transferred into a hormone-free medium, the majority of the oocytes matured during the second incubation period. Thus, it is possible that a temporal rise in follicular cAMP, such as the LH-induced cAMP surge, permits the resumption of meiosis, whereas continuous exposure of the oocyte to high cAMP concentrations is inhibitory. The inhibitory action of some phosphodiesterase inhibitors may be explained by (a) long-term increase in follicular cAMP; (b) by unspecific effect of the drugs unrelated to their action as phosphodiesterase inhibitors (18). For instance, theophylline, at high concentration can be inhibitory to protein synthesis (19), a process shown to be essential for hormonal induction of oocyte maturation (5,23).

In conclusion, it appears that (i) the effect of LH on the resumption of meiosis is mediated by cAMP; (ii) the meiotic process is prevented by continuous exposure of the oocyte to high concentration of cAMP; and (iii) it is not clear whether LH induces maturation by a surge of cAMP in the germ cell itself, or indirectly through a rise in cAMP in some other follicular compartment.

## Protein Synthesis and the Resumption of Oocyte Maturation

A detailed account of oocyte protein synthesis during maturation is left for the expert treatment of Dr. P. Wassarman. Only one aspect of the involvement of protein synthesis in oocyte maturation will be dealt with here. The resumption of meiosis (GVB) by isolated mouse oocytes is not disturbed by the presence of puromycin (10 µg/ml) at a dose sufficient to suppress more than 95% of incorporation of [$^{35}$S]-methionine into TCA-insoluble material. Protein synthesis is needed for later stages of maturation beyond the circular bivalent stage (20,21,22). By contrast, LH-induced maturation of rat follicle-enclosed oocytes is blocked by cycloheximide or puromycin (5,23). This discrepancy in the behavior of the denuded and follicle-enclosed oocyte may indicate that LH exerts its action on the resumption of meiosis through a protein synthesized in follicle cells.

## Divalent Cations and Oocyte Maturation

Exposure of sea urchin (*Spisula solidissima*) oocytes to divalent ionophore A23187 resulted in GVB (24). High $Ca^{++}$ or $Mg^{++}$ ion concentration in the presence of the ionophore (25), or electrophoretic introduction of $Ca^{++}$ ions (26) induced the resumption of meiosis of *Xenopus laevis* oocytes. We examined the possible involvement of divalent cations in the maturation of isolated and follicle-enclosed rat oocytes by the use of the cationophore A23187 and of $Ca^{++}$ and $Mg^{++}$-deficient media (Tsafriri and Bar-Ami, submitted).

The spontaneous maturation of isolated oocytes was unaffected by the presence of the ionophore ($10^{-7}$ to $10^{-5}$M) or of $Ca^{++}$ or $Mg^{++}$-deficient medium with or without the addition of EDTA or EGTA (1 mM).

Addition of cationophore A23187 ($10^{-5}$M) to follicle-enclosed oocytes cultured in hormone-free Eagle's medium induced the maturation in 61% of the oocytes (Fig. 2A). In calcium deficient Eagle's medium, the follicle-enclosed oocytes also matured (66% GVB) without hormone supplementation. This maturation of follicle-enclosed oocytes in $Ca^{++}$-deficient medium was unaffected by the addition of the ionophore ($10^{-7}$ or $10^{-5}$M)(Fig. 2B). Addition of LH to $Ca^{++}$-deficient medium resulted in a further increase in the rate of maturation of the follicle-enclosed oocytes (to 86% GVB) and addition of EDTA or EGTA (1 mM) was without effect on the resumption of meiosis, though an exceptionally high rate of degeneration (up to 35%) was observed (data not shown). Thus, divalent cations, possibly calcium, seem to be involved in the regulating hormonal mechanisms of meiotic maturation of mammalian oocytes, as they are involved in the maturation of sea urchin and *Xenopus laevis* oocytes. We were unable to define whether the meiosis-inducing action of LH or of the ionophore depend upon the presence of $Ca^{++}$ in the external medium, since both follicle-enclosed as well as isolated oocytes matured spontaneously in $Ca^{++}$-deficient medium. Maturation occurred in this medium even in the presence of EGTA or EDTA. Nevertheless, the observed independence of maturation upon the presence of $Ca^{++}$

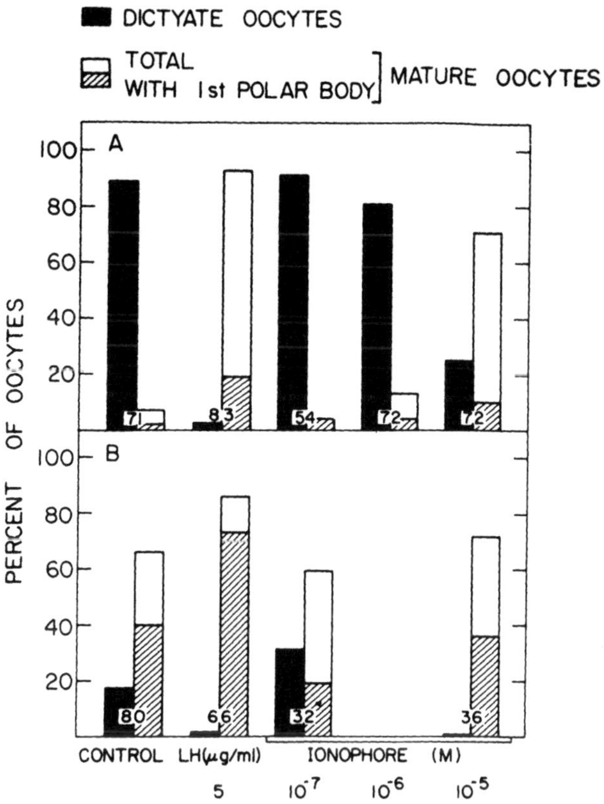


Fig. 2. Effect of divalent cationophore A23187 and of $Ca^{++}$-deficient medium on the maturation of follicle-enclosed oocytes. Follicles explanted on the morning of the day of proestrus (4) were cultured for 18-20 h in: A. normal Eagle's medium; B. $Ca^{++}$-deficient medium. The number of oocytes is indicated at the base of columns. Dictyate oocytes, containing intact germinal vesicle; mature oocytes, all oocytes beyond the stage of germinal vesicle breakdown without signs of degeneration. The percentage of degenerated oocytes is not shown (from Tsafriri and Bar-Ami, 1978, submitted).

in the medium and the induction of maturation by the ionophore may be explained by the involvement of internally sequestered calcium, rather than transmembrane transport of the cation, in the regulation of meiosis. Further support for this assumption comes from the use of agents believed to interfere with transmembrane transport of $Ca^{++}$, such as $La^{+++}$ and verapamil. Preliminary results indicate that neither verapamil ($5 \times 10^{-5}$M) nor $LaCl_3$ (1 mM) inhibited the LH or ionophore-induced maturation of follicle-enclosed oocytes. Neither of these drugs induced maturation by itself.

In the preceding Ovarian Workshop, Biggers and Powers (27) presented evidence implicating changes in membrane permeability and resulting membrane potential in the spontaneous maturation of mouse oocytes. Similarly, a change in the gradient of various cations across the cell membrane was implicated in the regulation of meiosis in *Xenopus laevis* oocytes (28). Further studies are required in order to evaluate the importance of changes in intracellular distribution and transmembrane movement of various cations in the regulation of meiosis.

## The Regulation of Oocyte Maturation by Follicular Constituents

Pincus and Enzmann (1) originally suggested that the maturation of the follicular oocyte is prevented either by a substance produced by follicle cells which inhibits nuclear maturation directly or indirectly by keeping the oocyte in a "nutritional state" wherein maturation is impossible. Foote and Thibault (6) presented evidence that oocyte contact with granulosa cells prevents the resumption of meiosis by porcine oocytes in culture. Maturation of porcine oocytes cocultured with porcine granulosa cells was inhibited in a dose dependent manner (3). This inhibition, however, was not reversible by LH. Addition of porcine follicular fluid (FFl) or a low molecular weight fraction thereof (PFFl) to porcine oocytes cultured within their cumulus inhibited spontaneous maturation, whereas porcine serum or serum constituents was without effect (7,29). This inhibitory action of FFl or PFFl was overcome by addition of LH (7,29) or reversed by transfer of oocytes into a PFFl-free medium (30). Partial purification and characterization of the inhibitor was achieved by the use of Sephadex G-25 and paper electrophoresis. The available data are consistent with the suggestion that the inhibitor is a peptide of about 2000 daltons (29). A similar inhibitor from bovine follicular fluid inhibits the maturation of hamster oocytes (31). Channing *et al.* demonstrated that also human follicular fluid or its low molecular weight fraction inhibit the maturation of porcine oocytes in culture (32).

Recently it was demonstrated that porcine FFl (50%) and PFFl (85 μg protein/ml) also inhibit the maturation of rat oocytes cultured within their cumuli (33). Furthermore, these studies demonstrated a change in the sensitivity of rat oocytes to the inhibitor. Oocytes collected from preovulatory follicles, i.e. 44 h following PMSG-administration, are less sensitive to the inhibitory action of PFFl than oocytes harvested from animals 20 h following PMSG treatment. The inhibition of rat oocyte maturation by FFL or PFFl is partially overcome by LH (5 μg/ml).

Immunization of rabbits (34) or rats (unpublished observations) with PFFl conjugated to BSA produced an antiserum able to neutralize

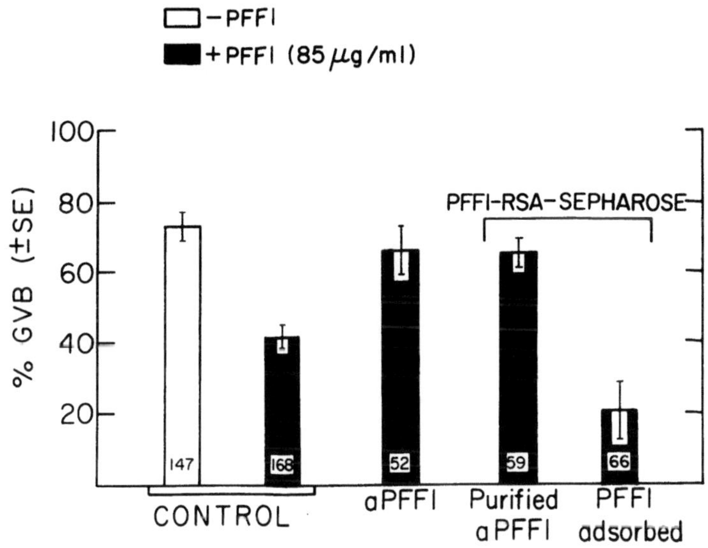


Fig. 3. Effect of purified antibody to PFF1 on maturation of isolated rat oocytes. Oocytes were isolated 20 h after PMSG-administration to 26-day-old rat and cultured within their cumuli for 6 h (33). The ability of the antiserum to PFF1 (aPPF1), antiserum to PFF1 adsorbed on PFF1-RSA-sepharose column (PFF1-adsorbed) and of purified antibodies to PFF1 to neutralize the inhibitory action of PFF1 was tested [from Tsafriri, Weinstein *et al.* (35) in press, by permission].

the inhibitory action of PFF1 on rat oocyte maturation. Furthermore, the antibodies were purified by affinity chromatography and the inhibitory action of PFF1 was neutralized only by the pure antibody but not by the PFF1-adsorbed serum (Fig. 3)(35).

Coculture of freshly isolated granulosa cells (from $10^4$ to $10^7$ cells per culture) and oocytes did not prevent spontaneous maturation. On the other hand, maturation was inhibited when the oocytes were added to granulosa cells previously cultured for 24 h. Addition of LH (5 μg/ml) at the beginning of the oocyte incubation abolished this inhibitory effect of granulosa cells (Fig. 4)(34). Simultaneous presence of granulosa cells and oocytes is not obligatory for the inhibitory effect: a medium in which granulosa cells were cultured previously for 48 h also inhibited the resumption of meiosis. Addition of LH to oocytes cultured in such granulosa cell-conditioned medium failed to overcome the inhibition or did so only partially (Fig. 5)(34).

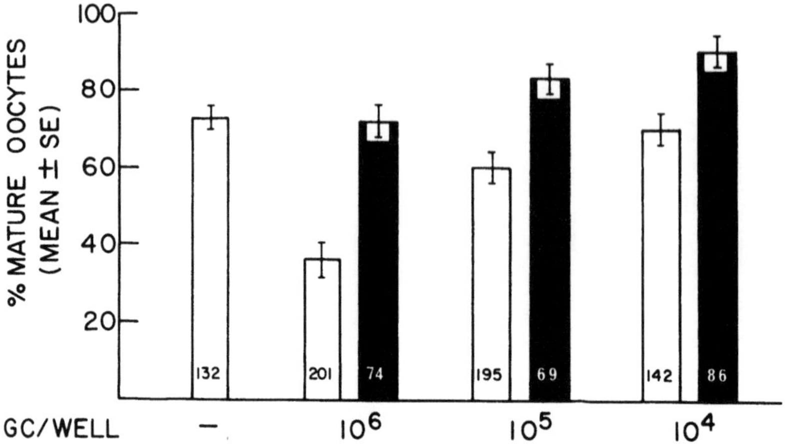


Fig. 4. Effect of coculture of rat granulosa cells and oocytes upon oocyte maturation. Granulosa cells (GC; number indicated) were cultured for 24 h in 0.2 ml Eagle's medium on Falcon Microtest II plates. Freshly isolated oocytes (10-15) were then added to the culture alone (open columns) or together with 5 μg/ml LH (black columns). Germinal vesicle breakdown was scored following a 6 h culture (mature oocytes). The number of oocytes cultured is shown on the columns; vertical brackets, ± S.E.M.) [from Tsafriri (34) by permission].

This dependence of the meiosis-inducing action of LH upon the presence of granulosa cells is compatible with the notion that this effect of LH on maturation is exerted through the mediation of granulosa cells.

## CONCLUDING REMARKS

Comparison of germinal vesicle breakdown in isolated oocytes maturing spontaneously with LH-induced maturation of follicle-enclosed oocytes reveals a delay of about 1 h in the completion of the process and the need for protein synthesis in the latter model. These differences suggest that by dislodging the oocyte from the follicle it is released from the physiologic mechanism(s) keeping meiosis in abeyance. Within the preovulatory follicle, LH induces the resumption of meiosis, probably through the mediation of granulosa cells, by a process that includes protein synthesis and lasts at least 1 h. This triggering action of LH is mediated, at some stage, by a temporal increase in follicular cyclic AMP. Whether this surge of cAMP is limited only to granulosa cells or includes the germ cell also still remains to be elucidated.

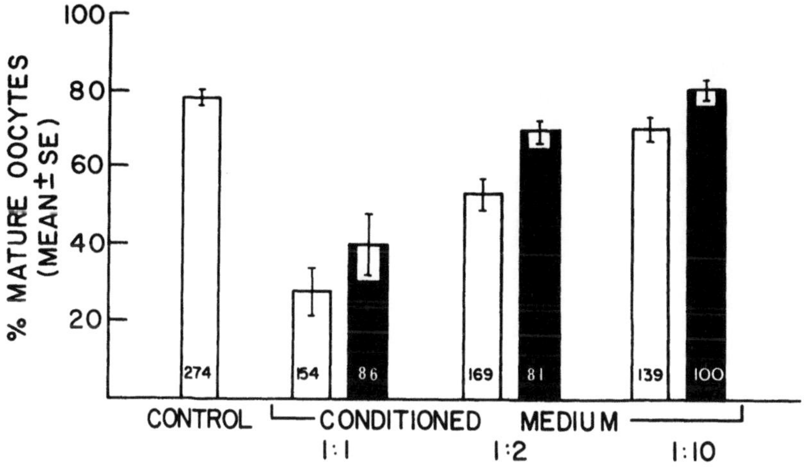


Fig. 5. The effect of granulosa cell conditioned medium upon the maturation of rat oocytes. Granulosa cells ($10^6$/ml) were cultured for 48 h in Eagle's medium. The medium was then withdrawn and sterilized by filtration through millipore membrane (0.22 μm). Oocytes were cultured for 6 h in the conditioned medium diluted with fresh medium as shown. Open columns, no hormone added; black columns, LH (5 μg/ml) added. The number of oocytes cultured is shown on the columns; vertical brackets, ± S.E.M. [from Tsafriri (34) by permission].

Coculture of follicular constituents with isolated oocytes provided experimental support for the early suggestion that resumption of the meiotic process is held in abeyance within the mature follicles by a factor produced by granulosa cells and accumulated in the follicular fluid. Furthermore, maturation *in vitro* of hamster follicle-enclosed oocytes was obtained either by addition of LH or by washing the explanted follicles (31). Nevertheless, the exact mechanism by which this inhibitor acts on maturation is not known: it may act directly on the oocyte, through the mediation of cumulus cells or by preventing some process obligatory for the resumption of meiosis.

Though oocytes obtained by spontaneous or by hormone-induced maturation cannot be distinguished morphologically, the developmental potential of oocytes matured spontaneously has been questioned, at least in some species. Even assuming that spontaneous maturation includes all the oocytic changes inherent in the normal meiotic process, there is little doubt that in spontaneous maturation some of the regulatory events which lead to the resumption of meiosis following the LH-triggered maturation are bypassed. From this aspect, spontaneous maturation of the isolated oocyte may be regarded as an artifact. Nevertheless, important principles may be established by use of artifactual model systems. Examples are the studies demonstrating the necessity of pyruvate for the maturation of denuded mouse oocytes, the definition of optimal media for oocyte culture (36) and the descriptions of changes in protein synthesis within the oocyte (22). When the aim is to disclose physiological mechanism(s), critical evaluation is required to discriminate between artifacts inherent in the model and normal processes. The coculture studies with isolated oocytes appear to withstand such critical evaluation: the dependence of maturation upon the normal hormonal stimulus is restored; the inhibitory activity of porcine follicular fluid is limited to a specific fraction thereof, and this inhibition is abolished by a purified antibody to this fraction. Collectively, these findings support the view that meiosis is prevented in the preovulatory follicle by a local factor, produced by granulosa cells.

Our understanding of the mechanisms involved in the maturation of both mammalian and lower vertebrate oocytes is only fragmentary at the present. However, some apparent differences in the control of maturation in mammals and amphibians were described: amphibian oocytes do not mature spontaneously in culture; protein synthesis is obligatory for GVB in amphibian oocytes; the meiosis inducing action of gonadotropins in amphibians is mediated by steroids synthesized by follicle cells; inhibition of maturation by follicle cells has been demonstrated only in mammalian but not amphibian oocytes. It is quite possible that when more details will be available we may find the explanations for the apparent inconsistencies and differences. In the meantime two factors should be stressed: (a) when the behavior of the amphibian oocyte is compared to that of the mammalian follicle-enclosed oocyte, rather than to isolated mammalian germ cells, some of the inconsistencies disappear; (b) the ovary of mature polyestrous mammals contains a number of fully grown oocytes competent to mature upon isolation, which exceeds the number of ova released at ovulation at each cycle. On the other hand, in the amphibians investigated there is only one ovarian cycle each year and the fully grown oocytes are laid in one clutch. It is possible that some of the differences between mammals and lower vertebrates in the mode of control of meiosis may be related to this evolutional divergence in reproductive behavior, the mammalian ovary having acquired additional intra-ovarian control mechanisms to restrict the number of eggs ovulated at each cycle.

## ACKNOWLEDGEMENTS

I am grateful to my colleagues Mr. S. Bar-Ami, Drs. Y. Weinstein, C.P. Channing, S.H. Pomerantz and T. Hillensjö for permission to quote unpublished results and to Dr. H.R. Lindner for his continuous interest in these studies and for critical reading of the manuscript. The studies reported were supported in part by the World Health Organisation, the Ford Foundation and the Israel-USA Binational Science Foundation, Jerusalem, Israel. I thank Mrs. A. Tsafriri and Mrs. M. Schattner for their able technical assistance and Mrs. M. Kopelowitz for typing the manuscript.

## SUMMARY

Maturation of mammalian oocytes is studied mainly in two dissimilar *in vitro* models: isolated oocytes maturing spontaneously in culture, and hormone-induced maturation of follicle-enclosed oocytes. In this discussion the following aspects of maturation *in vitro* in the two aforementioned models were compared: timing of germinal vesicle breakdown (GVB), involvement of cyclic nucleotides, protein synthesis and divalent cations. A third approach to the study of oocyte maturation *in vitro*, namely oocyte co-culture with follicular constituents was adopted in order to test the role of follicular components in the control of the resumption of meiosis. Such studies demonstrated an inhibitory action of granulosa cells, granulosa cell-conditioned medium and of follicular fluid upon the spontaneous maturation of the co-cultured oocytes. Furthermore, addition of LH to co-cultures of oocytes and granulosa cells induced resumption of meiosis.

Although oocytes obtained by spontaneous or by hormone-induced maturation cannot be distinguished morphologically, the developmental potential of oocytes matured spontaneously has been questioned, at least in some species. Furthermore, analysis of the kinetics of spontaneous maturation suggests that oocytes dislodged from their follicles escape physiologic mechanisms ensuring meiotic arrest and skip some of the regulatory steps involved in the normal hormonal triggering of maturation. Oocyte co-culture with granulosa cells offers a close approximation to physiological conditions in that meiotic maturation depends on hormonal stimulation, while the system permits the application of separate treatments to the oocyte-cumulus complex and granulosa cells. As in any *in vitro* system, critical evaluation is required to discriminate between artifacts inherent in the model and physiological processes.

## REFERENCES

(1) G. Pincus and E.V. Enzmann. J.exp.Med. 62(1935)665.

(2) A.W. Schuetz. Biol.Reprod. 10(1974)150.

(3) C.P. Channing and A. Tsafriri. Metabolism 26(1977)413.

(4) A. Tsafriri, H.R. Lindner, U. Zor and S.A. Lamprecht. J. Reprod.Fert. 31(1972)39.

(5) H.R. Lindner, A. Tsafriri, M.E. Lieberman, U. Zor, Y. Koch, S. Bauminger and A. Barnea. Rec.Progr.Horm.Res. 30(1974)79.

(6) W.D. Foote and C. Thibault. Ann.Biol.anim.Bioch.Biophys. 9(1969)329.

(7) A. Tsafriri and C.P. Channing. Endocrinology 96(1975)922.

(8) A. Tsafriri and P.F. Kraicer. J.Reprod.Fert. 29(1972)387.

(9) J.P.W. Vermeiden and G.H. Zeilmaker. Endocrinology 95(1974)341.

(10) C. Magnusson, T. Hillensjö, A. Tsafriri, R. Hultborn and K. Ahrén. Biol.Reprod. 17(1977)9.

(11) J.M. Marsh. Biol.Reprod. 14(1976)30.

(12) T. Hillensjö. Acta Endocr. 82(1976)809.

(13) T. Hillensjö, C. Ekholm and K. Ahrén. Acta Endocr. 87(1978)377.

(14) U. Zor, S.A. Lamprecht, A. Tsafriri, S.H. Pomerantz, Y. Koch and H.R. Lindner. Israel J.Med.Sci. 8(1972)1774.

(15) K.W. Cho, S. Stern and J.D. Biggers. J.Exp.Zool. 187(1974)383.

(16) C. Magnusson and T. Hillensjö. J.Exp.Zool. 201(1977)139.

(17) P.M. Wassarman and E. Turner. J.Exp.Zool. 196(1976)183.

(18) M.M. Appleman, W.J. Thompson and T.R. Russel, in: Advances in Cyclic Nucleotide Research, Vol. 3, eds. P. Greengard and G.A. Robinson (Raven Press, New York, 1973) p. 65.

(19) I.D.K. Halkerston, M. Feinstein and O. Hechter. Proc.Soc. Exp.Biol.Med. 122(1966)896.

(20) S. Stern, H. Rayyis and J.F. Kennedy. Biol.Reprod. 7(1972)341.

(21) P.M. Wassarman, W.J. Josefowicz and G.E. Letourneau. J.Cell.Sci. 22(1976)531.

(22) R.M. Schultz and P.M. Wassarman. J.Cell.Sci. 24(1977)167.

(23) A. Tsafriri, M.E. Lieberman, A. Barnea, S. Bauminger and H.R. Lindner. Endocrinology 93(1973)1378.

(24) A.W. Schuetz. J.Exp.Zool. 191(1975)433.

(25) W.J. Wasserman and Y. Masui. J.Exp.Zool. 193(1975)369.

(26) M. Moreau, M. Doree and P. Guerrier. J.Exp.Zool. 197(1976)443

(27) J.D. Biggers and R.D. Powers, in: Ovarian Follicular Development and Function, eds. A.R. Midgley and W.A. Sadler (Raven Press, New York, in press).

(28) E. Baltus, J. Hanocq-Quertier, A. Pays and J. Brachet. Proc.Natl.Acad.Sci.U.S.A. 74(1977)3461.

(29) A. Tsafriri, S.H. Pomerantz and C.P. Channing. Biol.Reprod. 14(1976)511.

(30) C.P. Channing, S.L. Stone, A.S. Kripner and S.H. Pomerantz, in: Novel Aspects of Reproductive Physiology, eds. J. Wilkes and L. Spielman (in press).

(31) R.B.L. Gwatkin and O.F. Andersen. Life Sciences 19(1976)527.

(32) C.P. Channing, T. Hillensjö and F.W. Schaerf, in: Clinics in Endocrinology and Metabolism, eds. M. Lipsett and G.T. Ross (Saunders, Philadelphia, in press).

(33) A. Tsafriri, C.P. Channing, S.H. Pomerantz and H.R. Lindner. J. Endocr. 75(1977)285.

(34) A. Tsafriri. Ann.Biol.anim.Bioch.Biophys. 18(1978)523.

(35) A. Tsafriri, Y. Weinstein, S. Bar-Ami, C.P. Channing, S.H. Pomerantz and H.R. Lindner, in: Recent Advances in Steroid Research, ed. C. Conti (Academic Press, New York, 1978, in press).

(36) J.D. Biggers, D.G. Whittingham and R.P. Donahue. Proc.Natl. Acad.Sci.U.S.A. 58(1967)560.

DISCUSSION - CHAIRPERSON - A.W. SCHUETZ

K. McNATTY: In collaboration with Dr. Diane Smith, we have been looking at how granulosa cells and/or steroids may influence oocyte maturation in vitro. And the first point I would like to make is in the human, unless the oocyte is of a certain size, 112 microns, it won't undergo meiosis in vitro, and secondly, the oocyte doesn't seem to resume meiosis in vitro unless it has been exposed to some estrogen in vivo. The absolute level of estrogen isn't important, but the fact is that there must be more estrogen than such androgens as androstenedione, testosterone or dihydrotestosterone. An interesting point is that it may be that the granulosa cells are influencing the oocyte in this respect, because in the human, if we assume that the leakage of steroid out of the follicle is slowed, you can say that almost 100% of the estrogen is derived from the granulosa cells, and at least 50% of the androgen. So in this way, there may be some influence of the steroids produced by the granulosa cells on the human oocyte in vitro. Thank you.

A. TSAFRIRI: I wish to emphasize that in our studies with follicular fluid it was charcoal extracted so that the possibility of steroid effect was very much reduced. Furthermore, in some of our experiments with the low molecular weight fraction of follicular fluid it was ether extracted prior to addition to the cumulus enclosed oocyte and, therefore, its inhibitory effect on oocyte maturation cannot be attributed to steroids.

J. MARSH: I would just like to make a comment that the data you showed of Hillensjö on the two types of incubation with dibutryl cyclic-AMP reminds me of the changes in endogenous cAMP in the follicle after LH treatment. In the rabbit, and as has been shown by others in the rat, cyclic AMP rises very rapidly after the LH surge and then falls off. We proposed in our early studies that this fall-off might be related to ovulation. We have never proven or disproven this hypothesis but perhaps it's related not to ovulation, but to oocyte maturation.

A. TSAFRIRI: I agree. My argument rests in part upon this time-course of the rise in cAMP within the preovulatory follicle.

# ACTION OF PORCINE FOLLICULAR FLUID OOCYTE MATURATION INHIBITOR IN VITRO: POSSIBLE ROLE OF THE CUMULUS CELLS

T. Hillensjö, A.S. Kripner, S.H. Pomerantz, and
C.P. Channing
Departments of Physiology and Biochemistry,
University of Maryland School of Medicine
Baltimore, Maryland 21201

## INTRODUCTION

Evidence has been presented in the pig that an oocyte maturation inhibitor (OMI) may be responsible for keeping the oocyte in meiotic arrest in the first prophase (1). This inhibitor is present in the fluid of antral porcine follicles. Partial purification has revealed it to be a small polypeptide with a molecular weight in the order of 2000 (2). This inhibitor prevents the spontaneous resumption of meiosis in cultured cumulus-enclosed porcine oocytes. Its inhibitory effect can be overcome by luteinizing hormone (LH) (3) and it can be reversed by removing the factor from the culture medium (4).

In the above mentioned studies the effect of OMI has been analyzed using cumulus-enclosed oocytes cultured in a complex medium containing serum (1). More recently it has been shown that isolated porcine oocytes mature as well in a chemically defined minimal culture medium (5). Furthermore, it has been demonstrated that mechanically denuded oocytes (stripped of surrounding cumulus cells) have the capacity to mature spontaneously, although less efficiently compared to cumulus-enclosed oocytes (5).

In an attempt to study the mode of action of OMI we have compared the effect of porcine follicular fluid (PFFl) upon the spontaneous maturation of porcine oocytes cultured either in a complex, serum containing medium or in a defined, minimal culture medium. Cumulus-enclosed as well as denuded oocytes were tested.

The viability of the cumulus cells during the culture was assessed by morphological examination and by analysis of progesterone formation. The present study demonstrates that cumulus cells appear to be required for the inhibitory action of OMI since it only works in cumulus-enclosed, not in denuded oocytes.

## MATERIALS AND METHODS

Porcine ovaries were obtained from a local slaughter house. Oocytes were harvested from medium-sized (3-5 mm) follicles and cultured within 3 h of slaughter. For each experimental treatment 4-5 x 15 oocytes were cultured in wells of micro-test plates II (Falcon) containing 0.2 ml of culture medium with or without PFFl fractions. Each experiment was repeated on two or more occasions. Only oocytes surrounded by a dense layer of cumulus cells, of similar size, were selected for the experiments. They were cultured intact (cumulus-enclosed) or after removal of the cumulus cells (denuded). Complete removal of the cumulus cells was achieved by forcing the cell-complexes through 3 sets of thin-bore glass capillaries controlled by mouth pressure (6).

The culture medium was modified TC 199 containing 15% pig serum (199A; 1) or defined, minimal culture medium containing dextran T70 (BMOC; 5). Two pools of PFFl, collected separately, were tested. The fluid was filtered through a PM-10 membrane (Amicon) to obtain the low molecular weight fraction (lower than 10,000). The filtrates were lyophilized and re-dissolved in water at a 50-fold concentrate of the native PFFl as described (2). Part of the PM-10 filtrates were used for further purification by Sephadex G25 column chromatography and the fractions corresponding to Peak A (4) used in the present experiments. The fractions were diluted in culture medium and sterilized by Milipore filtration immediately before the experiment. The cultures were incubated at 37° C under 5% $CO_2$ in air for 42-44 h.

At the end of culture the cumulus cells were examined by an inverted microscope to determine their growth pattern. The oocytes were then fixed in ethanol-acetic acid (3:1) and stained with 2% aceto-orcein (1). Classification of oocytes was made according to Tsafriri and Channing (1) into one of three categories: (i) oocytes with an intact nuclear membrane were considered immature; (ii) oocytes which had matured beyond the germinal vesicle stage were classified as mature; (iii) oocytes with shrinkage or fragmentation of the ooplasm were classified as degenerated. In the case of the cumulus-enclosed oocytes this fraction was consistently below 2% and is therefore not shown in the figures.

Medium levels of progesterone were analyzed by radioimmunoassay (RIA) using a highly specific antiserum generated against the 11-OH-progesterone BSA conjugate (kindly donated by Dr. Gordon Niswender). Progesterone was assayed in diluted, unextracted aliquots of the media as described (7). The sensitivity was 0.1 ng progesterone per culture. Medium blanks were consistently lower than this value.

The number of cells in an average-sized oocyte-cumulus complex was determined in representative samples taken at the start of the culture as well as after two days of culture. Fifteen complexes were incubated in citric acid and the released nuclei stained by crystal violet and counted in a hemocytometer. The number of cells per cumulus was approximately 2100, with an interexperimental variation of 23% (C.V.). No significant difference in cell number was seen after 2 days of culture.

Statistical analysis was performed using Duncan's Multi-range test or, when two groups were compared, Student's t-test.

## RESULTS

The maturation of cumulus-enclosed oocytes was comparable in 199 A and BMOC ( Fig 1 ). Of the maturing oocytes 50-75% had extruded the first polar body (not shown).

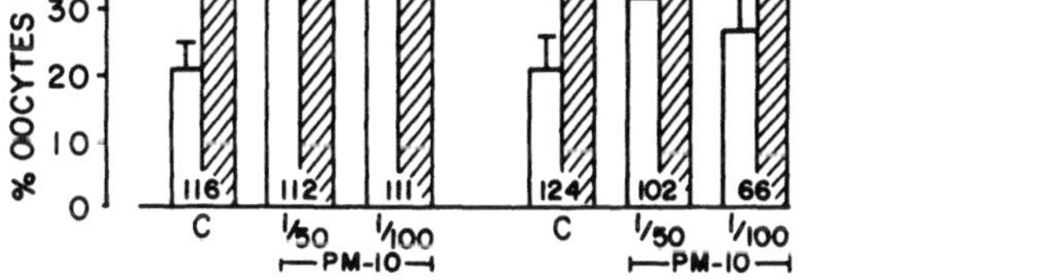


Figure 1. Cumulus-enclosed oocytes were cultured in the absence or presence of the PM-10 filtrate of PFFl. Significant inhibition of maturation was found only when 199 A was used ($p<0.01$).

When cultured in 199 A the cumulus cells differentiated into fibroblast-like cells containing sparse granules, attached to the surface and secreted significant amounts of progesterone ( Fig 2 ). When BMOC was used the cumuli did not attach but remained loosely arranged around the oocytes and nondetectable amounts of progesterone were secreted ( Table 1 ).

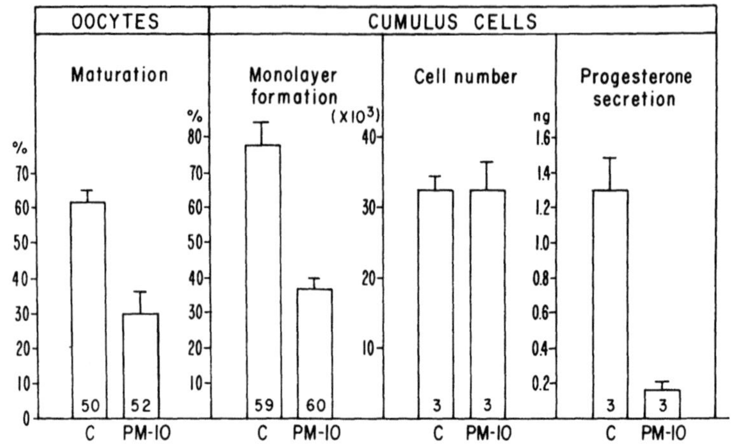


Figure 2. Cumulus-enclosed oocytes were cultured in 199 A in the absence and presence of the PM-10 filtrate of PFF1 in 1/50 dilution.

Mean + SE. The numbers on the bars of the two left panels indicate the total number of oocytes examined for maturation or cumuli examined for monolayer formation. Cell number and progesterone secretion expressed per culture (n=3) each consisting of 15 oocyte-cumulus complexes. The effect of the inhibitor was significant (p<0.01) except on the cell number.

TABLE 1. Effect of PM-10 filtrate of PFF1 on progesterone secretion by cumulus cells (ng/culture/day)

| | 199 A | BMOC |
|---|---|---|
| Control | 1.93 ± 0.23 (8) | <0.1 (10) |
| PM-10 1/50 | 0.12 ± 0.02 (7)** | <0.1 (10) |
| PM-10 1/100 | 0.27 ± 0.05 (6)** | ------- |

Mean ± SE. (n) indicates number of cultures, each including 15 cumuli (about 30,000 cells). ** p<0.01 vs control.

The maturation of denuded oocytes was lower than that of cumulus-enclosed in both types of media and a high proportion degenerated, particularly in 199 A (Fig 3).

The low molecular weight fraction of PFF1 (PM-10 filtrate) inhibited oocyte maturation and cumulus differentiation when 199 A was used as the culture medium (Figs 1 and 2, Table I). The partially purified OMI (Sephadex peak A fraction) also inhibited oocyte maturation and cumulus differentiation in terms of progesterone secretion. (Table II).

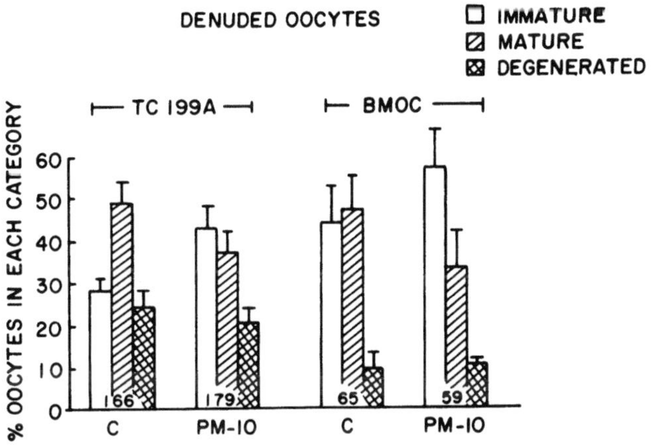


Figure 3. Denuded oocytes were cultured in the absence or presence of the PM-10 filtrate of PFF1 in 1/50 dilution. There was no significant effect on maturation by the inhibitor.

Table II. Effect of Sephadex peak A fraction of PFF1 on oocyte maturation and cumulus cell progesterone secretion

| | Mature oocytes % | Cumulus progesterone ng/culture/day |
|---|---|---|
| Control | 52 ± 3 (148) | 2.14 ± 0.23 (11) |
| peak A 1/20 | 21 ± 2 (124)** | <0.10 (11)** |
| peak A 1/80 | 30 ± 4 (113)** | 0.62 ± 0.17 (5)** |
| peak A 1/320 | 53 ± 6 (61) | 1.74 ± 0.34 (5) |

Cumulus-enclosed oocytes were cultured in 199 A in the absence and presence of various dilutions of Sephadex peak A fraction. Mean ± SE. (n) indicates total number of oocytes examined, or total number of cultures analyzed for progesterone (each culture normalized per 15 oocyte-cumulus complexes)** $p<0.01$ vs control.

There was no effect of the low molecular weight fraction on maturation of cumulus-enclosed oocytes cultured in BMOC. (Fig 1). Maturation of denuded oocytes was not inhibited by the PM-10 filtrate in either culture media (Fig 3).

## DISCUSSION

The two types of culture media used in the present study, the complex 199A, and the minimal, BMOC, differ widely in chemical composition (1, 5). Nevertheless, both media supported oocyte maturation equally well. Only the complex medium, however, supported cumulus cell function in terms of morphological differentiation and progesterone secretion. It is well established that defined minimal culture media can support oocyte maturation *in vitro* in several species (8). On the other hand it is known that more enriched media, with or without serum, are required for prolonged maintenance of granulosa cells in culture (9).

The low molecular weight fraction of PFF1 inhibited oocyte maturation only when cumulus-enclosed oocytes were cultured in 199A. Under these conditions the morphological differentiation of the cumulus cells into fibroblast-like cells and the progesterone secretion were inhibited by PFF1. Also the more purified Sephadex peak A fraction had this effect. Additional studies, reported elsewhere (7), have shown that when various fractions of PFF1 are tested (native PFF1 as well as partially purified Sephadex fractions) there is a close correlation between the meiosis-inhibiting activity and the progesterone inhibiting activity. Generally, the progesterone secretion has been affected by lower doses of the fractions than that required to inhibit meiosis. The PFF1 inhibition of progesterone secretion will therefore be followed as an additional test in further purification steps aimed at isolating OMI. A previous study has shown that the PFF1 inhibition of meiosis is reversible (4). Similarly, the inhibition of cumulus progesterone secretion is reversible (7). The finding that denuded oocytes, in contrast to cumulus-enclosed, did not respond to the inhibitor has also been found using mouse oocytes (Tsafriri, personal communication).

Progesterone or related steroids have been implicated in the resumption of meiosis in amphibia (10). There are, however, few indications that progesterone serves as a trigger of meiosis in mammals. In contrast dissociation between meiosis and steroidogenesis has been reported for cultured preovulatory rat follicles (11). In the pig oocyte such a dissociation also has been found: aminoglutethimide (0.01-1 mM) inhibited cumulus cell progesterone secretion without impairing spontaneous oocyte maturation (7). In the present study the rate of oocyte maturation was high in the minimal culture medium although progesterone levels remained non-detectable. We therefore believe that the inhibitory effect by PFF1 on cumulus progesterone may merely be an indicator of the functional state of the cells rather than being responsible for the arrest of meiosis.

In summary, PFF1 appears to inhibit oocyte maturation only when the oocyte is surrounded by cumulus cells capable of secreting progesterone. In parallel to inhibiting meiosis PFF1 inhibits cumulus progesterone synthesis. Further studies are required to elucidate the physiological significance of these findings. It is best concluded from these studies that the inhibitory action of OMI upon porcine oocyte maturation is mediated by the surrounding cumulus cells. It is suggested that OMI inhibition of cumulus cell progesterone secretion, cumulus cell differentiation (partial luteinization) and oocyte maturation all reflect actions aimed at keeping the oocyte-cumulus cell mass in an immature, "arrested" state.

## SUMMARY

To study the mode of action of porcine follicular fluid oocyte maturation inhibitor isolated cumulus-enclosed or mechanically denuded pig oocytes were used. Two types of culture media were employed, a complex containing 15% pig serum (TC199A) and a defined, minimal medium (BMOC). The maturation of cumulus-enclosed oocytes was comparable in the two types of culture media, but only in the complex medium did the cumulus cells remain functional in terms of morphology and progesterone secretion. The low molecular weight portion of follicular fluid partially inhibited oocyte meiosis and cumulus progesterone secretion in 199A. No inhibition of oocyte maturation was seen when follicular fluid was added to BMOC. Since denuded oocytes did not respond to follicular fluid in either culture medium, it is suggested that the cumulus cells may mediate the action of the follicular fluid oocyte maturation inhibitor.

## Acknowledgements

This work was supported by grants from the Ford Foundation (760-0530 to Cornelia P. Channing) and the Swedish Medical Research Council (B78-14F-5158-01, B78-14R-5158 to T. Hillensjö). We are thankful to Dr. R. McGaughey for supplying the chemically defined medium and to Dr. A. Tsafriri for valuable suggestions. For technical assistance we thank Ms. Sandra Fowler.

## REFERENCES

(1) A. Tsafriri and C. P. Channing Endocrinology 96 (1975) 922.

(2) A. Tsafriri, S. H. Pomerantz and C. P. Channing Biol. Reprod. 14 (1976) 511.

(3) A. Tsafriri, S. H. Pomerantz and C. P. Channing In: Ovulation in the human, (Academic Press, New York, 1976) 31.

(4) S. L. Stone, S. H. Pomerantz, A. S. Kripner and C. P. Channing Biol. Reprod. 1978 (in press).

(5) R. W. McGaughey Exp. Cell Res. 109 (1977) 25.

(6) T. Hillensjö, L. Hamberger and K. Ahren Acta Endocr. (Kbh) 78 (1975) 751.

(7) T. Hillensjö, C. P. Channing,,S. H. Pomerantz, and A. S. Kripner. In Vitro (submitted)

(8) R. G. L. Gwatkin In Vitro 8 (1972) 59.

(9) C. P. Channing and F. Ledwitz-Rigby In: Methods in Enzymology, Vol. 39, eds. J. G. Hardman and B. W. O'Malley (Academic Press, New York, 1975) 183.

(10) A. W. Schuetz Biol. Reprod 10 (1974) 150.

(11) M. E. Lieberman, A. Tsafriri, S. Bauminger, W. P. Collins, K. Ahren and H. R. Lindner Acta Endocr. (Kbh) 83 (1976) 151.

DISCUSSION - CHAIRPERSON - A.W. SCHUETZ

B.W. RIGBY: Dr. Hillensjö, I was interested in your PM-10 effect on progesterone secretion in the cumulus. We've been trying to purify the luteinization inhibitor and in our experiments, a low molecular weight fraction of serum is equally inhibitory as follicular fluid--what was your dummy run through the column? Was it serum or just buffer?

T. HILLENSJÖ: The dummy peak was prepared from a region of the column that did not have any oocyte maturation inhibitory activity.

B.W. RIGBY: Have you looked at a comparison between the less than 10,000 molecular weight fractions of serum and follicular fluid on progesterone secretion?

T. HILLENSJÖ: We haven't done very much on pig serum regarding progesterone secretion. Human follicular fluid has similar inhibitory effects on oocyte maturation and on progesterone secretion. The low molecular weight fraction of human serum, however had no inhibitory effect on oocyte maturation or on progesterone secretion. In the case of the inhibitor in pig follicular fluid, previous work by Tsafriri and Channing has clearly shown that the serum, whether you test native serum or the low molecular weight fraction, does not have any inhibitory effect on oocyte maturation. Thus far however, we haven't really looked into the effects of the low molecular fraction of pig serum on progesterone secretion, but it might be a good idea.

B.W. RIGBY: We have seen an inhibitory effect of the low molecular weight fraction of pig serum upon progesterone secretion by cultured porcine granulosa cells harvested from large porcine follicles.

Relationship Between Steroidogenesis and Oocyte Maturation

in Rat Graafian Follicles Cultured in vitro

C. Readhead*, M. H. Kaufman+, A. W. Schuetz°, and G. E. Abraham++
*Physiological Laboratory, University of Cambridge Cambridge, England
+Department of Anatomy, University of Cambridge Cambridge, England
°The Johns Hopkins University, Baltimore, Maryland
++Department of Obstetrics and Gynaecology, UCLA Los Angeles, California

ABSTRACT

Rat Graafian follicles were dissected from ovaries during the morning of proestrus and incubated invarious media. After an 8 h incubation period, ovarian follicles were collected and analyzed, by radioimmunoassay, for progesterone, 17α-hydroxyprogesterone, androstenedione, estradiol-17β, and dihydrotestosterone. Following incubation in proestrus (a.m.) serum, preovulatory follicles secreted all these steroids and the output of progesterone, 17α-hydroxyprogesterone and androstenedione was markedly stimulated following the addition of LH to the media.

Follicle-enclosed oocytes resumed meiosis when incubated in proestrus serum without added LH. However, germinal vesicle breakdown (GVBD) under these conditions was considerably slower than that observed following addition of LH to the medium. "Spontaneous" maturation of oocytes in proestrus serum was associated with increased production of steroids. Oocyte maturation and steroidogenesis both occurred following incubation of ovarian follicles in serum obtained from hypophysectomized rats. Addition of LH to the media markedly accelerated these responses. These results suggest that oocyte maturation and steroidogenesis also occur in cultured ovarian follicles in the absence of gonadotrophic hormones. The relationship between steroidogenesis and/or spontaneous oocyte maturation are discussed.

## INTRODUCTION

During the afternoon of proestrus in the rat, high levels of LH are found in the blood (1) and a few hours later several oocytes resume meiosis (2). 8 h after the LH surge, the follicles rupture liberating the oocyte. In order to study the interaction between LH, the follicular cells and the oocyte, several workers have isolated Graafian follicles prior to the LH surge and measured their response to the LH *in vitro* (3, 4, 5, 6).

Previous work has shown that follicle-enclosed oocytes will only mature *in vitro* in the presence of LH (3, 4) but that oocytes released from follicles will mature "spontaneously" (7, 8). LH also has a stimulatory effect on follicular steroidogenesis (5, 6, 9, 10) probably through its action on cyclic AMP (11).

The experiments described in this paper were designed to examine the pattern of steroids produced by rat Graafian follicles *in vitro* and to determine the effect of LH on steroidogenesis and oocyte maturation. The follicles were cultured in proestrus (a.m.) serum in order to mimic as closely as possible their environment prior to the LH surge. In some instances hypophysectomized rat serum was used as a culture medium to eliminate the effect of endogenous gonadotrophins on steroidogenesis and oocyte maturation.

## MATERIALS AND METHODS

### Culture of Graafian Follicles

Large rat Graafian follicles (>0.8 mm) were isolated during the morning of proestrus and cultured as described previously (12). Briefly, after dissection, the follicles were pooled and 4-6 were transferred to a glass roller bottle containing 1 ml of either proestrus (a.m.) rat serum or hypophysectomized rat serum. The incubation medium was equilibrated with 66% $O_2$, 34% $N_2$. The stoppered bottles were rolled continuously on cylindrical rollers at 37°C. After an 8 h incubation, the medium was aspirated and frozen at -20°C for steroid assay at a later date, and the follicles were transferred to a petri dish for oocyte examination. Oocytes were classified into three morphologically distinct groups depending on whether the germinal vesicle was present (GV+) or had broken down (GVBD) or that the germinal vesicle had broken down and a polar body was present (GVBD+PB). Oocytes which had undergone germinal vesicle breakdown were termed mature.

### Assay of Steroids

Steroids in the incubation medium were extracted with diethyl ether and fractionated on celite columns (13) and the concentrations of progesterone (14), 17α-hydroxyprogesterone (15), estradiol-17β (16), androstenedione (17) and dihydrotestosterone (18) were measured by radioimmunoassay.

## RESULTS

### Follicles Incubated in Proestrus (a.m.) Rat Serum

After an 8 h incubation in proestrus (a.m.) serum, 65% of the follicle-enclosed oocytes had matured in the absence of added LH (Table I). The endogenous levels of LH in proestrus (a.m.) serum of these rats is 47.6 ± 15.0 ng/ml (Readhead and Abraham, unpublished observations). In cultures with added LH (10 μg oLH), 93% of the oocytes matured, half of the matured oocytes had polar bodies.

At the end of the incubation period, levels of progesterone, 17α-hydroxyprogesterone, estradiol-17β, androstenedione and dihydrotestosterone in the medium had increased (Fig. 1, Schuetz et al., unpublished). The addition of LH to the medium stimulated the follicular output of progesterone, 17α-hydroxyprogesterone and androstenedione, however, it did not stimulate the output of estradiol-17β or dihydrotestosterone.

### Follicles Incubated in Hypophysectomized Rat Serum

Follicles incubated in hypophysectomized rat serum show a similar pattern of steroid output as those grown in proestrus serum. The levels of progesterone, 17α-hydroxyprogesterone and androstenedione in the medium increase after the addition of LH (Fig. 2, Schuetz et al., unpublished). The response of estradiol-17β to LH differs in the different media. Follicles cultured in hypophysectomized serum increase their output of estradiol-17β in response to LH whereas those cultured in proestrus serum do not.

After an 8 h incubation, 50% of the follicle-enclosed oocytes mature in response to LH (Table I).

## DISCUSSION

Previous work has shown that follicle-enclosed rat oocytes mature in vitro after LH is added to the medium (3). The data presented here confirm this finding and in addition demonstrate for the first time that follicle-enclosed oocytes also mature "spontaneously" in the absence of gonadotrophins. However, germinal vesicle breakdown (GVBD) under these conditions is rather slow (8 h) compared to GVBD in response to LH (2.5 h) (12).

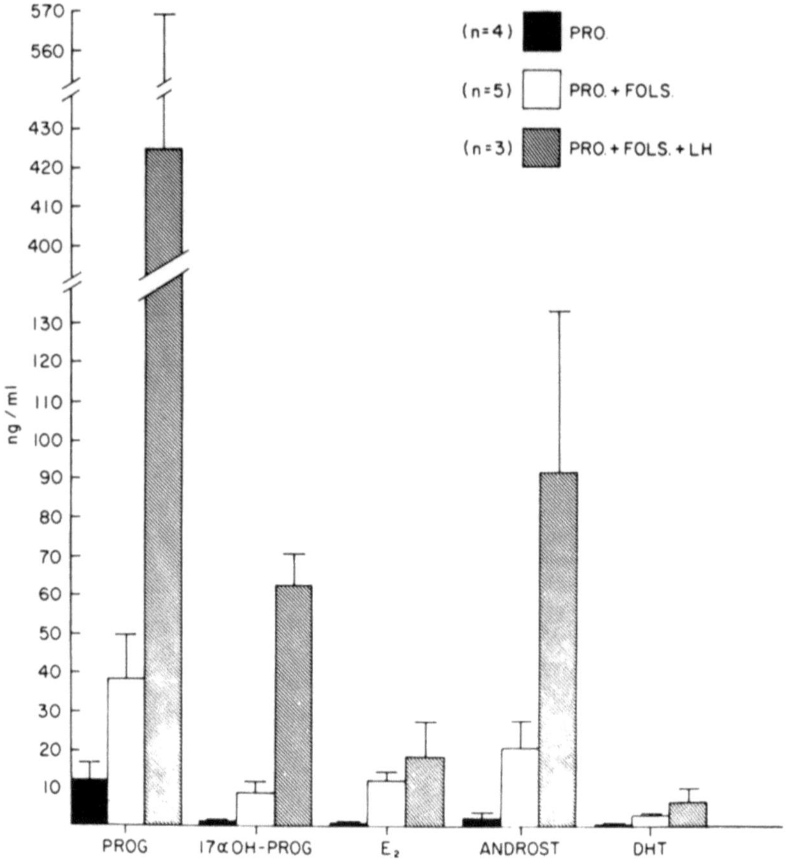


Figure 1. Output of steroids (ng/ml medium) by rat Graafian follicles after an 8 h incubation in proestrus (a.m.) serum, with and without added LH (10 μg/ml).

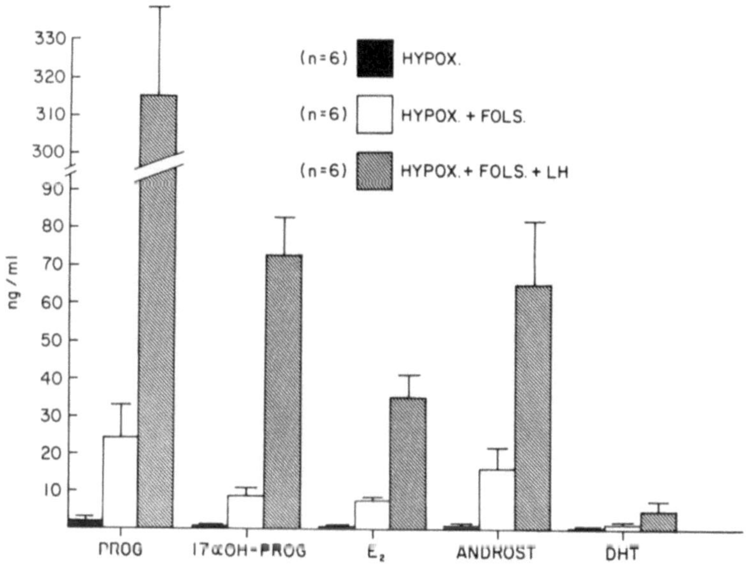


Figure 2. Output of steroids (ng/ml medium) by rat Graafian follicles after an 8 h incubation in hypophysectomized rat serum, with and without added LH (10 µg/ml).

TABLE I

OOCYTE MATURATION IN FOLLICLES INCUBATED

IN PROESTRUS (a.m.) OR HYPOPHYSECTOMIZED RAT SERUM

| Incubation Medium | Total No. Oocytes | GV+ | GVBD | GVBD + PB | Degenerate or Lost Oocytes | % Maturation |
|---|---|---|---|---|---|---|
| Proestrus Serum | 26 | 8 | 13 | 4 | 1 | 65 |
| Proestrus Serum + LH (10 μg/ml) | 15 | 0 | 7 | 7 | 1 | 93 |
| Hypophysectomized Serum | 34 | 14 | 17 | 0 | 3 | 50 |
| Hypophysectomized Serum + LH (10 μg/ml) | 35 | 0 | 19 | 15 | 1 | 97 |

(Schuetz, Kaufman, Abraham and Readhead, unpublished).

GV+ germinal vesicle present

GVBD germinal vesicle breakdown

GVBD + PB germinal vesicle breakdown and the presence of the 1st polar body

"Spontaneous" oocyte maturation also occurs in oocytes liberated from their follicles (7, 8). This is believed to be due to the oocytes escaping from an inhibitor of maturation present in the follicle. A polypeptide which will inhibit "spontaneous" oocyte maturation has been found in porcine follicular fluid (19) and in hamster and bovine follicular fluid (20).

The mechanism for the resumption of meiosis by oocytes is not fully understood. In amphibians steroids trigger oocyte maturation (21) and progesterone-induced maturation is under translational control since cyclohexamide inhibits maturation (22, 23, 24). In mammalian oocytes, no direct association between steroidogenesis and oocyte maturation has been established, and various conflicting pieces of evidence regarding the subject have been described. Lieberman et al. (25) showed that inhibitors of steroidogenesis, such as cyanoketone or aminoglutethimide do not block LH-induced maturation in follicle-enclosed rat oocytes _in vitro_. In contrast to this, pig oocyte maturation _in vivo_ is associated with elevated levels of estrogens in the follicular fluid (26). Rat Graafian follicles have the capacity to produce progesterone, 17α-hydroxyprogesterone, androstenedione, estradiol-17β and dihydrotestosterone in the absence of gonadotrophins, but LH greatly stimulates the production of progesterone, 17α-hydroxyprogesterone and androstenedione. The data presented here shows that oocyte maturation whether "spontaneous" or LH-induced is always associated with increased levels of steroids in the medium. Thus, it is tempting to suggest that steroids play a role in initiating the resumption of meiosis in rat oocytes.

## REFERENCES

(1) T. A. Daane and A. F. Parlow. Endocrinology 88 (1971) 653.

(2) A. Tsafriri and P. F. Kraicer. J. Reprod. Fert. 29 (1972) 387.

(3) A. Tsafriri, H. R. Lindner, U. Zor, and S. A. Lamprecht. J. Reprod. Fert. 31 (1972) 39.

(4) R. M. Moor. J. Reprod. Fert. 32 (1973) 545.

(5) W. J. LeMaire and J. M. Marsh. J. Reprod. Fert. Suppl. 22 (1975) 53.

(6) E. V. YoungLai. J. Endocr. 71 (1976) 167.

(7) G. Pincus and E. V. Enzmann. J. Exp. Med. 62 (1935) 665.

(8) R. G. Edwards. Nature 208 (1965) 349.

(9) H. R. Lindner, A. Tsafriri, M. E. Lieberman, U. Zor, Y. Koch, S. Bauminger, and A. Barnea. Rec. Prog. Horm. Res. 30 (1974) 79.

(10) R. F. Seamark, R. M. Moor, and J. E. A. McIntosh. J. Reprod. Fert. 41 (1974) 143.

(11) J. M. Marsh. Biol. Reprod. 14 (1976) 30.

(12) C. Readhead, M. H. Kaufman, and A. W. Schuetz. Biol. Reprod. (submitted).

(13) G. E. Abraham. Path. Biol. 23 (1975) 889.

(14) G. E. Abraham, R. S. Swerdloff, D. Tulchinsky, and W. D. Odell. J. Clin. Endoc. Metab. 32 (1971) 619.

(15) G. E. Abraham, R. S. Swerdloff, D. Tulchinsky, K. Hopper, and W. D. Odell. J. Clin. Endoc. Metab. 33 (1971) 42.

(16) G. E. Abraham, K. Hopper, D. Tulchinsky, R. S. Swerdloff, and W. D. Odell. Analyt. Letters 4 (1971) 325.

(17) G. E. Abraham and Z. H. Chakmakjian. J. Clin. Endoc. Metab. 37 (1973) 581.

(18) G. E. Abraham, F. S. Manlimos, M. Solis, and A. C. Wickman. Clin. Biochem. 8 (1975) 374.

(19) A. Tsafriri and C. P. Channing. Endoc. 96 (1975) 922.

(20) R. B. L. Gwatkin and O. F. Andersen. Life Sciences 19 (1976) 527.

(21) A. W. Schuetz. Biol. Reprod. 10 (1974) 150.

(22) A. W. Schuetz. J. Exp. Zool. 166 (1967) 347.

(23) W. Wasserman and Y. Masui. Biol. Reprod. 11 (1974) 133.

(24) K. Drury and S. Schoderet-Slatkine. Cell 4 (1975) 269.

(25) M. E. Liebermàn, A. Tsafriri, S. Bauminger, W. P. Collins, K. Ahren, and H. R. Lindner. Acta Endocr. 83 (1976) 151.

(26) R. H. F. Hunter, B. Cook, and T. G. Baker. Nature 260 (1976) 156.

DISCUSSION - CHAIRPERSON - A.W. SCHUETZ

**H. LINDNER:** I am a little puzzled by these results because, as you may be aware, we do observe the LH effect on oocyte maturation in vitro in the presence of inhibitors of steroidogenesis, such as aminoglutethimide or cyanoketone, which obliterate the stimulatory effect of LH on the synthesis of hormonal steroids and even suppress follicular steroid accumulation (progesterone, androgen and estrogen) to a level well below that seen in controls cultured without LH [Lieberman, Tsafriri, Bauminger, Collins, Ahrén and Lindner, Acta Endocrinologica, 83, 151 (1976)]. Under these conditions, oocyte maturation does take place in response to LH, whereas in follicles cultured in medium devoid of LH the oocytes persist in the dictyate state throughout 24 hours of culture [Lindner et al., Recent Progress in Hormone Res., 30, 79 (1974)] despite continued steroidogenesis at the basal rate. I wonder why the oocytes behave differently under your culture conditions. Could it be that the submerged culture in rotating tubes which you used causes detachment of the cumulus mass from the oocyte, - perhaps by shear or in some way, - so that the oocyte in effect becomes sequestrated within the follicle? We observed that the delicate cytoplasmic projections of the corona cells across the zona pellucida form gap junctions with the oolemma [Amsterdam, Josephs, Lieberman and Lindner, J. Cell Sci. 21, 93 (1976)] and proposed that these specialized junctions may be important in maintaining meiotic arrest [Lindner et al., J. Reprod. Fert. 51, 215 (1977)]. Perhaps under your culture conditions something happens to these gap junctions so that they become detached or functionally incompetant, thereby releasing the oocyte from an inhibitory action exerted by the follicle cells and thus triggering meiosis?

C. READHEAD: Could I answer the first part about steroidogenesis. I am aware of your work on this and if I remember the data correctly, progesterone was depressed, but there was still quite substantial amounts of androstenedione present, in the medium. The medium steroid concentrations only partially reflect what really is inside the follicles. What is outside is usually much lower than what presumably is inside.

H. LINDNER: Well, these agents do inhibit entirely the steroidogenic effect of LH, so I don't think we can explain the effect of LH by stimulation of either progesterone or androstenedione production by the follicle. We have measured steroid accumulation in both the medium and the follicular tissue, and also demonstrated suppression of 3β-hydroxysteroid dehydrogenase throughout the follicle by cyanoketone. Nevertheless, you get the LH effect on ovum maturation which is not seen in LH-free controls.

A. SCHUETZ: We would now like to open this session up for general discussion of the topic of oocyte maturation. Two people have indicated they want to show one or two slides. In view of

Dr. Lindner's comments it would seem appropriate if we first ask if anyone has information relating to the question of whether steroidogenic or nonsteroidogenic mechanisms regulate meiosis in the mammal?

T. HILLENSJÖ: I might just add that in collaboration with Dr. Dekel and Dr. Ahrén, similar observations as those Dr. Lindner described have been made in Göteborg independently showing that inhibitors of steroidogenesis do not affect LH-induced resumption of maturation in follicle-enclosed rat oocytes.

A. SCHUETZ: Although we say that we inhibit steroidogenesis with various inhibitors, generally only one or two steroids are measured. I realize that the evidence toward the action of some of these compounds is fairly convincing--however, in many cases, it has not really been established that all rather than selected aspects of steroidogenic pathways have been inhibited.

H. LINDNER: Can you accelerate oocyte maturation in your system by adding exogenous steroids, such as progesterone, to make it comparable to the kinetics you get with LH? We were unable to induce or accelerate maturation of follicle-enclosed rat oocytes in organ cultures by adding various steroids, including progesterone, 20α-dihydroprogesterone and oestradiol to the medium over a wide range of concentrations, whereas in the amphibian oocyte you clearly can. There does seem to be a fundamental difference.

A. TSAFRIRI: I want to remind you that aminoglutethimide inhibits 20α-hydroxylase and/or 20:22 lyase and hence blocks steroidogenesis at an early step, so that it seems to me very unlikely that use of aminoglutethimide caused accumulation of some unspecified steroid which induces oocyte maturation. May I ask Dr. Readhead what was the volume of the medium used in your cultures?

C. READHEAD: The volume was 1 ml.

A. TSAFRIRI: Is it possible that in your rotating culture system the inhibitor of meiosis was washed out from the follicles and probably that is the reason why most of the oocytes escaped meiotic arrest?

C. READHEAD: We have some recent data that I haven't shown here which indicates that steroid concentrations in the follicular fluid are tremendously high. If, as you suggest, there's leakage out of the follicles, I would assume you wouldn't get such a huge accumulation of steroids in the follicular fluid.

**A. TSAFRIRI:** I was talking about the inhibitor of meiosis within follicular fluid, and you cannot compare leakage rates of different molecules. Each substance may have a completely different rate of passage through follicular wall.

**K. McNATTY:** I would just like to make a point from our data. If the oocyte is not exposed to steroid in vivo, it doesn't seem to have the ability to resume meiosis in vitro, however, the steroids themselves, when added in vitro don't seem to block meiosis at all!

N. DEKEL: To study the paradoxical situation which seems to be present in the preovulatory maturing follicle in which the oocytes do mature in presence of high levels of cyclic-AMP, we tried to analyze the ability of gonadotropins to overcome the inhibition caused in vitro by cyclic-AMP, and induce oocyte maturation. We found that LH and FSH could induce meiosis in oocytes that were maintained in vitro, in the immature stage.

Rat oocytes enclosed in their cumuli were cultured in medium containing 0.2 mM of MIX, and maintained in the immature stage. 95% of them contained the germinal vesicle structure after 4 hours of incubation. The oocytes were divided into three groups, one group was continuously incubated in the presence of MIX, in the other group, LH was added to the medium, in presence of MIX; the third group of oocytes was transferred to an inhibitor-free medium. We found that oocyte maturation was induced either by removal of MIX or by addition of LH in the presence of MIX. There was a difference in timing of germinal vesicle breakdown between the two groups. The time curve representing maturation in oocytes that were released from the inhibitory medium is parallel to the time curve that was obtained when oocytes were released from the inhibitory environment of the follicular fluid, while oocyte maturation induced by LH appeared to be slower.

Our proposed model on the LH effect on oocyte maturation in this system is based on the possibility that the cumulus cells act as mediators in transferring an inhibitor to the oocyte. This inhibitor is cyclic-AMP in the in vitro system, or any other substance which is present in the follicular fluid. It has been shown that gap-junctions between the cumulus cells and the oocytes in the preovulatory stages are very common. Administration of hCG causes breakdown of the gap-junctions and interruption of the communication between the cumulus cells and the oocytes (Gilula _et al._, 1978, _J. Cell. Biol._ _78_: 58). We think that the interruption of communication between the cumulus and the oocytes might have a crucial role in permitting maturation of the oocyte in the preovulatory follicular environment, in which cyclic-AMP levels are high.

K.J. CATT: One general point that should be made about the use of MIX and similar compounds to increase cyclic AMP levels in intact cell systems is that these agents have several other actions than inhibition of phosphodiesterase. They block steroidogenesis and also protein synthesis when present in high concentration, so the levels employed to inhibit phosphodiesterase should be selected to avoid these additional effects.

DR. SWANSON: Just a comment on the steroids. We have injected some antiserum to BSA-conjugated progesterone, and have found no effect on the maturation of the ovum, but have blocked ovulation with it. This antiserum was raised in a sheep against BSA-conjugated progesterone. In relation to some of the data that was presented here, that might be of interest.

We also have shown that the ovum will stop at Metaphase-II, as some of those have presented here this morning.

R. SHALGI: I would like to add to the presentations regarding oocyte maturation induced by LH. In the work done with Dr. Dekel, we were trying to evaluate maturation of oocytes by testing the fertilizability. Rat oocytes were cultured with cumulus cells and were allowed to mature in vitro spontaneously, in the presence of LH. Mature oocytes from both groups were transferred to oviducts of mated females where they were tested for fertilization. We found that there is a three-fold increase in sperm penetration and fertilization within oocytes that have matured in vitro in the presence of LH compared to those matured without LH in the culture medium.

G. ERICKSON: Dr. Wasserman, you presented data suggesting that the ability of the oocyte to undergo meiotic maturation occurs after the oocyte has completed a significant portion of its ultimate growth. As you know we have published data which is consistent with your observations. However, I have been puzzled by the observations reported by Professor Ono, (Cytogenetics 3: 324, 1964), that some embryonic oocytes which do not become associated with presumptive granulosa cells during primordial follicle formation proceed beyond diplotene and enter diakinesis of an abortive Metaphase I. Could this suggest that the capacity of a mammalian oocyte to continue or resume the meiotic sequence indeed occurs very early during the initiation of meiosis in the embryo and that the meiotic sequence is inhibited at this point only if the oocyte becomes incorporated into a primordial follicle?

P. WASSERMAN: I was not aware of those very interesting results. They do suggest that there perhaps is some sort of inhibition by the follicle cells. It is difficult to reconcile certain results with their observations, since as you know, if you take out small denuded oocytes from the ovary, they do not resume meiosis in vitro.

S. SEGAL: I have a question for Dr. Tsafriri. Is there any situation that you know of in which a dissociation between cumulus cells and the oocyte or a breaking of their gap-junctions does not immediately result in a resumption of the meiotic process?

A. TSAFRIRI: From ultrastructural studies in which the timing of the disruption of cumulus/oocyte gap-junctions was studied it appears that germinal vesicle breakdown precedes this change in cumulus/oocyte association (Szollosi et al., Ann. anim. Bioch. Biophys. 18: 511, 1978). But of course, one may say that physiological dissociation takes place before any morphological changes can be detected. I think that much more research is needed in order to understand the role of cumulus/oocyte association in meiotic maturation.

A. SCHUETZ: I would like to raise the question of whether in all these discussions this morning, we are in fact talking about different follicles or oocytes and different types of maturation (spontaneous maturation and/or induced maturation). It is my view that most oocytes obtained from mouse ovaries undergo spontaneous maturation in culture and do not resemble preovulatory oocytes in the sense that the cumulus cells are very loose and easily detached.

Am I wrong in that assumption?

A. TSAFRIRI: In this regard there is a variation among mammalian species. When follicular oocytes are collected in the mouse you will find many denuded oocytes, whereas in the rat and pig most of the oocytes dislodged from the follicles will be enclosed within the cumulus. Furthermore, if you culture these few rat or pig oocytes which were found denuded - most of them will degenerate. To answer your question: rat follicle-enclosed oocytes cultured in the presence of LH undergo maturational changes and cannot be distinguished morphologically from in vivo matured cumuli and oocytes.

A. SANYAL: Several years back, it was shown in the rabbit treated with hCG that the steroid levels in the follicular fluid fall before ovulation. So the conclusion was that the leaching of the steroid from the follicular fluid may be important before ovulation. If this process is related to oocyte maturation it seems very confusing now. If anybody has an answer to this question in the context of ovulation and the maturation of the oocyte.....

A. TSAFRIRI: Since germinal vesicle breakdown occurs earlier than the decrease in steroid levels you cannot assume that the latter causes germinal vesicle breakdown.

QUESTION: Do you know how many hours before ovulation GV breaks down in the rabbit?

A. TSAFRIRI: In the rabbit germinal vesicle breakdown occurs 3-4 hours post coitum or hCG treatment.

QUESTION: Thus, in rabbits you have to give hCG to induce ovulation, so following hCG, this leaching occurs. Obviously, in the rabbit there is no GV breakdown unless you give hCG. So it is quite possible that leaching of the steroid might influence the GV breakdown?

A. TSAFRIRI: But what I am trying to say is that germinal vesicle breakdown occurs before the steroid levels in the follicle are reduced; on the contrary in response to hCG there is first an increase in progesterone synthesis within the preovulatory follicle and it will be reduced only later.

S.K. BATTA: In the presentation by Dr. Tsafriri, he alluded to the role of calcium in LH-induced germinal vesicle breakdown. Dr. Knudsen and I have determined the concentration of calcium in oocytes obtained from the immature female rat after treatment with PMSG, and we found that after the LH surge occurs there is an increase in calcium levels in the oocytes which slightly precedes the total germinal vesicle breakdown. I would think that this observation would support what Dr. Tsafriri found.

CHAIRPERSON'S COMMENTS - In Manuscript

THE SOMATIC-GERM CELL COMPLEX: INTERACTIONS AND TRANSFORMATIONS IN RELATION TO OOCYTE MATURATION

Allen W. Schuetz

Dept. of Population Dynamics
Johns Hopkins School of Hygiene
615 N. Wolfe Street
Baltimore, Md. 21205

In recent years considerable interest has been focused on elucidating the mechanisms which control the multifaceted processes involved in reinitiation and mediation of oocyte differentiation from the stage of meiotic prophase I arrest. Although detailed information is incomplete, considerable controversy exists concerning the specific and unique nature of these mechanisms in mammalian as contrasted to non-mammalian species (1,2). In general these differences relate to the following: 1) whether oocyte maturation is under the positive or negative external control of specific molecules, 2) the chemical nature of these molecules, and 3) the nature and role of somatic-germ cell interactions as contrasted to the independent functions of these cells.

Extensive experimental analyses in a number of invertebrates and non-mammalian vertebrates indicate that somatic-germ cell interactions within the ovary play a crucial role in controlling meiosis reinitiation (1,3,4,5). Results of these studies suggest that factors (hormones) produced by somatic cells present in the microenvironment act directly on the oocyte to reinitiate the meiotic process. These factors cause complex internal biochemical mechanisms to be activated or initiated and differentiation to proceed. Thus oocyte maturation may be considered to be under the direct, positive control of these regulatory molecules (6,7, 8,9). Important features of such a positive control mechanism are illustrated in figure 1 (type I). In this case an increasing level of the regulating hormone causes meiosis to occur in an increasing number of adequately developed oocytes.

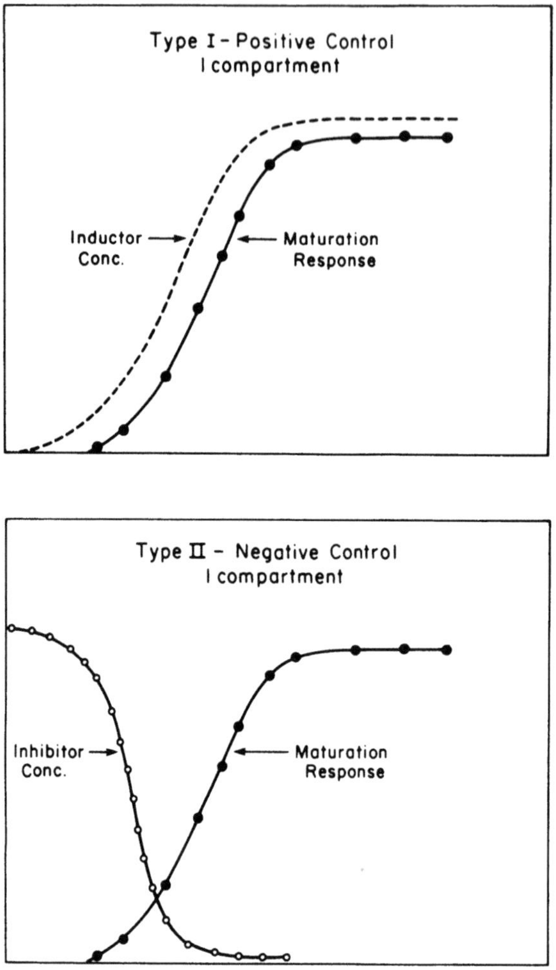


Figure 1. Diagrammatic illustration of postulated relationship between a single regulatory factor (inductor (type I) or inhibitor (type II) ) concentration and reinitiation of meiotic maturation.

In contrast to this mechanism various data in mammals indicate that meiotic maturation is controlled by negative or inhibitory factors present within the ovarian follicle granulosa cells and follicular fluid (10, 11, 12). In this case the inhibitory factors are thought to prevent the oocyte from fulfilling its inherent capability of undergoing maturation. Therefore, decreasing levels of the inhibitory molecules are a necessary prerequisite for reinitiation of meiotic maturation in adequately differentiated oocytes. (Fig. 1 - Type II) Significantly, however, these

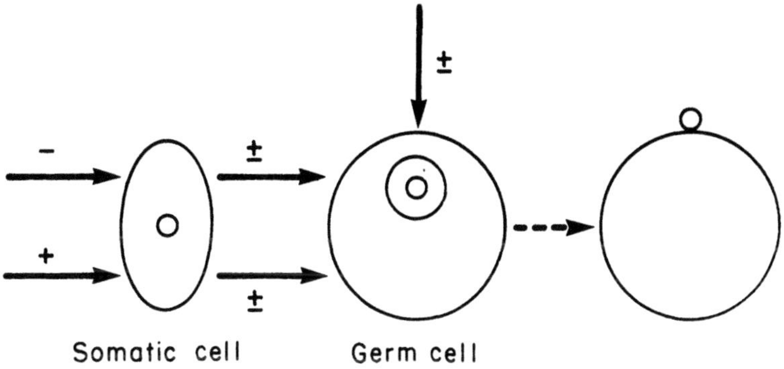


Figure 2. Diagrammatic illustration of various alternative paths by which a single postulated stimulatory (+, positive control) or inhibitory (-, negative control) factor may act to directly regulate meiosis reinitiation in individual germ cells (vertical arrow) or indirectly in cell preparators (horizontal) containing 2 interacting cell types (germinal and somatic).

conclusions are based on data obtained using isolated oocytes incubated in the presence of somatic cells. Thus the question arises as to whether these inhibitors act on the oocyte rather than the somatic cells (Figure 2). In such a two compartment system various alternative regulatory mechanisms are feasible. In particular if inhibitory substances act to suppress the synthesis and/or release of inductor molecules produced by the somatic cells, the subsequent effects of these inductor substances would be similar to those described in Figure 1 (type I) and therefore would be part of a positive control mechanism (Fig. 3).

Because of the fundamental importance of the somatic-germ cell interactions in defining the nature of meiotic regulation, a short discussion of this topic appears warranted. Considerable attention has been given to apparent differences which exist in the regulation of meiosis in mammalian as opposed to non-mammalian species. However, our comparative studies, as well as those of others, indicate a remarkable degree of basic similarity in the cellular and molecular events comprising maturation in echinoderm, amphibian, and mammalian species (1,4,5). During the course of growth and differentiation, oocytes of starfish, frogs and mammals remain encapsulated by and in contact with a small group of somatic cells. The nature of this association is so intimate that both of these cellular components are released from the ovary or ovarian follicle as a single unit, the somatic-germ cell complex. In mammals this complex is comprised of the oocyte and its surrounding cumulus cells. In all these species this complex undergoes

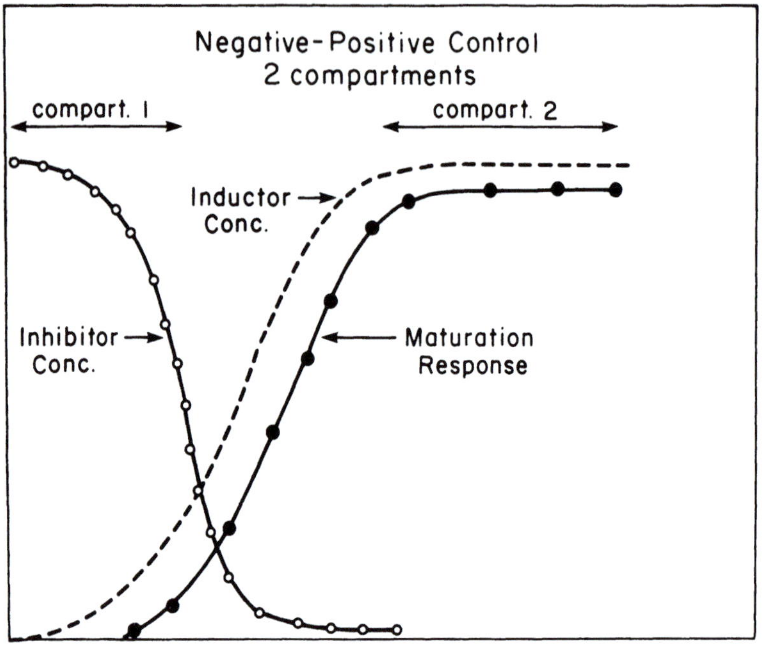


Figure 3. Diagrammatic illustration of postulated relationship between regulatory factors acting in a system containing 2 cell types. The effects of the inhibitor are shown to be on somatic cells (compartment 1) and their secretion of a hypothetical inductor of meiosis in oocytes (compartment 2). Compare to illustration in Figure 2.

similar morphological and physiological changes in association with the oocyte maturation process. Somatic cells do not partici pate in embryogenesis and are thus typically lost from the oocyte at select stages of their differentiation.

In the starfish ovary, the gonadotrophic hormone which initiates maturation also stimulates the detachment of follicle cells from the oocyte. Furthermore, detachment and separation of somatic cells from the oocyte are produced by the starfish meiosis inducing hormone, 1 methyladenine when it is added either to the ovary or to isolated oocyte-follicle cell complexes in vitro (5,14). Additional evidence indicates that these same starfish follicle cells are sites of gonadotrophic hormone action, major sources of 1-methyladenine, and involved in the mediation of both induced and spontaneous maturation (3,4,5).

Surrounding the immature ovarian oocyte of the frog is a single layer of follicle cells which is closely applied to the exterior surface of the extracellular vitelline membrane. Macrovilli project from the follicle cells, penetrate the vitelline membrane and come in direct and firm contact with the plasmalemma of the oocyte (1,15). Following manual microdissection of individual follicles one obtains an oocyte with the single continuous layer of attached follicle cells. These somatic cells remain in place for extended periods of time in culture. However, the follicle cell-oocyte linkage as well as the follicle cells and oocyte themselves undergo drastic alterations following exposure to the gonadotrophic or steroidal (progesterone) hormones which initiate meiotic maturation. Physical detachment and separation of these 2 cell types are associated with retraction of follicle cell macrovilli from the oocyte and the formation of loose clumps of cells on the vitelline membrane. Simultaneously oocyte microvilli retract from the inner portion of the vitelline membrane and the gamete surface becomes considerably flattened. Although the present evidence indicates that gonadotrophic hormone effects are mediated through steroid hormones it is unclear whether any or all of these changes are the result of direct or indirect actions of the steroids (1,5,16). Similar changes in the oocyte surface are also seen when mature oocyte cytoplasm containing maturation promoting factor (MPF) is injected into immature oocytes and meiosis is reinitiated. Such results suggest that the final signal mediating the disengagement of somatic and germ cells arises internal to the oocyte. Inhibitors of protein synthesis prevent these hormonally induced changes (6,15,17).

In mammals, as in the other species mentioned, the cumulus cells which surround and interdigitate with the oocyte through the zonapellucida undergo dramatic changes in morphology and function during the course of in vivo and in vitro maturation. (18, 19,20). Somatic and germ cell disengagement are associated with the hormonally induced meiotic maturation process and are correlated with alterations in the gap junctions which connect cumulus cells together as well as to the oocyte (20,21,22,23). Our studies also show that during the maturation process drastic changes occur in the sensitivity of the cumulus cells to dissociation by hyaluronidase. Cumulus cell complexes surrounding immature oocytes are totally insensitive to hyaluronidase dispersing activity whereas cumulus cells associated with ovulated oocytes are rapidly dissociated (24). Furthermore, onset of cumulus cells sensitivity to hyaluronidase occurred within the ovary following gonadotrophic hormonal stimulation and was closely associated with the breakdown of the germinal vesicle in these oocytes.

Thus, the evidence obtained in all these species including mammals, indicated that interruption of somatic cell contact with the oocyte is closely linked with ovulation or spawning and reinitiation of meiotic maturation. Data discussed further indicate that follicle cells; in echinoderm and amphibian species, secrete endocrine products which are directly involved in triggering meiosis. Although whole mammalian follicles have been shown to have extensive endocrine functions, the question of whether the somatic-oocyte complex is similarly active has been largely ignored. We have examined this question by incubating isolated rat cumulus masses in rat serum (25 ). Following an eight hour incubation period in roller tubes, the media were analyzed by radioimmunoassay for progesterone, estradiol and prostaglandin F (Table 1). Results show that large quantities of

Table I

Endocrine activity of rat cumulus masses cultured *in vitro*. Secretion of progesterone, estradiol and prostaglandin.

| Tube Number | No #. Cumulus Masses | ng/ml Prog. | ng/ml Estradiol | ng/ml PGF |
|---|---|---|---|---|
| 1 | 13 | 80.4 | | 1.75 |
| 2 | 10 | 96.2 | | 1.93 |
| 3 | 68 | $>$186.0 | .10 | 2.50 |
| 4 | 72 | $>$194.0 | .10 | 2.70 |
| 5 | 0 | 10.1 | .10 | .65 |

(Schuetz and Dubin, unpublished data).

progesterone, smaller increases of prostaglandin, and unchanged levels of estradiol were present in media incubated with cumulus masses. Although the results show that cumulus cells secrete hormones it remains to be established whether these are the only important molecules or whether other products or activities of the cumulus cell-oocyte complex play a significant role in the meiotic process. Furthermore do these or other products act as inductors or inhibitors of oocyte differentiation during the course of induced or spontaneous meiotic maturation? Present evidence concerning the role of steroids in mammalian oocyte maturation remains equivocal (2,26, 27, 28, 29).

## Acknowledgments

Research support by grants from NIH and NSF and the Population Council.

## References

1. A. W. Schuetz. Biol. Reprod. 10 (1974) 150.

2. H. R. Lindner, A. Tsafriri, M.E. Lieberman, W. Zor, Y. Koch, S. Bauminger and A. Barnea. Recent Prog. Hormone Res. 30 (1974) 79.

3. A. W. Schuetz and J. G. Cloud. Ann. Biol. Anim. Biochem. Biophys. 18(2B) (1978) 493.

4. P. Guerrier, M. Moreau and M. Dorée. Ann. Biol. Ann. Bioch. Biophys. 18(2B)(1978) 441.

5. H. Kanatani in: Oogenesis ed. J. D. Biggers and A. W. Schuetz (University Park Press, Baltimore, 1972) p. 459.

6. Y. Masui and C. L. Markert. J. Exp. Zool. 177 (1971) 129.

7. J. Maller and E. G. Krebs. J. Biol. Chem. 252 (1977) 1712.

8. W. J. Wasserman and Y. Masui. Science 191 (1976) 1266.

9. A. W. Schuetz. Ann. N. Y. Acad. Sci. 286 (1976) 408.

10. A. Tsafriri and C. P. Channing. J. Reprod. Fert. 43 (1975) 149.

11. A. Tsafriri,.S. H. Pomerantz and C. P. Channing. Biol. Reprod. 14 (1976) 511.

12. A. Tsafriri. Ann. Biol. Anim. Bioch. Biophys. 18(2b) 1978. 523.

13. A. Tsafriri and C. P. Channing. Endocrinol. 96(1975) 922.

14. A. W. Schuetz. Gen. Comp. Endocrinol. 12(1967) 209.

15. A. W. Schuetz. Biol. Reprod. 6 (1972) 67.

16. L. D. Smith in Biochemistry of Animal Development. Vol. 3 eds, R. Weber (Academic Press, New York) 1975.

17. A. W. Schuetz, T. G. Hollinger, R. A. Wallace and Samson, D. A. Develop. Biol. 58(1977) 428.

18. T. Hillensjö, N. Dekel and K. Ahrén. Acta physiol. scand. 96 (1976) 558.

19. S. Stoklosowa and M. Szoltys. Ann . Biol. Anim. Bioch. Biophys. 18(b) (1978) 503.

20. D. Szollosi, M. Gerard, Y. Ménézo and C. Thibault. Ann. Biol. Anim. Bioch. Biophys. 18(2B) (1978) 511.

21. Albertini, D. F. and E. Anderson. J.Cell Biol. 63(1974) 234.

22. Amsterdam, A., R. Joseph, M.E. Lieberman and H. R. Lindner. J. Cell Sci. 21(1976) 93.

23. E. Anderson and D. F. Albertini. J. Cell Biol. 71(1976) 680.

24. A. W. Schuetz and W. Swartz. SSR Abstr. (1976).

25. A. W. Schuetz and N. Dubin. Unpublished data.

26. R. M. Moor and A.O. Trounson. J. Reprod. Fert. 49(1977)101.

27. R. M. Moor. Ann. Biol. Anim. Bioch. Biophys. 18/2B(1978)477.

28. M.E. Lieberman, A. Tsafriri, S. Bauminger, W.P. Collins, K. Ahrén and N.A. Lindner. Acta endocr. (Kbh) 83(1976)151.

29. C. Readhead, M.H. Kaufman and A. W. Schuetz. Submitted for publication.

# RAPID OOCYTE DEATH PRIOR TO AND AT PUBERTY IN NEONATALLY THYMECTOMIZED MICE

Teruto Sakakura, Osamu Taguchi, Akinori Kojima and Yasuake Nishizuka

Laboratory of Experimental Pathology, Aichi Cancer Center Research Institute, Nagoya, 464, Japan

## INTRODUCTION

It is well established that the thymus plays a critical role in the development of immune systems, especially T-cell participating function. The hypothesis that the thymus is involved in the development of the neuroendocrine system in mammals and that, conversely, neuroendocrine regulation dominates thymic functions, is proposed(1, 2). Based on or not basing on such a hypothesis, experimental data have been accumulating that indicate some interrelation between the thymus and gonads. In 1969, Nishizuka and Sakakura described that thymectomy(Tx) in the mouse at 2-4 days of age leads to complete loss of oocytes, follicles and corpora lutea(3). This ovarian dysgenesis is noticed at 6-9 weeks of age in Tx mice and occurs under thymic regulatory functions and independently of wasting disease(4,5). Similar results, but to a lesser extent, were subsequently reported in mice(6) and in rats(7).

The nature and mechanism of post-Tx ovarian failure is still a matter of conjecture. Recently, Lintern-Moore(8) reported that alterations in the number and size distribution of ovarian follicles are initiated in Tx rats by the 10th day of life, and suggested that the presence of an intact thymus is necessary for the regulation of primordial follicular number and the initiation of follicular growth, connected in some way with pituitary function.

In this communication, we will describe histological sequences which ended in complete deletion of oocytes and follicles after neonatal Tx in a hybrid strain of mice, and show that a rapid death of all oocytes occurred prior to and at puberty. Further, we will discuss the possibility of the autoimmune mechanism in the developmental process of post-Tx ovarian failure in the light of our parallel experiments designated from an immunological aspect.

## MATERIAL AND METHODS

Female mice of (C57BL/6Cr X A/JCr)$F_1$ hybrid (B6A), raised under conventional conditions in an air-conditioned room ($24 \pm 2$°C; 14 h light: 10 h dark) of our laboratory, were used throughout the experiment. The thymus was removed on Day 3, or Day 7 in a few cases (the day of birth: Day 0), as described previously(4). A corresponding number of littermates received sham-operation and served as controls. Both groups were weaned on Day 24, and fed a commercial pellet food (Oriental Yeast Co. Ltd., Tokyo) with tap water *ad lib.* The animals which received Tx and sham-Tx were killed on Day 4, 10, 20, 24, 30, 60, and 90; some intact animals were killed on Day 0. The pituitary, ovaries, adrenals, spleen, mesenteric lymph nodes, salivary glands were weighed and fixed immediately in Bouin's fixative. The thyroids and stomach were preserved also for histological study. At autopsy, completeness of Tx was assessed grossly and examined histologically when necessary.

For the purposes of the present study, the right ovary from each mouse was embedded in paraffin, serially sectioned at 6 μm, and stained with hematoxylin and eosin. Total number of oocytes was counted at the level of nucleolus in every third section in each ovary. Histological changes of the left ovary were checked in each case to compare to those of the right ovary, and sections stained with periodic acid Schiff technique and Sudan III were examined when necessary.

## RESULTS

The total number of oocytes in each animal is illustrated in CHART 1. In the present observation no differential counting of primordial and growing follicles was made. In all cases, no significant difference in histological patterns was found between the right and left ovary.

In the mouse ovary, rapid and continuous reduction of small oocytes occurs regularly during infancy and juvenile period. Apparently, two ways of eliminating small oocytes are present: lysis or phagocytosis and passed through the surface epithelium of the ovary (9). It seemed that this phenomenon was not disturbed by neonatal Tx, and a remarkable but normally occurring decline of oocyte numbers was observed equally in both Tx and non-Tx groups by Day 20. Up to this age, no abnormal pattern of folliculogenesis was noticed in Tx mice, numerous growing follicles of medium sizes, Type 3b - Type 5a, and of large sizes, mostly Type 5b (according to Pedersen and Peters classification(10)), was found (FIGS. 1a,1b). Of interest was that on Day 24, infiltration of lymphocytes was first noticed in many sections inside and around follicles of medium and large sizes, and resulted in acute death of oocytes located in affected follicles, accompanied with severe destruction of multi-layers of growing follicles (FIGS. 2a,2b). However, reduction of the total oocyte number was not induced so much at this age in Tx mice compared with that in non-Tx mice, probably because lymphocyte infiltration did not spread out into the peripheral cortical region of the ovary where many small and primordial follicles are lodged. On Day 30, lymphocyte infiltration became very prominent and intermingled with few but a definite number of plasma cells (FIGS. 3a, 3b). Remarkable decrease in oocyte numbers, obviously far below the normal level, was found in some cases. With advance of age, follicle destruction by cell infiltration became more intensive and all oocytes disappeared rather rapidly by Day 60, neither oocyte nor follicle was recognized in many cases. The ovary was occupied mostly by interstitial cells and reduced its weight (FIGS. 4a,4b). In B6A hybrid mice, vaginal opening occurs usually on Day 45 - Day 55. It can be said therefore that rapid and complete disappearance of all oocytes is taken place prior to and at puberty

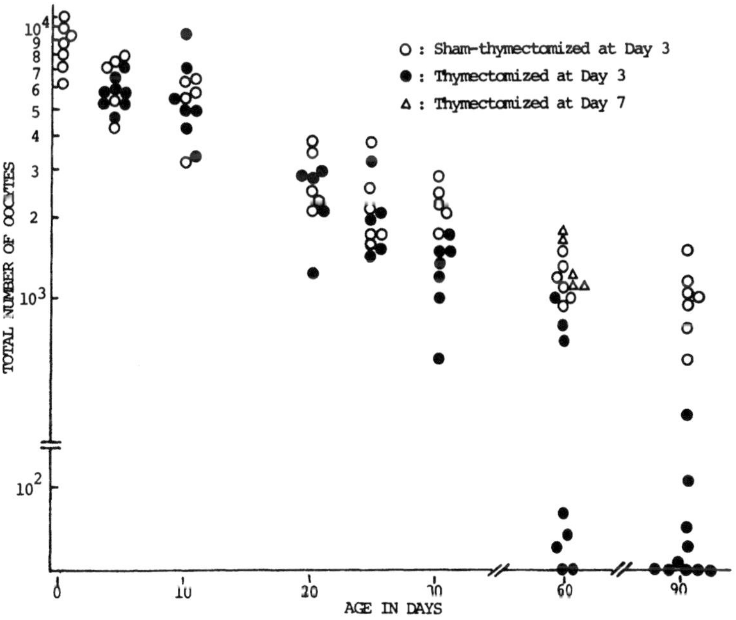


CHART 1. Total number of all oocytes per ovary on Day 0, 4, 10, 20, 24, 30, 60, and 90 in (C57BL/6Cr X A/JCr)$F_1$ mice after thymectomy(Tx) on Day 3, Day 7, and after sham-Tx.

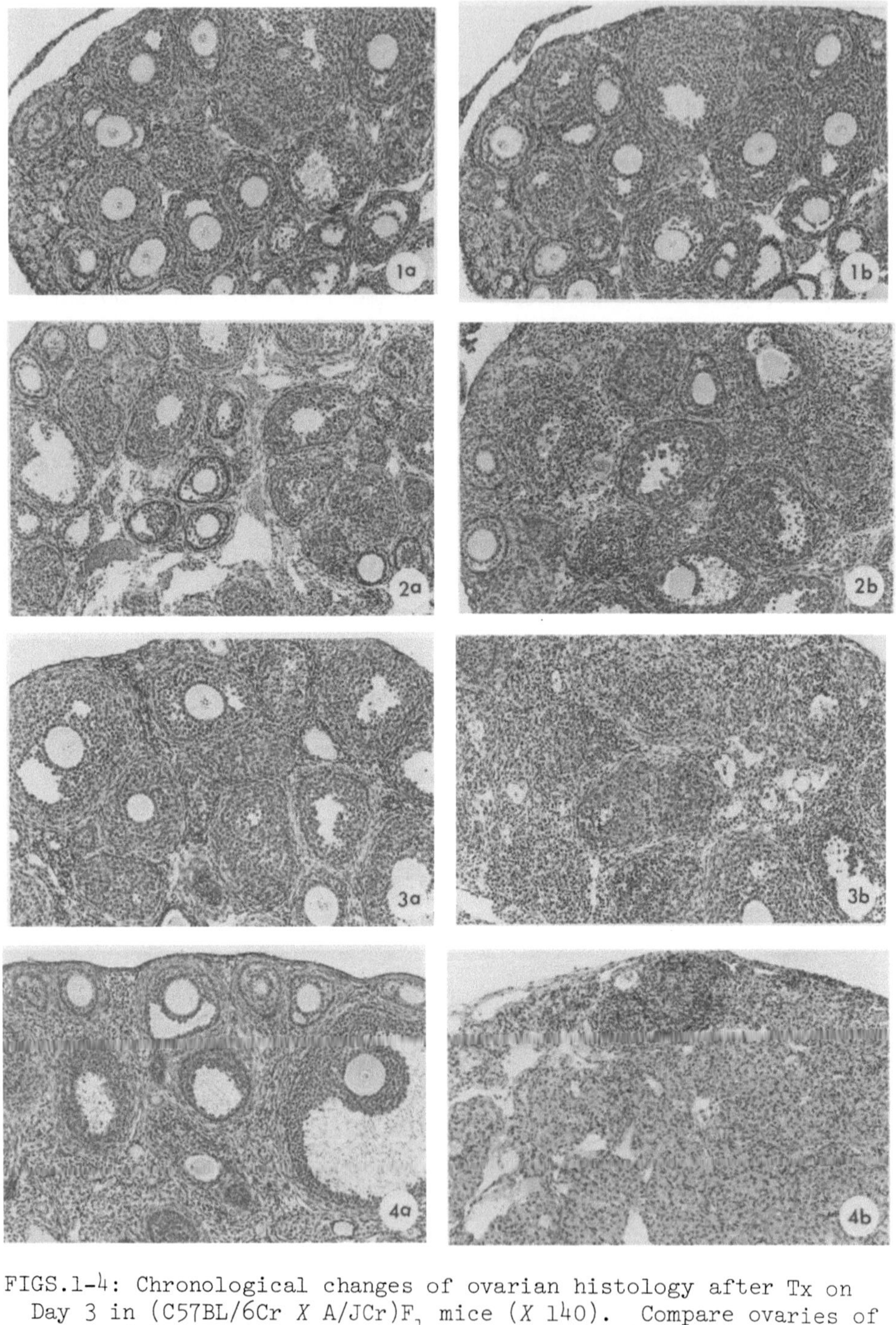


FIGS.1-4: Chronological changes of ovarian histology after Tx on Day 3 in (C57BL/6Cr *X* A/JCr)$F_1$ mice (*X* 140). Compare ovaries of Tx mice[b] with normal ovaries of non-Tx mice[a] of corresponding ages. FIG.1: Day 20; FIG.2: Day 24; FIG.3: Day 30; FIG.4: Day 60.

after Tx on Day 3. However, massive oocyte death was not observed after Tx on Day 7 (CHART 1). This result confirms our previous study which demonstrated that the critical age for effective Tx for induction of ovarian dysgenesis falls into a rather short period, 2-4 days after birth(3,4).

## DISCUSSION

The present chronological observation on the mouse ovary after neonatal Tx indicates that Tx at Day 3 but not at Day 7 results in rapid, and frequently, complete disappearance of all oocytes prior to and at puberty. It is likely that the process of acute oocyte death is directly caused by massive lymphocyte infiltration mixed with plasma cells inside and around follicles. This result clearly coincides with our previous data that neonatal Tx in the mouse causes early and/or complete sterility in adult life(11).

As for the basic nature of such ovarian dysgenesis after Tx, our extensive parallel studies may indicate so far its uatoimmune nature (O.Taguchi, et al., unpublished data; A.Kojima, et al., unpublished data). The following are fundamental findings by which we could support this hypothesis:

i) Severe lymphocyte infiltration with plasma cells in the ovary as described above;

ii) Success in detection of anti-oocyte antibody or antibodies in the sera taken from Tx mice with ovarian dysgenesis;

iii) Successful passive transfer of ovarian dysgenesis into normal infant isogeneic mice by intraperitoneal injections of certain numbers of the spleen cells prepared from Tx mice with ovarian dysgenesis; failure in passive transfer by injections of the spleen cells from the same source, but pretreated with anti-theta serum before injections to recipients;

iv) Co-existence of organ-localized autoimmune diseases in the thyroid(12) and gastric mucosa in Tx females, and in Tx males prostate and testis as well.

Auto-antibody or antibodies against oocytes were demonstrated by both indirect immunofluorescence and horseradish peroxidase immune histochemical techniques. It appears so far that antibody or antibodies in sera can react with the ooplasma consistently and with zona pellucida with fewer oocytes, located in growing follicles of medium and large sizes. These data may suggest that antigen or antigens exist and appear with follicle and/or oocyte maturation. Also, it would be possible to presume that massive infiltration of lymphocytes inside and around follicles indicates immune reactions and reflects their preponderant sites.

## SUMMARY

Neonatal thymectomy(Tx) at Day 3 in (C57BL/6Cr *X* A/JCr)$F_1$ mice results in rapid and frequently complete death of all oocytes prior to and at puberty. Chronological examination of ovarian histology reveals that massive infiltration of lymphocytes inside and around growing follicles, probably somehow connecting with follicular and/or oocyte maturation, is a direct process leading to acute loss of oocytes. This process was first noticed at Day 24 and increased its grade and extent rather rapidly thereafter, and reached its maximum intensity around puberty, Day 50 - Day 60. Our parallel studies strongly suggest so far, autoimmune mechanism of the developmental process of the ovarian destruction after neonatal Tx performed at the critical age.


## ACKNOWLEDGEMENT

This work was supported by Contract NO1-CP-55650 between Dr. Y.Nishizuka, Aichi Cancer Center Research Institute and National Cancer Institute, USA.

## REFERENCES


(1) W.Pierpaoli, H.G.Kopp and E.Bianchi. Clin. exp. Immunol. 24 (1976)501.
(2) H.Besedovsky and E.Sorkin. Clin. exp. Immunol. 27(1977)1.
(3) Y.Nishizuka and T.Sakakura. Science 166(1969)753.
(4) Y.Nishizuka and T.Sakakura. Endocrinol. 89(1971)886.
(5) T.Sakakura and Y.Nishizuka. Endocrinol. 90(1972)431.
(6) S.Lintern-Moore and I.Norbaek Sorensen. Mech. Aging Dev. 5 (1976)235.
(7) M.Hattori. Endocrine Function of the Thymus, Ph.D. Thesis, Australian National University (Nov. 1977).
(8) S.Lintern-Moore. Biol. Reprod. 17(1977)155.
(9) H.Peters. Acta endocrinol.(kbh.) 62(1969)98.
(10) T.Pedersen and H.Peters. J. Reprod. Fertil. 17(1968)555.
(11) A.Kojima, T.Sakakura, Y.Tanaka and Y.Nishizuka. Biol. Reprod. 8(1973)358.
(12) A.Kojima, Y.Tanaka-Kojima and Y.Nishizuka. Lab. Invest. 34 (1976)550.

# MATURATION AND ATRESIA OF HUMAN OOCYTES *IN VIVO*

Mrinal K. Sanyal

Laboratory of Environmental Toxicology

National Institute of Environmental Health Sciences
P.O. Box 12233, Research Triangle Park, N.C. 27709

A small number of oocytes mature and ovulate during the entire reproductive life of women, while a large population of oocytes degenerate during follicular development in the adult human ovary (1). A broad light microscopic survey of the cytological feature of oocytes randomly collected from ovaries of patients without apparent ovarian pathology, revealed that the follicular oocytes could be classified into four different types (2). Briefly, these oocyte types were (i) oocytes with nuclei containing diffused or filamentous chromatin (Type 1), (ii) mature oocytes with metaphase chromosomes or first polar bodies (Type 2), (iii) oocytes with clumped chromatin within the nuclei (Type 3), and (iv) degenerated oocytes (Type 4). The percentages of oocytes of each type during different reproductive states is given in Table 1.

The total number of "normal" oocytes with diffused chromatin (Type 1) and mature oocytes (Type 2) during the proliferative phase of the menstrual cycle was similar to the number of "normal" oocytes (Type 1) of the secretory phase. More than 50% of the oocytes were of Types 3 and 4 during both phases of the menstrual cycle, presumably constituting the "abnormal" population of follicular oocytes. The ratio of "normal" to "abnormal" oocytes is markedly altered in the ovaries of patients with polycystic ovarian disease, or in ovaries obtained from steroid treated patients. The number of "abnormal" follicular oocytes increased sharply in these ovarian samples. The incidence of ovulation is rare during these pathological states. It seems likely that clumping of the nuclear material may be an event associated with atresia of oocytes in the adult human ovary.

TABLE 1

Percent oocytes of different types during various reproductive states.*

| Reproductive States (No. of oocytes) | Type 1 (Diffused chromatin) | Type 2 (Mature oocytes) | Type 3 (Clumped chromatin) | Type 4 (Degenerated oocytes) |
|---|---|---|---|---|
| Proliferative phase (87) | 35 | 6 | 45 | 14 |
| Secretory phase (64) | 44 | - | 40 | 16 |
| Polycystic ovary (43) | 15 | - | 70 | 15 |
| Steroid treated (14) | 7 | 7 | 86 | - |

*Adapted from Sanyal *et al*. (1976).

Sections or whole mount preparations of oocytes show that a substantial percentage of oocytes had nuclei containing clumps of chromatin (Types 3 & 4). The clumps stained deeply with aceto-orcein or with osmium tetroxide and toluidine blue. They were irregular in shape and frequently were associated with the nucleolus. The nuclear shapes in these oocytes were not uniform and often the cytoplasm was retracted from the zona pellucida (Fig. 1).

The cytologically distinguishable features in oocytes approaching maturation or atresia and the relationships of these processes with the characteristics of the ovarian follicles are poorly understood. The effect of the follicular environment in inducing maturation or atresia seems to be of primary importance. It is conceivable that under an unfavorable milieu, meiosis may not be resumed and the chromatin material forms clumps around the nucleolus. The processes associated with condensation of chromatin material into filamentous chromosomes from the diffused state in the normal ovum or clumping of the nuclear materials leading to eventual degeneratation of oocytes are areas of research which deserve considerable attention.

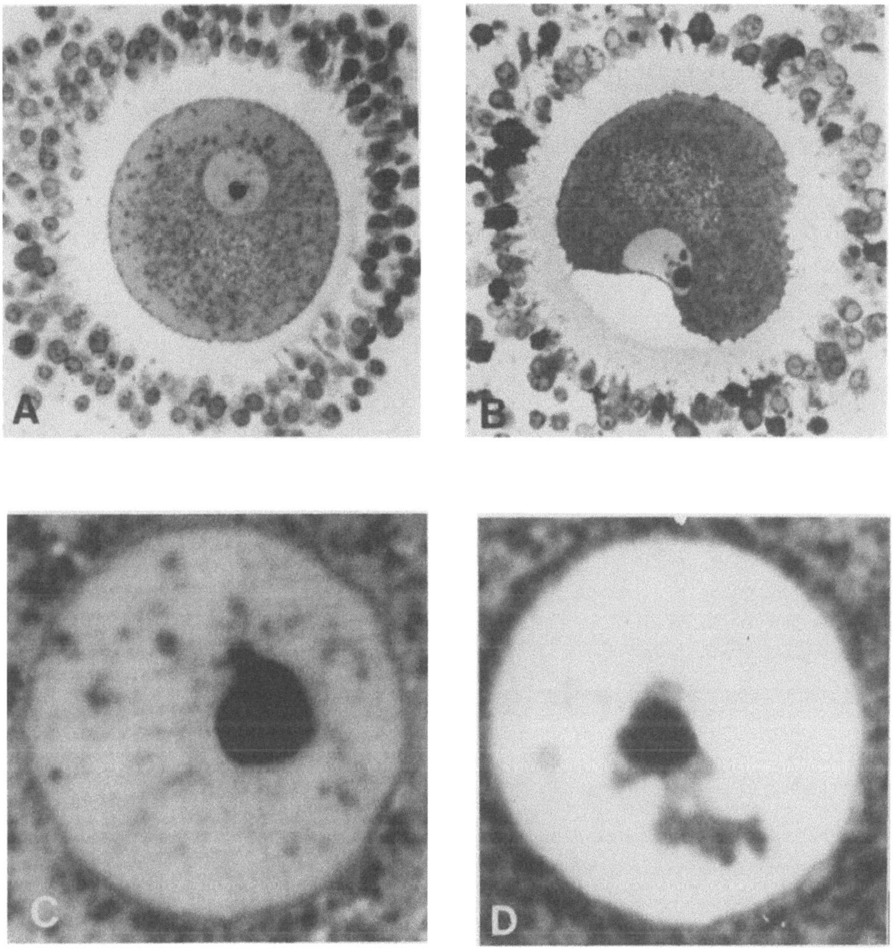

Fig. 1: Sections of oocytes. A. "Normal" oocyte with round nucleus and diffused chromatin. B. "Abnormal" oocyte with cytoplasm retracted from the zona pellucida. (A & B approximately X 500). C. Magnified nucleus of a "normal" oocyte showing diffused chromatin. D. Clumped chromatin around the nucleolus within the nucleus of an "abnormal" oocyte. (C & D approximately X 2000).

## References

(1) L. L. Franchi and T. G. Baker, in: Human Reproduction-Conception to Contraception, ed. E. S. E. Hafez and T. N. Evans (Harper & Row, New York, 1973) p. 53.

(2) M. K. Sanyal, M. L. Taymor and M. J. Berger. Fertility and Sterility 27 (1976) 501.

SESSION NUMBER: V

TITLE: FOLLICULAR NONSTEROIDAL REGULATORS

PRINCIPAL SPEAKER: Cornelia P. Channing

INVITED DISCUSSANT: Florence Ledwitz-Rigby

PRESENTED ABSTRACTS:

10.) S.C. Chappel, T. Acott and H.G. Spies. Inhibition of In Vivo and In Vitro Follicle-Stimulating Hormone Secretion by Steroid-Free Ovarian Extracts.

11.) J.R. Lorenzen and N.B. Schwartz. The Differential Ability of Porcine Follicular Fluid to Suppress Serum FSH in Female Rats From 6 Days of Age to Adulthood.

12.) N.C. Darga and L.E. Reichert, Jr. Evidence for the Presence of a Low Molecular Weight Follitropin Binding Inhibitor in Bovine Follicular Fluid.

CHAIRPERSON: Neena B. Schwartz

FOLLICULAR NON-STEROIDAL REGULATORS

Cornelia P. Channing

Department of Physiology
University of Maryland, School of Medicine
660 W. Redwood Street
Baltimore, Maryland 21201

## INTRODUCTION

Chemical messengers synthesized both within and without the ovary control follicular and oocyte maturation and ovulation. The extraovarian messengers, whose roles are better defined, are the pituitary gonadotropins-luteinizing hormone (LH) and follicle stimulating hormone (FSH). Other hormones-insulin, cortisol, thyroxin, and prolactin have poorly defined "permissive" functions. Intraovarian chemical messengers include ovarian steroids (estrogen, progesterone, and androgens) and polypeptide regulators such as oocyte maturation inhibitor (OMI), luteinization inhibitor (LI) ovarian inhibin (folliculostatin), luteinizing hormone binding inhibitor (LHRBI), and a possible follicular fluid luteinizing promoting substance.

The roles of some of the non-steroidal regulators will be discussed here. Actions of LH and FSH upon the ovary have been discussed elsewhere (1,2) and are further summarized in other chapters in this volume.

### Review of Follicular Maturation

During follicular maturation LH and FSH stimulate follicular growth and maturation beyond the antral stage, stimulate estrogen secretion by the theca of the primate follicle, and stimulate granulosa cell growth. Discussion here will be restricted to post antral stages of follicular development. During follicular growth, granulosa cells divide and acquire FSH receptors (1,3,4) and as follicular maturation progresses, FSH acts to stimulate the appearance of LH receptors (2-9) in the granulosa cells. There is

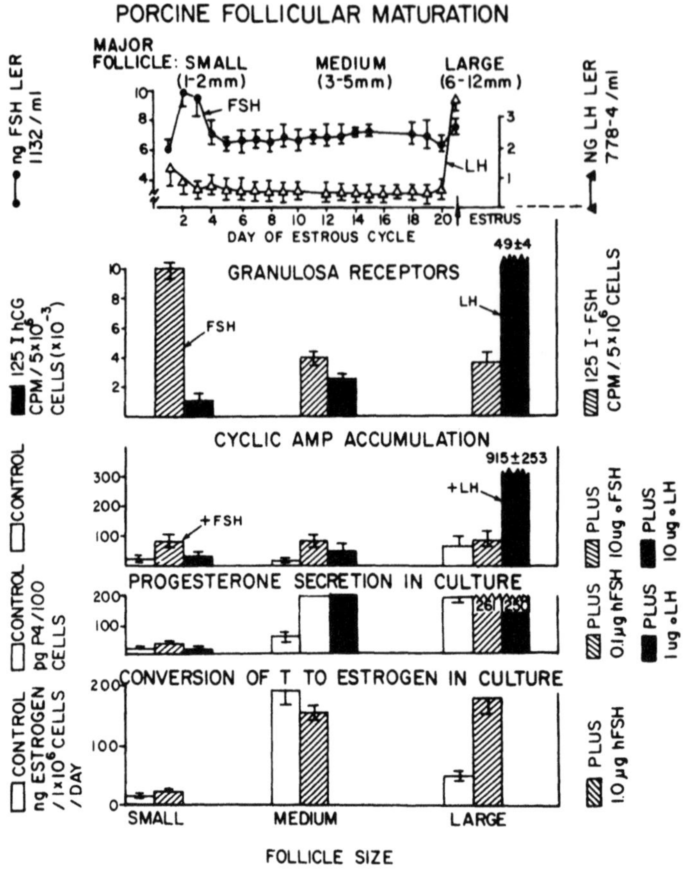


FIGURE 1. Changes occurring in porcine granulosa cells during follicular maturation. Data on the top represent serum FSH and LH levels occurring throughout the porcine estrous cycle. Levels of FSH are adapted from the data of Rayford, et. al. (14) and levels of serum LH are adapted from the data of Niswender, et. al. (15). Data on granulosa FSH binding are taken from Nakano, et. al. (6) and data on hCG binding are taken from Channing and Kammerman (16,17). Data represent the mean ± SEM of 6-15 observations. Data on the ability of FSH and LH to stimulate intracellular cyclic AMP accumulation represent the mean ± SE of 6-11 observations and are taken from Lindsey and Channing (10,11). Observations on the effect of human FSH (LER 1801-3) and 1.0 µg ovine upon progesterone secretion are taken from the data of Thanki and Channing (12,13) and Channing (18,19), respectively and represent the mean ± SE of 6-12 observations on granulosa cells cultured for 4 days. Data on the effect of follicular maturation upon the ability of granulosa cells to convert testosterone to estrogen was adapted from the findings of Anderson, Schaerf and Channing (20).

a later gradual loss of granulosa cell FSH receptors and FSH responsiveness in terms of ability to stimulate cyclic AMP accumulation (6,10,11) and progesterone secretion (12,13) (Figure 1). Immediately prior to ovulation the preovulatory surge of LH brings about resumption of meiotic maturation of the oocyte in the preovulatory follicle so that the immature (dictyate) oocyte completes maturation to the first polar body stage (reviewed in 1). Granulosa cells increase at least 100-fold in their numbers of LH/hCG receptors and in their ability to respond to LH (16,17) by accumulation of cyclic AMP (10,11). They also increase in ability to luteinize if removed from the follicle and cultured (18,19). All of these changes in the follicle and the component cell types are probably regulated by intraovarian factors. This is necessary because LH and FSH enter into follicles of various sizes and states of maturation (21-24). Yet during the primate menstrual cycle only one follicle matures and ovulates and only one oocyte matures. It is proposed that a substance in the non-maturing follicle keeps them from responding to LH and FSH. These factors will be discussed. The earlier literature prior to 1976 has been discussed elsewhere (1).

## Oocyte Maturation Inhibitor

In 1975 Tsafriri and Channing reported that addition of porcine follicular fluid or granulosa cells could inhibit the spontaneous resumption of meiosis in isolated, cumulus-enclosed porcine oocytes (25,26). Addition of denuded (granulosa cell-free) porcine theca or human foreskin fibroblasts did not exert an inhibitory action upon the oocytes (25). Tsafriri, Pomerantz and Channing carried out a partial purification of the porcine follicular fluid oocyte maturation inhibitor (OMI) which included Amicon PM-10 membrane filtration and Sephadex G-25 chromatography (27). More recently an additional paper electrophoresis step has led to a 2500-fold purification of OMI from follicular fluid. The OMI activity is destroyed by treatment of follicular fluid with trypsin, so it probably is a peptide (27). Chromatography on Sephadex G-25 and analysis of each eluate in the porcine oocyte assay demonstrated that there were two peaks of OMI activity; one corresponding to a molecular weight of nearly 2000 daltons and a second corresponding to a molecular weight of about 1000 daltons (Figure 2) (28,29). These two peaks were reproducibly obtained from 3 pools of porcine follicular fluid. There was more OMI activity in fluid obtained from small (1-2 mm) follicles compared to fluid obtained from large (6-12 mm) porcine follicles (28). When OMI material obtained in Peak A was pooled and rechromatographed on Sephadex G-25, it ran in the same region in the second chromatogram.

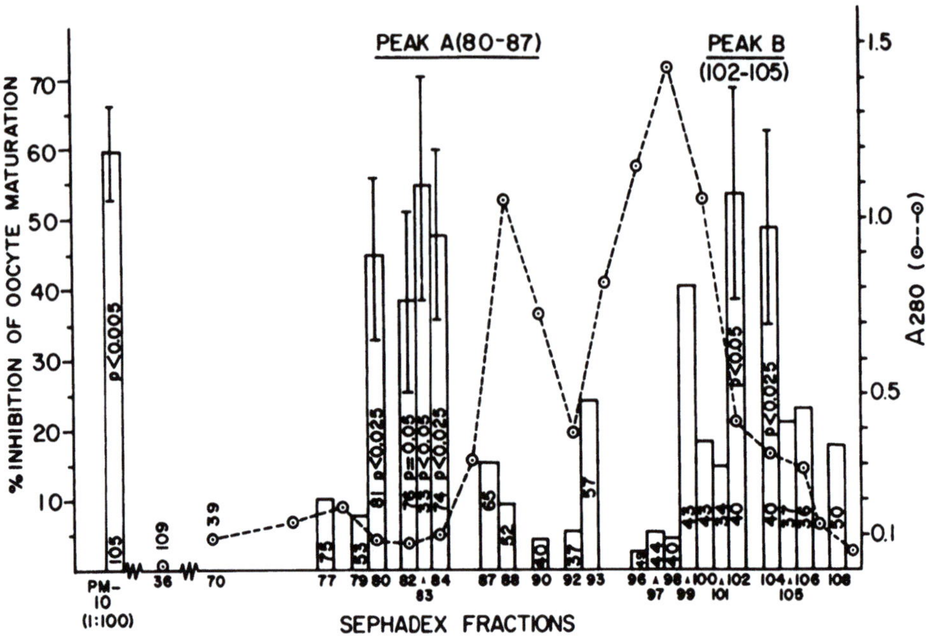


FIGURE 2. Fractionation of low molecular weight fraction of porcine follicular fluid on Sephadex G-25. The absorbency at 280 nm (o--o) was determined on the original fractions (15.7 ml). The fractions were lyophilized, redissolved in 2 ml $H_2O$ and the oocyte inhibitory activity assayed at a dilution of 1:100. The numbers in the lower parts of the bars are the numbers of oocytes for each test. The bars show ± SE. Inhibition values without error bars were not significantly different from controls. (Taken from 28 with permission from The Biology of Reproduction.)

The OMI material eluting in Peak A and Peak B inhibited oocyte maturation and cumulus cell progesterone in cultured porcine cumulus-enclosed oocytes (Figure 3).

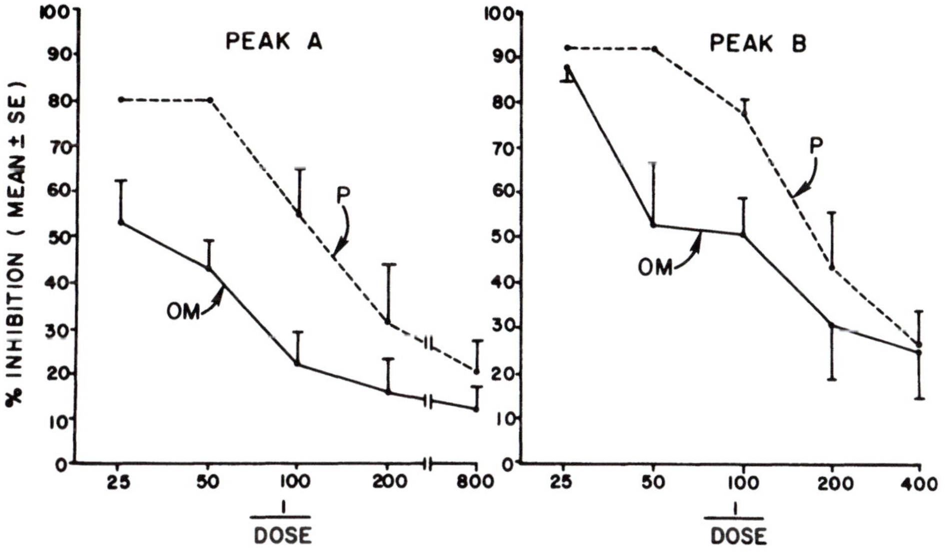


FIGURE 3. Effect of Sephadex G-25 fractions upon oocyte maturation and cumulus cell progesterone secretion. Cumulus-enclosed oocytes were cultured in TC 199A containing 15% pig serum for 2 days followed by examination of the oocytes to reveal maturational stage. The progesterone content in each culture was analyzed by radioimmunoassay. Various dilutions of Sephadex peak A (left panel) and peak B (right panel) fractions were tested. The amount of peptide in these two peaks were measured by fluorescamine assay with β-MSH as standard and found to correspond to 11 mg/ml in the 1/25 dilution of peak A and 2.1 mg/ml in the 1/25 dilution of peak B. Percent inhibition of oocyte maturation or progesterone secretion was calculated as follows: (Value of control-value of inhibitor/value of control) x 100. When high concentrations of inhibitors were used the progesterone levels were lower than the detection limit of the assay (0.1 ng/culture) and the values shown for these groups (the points without SE) Therefore represent the minimal inhibition.
(Hillensjö, Pomerantz and Channing, unpublished observations).

Inhibition of progesterone secretion occurred in each tube from the Sephadex column in which inhibition of oocyte maturation occurred. The inhibition of both oocyte maturation and progesterone secretion was dose dependent. Both fractions also inhibited outgrowth of the cumulus cells. Thus inhibition of progesterone, oocyte maturation and cumulus outgrowth occurred together, representing a generalized action of the OMI to keep the cumulus oocyte complex "immature". The inhibitory actions were reversed after removal of the OMI and replenishment of the medium with control medium (Figure 4 and 29,30).

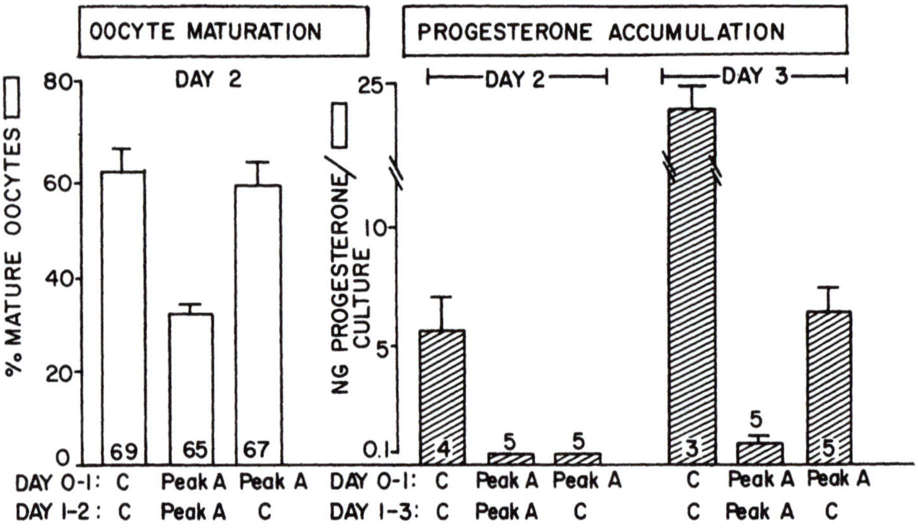


FIGURE 4. Reversibility of inhibitory action of OMI upon porcine oocyte maturation and cumulus cell progesterone secretion. Cumulus-enclosed oocytes were cultured for 1 day in TC 199A containing 15% pig serum in the absence or presence of partially purified OMI (Sephadex peak A; 5 mg/ml). The media were then changed as indicated and the culture continued for 1 or 2 additional days. Oocyte maturation was studied in cultures terminated after 2 days and progesterone secretion was examined in cultures terminated after 2 or 3 days. When the inhibitor-containing medium was changed into control medium (designated C) significant reversal of oocyte maturation occurred ($p<0.01$). A reversal of inhibition of progesterone was seen after 3 days of culture ($p<0.01$). (Hillensjö, Kripner and Channing, unpublished observations).

Hillensjö et. al. observed that the inhibitory action of OMI *in vitro* requires the presence of cells surrounding the oocyte since the OMI could not inhibit the spontaneous maturation of denuded oocytes or oocytes grown in serum-free media which did not support long term growth of the cumulus cells.

In further support of an important role of cumulus cells in mediating the inhibitory action of OMI upon oocyte maturation is the finding of Alexander Tsafriri who observed that the low molecular weight fraction of porcine follicular fluid (OMI) inhibited spontaneous maturation of cultured cumulus-enclosed but not denuded mouse oocytes (Tsafriri, unpublished observations).

The mechanism of inhibitory action of OMI on cumulus cell progesterone secretion and cumulus outgrowth is not known and further purification will be necessary to state that the inhibitory effects on oocyte maturation and cumulus differentiation are inherent in the same molecule of the partially purified OMI. It does not appear to be acting at the level of the LH/hCG receptor since addition of the low molecular weight fraction of porcine follicular fluid does not inhibit binding of the $^{125}$I-hCG to porcine granulosa cells (Anderson and Channing, unpublished observation). It may interfere with FSH binding since addition of the low molecular weight fraction of porcine follicular fluid can inhibit FSH stimulation of cyclic AMP accumulation (31). Studies carried out by Darga and Reichert indicate that the same low molecular weight fraction of porcine follicular fluid can inhibit FSH binding to bovine granulosa cells (32). The low molecular weight fraction of serum also inhibited FSH binding to granulosa cells to a comparable degree (32). However, further purification of both the serum and follicular fluid inhibitor will be needed to determine whether they are the same.

## Luteinization Inhibitor

In 1973 Ledwitz-Rigby, Stetson and Channing (33) reported that addition of 50% fluid obtained from small but not large porcine follicles or serum inhibited spontaneous luteinization of cultured porcine granulosa cells harvested from large follicles. It also inhibited LH stimulation of cyclic AMP in these cells. Later they extended the observation (34) and found that the follicular fluid inhibited progesterone secretion by cultured granulosa cells from large porcine follicles. Bernard made a similar observation with bovine follicular fluid in cultured rat granulosa cells (35). The luteinization inhibitor in small porcine follicular fluid also brought about a decrease in LH/hCG receptors after 2-4 days of culture of granulosa cells from small porcine follicles (7,8). Addition of excess hFSH plus insulin, thyroxin and cortisol can overcome the follicular fluid inhibition of induction of LH/hCG receptors. This observation would support the concept of antagonism between follicular fluid inhibitors and pituitary gonadotropins. Recent advances in the actions of porcine follicular fluid upon granulosa cell luteinization are reported elsewhere in this workshop (36).

## Ovarian Inhibin (Folliculostatin)

In some species, namely the pig (14) and rat (37,38), there is a secondary rise in FSH after ovulation. In the rat within 3-4 hours after castration, there is a prompt rise in FSH with no concurrent rise in LH until 1-2 days later (37,38). The early post-castrational rise in FSH is not suppressable with estrogen. Such observations led to the speculation that the ovary was secreting a non-steroidal substance which acted to inhibit FSH secretion. This substance could be lost at ovulation, accounting for the postovulatory rise in FSH. In support of the concept of a follicular fluid FSH supressing substance (ovarian inhibin) DeJong and Sharp observed that injection of bovine follicular fluid into castrate male rats could inhibit FSH (39). Marder, Channing and Schwartz (49) observed that injections of as little as 150 μl charcoal-treated porcine follicular fluid could inhibit FSH in intact and short-term castrated adult female rats. They also observed that 0.5 ml porcine follicular fluid, given at 3:45 P.M. and 5:30 P.M. could inhibit the secondary surge (post-ovulatory) of FSH observed at 4:00 A.M. on the morning of estrus as well as the LH induced FSH surge in Nembutal-blocked proestrous intact rats (41). If the injection of porcine follicular fluid was advanced to 11:00 A.M. or 12:00 noon, the primary FSH surge was blocked (42,43).

Lorenzen, Schwartz and Channing (44,45) observed that there was more follicular inhibin activity in fluid obtained from small compared to large porcine follicles and that the active principle was trypsin-labile and greater than 10,000 daltons in size based on fractionation on an Amicon PM-10 membrane.

The likely site of inhibitory action of the ovarian inhibin is at the level of the anterior pituitary since addition of charcoal-treated porcine follicular fluid can inhibit pituitary cultures from secreting FSH both under basal conditions and in the presence of luteinizing hormone releasing factor (LRF)(46,47). DePaolo and his colleagues also observed that the porcine follicular fluid inhibited LRF-stimulated FSH secretion but not LH secretion in the intact Nembutal-blocked LRF infused rat (43). Employing monolayer cultures of rat anterior pituitaries Shander et. al. (47) observed that as little as 2% porcine follicular fluid, given for 2 hours followed by a post incubation with LRF or control medium could inhibit FSH secretion over a subsequent 4 hour period of incubation. Addition of as little as 0.1% porcine follicular fluid for a longer time, 24 hours, also inhibited FSH but not LH secretion by the pituitary cultures.

The inhibitory action of porcine follicular fluid upon serum FSH levels is not limited to the rat. Recently Channing, Anderson and Hodgen observed that I.P. injections of charcoal-treated porcine follicular fluid could inhibit serum FSH levels in both the castrate and the intact rhesus monkey (48). In a series of 4 castrate monkeys subcutaneous administration of either 5 or 15 ml of porcine follicular fluid in two injections 6 hours apart led to an 83, 60, 28 and 14% decline in serum FSH levels after 24-30 hours. There was less than 5% decline in serum LH levels in the same monkeys. When four intact monkeys were given 4 ml intraperitoneal injection of porcine follicular fluid every 8 hours from days 1 to 4 of the cycle there were declines of 87, 85, 52 and 14% in serum FSH within 24-36 hours of treatment. There was less than a 5% decline in serum LH levels in these follicular fluid treated monkeys. In one monkey if porcine follicular fluid was given for a 4 day period prior to and during the expected midcycle surge of FSH the FSH surge was abolished. In another monkey if the follicular fluid was given on 4 days starting 1/2 day after the initiation of the midcycle FSH surge the duration of the FSH surge was shortened. Administration of the fluid early in the cycle caused the follicle at midcycle to contain less than 10% of the normal number of granulosa cells. Figure 5 depicts the time course of the inhibitory effect of porcine follicular fluid upon serum FSH and LH in a cycling and a castrate rhesus monkey.

Inhibin activity has also been found in human follicular fluid by Batta and his colleagues (52).

The significance of an ovarian inhibin in physiological control of serum FSH may be great. Sherman and Korneman observed that there was a rise in serum FSH prior to LH in women just before the onset of menopause in spite of relatively normal serum estrogen levels (53). This observation could be explained by a decrease in ovarian inhibin. The possibility exists that the follicle may make excessive inhibin in polycystic ovarian disease since in the syndrome there is diminished FSH, normal or elevated LH levels, and non-ovulating cystic follicles (54). The fact that there can be ovulation in follicular fluid-treated rats who do not have preovulatory FSH surge is evidence for the lack of requirement of preovulatory FSH for follicular rupture and oocyte maturation. The effect of administration of follicular fluid one estrous cycle prior to the one observed may give insight to the role of early exposure of the follicle to FSH. The monkey

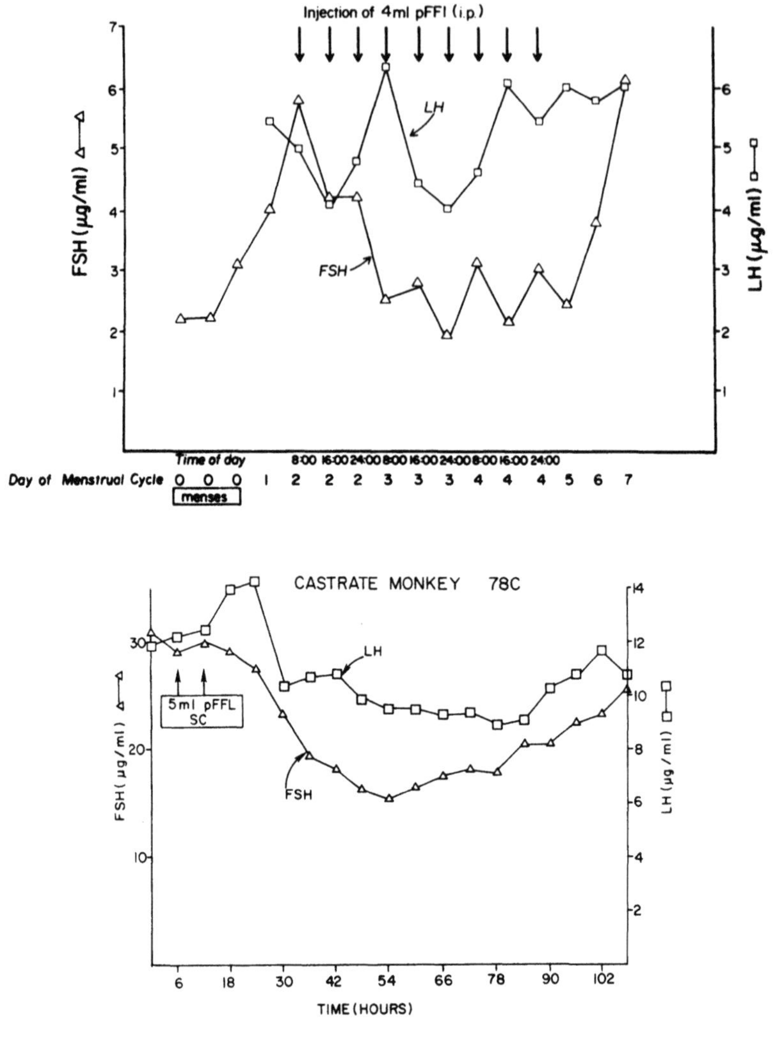


FIGURE 5. Effect of charcoal treated porcine follicular fluid upon LH and FSH in an intact (top panel) and a long term castrate (bottom panel) adult rhesus monkey. The intact rhesus monkey was given 4 ml porcine follicular fluid (ip) every 8 hours for a period of 4 days from day 1 to 4 of the menstrual cycle. Serum LH was measured by the radioimmunoassay method of Niswender (49) using crude rhesus monkey pituitary extract (LER 1909-2) as a standard; serum FSH was measured by the radioimmunoassay method of Hodgen et. al. (50) using LER-1909-2 as a standard. Methods for administration of the porcine follicular fluid (48) and monkey blood sampling (51) are described elsewhere. The castrate monkey was given two injections of 15ml follicular fluid (sc) at the start of the experiment.

studies indicate that early follicular deprivation of FSH is deleterious to the later growth of the follicle and of the granulosa cells. Miller and Ginther have made a similar observation in the sheep (55). Since whole follicular fluid was used in the monkey and sheep studies and not purified inhibin additional inhibitory actions exerted by the follicular fluid directly upon the ovary cannot be ruled out. Chemical purification of ovarian inhibin will be required for a more rigorous evaluation of its physiological role. Ovarian inhibin will also provide a useful means to selectively inhibit FSH secretion without inhibiting LH so that the physiological role of FSH may be studied.

### Luteinizing Binding Inhibitor (LHRBI)

In 1976 Yang, Saaman and Ward observed that aqueous extracts of ovaries from both pseudopregnant and pregnant rats but not extracts of ovaries from immature rats inhibited binding of LH to corpus luteum receptors (56). Sakai, Engel and Channing (57) found a similar LH/hCG binding inhibitor in aqueous extracts of frozen porcine corpus luteum. The inhibitory factor(s) was absent from non-luteal ovarian tissue, lung or heart issue. Tucker, Kumari and Channing observed that the inhibitor was present in greater amounts of old compared to young porcine corpus luteum tissue (58).

## CONCLUSIONS

The precise interaction between the extraovarian hormones (gonadotropins and steroids) and intraovarian steroid hormones and follicular non-steroidal regulators in control of follicular function remains the subject of future study. A diagrammatic sketch of follicular nonsteroidal regulator action is shown in Fig. 6.

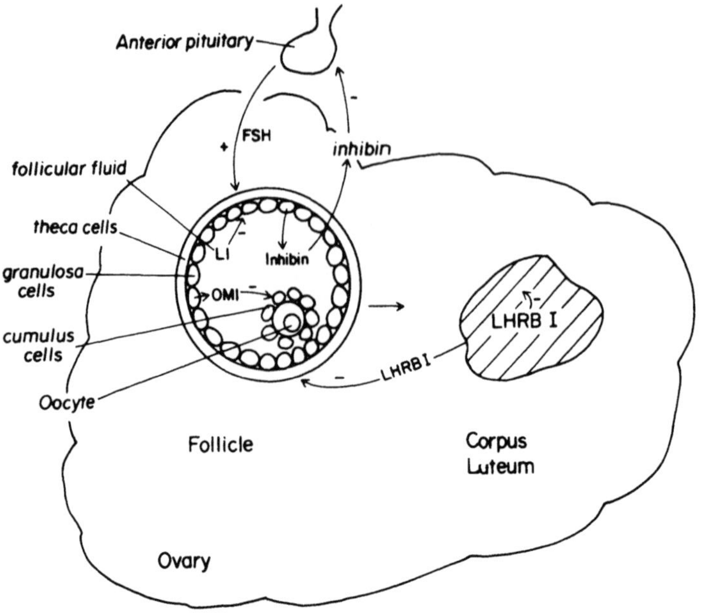


FIGURE 6. Diagramatic Sketch of Follicular Non-steroidal Ovarian Regulators.

Isolation of intraovarian follicular inhibitors does not imply an inherent central role in physiological function of the follicle. Further experiments relating follicle physiology to the levels of these inhibitors (which should be purified chemically and subjected to proper radioimmunoassay measurements) will answer some of these questions. They also have potential as contraceptive agents since they may control important ovarian events and ideally may have no side affects outside the ovary.

## SUMMARY

Porcine, as well as human follicular fluid contains at least three nonsteroidal ovarian regulators: an oocyte maturation inhibitor (OMI), a luteinization inhibitor (LI) and an ovarian inhibin ("folliculostatin"). OMI is a polypeptide of about 2000 daltons in size, is secreted by the follicular granulosa cells, and acts to keep the oocyte in the arrested dictyate state. Its inhibitory action can be reversed by removal of it or addition of gonadotropins. Luteinization inhibitor is present in greater amounts in the fluid of small follicles as compared to large follicles and acts to keep granulosa cells from luteinizing in terms of morphology and progesterone secretion. It acts to inhibit maturation of LH receptors and acute LH stimulation of cyclic AMP accumulation.

Ovarian inhibin is a follicular fluid polypeptide >10,000 daltons in size which acts to decrease blood levels of FSH. Its probable site of inhibitory action is at the level of the anterior pituitary gland since it can act to inhibit cultured anterior pituitary cells in their ability to secrete basal as well as LHRH stimulated release of FSH. It can also inhibit LHRH stimulation of FSH secretion in the rat in vivo. Charcoal treated porcine follicular fluid can inhibit FSH levels in the intact as well as the castrate adult female rhesus monkey. Administration of porcine follicular fluid early in the menstrual cycle leads to a decreased number of granulosa cells appearing in the follicle present at mid cycle.

The porcine and rat corpus luteum contains a fourth ovarian non-steroidal regulator a luteinizing hormone receptor binding inhibitor (LHRBI) which acts to inhibit binding of LH or hCG to its receptor. This is a polypeptide which exists in a low molecular weight from ∿3800 daltons and in a high molecular weight form >10,000 daltons. This LHRBI increases in the porcine corpus luteum as it gets older and may play a role in decline of luteal function occurring late in the luteal phase. It also inhibits LH stimulation of progesterone secretion.

## Acknowledgements

This work was supported by grants from NIH (HD 08834) and the Ford Foundation (760-0530).

The discussions and able collaboration of Drs. Seymour Pomerantz, Torbjorn Hillensjö, Larry Anderson, Gary Hodgen and Satish Batta and of Mr. Frederick Schaerf, Mrs. Saundra Tucker Mrs. Alison Kripner and Ms. Sandra Fowler is appreciated.

## REFERENCES

(1) C.P. Channing and A. Tsafriri. Metabolism 26(1977)413.

(2) J. Marsh. Biol. Reprod. 14(1976)30.

(3) R.J. Ryan and C.Y. Lee. Biol. Reprod. 14(1976)16.

(4) J.S. Richards and A.R. Midgley, Jr. Biol. Reprod. 14(1976)82.

(5) A.J. Zeleznik, A.R. Midgley, Jr. and L.E. Reichert, Jr. Endocrinology 95(1974)818.

(6) R. Nakano, T.A. Kahori, K. Katayama, and S. Tojo. J. Reprod. Fert. 51(1977)23.

(7) C.P. Channing, L.D. Anderson and S.K. Batta, in: Clinics in Obstetrics and Gynecology, Vol. 5, ed. R. Tyson (Saunders, London, 1978) in press.

(8) L.D. Anderson and C.P. Channing. Fed. Proc. 37(1978)283, (Abstract 381).

(9) C.P. Channing. Proc. Soc. Exper. Biol. Med. 149(1975)238.

(10) A.M. Lindsey and C.P. Channing, in: Workshop on Ovarian Function ed. A.R. Midgley, Jr. and W. Sadler (Raven Press, New York, 1978) in press.

(11) A.M. Lindsey and C.P. Channing. Biol. Reprod. (1978) Submitted for publication.

(12) K.H. Thanki and C.P. Channing. Endo. Res. Comm. 3(1976)319.

(13) K.H. Thanki and C.P. Channing. Endocrinology (1978) in press.

(14) P.L. Rayford, H.J. Brinkley, E.P. Young and L.E. Reichert. J. Anim. Sci. 39(1976)348.

(15) G.D. Niswender, L.E. Reichert and D.R. Zimmerman. Endocrinology 87(1970)576.

(16) C.P. Channing and S. Kammerman. Endocrinology 92(1973)531.

(17) C.P. Channing and S. Kammerman. Biol. Reprod. 10(1974)179.

(18) C.P. Channing. Endocrinology 87(1970)49.

(19) C.P. Channing. Endocrinology 87(1970)156.

(20) L.D. Anderson, F.W. Schaerf and C.P. Channing, in: Workshop on Ovarian Follicular and Corpus Luteum Function. ed. C.P. Channing, J. Marsh and W. Sadler (Plenum Press, New York, 1978) in press.

(21) C.N. Sakai and C.P. Channing. Fed. Proc. 37(1978)723 (Abstract 2676).

(22) C.N. Sakai and C.P. Channing. Endocrinology (1978) submitted.

(23) C.N. Sakai and C.P. Channing. Endocrinology (1978) submitted.

(24) K. McNatty and R.S. Sawers. J. Endocrinology 66(1975)391.

(25) A. Tsafriri and C.P. Channing. Endocrinology 96(1975)922.

(26) A. Tsafriri and C.P. Channing. J. Reprod. Fert. 43(1975)149.

(27) A. Tsafriri, S.H. Pomerantz and C.P. Channing. Biol. Reprod. 15(1976)235.

(28) S. Lipford-Stone, A. Schwartz-Kripner, S.H. Pomerantz and C.P. Channing. Biol. Reprod. (1978) in press.

(29) C.P. Channing, S. Lipford-Stone, A.S. Kripner and S.H. Pomerantz, in: Novel Aspects of Reproductive Physiology C.H. Spilman and J.W. Wilks (Spectrum Press, New York, 1978) p. 37.

(30) T. Hillensjö, A.S. Kripner, S.H. Pomerantz and C.P. Channing, in: Workshop on Ovarian Follicular and Corpus Luteum Function. ed. C.P. Channing, J. Marsh and W. Sadler (Plenum Press, New York, 1978) in press.

(31) C.P. Channing, T. Hillensjö and F.W. Schaerf, in: Clinics in Endocrinology and Metabolism ed. M. Lipsett and G. Ross (Saunders, London, 1978) in press.

(32) N. Darga and L.E. Reichert, in: Workshop on Ovarian Follicular and Corpus Luteum Function, ed. C.P. Channing, J. Marsh and W. Sadler. Plenum Press, New York (in press) 1968.

(33) F. Ledwitz-Rigby, M. Stetson and C.P. Channing. Biol. Reprod. 9(1973)94 (Abstract 85).

(34) F. Ledwitz-Rigby, B. Rigby, V.L. Gay, M. Stetson, J. Young and C.P. Channing. J. Endocr. 74(1977)175.

(35) J. Bernard. Compt. Rendu des Sci. Soc. Biol. 167(1973)882.

(36) F. Ledwitz-Rigby and B.W. Rigby, in: Workshop on Ovarian Follicular and Corpus Luteum Function. ed. C.P. Channing, J. Marsh and W. Sadler (Plenum Press, New York, 1978) in press.

(37) C.S. Cambell and N.B. Schwartz. J. Toxicol. Environ. Health (1978) in press.

(38) S. Chappel and C.A. Barraclough. Endocrinology 98(1976)927.

(39) F.F. DeJong and R.M. Sharpe. Nature 263(1976)71.

(40) M.L. Marder, C.P. Channing and N.B. Schwartz. Endocrinology 101(1977)1639.

(41) N.B. Schwartz and C.P. Channing. Proc. Nat. Acad. Sci. U.S.A. 74(1977)5721.

(42) J. Hoffman and N.B. Schwartz. Abstract presented at Society for Study of Reproduction 1978 Annual Meeting.

(43) L. DePaolo, P.M. Wise, L.D. Anderson, C.A. Barraclough and C.P. Channing. Abstract presented at the 1978 Annual Meeting of the Society for the Study of Reproduction.

(44) J. Lorenzen, N.B. Schwartz and C.P. Channing. Biol. Reprod. (1978) in press.

(45) J. Lorenzen, N.B. Schwartz and C.P. Channing. Fed. Proc. 1978 (Abstract).

(46) L. Lagace, F. Labrie, N.B. Schwartz, J. Lorenzen and J.H. Dorrington. Abstract in Program of 1978 Annual Meeting of The Endocrine Society.

(47) D. Shander, L.D. Anderson, C.A. Barraclough and C.P. Channing. in: Workshop on Ovarian Follicular and Corpus Luteum Function. ed. C.P. Channing, J. Marsh and W. Sadler (Plenum Press, New York (1978) in press.

(48) C.P. Channing, L.D. Anderson and G. Hodgen, in: Workshop on Ovarian Follicular and Corpus Luteum Function. ed. C.P. Channing, J. Marsh and W. Sadler (Plenum Press, New York, 1978) in press.

(49) Niswender, G.D., S.E. Monroe, W.D. Peckham, A.R. Midgley, E. Knobil, and L.E. Reichert. Endocrinology 88(1971)1327.

(50) G. Hodgen, J.W. Wilks, J.L. Vaitulaitis, H.C. Chen, H. Papkoff and G.T. Ross. Endocrinology 99(1966)137.

(51) C.P. Channing and S.P. Coudert. Endocrinology 98(1976)568.

(52) S.K. Batta, A.C. Wentz, J. Sulewski. Abstract presented at the 1978 Annual Meeting of the Society for The Study of Reproduction.

(53) B.M. Sherman and S.G. Kormeman. J. Clin. Endocr. Metab. 42(1976)629.

(54) D.T. Baird, C. Corber, D.W. Davidson, W.M. Hunter, E.A. Michie and P.F.A. VanLock. J. Clin. Endocr. Metab. 45(1977)798.

(55) K. Miller, J.K. Crister and O.J. Ginther. Abstract presented at the 1978 Annual Meeting of the Society for The Study of Reproduction.

(56) K. Yang, N.A. Saaman and D.N. Ward. Endocrinology 98(1976)233.

(57) C.N. Sakai, B. Engel and C.P. Channing. Proc. Soc. Exper. Biol. Med. 155(1977)373.

(58) S. Tucker, L. Kumari and C.P. Channing, in: Workshop on Ovarian Follicular and Corpus Luteum Function. ed. C.P. Channing, J. Marsh and W. Sadler (Plenum Press, New York, 1978) in press.

DISCUSSION - CHAIRPERSON - N.B. SCHWARTZ

J. FINDLAY: Could Dr. Channing briefly please summarize what the stability of some of these non-steroidal factors is when they are purified. Secondly, can they be detected in ovarian vein blood or peripheral blood where they should be present if they are to have a physiological role controlling pituitary function.

C.P. CHANNING: We have provided Dr. Darrell Ward with samples of charcoal treated porcine follicular fluid which have been kept either at room temperature, or 4°C or frozen for 4 days. Subsequently the samples were frozen and sent to Neena Schwartz and Janice Lorenzen who assayed them for inhibin activity. Surprisingly the period of standing at room temperature failed to alter inhibin activity! Further studies on the heat stability of more purified inhibin preparations are currently being carried out by Drs. Lorenzen, Schwartz and Ward. If charcoal treated porcine follicular fluid is heated to 60°C for 1 hour it retains activity whereas after heating to 80°C for 1 hour, activity is destroyed (Lorenzen, Schwartz, Pomerantz and Channing, Biol. Reprod. in press, 1978).

J. WEISZ: Two questions, Dr. Channing. In your slide on the follicular fluid in monkeys, it appeared to me that two monkeys responded beautifully, and two monkeys didn't do so well. I wonder whether the degree of inhibition in the latter two was statistically significant? I wonder also whether the degree of difference in follicular development correlated with these differences in suppression, i.e., effectiveness of follicular fluid in that particular monkey. My second question relates to your LH binding inhibiting factor. Is this like the one described by Dr. Ward and coworkers? That is, something that appears on storage of the corpus luteum, or are you able to obtain this from the fresh corpus luteum? This may be of importance in trying to determine whether we are dealing with a useful molecule for experimentation or something that the body really uses for its own purposes.

C.P. CHANNING: We did not experiment on enough monkeys yet to obtain adequate data to enable us to statistically correlate the degree of inhibition of follicular development with the percent inhibition of serum FSH. More animals are being examined and an answer should be forthcoming next year. The one animal (C-1) out of the 4 examined who did not show much depression in serum FSH after injection of follicular fluid had a high basal FSH level and was perimenopausal. All of the other 3 animals with 60-80% inhibition of serum FSH had complete loss of granulosa cells or no detectable follicles in the expected preovulatory time of that menstrual cycle, demonstrating that this was probably more than the maximal dose which led to a decrease in FSH which was essential for follicular growth. Studies with various doses (i.e., submaximal doses) of follicular fluid will be carried out to further study the relation between the degree of depression of FSH and inhibition of follicular growth,

J. WEISZ: The second part of my question concerned the morphological appearance of the follicles. How did this correlate with the suppression of FSH? This may provide a clue whether we are only dealing with a substance that suppresses FSH, or also with some other more direct effect on the developing follicles.

C.P. CHANNING: The question is very meaningful and at this stage of our studies it is impossible to definitively state if the inhibition of follicular function and granulosa cell growth was due to the follicular fluid induced decrease in serum FSH or due to some other component in the follicular fluid. In all instances there was a decrease in granulosa cell growth and viability as well as oocyte viability with or without a decrease in thecal size. This most likely is indicative of an inhibition of FSH since we have shown that FSH added to monkey granulosa cell cultures, but not to thecal cultures, can stimulate progesterone secretion (C.P. Channing, Manuscript in preparation, 1978).

The LHRBI isolated by us from the porcine corpus luteum is most likely similar to that isolated by Yang and Ward from rat corpus luteum. In both the luteinized rat ovary and porcine corpus luteum LHRBI activity increases upon storage in the freezer. More recently Yang and Ward (unpublished observations) found that LHRBI could be extracted from fresh rat and pig corpus luteum tissue with Triton X-100 demonstrating that the freezing is not obligatory for recovery of LHRBI from corpus luteum. It is possible that the LHRBI is bound up with lipid which may act to stabilize it since treatment of porcine corpus luteum extracts with ether lead to destruction of the LHRBI activity (Tucker and Channing, Manuscript in this Workshop).

N.B. SCHWARTZ: I would like to take a shot at Dr. Weisz' question from the standpoint of the rat. Dr. Channing showed our data on the ovariectomized metestrous rat which has been a very good dose-dependent model for folliculostatin activity--we're clearly getting a good response in the absence of the ovary.

But we also do see in rats, particularly those without ovaries, some variability which we cannot quite explain.

# THE INFLUENCE OF FOLLICULAR FLUID ON PROGESTERONE SECRETION BY PORCINE GRANULOSA CELLS IN VITRO

Florence Ledwitz-Rigby and Brian W. Rigby

Department of Biological Sciences
Northern Illinois University
DeKalb, Illinois 60115

Follicular fluid, once considered to be an ultrafiltrate of serum, has been found to contain many components which make it a unique biological fluid (1,2,3,4,5,6,7). The contents of follicular fluids have been reported to change with the phases of the estrous or menstrual cycles (8,9,10) and with the size of the follicles (4,11). Recently a number of reports have indicated that components of follicular fluid are capable of influencing the function of granulosa cells (11,12,13,14) oocytes (15) follicular growth initiation (16) whole ovaries (17) the hypothalamic-pituitary axis (18) and Yoshida Ascites cells (19). Many of these observations have been made with fluid from one species and cells of another. Thus, follicular fluid may have numerous important regulatory roles in mammalian reproductive function.

We have concentrated our studies on examining the ability of porcine follicular fluid to influence luteinization of porcine granulosa cells in vitro. We have found that the addition of follicular fluid to culture media has striking effects on granulosa cell morphology, progesterone secretion and LH stimulation of cyclic-AMP accumulation (11,12,20). There appear to be at least two luteinization influencing factors in porcine follicular fluid. One has been observed to enhance progesterone secretion by granulosa cells from small, immature antral (1-2mm diameter) follicles (20). The other inhibits progesterone secretion by granulosa cells from medium (3-5mm) and large preovulatory (6-12mm) follicles (11,21). The inhibitor is found predominantly in fluid from small follicles (11), whereas, the stimulator is most active in fluid from large follicles, but also exists in the fluid from small follicles (20).

Follicular Fluid Stimulation of Progesterone Secretion by Granulosa Cells from Immature Follicles.

Granulosa cells from small antral follicles secrete little progesterone when cultured in tissue culture media supplemented with serum (22). Thanki and Channing (23) have demonstrated that progesterone secretion by such cells can be enhanced with the addition of insulin, thyroxine, cortisol, FSH and LH to the media. These hormones, however, do not bring about complete luteinization in the presence of serum. It is possible that the maturation of granulosa cells in vivo requires the influence of a local ovarian factor to allow them to fully respond to the hormones mentioned above.

We have cultured granulosa cells from small follicles in monolayer for six days as previously described (11,12,24). The cells were incubated in either: a) 50% adult female porcine serum: 50% TC 199 (50% Serum), b) 50% fluid from small follicles: 50% TC 199 (50% SFFl) or c) 50% fluid from large follicles: 50% TC 199 (50% LFFl). All treatments were triplicated within each experiment. Culture media were changed every two days and assayed for progesterone content by radioimmunoassay as previously described (11). The morphology of the live unstained cells was examined with phase and Hoffman modulation microscopy. This same methodology was also employed for the studies on granulosa cells from medium-sized and large follicles described in the next section.

Granulosa cells cultured in 50% LFFl secreted substantially more progesterone during all six days of culture than cells cultured in 50% SFFl or 50% Serum in all twelve experiments conducted. Over the first two days of culture granulosa cells incubated in 50% LFFl secreted 14.1 ± 4.0 fold more progesterone than cells cultured in 50% Serum (n=12). On days 3 and 4 of culture, progesterone secretion was enhanced 46.7 ± 20.9 fold (n=6) and 28.1 ± 11.8 fold on days 5 and 6 (n=6). The ability of 50% SFFl to enhance progesterone secretion over that observed in cultures incubated in 50% Serum was less dramatic and took longer to exert its influence. When granulosa cells were incubated in 50% SFFl for 2 days, progesterone secretion was enhanced in only 6 out of 12 experiments. In those experiments in which progesterone secretion was stimulated, cells incubated in 50% SFFl secreted 5.1 ± 2.3 fold more progesterone than cells incubated in 50% Serum (n=6). However, when the experiments were continued up to six days, progesterone secretion was enhanced on days 3 through 6 in all experiments including ones in which no stimulatory effect had been observed over the first two days. Cells incubated in 50% SFFl secreted 1.8 ± 0.2 fold as much progesterone as cells incubated in 50% Serum on days 3 and 4, and 4.5 ± 2.1 fold as much on days 5 and 6 (n=6).

We have also observed that both SFFl and LFFl enhanced progesterone secretion in response to FSH and LH by granulosa cells from small follicles. Granulosa cells were cultured in monolayer for six days in 50% Serum, 50% SFFl or 50% LFFl in the presence or absence of 100 ng NIH-LH-S19 and/or 2 μg NIH-FSH-S11/ml. The results of a typical experiment (one out of five) are shown in Figure 1. FSH and LH alone or in combination significantly enhanced progesterone secretion by granulosa cells in all three culture media on all six days of culture. Progesterone secretion was always greatest in cultures containing LFFl. Cells cultured in 50% SFFl were similar in response to the gonadotrophins to the cells cultured in 50% Serum on the first two days of culture, but by days three through six secreted significantly more progesterone in response to LH and/or FSH. The relative stimulation of progesterone secretion by the gonadotrophins over that secreted in the absence of exogenous hormones was greater for cells incubated in SFFl than in LFFl. By day four of culture, granulosa cells incubated in 50% SFFl or 50% LFFl in the presence or absence of gonadotrophins were primarily epithelioid-like (approximately 80% of the cells) whereas the cells incubated in 50% Serum were primarily fibroblast-like (approximately 80% of the cells).

Follicular Fluid Inhibition of Progesterone Secretion by Granulosa Cells from Medium-sized and Large Preovulatory Follicles.

Granulosa cells from medium sized (3-5mm diameter) porcine follicles will luteinize in vitro in the presence of FSH or LH. Granulosa cells from large (6-12mm diameter) follicles which have been exposed to the preovulatory surge of gonadotrophins in vivo will luteinize in vitro without requiring further exposure to gonadotrophins (22). In sharp contrast to the response of granulosa cells from immature follicles to follicular fluids, granulosa cells from both medium and large follicles secreted less progesterone when cultured in 50% SFFl than when cultured in 50% serum. LFFl no longer exerted a highly stimulatory action but was merely as permissive of progesterone secretion as serum (Table 1).

In addition to depressing progesterone secretion, SFFl depressed LH stimulation of cyclic AMP accumulation and prevented the morphological luteinization of granulosa cells from large follicles. Here again, LFFl was at least as permissive as serum in both these actions (11,12). Incubations of granulosa cells from large follicles in 50% SFFl lasting 18 hours or less were ineffective in suppressing LH stimulation of cyclic AMP accumulation. 24 hour incubations significantly decreased LH stimulated cyclic AMP accumulation in only 50% of the experiments conducted but 48 hour incubations were always effective, suppressing LH stimulated cyclic AMP accumulation to 29.0 ± 2.6% of control levels (n=57) (11). Similarly progesterone secretion by granulosa cells from large follicles was not depressed over the first 24 hours of culture in 50% SFFl but was only 14.0 ± 3.0% (n=3) that secreted by cells

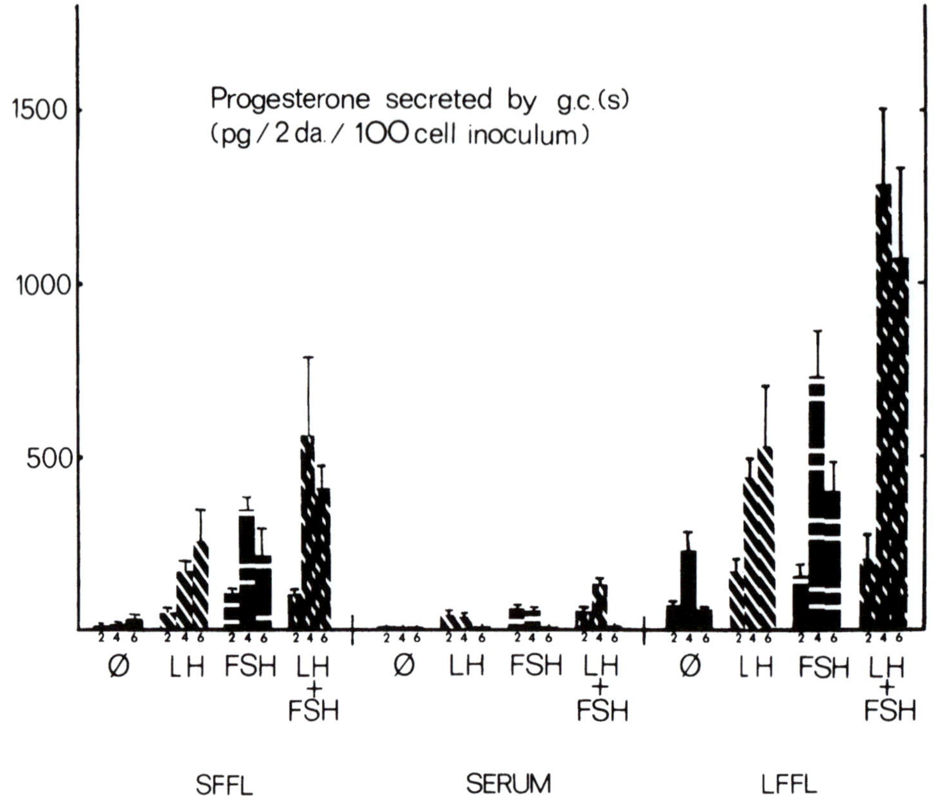


Figure 1. Progesterone secretion by granulosa cells from small (1-2 mm) porcine follicles during 6 day monolayer cultures in 50% TC 199: 50% fluid from small follicles (SFF1), 50% fluid from large (6-12 mm) follicles or 50% serum with or without 2 μg NIH-F-S11/ml and/or 100 ng NIH-LH-S19/ml. Each bar represents the mean of three cultures. The lines indicate the standard errors of the means.

TABLE I The influence of follicular fluids on progesterone secretion by granulosa cells from medium and large follicles

| | | CULTURE MEDIUM | | |
|---|---|---|---|---|
| | | 50% SFF1 | 50% LFF1 | 50% Serum |
| | | PROGESTERONE pg/100 cells/48 hours | | |
| Exp. # | SIZE FOLLICLES CELLS OBTAINED FROM | | | |
| 1 | medium | 0.2±0.02[a] (3) b | 1.6±0.5 (3) | 1.2±0.08 (3) |
| 2 | medium | 7.2±0.5 (3) | 14.9±1.0 (3) | 13.4±0.4 (3) |
| 3 | large | 11.2±0.4 (3) | 23.3±0.9 (3) | 24.7±0.2 (3) |
| 4 | large | 4.6±1.5 (3) | 9.3±0.4 (3) | 9.9±0.6 (3) |

a-means ± s.e.m.
b-number of cultures

incubated in 50% LFF1 during hours 25 to 48 of incubation (25).

Morphometric analysis of the ultrastructure of granulosa cells from large follicles following 48 hour monolayer cultures in 50% SFF1, 50% LFF1 or 50% serum indicated that cells cultured in 50% SFF1 were viable but less differentiated with respect to intracellular machinery for progesterone secretion than granulosa cells cultured in 50% LFF1 or 50% Serum. Cells incubated in 50% SFF1 contained significantly less lipid volume ($p < 0.0025$), mitochondrial volume ($p < 0.0005$) and lysosomal-appearing bodies ($p < 0.0005$) than cells cultured in 50% LFF1 or 50% Serum. Whorled smooth endoplasmic reticulum was found five times more frequently and in larger configurations (up to 3.5 μ in cross-sectional diameter) in cells cultured in the presence of LFF1 or Serum than in SFF1. The granulosa cells exposed to SFF1 had significantly less mitochondrial membrane area ($p < 0.005$) and less smooth endoplasmic reticulum ($p < 0.05$) than cells exposed to LFF1 or Serum. Rough endoplasmic reticulum, however, was significantly greater in cells incubated in SFF1 ($p < 0.01$) or Serum ($p < 0.01$) as compared with cells incubated in LFF1. LFF1 treated cells contained more lipid than Serum treated cells ($p < 0.05$) (12). These last two observations suggest that cells incubated in 50% LFF1 may undergo the morphological changes of luteinization somewhat faster than cells incubated in 50% Serum. By day six of culture, granulosa cells grown in the presence of either Serum or LFF1 appeared well luteinized at the light microscopic level whereas the cells grown in SFF1 did not (11).

The ability of LH to stimulate progesterone secretion by granulosa cells from medium sized follicles was reduced during the first 48 hours of monolayer culture in 50% SFF1 as compared to cells cultured in 50% LFF1 or 50% Serum. The minimal concentration of LH (100 ng/ml) which elicited a maximal response, was not different, however, for cells incubated in the presence of SFF1, LFF1 or Serum (Table 2).

When the granulosa cells from medium sized follicles were preincubated for two days in 50% SFF1 or 50% LFF1 and then cultured in 10% Serum in TC 199 containing 100 ng NIH-LH-S19/ml on days 3 and 4 the cells preincubated in 50% SFF1 secreted less progesterone than did the cells preincubated in 50% LFF1 (Figure 2). Similarly the ability of the phosphodiesterase inhibitor, 3-isobutyly-1-methylxanthine (MIX) to stimulate progesterone secretion was decreased in cells exposed to SFF1 as compared to those exposed to LFF1. Interestingly however, when granulosa cells from medium sized follicles, which had been preincubated in 50% SFF1 for two days were cultured in the presence of both 100 ng LH/ml and 0.2 mM MIX, they secreted as much progesterone as cells which had been preincubated in 50% LFF1 and then exposed to the combination of LH and MIX (Figure 2). This observation suggests that LH is fully capable of stimulating progesterone secretion in SFF1 treated granulosa cells

TABLE 2. The Influence of Follicular Fluids on LH stimulation of Progesterone Secretion by Granulosa Cells from Medium-sized Follicles during the first 48 hours of culture.

| LH added (ng NIH-LH-S19/ml) | CULTURE MEDIUM 50% SFF1 | 50% LFF1 | 50% Serum |
|---|---|---|---|
| | PROGESTERONE pg/100 cells/48 hours) | | |
| none | 0.18±0.02[a] (3) [b] | 1.56±0.05 (2) | 1.24±0.08 (3) |
| 10 | 0.38±0.02 (3) | 5.87±1.60 (3) | 2.29±0.23 (3) |
| 100 | 1.19±0.18 (3) | 28.27±2.75 | 31.32±2.10 (3) |
| 1000 | 1.51±0.27 (3) | 28.69±2.33 (2) | 31.27±0.80 (3) |
| 10,000 | 1.21±0.15 (2) | ----- | 22.73±1.78 (2) |

a-means ± s.e.m. b-number of cultures

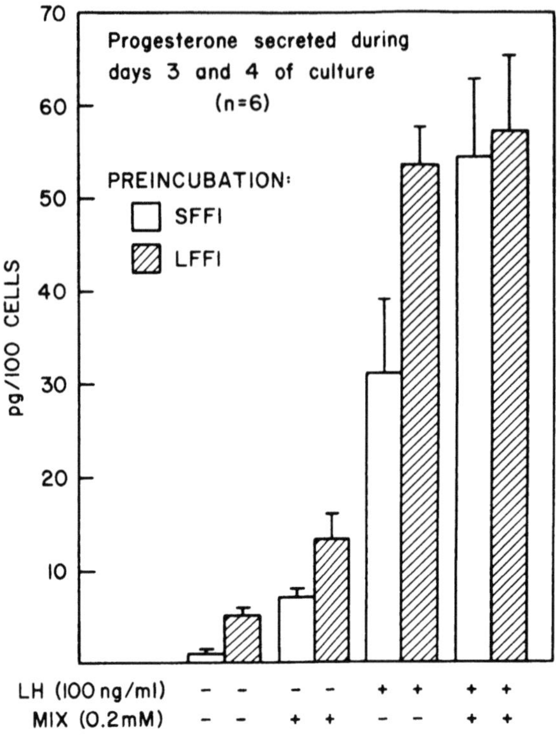


Figure 2. Progesterone secretion during days 3 and 4 of culture by porcine follicles following 48 hour preincubations in 50% TC 199: 50% fluid from small (1-2 mm) follicles (SFF1) or 50% fluid from large (6-12 mm) follicles. On days 3 and 4 of culture the granulosa cells were incubated in 90% TC 199: 10% serum with or without 100 ng NIH-LH-S19/ml and/or 0.2 mM 3-isobutyl-1-methyl xanthine (MIX). Each bar represents the mean of six cultures from two separate experiments. The lines indicate the standard errors of the means.

in the presence of a phosphodiesterase inhibitor (21). We have previously reported that either theophylline or MIX can restore LH stimulated cyclic AMP accumulation by granulosa cells from large follicles preincubated in 50% SFF1 to approximately 50% of control levels (26). In combination, these observations suggest that the defect in the ability of LH to stimulate cyclic AMP levels is at least partially responsible for the defect in the ability of LH to stimulate progesterone secretion.

Preliminary attempts to characterize the follicular fluid inhibitor of luteinization have shown that it:

1) was not an unbound steroid, as follicular fluids extracted of 98% of their steroids with charcoal or petroleum ether were as capable of altering luteinization of granulosa cells from large follicles as unextracted fluids (11),

2) was not likely due to the Prolactin content of the follicular fluids as concentrations of Prolactin of up to 50 ng NIH-P-S12/ml were found to enhance progesterone secretion by porcine granulosa cells in vitro (11) and measurement of porcine follicular fluid content of Prolactin using rat Prolactin as a standard indicated that the Prolactin contents of LFF1 and SFF1 did not differ (27),

3) was not dialysable (11,13),

4) had a molecular weight of greater than 100,000 based on retention by amicon filter Xm 100A (28),

5) was stable at $60^0$ (11) but not at $80^0$ (13),

6) exhibited the same activity whether the fluid was collected within minutes after slaughter or after a few hours (29) and

7) was stable after storage at $-20^0$ for up to two years and several freeze-thaws (25).

## CONCLUSIONS AND A WORKING HYPOTHESIS

As the follicles mature, there are dramatic changes occuring in both the composition of the follicular fluid and the ability of the granulosa cells to respond to numerous hormones (30,31,32) and to the follicular fluid luteinization factors which we are currently characterizing. We have hypothesized that for follicular development to proceed to the preovulatory state the granulosa cells and the fluid must mature in a coordinated manner. If the cells mature to the point of responding to the luteinization inhibitory factor found in fluid from small follicles before the fluid changes into a more permissive or even stimulatory environment for luteinization, further follicular development might be inhibited and atresia might result (33). The factors controlling the changes of the luteinization influencing activities of follicular fluid are unknown but are most likely related in some way to the factors currently known to influence atresia in vivo, such as the concentrations of FSH, estrogens and androgens (34,35,36).

## SUMMARY

Porcine follicular fluid contains at least two factors which can modify porcine granulosa cell function in vitro. One factor, found primarily in fluid from 1-2 mm follicles inhibits LH stimulated cyclic AMP accumulation, basal and LH stimulated progesterone secretion, and morphological luteinization of granulosa cells from medium sized (3-5 mm) and large (6-12 mm) follicles. Following inhibition by the follicular fluid luteinization inhibitor, incubation with both LH and a phosphodiesterase inhibitor partially restores cyclic AMP accumulation and fully restores progesterone secretion. This suggests that the defect induced by the luteinization inhibitor is primarily an alteration of cyclic nucleotide metabolism. Another factor found primarily in fluid from large preovulatory follicles stimulates basal and gonadotrophin stimulated progesterone secretion by granulosa cells from small antral follicles. We hypothesize that it may be necessary for the fluid within an individual follicle to shift from being inhibitory of luteinization to being permissive or stimulatory for the granulosa cells to mature rather than become atretic.

REFERENCES

(1) R.G. Edwards. J. Reprod. Fertil. 37(1974)189.

(2) A.W. Scheutz and A. Anisowicz. Biol. Reprod. 11(1974)64.

(3) R. Shalgi, P. Kraicer, A. Rimon, N. Pinto and N. Soferman. Fertil. Steril. 24(1973)429.

(4) R.W. McGaughey. Biol. Reprod. 13(1975)147.

(5) P. Cerletti and L. Zichella. Clin. Chim. Acta. 6(1961)581.

(6) S.S.C. Chang, J.D. Jones, R.D. Ellefson and R.J. Ryan. Biol. Reprod. 15(1976)321.

(7) M. Yanagashita, D. Rodbard and V.C. Hascall. Fed. Pro. Fed. Am. Soc. Exp. Biol. 37(1978)1776.

(8) R.V. Short. J. Endocrinol. 22(1961)153.

(9) K.P. McNatty, W.M. Hunter, A.S. McNeilly and R.S. Sawers. J. Endocrinol. 64(1975)555.

(10) H. Eiler and A.V. Nalbandov. Anat. Rec. 175(1973)313.

(11) F. Ledwitz-Rigby, B.W. Rigby, V.L. Gay, M. Stetson, J. Young and C.P. Channing. J. Endocrinol. 74(1977)175.

(12) J.S. Alexander, B.W. Rigby and F. Ledwitz-Rigby. Biol. Reprod. 18(1978)in press.

(13) J. Bernard. C.R. Seances Soc. Biol. Ses. Fil. 167(1973)882.

(14) J. Bernard and A. Psychoyos. J. Reprod. Fertil. 49(1977)355.

(15) A. Tsafriri and C.P. Channing. Endocrinology 96(1975)922.

(16) H. Peters, A.G. Byskov and M. Faber, in: The Development and maturation of the ovary and its functions; Int. Cong. Ser. No. 267, ed H. Peters. (Excerpta Medica, Amsterdam, 1973).

(17) Kraiem and B. Lunenfeld. 58th Annual Meeting of the Endocrine Society (1976) Abstract #169.

(18) N.B. Schwartz and C.P. Channing. Proc. Nat. Acad. Sci. U.S.A. 74(1977)5721.

(19) G.P.M. Moore, S. Lintern-Moore, H. Peters, A.G. Byskov, M. Anderson and M. Faber. J. Cell Physiol. 86(1975)882.

(20) F. Ledwitz-Rigby and B.W. Rigby. J. Cell Biol. 75(1977)178a.

(21) F. Ledwitz-Rigby. 11th Annual Meeting of the Society of the Study of Reproduction (1978) Abstract #

(22) C.P. Channing. Endocrinology 87(1970)156.

(23) K.H. Thanki and C.P. Channing. Endocr. Res. Commun. 3(1976)319.

(24) C.P. Channing and F. Ledwitz-Rigby, in: Methods in Enzymology Vol. 39, Part D, ed. J.G. Hardman and B.W. O'Malley (Academic Press, New York, 1975) p. 183.

(25) F. Ledwitz-Rigby. Unpublished observations.

(26) F. Ledwitz-Rigby. Seventh Annual Meeting of the Society for the Study of Reproduction (1974) Abstract # 16.

(27) Assay Courtesy of N.B. Schwartz.

(28) B.W. Rigby. Unpublished observations.

(29) J.S. Alexander, F. Ledwitz-Rigby and B.W. Rigby. Unpublished observations.

(30) C.P. Channing and F. Ledwitz-Rigby, in: The Physiology and Genetics of Reproduction, Vol II, ed. E. Coutinho and F. Fuchs (Plenum Press, New York, 1974) p. 353.

(31) S. Kammerman and C.P. Channing. Biol. Reprod. 10(1974)1215.

(32) H.J. Rajaniemi, A.R. Midgley, Jr., J.A. Duncan and L.E. Reichert, Jr. Endocrinology 101(1977)898.

(33) F. Ledwitz-Rigby, in: Ovarian Follicular Development and Function, eds. A.R. Midgley and W.A. Sadler (Raven Press, New York, 1978).

(34) G. Greenwald. J. Reprod. Fertil 2(1961)351.

(35) S.M. Harman, J.P. Louvet and G.T. Ross. Endocrinol. 96(1975)1145.

(36) J.P. Louvet, S.M. Harman, J.R. Schreiber and G.T. Ross. Endocrinol. 97(1975)366.

Acknowledgements: The authors gratefully acknowledge: Jane Glaser for drawing and photographing the figures, and Christine Mlot, John Hauwiller, Patricia Barnes, Bernadine Upshaw and Shelita Young for technical assistance. The study on Ultrastructure of Porcine Granulosa cells following Incubation with Follicular Fluids, was done by James S. Alexander in partial fulfillment of the M.S. degree at N.I.U.

These studies have been supported in part by: The Ford Foundation, The Population Council, The American Cancer Society - Illinois Division, the Pan American Health Organization and Northern Illinois University.

# INHIBITION OF IN VIVO AND IN VITRO FSH SECRETION BY STEROID-FREE HAMSTER OVARIAN EXTRACTS

S. C. Chappel, T. Acott and H. G. Spies
Reproductive Physiology
Oregon Regional Primate Research Center
505 N.W. 185th Avenue
Beaverton, Oregon 97005

We designed experiments to determine if a nonsteroidal substance capable of inhibiting the secretion of follicle-stimulating hormone (FSH) exists in the hamster ovary. Proestrous hamster ovaries were homogenized, centrifuged, and treated with dextran-coated charcoal for removal of steroids. Charcoal-treated ovarian extract (CTOE) was injected intraperitoneally at hourly intervals into phenobarbital-treated (phen-treated) proestrous hamsters immediately after ovariectomy at 1300. The CTOE inhibited the selective rise in serum FSH that occurred in phen-treated, ovariectomized (ovx) controls.

To determine the biochemical nature of this ovarian substance, we performed a series of experiments with a hamster pituitary cell culture system as the assay. Gonadotropin secretion was stimulated by addition of $10^{-8}$ M luteinizing-hormone-releasing hormone (LHRH) to the incubation medium. The presence of FSH-inhibiting activity was determined by the degree of inhibition of LHRH-induced FSH secretion after addition of test substances. The CTOE was separated with a Millipore molecular separator into two fractions, $<$ and $>$ 10,000 daltons. The material with a mol wt $>$ 10,000 daltons contained the FSH-inhibiting activity. This high-molecular-weight fraction neither bound to DEAE-cellulose nor was retained by a concanavalin A (Con A)-agarose column. Molecular sieving of this material on Sephadex G-75 gave a mol wt estimate of $<$ 30,000 daltons. The partially purified material exerted little or no inhibitory influence on LHRH-induced luteinizing hormone (LH) secretion. Thus, the mol wt of hamster "folliculostation" appears to be between 10,000 and 30,000 daltons. It contains an overall neutral or positive charge and does not contain the carbohydrate groups that would attach to the lectin (Con A).

## Introduction

Bilateral ovariectomy of rats (1,2) and hamsters (3) results in an immediate elevation in peripheral concentrations of FSH but not LH. This rise in serum FSH levels is only partially suppressed by administration of physiologic amounts of estradiol ($E_2$) (1,3). Some other ovarian substance may exert inhibitory effects on pituitary FSH secretion.

Since 1932, many investigators have proposed the existence of a nonsteroidal secretory product of the testes ("inhibin") that selectively suppresses the release of pituitary FSH (4). "Inhibin-like" material has been demonstrated in seminal fluid (5), testicular extracts (6,7), rete testes fluid (8), extracts of spermatozoa (9) and cultured Sertoli cells (10). Recently, the presence of such a factor was reported in bovine follicular fluid (11). Schwartz and Channing (12) have shown that an FSH-inhibiting substance (folliculostatin) exists in porcine follicular fluid.

Using acutely ovx proestrous hamsters and hamster pituitary cells in culture, we performed the studies described below to determine if folliculostatin exists within hamster ovaries. We also have preliminary data on the physical characteristics of this substance.

## Materials and Methods

### In Vivo Studies

Adult female hamsters (*Mesocricetus auratus*) were caged in groups of five and maintained on a regimen of 14 h of light and 10 h of dark (lights on from 0500 to 1900) at 23°C. All animals exhibited at least three 4-day vaginal estrous cycles before they were assigned to one of the four groups below on the morning of proestrus.

Group 1. Proestrous controls. At hourly intervals from 1300 to 1800, five proestrous hamsters were decapitated and trunk blood was collected. Blood in this and all other groups was allowed to clot at 4°C. The serum was removed, and each sample was snap-frozen and stored until the day of the radioimmunoassay (RIA).

Group 2. Proestrous hamsters treated with phenobarbital. At 1245, 30 proestrous hamsters were treated with phen (100 mg/kg of body weight [BW] subcutaneously [sc]). At hourly intervals from 1300 to 1800, five phen-treated hamsters were decapitated and trunk blood was collected and processed as described for group 1.

Group 3. Phenobarbital-treated, ovariectomized hamsters. Thirty proestrous hamsters were treated with phen (100 mg/kg of BW, sc) and bilaterally ovx at 1245. At hourly intervals from 1300 to 1800, five hamsters were decapitated and trunk blood was processed as described for group 1.

Group 4. Phenobarbital-treated, ovariectomized hamsters given ovarian extract or charcoal-treated ovarian extract. Sixty proes-hamsters were subjected to the experimental manipulations described for group 3. At hourly intervals after ovariectomy until the time of sacrifice, 30 hamsters so treated each received an injection (1 ml, ip) of ovarian extract (OE); the remaining 30 animals received hourly injections (1 ml, ip) of charcoal-treated OE (CTOE). Animals from the OE-treated, ovx and the CTOE-treated, ovx groups were decapitated at hourly intervals for 5 h after ovariectomy. Trunk blood was processed as described above.

Preparation of ovarian extract. Three hundred proestrous hamster ovaries were collected between 1100 and 1300 and homogenized in 150 ml of phosphate-buffered saline (PBS). The OE was centrifuged at 3,000 x g for 30 min. The supernatant was divided into two equal volumes. One volume was divided into 75 1-ml aliquots. The remaining 75 ml of OE were combined with 200 mg of Norit (charcoal) and 20 mg of dextran, vortexed and incubated at 4°C for 30 min. This CTOE was centrifuged again for 30 min and divided into 75 aliquots of approximately 1 ml each. All aliquots were snap-frozen and stored until use. Progesterone (13) and $E_2$ (14) concentrations in OE and CTOE were determined by the appropriate RIA.

## In Vitro Studies

Forty milliliters (80 proestrous hamster ovaries) of CTOE were prepared as described above. The CTOE was separated into two fractions, one > and one < 10,000 daltons, with a Millipore molecular separator. These two fractions were collected, dialyzed against PBS, diluted to a final volume of 4 ml each, snap-frozen and stored until the day of the *in vitro* assay.

The high-molecular-weight material of an additional 40 ml of CTOE (80 proestrous hamster ovaries) was applied to a DEAE-cellulose anion-exchange column. The material that was bound to the resin was eluted from the column with 500 mM NaCl. Both the bound and unbound fractions were dialyzed against PBS and concentrated to a final volume of 4 ml. In the third and fourth column studies, 40 ml of CTOE (mol wt > 10,000 daltons) was divided into two fractions and applied to either a Con A - agarose or a Sephadex G-75 column. Materials not bound to the Con A column were collected. Those that were bound to the Con A column were eluted with α-D-methyl mannoside (100 mM) and collected. These two fractions were dialyzed against

PBS and concentrated to a volume of 4 ml each. The second half of the CTOE was applied to a Sephadex G-75 column. Fractions were pooled to form two groups of material, one > and one < 30,000 daltons. These two pools were dialyzed against PBS and concentrated to 4 ml each. All fractions were snap-frozen and stored until the day of the *in vitro* assay.

All column samples described above were assayed for folliculostatin activity in a hamster pituitary cell monolayer culture. Anterior pituitary cells (0.5 x $10^6$/dish) were cultured for 4 days at 37°C as previously described for rat cells (15). The incubation medium, Dulbecco modified Eagle's medium (DMEM), was supplemented with 15% fetal calf serum (FCS), 3% horse serum (HS), penicillin (50 U/ml) and streptomycin (50 μg/ml). On the 5th day, medium was collected. The cells were washed three times with DMEM without FCS or HS and then suspended in 2.5 ml of a test medium. The test medium contained either 1.5 ml of DMEM plus 1 ml of PBS; 1.5 ml of DMEM in which $10^{-8}$ M LHRH had been dissolved plus 1 ml of PBS; or 1.5 of DMEM in which $10^{-8}$ M LHRH had been dissolved plus 1 ml of the test substance obtained from the various column preparations. After 4 h, the test medium was collected and centrifuged; the supernatant was snap-frozen and stored for RIA. Folliculostatin activity was determined by comparison of the percent inhibition of the LHRH-induced FSH rise versus that in PBS-treated, LHRH-treated cells.

The FSH (16) and LH (17) concentrations in the serum obtained in the *in vivo* studies and in the medium obtained from the *in vitro* studies were determined by the appropriate RIA. The within- and between-assay variabilities for the RIAs used to measure FSH were 7 and 9.5% and for LH, 4.3 and 7.5%, respectively. Within- and between-assay variability was calculated by a comparison of a standard pool of ovx hamster sera measured in each RIA. Statistical differences between control and treatment groups were determined with the Student's *t* test.

## Results

### In Vivo Studies

In proestrous hamsters (Fig. 1), serum concentrations of LH and FSH reached peak levels by 1600 (1818 ± 249 ng/ml and 688 ± 98 ng/ml for LH and FSH, respectively) and returned to basal serum levels by 1800 (302 ± 209 ng/ml and 206 ± 18 ng/ml for LH and FSH, respectively).[1] Injection of phen at 1245 of proestrus inhibited the rise in serum gonadotropins that otherwise would have occurred (Fig. 1). Serum LH and FSH levels remained low throughout the afternoon of proestrus (< 40 ng/ml and < 150 ng/ml for LH and FSH, respectively).

---

[1] Data from Fig. 1 and 2a taken from reference 3.

Proestrous hamsters treated with phen at 1245 and ovariectomy immediately thereafter exhibited a rise in serum FSH within 2 h (Fig. 2A). Elevated serum levels of FSH (810 ± 161 ng/ml) were observed within 5 h. Serum LH levels after ovariectomy remained at basal levels (< 40 ng/ml) throughout the 5-h sampling interval.

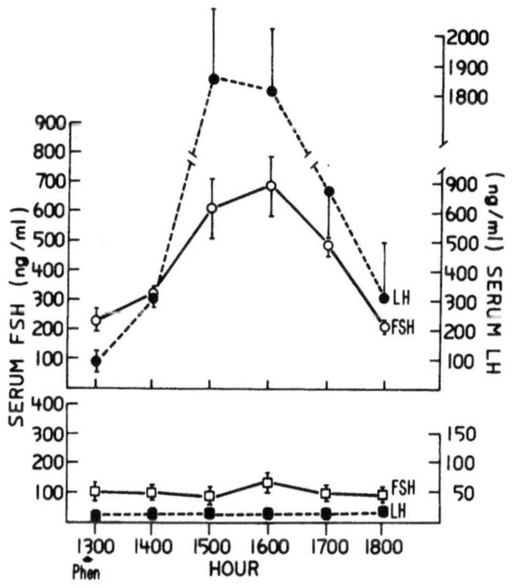


Fig. 1. Serum concentrations of LH (shaded) and FSH (open) on the afternoon of proestrus in control (circles) and phenobarbital-treated (squares) hamsters. Vertical bars indicate the SEM of 5 animals per point.

Hourly injections of OE suppressed the expected rise in serum FSH 3 h after ovariectomy (102 ± 16 ng/ml and 492 ± 53 ng/ml for OE-treated ovx and ovx controls, respectively [$p < 0.01$]). (Fig. 2B). The LH levels also remained basal (< 40 ng/ml) during the first 3 h as they did in group 3 controls. A delayed and attenuated rise in both gonadotropins was observed within 4 to 5 h post ovariectomy (402 ± 38 ng/ml and 136 ± 51 ng/ml of FSH and LH, respectively). Each milliliter of OE had 3.94 ± 0.38 ng/ml of $E_2$ and 3.30 ± 0.73 ng/ml of progesterone. As observed in the OE-treated ovx group, hourly injections of CTOE, which contained 6 ± 2 pg/ml of $E_2$ and 7 ± 3 pg/ml of progesterone (levels comparable to assay blanks) were capable of suppressing the ovariectomy-induced serum FSH rise (Fig. 2C). The serum levels of this gonadotropin dropped from 186 ± 35 ng/ml (immediately before ovariectomy) to 23 ± 20 ng/ml

2 h post ovariectomy. Serum FSH levels began to rise from basal concentrations after 3 h and reached a level of 457 ± 33 ng/ml within 5 h, in spite of the continued injection of CTOE. Serum LH concentrations in these animals remained basal (approximately 50 ng/ml) for 5 h, after which a small elevation (109 ± 60 ng/ml) was observed.

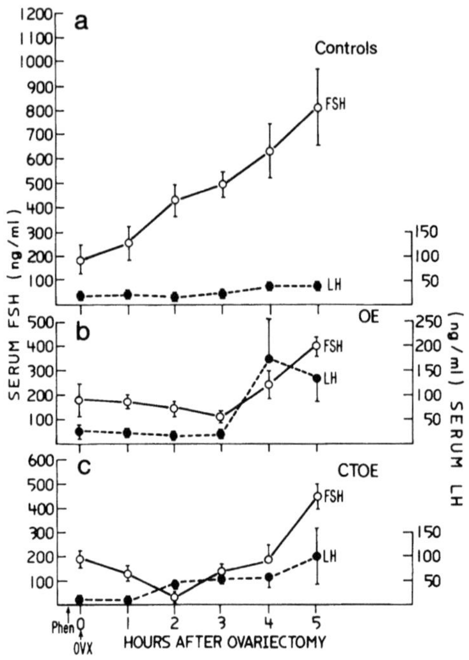


Fig. 2. Changes in serum levels of LH (shaded circles) and FSH (open circles) at hourly intervals after ovariectomy (ovx) (panel a); after ovx plus hourly injections of ovarian extract (OE) (panel b); or after ovx plus hourly injections of charcoal-treated ovarian extract (CTOE) (panel c). Phen = phenobarbital.

In Vitro Studies

Addition of LHRH ($10^{-8}$ M) to the pituitary cell culture medium (DMEM) increased the amount of FSH ($p < 0.02$) and LH ($p < 0.005$) released within 4 h (132 ± 2% and 148 ± 5% versus 100 ± 7% and 100 ± 15%, for FSH and LH concentrations in the medium of unstimulated [PBS-treated] control cells) (Fig. 3). Addition of CTOE with a mol wt < 10,000 daltons had no inhibitory effect on the amount of LHRH-induced FSH or LH release compared to the gonadotropin release in LHRH-plus-PBS-treated control cells. Addition of

CTOE with a mol wt > 10,000 daltons suppressed ($p < 0.001$) the amount of FSH released into the medium (96 ± 11% of the FSH concentration in the PBS-treated control medium) when compared to a 132 ± 2% rise in the FSH concentration of medium collected from LHRH-plus-PBS-treated cells. The LHRH-induced LH release in medium from CTOE-treated (> 10,000 daltons) cells was similar to that observed in the LHRH-PBS-treated group (152 ± 6% rise versus 148 ± 5% rise in medium from LHRH-plus-PBS-treated control cells).

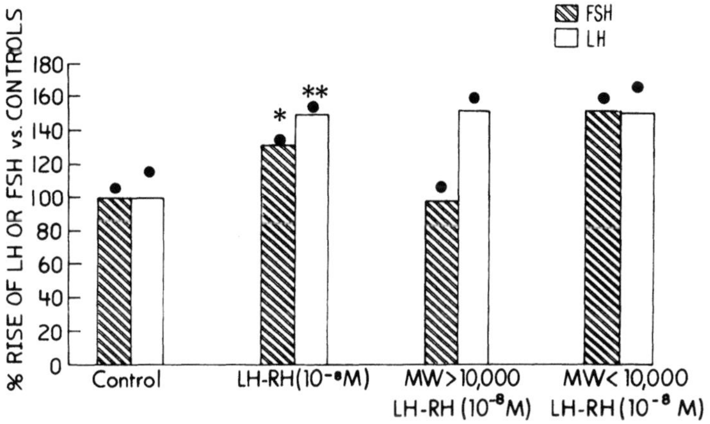


Fig. 3. The effects of LHRH and LHRH plus charcoal-treated ovarian extract (CTOE) > and < 10,000 daltons on the secretion of LH and FSH by cultured anterior pituitary cells. The dots in this and the following figures represent the SEM of 4 dishes per group. * = $P < 0.05$; ** = $p < 0.01$; MW = molecular weight.

When CTOE (> 10,000 daltons) was placed on a DEAE anion-exchange column, the unbound material exhibited a potent inhibitory effect ($p < 0.02$) on the amount of FSH (but not LH) released into the medium (44 ± 25% of the amount of FSH released into the medium of PBS-treated cells compared to a 129 ± 18% elevation in medium FSH concentrations of LHRH-plus-PBS-treated cells (Fig. 4). In this anterior pituitary cell preparation, LHRH did not provoke a significant increase in the FSH concentration of the medium from the control (LHRH-plus-PBS-treated) group (129 ± 18% elevation [$p < 0.2$] versus 101 ± 10% increase in the FSH concentration of medium from PBS-treated control cells). However, the basal FSH secretory rate in this experiment was profoundly affected by DEAE-unbound CTOE (> 10,000 daltons). The LHRH-induced LH rise was affected by neither the DEAE-unbound nor the DEAE-bound material (146 ± 14% and 186 ± 3% rises, respectively). LHRH-plus-PBS treatment elicited a 186 ± 9% rise in medium LH levels versus that of PBS-treated controls (100 ± 2%).

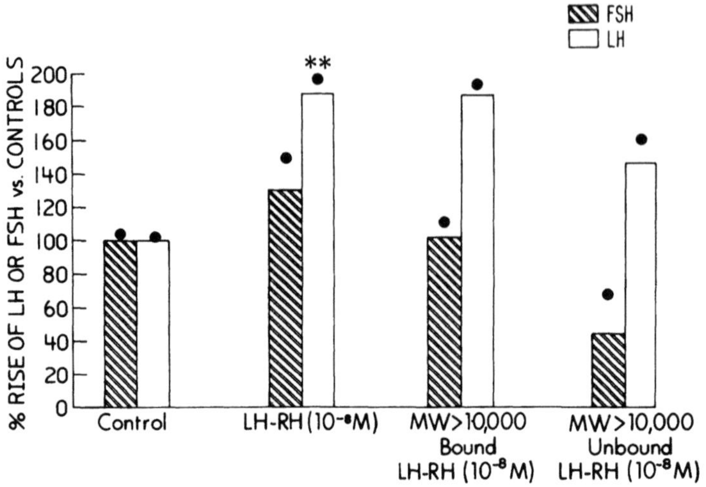


Fig. 4. The effects of LHRH and charcoal-treated ovarian extract (CTOE) (> 10,000 daltons) that was bound or unbound to a DEAE - anion-exchange resin on the secretion of LH and FSH by cultured anterior pituitary cells. MW = molecular weight.

Figure 5 illustrates that CTOE not bound to the Con A column exerted an inhibitory effect ($p < 0.001$) on the FSH secretion in the pituitary cell culture system (82 ± 5% medium concentration increase compared to 256 ± 9% increase in medium concentrations for LHRH-plus-PBS-treated control cells). The FSH-inhibiting material not bound to the Con A column exerted no inhibitory effect on the LHRH-induced LH rise.

Finally, separation of CTOE > and < 30,000 daltons revealed that the FSH-inhibiting activity was present in the < 30,000 dalton fraction. This material, when combined with DMEM and added to the cell culture system, suppressed ($p < 0.001$) the LHRH-induced FSH release (67 ± 5% compared to 256 ± 9% rise in FSH for LHRH-plus-PBS-treated control cells). In this group, the LHRH-induced LH release was unaffected (140 ± 16% versus 180 ± 30% for the LHRH-treated controls) in the presence of CTOE. The > 30,000 dalton CTOE fraction exerted no inhibitory effect on the LHRH-induced FSH rise.

## Discussion

These studies demonstrate that hourly injections of the equivalent of two proestrous hamster ovaries are capable of suppressing the ovariectomy-induced rise in serum FSH in proestrous, phen-treated hamsters. The OE contains large amounts of $E_2$ and progesterone. Proestrous, phen-treated, ovx hamsters given OE (ip) exhibited a

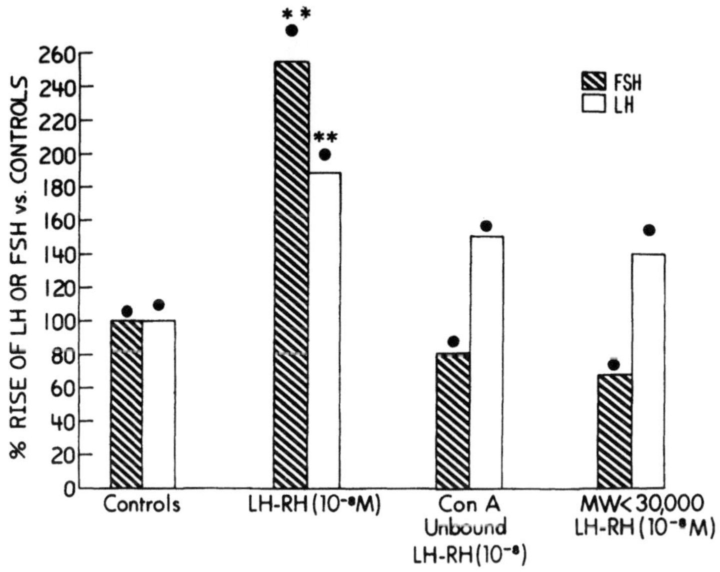


Fig. 5. The concentrations of LH and FSH in the incubation medium after treatment with: LHRH; LHRH plus charcoal-treated ovarian extract (CTOE) that was not bound to a concanavalin A (Con A) resin; or CTOE with a mol wt (MW) < 30,000 daltons.

delayed and attenuated rise in both gonadotropins. Siegel et al. (18) observed that administration of progesterone can partially override the inhibitory effects of an injection of phen in proestrous hamsters. Our results with OE may be due to progesterone contamination.

Hourly injections of steroid-free OE (CTOE) also suppressed the ovariectomy-induced FSH rise in phen-treated hamsters. The inhibitory material present within the CTOE appeared as effective as the steroid-containing OE in the suppression of ovariectomy-induced FSH secretion. Similar results have been reported with rete testes fluid (8) and porcine follicular fluid (19). However, the late rise in serum FSH indicated that the mechanisms controlling the ovariectomy-induced FSH release can overcome the inhibitory effects of CTOE injections. We have no explanation for this phenomenon.

We feel that these *in vivo* data suggest the presence of an "inhibin-like" material in the hamster ovary. Furthermore, the demonstration that this substance is capable of depressing the LHRH-induced FSH rise in an *in vitro* pituitary cell culture system suggests a pituitary site of action. Our *in vitro* studies have shown this substance to be of a mol wt between 10,000 and 30,000 daltons. These results are similar to those observed by others (10,11). Our studies also suggest that this material has an overall neutral or positive charge since it was not bound to a DEAE-cellulose anion-exchange resin. It appears that this FSH-inhibiting substance does not contain mannose or glucose residues that would allow it to bind to lectin (Con A).

To our knowledge, these studies are the first to demonstrate the existence and activity, both in vivo and in vitro, of an "inhibin-like" substance present within the hamster ovary. Because $E_2$ has been shown to exert an inhibitory effect on gonadotropin secretion at the hypothalamic (20) and pituitary (21) levels, this nonsteroidal material may act synergistically with $E_2$ (at the level of the anterior pituitary gland) to regulate the secretion of FSH.

## Acknowledgments

The work described in this article, Publication No. 966 of the Oregon Regional Primate Research Center, was supported by NIH Grants RR-00163, HD-07602, and HD-05969.

## References

1. C.S. Campbell, N.B. Schwartz and M. Gorski Firlit. Endocrinology 101 (1977) 162.

2. S.C. Chappel and C.A. Barraclough. Endocrinology 101 (1977) 24.

3. S.C. Chappel, R.L. Norman and H.G. Spies. Biol. Reprod. (1978) in press.

4. D.R. McCullagh. Science (Wash.) 76 (1932) 19.

5. P. Franchimont, S. Chari, M.T. Hagelstein and S. Duraiswami. Nature 257 (1975) 402.

6. E.J. Keogh, V.W.K. Lee, G.C. Rennie, H.G. Burger, B. Hudson and D.M. de Kretser. Endocrinology 98 (1976) 997.

7. N.R. Moudgal, S.G. Nandini and H. Lipner, in Hypothalamus and Endocrine Function, ed. F. Labrie, J. Meites and G. Pelletier (Plenum Press, New York, 1976) p. 495 (Curr. Top. Mol. Endocrinol.).

8. B.P. Setchell and F. Jacks. J. Endocrinol. 62 (1974) 675.

9. G. Lugaro, M.M. Casellato, G. Mazzola, G. Fachini and G. Carrea. Neuroendrocrinology 15 (1974) 62.

10. A. Steinberger and E. Steinberger. Endocrinology 99 (1976) 918.

11. F.H. de Jong and R.M. Sharpe. Nature 263 (1976) 71.

12. N.B. Schwartz and C.P. Channing. Proc. Natl. Acad. Sci. (U.S.A.) 74 (1977) 5721.

13. J.A. Resko, R.L. Norman, G.D. Niswender and H.G. Spies. Endocrinology 94 (1974) 128.

14. J.A. Resko, J.G. Ploem and H.L. Stadelman. Endocrinology 97 (1975) 425.

15. L.K.L. Tang and H.G. Spies. Endocrinology 96 (1975) 349.

16. J.D. Bast and G.S. Greenwald. Endocrinology 94 (1974) 1295.

17. C.A. Blake, R.L. Norman and C.H. Sawyer. Biol. Reprod. 8 (1973) 299.

18. H.I. Siegel, J.D. Bast and G.S. Greenwald. Endocrinology 98 (1976) 48.

19. M.L. Marder, C.P. Channing and N.B. Schwartz. Endocrinology 101 (1977) 1639.

20. J.M. Davidson, in Frontiers in Neuroendocrinology, ed. W. Ganong and L. Martini (Oxford University Press, New York, 1969) p. 343.

21. A. Negro-Vilar, R. Orias and S.M. McCann. Endocrinology 92 (1973) 1680.

DISCUSSION - CHAIRPERSON - N.B. SCHWARTZ

A. GOODMAN: In one slide, which depicted the effects of the extract after DEAE Sepharose chromatography, it appeared that there was some suppression of LH, as well as FSH. Was that not significant?

S.C. CHAPPEL: There was, compared to the LHRH-treated controls, a bit of LH suppression, and admittedly, the extracts that we're using are not purified. We may also have had some kind of enzymatic activity that partially degraded the signal, in this case, LHRH. In other studies we have seen that the inhibin is able to suppress the basal secretion rate of FSH in cultures that do not have LHRH added. You are right, there was a minor suppression of LH, but certainly not to the extent of that observed for FSH.

QUESTION: (Kansas City) I wonder if you were to take the extracts of the hamster ovaries in the evening of the proestrus, and inject your extracts using the same model you currently use, would you get the same results? It would be interesting to see if those ovaries obtain inhibin after they had been exposed to the LH surge.

S.C. CHAPPEL: Right, that's a good question, and that happens to be what we're doing right now. We are obtaining ovaries from different stages of the estrous cycle and measuring for inhibin content. We would expect that the levels may change and perhaps be inversely related to FSH concentrations in the serum.

N.B. SCHWARTZ: Why would you expect them to be inversely related?

S.C. CHAPPEL: I assume that it would be like comparing pituitary LH concentrations to serum concentrations of LH, I don't know. But any relationship between serum FSH and ovarian inhibin would be interesting.

N.B. SCHWARTZ: We're doing the same experiment and we agree with you that it would be of interest.

H. LIPNER: I would like to first make a comment. As I understand McCullough's 1932 paper, he was referring to inhibin as a water-extractable substance from the testes which had an effect directly on pituitary, and he wasn't distinguishing between an action on FSH or LH.

S.C. CHAPPEL: You're right, I got overexcited!

H. LIPNER: I wonder what your protein concentration is in the material that you've run through your concanavalin-A column, because there is always the possibility that you've got sufficient glycoproteins in your material since most of the protein that is present in follicular fluid was derived from the plasma and I assume that most of the protein you've got in your extract is also derived from the plasma. Therefore, there is probably a large concentration of glycoprotein present which could swamp out the concanavalin-A effect, and you wouldn't be able to say whether or not what you're looking at is a glycoprotein. Thirdly, I wondered what the pH was when you ran your material through your DEAE column.

S.C. CHAPPEL: The DEAE column was run at a basic pH, I believe. I'm sorry, I did not run the columns, the second author did. It was, I believe, pH 8. Your question about protein concentrations: it should have been done and it wasn't.

H. LIPNER: Chari, in her recent paper describing the apparent molecular weight and amino acid composition, suggested that the inhibin from the testis is probably an acidic molecule.

S.C. CHAPPEL: Right.

H. LIPNER: Therefore, you may have been at the wrong pH, as far as getting binding on DEAE Sepharose...

S.C. CHAPPEL: Right, since we observed that the activity did not bind to DEAE, we have started to use cation-exchange. Hopefully, it will bind. Also in that paper, I believe that Chari found a molecular weight of 19,000 daltons which goes along well with what we found here. But you're right, the next step is to use cation-exchange chromatography to purify it further.

# THE DIFFERENTIAL ABILITY OF PORCINE FOLLICULAR FLUID TO SUPPRESS SERUM FSH IN FEMALE RATS FROM 6 DAYS OF AGE TO ADULTHOOD

Janice R. Lorenzen and Neena B. Schwartz

Department of Biological Sciences, Northwestern University, Evanston, Illinois 60201

It has been speculated that there is a non-steroidal ovarian hormone which acts as a negative feedback signal for FSH (1). Recent data from our own laboratory (2,3,4) as well as others (5, 6,7) suggest that this hormone, folliculostatin or female inhibin, can be found in charcoal extracted porcine or bovine follicular fluid. In the adult (67-80 days old) female rat ovariectomized (ovax) at 0800h on the day of metestrus, serum FSH is elevated by 9h post-ovax (1700h). As little as 150 µl of porcine follicular fluid (PFF) injected at 1130h (3.5h post-surgery) can suppress this rise (3). Furthermore, this same amount of PFF injected into a sham-operated adult will suppress the basal levels of FSH (3). Since FSH has been implicated in the neonatal development of the ovary (8,9,10,11,12), it was of interest to determine if PFF could selectively suppress serum FSH in the neonatal and prepuberal female rats. Thus, the purpose of the following experiments performed in neonatal and prepuberal rats was to see if PFF could suppress FSH without altering LH and thus lead to a model for investigating the role of FSH in neonatal ovarian development.

Female rats (26 days of age) were ovariectomized or sham-ovax at 0800h. At 1130h, 5.5h before sacrifice, they were injected intraperitoneally with either 250 µl PFF or 250 µl of charcoal extracted porcine serum (PS). At 1700h the rats were sacrificed by decapitation. The data are presented in Table 1. The first observation is that FSH has not yet risen much above basal levels by 9h post-ovax (compare the sham/PS group to the ovax/PS group). Secondly, PFF had no effect on this post-ovax 9h non-rise, but did suppress FSH in the sham-operated animals. These results are in contrast to the adult female, but similar to what we find in the

Table 1. The Effect of Follicular Fluid on Serum FSH and LH in Ovariectomized and Sham-operated 26 Day Old Female Rats

| | Fluid | n | Body Weight (g) | LH (ng/ml) | FSH (ng/ml) |
|---|---|---|---|---|---|
| OVAX | PFF[1] | 4 | 62±4 | <1.0 | 314±44 |
| | PS | 3 | 62±4 | <1.0 | 417±49 |
| SHAM | PFF[1] | 4 | 61±5 | <1.0 | 147±6* |
| | PS | 3 | 60±8 | <1.0 | 336±22 |

[1]Follicular fluid was collected from medium sized (3-5 mm in diameter) porcine follicles.
*Significantly ($p<0.05$) different from PS treated controls.

adult castrate male given PFF under the same protocol (13). The results suggest that in the 26 day old female something is happening during the 9h after removal of the ovaries which inhibits the action of folliculostatin, or that the surgical stress in the presence of the ovaries enhances the effect of folliculostatin. LH had not risen by 9h post-ovax either, and, thus, any effect PFF might have had on this hormone could not be assessed.

In the younger ages tested, the animals were not surgically treated in order to assess the ability of PFF to reduce FSH in the presence of the ovaries. If this were possible it could lead to the ultimate goal of suppressing FSH during development and learning what derangements, if any, in adult ovarian function would be made manifest. Intact male and female 21 day old rats were treated with 250 µl PFF or PS at 1130h and sacrificed, as before, at 1700h. At this age, the high levels of FSH seen in the neonatal female have decreased. In the male, serum FSH has remained virtually unchanged from birth until weaning (14). PFF treatment at this age resulted in a significant ($p<0.05$, ANOVA) decrease in basal FSH levels in both sexes (Table 2); however, FSH did not decrease to the levels seen in the PFF suppressed adult metestrous sham-operated rat (140±9 ng/ml (3)).

Intact neonatal female rats, age 6 and 12 days, were treated intraperitoneally at 1130h with either 200 µl of PS or a high molecular weight (>10,000) fraction of PFF which has been shown to contain folliculostatin (15). Blood was collected by decapitation at 1700h and pooled for hormone determinations (3-8 animals/pool at 6 days; 2-4 animals/pool at 12 days). The data are presented in Table 3. PFF had no effect on LH (occasional high values were seen in the females as reported previously (16,17)), yet sup-

Table 2. The Effect of Follicular Fluid on Serum FSH and LH in Intact Male and Female 21 Day Old Rats

| | Fluid | n | Body Weight (g) | LH (ng/ml) | FSH (ng/ml) |
|---|---|---|---|---|---|
| FEMALE | PFF | 6 | 37±2 | 4.0± 2.1 | 667±126 |
| | PS | 5 | 38±2 | 18.8±16.9 | 911±226 |
| MALE | PFF | 6 | 37±1 | 3.0±0.7 | 390±29 |
| | PS | 5 | 40±2 | 1.9±0.2 | 510±57 |

PFF has a significant ($p<0.05$; one way analysis of variance) effect on FSH levels.

Table 3. The Effect of Follicular Fluid on Serum FSH and LH in Intact 6 and 12 Day Old Female Rats

| Age (days) | Fluid | Volume (μl) | Total n | Body Weight(g) | Pool n | LH (ng/ml) | FSH (ng/ml) |
|---|---|---|---|---|---|---|---|
| 6 | PFF[1] | 200 | 33 | 11±0 | 5 | 4.6±1.2 | 712±41* |
| | | 100 | 30 | 12±0 | 6 | 3.3±1.1 | 1178±53* |
| | PS | 200 | 35 | 12±0 | 5 | 3.1±0.3 | 978±47 |
| | | 100 | 34 | 11±0 | 7 | 2.6±0.6 | 1415±81 |
| 12 | PFF[1] | 200 | 13 | 25±1 | 5 | 4.6±2.1 | 718±47 |
| | | 100 | 21 | 21±1 | 6 | 2.7±0.7 | 1091±87* |
| | PS | 200 | 12 | 21±0 | 5 | 2.7±0.9 | 870±78 |
| | | 100 | 24 | 20±0 | 7 | 8.3±3.2 | 1580±170 |

[1]A fraction of PFF of molecular weight >10,000 was used.
*$p<0.05$ compared to appropriate PS control.

pressed FSH in all cases when compared to the identical PS dose, except in the 12 day old pups given 200 μl. The effect appeared to be dose dependent in that the larger volume resulted in lower FSH values. Surprisingly, the PS injections manifested a dose response relationship also. Our first thought was that this represented a volume or osmotic "stress" effect. If the 200 μl dose increased intraperitoneal osmotic pressure, thus increasing extracellular fluid volume, one would expect the weights of the peritoneal organs (e.g. ovaries and uterus) to reflect this change. However, the tissue weights (Table 4) do not support this hypothesis, since the trend in increasing wet weights does not follow the trend in decreasing FSH. Only the 200 μl dose in the 6 day old rat had any significant effect on ovarian wet weight, while

Table 4. The Effect of Follicular Fluid on Ovarian and Uterine Wet Weight in Intact 6 and 12 Day Old Rats

| Age (days) | Fluid | Volume (µl) | Ovarian Wet Weight (mg) | Uterine Wet Weight (mg) |
|---|---|---|---|---|
| 6 | PFF[1] | 200 | 2.16±0.10 (33)[2] | 2.74±0.21 (26) |
| | | 100 | 2.07±0.08 (10) | 2.88±0.11 (10) |
| | PS | 200 | 2.51±0.19 (34) | 3.01±0.28 (25) |
| | | 100 | 1.75±0.14 (11)* | 2.37±0.26 (11) |
| 12 | PFF[1] | 200 | 8.81±0.90 (13) | 14.47±1.19 (12) |
| | | 100 | 5.44±0.15 (4) | 9.43±0.70 (4)* |
| | PS | 200 | 7.38±0.66 (12) | 13.95±1.06 (12) |
| | | 100 | 5.29±0.14 (4) | 8.95±0.77 (3)* |

[1]A fraction of PFF of molecular weight >10,000 was used.
[2]n = ( ).
*Significant ($p<0.05$) from 200 µl dose.

the uterine wet weight was only affected by the higher dose in the 12 day old rat. Furthermore, the PS dose effect appears to be greater in the older rat. If this was due to a volume "stress," one would expect the stress to be greater in the younger, smaller rat.

A second possibility is that we have found FSH suppressing activity in PS. The PS used in these experiments was acquired from Grand Island Biological Company (Grand Island, NY) which collects serum from slaughterhouse male and female 6 months old, probably non-castrated, pigs (GIBCO, personal communication). If folliculostatin (inhibin) is an endogenous feedback signal for FSH, one would expect its presence in serum. Using factorial

Table 5. Bioassay of Folliculostatin in Porcine Follicular Fluid and Porcine Serum[1]

| Age (days) | Fluid | λ, Index of Precision | Relative Potency | Range of Potency |
|---|---|---|---|---|
| 6 | PFF | 0.085 | 1.20 | 1.11→1.35 |
| | PS | | 1.00[2] | |
| 12 | PFF | 0.148 | 1.22 | 1.01→1.80 |
| | PS | | 1.00[2] | |

[1]Bioassay statistics by the method of Bliss (18).
[2]Arbitrarily designated as standard (potency = 1.00).

assay technique for two doses (18), the relative potency of PFF, assuming PS had a potency of 1.00, was determined in both the 6 and 12 day old rat (Table 5). The dose-response lines are parallel in both ages suggesting very strongly the presence of folliculostatin in PS. It is not surprising that this effect is not seen in the adult metestrous rat (3), since the ratio of PS injected (610 µl at maximum) to body weight (200-300g) is much smaller than that used in the neonate (200 µl PS in an 11g rat). One would have to inject 4 mls of PS into a 200g rat to equal the dose/BW achieved in the 6 day old pup.

In summary, PFF (or its active agent, folliculostatin) can suppress serum FSH in intact or sham-operated females of all ages tested; 1) when FSH is high in the neonatal period, 2) after FSH has fallen post-weaning, but prepuberally, as well as 3) in the adult. It appears to have no effect on LH at any age. This is in contrast to the effects in males where PFF administration with similar protocols has seldom suppressed FSH in the adult; however, in the immature male (21 days old (Table 2) and 34 days old, during the time FSH is peaking) PFF can suppress FSH (13). These results suggest that we now have the capability to partially suppress serum FSH in the female at any age without altering other gonadotrophic functions. This, in turn, should lead to an increased understanding of the role of FSH in follicular development. Additionally, as an unexpected bonus of these investigations, we have uncovered the suggested presence of folliculostatin (inhibin) in porcine serum, indicating the possibility of using the 6 and/or 12 day old female rat as a sensitive bioassay for serum levels of folliculostatin.

Acknowledgements. We wish to thank John Gesme and William L. Talley for excellent technical assistance; Dr. C.P. Channing for supplying the follicular fluid; the National Institutes of Health, Endocrinology Study Section, for generous supplies of gonadotropic hormones used as standards for assays and for the FSH kit; Dr. L. E. Reichert, Jr. for the LH used for radioiodination; and Dr. G.D. Niswender for contributing the antiserum for the radioimmunoassays. Supported, in part, by U.S. Public Health Service Grant HD 07504.

References

1. C.S. Campbell and N.B. Schwartz. J.Toxicol. Environ. Health 3(1977)61.

2. N.B. Schwartz and C.P. Channing. Proc. Nat'l. Acad. Sci.74 (1977)5721.

3. M.L. Marder, C.P. Channing and N.B. Schwartz. Endocrinology 101(1977)1639.

4. J.R. Lorenzen and N.B. Schwartz, in: Novel Aspects of Reproductive Physiology, ed. C.H. Spilman and J.W. Wilks (Halsted Press, New York, 1978) p. 339.

5. F.H. deJong and R.M. Sharpe. Nature 263(1976)71.

6. C.R.N. Hopkinson, G. Sturm, E. Daume, E. Fritze and C. Hirschhauser. IRCS Med. Sci. 3(1975)588.

7. R. Welschen, J. Dullaart, and F.H. deJong. Biol. Reprod. 18(1978)421.

8. C.H. Anderson, N.B. Schwartz, L.G. Nequin and C.A. Ely. Fertility and Sterility 27(1976)47.

9. B. Hardy, D. Danon, A. Eshkol and B. Lunenfeld. J. Reprod. Fert. 36(1974)345.

10. H. Peters, A.G. Byskov, S. Lintern-Moore and M. Faber. J. Reprod. Fert. 35(1973)619.

11. J.S. Richards and A.R. Midgely, Jr. Biol. Reprod. 14(1976) 82.

12. J. Th. J. Uilenbroek, E.A. deWolff-Exalto and R. Welschen. Ann. Biol. Anim. Bioch. Biophys. 16(1976)297.

13. J.R. Lorenzen, G.H. Dworkin, N.B. Schwartz and C.P. Channing. Prog. 60th Meeting Endo. Soc. (1978)97.

14. K.D. Dohler and W. Wuttke. Endocrinology 97(1975)898.

15. J.R. Lorenzen, N.B. Schwartz and C.P. Channing. Biol. Reprod. (in press).

16. K.D. Dohler and W. Wuttke. Acta Endocrin. 83(1976)269.

17. P.C.B. MacKinnon, J.M. Mattock and M.B. terHaar. J. Endocrin. 70(1976)361.

18. C.I. Bliss. Drug Standards 24(1956)33.

DISCUSSION - CHAIRPERSON - N.B. SCHWARTZ

A. GOODMAN: On one of your slides Dr. Lorenzen, perhaps that for showing data obtained in 12 day female rats, LH appeared to be suppressed after treatment with pig serum, but not after the follicular fluid. Is that correct?

J. LORENZEN: No. We do see that high doses of follicular fluid in adult female and male rats will suppress LH as well as FSH but that much less follicular fluid is required to suppress FSH compared to LH.

A. GOODMAN: Did you show the corresponding slide for the 6-day old female?

J. LORENZEN: Yes.

A. GOODMAN: And there were no differences?

J. LORENZEN: LH levels were not suppressed. There was not as much variation.

S.K. BATTA: It is presumed you were using the PM filtrate for the follicular fluid...

J. LORENZEN: The PM-10 retentate (> 10,000 daltons in molecular weight) of follicular fluid.

S.K. BATTA: I was wondering, have you tried the PM-10 retentate of porcine serum?

J. LORENZEN: No, I haven't done that. That's a good suggestion.

H. LIPNER: After we saw that paper by Dolan, we were also fascinated by the possibility of generating a bioassay using the 15-day-old female rat. Unfortunately, in our hands, using very large doses of porcine follicular fluid, a follicular fluid which gives us an identical curve to one that you have shown in adult female rats, we failed to get any suppression in 15 day old female rats. We are also rather interested in the way you were bracketing the age group with which we worked.

J. LORENZEN: If you look closely at the data I presented, that ability of follicular fluid to suppress FSH falls from 6 to 12 to 21 days of age. It wasn't as good at days 12 and 21, as it was at days 6 and 26 in the adult, so there may be some time where the effect fades away and then comes back.

# EVIDENCE FOR THE PRESENCE OF A LOW MOLECULAR WEIGHT FOLLITROPIN BINDING INHIBITOR IN BOVINE FOLLICULAR FLUID

Nina C. Darga and Leo E. Reichert, Jr.

Department of Biochemistry

Emory University School of Medicine, Atlanta, GA 30322

The interaction of FSH with granulosa cell receptors has previously been studied by Nimrod et al. (1), Louvet and Vaitukaitis (2) and Lee and Takahashi (3) in the rat and by Zeleznik et al. (4) in the porcine. In this report, we describe the binding of FSH to bovine granulosa cells and summarize studies suggesting the presence in bovine follicular fluid of a low molecular weight inhibitor of FSH binding (FSH-BI) to bovine granulosa cell receptors.

## BINDING OF FSH TO BOVINE GRANULOSA CELL RECEPTORS

The binding assay used in these studies has been described in detail elsewhere (5) and utilizes radioiodinated, biologically active human follitropin ($^{125}$I-hFSH) as the ligand and granulosa cell suspensions from bovine ovarian follicles as the receptor source. Iodination was carried out by a modified chloramine T method (6).

A series of studies were performed to determine the binding of $^{125}$I-hFSH to bovine granulosa cells from follicles at various stages of development, as judged by size. Small sized follicles were those less than 3 mm in diameter, medium sized follicles were between 3-6 mm in diameter, and large sized follicles were those having a diameter greater than 6 mm (but not cystic). The results, shown in Table 1, indicated a slight increase in $^{125}$I-hFSH binding as follicle size increased. These observations are similar to those of Nimrod et al. (1) who reported that, in the rat, large follicles exhibited a greater degree of binding than did medium follicles. These same workers observed no difference in FSH binding to granulosa cells from medium or large follicles

TABLE 1. Binding of $^{125}$I-hFSH to granulosa cells from bovine follicles at various stages of maturation.

| Size of follicle | No. of expts. | Cells x $10^6$ per 0.1 ml | % Specific binding | Moles x $10^{-14}$ of $^{125}$I-hFSH bound to $10^6$ cells |
|---|---|---|---|---|
| A) Small | 4 | 2.35 ± 0.18* | 21.8 ± 2.05* | 0.27 ± 0.05* |
| B) Medium | 4 | 2.45 ± 0.06 | 25.2 ± 2.05 | 0.31 ± 0.07 |
| C) Large | 4 | 2.29 ± 0.22 | 30.1 ± 1.23 | 0.40 ± 0.05 |

* = Standard error

in the rabbit. However, Zeleznik et al. (4) reported that in the porcine, granulosa cells from small follicles exhibited the greatest binding capacity for rat FSH, and that this binding capacity decreased as follicle size increased.

## INHIBITION OF $^{125}$I-hFSH BINDING TO GRANULOSA CELLS BY FOLLICULAR FLUID

Reports of various types of inhibitors in follicular fluid are not new. Tsafriri et al. (7) described a porcine follicular fluid inhibitor of rabbit oocyte maturation. A luteinization inhibitor has been observed in rabbit follicular fluid by El Fouly et al. (8). Channing and Kammerman et al. (9) have shown that LH-induced cyclic AMP production in porcine granulosa cells can be inhibited by fluid from small or medium follicles.

Gonadotropin binding inhibitors have also been described in the recent literature. Yang et al. (10) observed an LH-binding inhibitor in aqueous extracts of ovaries from pseudopregnant or pregnant rats. Inhibitors of FSH binding to rat testicular receptors have recently been reported in human serum (11,12) and in rat testis extracts (13). Due to the similarities between serum and follicular fluid, it seemed of interest to determine whether inhibitors of FSH binding to ovarian receptors might also exist in follicular fluid. As shown in Figure 1, $^{125}$I-hFSH binding to bovine granulosa cell receptors was inhibited by bovine follicular fluid, in a dose-dependent manner.

An attempt was made to quantitate the degree of binding inhibition by fluid from follicles at various stages of development (as judged by size). The concentration of FSH-BI (ng hFSH equiv./ml FF) was expressed in terms of highly purified hFSH and based on the amount of fluid required for 50% binding inhibition of $^{125}$I-hFSH to granulosa cell receptor. For example, if a fraction possessed 10 ng hFSH equiv./ml FF, then 1 ml of follicular fluid inhibited binding of $^{125}$I-hFSH to receptor to the same degree as 10 ng of unlabelled hFSH. As seen in Table 2, in each series of experiments, a 2 to 3-fold increase in FSH-BI concentration was observed in fluid from large follicles compared to fluid from

TABLE 2. FSH binding inhibition activity of fluid from bovine follicles at different stages of maturation.

| Series | Size of follicle | No. of folli-cles | ul FF/ follicle | Ng hFSH equiv./ ml FF* | Ng FSH equiv./ follicle |
|---|---|---|---|---|---|
| A) | Small | 125 | 5.0 | 604 (513-701) | 3.00 |
| | Medium | 238 | 32.7 | 1190 (1067-1329) | 38.9 |
| | Large | 35 | 258.6 | 1244 (1124-1378) | 321.7 |
| B) | Small | 179 | 7.6 | 270 (174-374) | 2.10 |
| | Medium | 231 | 36.4 | 675 (534-833) | 24.6 |
| | Large | 46 | 434.0 | 756 (611-917) | 328.1 |

*Results expressed in terms of binding inhibition obtained with pure hFSH. Estimate based on amount of FF required for 50% inhibition of binding of 1 ng $^{125}$I-hFSH to receptors on granulosa cells from large follicles.

small follicles. When the average fluid volume per follicle was considered, the total amount of FSH-BI per follicle increased over 100-fold as follicular size increased.

SOME PROPERTIES OF THE FOLLICULAR FLUID BINDING INHIBITOR

Initially, the possibility was considered that binding inhibition might be due to the presence of endogenous FSH in follicular fluid. The FSH-BI activity of follicular fluid was markedly reduced, by about 70% after dialysis against deionized water using Spectrapor #1 membrane as the dialysing barrier. This membrane passes molecules of 8000 daltons or less. FSH-BI activity was detected in buffer-reconstituted lyophilized dialysates, which inhibited $^{125}$I-hFSH binding to granulosa cell receptor in a dose-dependent manner (Figure 1). The finding that FSH-BI passed

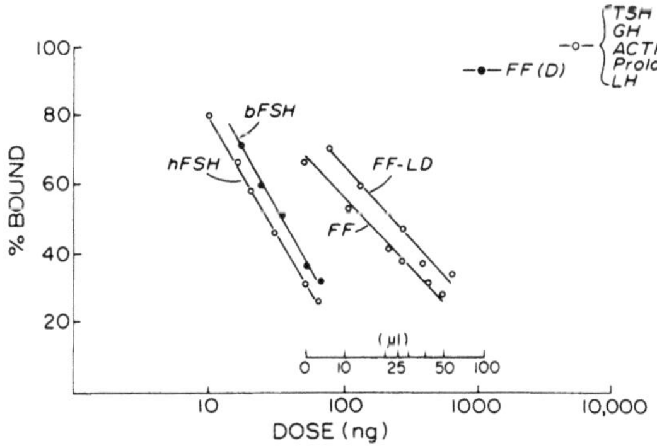


Figure 1. FSH=follitropin, b=bovine, h=human, FF=follicular fluid, LD=lyophilized Spectrapor #1 dialysate, FF-D=fraction not passing Spect. #1 membrane on dialysis.

through the Spectrapor #1 membrane indicated that the FSH binding inhibition previously noted in follicular fluid could not be completely due to endogenous pituitary FSH.

In earlier experiments (18), we have shown that in addition to passing Spectrapor #1 membrane on dialysis, FSH-BI activity will also pass Amicon UM2 membrane (MW exclusion 1000 daltons) on ultrafiltration. In the present experiments, both buffer reconstituted lyophilized dialysates and charcoal adsorbed follicular fluid from medium and large bovine follicles were ultrafiltered through Amicon UM-05 membrane (MW retention 500 daltons). Most (80%) of the FSH-BI activity originally present passed the membrane and was recovered in the filtrate. The UM-05 filtrate was lyophilized, reconstituted with deionized water (soluble) and sieved through a column of Sephadex G-10 (MW exclusion 700 daltons) in deionized water. Scanning receptor assays indicated FSH-BI activity in two regions of the elution profile, Tubes 52-61 (Fraction A) and Tubes 95-102 (Fraction B) (Figure 2).

Fraction B was strongly retarded (Ve/Vo tube 95 = 1.92), had a relatively low level of FSH-BI activity (1000 mcg inhibits $^{125}$I-hFSH binding by 50%) and retained FSH-BI activity after incubation at 100 C for 24 hours in the presence or absence of 6N HCl.

Fraction A was the most active with regard to FSH-BI activity (50 mcg inhibits $^{125}$I-hFSH binding by 50%) and emerged about the Vo of the column, although apparently slightly retarded (Ve/Vo = 1.10, mean of 3 experiments).

In earlier studies it has been shown that FSH-BI activity of follicular fluid or Spectropor #1 or UM-05 membrane dialysates was not lost following adsorption by charcoal or extraction by diethyl ether. Fraction A (Figure 2) did not loose FSH-BI activi-

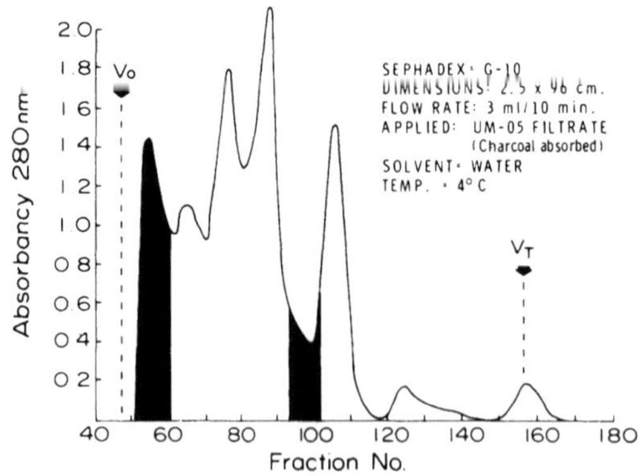


Figure 2. Gel filtration of FSH-BI through Sephadex G-10.

ty after adsorption with charcoal and, in preliminary experiments, did not seem to loose FSH-BI activity after adsorption with conconavalin A. FSH-BI activity of Fraction A was lost after rather severe heat treatment (100 C for 24 hours) although more careful studies of thermal stability have not yet been performed. In addition to having significant absorbancy at 280 nm (Figure 2), Fraction A released amino acids after acid hydrolysis (6N HCl, 105 C, 24 hours) as determined by the TNBS reaction.

The chemical nature of the FSH-BI component of Fraction A is not known, although on the basis of studies described above, it would not seem to be steroid or nucleotide in nature, nor to contain carbohydrate of the type having an affinity for conconavalin A. The evidence for a peptide nature to the FSH-BI is as yet inconclusive. A practical problem in purification and characterization of the FSH-BI is the low and seemingly variable level of inhibitory activity in follicular fluid, about 1 to 5 mg of Fraction A equivalent inhibitor fraction per 100 mls of FF, together with the relative insensitivity of the inhibitor assay at its present stage of development. Also, before a physiologic role can be assigned to the FSH-BI, evidence for effectiveness in *in vivo* experiments is required.

It has been reported, both *in vivo* and *in vitro*, that FSH can induce binding sites for hCG on rat and porcine granulosa cells (14,15,16). Richards et al. (17) have suggested autoregulation, coordinate regulation and heteroregulation as possible mechanisms of hormone regulation of receptors. The studies described in this paper suggest another possible mechanism by which hormone-receptor interactions might be regulated. Investigation is currently underway to further purify and characterize the follicular fluid FSH-BI and to assess its possible physiologic significance.

## ACKNOWLEDGEMENT

This study was supported by USPHS Grant HD-08228 (LER) and Training Grant T32-HD-07121 (NCD). We are grateful for the expert technical assistance of Ms. Nancy Shih in various phases of this study.

## REFERENCES

(1) A. Nimrod, G.F. Erickson and K.J. Ryan. Endocrinology. 98 (1976)56.

(2) J.P. Louvet and J.L. Vaitukaitis. Endocrinology. 99(1976) 758.

(3) C.Y. Lee and H. Takahashi. Endocrinology. 101(1977)869.

(4) A.J. Zeleznik, C. Desjardins, A.R. Midgley, Jr. and L.E. Reichert, Jr. Fed. Proc. 33(1974)213.

(5) N.C. Darga and L.E. Reichert, Jr. Biol. Reprod. (1978) (In Press).

(6) L.E. Reichert, Jr. and V.K. Bhalla. Endocrinology. 94(1974) 483.

(7) A. Tsafriri, S.H. Pomerantz and C.P. Channing. Biol. Reprod. 14(1976)511.

(8) M.A. El Fouly, B. Cook, M. Nekola and A.J. Nalbandov. Endocrinology. 87(1970)288.

(9) C.P. Channing and S. Kammerman. Endocrinology. 92(1973)531.

(10) K.P. Yang, N.A. Samaan and D.N. Ward. Endocrinology. 98 (1976)233.

(11) L.E. Reichert, Jr., in: Novel Aspects of Reproductive Physiology, ed. C.H. Spilman and J.W. Wilks (J. Wiley and Sons, New York, 1978) p. 355.

(12) L.E. Reichert, Jr., H. Abou-Issa, E.B. Carter and N.L. Shih. Endocrinology. 100(1977)A-22.

(13) L.E. Reichert, Jr. and H. Abou-Issa. Biol. Reprod. 17(1977) 614.

(14) A.J. Zeleznik, A.R. Midgley, Jr. and L.E. Reichert, Jr. Endocrinology. 95(1974)818.

(15) C.P. Channing. Proc. Soc. Exptl. Biol. Med. 149(1975)238.

(16) A. Nimrod, A. Tsafriri and H.R. Lindner. Nature. 267(1977) 632.

(17) J.S. Richards, J.J. Ireland, M.C. Rao, G. Bernath, A.R. Midgley, Jr. and L.E. Reichert, Jr. Endocrinology. 99(1976) 1562.

(18) N.C. Darga and L.E. Reichert, Jr. Endocrinology. 102(1978) A-574.

DISCUSSION - CHAIRPERSON - N.B. SCHWARTZ

**B.F. RICE: I wonder if you could tell us how specific this effect is; is the inhibitor specific for ovarian tissue or can these effects be observed with tissues from other sources?**

**N. DARGA: There does seem to be an FSH binding inhibitor in human serum and in rat testicular tissues.**

A.T. WILLIAMS: Using porcine follicular fluid I found no significant effect on serum LH levels in vivo in the rat. In a bioassay system comparable to that of Dr. Schwartz', we obtained a dose-response curve for FSH that paralleled very well with that found in Dr. Schwartz' lab.

In all the work that I'll present here, the model is the same that was presented by Dr. Schwartz' lab, that is, the metestrous adult female rat is ovariectomized at 8 am in the morning, and injected at 11:30 am, and blood is collected at five in the afternoon. All test material was charcoal-treated porcine follicular fluid (PFFl) or serum (PS).

It is well established that follicular fluid has a high concentration of plasma derived protein and that albumin is present in extremely large concentrations. We decided to remove the albumin and reduce the total protein concentration by 60 to 80% with Affi-gel blue (a Cibacron dye) to which albumin binds very tightly. The residual material placed on a column of G 150 Superfine Sephadex. The active factor was found in these fractions eluting in the molecular weight region corresponding to 10,000 and 35,000 daltons.

Since the protein concentration and distribution in pFFl is very similar to that found in plasma we decided to remove the plasma proteins immunochemically. Serum obtained from barrow blood (male pigs castrated at birth) was used to generate an antiserum. Treatment of follicular fluid with barrow-serum antiserum resulted in the removal of most of the plasma protein.

We generated an antibody against all the plasma proteins in the porcine follicular fluid, except those that are endogenous to the follicle. This resulted in removal of about 99.7 to 99.9% of the protein in the follicular fluid. The residual material was run on a preparative isoelectric focusing column between pH 3 and 10. (Fig. 1).

The linear pH gradient is shown on the left ordinate as a function of the elution volume indicated on the abscissa, and is represented by the curve in the middle; 7 pools were formed with the pH boundaries indicated at the top. The assays of that mate-

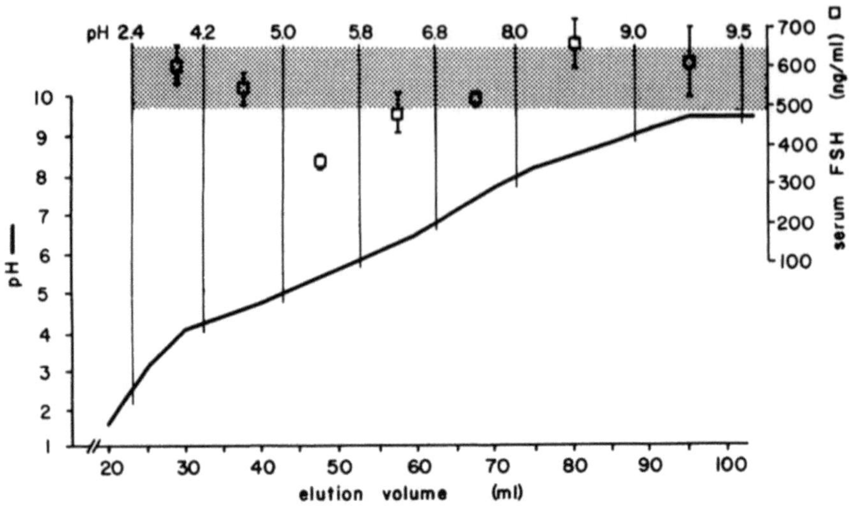


FIG. 1 Isoelectric focusing of pFF1 after immunoadsorption with castrate male pig serum antiserum. See discussion. (A. T. Williams).

rial are indicated by the FSH levels represented by the open squares. You can see that the fractions having a pH between 5 and 5.8 have a statistically significant depression in FSH, whereas, the other values fall in the shaded control area. We feel that the combination of the removal of a large amount of protein initially by Sephadex G 150 Superfine followed by antiserum absorption and finished with isoelectric focusing is a very fine method for the purification of pFF1. Thank you.

L.D. ANDERSON: In addition to some of the experiemnts that Dr. Channing has shown in her original talk tonight, and to Dr. Chappel's work, that looked very nice, we have additional data demonstrating that porcine follicular fluid can cause a decrease in secretion of FSH by rat pituitary cells in vitro. Details are shown in a manuscript elsewhere in this workshop (Shander et al.). As a control we used 2% pig serum that was charcoal-treated and as an experimental fluid we used 2% pig follicular fluid from medium-sized follicles that was charcoal treated.

Addition of 0.5% and 2% pig follicular fluid for 24 hours and an addition of 2% pig follicular fluid for only 2 hours brought about a significant decrease in FSH secretion by rat pituitary cells cultured according to the method of Vale et al.

Addition of 0.5% and 2% of porcine follicular fluid for 24 hours with incubation or preincubation with or without 2 x $10^{-10}$M LHRH inhibited the LHRH stimulation of FSH secretion. Preincubation with 0.5% follicular fluid for only 2 hours failed to inhibit LHRH stimulation of FSH, and at a higher concentration of LHRH we see the same pattern. It seems as though it takes at least 2 hours of preincubation to have this effect of pig follicular fluid on rat pituitary cells in vitro. Pig serum seems to have no effect, if any, as compared to not being added to the Delbuccos medium. During our experiments we preincubate for 24 hours, then wash the cells 2-3 times with plain media (nothing added) and then add back only LHRH so that we have nothing else present with the LHRH at that time. Thank you.

I would like to answer a question posed earlier by Dr. Rigby. We added follicular fluid from small pig follicles to granulosa cells obtained from medium-sized follicles and mixed it with FSH and testosterone. This resulted in a stimulation of progesterone secretion. Our culture conditions differed from yours; you added LH and not FSH. Your experiments included previous incubation with follicular fluid and then a challenge with LH, while in our experiments the follicular fluid was present with FSH and testosterone, therefore you obtained an inhibition of progesterone secretion we saw a stimulation. These differences in culture conditions could account for our differing data.

L. AINSWORTH: I would like to change the topic somewhat and summarize some of our recent work on another group of non-steroidal regulators of ovulation, prostaglandins (PG). This work has been carried out in collaboration with Dr. G.J. Marcus, Animal Research Institute, Ottawa, Ont., Dr. B.R. Downey, Macdonald College, Ste. Anne de Bellevue, Que., and Drs. B.K. Tsang and D.T. Armstrong, University of Western Ontario, London, Ont. Canada. We use prepubertal gilts treated with 750 IU PMSG(IM) at time zero followed by 500 IU hCG(IM) 72 hr later, to induce follicular growth and ovulation, as the experimental model. With this treatment, animals ovulate about 42 hr after HCG. Using this model we have demonstrated previously that ovulation is preceded by a marked elevation of follicular fluid PGF levels (L. Ainsworth, R.D. Baker and D.T. Armstrong, Prostaglandins, 9: 915, 1975). These studies have been extended to investigate further the role of prostaglandins in the ovulatory process by treatment of PMSG/hCG treated gilts with 650 mg indomethacin (an inhibitor of PG synthesis) IM at 20, 24, 32 or at 24 and 72 hr after HCG treatment.

Tables 1 and 2 summarize the results of indomethacin treatment on ovulation and follicular fluid levels of PGE and PGF respectively, in groups of PMSG/hCG treated prepubertal gilts which were subjected to laparotomy at various times after hCG, for observation of the ovaries and collection of follicular fluid. In control gilts over 90% had ovulated by 48 hr after hCG treatment (Table 1). In contrast, irrespective of the time indomethacin was administered, ovulation in the treated groups was inhibited as indicated by absence of ovulation points, by failure to recover tubal ova and by recovery of oocytes from unruptured follicles at 48 and 72 hr after hCG treatment. In those animals from which it was possible to collect follicular fluid at the time of laparotomy, the follicular fluid levels of PGE and PGF were similar to those observed in a zero time control group and did not show the characteristic preovulatory rise observed in untreated controls at 38 hr after hCG treatment (Table 2). These results indicate that in-

TABLE 1

EFFECT OF INDOMETHACIN ON OVULATION IN PREPUBERTAL GILTS TREATED WITH PMSG/hCG TO INDUCE OVULATION

| NUMBER OF ANIMALS | TIME AFTER hCG (HR) | FOLLICLES OVULATED/TOTAL PREOVULATORY FOLLICLES |
|---|---|---|
| 6 | 48 | 0/100 |
| 6 | 72 | 3/56 |
| 6 | 96 | 9/100 |
| 6 | 120 | 17/74 |
| 8 | 48 (CONTROL) | 147/166* |

*One animal (0/14) did not ovulate

TABLE 2

EFFECT OF INDOMETHACIN ON FOLLICULAR FLUID PROSTAGLANDIN LEVELS IN PREPUBERTAL GILTS TREATED WITH PMSG/hCG TO INDUCE OVULATION

| NUMBER OF ANIMALS | TIME AFTER hCG (HR) | FOLLICULAR FLUID LEVELS | |
|---|---|---|---|
| | | PGE (NG/ML) | PGF (NG/ML) |
| 4 | 48 | $1.45 \pm 0.18$ | $0.42 \pm 0.07$ |
| 4 | 72 | $0.94 \pm 0.11$ | $0.19 \pm 0.03$ |
| 2 | 96 | $0.79 \pm 0.01$ | $0.06 \pm 0.003$ |
| 3 | 120 | $1.02 \pm 0.14$ | $0.26 \pm 0.04$ |
| 4 | 38 (CONTROL) | $176 \pm 26$ | $231 \pm 55$ |
| 4 | 0 (CONTROL) | $1.2 \pm 0.2$ | $0.16 \pm 0.02$ |

hibition of the preovulatory rise in follicular fluid prostaglandin levels prevents ovulation and provides further evidence for a physiological role for prostaglandins, acting at the level of the follicle, in the ovulatory process.

A. AMSTERDAM: We have heard from the fine talk of Dr. Ledwitz-Ribgy and the summary of Dr. Channing that the follicular fluid inhibits cyclic-AMP accumulation. But still there is a question whether this inhibition is exerted by inhibition of adenylate cyclase or by activation of phosphodiesterase activity. We have some evidence that follicular fluid exerts a direct effect on the adenylate cyclase system. We took purified rat ovarian plasma membranes and we incubated them with bovine follicular fluid. The LH-sensitive and hCG-sensitive adenylate cyclase were inhibited by bovine follicular fluid in a dose-dependent manner and so there is 50% inhibition of the enzyme using a concentration of 5 mg per ml. of the bovine follicular fluid as measured by Lowry protein (editors remark: 5 mg/ml bovine follicular fluid is equivalent to about 50% FFl). In contrast, the follicular fluid did not alter basal adenylate cyclase activity. The follicular fluid exerted only a slight inhibition of binding of $^{125}$I-hCG. We are unsure as to the nature of this follicular fluid adenyl cyclase inhibitor. Recently we have found heparin-like substances in the ovary, and have found that purified heparin can inhibit adenyl cyclase. Whether or not the inhibitor is heparin cannot be ascertained at the present moment. We set out to see if there was any correlation between follicular maturation and inhibition of adenylate cyclase. For these studies we took post-microsomal supernatant from ovaries different times after PMSG treatment. This represents basal activity while in the upper part at the left, you see two stars, and these represent the hormone-stimulated activity without having any fraction. From 24-36 hours after PMSG treatment addition of 2 mg per ml of post-microsomal supernatant abolishes the hormone-sensitive adenylate cyclase. There is a loss of this inhibitory activity after 36-48 hours.

We can observe a dose response curve with the bovine follicular fluid inhibition of adenyl cyclase and can see an effect with 5 and 1 mg/ml and see no effect with 0.5 mg/ml. So, in conclusion, we demonstrated a direct effect of the bovine follicular fluid on adenylate cyclase and also that there is an inverse relationship between follicular maturation and the inhibitory effect which may suggest that there is some physiological role in sensitization of the preovulatory follicle. Thank you.

A. GOODMAN: I would like to ask members of the panel if they would discuss explicitly what has so far remained as an implicit assumption underpinning these kinds of studies. Namely, is the follicular fluid compartment a "source," or "sink?" That is, does the presence of this FSH inhibiting substance in follicular fluid mean that it was, in fact, produced by the follicle, or was it merely trapped and concentrated there? Along this same line: the mere presence of this inhibitor in peripheral blood does not, by itself, reveal the follicle or the ovary as its source. To do so, its levels in ovarian venous blood would have to be demonstrably higher than those in the systemic circulation; and if I may ask Dr. Channing to discuss a subject near and dear to my heart (the selection of follicles), how does she envisage something which affects systemic FSH levels would regulate the choice of the ovary that is going to ovulate, as well as the particular follicle in the ovary that is going to ovulate, in a species like the monkey, which is monotonous.

N.B. SCHWARTZ: I think what you are saying is correct. We don't know whether the ovary is a sequestering place for the FSH inhibiting substance. We have already done the experiment that you are suggesting with respect to ovariectomizing animals or looking at different stages of the cycle; we're looking in the serum, as Dr. Chappel will be looking in the ovary. We will let you know when the FSH radioimmunoassay is done.

C.P. CHANNING: It is of course, important to measure the material in the ovarian vein versus the peripheral circulation. Mr. Lou DePaolo, Dr. Barraclough and his colleagues are currently measuring inhibin in ovarian venous blood of the rat throughout the rat estrous cycle and should have some answers shortly.

The question of how a follicle is chosen, is an extremely complex one, and maybe inhibin is involved. Until we can measure it, under different conditions, we really can't say what its critical role is in choice of a follicle. We do know however, that a lack of ovarian inhibin can explain the rise in FSH prior to menopause. Sherman and Korenman (J. Clin. Endocr. Metab. 42 (1976) 629) have observed a rise in FSH which occurs before the rise in LH prior to menopause. At this time there is a decrease in the number of follicles and probably also inhibin. This would indicate a physiological role of inhibin in controlling FSH secretion. Maybe at a future workshop we will know more about the control of follicular maturation and the way in which inhibin and FSH interact in this process.

J. WEISZ: I have been bothered throughout by the use of charcoal extraction and the measurement of the levels of a couple of steroids in the charcoal-extracted material, as a means of excluding the presence of steroid in material that is being used. For example, as far as I know, the Amicon-filter actually retains steroid through adsorption. Considering that we are starting with enormous concentrations of steroids in the follicular fluid, a small percentage left behind and then concentrated at any of the steps used could result in steroid being present in effective concentrations in the test material. Now, I don't want to be misunderstood. I'm not saying that what we have seen is necessarily a steroid; but if there are steroids left behind and different amounts are getting concentrated in different fractions, we will further complicate what is already a complicated enough situation to analyze. I would suggest that fairly routinely and, as different new approaches are introduced, people add steroids, large quantities of radioactive material, and check the different fractions for radioactivity. Furthermore, it might be worth checking not only the unconjugated but also the conjugated steroid.

J. LORENZEN: Two points to answer that question. One, the charcoal extraction is done after the fluid is concentrated on the filter, and secondly, we will do the experiment to add the steroids back to see if we get the same results.

N.B. SCHWARTZ: However, the steroid levels in the charcoal-treated follicular fluid are detectable in our assays, and are actually lower than the blood values of the animals into which we're injecting the fluid, which suggests to me that we're not really dealing with a steroid problem.

F. LEDWITZ-RIGBY: I have added μg concentrations of both progesterone and estrogen to the follicular fluids and have not been able to see any reversal of inhibition.

G. ERICKSON: In connection with the previous question regarding the origin of follicular inhibin, I would like to present some data from experiments in our laboratory that were designed to test the hypothesis that granulosa cells produce "inhibin". Granulosa cells, collected from preovulatory follicles of adult rats on the morning of proestrus and from pre-antral follicles of immature hypophysectomized animals, were cultured in vitro ($2 \times 10^3$ to $2 \times 10^5$ viable cells) for 3 days in McCoy's medium containing charcoal-treated 10% horse and 2.5% fetal calf sera. The spent media from granulosa cell cultures were then extracted twice with charcoal to remove steroid hormones and subsequently diluted 1:1 with fresh culture medium to replace nutrients that were lost during the initial culture and subsequent charcoal treatment. To assay the spent granulosa cell media for its FSH suppressing activity it was used as a medium in a 3 day culture of collagenase-dispersed pituitary cells ($1 \times 10^6$ viable cells/dish) from adult ovariectomized rats. At the end of the culture period, the amount of FSH and LH released by the pituitary cells was measured by radioimmunoassay. Control cultures of pituitary cells released approximately 2.5 μg of FSH during the 3 day period. Spent media from cultures of granulosa cells obtained from preovulatory follicles inhibited FSH secretion by pituitary cells. The amount of inhibition increased with increased numbers of cultured granulosa cells and ranged from

14% inhibition with $2 \times 10^3$ cells to a maximum of 61% inhibition by $2 \times 10^5$ cells. In contrast, the amount of LH released (about 30 μg) by the control pituitary cells was suppressed only by spent media from the highest concentrations of granulosa cells ($2 \times 10^5$ cells, 19% inhibition). Comparable results were found with cultured granulosa cells from immature hypophysectomized rats. These data indicate that the granulosa cell is a source of follicular inhibin and that the anterior pituitary is a site of ovarian inhibin action. In addition, the results suggest that the granulosa cell acquires the capacity to secrete "inhibin" early in follicle development. These findings are not inconsistent with data presented earlier in this session that an FSH-inhibiting substance is present in the follicular fluid from small, medium and large antral follicles.

D.W. SCHOMBERG: Some wise old anonymous sage once remarked that it's amazing how if the only tool one has is a hammer, with time eventually the whole world begins to look more and more like a nail....

I have a question for Dr. Ledwitz-Rigby. In our work with androgen-stimulated progesterone synthesis by granulosa cells, we were sensitized to the possibility that the significance of the observations (with respect to continued follicle development) on cells from pre-antral and moderately differentiated follicles might be minimal because the majority of these follicles are destined to become atretic. So, the questions to you is, is it too heretical to entertain the idea that the inhibitory activity you're seeing is coming from those follicles which are in the process of atresia since no distinction is made between those destined to ovulate and those which are in the process of atresia when the follicular fluid is harvested?

F. LEDWITZ-RIGBY: I have collected follicular fluid from two different groups of small follicles. One group was found only on ovaries that had the large follicles present already; we presumed that the majority of these small follicles were atretic. The other group of small follicles was on ovaries that had only small follicles; at least some of those follicles were not atretic, although obviously, some would be. There was no difference in the inhibitory actions of these two different fluids from small follicles. However, I think that the inhibitor probably does have a major role in deciding which follicles are going to become atretic. I would love to know what controls the production of the inhibitor; it could very well be many of the factors that have already been reported to influence the rate of atresia. Perhaps the failure to eliminate this inhibitor may doom certain follicles to become atretic. Does that answer your question?

D.W. SCHOMBERG: It's a difficult question because one can't a priori select the follicles.

M. RAJ: Much has been said on the possible use of these inhibitors in the future in some form or other for contraceptive purposes. This prompts me to ask whether anyone has any data on the chronic effects of these inhibitors on reproductive processes, for example: ovum maturation, folliculogenesis. We know that chronic inhibition of FSH using specific antiserum affects follicular maturation in the hamster (Moudgal et. al. Rec. Prog. Horm. Res. 30, 55, 1974), and first wave of spermatogenesis in the rat (Madhwa Raj and Dym, Biol. Reprod. 14: 489, 1976). I might point out that at least this kind of data is being generated now, using inhibin in the male. Dr. deJong presented some of this data in the recent testes workshop. Is there any data available in the female on this subject? If not, would anybody like to speculate on the chronic effects of inhibiting FSH using inhibin in the female?

C.P. CHANNING: Maybe I can answer this. In very preliminary studies in the monkey, we gave some of the monkeys follicular fluid over a 4-day period. We examined them 1, 2, and 4 cycles later, and in two of the monkeys, 2 cycles later they had an ovulatory cycle. So it looks like its effects are reversible and not long-lasting after the fluid is given. We have not given follicular fluid to monkeys for more than 4 days or to young animals, so we cannot actively assess long term effects of follicular fluid on fertility yet experiments are underway. The effects on oocyte maturation in vitro indicates that the effect of OMI is reversible. If cultured cumulus-enclosed oocytes are cultured for 24 hours in the presence of OMI followed for an additional culture period under control conditions, the oocytes mature and the cumulus cells resume their ability to secrete progesterone (Hillensjö et. al., In Vitro, in press, 1978).

QUESTION: I would like to direct a question to Dr. Lorenzen, which she can speculate about if she wants, and that is, relating what you have done to the control of FSH secretion in the prepubertal rat. There is a rise in FSH occurring fairly early on and then a rather precipitous drop around day 15 to day 18 of life. Does anyone in the audience, including Dr. Lorenzen, have any information about when the ovary starts making some of these nonsteroidal controlling substances? Certainly they seem to be concentrated in small follicles, and young ovaries contain lots of small follicles. Is there any information about when they're made, or when they might be released into the circulation?

J. LORENZEN: I don't know the answer to that -- if we had good antibodies to the active component we could measure it in small amounts. You do not get a lot of blood from a 6-day old rat. I would guess it doesn't come out until the FSH starts to fall. The follicles in the younger rats have no antra, you can't pull out any follicular fluid, even if they were big enough, because they're just solid cells.

J. FINDLAY: I would like to make a comment about atresia, and ask a question of Dr. Ledwitz-Rigby. As I mentioned this morning it's almost certain that all the follicle class sizes that you look at will have a fair proportion of them atretic. So you're really stuck with that problem of not knowing whether you've got atretic ones or not.

I think it's also pertinent here to point out that Bob Moore and Mary Hay have shown that there is a decrease in the blood flow into the theca of sheep follicles associated with atresia. This begs a question, then, as to what causes the decrease in blood flow, but I don't think we should lose sight that there may be some extrafollicular mechanism involved.

My question is: have you measured any other steroids, besides progesterone in the studies you have done, as it is well known that there is a differentiation of steroidogenesis within the granulosa cell?

F. LEDWITZ-RIGBY: No, the only steroid I have measured is progesterone.

C.P. CHANNING: May I add, in answer to your question about atretic versus viable follicles, the mare is an ideal animal for which to study the mechanism of atresia. In some studies with Drs. Condon, Ganjam and Dr. Kenny, in New Bolton Center, we have harvested atretic and viable follicles and histologically characterized each one. Estrogen, progesterone and testosterone in follicular fluid declines during atresia. Presently we are in the process of measuring inhibin, luteinization inhibitor and OMI of follicles and I think this is the only way you're going to answer some of these questions.

N.B. SCHWARTZ: Dr. Midgley.

A.P. MIDGLEY: A couple of comments, and a question or two. One is, I was intrigued by the presentation of Erickson. I thought that was beautiful data, and I want to go on record as such. It is of interest that cultured granulosa cells can secrete inhibin.

Regarding the LHRBI (which reminds me an awful lot of the name that is given to the Institute in Harvard), you made the interesting comment that this substance appears to be present in fresh tissue if one uses a detergent extraction. This suggests to me that it's present in the membranes someplace in the cell, and if that is the case, how can it be regulatory?

C.P. CHANNING: We don't know enough about it yet in order to answer your question. We haven't examined a membrane fraction vigorously, and I think at this stage in the game, it's too premature even to speculate.

A.P. MIDGLEY: Fair enough. I would like to ask--Florence a question. I noted that in many of the slides--I may have missed this--it seems to me you're representing the data as progesterone per 100 cells inoculated. There's a problem here when one looks at 2 and 4 and 6 days later, wondering about how many cells are still there. Have you thought about ways of trying to express the data as progesterone production rates after the 2 - 6 day period of time per viable cell and preferably per identifiable viable cell that happens to be a granulosa cell?

F. LEDWITZ-RIGBY: Yes, we've just had two undergraduate honor students, Marta Voytovich and Mark Hurt, look at the rate of cell growth in 50% fluid from large follicles, fluid from small follicles, or serum. There was a decline in the number of cells between days 0 and 6, after which there was a very rapid increase in cell number in all three culture media. We could not find any statistically significant difference between either the rate of the decline or of the cell multiplication that was consistent, between any of the three different culture media. We did have one intriguing experiment where the cells cultured in fluid from small follicles were actually multiplying faster than the others, but this was not a consistent finding.

N.B. SCHWARTZ: I would like to try a little Greenwaldian speculation about some of the things we've been hearing about today, trying to connect them with some of the things we heard yesterday.

Essentially, we have been discussing intraovarian regulation. This is unusual for a group of endocrinologists to be talking about because until a few years ago, most of what we discussed was hormones made by the ovary to go someplace else. The one material that we have talked about that has not been called an "intraovarian" signal is the folliculostatin, or female inhibin. As has been pointed out, we do not really know yet that it is in the blood, although we feel that it does have a physiological role. I would like to tell you what I think that role is. Yesterday evening, there was speculation about the recruitment of follicles in animals with short cycles. I do not remember whether it was Dr. Richards or someone else who said that it is usually considered that the proestrous surge of LH and FSH are responsible for that recruitment. I would like to discuss data that we have on the effects of folliculostatin on the primary surges of LH and FSH, and on the secondary surge of FSH, and then tell you about some recent observations made by a graduate student on follicular recruitment.

As presented in the 1978 Federation Proceedings (abstract 2686) Hoffmann and I have injected 0.5 ml of this material at 1200 h of proestrous, and we found that there is good suppression of FSH at 1700 h, and at 1830 h.

When we inject 0.5 ml of FFl at 1545 h and at 1830 h of proestrous, after the primary surges, we completely suppress FSH at 0400 h estrus.

David Hoak, a graduate student in the laboratory has recently redone these experiments. Instead of just looking at the hormone levels, he has counted crops of follicles in the ovary the next morning (estrus). Normally, there is a cohort of follicles which comes into the ovary on the morning of estrus, which appears to be that crop of follicles which is ovulated at the next estrus. We have now counted that cohort (the cohort over 350 microns) which simply is not present at proestrous and then is recruited into the ovary by the next morning. When we block the primary FSH surge by giving early pFFl, we do not block completely the recruitment of that cohort. But if we block the secondary FSH surge, we do block the recruitment of the cohort! So, while the secondary FSH surge is not very high, it is prolonged for about 10 hours, and that 10 hours of FSH apparently is enough to move one cohort of follicles from below 350 microns to above it, and we believe then to constitute the next crop to be ovulated. Thus, we think there is a physiological role, at least for the secondary FSH surge. We have succeeded in blocking it with this very interesting tool, and by implication, we also believe that a change in concentration of folliculostatin may normally be responsible for the recruitment of FSH in a short-cycle animal for the next cycle.

INHIBITION BY FOLLICULAR FLUID OF LUTEINIZING HORMONE-SENSITIVE ADENYLATE CYCLASE IN OVARIAN PLASMA MEMBRANES

A. Amsterdam, M. Shemesh* and Y. Salomon

Department of Hormone Research, The Weizmann Institute of Science, Rehovot, and *Kimeron Veterinary Institute Beit-Dagan, Israel

## INTRODUCTION

The regulation of ovarian function by luteinizing hormone (LH) is mediated by a hormone-sensitive adenylate cyclase (AC). The respective hormone receptors are located on the outersurface of the cell membrane of the follicular cells (for review see ref. 1). Since hormone binding takes place in the extracellular *milieu*, one should consider the possibility that extracellular factors may affect hormone-receptor interactions and consequently influence the cellular response to the hormone.

The role of extracellular components which accumulate in the follicular antra (i.e. follicular fluid, FF) is not yet clear. It has been suggested that degradation of macromolecular components present in FF shortly before ovulation will lead to an elevation of the intrafollicular pressure and thus facilitate follicular rupture (2,3). There is still some debate on the question of whether such a mechanism is really operative. It was recently demonstrated that porcine FF inhibits ovum maturation *in vitro* (4). Furthermore, porcine FF inhibits granulosa cell luteinization *in vitro* as judged by morphological transformation, progesterone secretion and accumulation of 3'5'-cyclic AMP (cAMP) in response to LH (5). Follicular fluid was shown by us to contain heparin-like substances (6-8) and heparin was shown to inhibit LH-sensitive AC in ovarian plasma membrane preparations (9). In this report we present evidence that FF contains a potent inhibitor of LH sensitive AC and show that the activity of the intraovarian inhibitor decreases with follicular maturation. Intraovarian inhibitor(s) are suggested to play a role in modulating hormonal responsiveness of the ovary.

## THE INHIBITORY EFFECT OF BOVINE FF ON LH/hCG SENSITIVE AC ACTIVITY AND ON $^{125}$ I-hCG BINDING TO OVARIAN PLASMA MEMBRANES

Adenylate cyclase activity was determined using purified ovarian plasma membrane (PM) preparations. Ovaries were obtained from 25 d.o. rats 48 hrs after administration of pregnant mare serum gonadotropin (PMSG, 15 i.u./rat)(10). Bovine FF was obtained from medium to large follicle at midcycle.

Shown in Figure 1 is an experiment in which AC activity was determined in the presence of increasing concentrations of FF. It can be seen that LH sensitive AC is inhibited by FF in a dose-dependent manner. Fifty percent inhibition was reached at a concentration of FF of 3 mg protein/ml while at 12 mg/ml almost complete inhibition was achieved. Stimulation of AC by hCG was similarly affected by FF. It should be noted that the highest concentration of FF used is equivalent to one-seventh of the original FF concentration, suggesting that the effects observed may be of physiological significance. In contrast to the intensive effect of the FF on hormone stimulable AC, it was found that fluoride activated enzyme was relatively resistant to the inhibitor. It is generally accepted that fluoride ions exert their effect on AC by direct interaction with its catalytic component. The inhibition of basal AC activity may be explained by an inhibitory effect of FF on a component of AC activity stimulated by residues of PMSG bound to these membrane preparations. The inhibitory activity of the bovine FF was recovered in the void volume of a Sephadex G-10 column, suggesting that the molecular weight of the inhibitor is in excess of 700 Daltons.

It is shown in Fig. 1B that inhibition of $^{125}$I-hCG binding to ovarian plasma membranes by FF is 18% at 3 mg FF protein/ml and 37% at 12 mg/ml. This inhibition was small when compared to the effect seen on the hormone stimulable AC activity (50% inhibition at 3.0 mg/ml).

The inhibition by FF of LH sensitive AC was found to be independent of LH concentrations. Bovine FF (2.5 mg protein/ml) inhibited about half of the AC activity stimulated by LH in the concentration range 0.1 - 100 nM (Fig. 2). This result indicates that inhibition of AC does not involve direct interaction of the hormone with components present in FF.

Inhibition of hormone binding is unlikely to account for the observed effects of FF on the AC system. This is inferred from the relatively small effect of FF on $^{125}$I-hCG binding. The resistance of the fluoride-activated enzyme to the inhibitory action of the FF renders unfavorable the possibility that the FF interferes strictly with catalytic activity. It is therefore suggested that the putative inhibitor present in FF may interfere with the coupling of the

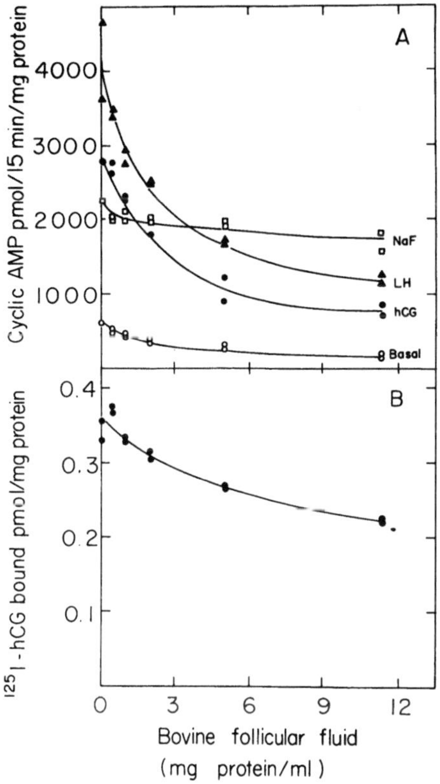


Fig. 1. The effect of bovine FF on AC activity and $^{125}$I-hCG binding in purified ovarian plasma membranes: Fig. 1A. Adenylate cyclase activity unstimulated (Basal) or stimulated by NaF (10 mM), LH ($10^{-7}$ M) or hCG ($10^{-8}$ M) was determined in the presence of increasing concentrations of bovine follicular fluid. Each assay contained 3 µg of PM protein. All other details were as described elsewhere (11). Fig. 1B. $^{125}$I-hCG binding was determined at increasing concentrations of bovine follicular fluid. Each assay system contained 10 µg of ovarian PM. $^{125}$I-hCG concentration was 1.7 nM (from Amsterdam, Shemesh, Riesel, Mintz and Salomon, 1978, submitted).

receptor to the catalytic unit either by restricting the relative mobility of these components or by inducing conformational changes which will not allow normal interaction between these membrane components. We have recently reported that heparin like substances are present in the rat ovary (6-8). We also found that LH-sensitive AC is strongly inhibited by heparin, which primarily interferes with LH-stimulation but has only a small effect on stimulation of the enzyme by NaF (9). Therefore, ovarian sulfated mucopolysaccharides

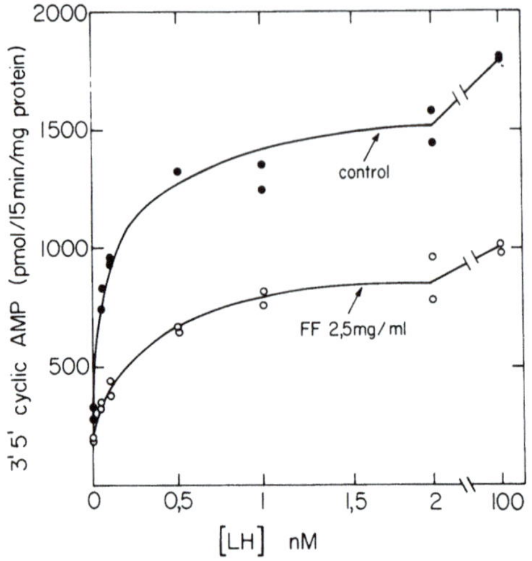


Fig. 2. The effect of bovine follicular fluid on the activity of LH sensitive AC at various LH concentrations: AC activity was determined at the indicated concentrations of LH in the absence (•) or in the presence (o) of bovine FF (2.5 mg protein/ml) in the reaction mixture. Each assay system contained 3 μg of PM protein. All other details were described elsewhere (11) (from Amsterdam, Shemesh and Salomon, in preparation).

such as heparin may be considered as natural modulators of hormonal control in the ovary.

## CHANGES IN INHIBITOR ACTIVITY DURING FOLLICULAR MATURATION

In searching for physiological role of the intra-ovarian factor which inhibits LH-sensitive AC we dediced to test whether increased responsiveness to LH observed with follicular maturation (10) may be explained in part by a decrease in inhibitor activity. Treatment with PMSG at the dose level chosen caused synchronized growth of a large group of follicles: 24 hr after injection many small and medium sized antral follicles were observed, whereas after 48 hrs most of the ovarian volume was occupied by large antral follicles. Incubation of plasma membranes with the $10^5$ x g supernatant fraction obtained from rat ovarian homogenates 24-36 hrs after injection of PMSG abolished LH-stimulable AC activity at supernatant protein concentrations of 2 mg/ml. 42 and 48 hrs after injection of PMSG the degree of inhibition was reduced to 67% and 33% respectively (Fig. 3). A progressive decline of inhibitory activity with time after PMSG injection was also evident using a concentration of 1.0 mg/ml of the

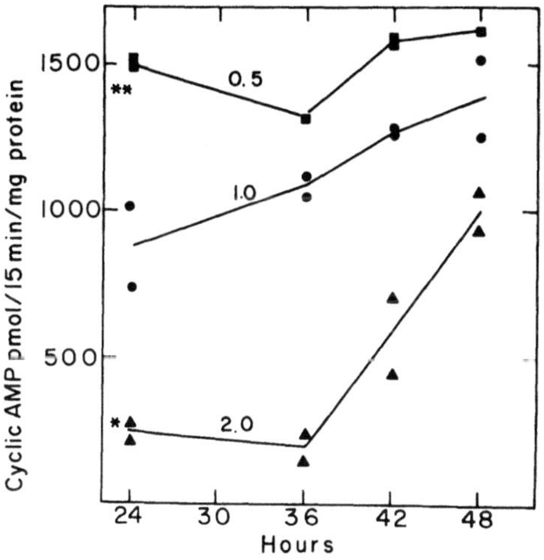


Fig. 3. The effect of the soluble fractions from ovarian homogenates obtained from PMSG-treated rats on LH-sensitive AC activity: Ovarian post-microsomal supernatants were prepared at the indicated times after PMSG (15 i.u./rat) injection. These preparations at a final concentration of 0.5, 1.0 and 2.0 mg/ml were included in standard adenylate cyclase assay. Enzyme activity was determined in the presence of $10^{-7}$ M LH. Each assay system contained 3.0 µg of membrane protein. All other details were as described elsewhere (11). Indicated are Basal (*) and LH stimulated activity (**) in the absence of FF (from Amsterdam, Shemesh, Riesel, Mintz and Salomon, 1978, submitted).

soluble fraction. At 0.5 mg protein/ml no inhibition of the enzyme was detected at any time. The progressive decrease in inhibitor activity shortly before the LH surge may thus be of physiological importance. The mechanism which controls the level of the inhibitor is not yet understood. Early findings of Zachariae suggested degradation of high molecular weight components of the FF close to ovulation (2). Furthermore, we have recently found that the biosynthesis of sulfated mucopolysaccharides is inhibited by progesterone and LH (6-8).

## CONCLUDING REMARKS

The exact physiological role of the macromolecular components which accumulate in the follicular antra is not yet clear. In this paper we present evidence that follicular fluid can inhibit specifically the LH-sensitive AC suggesting that it may play a role in controlling hormone action in the ovary. The observation that ovarian inhibitor activity declines concomitantly with follicular maturation supports the contention that elimination of inhibitor(s) may be part of the preconditioning of the follicle for subsequent sensitization to LH.

## ACKNOWLEDGEMENT

We thank Dr. H.R. Lindner for helpful discussions and Mrs. M. Kopelowitz for typing the manuscript. This work was supported in part by a grant to H.R.L. by the Ford Foundation and Population Council, Inc., New York and partly by grants to Y.S. and A.A. by the U.S.-Israel Binational Science Foundation (BSF), Jerusalem. A.A. is the incumbent of the Gestetner Career Development Chair. Y.S. is the incumbent of the Charles W. and Tillie Lubin Career Development Chair.

## REFERENCES

(1) K.J. Catt and M.L. Dufau. Ann.Rev.Physiol. 39(1977)529.

(2) F. Zachariae. Acta Endocr.Suppl. 47(1959)33.

(3) D. Rodbard. J.Clin.Endocrinol. 38(1968)849.

(4) A. Tsafriri and C.P. Channing. Endocrinology 96(1975)922.

(5) F. Ledwitz-Rigby, B.W. Rigby, V.L. Gay, M. Stetson, J. Young and C.P. Channing. J.Endocr. 74(1977)175.

(6) H. Gebauer, Y. Koch, H.R. Lindner and A. Amsterdam. V. Int'l. congr. of Endocrin., Hamburg. Abstr. 811(1976)335.

(7) H.R. Lindner, A. Amsterdam, Y. Salomon, A. Tsafriri, A. Nimrod, S.A. Lamprecht, U. Zor and Y. Koch. J.Reprod.Fert. 51(1977)215.

(8) H. Gebauer, H.R. Lindner and A. Amsterdam. Biol. Reprod. 8(1978)350.

(9) Y. Salomon and A. Amsterdam. FEBS Lett. 83(1977)263.

(10) Y. Salomon, A. Yanovsky, Y. Mintz, Y. Amir and H.R. Lindner. J.Cyclic Nuc.Res. 3(1977)163.

(11) Y. Mintz, Y. Amir, A. Amsterdam, H.R. Lindner and Y. Salomon. Mol.Cell. Endocrin.(1978) in press.

# INHIBITORY EFFECT OF CHARCOAL-TREATED PORCINE FOLLICULAR FLUID UPON SERUM FSH LEVELS AND FOLLICULAR DEVELOPMENT IN THE RHESUS MONKEY

Cornelia P. Channing[1], Larry D. Anderson[1] and Gary D. Hodgen[2]
[1]Dept. of Physiology, University of Maryland School of Medicine, 660 W. Redwood Street, Baltimore, Maryland 21201
[2]Endocrinology Section, Endocrinology and Reproduction Research Branch, National Institute of Child Health and Human Development, NIH, Bethesda, Maryland 20014

## INTRODUCTION

Porcine follicular fluid (pFFl) has been shown to contain a nonsteroidal FSH suppressing substance active in castrated (1) and intact (2) adult female rats. However, the follicular fluid did not inhibit serum LH. The role of physiological levels of FSH in controlling follicular development and ovulation in the rhesus monkey is not known, since there are no methods available other than immunization against FSH which can selectively and reversibly remove FSH without removing LH at the same time. Based on the findings in the rat, it seemed feasible that pFFl might also depress serum FSH levels in the rhesus monkey. In order to examine the role of the FSH secreted early in the menstrual cycle in the rhesus monkey, a series of injections of pFFl have been employed to depress serum FSH early in the menstrual cycle. Serum FSH, LH, estrogen and progesterone levels were measured during the treated and control cycle to monitor the effects of the pFFl. In addition, the dynamics of using pFFl to suppress peripheral serum FSH levels were examined in both postmenopausal and long term castrated female monkeys, where sustained hyper-secretions of both FSH and LH may dramatize the selective suppression of circulating FSH by pFFl.

## MATERIALS AND METHODS

### Cycling monkeys and treatments

Ten intact normally cycling adult female rhesus monkeys were checked daily for vaginal bleeding and housed and fed as previously described (3,4). Each monkey was allowed to go through two regular menstrual cycles prior to use. During the third cycle, designated as the "control cycle", the monkey was bled daily by femoral puncture while under light Ketamine anesthetic (4). On the anticipated last day of the luteal phase and during the first 2-4 days of the new menstrual cycle (total of 4 days) four monkeys were given 4 ml charcoal-treated porcine follicular fluid intraperitoneally every 8 hours at 8:00 A.M., 4:00 P.M. and 12:00 midnight. Blood samples were taken at each treatment time and daily thereafter for the duration of the "treated" menstrual cycle. On day 10-12 of the treated cycle, the monkey was anesthetized with Nembutal (30mg/kg) and laparotomized. The ovary was exposed and the diameter of the largest follicle was measured (5). Granulosa cells and follicular fluid were aspirated from that follicle. All the granulosa cells recovered from that follicle were cultured for 2 days in 15% male monkey serum in tissue culture medium 199 (6). After 2 days in the culture, the cells were counted and the progesterone content of the spent culture medium determined by radioimmunoassay using a specific progesterone anti-11 OH progesterone BSA conjugate (7). Four control monkeys were given porcine serum instead of porcine follicular fluid.

### Menopausal and castrated monkeys and treatments

Two postmenopausal monkeys (8) and four adult female monkeys castrated for more than 2 years were given 10 or 30 ml sc of charcoal-treated pFFl over 6 hours. Blood was obtained from the femoral vein at 6 hour intervals beginning 12 to 24 hours before treatment and extending from 72-96 hours beyond pFFl injections.

### Porcine follicular fluid

Porcine follicular fluid was collected from a pool of small (1-2mm), medium (3-5mm) and large (6-12mm) porcine follicles with 30% coming from small, 50% from medium and 20% coming from large follicles (9). The fluid was centrifuged at 1000 x G to remove granulosa cells, frozen and later charcoal-treated to remove steroids, as outlined previously (10). The follicular fluid pool used in the present studies contained 49,200 pg/ml estrogen; 16,000pg/ml progesterone before charcoal treatment and 50pg/ml estrogen and 30 pg/ml progesterone after charcoal treatment.

Hormone measurement

Serum estrogen and progesterone were measured by radioimmunoassay in diethyl ether and petroleum ether extracts, respectively, as detailed previously (5). Serum LH and FSH were measured by radioimmunoassay as described elsewhere (11,12).

RESULTS

Administration of 4 ml of pFFl every 8 hours to 4 intact cycling monkeys early in the menstrual cycle ultimately exerted 87, 85, 52 and 14% inhibitory effect upon serum FSH. (Table 1). The FSH suppression was evident within about 12 hours. The FSH levels returned to preinjection control levels within 1-3 days after cessation of treatment. Serum LH was essentially unaltered (Figure 1,2). Among the 4 monkeys given pFFl early in the menstrual cycle, there was no preovulatory estrogen surge (Figure 2) and laparotomy at midcycle during the treatment revealed that the follicle was smaller than normal and contained few granulosa cells ( <1 x $10^6$ cells ) compared to serum-treated or

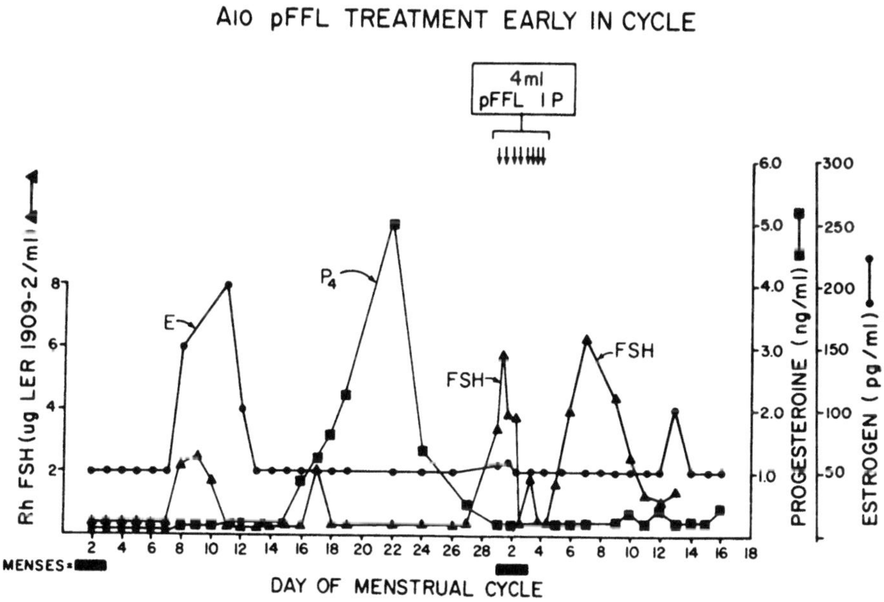


FIGURE 1. Temporal course of effect of porcine follicular fluid upon serum FSH and LH in the rhesus monkey. The fluid was given intraparitoneally in a dose of 4 ml every 8 hours from day 1 to 4 of the menstrual cycle of an intact monkey and serum samples taken every 8 hours during the treatment period and daily thereafter.

TABLE 1. Effect of charcoal treated porcine follicular fluid upon serum FSH and LH in the monkey.

| Monkey Number | Serum FSH (μg LER 1909-2/ml) | | | | Serum LH (μg LER 1909-2/ml) | | | |
|---|---|---|---|---|---|---|---|---|
| **Intact Monkeys:** | | | | | | | | |
| No. | Before | During | After | % Inhibition | Before | During | After | % Inhibition |
| A10 | 4.6 | 0.6 | 6.5 | 87 | | | | |
| A11 | 2.7 | 0.73 | 3.0 | 85 | 7.8 | 9 | 10 | 0 |
| A21 | 3.8 | 2.4 | 4.1 | 52 | 5.4 | 4-6 | 6 | 0 |
| C1 | 6.6 | 5.75 | 14 | 14 | | | | |
| **Castrate Monkeys:** | | | | | | | | |
| C-467 | 25 | 10 | ---- | 60 | 12 | 12 | ---- | 0 |
| C-142 | 36 | 26 | ---- | 28 | 21 | 9.3 | ---- | 8 |
| 78C | 29.5 | 15 | ---- | 83 | 12 | 9.3 | 10.8 | 22 |
| 600-D | 35.5 | 16.5 | ---- | 54 | 12 | 11.3 | 10.5 | 6 |

The 4 intact monkeys were given 4 ml intraperitoneal injection of follicular fluid every 8 hours from day 1 to 4 of the menstrual cycle the 4 long term castrated monkeys were given two subcutaneous injections of follicular fluid of either 5 ml (in the case of monkey 78C) or 15 ml (monkey 600-D, C-142, C-467) and sampled for blood every 6 hours for 2-4 days thereafter.

TABLE 2. Effect of porcine follicular fluid administered in 4 ml doses every 8 hours from day 1 to 4 of the menstrual cycle upon follicular size and recovery of granulosa cells at midcycle.

| Monkey Treated with pFF1 | Largest Follicle Present | |
|---|---|---|
| | Follicle size (mm) | Number of Granulosa cells |
| $A_{11}$ | 11 | $\sim 0.18 \times 10^6$ |
| $C_1$ | 4 | $0.002 \times 10^6$ |
| $A_{10}$ | <2 | No Visible Follicle |
| $A_{21}$ | 6 | Too Few To Count |
| Serum Treated Control | | |
| $C_4$ | 4 | $1.4 \times 10^6$ |
| $C_6$ | 11 | $3.4 \times 10^6$ |

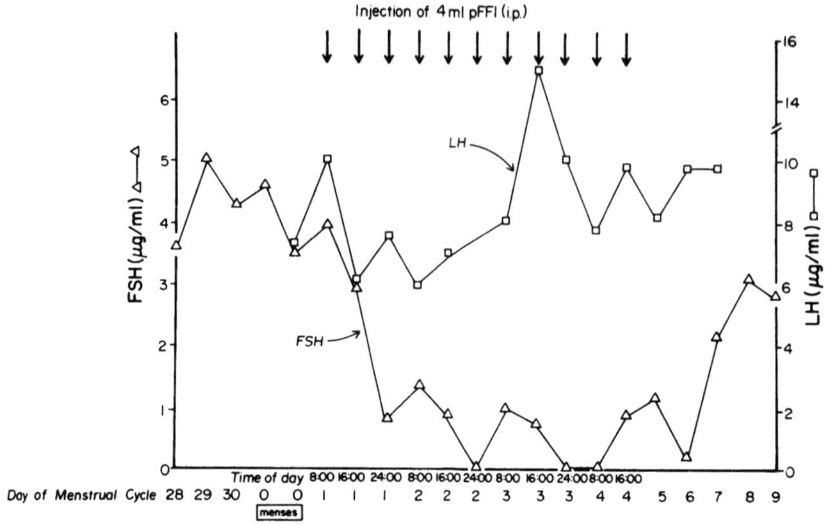


FIGURE 2. Temporal course of inhibitory action of procine follicular fluid upon serum FSH in a rhesus monkey given porcine follicular fluid every 8 hours for 4 days during the early follicular phase of the menstrual cycle.

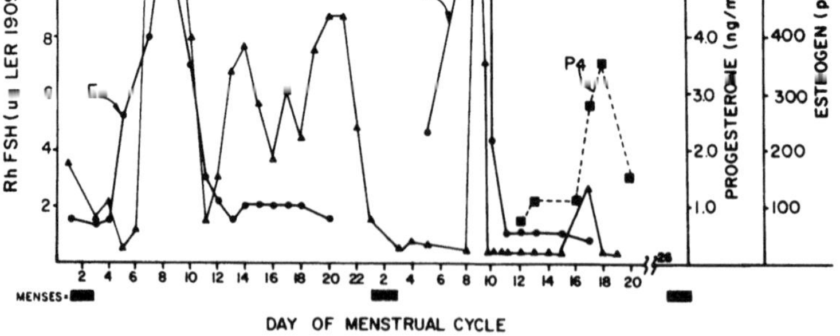


FIGURE 3. Temporal course of inhibitory action of porcine follicular fluid upon serum FSH in a rhesus monkey given porcine follicular fluid every 8 hours for 4 days during the mid-follicular phase of the menstrual cycle.

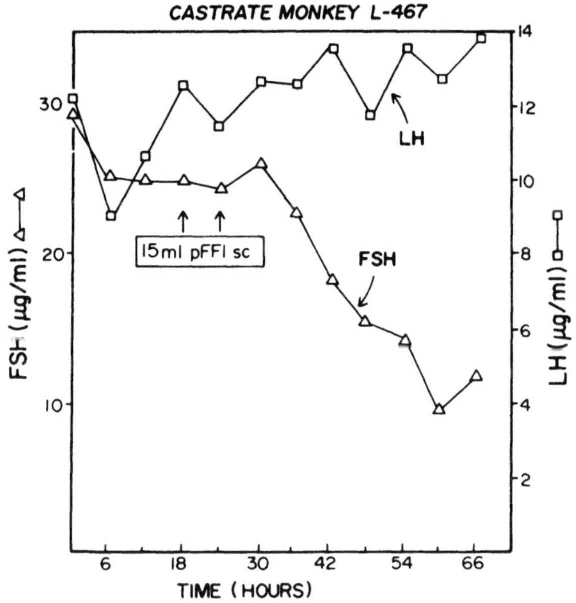


FIGURE 4. Temporal course of inhibitory action of porcine follicular fluid upon serum LH and FSH in a castrate monkey. The fluid was given subcutaneously in two 5 ml doses 6 hours apart followed by blood samples taken every 6 hours thereafter for 3 days. A long term castrate female monkey was used for this study. Serum LH and FSH levels were measured in each blood sample.

untreated control monkeys whose follicles contained 2-50 x $10^6$ granulosa cells at midcycle (Table 1). Two monkeys were given pFFl for four days at midcycle: in one case, the treatment was started 1 day prior to the expected midcycle gonadotropin surge of FSH. The midcycle FSH surge was delayed until after cessation of pFFl treatment. In the other monkey, in which treatment was started 8 hours after the onset of the FSH surge, apparently the FSH surge was abbreviated to about 1/3 of normal (Figure 3).

Administration of pFFl sc every 6 hours in two doses of either 5 or 15 ml to four long term castrated adult female monkeys led to a progressive decline of serum FSH levels by as much as 83, 60, 28 and 14% within 24-30 hours after the initial injection. The decline started within 18 hours and was maintained for about a 3 day period. FSH levels returned to pretreatment levels by 96 hours. Serum LH was not decreased significantly (Figure 4). Similarly this treatment regimen selectively suppressed circulating FSH levels in two postmenopausal monkeys, without changing LH levels.

## DISCUSSION

These findings demonstrate that charcoal extracted pFFl contains a substance which can cause a selective decrease in blood FSH levels in the intact, castrated and postmenopausal rhesus monkey. Exposure to the pFFl for 4 days early in the follicular phase of the cycle led to abnormal granulosa cell development in the treated cycle, demonstrating that appropriate exposure of the growing follicle to FSH early in the menstrual cycle may be essential for its eventual normal development in the late follicular phase. It is also possible that the pFFl has a direct inhibitory action upon the follicle unrelated to its effects on serum FSH levels.

Under conditions of hyper-secretion of FSH and LH in castrated and postmenopausal monkeys, pFFl acted by suppressing serum FSH without significant effects on LH; yet, the nature of this pFFl reduction of circulating FSH seemed to plateau after about 24 hours, before recovery to pretreatment levels during the next 36-48 hours.

## SUMMARY

Injection of charcoal treated porcine follicular fluid into 4 intact, 4 long term castrate and 2 postmenopausal rhesus monkeys brought about a selective decline in serum FSH levels. Luteinizing hormone levels were not significantly decreased. Intraperitoneal injection of 4 ml porcine follicular fluid every 8 hours for the first 4 days of the menstrual cycle led to a decrease in follicle and granulosa cell growth observed at that midcycle. Treatment with porcine serum did not exert this effect.

## Acknowledgements

This research was supported by NIH grant HD 08834 and Ford Foundation grant 760-0530. The expert assistance of Ms. Sandra Fowler and Mr. Charles Turner is gratefully acknowledged.

REFERENCES

(1) M.L. Marder, C.P. Channing and N.B. Schwartz. Endocrinol. 101(1977)1639.

(2) N.B. Schwartz and C.P. Channing. Proc. Nat. Acad. Sci. 74(1977)5721.

(3) C.P. Channing. Endocrinol. 87(1970)156.

(4) C.P. Channing, B. Engel, S. Fowler and K. Vitek. Proc. Soc. Exper. Biol. Med. 155(1977)615.

(5) C.P. Channing and S.P. Coudert. Endocrinol. 98(1976)568.

(6) C.P. Channing. Endocrinol. 94(1974)1215.

(7) C.P. Channing, V. Tsai and D. Sachs. Biol. Reprod. 15(1976)235.

(8) G.D. Hodgen, A.L. Goodman, A. O'Connor and D.K. Johnson. Amer. J. Obs. Gyn. 124(1977)581.

(9) C.P. Channing and F. Ledwitz-Rigby. In: Methods in Enzymology (J.G. Hardman and B.W. O'Malley, eds.) Vol 39 Part 2 Academic Press, New York (1975) pp. 83-230.

(10) F. Ledwitz-Rigby, B.W. Rigby, V.L. Gay, J. Young, M. Stetson and C.P. Channing. J. Endocrinol. 74(1977)175.

(11) G.D. Hodgen, J.W. Wilks, J.L. Vaitukaitis, H.C. Chen, H. Papkoff, and G.T. Ross. Endocrinol. 99(1976)137.

(12) G.D. Niswender, S.E. Monroe, W.D. Peckham, A.R. Midgely Jr. E. Knobil and L.E. Rechert Jr. Endocrinol. 88(1971)1327.

# FOLLICULAR FLUID INHIBITION OF OVULATION

Kurt F. Miller, John K. Critser and O. J. Ginther

University of Wisconsin

Madison, Wisconsin 53706

Summary Ovulation was inhibited in the bovine following treatment with whole bovine follicular fluid. Ovulation was not inhibited following treatment with either steroid free or protein denatured follicular fluid. However treatment with the combination of steroid free follicular fluid and protein denatured follicular fluid was as effective as whole follicular fluid in inhibiting ovulation. Although steroid free follicular fluid did not inhibit ovulation, two of three animals which recieved this treatment did not form normal corpora lutea, as indicated by palpation and the occurance of ovulation 5 and 6 days after the initial post-treatment ovulation.

## INTRODUCTION

Steroid free follicular fluid (FF) has been demonstrated to contain a substance which has an inhibitory effect on circulating concentrations of FSH. Bovine FF in castrated male rats (1), bovine or porcine FF in ovariectomized (OVX) rats (2), porcine FF in OVX or intact female rats (3), and equine FF in OVX mares (4) significantly decreased circulating levels of FSH. It has also been reported that administration of whole FF causes inhibition of follicle growth (immature mouse, 5, cycling ewe, 6) and ovulation (cycling ewe, 6). In the intact female the active FF substance does not occur in a steroid free milieu, but rather in combination with the steroids produced by the follicle. Indeed, Marder et al (3) report that sham operated rats were more sensitive to steroid free FF than were OVX rats. Therefore, two studies in the bovine were designed to determine, first the effect of bovine FF on ovulation, and secondly which category of molecular compounds possessed the inhibitory activity.

## MATERIALS AND METHODS

All animals were nulliparous Holstein heifers weighing from 350 to 500 kg. The FF was obtained by aspiration of all grossly visable follicles from cow ovaries obtained at a local slaughterhouse. All FF was stored at -20 C until immediatly before injection.

Experiment 1 Six heifers, on day 8 - 12 of diestrus, were injected, IM, with 30 mg of Prostaglandin $F_{2\alpha}$ ($PGF_{2\alpha}$) to cause luteal regression and synchronized follicular development. They were assigned randomly to either a control or a treated group. Controls recieved 10 ml of saline, and the treated animals recieved 10 ml of whole FF. Injections were given, IV, twice daily for three days, and were begun at the time of the $PGF_{2\alpha}$ injection. All animals were palpated twice daily to confirm luteal regression and to determine the time of ovulation.

Experiment 2 Fifteen heifers, on day 8 - 13 of diestrus, were injected, IM, with 30 mg $PGF_{2\alpha}$ and assigned randomly to one of five groups. Groups and treatments were:

Group C - control (saline).

Group HFF - heat treated FF (10 ml FF heated to 100 C for 2 min).

Group CFF - charcoal extracted FF (10 ml FF extracted with 50 mg charcoal per ml FF for 1 hr at 20 C followed by centrifugation at 5,000 X g and 4 C for 45 min).

Group HFF + CFF - heat treated FF and charcoal extracted FF (10 ml of each).

Group WFF - whole FF (10 ml).

Treatments were given, IM, in a total volume of 20 ml, twice daily for three days begining at the time of the $PGF_{2\alpha}$ injection. All heifers were palpated daily to confirm luteal regression and to determine the time of ovulation.

## RESULTS

Experiment 1 There was no difference in luteal regression between the two groups as measured by corpus luteum (CL) diameter. The diameter (mm) of CL 48 hr after $PGF_{2\alpha}$ treatment was 10.0 ± 1.2 (mean ± SEM) for the control and 11.3 ± 1.8 for the treated group. However, treatment with FF significantly ($p < .001$) lengthened the interval from $PGF_{2\alpha}$ administration to ovulation (Figure 1).

Experiment 2 As in the previous experiment, there were no differences in the amount of luteal regression among groups. The diameter (mm) of the CL at 72 hr after $PGF_{2\alpha}$ administration was 11.3 ± .3, 12.3 ± 1.5, 11.0 ± .6, 13.0 ± .6, and 12.7 ± .7 for

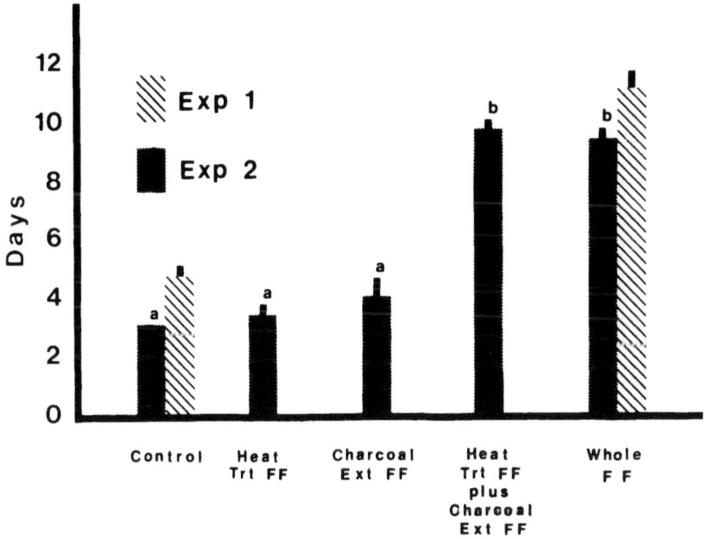


Figure 1 Days (+SEM) from $PGF_{2\alpha}$ administration to ovulation in diestrous heifers treated with follicular fluid or fractions thereof. See text for explanation of fractions and treatment regimens.

a,b means with different superscripts in Experiment 2 are different at the .01 level.

groups C, HFF, CFF, HFF + CFF, and WFF respectivly. Analysis of variance indicated an effect ($p < .001$) of treatment on interval to ovulation. Examination of the means by use of Duncan's new multiple range test (7) indicated that intervals to ovulation for groups C, HFF, and CFF were not different from each other but were different from those of groups HFF + CFF and WFF (Figure 1). It was noted that two of three heifers in group CFF did not develop normal CL subsequent to this ovulation, based on palpation, and ovulated for a second time five and six days after the initial ovulation.

## DISCUSSION

Neither steroid free FF (group CFF) nor protein denatured FF (group HFF) increased the interval to ovulation over that of untreated controls (group C). However the combination of both fractions (group HFF + CFF), presumably with unaltered proteins from the charcoal extracted FF and steroids from the heat treated FF, restored the inhibitory activity seen in whole FF (group WFF).

Therefore it is concluded that the ovulation inhibiting activity demonstrated in bovine FF is due to a combination of a protein factor and a steroid factor. The mechanism of action is not known. It may be through an effect on the hypothalamic-pituitary axis, to alter gonadotropin release, or it may be a direct effect at the level of the ovary or a combination of these two mechanisms.

Steroids in FF could exert a local inhibitory effect on the ovary. However the present results do not support a local effect due solely to steroids, since group HFF, which should have unaltered steroids, was not different from group C. That heifers in groups HFF and CFF ovulated at the same time as those in group C suggests normal LH profiles. Delay of ovulation in the groups with both steroid and protein present (group WFF and group HFF + CFF) may be due to suppression of both FSH and LH. Suppression of LH in addition to FSH is supported by data from mares (4) showing that plasma levels of FSH and initially LH decreased following IV injection of whole FF.

If steroid free FF selectivly suppresses only FSH in the bovine as it does in other species (1,2,4) ovulation would still be expected to occur, as it did in group CFF. However, since FSH has been shown to play a role in the development of the ability of granulosa cells to luteinize (8,9 for review) luteal function may be disrupted. The unexpected observation that two of three heifers in group CFF ovulated but did not form normal CL and ovulated again five and six days later is consistent with this hypothesis. A similar condition of inadequate luteal function attributed to an abnormal FSH / LH ratio prior to ovulation has been reported in humans (10,11) and in the rhesus monkey (12). The fact that heifers in group WFF and group HFF + CFF appeared to have normal length post-treatment estrous cycles with apparently adequate luteinization is attributable to the relativly long time between treatment and ovulation. This interval, subsequent to inhibition of ovulation at the normal time, allowed for the re-establishment of normal gonadotropin patterns and normal ovulation and luteinization in groups WFF and HFF + CFF, whereas in group CFF ovulation occurred at the expected time but with a deficit in FSH and subsequently a deficit in luteinization.

In summary, bovine FF had an inhibitory effect on ovulation in the bovine, which required both protein and steroid factors. This may have been either a direct effect at the ovarian level or, more likely, may have been mediated via the hypothalamus and / or the pituitary.

## REFERENCES

1. F. H. DeJong and R. M. Sharpe. Nature 263:71, 1976.
2. R. Welschen, W. P. Hermans, J. Dullart and F. H. DeJong. J. Reprod. Fert. 50:129, 1977.
3. M. L. Marder, C. P. Channing and N. B. Schwartz. Endocr. 101:1639, 1977.

4. K. F. Miller and O. J. Ginther. Federation Proceedings 37:285, 1978.
5. H. Peters, A. G. Byskov and M. Faber. in: The development and maturation of the ovary and its functions, ed. H. Peters (Exerpta Medica, Amsterdam, 1973).
6. K. F. Miller, J. K. Critser and O. J. Ginther. abstract for Society for the Study of Reproduction, 1978.
7. R. G. D. Steel and J. H. Torrie. Principles and Proceedures of Statistics (McGraw-Hill, New York, 1960) p. 107.
8. A. J. Zeleznik, P. L. Keyes, K. M. Menon, A. R. Midgley and L. E. Reichert. Am. J. Physiol. 233:E229, 1977.
9. C. P. Channing and A. Tsafriri. Metabolism 26:413, 1977.
10. B. M. Sherman and S. C. Korenman. J. Clin. Endocr. Metab. 38:89, 1974.
11. B. M. Sherman and S. G. Korenman. J. Clin. Endocr. Metab. 39:145, 1974.
12. J. W. Wilkes, G. P. Hodgen and G. T. Ross. J. Clin. Endocr. Metab. 43:1261, 1976.

# MODULATION OF PITUITARY RESPONSIVENESS TO LHRH BY PORCINE FOLLICULAR FLUID: TIME AND DOSE- DEPENDENT EFFECTS

Douglas Shander, Larry D. Anderson, Charles A. Barraclough and Cornelia P. Channing

Department of Physiology, University of Maryland School of Medicine
660 W. Redwood St., Baltimore Maryland 21201

## INTRODUCTION

The injection of porcine follicular fluid (pFFl) can suppress both the rise in serum FSH that occurs in rats after ovariectomy (1) and FSH surges that occur in rats during proestrus (1600 h and 1900 h) and estrus (0800 h and 1100 h)(2). Furthermore, it has been shown that pFFl inhibits FSH secretion after LHRH infusion in phenobarbital-treated, proestrus rats (2), suggesting that pFFl exerts its suppressive effects at the level of the pituitary gland. Previously, it has been shown in vitro, using monolayers of pituitary cells, that non-steroidal substances of gonadal origin can directly inhibit gonadotropin responses to LHRH (3). Thus, the addition of rete testes fluid (free of steroids) to pituitary cells in a monolayer culture system resulted in an inhibition of both LH and FSH secretion in response to LHRH. The above findings prompted the present investigation which was designed to study the effects of charcoal-treated pFFl on both basal and LHRH-stimulated secretion of FSH and LH from rat anterior pituitary cells cultured in monolayer.

## MATERIALS AND METHODS

Anterior pituitary glands were obtained from adult, female Sprague-Dawley rats (ARS/Madison, Wisconsin) and treated with

0.1% hyaluronidase (Sigma; St. Louis, Mo.), 0.35% collagenase (Sigma; St. Louis, Mo.) and 0.25% viokase (Gibco) at 37C. The dispersed cells were counted using a hemocytometer and viable cells were determined by their ability to exclude 0.066% trypan blue. The cells were diluted to a concentration of $6 \times 10^5$ viable cells/ml. in Dulbecco Modified Eagles Medium (DMEM) containing 10% rat, 3% horse, 2.5% fetal calf serum and fungizone (2.5 μg/ml); and the cells were cultured in a monolayer system, as described by Vale, et al (4). Cell attachment was completed 3 days after enzymatic dispersal and the media was replaced with DMEM containing estradiol ($1 \times 10^{-8}$ M) and progesterone ($1 \times 10^{-7}$M) ± either charcoal-treated pig serum (2%) or charcoal-treated pFFl (2% or 0.5%; collected from medium antral follicles, 3-5 mm diameter) for 2, 24, or 26 hours. Following this preincubation time period with either pig serum or pFFl, synthetic LHRH ($1.9 \times 10^{-10}$ M or $1.9 \times 10^{-8}$ M; Beckman, Palo Alto, Ca.) was added for 4 hours; the amount of FSH and LH secreted into the media was determined by radioimmunoassay as previously described for FSH (5) and LH (6). Data are expressed in terms of NIAMDD-rat-LH-RP-1 and rat-FSH-RP-1. The cells were then fixed in 10% formalin in water and stained with hematoxylin-oil red O for morphological examination and cell counting.

## RESULTS

The presence of 2% pFFl for 26 hours decreased both basal and LHRH-stimulated, FSH secretion from pituitary cells by 30-50% ($p<0.01$) as compared to cells incubated with 2% pig serum. The amount of LH secreted by pituitary cells was not affected by pFFl except with the lower dose of LHRH, $p<0.01$ (Figure 1).

The amount of LH and FSH secreted into the medium by pituitary cells treated with 2% pig serum did not differ from that found by cells incubated without pig serum (data not shown). Similar to the data in Figure 1, the incubation of pituitary cells with 2% or 0.5% pFFl for 24 hours (Figures 2 and 3) decreased the amount of LH and FSH secreted in response to the lower dose of LHRH ($p<0.01$); only FSH secretion was decreased in response to the high concentration of LHRH, $p<0.01$, (LH data not shown). Basal secretion of FSH was decreased 40-60% below control levels with incubation of 2% pFFl for 24 hours and 0.5% pFFl for 24 and 2 hours ($p<0.01$), while LH was decreased 35% with 2% pFFl for 24 hours ($p<0.01$) and to a lesser magnitude (20-25%) with 0.5% pFFl for 24 and 2 hours ($p<0.10$). However, incubation with 0.5% pFFl for only 2 hours had no effect on suppressing the LHRH stimulation of FSH and LH.

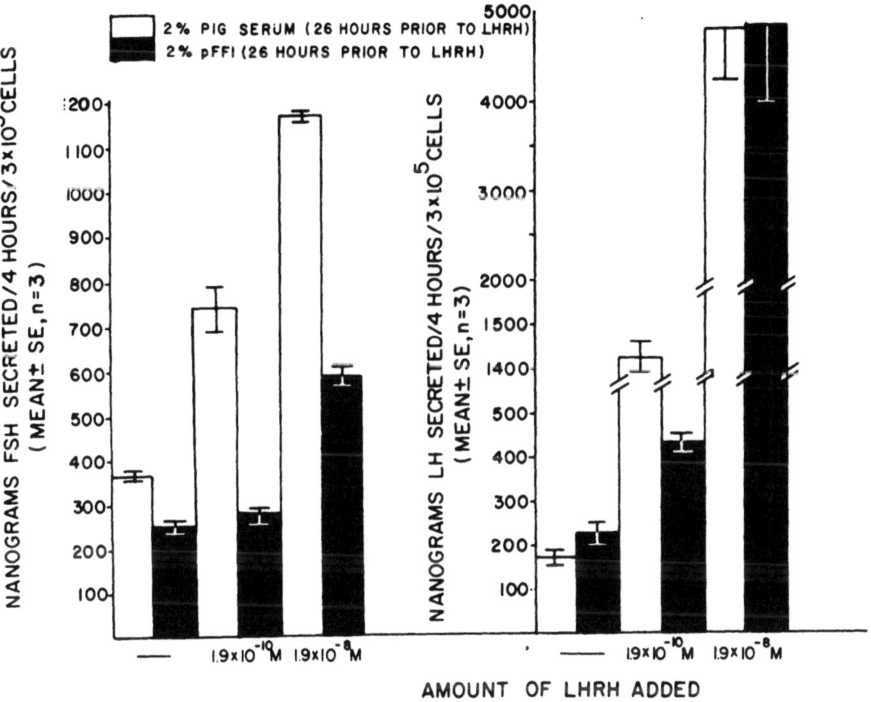


Figure 1. Influence of pFF1 on FSH and LH secretion by rat anterior pituitary cells in response to LHRH. The pituitary cells were prepared and treated as described in the Materials and Methods section. Pig serum (2%) or pFF1 (2%) were present for 24 hours, after which the media was changed and replaced with medium containing either pig serum of pFF1 in the same concentrations as the preceeding 24 hours. After 2 hours of incubation, either LHRH (50μl) or DMEM (50μl) was added so that the cells were exposed concurrently to LHRH and either pig serum or pFF1. After an additional 4 hour incubation, the media was removed and the amount of FSH and LH secreted was determined by RIA. The data represents the mean ± S.E. from triplicate cultures.

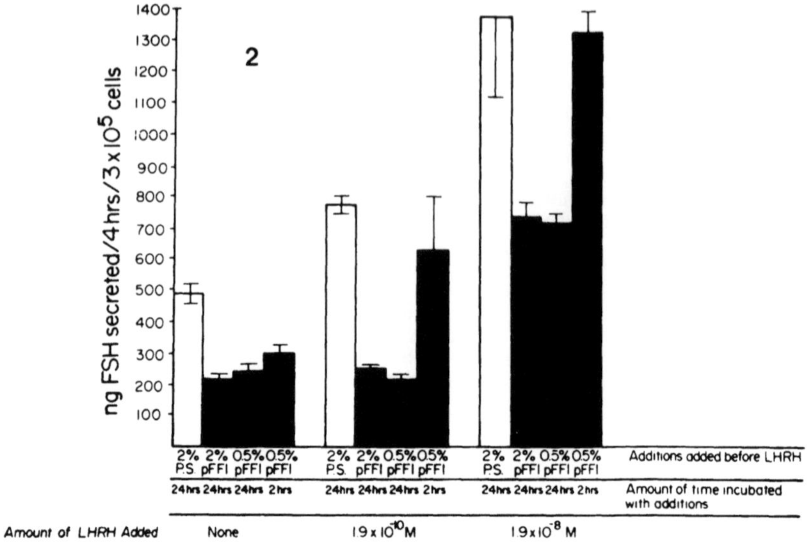


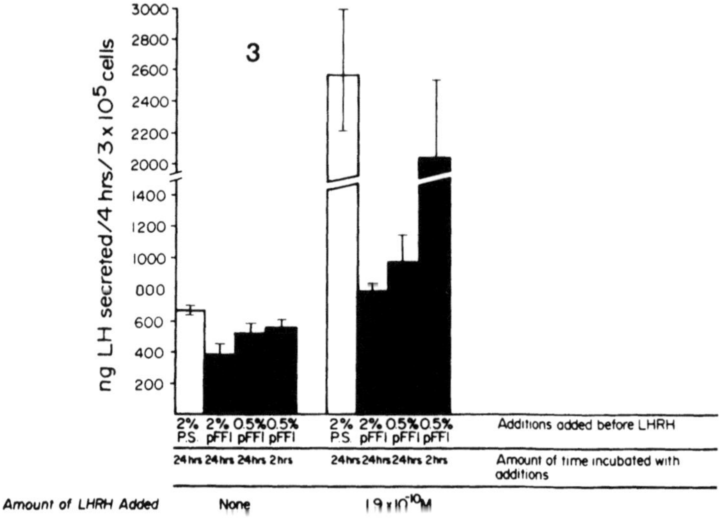


Figures 2 and 3. Effect of dose and incubation time with pFF1 upon FSH and LH secretion from rat anterior pituitary cells in response to LHRH. The pituitary cells were prepared and treated as described in the Materials and Methods section. Pig serum (2%) or pFF1 (2% or 0.5%) was present for either 24 or 2 hours. The cells were then washed three times with DMEM and replaced with DMEM $\pm$ LHRH (Note that pig serum or pFF1 was not present during the LHRH exposure as in Figure 1). After 4 hours, the media was removed and the amount of FSH and LH secreted during this 4 hour interval was determined by RIA. The data represents the mean $\pm$ S.E. from triplicate cultures.

## DISCUSSION

These experimental results indicate that porcine follicular fluid (pFFl) can suppress both basal and LHRH-stimulated secretion of FSH from rat anterior pituitary cells, and in some instances it also can suppress LH secretion. Furthermore, the degree of inhibitory action of pFFl is quantitatively and temporally cumulative since both an increase in the dose and length of exposure to pFFl results in an increased inhibition of pituitary responsiveness to LHRH. This data supports the concept that non-steroidal substances of gonadal origin play an important role in the regulation of pituitary gonadotropin secretion. Presently, it is not known whether this substance has additional effects at the level of the hypothalamus, or whether steroids interact with the action of this substance.

Recent observations have suggested that the granulosa cell of the ovary is the possible site of secretion for this 'ovarian inhibin' (7). Presently, investigations are being carried out to study the mechanism(s) by which ovarian inhibin secretion is regulated, and whether there are changes in ovarian inhibin present in serum which can be correlated with changes in pituitary sensitivity to LHRH. In addition, it has been shown that the inhibitory activity of pFFl is restricted to material having a MW greater than 10,000 (8,9). Preliminary observations have shown that a substance, >10,000 MW, is secreted by porcine granulosa cells *in vitro* which has the ability to inhibit FSH secretion from rat anterior pituitary cells *in vitro* (10).

## SUMMARY

The results of these experiments indicate porcine follicular fluid (pFFl) from medium-sized, antral follicles has the ability to suppress both basal and LHRH-stimulated FSH secretion from rat anterior pituitary cells, *in vitro*. The degree of inhibitory activity is quantitatively and temporally cumulative since both an increase in the dose and length of exposure to pFFl results in an increased degree of inhibition of pituitary responsiveness to LHRH. This data supports the concept that non-steroidal substances of gonadal origin play an important role in the regulation of pituitary gonadotropin secretion.

## Acknowledgements

Supported in part by Ford Foundation grant #760-0530 (to CPC) and NIH grant HD-02138 (to CAB) and by NIH Training Grant HD 02138 (LDA, DS).

## REFERENCES


(1) Marder, M.L., C.P. Channing and N.B. Schwartz Endocrinology 101(1977)1639.

(2) DePaolo, L.V., P.M. Wise, L.D. Anderson, C.A. Barraclough and C.P. Channing Society for the Study of Reproduction Abstract # 182 (1978).

(3) Baker, H.W.G., W.J. Bremmer, H.G. Burger, D.M. De Krester, Ausma Dulmanis, L.W. Eddie, B. Hudson, E.J. Keogh, V.W.K. Lee and G.C. Rennie Rec. Prog. Horm. Res., 32(1976)429.

(4) Vale, W., G. Grant, M. Amoss, R. Blackwell and R. Guillemin Endocrinology 91(1972)562.

(5) Chappel, S.C. and C.A. Barraclough Endocrinology 98(1976)927.

(6) Niswender, G.D., A.R. Midgley, Jr., S.E. Monroe and L.E. Reichert, Jr. Proc. Soc. Exp. Biol. Med. 128(1968)807.

(7) Erickson, G.F. and A.J.W. Hsueh Endocrine Society Abstract #540 (1978).

(8) Chappel, S.C., T. Acott and H.G. Spies in Workshop on Ovarian Follicular and Corpus Luteum Function (C.P. Channing, J. Marsh and W. Sadler, eds.) Plenum Press, New York (1978).

(9) Lagaće, L., F. Labrie, N.B. Schwartz, J. Lorenzen, and J.H. Dorrington Endocrine Society Abstract #727 (1978).

(10) Anderson, L.D., D. Shander, C.A. Barraclough and C.P. Channing, unpublished observations.

# ISOLATION AND PRELIMINARY CHARACTERIZATION OF INHIBIN-f

A. Trent Williams, Michael E. Rush and H. Lipner

Department of Biological Sciences
Florida State University
Tallahassee, Florida 32306

Although a gonadal regulation of hypophysial gonadotrophin secretion is established, the nature of this feedback is not fully known. In certain pathological conditions (1) or after selective destruction of the germinal epithelium (2), FSH levels are elevated whereas LH levels are not. These observations suggest that the secretion of FSH is under the control of a factor other than that controlling LH secretion (2). Recent evidence in the female (3) also supports the concept of the dissociation of secretion of FSH and LH.

Since neither testosterone nor estradiol alone can return elevated FSH levels to baseline (2,3), FSH secretion is probably under a dual negative feedback control, consisting of steroidal and non-steroidal gonadal factors. There is now overwhelming evidence to support the existence of a non-steroidal, water-soluble gonadal factor which selectively inhibits FSH secretion. This factor has been named inhibin. Inhibin-like activity is found in testicular fluid (2) and in ovarian follicular fluid (3).

This report describes our preliminary isolation and characterization of porcine ovarian inhibin ($_{p}I_{f}$).

## MATERIALS AND METHODS

### Preparation of Porcine Follicular Fluid (pFF)

Porcine ovaries were collected over a normal 8 hour work day at a local meat packing plant and were immediately placed on ice.

Approximately 600 ml of follicular fluid were aspirated from 1000 ovaries at 4C. The cellular contaminants were removed by centrifugation. Charcoal (Norit A; 30 mg/ml) was added to remove steroids and the slurry was stirred for 16 hours at 4C. The charcoal was removed by centrifugation and filtration. The biological activity was demonstrated by bioassay (Fig. 2) and was designated pFF (Fig. 1).

Albumin was removed from pFF (Fig. 1) by affinity to Cibacron Blue F3Ga bound to agarose beads (Affi-Gel Blue, AGB) in 0.02 M phosphate buffer (Buffer A) at pH 7.2. The AGB-pFF mixture was gently shaken for 1 hour at room temperature, suctioned, filtered, and was washed with 1 bed volume of Buffer A. The resulting solution was concentrated with hollow fiber dialysis and either treated with polymerized sheep barrow serum antiserum or sheep barrow serum immunoglobulin immobilized on sepharose. An alternative and preferential procedure was to fractionate the solution on Sephadex G-150 and select the fractions in the 10,000 to 35,000 mwt range for purification.

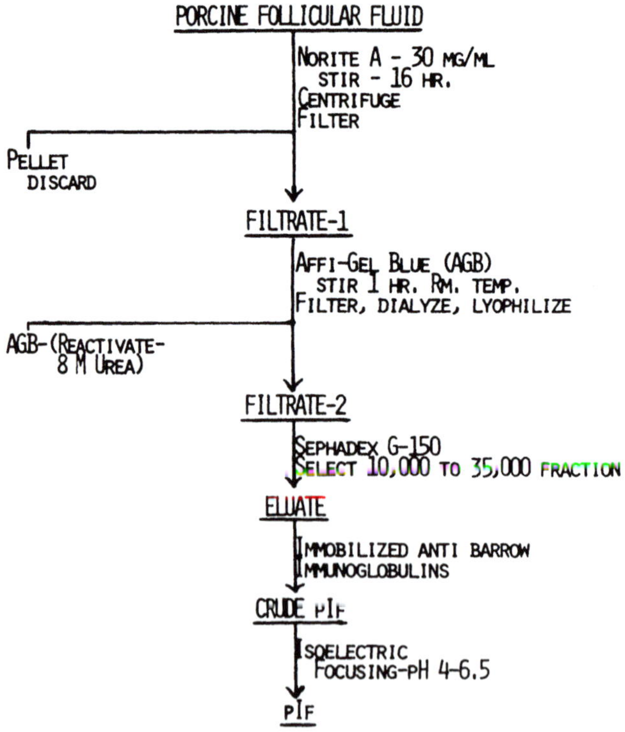


Fig. 1. Flow chart for isolation of porcine ovarian inhibin ($_pI_f$).

## Preparation of Barrow Serum Antiserum

Barrow (orchidectomized boar) blood was obtained at the packing plant. The serum was separated and stored at -20C until used. Barrow serum in complete Freunds' adjuvant (1:1, v:v) was injected at multiple intradermal sites into a 1 year old female sheep. The injections were repeated weekly for 4 weeks. A test bleeding proved immunologically active in a double-diffusion plate test. The sheep was bled at weekly intervals for 1 month and then exsanguinated. Immunoglobulins were isolated by fractional precipitation with 35% saturated ammonium sulfate at pH 7.3.

The isolated immunoglobulins were either attached to cyanogen bromide activated agarose or to the N-hydroxysuccinamine ester of succinylated aminoalkyl agarose gel beads (Affi-Gel 10). Subsequently, antisera to barrow serum were prepared in rabbits according to the previously mentioned procedure.

## Bioassay

The metestrus castrate rat assay (3) was used to bioassay the biological activity of pFF during the various stages of purification of pIf. Rats, ovariectomized at 0800 hours on the day of metestrus, were administered the test substance at 1130 hours IP and exsanguinated at 1700 hours. The serum was frozen until the time of analysis for FSH and LH (Fig. 2).

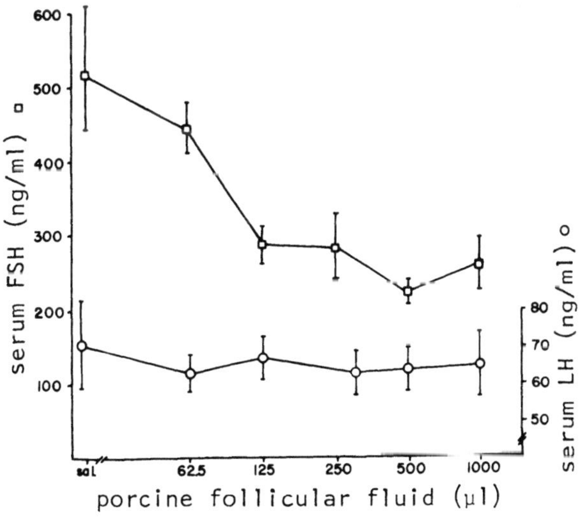


Fig. 2. Dose dependent suppression of FSH by porcine follicular fluid (pFF). LH over the entire dose range is unaffected.

### Radioimmunoassay of FSH and LH

Both FSH and LH were assayed using the rat radioimmunoassay kits generously supplied by the National Institute for Arthritis and Metabolic Diseases. Intra- and interassay variabilities were less than 10%.

### Protein Determination

The concentrations of protein at all stages of the isolation were determined by the Folin-Cicolteau method. Bovine serum albumin was used as the standard.

### Polyacrylamide Gel Electrophoresis (PAGE)

The removal of albumin was demonstrated with PAGE performed at pH 8.9. The sample was dissolved in 0.01 M Tris-HCl buffer and 10% glycerol and layered in the troughs of a vertical gel slab.

## RESULTS

### Biological Activity of pFF

Porcine follicular fluid caused a dose related depression of plasma FSH from castrate to intact animal levels over a dose range of 62.5 μl to 125 μl. Plasma LH levels were unaffected even at the 1000 μl dose. This bioassay was used to test for biological activity in the fractions obtained at various stages of the iso-electric procedure.

### Isolation of pIf

The procedure used in the isolation of pIf is diagrammed in the flow chart in Figure 1. The effectiveness of the procedures employed in the removal of protein are indicated in Table 1. AGB removed between 60 and 80% of the protein from the pFF. Gel filtration on a Sephadex G-150 column resulted in a further sharp reduction of protein leaving less than 1% of the original protein and a biologically active material. Based on the use of protein markers of known molecular weight the active fraction was found in the range 10,000 to 35,000 Daltons (Fig. 3). In an alternative procedure we subjected pFF to AGB and then to polymerized sheep antibarrow serum. In this approach 87% of the initial protein content was removed. The resulting material was subjected to isoelectric focusing over a pH range of 3 to 10.

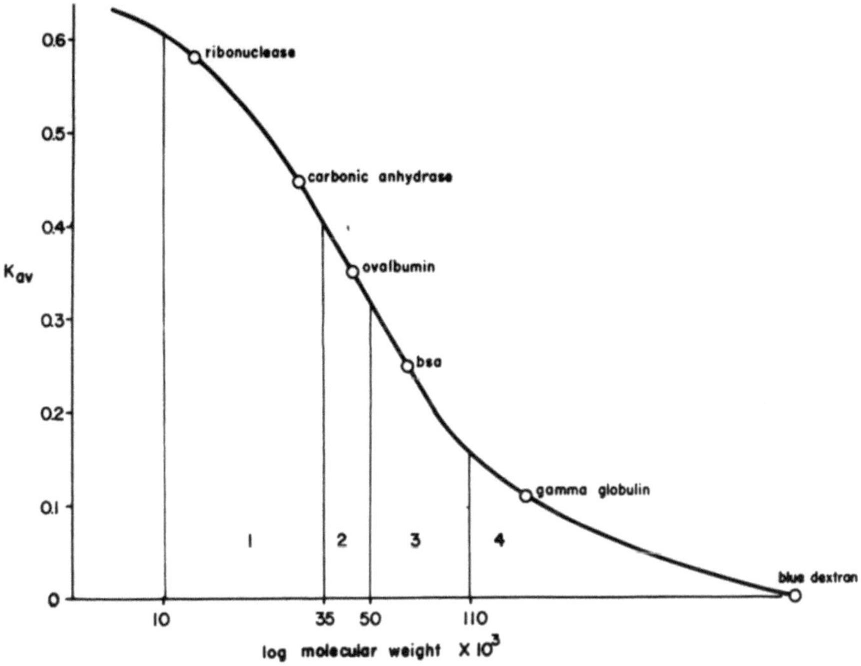


Fig. 3. Approximation of the molecular weight of $_pI_f$ based on a Sephadex G-150 column. Biological activity was present in fraction IV.

The biologically active material appeared in the range of pH 5 to 5.8 (Fig. 4).

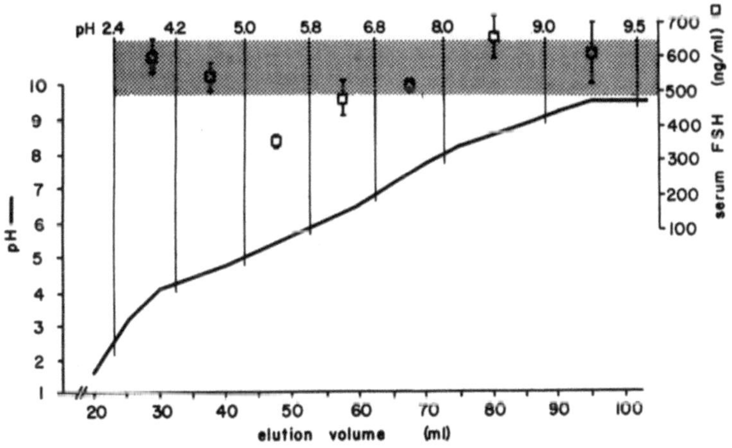


Fig. 4. Preparative isoelectric focusing of partially purified pFF--Ampholine pH range 3-10. Only the factor(s) focusing in the pH 5.0-5.8 range significantly suppressed FSH.

Table 1. Effect of different procedures on removal of protein from porcine follicular fluid (pFF)

| Treatment | Conc. of Protein mg/ml | % Decrease in Protein | Conc. of Protein mg/ml | % Decrease in Protein |
|---|---|---|---|---|
| | 1st Procedure | | 2nd Procedure | |
| Native pFF | 52 | - | 68 | - |
| Ether Extraction | 52 | - | | |
| Carbon Adsorption | | | 50 | 26 |
| AGB | 20.5 | 60 | | |
| Polymerized Barrow Serum A/S | 3.5 | 83 | | |
| | 3rd Procedure | | 4th Procedure | |
| Carbon Adsorption | 45 | | 57 | - |
| AGB | 9.6 | 79 | 21 | 64 |
| Sephadex G-150 MW 10,000-35,000 | | | .68 | 98.8 |
| Barrow Serum A/S | 5.8 | 97 | | |

## DISCUSSION

The isolation of a highly active porcine ovarian inhibin ($_pI_f$) has proven to be remarkably simple. The methods depend on our recognition that the largest amount of follicular fluid protein is derived from the plasma (4). Affinity chromatography with AGB effecting the removal of serum albumin resulted in between 60-80% reduction in protein content. Affinity chromatography with immunobilized antibarrow serum resulted in removal of much of the remaining plasma dervied protein. By interposing a gel filtration step between the two affinity chromatography steps the capacity of the system was augmented, and the final protein content at this stage reduced by 99.7-99.9%.

The data suggest that a specific substance present in follicular fluid possesses inhibin-like activity. This material has a molecular weight in the range of 10,000 to 35,000 Daltons and an isoelectric point range of pH 5-5.8. These limited physical constants suggest a similarity to the male inhibin described by Chari et al (5). As larger quantities of inhibin become available more precise information about its physial and chemical properties, as well as its constituent molecules, will become available.

## ACKNOWLEDGEMENTS

The authors wish to thank the Department of Biological Science at FSU and The Population Council for the support that made this work possible. We appreciate the technical assistance of Ms. Carolyn McRae and the secretarial assistance of Mrs. Linda F. Derby.

## BIBLIOGRAPHY

(1) H.W.G. Baker, W.J. Bremner, H.G. Burger, D.M. deKretser, A. Dulmanis, L.W. Eddie, B. Hudson, E.J. Deogh, V.W.K. Lee, and G.C. Rennie. Recent Progr. Horm. Res. 32 (1976) 429.

(2) B.P. Setchell and S.J. Main. Bibliogr. Reprod. 24 (1974) 245.

(3) M.L. Marder, C.P. Channing, and N.B. Schwartz. Endocrinology. 101 (1977) 1639.

(4) H. Lipner. in: Handbook of Physiology, Endocrinology II, Part 1, eds R.O. Greep and E.B. Astwood (American Physiological Society, Washington, D.C., 1973) p. 409.

(5) S. Chari, S. Duraiswami, and P. Franchimont. Acta Endocrinol. 87 (1978) 434.

# EVIDENCE THAT THE PITUITARY GLAND IS THE SITE OF INHIBITORY ACTION OF PORCINE FOLLICULAR FLUID UPON FSH SECRETION IN THE RAT

Phyllis M. Wise, Louis V. DePaolo, Larry D. Anderson,
Cornelia P. Channing, and Charles A. Barraclough
Department of Physiology
University of Maryland
Baltimore, Md. 21201

## ABSTRACT

This study has examined the effects of porcine follicular fluid (PFF) on the adenohypophyseal secretion of LH and FSH following LHRH treatment. The animal model selected for study was the proestrous rat in which preovulatory LH and FSH surges were blocked with phenobarbital. To elicit the release of gonadotropins from this preparation, two different i.v. LHRH infusion protocols were used. The first elicited a proestrous-like LH and FSH surge while the second infusion schedule evoked the predominant release of FSH with only minor changes in LH. PFF suppressed LHRH-induced FSH secretion in both preparations but had only negligible effects on LH secretion. Interestingly, the suppressive effects of PFF were variable and depended upon the mode in which LHRH was presented.

## INTRODUCTION

Several investigators have proposed that the male and female gonads produce a non-steroidal substance which selectively inhibits the secretion of FSH. Water soluble testes extracts (1-3), rete testis fluid (1,4), seminal plasma (5), medium from cultured Sertoli cells (6), and extracts of spermatozoa have been found to suppress FSH without affecting LH. More recently an inhibin-like material has been found in bovine and porcine follicular fluid (7,8) and in the medium from cultured granulosa cells (9). Preliminary purification of the non-steroidal substance in PFF indicates that it is similar to that found in the male (10). The testicular inhibin is effective in the female (1) and the ovarian inhibin is effective in the male (11).

The present experiments were designed to test whether PFF altered the pituitary secretion of LH and FSH in response to LHRH.

## MATERIALS AND METHODS

Adult female Sprague-Dawley rats (ARS/Madison, Wisconsin) were maintained in temperature (22-24C) and light-controlled (lights on at 04.00 h off at 18.00 h) rooms. Only those rats which had exhibited at least 2 consecutive 4-day vaginal estrous cycles were used in these experiments.

Between 09.00 h and 11.00 h on proestrous, rats were anesthetized with ether and polyethylene cannulae (o.d. 0.038 in) were inserted into the external jugular vein and the femoral artery. This surgical procedure required about 10 min. At 11.00 h animals were given either 1 ml of PFF or saline i.p. and at 12.15 h phenobarbital (75 mg/kg body wt) was injected i.p.

At 13.00 h LHRH (Beckman) was infused into the jugular vein whereas control phenobarbital-treated animals received phosphate-buffered saline (PBS). The LHRH was dissolved in 0.01M phosphate buffered saline, pH 7.4 and was infused according to two different protocols: (1) LHRH (100 ng/ml) was infused at a rate of 0.5 ml/h for 3 h. Blood (0.6 ml) was drawn from the femoral artery before the infusion was begun and again at 60, 120, 180, 240 and 300 min. (2) In a second group of rats, LHRH (100 ng/ml) was infused at a rate of 0.5 ml/h for 30 min, followed by a 30 min period during which the infusion was discontinued. At the end of this interval LHRH (10 ng/ml) was again infused at a rate of 0.5 ml/h for 210 min. Blood was collected before beginning the infusion and again at 30, 60, 90, 120, 150, 210 and 270 min. Plasma from all blood samples was frozen until time of assay.

Porcine follicular fluid was obtained from medium sized follicles (3-5 mm in diameter). It was pooled and treated as described in detail elsewhere (12). In brief, it was charcoal-treated (10% v:v) overnight, then centrifuged and filtered through a 0.22 micron cellulose filter several times until no charcoal remained on the filter. Aliquots were frozen until use. Such treatment removed virtually all steroids (steroid concentrations remaining in charcoal treated PFF were estradiol, 54 pg/ml; progesterone, 5 ng/ml). This pool of PFF has been shown to inhibit the proestrous and estrous rise in FSH _in vivo_ (12).

Plasma LH and FSH levels were measured by radioimmunoassay as described previously (13). The within assay variation was 6.0% and 4.2% and between assay variation 5.1% and 5.7% for the LH and FSH assays respectively.

The data were statistically analysed by the Duncan's Multiple range test.

## RESULTS

Phenobarbital completely blocked the proestrous surge of LH and FSH (Fig. 1). This dose of anesthesia does not affect pituitary responsiveness to LHRH (14).

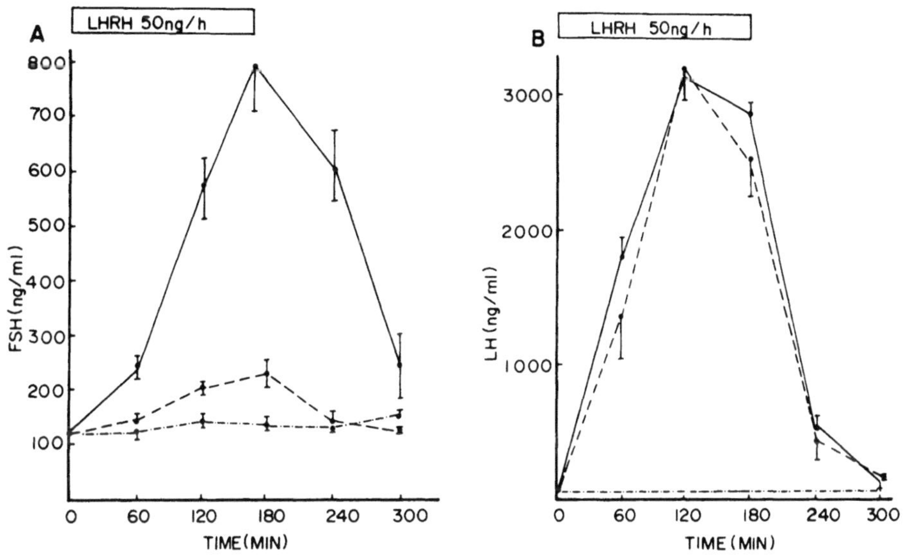


FIGURE 1. The effect of PFF on LHRH-induced rise of serum FSH (1A) and LH (1B) when LHRH was infused at a constant concentration (50 ng/h). Phenobarbital-blocked proestrous rats were treated with a) 1 ml saline at 11.00 h and infused with PBS for 3 h (•-·-·-•) (n=7); b) 1 ml saline at 11.00 h and infuse with LHRH for 3 h (•——•) (n=6); or c) 1 ml PFF at 11.00 h and infused with LHRH for 3 h (•- - -•) (n=6). Phenobarbital was injected i.p. at 12.15 h.

The effect of PFF on the LH and FSH response to a constant dose infusion of LHRH is shown in Fig. 1. LHRH, when infused at a concentration of 50 ng/h for 3 h stimulated the pituitary release of a proestrous-like surge of LH and FSH and confirm previous observations by Blake (14). When 1 ml PFF was injected 2 h prior to the beginning of LHRH infusion the peak FSH response was suppressed by 71.0% (Fig. 1). Inhibition was not complete as plasma FSH concentrations in PFF-treated animals still exhibited a significant response to LHRH when compared to PBS-infused controls. The LH response in PFF-treated females paralleled that of PBS-treated controls except at 60 min after the beginning of infusion. At this time LHRH-induced LH responses were slightly but significantly reduced ($p<.01$). The effect of PFF on the LH and FSH response to a low concentration of LHRH preceded by a short priming dose is shown in Figure 2.

When LHRH was infused at 2 different concentrations over a 270 min period, the pituitary response was very different. Plasma FSH concentrations rose gradually: The first significant response was observed at 90 min ($p<.01$) and by 270 min plasma concentrations (659±69 ng/ml) were comparable to those seen during the spontaneous proestrous surge. During this time the elevation in plasma LH was minor but significant. The peak concentrations (197±35 ng/ml) represent 6.5% of those seen during proestrous afternoon. A detailed explanation of this experimental protocol appears elsewhere (15). When PFF was injected into these animals 2 h prior to the beginning of the infusion, it completely inhibi-

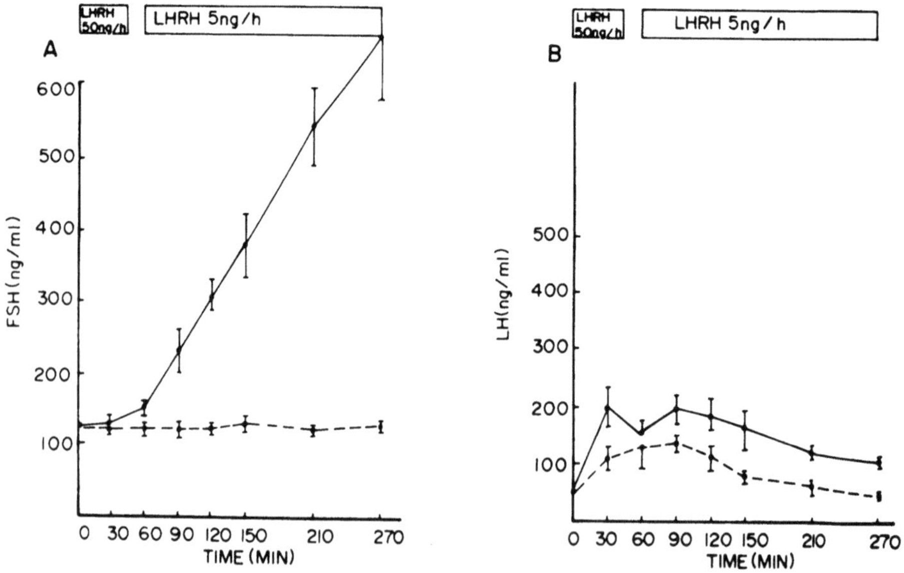


FIGURE 2. The effect of PFF on LHRH-induced FSH (2A) and LH (2B) when LHRH was infused at 2 different concentrations. LHRH was infused into phenobarbital-blocked proestrous rats at a concentration of 50 ng/h for 30 min followed by a 30 min respite. Then LHRH was infused at 5 ng/h for 210 min. Animals were treated with a) 1 ml saline at 11.00 h and infused with LHRH (•——•) (n=6) b) 1 ml PFF at 11.00 h and infused with LHRH (•- - - -•) (n=5). Phenobarbital was administered i.p. at 12.15 h.

ted the rise in FSH. Plasma levels were not significantly different from PBS-infused controls. Plasma LH response was partially inhibited by PFF: levels were significantly suppressed at all times except at 60 min.

## DISCUSSION

These results clearly indicate that the pituitary gland is the site of action of PFF. Whether it may also act at the hypothalamic level has not been examined; however, we can account for the inhibition without hypothesizing an additional site of action. A pituitary site of action has also been suggested by *in vitro* studies. Lagacé *et al.* (16) and Shander *et al.* (17) found that PFF suppressed LHRH-induced and basal secretion of FSH in a pituitary cell culture with a lesser effect on LH secretion. Whether it affects primarily FSH synthesis or secretion is not known, but its effects on the proestrous-estrous rise in FSH suggest that its primary action is to affect FSH secretion. DePaolo *et al.* (12) found that when PFF was administered to block the proestrous and/or the estrous rise in FSH it resulted in an unexpected rise in FSH after the effect of PFF had been dissipated. Perhaps this "rebound" in FSH secretion on estrus was the result of the accumulation of pituitary FSH within a releasable pool and its discharge on removal of inhibitory influences.

Interestingly the degree of FSH suppression was dependent upon the mode of infusion. The peak mean LHRH-induced FSH concentrations achieved by both infusion protocols were within the range observed during the endogenous proestrous FSH surge. Yet when LHRH was infused at a constant high concentration for a prolonged period of time, PFF inhibited only a portion of the LHRH-stimulated FSH secretion. In protocol 2, when one-tenth the concentration of LHRH was infused over a prolonged period of time and was preceded by a short priming dose of LHRH, PFF completely suppressed FSH secretion to baseline levels. It is possible that when LHRH is infused at a high concentration it may activate a higher number of receptor or second messenger sites than do lower concentrations of LHRH. However, both infusion protocols resulted in a comparable elevation in plasma FSH. The presence of PFF might only suppress a given number of these sites. Thus when the high concentration of LHRH was infused, sufficient extra receptors are still activated to allow some FSH secretion. However if the lower concentration of LHRH is infused a fewer number of receptors are activated, all of which can be blocked by the presence of PFF. Thus no rise in FSH can be observed.

The two infusion protocols had different effects on plasma LH. The LH secretion stimulated by the high concentration of LHRH was not appreciably affected by PFF. However, when only slight elevations in plasma LH were induced by LHRH, PFF significantly suppressed this gonadotropin's release except at 60 min. The difference in the effect of PFF on LHRH-induced LH secretion can be explained by the same mechanism as was proposed above for the differential effects on FSH secretion.

In summary, these experiments demonstrate that PFF changes the responsiveness of the pituitary gland to LHRH. It inhibits FSH secretion 70-100% depending upon the method of LHRH infusion. PFF has only a minor effect on LHRH-induced LH secretion.

## Acknowledgements

We would like to thank Joseph Connelly and Viktoria Reck-Malleczewen for their technical assistance and Gail Williams for typing this manuscript. Purified ovine LH (LER-1056-C2) was donated by Dr. L.E. Reichert and the ovine LH antiserum #15 was provided by Dr. G.D. Niswender.

## REFERENCES

(1) Baker, H.W.G., W.J. Bremner, H.G. Burger, D.M. deKretser, A. Dulmanis, L.W. Eddie, B. Hudson, E.J. Keogh, V.W.K. Lee, and G.C. Rennie. Rec. Progr. Horm. Res. 32(1976)429.

(2) Keogh, E.J., V.W.K. Lee, G.C. Rennie, H.G. Burger, B. Hudson, and D.M. deKretser. Endocrinology 98(1976)997.

(3) Nandini, S.G., H. Lipner, and N.R. Moudgal. Endocrinology 98(1976)1460.

(4) Setchell, B.P. and F.J. Jacks. J. Endocrinology 62(1974)675.

(5) Franchimont, P., S. Chari, M.T. Hagelstein, and S. Duraiswami.

(6) Steinberger, A. and E. Steinberger. Endocrinology 99 (1976) 918.

(7) Welschen, R., W.P. Herman, J. Dullaart, and F.H. DeJong. J. Reprod. Fertil. 50(1977)129.

(8) Marder, M.L., C.P. Channing, and N.B. Schwartz. Endocrinology 101(1977)1639.

(9) Erickson, G.F. and A.J.W. Hsueh. Endocrine Society Abstract #540 (1978).

(10) Lorenzen, J.R., N.B. Schwartz, and C.P. Channing. Fed. Proc. 37(1978)296.

(11) Lorenzen, J.R., G.H. Dworkin, N.B. Schwartz, and C.P. Channing. Endocrine Society Abstract #46 (1978).

(12) DePaolo, L.V., P.M. Wise, L.D. Anderson, C.A. Barraclough, and C.P. Channing. Submitted for publication.

(13) Chappel, S.C. and C.A. Barraclough. Endocrinology 98(1976) 927.

(14) Blake, C.A. Endocrinol. 98(1976)451.

(15) Wise, P.M., N. Rance, G.D. Barr and C.A. Barraclough. Submitted for publication.

(16) Lagacé, L., F. Labrie, N.B. Schwartz, J. Lorenzen, and J.H. Dorrington. Endocrine Society Abstract #727.

(17) Shander, D., L.D. Anderson, C.A. Barraclough, and C.P. Channing. Workshop on Ovarian Follicular and Corpus Luteum Function (C.P. Channing, J. Marsh and W.D. Sadler eds.) Plenum Press, New York in press, 1978.

SESSION NUMBER: VI

TITLE: CORPUS LUTEUM - LUTEINIZATION

PRINCIPAL SPEAKER: Landis Keyes

INVITED DISCUSSANT: Ken McNatty

PRESENTED ABSTRACTS:

13.) G.D. Niswender and M.A. Diekman. LH Receptors in the Ovine Corpus Luteum During the Estrous Cycle and Early Pregnancy.

14.) G.B. Sala, M.L. Dufau and K.J. Catt. Gonadotropin Action in Isolated Ovarian Luteal Cells: The Intermediate Role of Adenosine 3',5' Cyclic Monophosphate in Hormonal Stimulation of Progesterone Synthesis

CHAIRPERSON: John Marsh

# ESTROGEN ACTION IN THE CORPUS LUTEUM

P. Landis Keyes, Khe-Ching M. Yuh,
and Josephine B. Miller
Reproductive Endocrinology Program, Department of Pathology and the Department of Physiology, The University of Michigan, Ann Arbor, Michigan 48109

The search for unifying concepts and mechanisms concerning the hormonal regulation of the corpus luteum has been frustrated by the numerous reports of marked species differences in requirements for luteotropic hormones. Astwood's (1) observation that prolactin is luteotropic in the rat has been confirmed by many but a primary luteotropic role for prolactin in other species has not been found. Instead, LH seems to have a primary role (e.g., in the human female, 2) and in some species estradiol is luteotropic (3,4). To add a further disquieting note, it remains a mystery how the gonadotropic hormones maintain both steroidogenesis and cellular morphology. Prolactin is recognized for its luteotropic action in the rat (5), but the cellular mechanism of action is unknown. Rat luteal cells have LH receptors (6) and the corpora lutea are LH-dependent for several days during pregnancy (7,8), yet the respective roles of LH and prolactin in the regulation of the corpus luteum have not been defined. In those species in which estrogen is thought to act on the corpus luteum, the mechanism is unknown and the role of pituitary gonadotropic hormones is a matter of speculation. The research reported here will add some new ingredients to the ferment of research on the corpus luteum, in the hope that a fruitful new avenue of research might be realized and pursued.

In this communication, the luteotropic action of estradiol will be examined in two species - the rabbit and the rat. In the rabbit estradiol has been recognized for some time as the essential luteotropic hormone, since it maintains functional corpora lutea after hypophysectomy (3,9). The rabbit corpus luteum does not synthesize estrogen (10) nor does it appear to have aromatase activity (11). This explains the critical dependence of the corpus luteum upon estradiol secreted by the ovarian follicles (12,13). A luteotropic action of estradiol in the rat has been reported recently by several laboratories (4,14,15). The major source of estrogen for the rat corpus luteum may be the luteal tissue itself which is rich in aromatase activity (11). Although it is unclear whether the androgen precursor is normally derived from luteal tissue or from an extra-luteal source, the aromatase can be viewed as a system for maintaining optimal intraluteal estrogen concentrations. Our knowledge of the different pathways for access of estrogen to the corpora lutea of the rabbit and rat has permitted us to develop a different strategy for the study of luteal function in each species. Estradiol concentrations in blood and in corpus luteum have been manipulated to determine the effects of such changes on luteal progesterone synthesis or secretion and on luteal estradiol receptor. The results suggest that estradiol may be the ultimate luteotropic hormone in both species.

## STEROIDOGENIC EFFECT OF 17β-ESTRADIOL IN THE RABBIT: STIMULATION OF PROGESTERONE SYNTHESIS IN THE CORPUS LUTEUM.

Progesterone synthesis in the rabbit corpus luteum is critically dependent upon estradiol as shown by the rapid decline in serum progesterone after withdrawal of estradiol from the circulation (16) or after X-ray destruction of follicles (12,13) which synthesize estradiol (10). However, in these studies, the decline in progesterone secretion was antecedent to regression of corpora lutea which invariably occurred approximately 48 hours later. Since the decrease in progesterone could not be dissociated from luteal regression, evidence was still not at hand that estradiol could stimulate progesterone synthesis in vivo. Fuller et al. (17) have reported that luteal progesterone synthesis can be stimulated approximately 50 per cent by estradiol in vitro.

Figure 1 shows that estradiol can stimulate progesterone secretion by more than 200 per cent, provided estradiol is first withdrawn to initiate a decrease in the rate of progesterone secretion. During continuous treatment with estradiol (serum estradiol elevated 4- to 6-fold) for the first 10 days after ovulation, serum progesterone rises, reflecting in part the growth of the corpora lutea during this period (Fig. 1). These levels of serum progesterone are not significantly different from those we have observed in normal untreated pseudopregnant rabbits, suggesting that progesterone secretion is about maximal when it is sustained by estradiol at physiological concentrations; additional estradiol has little or no effect on progesterone secretion. The acute dependence of progesterone secretion upon estradiol is illustrated by the dramatic decline in serum progesterone during the next 24 hours after removal of the estradiol implant (Fig. 1).

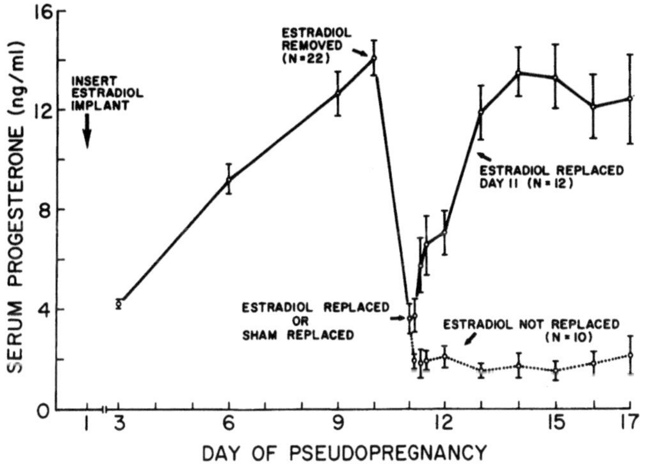


Fig. 1: The effect of removal and replacement of a Silastic capsule containing 17β-estradiol on serum progesterone in pseudopregnant rabbits. Day 1 = day after sterile mating; each data point is the mean ± SE.

Serum estradiol falls to basal values (3-4 pg/ml) within 2 hours after removal of the estradiol implant (16). With progesterone secretion reduced, replacement of the estradiol implant caused a significant increase in serum progesterone within 12 hours, and a sustained rise in serum progesterone over the next several days to values approximating those on day 10 (Fig. 1). In rabbits without estradiol replacement, serum progesterone continued to decrease, the corpora lutea regressed, and these animals returned to estrus. An explanation for the fall in progesterone after removal of estradiol on day 10 is that estradiol diffusing from the implant inhibits ovarian estradiol secretion (presumably by feedback inhibition of pituitary gonadotropins) to the point that the corpora lutea become dependent upon the exogenous estradiol. When the estradiol implant is removed, ovarian estradiol production is insufficient to prevent the demise of the corpora lutea (16).

The higher accumulations of progesterone in incubations of corpora lutea removed 12 hours after replacement of estradiol (Fig. 2) are consistent with higher rates of progesterone synthesis in these tissues. That the steroidogenic effect of estradiol was not mediated by decreased metabolism of progesterone to 20α-dihydroprogesterone is suggested by the greater per cent increase in accumulation of 20α-dihydroprogesterone compared to values in incubations of corpora lutea from rabbits without estradiol replacement (Fig. 2). It should be noted that the stimulation of progesterone accumulation/secretion at 12 hours was not accompanied by any change in luteal wet weight.

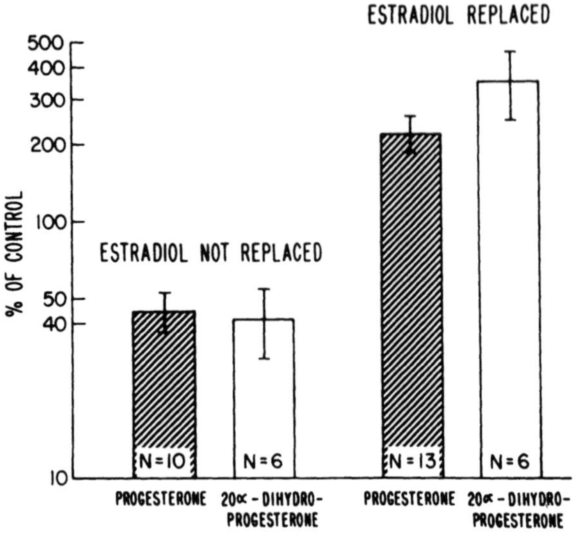


Fig. 2: The accumulation of progesterone and 20 α-dihydroprogesterone in medium from incubated corpora lutea removed from rabbits with estradiol replaced or not replaced on day 11 of pseudopregnancy. Corpora lutea were removed from one ovary of each rabbit and incubated; 12 h after replacement or sham-replacement of the estradiol implant, corpora lutea from the second ovary were removed and incubated. For each rabbit the accumulation of steroid during the second incubation is expressed as a per cent of the first or control incubation. The data are expressed as mean ± SE.

This is not surprising since a decrease in wet weight of corpora lutea is not observed until approximately 48 hours after estradiol withdrawal (18). Thus, the fall in progesterone during the first 24 hours after estradiol withdrawal and the subsequent increase stimulated by 12 hours of estradiol replacement are accomplished without detectable changes in luteal weight.

A role for LH in these responses cannot be excluded, since the rabbit corpus luteum has an LH-sensitive adenylate cyclase (19) and LH stimulates progesterone synthesis to a small extent in incubations of corpora lutea (20). However, it is difficult, conceptually, to invoke LH as the stimulatory agent that is responsible for the sustained rise in serum progesterone after replacement of estradiol, since estradiol suppresses the secretion of LH (21). It remains to be seen if LH has any significant role in the regulation of the corpus luteum, beyond its action in the initiation of luteinization.

To conclude, these data show: 1) that luteal progesterone synthesis is tightly coupled to estrogen action; 2) that decreasing progesterone synthesis and incipient regression of corpora lutea can be reversed by estradiol administration; and 3) that estradiol has a sustained steroidogenic effect on the corpus luteum that can be dissociated from changes in luteal macrostructure.

## NUCLEAR AND CYTOPLASMIC ESTROGEN RECEPTOR IN THE RABBIT CORPUS LUTEUM: EVIDENCE FOR TRANSLOCATION OF THE RECEPTOR TO THE NUCLEUS.

In the previous section, evidence was presented that the rabbit corpus luteum responds to estradiol, but little is known about the mechanism of action of estrogen in this tissue. The rabbit corpus luteum has a cytoplasmic macromolecule with properties resembling an estrogen receptor (22,23), and specific (diethylstilbestrol inhibitable) binding of [$^3$H]estradiol to nuclear fractions has been demonstrated (24). Assays have been developed for the quantification of cytoplasmic and nuclear estrogen receptor and conditions established for determination of both unoccupied and total binding sites permitting an estimation of occupied sites.

Preparation of cytoplasmic and nuclear fractions. Corpora lutea were removed from rabbits on day 11 of pseudopregnancy (11 days after hCG injection and ovulation), homogenized in TRIS-EDTA(TE) buffer (0.01M TRIS-HCl, 1.5 mM EDTA, pH 7.4 at 4 C), and the homogenate centrifuged at 800 x g for 10 min. to obtain the crude nuclear pellet. A small portion of the homogenate was saved for DNA determination (25). The nuclear pellet was washed 3 times and resuspended in the same buffer. The low speed supernatant was centrifuged at 105,000 x g for 60 min. to obtain the cytosol fraction.

Cytoplasmic and nuclear receptor assays. Cytoplasmic and nuclear preparations were incubated with 0.6 nM or 1 nM [$^3$H]estradiol-17β (115 Ci/mmol) in the presence or absence of 100-fold excess of diethylstilbestrol with a final volume of 500 μl. Specific estradiol-17β binding sites were calculated by subtraction of the nonspecific binding observed in the presence of excess diethylstilbestrol from total [$^3$H]estradiol-17β bound in the absence of diethylstilbestrol. Unoccupied nuclear and cytoplasmic receptor was determined by incubation at 0-4C for 2 hours. Total nuclear and cytoplasmic receptor was determined by incubation at 30C for 2 hours (nuclear) and 1 hour (cytoplasmic). Separation of bound and free estradiol in cytoplasm was accomplished by either charcoal-dextran adsorption (0.5 gm % Norit A and 0.05 gm % dextran T80 in TE buffer) or Millipore filtration. After incubation the nuclei were washed and centrifuged. The bound [$^3$H]estradiol-17β was then extracted overnight in 2 ml of absolute ethanol, and the residue was assayed for DNA (25).

The binding specificities are illustrated in Table 1. The cytoplasmic and nuclear receptor is highly specific for estrogens as non-estrogenic steroids did not compete with [$^3$H]estradiol for binding sites. Tables 2 and 3 show sedimentation coefficients and dissociation constants for the nuclear and cytoplasmic receptor.

Table 1: BINDING SPECIFICITY OF NUCLEAR AND CYTOPLASMIC RECEPTOR*

| Treatment | Specific [$^3$H]Estradiol Bound in Nucleus (% of control) | Specific [$^3$H]Estradiol Bound in Cytosol (% of control) |
|---|---|---|
| None (control) | 100 | 100 |
| Cortisol | 112 | 100 |
| Testosterone | 101 | 109 |
| Androstenedione | 104 | 106 |
| Progesterone | 104 | 101 |
| 17α-Hydroxyprogesterone | 85 | 128 |
| 20α-Dihydroprogesterone | 113 | 101 |
| Estrone | 6 | 0 |
| Estradiol | 6 | 6 |
| Estriol | 0 | 13 |

*Specific binding was determined by incubating with [$^3$H]estradiol in the absence or presence of 100-fold excess of diethylstilbestrol (control) plus 100-fold excess of different unlabeled steroids.

Table 2: SEDIMENTATION COEFFICIENTS*

| | |
|---|---|
| Nuclear receptor extracted with 0.4 M KCl, 10 mM Tris and 1.5 mM EDTA | 3.4 S |
| Cytoplasmic receptor in 10 mM Tris and 1.5 mM EDTA | 6.8 S |
| Cytoplasmic receptor in 0.4 M KCl, 10 mM Tris and 1.5 mM EDTA | 3.4 S |

*Nuclear receptor was incubated with 1 x $10^{-9}$ M [$^3$H]estradiol with or without 100-fold excess of diethylstilbestrol at 30 C for 2 h. At the end of incubation, the nuclear pellet was washed, extracted in one volume of 20 mM Tris, 3.0 mM EDTA, 0.8 M KCl, plus two volumes of 10 mM Tris, 1.5 mM EDTA, 0.4 M KCl, at 0-4 C for 1 h with vortexing every 10 min followed by centrifugation at 800 x g for 10 min. Cytoplasmic receptor was incubated with 1 x $10^{-9}$M [$^3$H]estradiol with or without 100-fold excess of diethylstilbestrol in the presence or absence of KCl at 0-4 C for 2 h. Sucrose density gradient centrifugation was conducted by the method of Martin and Ames (26) in a Beckman SW 50.1 rotor at 40,000 rpm, 0-4 C, for 18 h. Linear gradients consisted of 5-20% sucrose in 10 mM Tris, 1.5 mM EDTA with or without the presence of 0.4 M KCl. $^{14}$C-ovalbumin (3.6 S) and $^{14}$C-humanγ-globulin (7.0 S) served as marker proteins.

Table 3: DISSOCIATION CONSTANTS*

| | High Affinity | Low Affinity |
|---|---|---|
| Nuclear receptor | $8.9 \times 10^{-11}$M | $2.7 \times 10^{-8}$M |
| Nuclear receptor 20 min. after $E_2$ injection | $8.9 \times 10^{-11}$M | $3.2 \times 10^{-9}$M |
| Cytoplasmic receptor | $5.4 \times 10^{-11}$M | |
| Cytoplasmic receptor 20 min. after $E_2$ injection | $7.1 \times 10^{-11}$M | |

*Nuclear and cytoplasmic receptor before and after estradiol injection was incubated with $5 \times 10^{-11}$M to $2 \times 10^{-8}$M [$^3$H]estradiol in the absence or presence of 100-fold excess of diethylstilbestrol at 30 C for 2 or 1 h. The $K_D$ was derived from Scatchard plots.

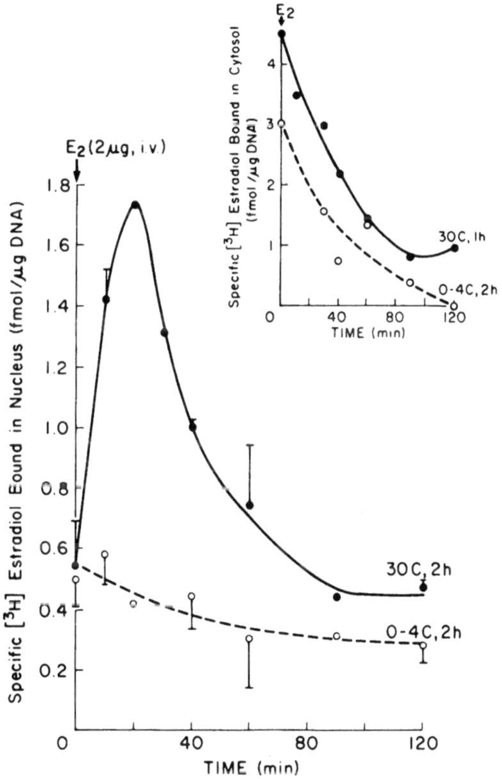


Figure 3: Changes in luteal cytoplasmic and nuclear estrogen receptor after i.v. injection of 17β-estradiol (2 μg) on day 11 of pseudopregnancy. Corpora lutea from one ovary were removed from each rabbit as control tissue ("time 0") before injection of estradiol. The solid line represents total receptor; the broken line, unoccupied receptor. Each point after estradiol injection represents duplicate determinations from 1 or 2 rabbits (mean ± SD).

The graph shows a rapid increase in total nuclear receptor, which peaks around 20 minutes followed by a steep decline in the receptor to approach prestimulation levels around 90 minutes after injection of estradiol. The graph for cytoplasmic receptor shows a concomitant depletion of cytoplasmic receptor. By 120 minutes, unoccupied cytoplasmic receptor was barely detectable. The increment in total nuclear receptor at 20 minutes is approximately the same as the amount of cytoplasmic receptor which is depleted, suggesting quantitative displacement of receptor from the cytoplasmic to nuclear compartment. The rapid decline in nuclear receptor after reaching a "peak," and the steady decline in cytoplasmic receptor during this period suggest that the nuclear receptor is being inactivated, rather than recycled to the cytoplasm. The time course for replenishment of cytoplasmic receptor and the possible regulation of this process by estradiol remain to be studied.

A constant finding in assays of nuclear receptor is the high percentage ($\sim$ 90%) of unoccupied binding sites in the nuclear fraction isolated from normal, untreated pseudopregnant rabbits (see values, "time 0", Fig. 3). If the assumption is made that cytoplasmic receptor reaches the nucleus as a receptor-steroid complex, the observation of unoccupied receptor in the nucleus is puzzling. A possible explanation for this has been found and the data are shown in Table 4.

Table 4: EFFECT OF REISOLATION OF NUCLEAR FRACTION ON NUMBER OF UNOCCUPIED BINDING SITES*

| | Specific [$^3$H]estradiol bound (fmol/μg DNA) | |
|---|---|---|
| | Unoccupied | Total |
| Nuclear receptor | 0.80 ± 0.14 | 0.88 ± 0.14 |
| $E_2$-treated nuclear receptor | 0.08 ± 0.04 | |
| $E_2$-treated nuclear receptor recombined with low speed supernatant and reisolated | 0.58 ± 0.13 | 0.83 ± 0.13 |

*A portion of the nuclear fraction was occupied with unlabeled estradiol. Excess unlabeled estradiol was washed out and the washed nuclear pellet was combined with the low speed supernatant and incubated at 0-4 C for 15 min with vortexing. The nuclear fraction was isolated again and specific [$^3$H]estradiol binding was determined as indicated in the text. The data are expressed as the mean ± SE.

During homogenization and isolation of a nuclear fraction, sites previously occupied with unlabeled estradiol become unoccupied. This was shown by first isolating a nuclear fraction from normal corpora lutea, treating this fraction with estradiol to completely occupy the sites, then recombining this nuclear fraction with a low-speed supernatant. When the nuclear fraction was again isolated the total binding sites were the same, but the majority of these sites had become unoccupied (Table 4). This suggests that unoccupied nuclear sites detected in the receptor assay may, in fact, represent sites that were originally occupied when the tissue was removed.

These data indicate that receptor translocation takes place in the rabbit corpus luteum, and they suggest that estrogen action in the corpus luteum may involve nuclear events. How these changes in receptor relate to progesterone synthesis remains to be determined.

A ROLE FOR INTRALUTEAL ESTROGEN IN THE MEDIATION OF LH ACTION ON THE RAT CORPUS LUTEUM.

At least three hormones have been designated as luteotropic in the rat: LH, prolactin and estradiol. This has led to considerable confusion about which hormone(s) are of primary importance and what function each hormone performs in the total regulation of the corpus luteum. Recently Gibori et al (14,27) have proposed that the period of LH-dependence for the corpus luteum during the first half of pregnancy (8) can be explained as a period of estrogen-dependence. The hypothesis was advanced that LH stimulates the synthesis of androgen which is aromatized to estrogen by the luteal aromatase (11), and that the maintenance of intraluteal estrogen is essential for normal rates of progesterone secretion (14,15,27). Corpora lutea were deprived of LH either by administration of LH antiserum or by hypophysectomy on day 10 of pregnancy and intraluteal estradiol concentrations were maintained by the administration of low amounts of testosterone via a Silastic capsule (27). The present experiments represent an extension of the studies of Gibori et al (27) to determine the changes in serum progesterone during the first 24 hours after the administration of LH antiserum when the major decline in serum progesterone occurs. Of particular interest was the effect of testosterone on serum progesterone and on intraluteal estradiol during this same period.

Figure 4 illustrates the rapid decline in serum progesterone that occurs after the administration of LH antiserum. These rats aborted within 48 hours and by day 15 the uteri were completely empty and ballooned, indicating a return to proestrus. This figure also shows the efficacy of testosterone in preventing the decline in progesterone.

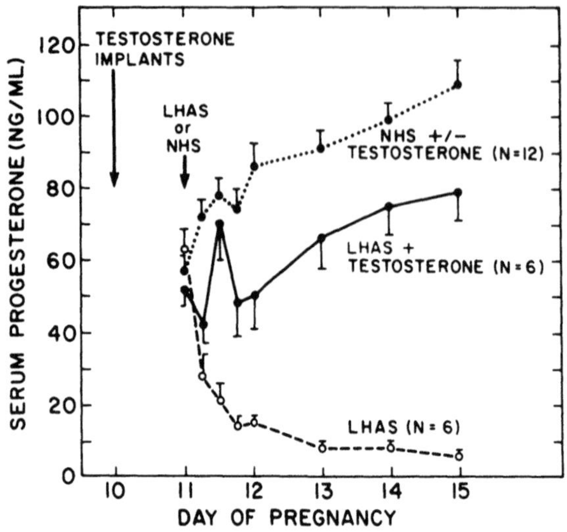


Fig. 4: Effect of testosterone on serum progesterone after injection of LH antiserum (LHAS) or normal horse serum (NHS) in pregnant rats. Two Silastic capsules (.078 in I.D.; .125 in O.D.; 0.5 cm filled length) filled with a mixture of crystalline testosterone and cholesterol (1:1) were placed subcutaneously on day 10. Progesterone values in rats treated with normal horse serum with or without testosterone were not different and therefore were pooled (top dotted line). Each point represents the mean ± SE.

This effect of testosterone was associated with the maintenance of normal luteal estradiol concentrations, while luteal estradiol concentrations dropped significantly within 6 hours in rats treated with LH antiserum alone (Fig. 5). Since serum progesterone had also decreased within 6 hours in rats treated with LH antiserum alone (Fig. 4) a cause and effect relationship between the decrease in intraluteal estradiol and that of serum progesterone cannot be seen. However, the continuation of progesterone secretion when intraluteal estradiol concentrations are maintained, either by administration of testosterone or by injection of large doses of estradiol (14) suggests an important role for estradiol. The distinct "peak" in serum progesterone 12 hours after injection of LH antiserum in rats also treated with testosterone (Fig. 4) is unexplained, but is of interest, because it is observed around the time that the corpora lutea would be expected to become first dependent upon the steroid. Serum progesterone in rats treated with LH antiserum and testosterone remains distinctly lower than controls throughout the experiment, and this may represent the permanent loss of a component of progesterone synthesis that is normally governed by LH. But it should be noted that fetal survival was lower in these animals (8 living embryos/rat vs 11 embryos/rat

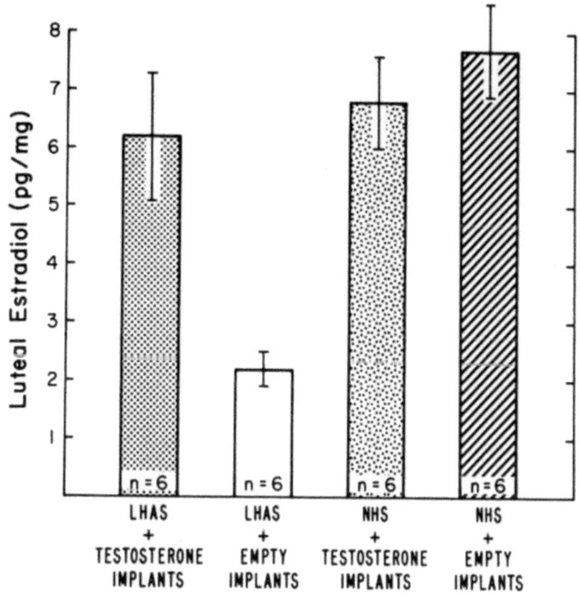


Fig. 5: Effect of testosterone on luteal estradiol concentrations after injection of LH antiserum (LHAS) or normal horse serum (NHS) on day 11 of pregnancy. Testosterone implants were placed sc on day 10 as described in Fig. 4. Corpora lutea were removed 6 hours after LHAS or NHS injection and estradiol measured in chromatographed extracts of luteal homogenates.

for normal horse serum treated rats with/without testosterone) and therefore there may have been less stimulation of the corpora lutea by placental lactogen.

Testosterone has not been measured in this study, but based upon the measurement of intraluteal estradiol, the serum concentrations of testosterone would be expected to be in the normal range. In a previous study using implants of similar size, serum immunoreactive androgen was 3.1 ng/ml while in control, pregnant rats without testosterone implants, the value was 1.8 ng/ml (15). The effectiveness of testosterone as a replacement hormone after the injection of LH antiserum has been interpreted as an action of estradiol produced within the corpus luteum by aromatization of testosterone. Another plausible interpretation is that testosterone directly stimulates the corpus luteum. An action of testosterone cannot be disproved readily, but the available evidence favors an action of estradiol. Estradiol alone is effective in preventing the effects of LH antiserum (14) and dihydrotestosterone, a nonaromatizable androgen, is ineffective as a replacement hormone (15,27).

## SUMMARY AND CONCLUSIONS

Figure 6 represents a working hypothesis for the hormonal regulation of the corpus luteum in the rabbit and rat, and is intended as a guide for further research. In the rabbit, LH initiates luteinization (28), and this is a sufficient stimulus for the corpora lutea to develop for several days and secrete progesterone. By the end of this period (approximately day 5 postcoitum), however, the corpora lutea develop an obligate dependence upon estradiol (24) which is produced by the developing ovarian follicles (10). In the absence of estradiol, progesterone synthesis begins to decrease, luteal development is arrested, and the corpora lutea die prematurely (24). As estradiol sustains the development of corpora lutea and progesterone synthesis, the inference can be made that LH (and FSH) indirectly supports the corpora lutea, through the stimulation of estradiol synthesis by the ovarian follicles. Only low blood concentrations of estradiol are required for a luteotropic effect (as opposed to local intra-ovarian concentrations) because estradiol from one ovary will maintain corpora lutea in a contralateral, irradiated ovary without follicles (13), or corpora lutea located beneath the kidney capsule (29,30). The mechanism of action of estrogen is unknown, but the interaction of estradiol with cytoplasmic receptor and the translocation of receptor to the nucleus suggest a classical steroid mode of action. The rapidity of the decline in progesterone synthesis after withdrawal of estradiol, and the stimulation of progesterone synthesis in vitro (17), and in vivo by estradiol indicate that steroidogenesis is acutely dependent on estradiol and responsive to estradiol. Those steps in the steroid pathway that are regulated by estrogen remain to be determined; evidence exists that estradiol acts at a step beyond the synthesis of cholesterol (17). Although it has been demonstrated that LH can directly stimulate adenylate cyclase (19) and progesterone synthesis (20), the significance of LH action in the sustained secretion of progesterone has not been determined. It seems clear that LH alone is not sufficient to maintain the corpus luteum, for when estradiol is withdrawn, LH is unable to prevent luteal regression (12,16). Regression of the corpus luteum may involve the gradual loss of estradiol receptor (22) possibly through the action of a luteolytic substance produced by the uterus (31). Withdrawal of estradiol from the circulation at the end of pseudopregnancy is not a necessary condition for a normal course of luteal regression. Corpora lutea regressed at the usual time (18 days post coitum) in the presence of estradiol maintained at elevated levels throughout pseudopregnancy (31).

Steroidogenesis in the rat corpus luteum appears to be different in at least two major respects from that in the rabbit. First, the rat corpus luteum is rich in aromatizing activity whereas the rabbit corpus luteum has no detectable aromatase.

Second, the rate of secretion of progesterone by the rat corpus luteum cannot be equated with progesterone synthesis, because of the well-established shift from progesterone secretion to the secretion of 20α-dihydroprogesterone after the withdrawal of luteotropic hormone (32). The secretion of 20α-dihydroprogesterone increases rapidly after the administration of LH antiserum (33), indicating that progesterone synthesis has not been stopped. In the rabbit we have observed no tendency of the corpus luteum to secrete more 20α-dihydroprogesterone after withdrawal of estradiol to initiate premature regression; both progesterone and 20α-dihydroprogesterone decrease. Therefore, a major criterion for the designation of a hormone as luteotropic in the rat must be its ability to block the induction of 20α-hydroxysteroid dehydrogenase activity and to maintain progesterone secretion.

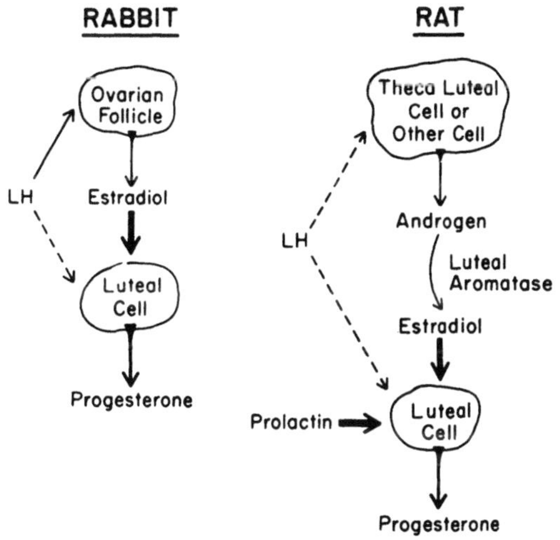


Fig. 6: Hypothesis for the hormonal regulation of the corpus luteum in the rabbit and rat. In both species, estradiol is proposed as the "ultimate" luteotropic hormone that is necessary for progesterone synthesis/secretion. The pituitary gonadotropins, LH and/or prolactin participate and are essential through actions to maintain estrogen responsiveness (receptor) or the availability of estrogen through the stimulation of estradiol synthesis or of androgen which can be aromatized. The two species differ in the source of estrogen for the corpus luteum: the rabbit corpus luteum lacks aromatase and is therefore totally dependent upon estradiol produced by the ovarian follicles; the rat corpus luteum is rich in aromatase and appears to produce its own estrogen either from an androgen precursor produced within the corpus luteum (theca luteal cells?) or from an extra-luteal cell, such as the theca interna of follicles.

The rat corpus luteum has receptor for prolactin and LH (6) and for estradiol (34). The action of prolactin during early luteinization is necessary for the acquisition of LH receptor (6), and prolactin also promotes an increase in luteal cell content of estrogen receptor (35). There is general agreement that prolactin is indispensable for progesterone secretion (8), and that as the prolactin surges wane toward the middle of pregnancy (36), placental lactogen must be present to maintain progesterone secretion and promote further luteal growth (8). But prolactin appears to act in concert with another hormone to maintain optimal luteal function and growth (8, 37). From the results reported here and from recent investigations (14, 27), we suggest that this hormone is estradiol (or another estrogen) which is produced within the corpus luteum and is also indispensable for normal rates of progesterone secretion. The requirement for LH during days 8 through 11 of pregnancy can be explained as an action to maintain intraluteal estradiol concentrations by stimulating the synthesis of androgen substrate that is accessible to the luteal aromatase. The cellular origin of androgen precursor has not been identified, but one possibility is the luteinized theca interna cell, and another is the ovarian follicles. Whatever the source of androgen, it is normally available to the corpora lutea in limited amounts, because the intraluteal estradiol concentration can be greatly increased in vitro (11) or in vivo (15, 27) by incubation with testosterone or by administration of testosterone. It is of interest to note that elevation of intraluteal estradiol above normal concentrations does not cause an elevation in serum progesterone, suggesting that the steroidogenic effect of estradiol is maximal at normal, physiological concentrations (15). After day 11 of pregnancy, the corpus luteum becomes independent of pituitary LH (38), and placental lactogen assumes the role of prolactin but the corpora lutea remain dependent upon estradiol (15, 39). The absence of a requirement for LH might be due to the availability of sufficient androgen in the circulation derived from the conceptus or to the activity of a placental "LH-like hormone" (40).

It is appropriate to explore briefly the implications of this work for an understanding of human luteal function. The human corpus luteum synthesizes estrogen (41) and the monkey corpus luteum readily aromatizes androgen (42). In view of the requirement for LH to sustain the corpus luteum (2) an intriguing possibility is that intraluteal estrogen may mediate the action of LH. A role for intraluteal estrogen in the primate corpus luteum has been proposed, but the evidence indicates that estrogen acts to inhibit the secretion of progesterone, at least in the latter half of the luteal phase (43). These observations offer the possibility that the human corpus luteum has a "built-in" mechanism to curtail the secretion of progesterone and terminate the luteal phase, and that this mechanism might be LH-dependent.

ACKNOWLEDGEMENTS

We thank Dr. W. Hansel and Dr. R.B. Snook for the LH antiserum and Russell M. Possley for technical assistance. This research was supported by NIH grant HD-07127 and by a Program Project grant NIH-HD-08333.

SUMMARY

The luteotropic action of estradiol has been studied in the rabbit and rat, and it is proposed that in both species, estradiol may be the "ultimate" luteotropic hormone. The acute dependence of the rabbit corpus luteum upon estradiol is illustrated by the rapid decline in serum progesterone after withdrawal of estradiol (removal of Silastic implant containing 17β-estradiol) and by the restoration of serum progesterone to normal values when the estradiol implant is replaced 24 hours later. The identification and characterization of a cytoplasmic and nuclear estrogen receptor in the rabbit corpus luteum and the translocation of the cytoplasmic receptor suggest that estradiol may be acting as it does in other estrogen target tissues. The administration of LH antiserum in pregnant rats causes rapid decreases in luteal estradiol and serum progesterone concentrations, which can be prevented by the administration of low doses of testosterone. This luteotropic effect of testosterone is attributed to an action of estradiol which is formed within the corpus luteum via the luteal aromatase. The rabbit corpus luteum, which has no aromatase, is totally dependent upon estradiol produced by the ovarian follicles and LH is essential to maintain follicular estradiol synthesis. The rat corpus luteum, which is rich in aromatase activity, may produce its own estrogen from an androgen precursor synthesized within the luteal tissue. The essential role of LH may be the stimulation of the synthesis of androgen precursor. As a working hypothesis it is proposed that in these two species LH is necessary to stimulate the synthesis of estradiol which then acts to sustain progesterone secretion.

REFERENCES

1. E.B. Astwood. Endocrinology 28 (1941) 309.
2. R.L. VandeWiele, J. Bogumil, I. Dyrenfurth, M. Ferin, R. Jewelewicz, M. Warren, T. Rizkallah, and G. Mikhail. Recent Progr. Hormone Res. 26 (1970) 63.
3. J.M. Robson. J. Physiol. (London) 90 (1937) 435.
4. M. Takayama and G.S. Greenwald. Endocrinology 92 (1973) 1405.
5. M.S. Smith, M.E. Freeman, and J.D. Neill. Endocrinology 96 (1975) 219.
6. J.A. Holt, J.S. Richards, A.R. Midgley, Jr., and L.E. Reichert, Jr. Endocrinology 98 (1976) 1005.

7. H.G. Madhwa Raj and N.R. Moudgal. Endocrinology 86 (1970) 874.
8. W.K. Morishige and I. Rothchild. Endocrinology 95 (1974) 260.
9. J. Hilliard, H.G. Spies, and C.H. Sawyer, in: The Gonads, ed. K.W. McKerns (Appleton-Century-Crofts, New York, 1969) p. 55.
10. T.M. Mills and K. Savard. Endocrinology 92 (1973) 788.
11. D.J. Elbaum and P.L. Keyes. Endocrinology 99 (1976) 573.
12. P.L. Keyes and A.V. Nalbandov. Endocrinology 80 (1967) 938.
13. P.L. Keyes and D.T. Armstrong. Endocrinology 83 (1968) 509.
14. G. Gibori, R. Rodway and I. Rothchild. Endocrinology 101 (1977) 1683.
15. G. Gibori and P.L. Keyes. Endocrinology 102 (1978) 1176.
16. J.A. Holt, P.L. Keyes, J.M. Brown and J.B. Miller. Endocrinology 97 (1975) 76.
17. G.B. Fuller, B.M. Markaverich, and W.C. Hobson. Biol. Reprod. 18 (1978) 176.
18. M.E. Cohen, P.L. Keyes, S.S. Han and L.J. Kleinsmith. Biol. Reprod. 17 (1977) 16.
19. M. Hunzicker-Dunn and L. Birnbaumer. Endocrinology 99 (1976) 185.
20. J.H. Dorrington and R. Kilpatrick, in: The Gonads, ed. K.W. McKerns (Appleton-Century-Crofts, New York, 1969) p. 33.
21. L. Dufy-Barbe, B. Dufy and J.D. Vincent. Biol. Reprod. 18 (1978) 118.
22. C. Lee, P.L. Keyes, and H.I. Jacobson. Science 173 (1971) 1032.
23. R.S. Scott and P.I.C. Rennie. Endocrinology 89 (1971) 297.
24. J.B. Miller and P.L. Keyes. Endocrinology 102 (1978) 31.
25. K. Burton. Biochem. J. 62 (1956) 315.
26. R.G. Martin and B.N. Ames. J. Biol. Chem. 236 (1961) 1372.
27. G. Gibori, P.L. Keyes, and J.S. Richards. Endocrinology 103 (1978) 162.
28. P.L. Keyes, G.D. Canastar and J.B. Miller. Endocrinology 91 (1972) 197.
29. P.L. Keyes and D.T. Armstrong. Endocrinology 85 (1969) 423.
30. P. Rennie. Endocrinology 83 (1968) 314.
31. J.B. Miller and P.L. Keyes. Biol. Reprod. 15 (1976) 511.
32. W.G. Wiest and W.R. Kidwell, in: The Gonads, ed. K.W. McKerns (Appleton-Century-Crofts, New York, 1969) p. 295.
33. N.R. Moudgal, H.R. Behrman and R.O. Greep. J. Endocrinol. 52 (1972) 413.
34. J.S. Richards. Endocrinology 95 (1974) 1046.
35. G. Gibori, J.S. Richards, and P.L. Keyes, in: Workshop on Ovarian/Follicular and Corpus Luteum Function, ed. C.P. Channing, J. Marsh and W. Sadler (Plenum Press, New York, 1978).
36. M.S. Smith and J.D. Neill. Endocrinology 98 (1976) 696.
37. J. Hilliard. Biol. Reprod. 8 (1973) 203.
38. G. Pepe and I. Rothchild. Endocrinology 91 (1972) 1380.
39. G. Gibori, E. Antczak, and I. Rothchild. Endocrinology 100 (1977) 1483.

40. F. Haour, G. Tell and P. Sanchez. C.R. Acad. Sci. 282 (1976) 1183.
41. K. Savard, J.M. Marsh, and B.F. Rice. Recent Progr. Hormone Res. 21 (1965) 285.
42. R. Stouffer, L. Bennett and G. Hodgen. Proceedings, 60th Annual Meeting, The Endocrine Society, Miami, 1978. Abstract no. 370.
43. F.J. Karsch and G.P. Sutton. Endocrinology 98 (1976) 553.

# FOLLICULAR DETERMINANTS OF CORPUS LUTEUM FUNCTION IN THE HUMAN OVARY

Kenneth P. McNatty

Laboratory of Human Reproduction and Reproductive Biology

Harvard Medical School, 45 Shattuck Street, Boston 02115

## INTRODUCTION

The corpus luteum is a direct continuation of follicle development. It forms after ovulation from the haemorrhagic tissues of the ruptured follicle and it becomes the major source of progesterone circulating in plasma. The main site of progesterone synthesis in the human corpus luteum is presumed to be the granulosa-lutein cells: these cells secrete appreciable amounts of progesterone in tissue culture (~6 pg/cell/day) (1) and they are the most numerous of the various cell-types present.

It is frequently, but incorrectly, assumed that the follicle spontaneously differentiates into luteal tissue and secretes large amounts of progesterone after it has ovulated. However, ovulation *per se* does not ensure that a normal corpus luteum will form or that progesterone secretion will result (2). Indeed recent studies on the human ovary suggest that a follicle is unlikely to develop into a normal corpus luteum unless it is prepared for eventual luteal function before ovulation takes place (1, 3). Several hormones, such as FSH, oestradiol, androstenedione and prolactin play important roles in preparing the follicle for eventual luteal function, while LH, in addition to influencing steroidogenesis by the growing follicle, is ultimately responsible for transforming the follicle into a corpus luteum.

It is suggested that at least three criteria need to be fulfilled before the follicle can be transformed into a fully functional corpus luteum: (a) there must be sufficient numbers of granulosa cells in the follicle before ovulation because after this event granulosa-lutein cell proliferation is absent; (b) the follicle must have granulosa cells with the capacity to produce sufficient progesterone; and (c) the granulosa cells as well as the theca cells must be responsive to LH [i.e., the follicle must have a high receptor content for LH; see (4) for review].

This communication describes some of the endocrine events which take place during the growth of the human antral follicle. It then examines the various relationships between the endocrine events within the follicle and the viability, mitotic activity and steroidogenic capacity of its granulosa cells. It is suggested that these endocrine events influence eventual corpus luteum function.

## The developing antral follicle and its hormonal environment

As a small antral follicle develops from the early follicular phase into a large preovulatory follicle just before ovulation, there is a precise sequence of endocrine events which takes place within its antral fluid (Fig. 1). The endocrine events within this follicle differ from those in peripheral plasma and from those in most, if not all, other antral follicles in the same ovary. In this follicle FSH is always present. Furthermore, the follicular concentrations of FSH increase as this follicle develops despite falling levels of FSH in peripheral plasma. As the levels of FSH rise, there is a concomitant increase in the follicular concentrations of oestradiol which peak during the mid- to late follicular phase reaching values ~10,000 times higher than those in peripheral plasma. Throughout the entire period of antral follicle growth, the concentrations of androstenedione within this follicle remain lower than those of oestradiol. Indeed during the preovulatory growth stage the androstenedione concentrations within this follicle are lower than those within all other follicles. Thus of all the follicles in the ovary, the preovulatory structure contains the highest concentrations of oestradiol and the lowest concentrations of androstenedione. Within this healthy 'oestrogenic' follicle, the levels of prolactin are also low, and like those of androstenedione, they decline to their lowest levels in the follicle just before ovulation. Before the mid-cycle peak of LH in plasma, LH is barely detectable in antral fluid (Fig. 2). At mid-cycle, when the levels of LH in plasma are elevated, this hormone enters the follicle in appreciable amounts for the first time (Fig. 2). After the theca cells have been exposed to the high blood levels of LH, and after the granulosa cells have been exposed to this hormone within the follicle, the concentrations of progesterone in antral fluid increase (Fig. 2), while those of androstenedione and eventually those of oestradiol decline. Shortly before it ovulates, the preovulatory follicle contains relatively high concentrations of FSH, LH and progesterone, low levels of androstenedione, prolactin and declining concentrations of oestradiol (5). This sequence of endocrine events only occurs within a single antral follicle, as it develops during the follicular phase. Thus, the granulosa cells in this follicle are

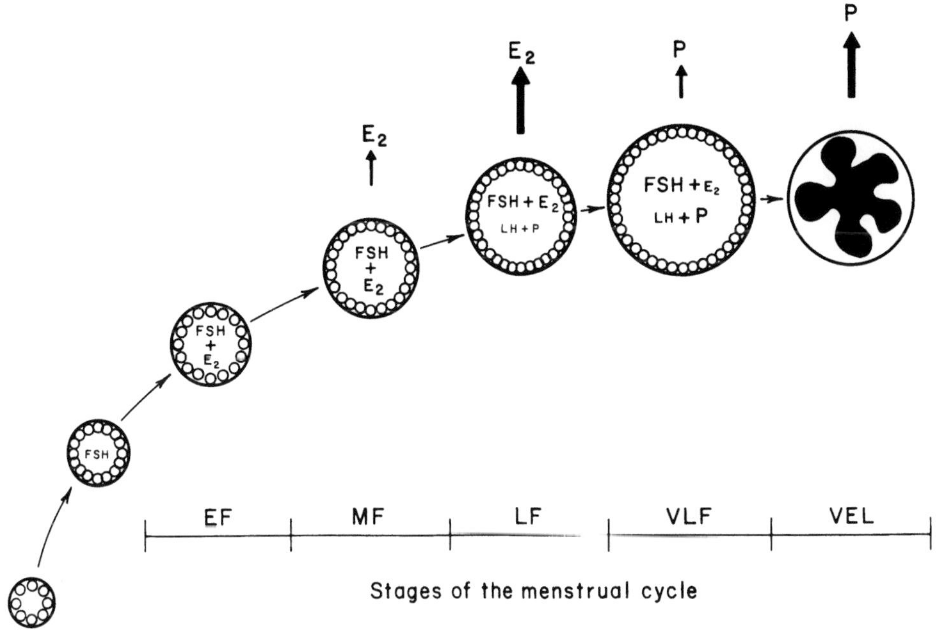


Fig. 1. Sequence of endocrine changes which take place within the developing antral follicle during the follicular phase of the menstrual cycle. The size of the lettering and the thickness of the arrows indicates the relative concentrations of these hormones in follicular fluid or in ovarian venous plasma. $E_2$ refers to oestradiol, P refers to progesterone. EF, MF, LF refer to early (patient still menstruating), mid-, and late follicular phase, respectively, and VEL refers to very early luteal phase.

exposed to a unique sequence of endocrine events before they differentiate into luteal cells.

## The effect of FSH on granulosa cell viability

*In vitro* evidence suggests that FSH prevents granulosa cells in antral follicles from undergoing degeneration. When human granulosa cells from antral follicles were grown for 10 days in a culture medium devoid of FSH, about 90% of the cells detached from the coverslip and died. In contrast, those cells which were cultured in the same medium, but with FSH added, remained viable and firmly attached to the coverslip (6). These findings are consistent with the suggestion that one of the major actions of FSH on the developing follicle is to sustain growth and prevent atresia (7). Without FSH support, it seems unlikely that a follicle could eventually develop into a functional corpus luteum.

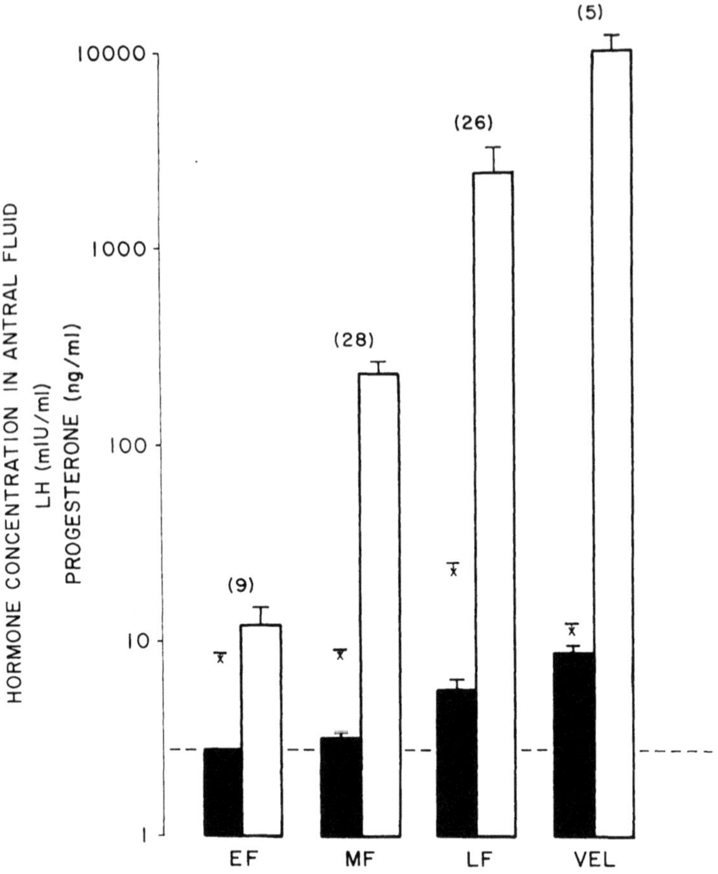


Fig. 2. Relationship between the concentrations of LH (black histogrammes) and progesterone (white histogrammes) in large antral follicles (≥ 8 mm diam.) and in recently ovulated follicles during the respective follicular phases and the very early luteal phase of the menstrual cycle. The mean concentrations of LH in plasma are also shown ($\bar{X}$). All results are expressed as means +1 S.E.M. The number of observations are in parentheses. For abbreviations see Fig. 1. The dotted line is the limit of detection of LH (2.8 mIU/ml). The limit of detection for progesterone is 10 ng/ml. Data from unpublished observations and from (3).

Relationships between granulosa cell number and the concentrations of hormones in antral fluid

During the menstruating to early follicular phase, the largest healthy antral follicle in the human ovary is approximately 4 mm in diameter and it contains about 1 million granulosa cells. Just before ovulation, some 10 to 15 days later, the follicle has increased in diameter to about 25 mm and it has accumulated about 50 million granulosa cells (5). Granulosa cell proliferation is rapid in the large antral follicle, it ceases after the LH-surge and is absent in the corpus luteum. The total number of granulosa cells in the preovulatory follicle is similar to the number of granulosa-lutein cells in the corpus luteum (8).

The hormones, FSH, oestradiol and androstenedione, have important roles in influencing the number of granulosa cells which accumulate in the developing antral follicle. In follicles which had detectable levels of FSH (>1.3 mIU/ml), there was a linear correlation between the number of granulosa cells present and the concentration of oestradiol in antral fluid [$p < 0.001$; (9)]. In follicles which had undetectable levels of FSH (<1.3 mIU/ml), the concentration of oestradiol remained low (≤ 200 ng/ml) irrespective of the number of granulosa cells that were present.

A close relationship was also found to exist between FSH and the levels of oestradiol and androstenedione in antral follicles. When FSH was detectable (> 1.3 mIU/ml), the concentrations of oestradiol exceeded those of androstenedione irrespective of the stage of the follicular phase, whereas the converse held true when FSH was not detectable (Table 1). In follicles devoid of granulosa cells, the levels of androstenedione exceeded those of oestradiol irrespective of whether FSH was present or absent within the follicle (9). Presumably, FSH influences the follicular levels of oestradiol by acting on the granulosa cells. *In vitro*, human granulosa cells in some antral follicles can be stimulated by FSH to metabolize androstenedione to oestradiol (10).

In order for the follicle to sustain a high rate of granulosa cell proliferation, it seems essential that the oestradiol concentrations within the follicle increase progressively. Furthermore, it also seems necessary that the oestradiol concentrations within the follicle exceed those of androstenedione. When antral follicles were found to have an oestradiol to androstenedione ratio greater than unity, the number of granulosa cells present was similar to the optimum number recorded. Conversely, in follicles where the oestradiol to androstenedione ratio was less than unity, the number of granulosa cells present was consistently lower than that in similar-sized 'oestrogenic' follicles (11). In tissue culture, human granulosa cells undergo their highest rate of mitoses provided they were: (a) harvested from follicles that contained FSH,

Table 1. Relationship between the concentrations of oestradiol-17β and androstenedione in antral fluid of non-cystic follicles* with and without FSH during the follicular phase of the menstrual cycle.

| FSH | Hormone | Concentrations (mean ± 1.S.E.M. ng/ml) | | |
|---|---|---|---|---|
| | | EF | MF | LF |
| + | $E_2$ | 440 | 2579 ± 41 | 2388 ± 49 |
| | Δ | 113 | 745 ± 68 | 141 ± 90 |
| | | (1) | (8) | (5) |
| - | $E_2$ | 69 ± 28 | 191 ± 89 | 143 ± 84 |
| | Δ | 520 ± 98 | 320 ± 51 | 676 ± 89 |
| | | (10) | (14) | (8) |

The limit of detection for FSH was 1.3 mIU/ml; + = follicles with detectable levels of FSH; - = follicles with undetectable levels of FSH; $E_2$ = oestradiol-17β; Δ = androstenedione. Number of observations in parentheses. For abbreviations see Fig. 1. *All follicles were ≥ 4 mm diameter and contained more than 50,000 granulosa cells (9).

high concentrations of oestradiol and low levels of LH (1); or (b) were grown in a medium with a similar endocrine environment to (a).

As the follicle develops during the follicular phase, it becomes the major ovarian source of both androstenedione and oestradiol. However, its capacity to secrete androstenedione exceeds that of oestradiol as judged by the respective levels of these hormones in ovarian venous plasma (Table 2). Although the developing follicle secretes more androgen than oestrogen its microenvironment must contain more oestrogen than androgen. Therefore, the presence of FSH within the follicle is critical for ensuring that the granulosa cells can metabolize the excess androgen to oestrogen. This ensures that the granulosa cells can continue to be exposed to an oestrogen-enriched environment. Moreover, as the follicle develops, thecal-derived androstenedione becomes an important prehormone for oestrogen biosynthesis within the follicle. It has been suggested that the entry of thecal-cell androstenedione into the FSH-enriched follicle promotes an intrafollicular, positive-feedback relationship between oestrogen production, granulosa cell proliferation, and FSH accumulation (9), thereby ensuring that a sufficient number of granulosa cells are present when the follicle is eventually transformed into a corpus luteum.

Table 2. Concentration of oestradiol and androstenedione in ovarian venous plasma (ng/ml; mean ± S.E.M.) during the follicular phase of the menstrual cycle.

| | Early Follicular | | Late Follicular | |
|---|---|---|---|---|
| | Right | Left | Large | Small |
| n | 3 | 3 | 4 | 4 |
| Oestradiol | 1.14 ± 0.55 | 1.01 ± 0.79 | 19.32* ± 6.19 | 0.65 ± 0.17 |
| Androstenedione | 11.95 ± 4.02 | 11.54 ± 4.13 | 59.40* ± 16.10 | 8.54 ± 2.41 |

n = number of pairs of measurements.
Samples were taken from: the right ovarian vein (right); left ovarian vein (left); vein draining an ovary containing a large follicle of ≥ 8 mm diameter (large); vein draining an ovary containing a small follicle of < 8 mm diameter (small). During the early follicular phase there were no follicles ≥ 8 mm diameter.
*Significantly different from the contralateral side ($p < 0.02$)(12).

Relationship between hormones in antral fluid and progesterone secretion by granulosa cells in vitro

As the antral follicle develops during the follicular phase, the capacity of its granulosa cells to synthesize progesterone or to respond to LH to secrete progesterone seems to depend upon their prior exposure to certain hormones in antral fluid (Table 3). The capacity of granulosa cells to synthesize progesterone, as judged by their response to dibutyryl cyclic AMP (dbcAMP) was greatest (i.e., 2.6 - 5.0 pg/cell/day) when the cells were harvested from follicles with FSH and oestradiol, and it was lowest (i.e., 0.65 pg/cell/day) when the cells were harvested from follicles devoid of FSH and with low levels of oestradiol (i.e.,<250 ng/ml). Moreover, it seems that the ability of an individual cell to secrete progesterone in response to dbcAMP increases progressively the longer the cell is present in a fluid environment enriched with FSH and oestradiol. In the human ovary, the majority of antral follicles contain low levels of FSH and/or oestradiol. Therefore, most of the follicles contain granulosa cells which do not have the capacity to synthesize large amounts of progesterone even when stimulated with dbcAMP (1). In the human ovary, the capacity of granulosa cells to synthesize progesterone is limited to those cells present in the one or two follicles which contain FSH and high concentrations of oestradiol in their antral fluids (1).

In the human ovary, granulosa cells in preovulatory follicles produce large amounts of progesterone about 24 to 36 hours before

Table 3. Relationship between the endocrine microenvironment of the human antral follicle and the average daily secretion of progesterone by granulosa cells exposed to dbdAMP, LH or LH-free media *in vitro*.

| IN VIVO | | IN VITRO | | |
|---|---|---|---|---|
| Hormones in antral fluid. | Stage of menstrual cycle. | Daily secretion rate of progesterone by granulosa cells [(pg/cell/day); mean (range)]. | | |
| | | [a]dbcAMP 0.1 mMol | [a]LH 30 mIU/ml | [b]LH-free media < 0.02 mIU/ml |
| -FSH $-E_2$ | F | 0.65 (0.40-0.88) | 0.57 (0.01-0.88) | 0.13 (0.04-0.21) |
| n | | 12 | 12 | 3 |
| +FSH $-E_2$ | F | 1.61 (0.04-4.00) | 0.55 (0.08-1.10) | --- |
| n | | 4 | 10 | |
| +FSH $+E_2$ | MF | 2.60 (0.40-6.00) | 0.04 (0.14-1.10) | 0.12 (0.09-0.18) |
| n | | 8 | 8 | 5 |
| +FSH $+E_2$ | LF | 5.00 (2.80-8.00) | 4.99 (2.80-8.00) | 0.85 (0.30-1.16) |
| n | | 5 | 16 | 5 |

$E_2$ = oestradiol-17β; dbcAMP = dibutyryl cyclic AMP; -FSH, <2.8 mIU/ml; $-E_2$, <250 ng/ml; +FSH, >2.8 mIU/ml; $+E_2$, >250 ng/ml.

a. Culture media contained 1.7 mIU/ml LH, 1.8 mIU/ml FSH and 5 ng/ml prolactin.

b. Culture media contained <0.02 mIU/ml LH, 2.6 mIU/ml FSH and 6 ng/ml prolactin.

F = any stage of follicular phase; MF = Mid-follicular phase; LF = Late follicular phase.

ovulation (3, 13). At this stage the granulosa cells produce more progesterone per unit time than at any earlier period of development. This increased output of progesterone occurs when the capacity of the cells to synthesize progesterone is maximal and this is also coincident with the time when the cells become responsive to LH stimulation and when LH enters the follicle in appreciable amounts for the first time (Fig. 1, Table 2). Presumably, the response of human granulosa cells to LH is related to the number of LH-receptors present in the cells (4). In vitro, granulosa cells only respond maximally to LH after the cells have been exposed for 8-10 days to a culture medium supplemented with serum and enriched with FSH and oestradiol (1).

These data suggest that as the healthy antral follicle develops during the follicular phase, there is a concomitant increase in the capacity of its granulosa cells to synthesize progesterone. However, the cells appear unable to express this capacity until the late follicular phase because it is not until this stage of differentiation that they become fully responsive to LH stimulation. These data also suggest that the steroidogenic capacity of an individual granulosa cell is always limited. Therefore before the LH surge there must be a coordinated increase in granulosa cell number and in the biosynthetic capacity of each cell so that the follicle can eventually develop into a fully functional corpus luteum.

## The effect of androstenedione on progesterone secretion by human granulosa cells

Androgens have been shown to stimulate progestin accumulation in rat and porcine granulosa cell cultures and to synergize with FSH in the stimulation of progestin accumulation (14, 15). In the human ovary, androstenedione is the major androgen present in human follicular fluid; dihydrotestosterone and testosterone are present at much lower concentrations. In antral follicles, the concentrations of androstenedione may vary from 100 to 1500 ng/ml. The lowest concentrations are usually found in healthy follicles and the highest in atretic or cystic follicles (11). In tissue culture, androstenedione at concentrations of 1000 ng/ml or more is mildly cytotoxic to granulosa cells. Moreover, at concentrations ranging from 100 to 1000 ng/ml, androstenedione either alone or in combination with FSH has no stimulatory effects on progesterone secretion by human granulosa cells when secretion was assessed on a per cell basis (10). These observations are consistent with those reported for rabbit granulosa cells (16).

## The effect of prolactin on follicular development and on progesterone secretion by granulosa cells

Throughout antral follicle growth, granulosa cells are probably always exposed to prolactin since this hormone has been detectable (i.e. >1.5 ng/ml) in all the follicles that have been examined. Normally, it seems that the prolactin levels in antral fluid are low unless the plasma levels of this hormone exceed 100 ng/ml (Table 4). During antral follicle growth, the mean concentrations of prolactin decline as the follicle approaches ovulation (3). After ovulation, the prolactin levels in luteal-tissue fluid progressively increase to reach values similar to those in plasma at the mid-luteal phase (17). Similar changes in prolactin concentrations occur in sheep follicles and corpora lutea (17). Why the prolactin levels decline to low levels at ovulation is not understood.

Table 4. Relationship between prolactin concentrations in peripheral plasma and in the fluid from human antral follicles.

| Prolactin concentration (ng/ml) | | |
|---|---|---|
| Peripheral plasma | Antral Fluid (mean ± 1 S.E.M.) | No. of Observations |
| >100 (101-300) | 33.7 ± 7.8 | 24 |
| 50 - 100 | 14.0 ± 4.7 | 47 |
| <50 | 14.2 ± 1.6 | 81 |

Peripheral blood was obtained immediately prior to ovariectomy. Antral fluid was collected within 3 hours of ovariectomy.

Prolactin appears to have at least three direct actions on follicle development and early luteal formation which may influence the eventual secretory activity of the corpus luteum:

1. During antral follicle growth, follicles with high concentrations of oestradiol ($\geq$ 1 μg/ml) have significantly lower concentrations of prolactin than follicles with low concentrations of oestradiol (<0.5 μg/ml) [high oestradiol, prolactin = 13.1 ± 1.6 ng/ml vs. low oestradiol, prolactin = 35.3 ± 4.7 ng/ml; $p < 0.01$; (5)]. When prolactin concentrations are elevated in plasma (i.e. >100 ng/ml) there are elevated levels of this hormone in antral fluid (Table 4) as well as low levels of oestradiol and a markedly reduced population of granulosa cells (11). Therefore, in women who have high blood levels of prolactin, antral follicle growth is impaired, presumably because the level of oestrogen within the follicle is insufficient to sustain the optimal proliferation rate of granulosa cells. This in turn may lead to the expression of inadequate luteal function.

2. When the prolactin concentrations in antral fluid exceed 25 ng/ml, they may inhibit the normal transformation of the follicle into luteal tissue (18). *In vitro* exposure of human granulosa cells to prolactin at concentrations in excess of 25 ng/ml prevents the accumulation of progesterone in tissue culture medium. This action of prolactin is reversible (8). This inhibitory action can also be

demonstrated on mouse follicular tissue (19), but the mechanism of action is not understood.

3. Prolactin has a luteotrophic action on human granulosa cells when present at physiological concentrations (18). When human granulosa cells were cultured in a media with a serum devoid of prolactin (<0.2 ng/ml), the accumulation of progesterone in the culture medium was ~60% lower than the control values. Suppression of prolactin during the follicular phase in normal cycling women leads to reduced progesterone output by the corpus luteum (20). These findings when considered together suggest that the granulosa-lutein cells require exposure to a certain level of prolactin in order to secrete maximal amounts of progesterone.

## SUMMARY

The corpus luteum is a direct continuation of preovulatory follicle development. Growth and differentiation of a follicle to the preovulatory stage depends on certain endocrine changes taking place within the follicle itself (5). These endocrine changes markedly influence the viability, the rate of proliferation, and the biosynthetic potential of granulosa cells. Therefore, the sequence of endocrine events within the developing antral follicle is likely to be of considerable importance for the formation and secretory activity of the corpus luteum.

If granulosa cells in developing antral follicles are exposed to certain hormones in the wrong sequence, the follicle may be prevented from developing into a normal corpus luteum. For example, if granulosa cells in antral follicles are not exposed to FSH, they are unable to generate high levels of oestrogen (Table 1). In the absence of oestrogen, granulosa cells do not increase in number. The same end result may also arise if the granulosa cells are exposed prematurely to elevated levels of LH or to high levels of prolactin: LH blocks granulosa cell mitoses even if the cells are exposed to oestrogen (1, 21); high levels of prolactin in antral fluid are associated with low levels of oestrogen and a reduced rate of granulosa cell proliferation.

In women, it is likely that a fully functional corpus luteum is formed from a follicle that was exposed to the most favourable sequence of endocrine changes during its antral stages of development.

## REFERENCES

(1) K.P. McNatty and R.S. Sawers. J. Endocrinol. 66(1975)391.

(2) W. Haresign, J.P. Foster, N.B. Haynes, D.B. Chrighton and G.E. Lamming. J. Reprod. Fertil. 43(1975)269.

(3) K.P. McNatty, W.M. Hunter, A.S. McNeilly and R.S. Sawers. J. Endocrinol. 64(1975)555.

(4) J.S. Richards and A.R. Midgley, Jr. Biol. Reprod. 14(1976)82.

(5) K.P. McNatty in: Gynaecological Endocrinology, Vol. 7, No. 3, ed. G.T. Ross and M.B. Lipsett, W.A. Saunders & Co., Ltd., U.K. 1978 (In press).

(6) K.P. McNatty, J.G. Bennie, W.M. Hunter and A.S. McNeilly in: Physiological Effects of Immunity Against Reproductive Hormones, ed. R.G. Edwards and M.H. Johnson, Cambridge University Press, Cambridge. 1975, p. 41.

(7) H. Peters, A.G. Byskov, R. Himelstein-Braw and M. Faber. J. Reprod. Fertil. 45(1975)559.

(8) K.P. McNatty. Ph.D. Thesis (1975) Edinburgh University, U.K.

(9) K.P. McNatty and D.T. Baird. J. Endocrinol. (1978) In press.

(10) K.P. McNatty, A. Makris, C. DeGrazia, R. Osathanondh and K.J. Ryan. (Unpublished data)

(11) K.P. McNatty (Unpublished data)

(12) K.P. McNatty, D.T. Baird, A. Bolton, P. Chambers, C.S. Corker and H. McLean. J. Endocrinol. 71(1976)77.

(13) D.T. Baird, T.G. Baker, K.P. McNatty and P. Neal. J. Reprod. Fertil. 45(1975)611.

(14) A. Nimrod and H.R. Lindner. Mol. Cell. Endocrinol. 5(1976)315.

(15) D.W. Schomberg, R.L. Stouffer and L. Tyrey. Biochem. Biophys. Res. Comm. 68(1976)77.

(16) G.F. Erickson and K.J. Ryan. Endocrinology 97(1975)108.

(17) K.P. McNatty. Acta Endocrinol. Suppl. 212 (1977) Abstract S21.

(18) K.P. McNatty, R.S. Sawers and A.S. McNeilly. Nature 250(1974) 653.

(19) K.P. McNatty, P. Neal and T.G. Baker. J. Reprod. Fertil 47 (1976)155.

(20) K.D. Schulz, W. Geiger, E. del Pozo and H.J. Künzig. Am. J. Obst. & Gyn. (1978) In press.

(21) F. Friedrich, P. Kemeter, H. Salzer and G. Breitenecker. Acta Endocrinol. 78(1975)332.

LUTEINIZATION DISCUSSION - J. MARSH - CHAIRPERSON

B.F. RICE: These are both very exciting observations, but I was just absolutely startled to see that the major secretory product of the developing follicle was androstenedione. I was also amazed to see that prolactin had such profound effects on the corpus luteum and on developing cells, when we and I'm sure others, have been unable to demonstrate any prolactin receptors in the human ovary.

I wonder if the two speakers would have some comments on those sort of discordant observations.

K. McNATTY: There is no evidence at the moment for the existence of prolactin receptors on human granulosa cells. Dr. R. Rollands and Osterlitz at Hershey have shown that porcine granulosa cells contain prolactin receptors.

L. KEYES: There are receptors for prolactin in the rat corpus luteum, but I am not sure whether receptors for prolactin have been demonstrated in rat granulosa cells.

B.F. RICE: My question specifically pertains to the human.

K. McNATTY: I might point out that in the human, there is a decline in follicular prolactin levels as the follicle develops despite the fact that there is no dramatic change in prolactin levels in plasma of women during the menstrual cycle. In the developing follicle, the prolactin concentrations within the follicle are lowest at the time of ovulation. These changing levels of prolactin also occur in sheep. It's interesting that this happens in these two species, because in the sheep there are certainly elevated levels of prolactin in plasma around the time of ovulation. Now, the fact that these changes do occur, and the fact that they do occur at the time when granulosa cells are undergoing luteinization and the fact that human granulosa cells do depend on some prolactin for steroidogenesis would implicate the possibility that there are receptors for prolactin in these cells.

B.F. RICE: What I asked is what is the evidence for prolactin doing something in the human follicle in terms of receptors and related cellular events?

K. McNATTY: Prolactin in some way may be altering FSH levels and estrogen levels within the follicle, and this may profoundly influence the cell number which is a critical determinant of corpus luteum function.

M. RAJ: We initially demonstrated that LH is luteotropic in the pregnant rat (Raj and Moudgal, Endocrinology 86, 874, 1970). In these studies antiserum to LH given at any time between days 8 and

11 of pregnancy caused abortion in the rat. Since this finding we have been intrigued as to the role of LH and prolactin in luteal maintenance. We have examined the effects of maintaining the feto-placental unit by dydrogesterone (a synthetic progestin) on luteal function. Rats were given LH antiserum or LH antiserum + dydrogesterone from day 8 through day 14 and autopsied. LH antiserum treatment led to abortion and lowering of ovarian weight. Luteal weights, progesterone content as well as concentration and plasma progesterone levels were lowered drastically with this treatment. All these effects of the LH antiserum were fully reversed when the feto-placental unit was maintained by the synthetic progestin. (refer to chapter in this volume). The decline in luteal progesterone production in the antiserum-treated group was associated with a decline in luteal estrogen content and concentration. Furthermore, the maintenance of the feto-placental unit and luteal progesterone production was associated with elevated intraluteal estradiol levels. These data seem to fit well with the elegant demonstration by Dr. Keyes on the role of estrogen in the luteotropic process.

R. KIRCHICK: Do estradiol implants in rabbits extend pseudopregnancy, and if not, why do you suppose the serum progesterone levels remain high until around day 17?

L. KEYES: The estradiol implant that we used in the rabbit does not extend pseudopregnancy, and this was puzzling until Josephine Miller showed that if you remove the uterus from these rabbits progesterone secretion is maintained indefinitely. It seems that there is something produced by the uterus which acts to kill the corpus luteum at the end of pseudopregnancy, and that withdrawal of estradiol from the circulation is not required for a normal course of luteal regression.

R. KIRCHICK: I was wondering why the serum progesterone is still elevated around Day 17 in rabbits with the implant replaced.

L. KEYES: The elevation of progesterone that we see late in pseudopregnancy in the rabbits with estradiol first withdrawn and then replaced is unexplained. But it could be that following the withdrawal of the estradiol implant, the ensuing decline in progesterone and estradiol causes an interruption of the luteolytic mechanism that normally exists in the uterus. When the estradiol implant is replaced the luteolytic factor is not present in high enough concentrations to prevent the restoration of progesterone secretion.

R. KIRCHICK: Did you use dihydrotestosterone in the rat experiments?

L. KEYES: Not in this study. Dr. Gibori has used it in some of her studies and dihydrotestosterone does not prevent LH antiserum-induced luteolysis.

R. KIRCHICK: Thank you.

NASS: As reported in our manuscript, in this book, we are interested in the short luteal phase, a recurrent event in the peripubertal rhesus monkey. Illustrated in the first slide (Fig. 2 of our manuscript) are the gonadotropin and steroid concentrations in the serum of peripubertal monkeys during both short and normal length luteal phases. Of particular interest to this discussion are the depressed FSH concentrations in the short luteal phase group on Days -7 and -8, which was followed by depressed estradiol concentrations on Days -4 through -2. One interpretation of these results is that the low FSH concentrations retarded the development of the preovulatory follicle, resulting in a reduced estradiol secretion by this follicle prior to ovulation. Dr. McNatty suggested in his presentation that without complete follicular maturation the resultant corpus luteum may not respond to LH with normal progesterone secretion. In these monkeys demonstrating short luteal phases, progesterone secretion was submaximal and showed a premature decline 5-7 days after ovulation.

The next two slides (Fig. 3, of our manuscript) show representative C.L. from short and normal luteal phases. The normal C.L. has a distinct, vascularized papilla and extensive subsurface luteal tissue. In contrast, the C.L. associated with a confirmed short luteal phase had a papilla with very little vascularity or subsurface luteal tissue.

Because of these hormonal and morphological characteristics, we propose that the short luteal phase may be a useful model for the investigation of follicular determinants of corpus luteum function. In addition, the relationship between nonsteroidal regulators of follicular growth and gonadotropin secretion may be important to an understanding of the short luteal phase.

T. MILLS: As with Dr. Keyes, studies in our laboratory have demonstrated estradiol binding activity in the corpus luteum of the rabbit during the early stages of pseudopregnancy. The results of these studies show that cytoplasmic receptor levels are significantly greater on Day 4 after injection than on Day 2 after injection of hCG to induce ovulation. The increased levels continue on Day 6. Furthermore, the receptor is shown to translocate into the nucleus within 30 minutes after injection of 100μg of estradiol. Nuclear receptors remain elevated at 60 minutes after estradiol injection. Appreciable receptor translocation can be demonstrated following estrogen administration on

Days 2, 4 and 6 of pseudopregnancy. The results of these studies confirm the previous report from this laboratory (Endocrinology 101:1744, 1977) that the cytoplasmic receptor is elevated as early as Day 4 of pseudopregnancy and support the conclusion that the rabbit corpus luteum contains an authentic receptor for estradiol which translocates to the nucleus in response to estradiol administration.

# LH RECEPTORS IN THE OVINE CORPUS LUTEUM DURING THE ESTROUS CYCLE AND EARLY PREGNANCY

G. D. Niswender and M. A. Diekman

Dept. Physiology & Biophysics, Colorado State University

Fort Collins, Colorado 80523

Luteinizing hormone (LH) appears to be the endocrine factor responsible for the stimulation of progesterone secretion in ovine luteal tissue (1,2,3). The action of LH is mediated, presumably, via binding to a hormone-specific membrane associated receptor (4) and a subsequent increase in synthesis of cAMP and steroidogenesis (5). However, the precise details of the mechanisms whereby LH regulates progesterone secretion are not known. Current concepts suggest that the primary factor regulating the quantity of progesterone secreted from the corpus luteum is the number of luteal receptors for LH which are occupied by this gonadotropic hormone. In sheep, serum concentrations of LH are at their lowest levels during the mid-luteal phase of the estrous cycle when progesterone secretion is maximal (6,7). Therefore, the total number of luteal receptors for LH or the affinity of the receptor for LH must increase during the mid-luteal phase of the cycle if the number of LH receptors occupied by endogenous hormone is to increase. Similarly, since progesterone secretion remains high during early pregnancy, even though serum levels of LH remain low, it seems likely that the total number of receptors occupied by LH will remain high during this reproductive state.

The present experiments were designed to determine the total number of luteal receptors for LH and the degree to which these receptors are occupied by endogenous hormone throughout the estrous cycle and early pregnancy in ewes. The affinity of the receptor for LH was also evaluated.

## MATERIALS AND METHODS

Western range ewes were checked visually for estrus twice daily using vasectomized rams. The first day a ewe stood to be mounted by the ram was designated as day 0 of the estrous cycle. Corpora lutea were collected surgically from 6 ewes each on days 4, 6, 10, 12, 14 and 16 of the cycle and from 18 ewes (6 pools of 3 corpora lutea each) on day 2. Jugular blood samples were collected just prior to collection of the luteal tissue, stored at 4 c for 16 hr, serum was harvested after centrifugation at 2000 x g for 30 min and progesterone concentrations were determined by radioimmunoassay (8).

The number of unoccupied receptors for LH was determined using $^{125}$I-hCG and Scatchard analysis (9). The progesterone concentration was also determined for each corpus luteum. The number of receptors occupied by endogenous LH was determined by elution of the hormone with formic acid (pH 2.3) and quantitation of the eluted LH (9). For calculation of the number of occupied receptors it was assumed that one LH molecule had been bound to one receptor.

Corpora lutea were also collected from six pregnant ewes each on days 12, 16 and 20. Serum and luteal concentrations of progesterone and the numbers of unoccupied and occupied LH receptors were determined as described above. One-way analysis of variance and Tukey's ω-procedure were used to test for statistical differences due to stage of the cycle or pregnancy (10).

## RESULTS

The data obtained from luteal tissues and sera collected during the estrous cycle are summarized in Table 1. Luteal weight and progesterone concentration and serum levels of progesterone increased from days 2 to 6, were maximal on days 10, 12 and 14 and then declined dramatically on day 16. A similar pattern was noted for both occupied and unoccupied luteal receptors for LH. The number of both unoccupied and occupied receptors increased on days 2, 4 and 6, were maximal on days 10, 12 and 14 and then dramatically declined on day 16. There were no differences in the affinity of the receptor for LH at any of the times studied. Luteal weight, serum and luteal progesterone concentrations, and the number of unoccupied and occupied luteal receptors for LH were all correlated significantly ($P < 0.01$) with each other throughout the estrous cycle.

No differences in serum concentrations of progesterone, luteal weight, or occupied and unoccupied luteal LH receptors were observed between days 12, 16 and 20 of pregnancy. Statistical comparisons were also made between luteal weight, serum and luteal progesterone concentrations and unoccupied and occupied luteal receptors for LH

TABLE 1

CORPORA LUTEA WEIGHT, SERUM AND LUTEAL PROGESTERONE, UNOCCUPIED AND OCCUPIED LUTEAL RECEPTORS FOR LH DURING THE ESTROUS CYCLE AND EARLY PREGNANCY

| Day of Cycle | CL Weight (mg) | Serum Progesterone (ng/ml) | Luteal Progesterone (μg/ml) | Unoccupied LH Receptors (x 10-11 M/CL)** | Occupied LH Receptors ($x 10^{-13}$ M/CL)** |
|---|---|---|---|---|---|
| 2 | 100 ± 17*[a] | .35 ± .11[a] | .72 ± .22[a] | .03 ± .01[a] | .11 ± .02[a] |
| 4 | 246 ± 28[b] | .90 ± .13[a] | 1.54 ± .38[a,b] | .05 ± .01[a] | .14 ± .02[a] |
| 6 | 344 ± 21[b,c] | 2.27 ± .28[b] | 2.69 ± .22[b] | .23 ± .02[b] | .16 ± .02[a] |
| 10 | 546 ± 41[e] | 3.04 ± .20[c] | 13.63 ± 1.56[c] | 1.12 ± .15[c] | .68 ± .05[b] |
| 12 | 512 ± 40[d,e] | 3.28 ± .28[c] | 11.45 ± 1.62[c] | .67 ± .17[c] | .63 ± .01[b] |
| 14 | 609 ± 36[e] | 3.38 ± .35[c] | 15.81 ± 2.14[c] | 1.22 ± .30[c] | .66 ± .05[b] |
| 16 | 420 ± 20[c,d] | .49 ± .17[a] | 3.31 ± 1.13[a,b] | .36 ± .12[b] | .16 ± .04[a] |
| Day of Pregnancy | | | | | |
| 12 | 622 ± 38*[e] | 3.07 ± .51[c] | 20.32 ± 1.25[d] | 1.07 ± .37[c] | .64 ± .04[b] |
| 16 | 490 ± 24[d,e] | 2.98 ± .52[c] | 14.55 ± 1.65[c] | 1.32 ± .16[c] | .66 ± .05[b] |
| 20 | 561 ± 48[e] | 3.75 ± .52[c] | 18.68 ± 1.48[c,d] | 1.33 ± .14[c] | .67 ± .03[b] |

** M equals moles

* Mean ± standard error of the mean (n=6)

[a,b,c,d,e] Means within a column which have a different letter superscript are significantly different ($P < 0.05$)

on day 12 of the estrous cycle and day 12 of pregnancy. There was no difference between any of the parameters tested except for luteal progesterone concentration ($P < 0.01$). There was no difference in the affinity of the receptor for LH when data from cycling and pregnant ewes were compared.

## DISCUSSION

Serum levels of LH are lowest during the luteal phase of the estrous cycle (6,7) and the affinity of the LH-receptor interaction did not change throughout the estrous cycle. Therefore, the increased number of receptors for LH appears to be the major factor regulating luteal progesterone secretion. Since the total quantity of luteal receptors for LH increased without a change in the affinity or serum levels of LH the law of mass action predicts that the number of receptors occupied by LH would increase during the mid-luteal phase of the cycle. The increased number of occupied receptors was highly correlated with the concentration of progesterone in serum suggesting a cause and effect relationship. The number of unoccupied luteal LH receptors per corpus luteum was 40-fold higher on day 10 than on day 2 of the estrous cycle. There was an increase of $3.3 \times 10^{10}$ molecules of LH bound to receptor per corpus luteum on day 10 even though the increase in occupied receptors per corpus luteum was only 6-fold. On day 10 of the cycle 0.6% of the total luteal receptors were occupied by LH. In Leydig cells only 1% of the 6000 receptor sites per cell need to be occupied to elicit maximum testosterone secretion *in vitro* (11).

The decline in luteal receptors for LH and progesterone secretion during regression of the corpus luteum does not imply a cause and effect relationship. It can not be elucidated from these data whether the decrease in luteal receptors precedes the decline in luteal function or whether the decreased function precedes a decline in receptor number.

Expressing the number of receptors per corpus luteum reflects relative changes in receptor number per granulosa-lutein cell since granulosa cells do not undergo secondary mitoses following ovulation (12). However, fibroblasts and endothelial cells continue to undergo mitoses after day 3 of the cycle (12) and therefore, the amount of DNA and/or protein being contributed by cell types other than the granulosa-lutein cells would not remain constant throughout the cycle. Thus, expressing the number of receptors per μg DNA or per μg membrane protein does not seem appropriate.

Niswender *et al.* (2) have demonstrated extreme heterogeneity of function within the corpus luteum *in vivo* including regional differences in the binding of $^{125}$I-hCG, progesterone concentrations and blood flow. Thus, if only a small fraction of the steroid-secreting cells in the corpus luteum are available to bind LH at any given time, in vivo, the number of occupied receptors per active cell could be drastically underestimated by averaging the quantity of LH bound over all luteal cells. The present study indicates that only 0.6% of the LH receptors are occupied by endogenous hormone during the mid-luteal phase of the cycle. However, if only a limited number of cells are actively secreting progesterone at a given time, the number of occupied receptors on these "active" cells could be much greater than 0.6%. Further studies are necessary to obtain meaningful data regarding the number of active luteal cells at a given point in time and the extent to which their LH receptors are occupied by LH.

There was no difference in the number of luteal LH receptors or affinity of the LH-receptor interaction between day 12 of the estrous cycle and days 12, 16 and 20 of pregnancy. Therefore, an increase in the number of luteal LH receptors or an increased affinity of LH binding does not appear to be the signal for maintaining the corpus luteum of pregnancy. Rather, data from the present study suggests that regression of the corpus luteum at the end of the cycle is an active lytic process and that pregnancy may simply prevent these luteolytic events or that another factor(s) (embryonic luteotropin?) is involved in maintenance of the corpus luteum of pregnancy.

## REFERENCES


(1) C.C. Kaltenbach, B. Cook, G.D. Niswender and A.V. Nalbandov. Endocrinology 82(1967)1407.

(2) G.D. Niswender, T.J. Reimers, M.A. Diekman and T.M. Nett. Biol. Reprod. 14(1976)64.

(3) K.R. Simmons, J.L. Caffrey, J.L. Phillips, J.H. Abel, Jr. and G.D. Niswender. Proc. Soc. Exp. Biol. Med. 152(1976)366.

(4) R.J. Ryan and C.Y. Lee. Biol. Reprod. 14(1976)16.

(5) A.W. Jordan, III, J.L. Caffrey and G.D. Niswender. Endocrinology (in press).

(6) G.D. Niswender, J.F. Roche, D.L. Foster and A.R. Midgley, Jr. Proc. Soc. Exp. Biol. Med. 129(1968)901.

(7) G.D. Niswender. Endocrinology 94(1974)612.

(8) G.D. Niswender. Steroids 22(1973)413.

(9) M.A. Diekman, P. O'Callaghan, T.M. Nett and G.D. Niswender. Biol. Reprod. (in press).

(10) R.G.D. Steel and J.H. Torrie. Principles and Procedures of Statistics (McGraw Hill Book Co., New York, 1960).

(11) C. Mendelson, M. Dufau and K. Catt, J. Biol. Chem. 250(1975) 8818.

(12) M.C. McClellan, M.A. Diekman, J.H. Abel, Jr. and G.D. Niswender. Cell Tiss. Res. 164(1975)291.

# GONADOTROPIN ACTION IN ISOLATED OVARIAN LUTEAL CELLS: THE INTERMEDIATE ROLE OF ADENOSINE 3', 5' CYCLIC MONOPHOSPHATE IN HORMONAL STIMULATION OF PROGESTERONE SYNTHESIS

G. B. Sala, M. L. Dufau and K. J. Catt

Endocrinology and Reproduction Research Branch, NICHD,

National Institutes of Health, Bethesda, Md. 20014

Recent studies in several endocrine target cells have revealed moderate to marked dissociations between steroidogenesis and cAMP formation during hormone action (1-3). In the corpus luteum, low concentrations of LH or hCG have been found to evoke full steroidogenic responses with little or no detectable change in cAMP formation (4). However, a rise in protein kinase activity has been observed in corpus luteum slices incubated with gonadotropin concentrations which produce a dose-related stimulation of progesterone synthesis, albeit in the absence of a discernible rise in cAMP (5). Such findings suggest that an increase in cAMP synthesis or turnover is stimulated by even the lowest doses of hCG that evoke a steroid response, and that measurement of this change should be possible under appropriate experimental conditions. A similar dissociation between steroidogenesis and cAMP during hormone action in the testis and adrenal has been recently resolved by optimization of incubation conditions and detailed analysis of cAMP levels in hormone-stimulated Leydig cells and and adrenal fasciculata cells (6,7). To evaluate the role of cAMP during hormone action in ovarian luteal tissue, we have analyzed the relationships between cAMP production and binding to protein kinase receptors, and stimulation of progesterone synthesis in isolated luteal cells.

For these studies, collagenase-dispersed luteal cells, prepared from ovaries of PMSG/hCG-treated immature rats, were preincubated for 2 hrs at 37C prior to stimulation with gonadotropin *in vitro*. Following incubation with increasing concentrations of hCG, the cells and medium were separated for assay of cAMP and progesterone. Aliquots of medium were assayed for extracellular cAMP (6,8) and progesterone (9) by radioimmunoassay. Cells were disrupted by sonication, and aliquots were assayed for intracellular cAMP and [$^3$H]cAMP binding capacity (6). Sonicates incubated with [$^3$H]cAMP were filtered through cellulose acetate membranes, which were analyzed for bound radioactivity as a measure of the number of free receptors on the regulatory subunit of protein kinase. Another aliquot was filtered directly and the membranes heated to extract receptor-bound cyclic AMP prior to radioimmunoassay (6). This fraction represented the endogenous cyclic AMP receptor protein corresponding to the occupied sites of the regulatory subunit.

Binding of [$^3$H]cAMP by the regulatory subunit of protein kinase.

In contrast to the results obtained in adrenal and Leydig cells (6,7), there was slower association (ka = $8.8 \times 10^5 M^{-1} min^{-1}$) and more rapid dissociation of cAMP from the regulatory subunit in luteal cell extracts. The dissociation rate constant (kd) was $1.5 \pm 0.16 \times 10^{-2} min^{-1}$ at 6C, and was slightly lower at 0°C ($1.1 \times 10^{-2} min^{-1}$). This property of the luteal cell protein kinase precluded the accurate measurement of the free cAMP receptors during binding studies with [$^3$H]cAMP due to exchange with the radioactive ligand during equilibration. Because of this, occupied receptors were measured by quantitation of cAMP bound to receptor protein, and not by the alternative method with labeled cAMP to measure 'empty' sites. The affinity constant (Ka) for cAMP binding was calculated by Scatchard plots from equilibrium binding studies at 6C to be $1.7 \pm 0.32 \times 10^8 M^{-1}$. The total binding capacity for cAMP derived from Scatchard and saturation plots ranged from 1 to 4 pmol per $10^6$ luteal cells.

Kinetic analysis of ovarian cell activation by gonadotropin.

The kinetics of cAMP production during incubation of luteal cells with the trophic hormone were determined at two hCG concentrations, a low dose ($10^{-12}$M) that was less than the $ED_{50}$, and a maximal steroidogenic dose ($2 \times 10^{-11}$M). Receptor-bound cAMP was increased significantly over the control incubation by the low dose ($10^{-12}$M hCG) at 15 to 30 minutes. With the maximal steroidogenic dose of hCG ($10^{-11}$M), significant increases were detected at 5 min and further increases were measured until 120 min.

The intracellular cAMP levels were significantly increased by the lower hCG dose. With the higher dose, significant increases were detected at 5 min and a 3-fold rise was observed at 120 min (Fig. 1, Left). Extracellular cAMP (Fig. 1, Right) was increased significantly by $10^{-12}$M hCG at 30 min over the control incubation. With the maximal steroidogenic dose ($10^{-11}$M), significant increases were detected at 5 min and further increases were

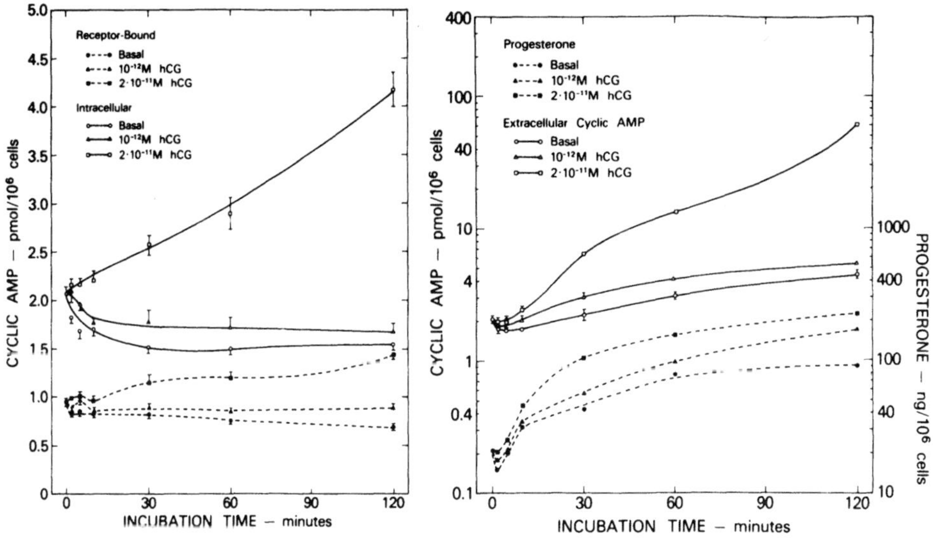


Figure 1. Changes in intracellular and receptor-bound cAMP (Left); and of extracellular cAMP and progesterone production (Right) in luteal cells stimulated with $10^{-12}$ and $2 \times 10^{-11}$M hCG for 2 to 180 min. Each point represents the mean $\pm$ SE of triplicate incubations.

measured until 120 min. Steroid values were significantly increased after 2 min by $10^{-11}$M hCG (Fig. 1, Right) and continue to rise to 6 µg at 120 min, a 30-fold increase over control levels. The low dose of hCG ($10^{-12}$M) stimulated progesterone production after 5 minutes, reaching a 60% increase at 30 min.

To examine the role of phosphodiesterase during acute stimulation by trophic hormone, the rate of cAMP formation and progesterone production was investigated during activation with a low dose of hCG ($10^{-12}$M) in the presence and absence of the phosphodiesterase inhibitor, 1-methyl-3-isobutyl xanthine (MIX). In control cells, extracellular cAMP values increased with time in the absence and presence of MIX. In the absence of MIX, a doubling of cAMP was observed, while in the presence of MIX the control values increased 7-fold at 120 min. In both cases, the low concentration of hCG significantly stimulated extracellular cAMP levels throughout the incubation period when analyzed by pooled variance and paired t test. Examination of intracellular cAMP levels showed no significant differences between control or stimulated values measured in the presence or absence of MIX between 30 and 120 min. However, at earlier times the presence of MIX had a small but significant effect upon the stimulated groups ($p < 0.01$). During measurement of cAMP bound to its receptor protein, there was no difference in values for control and hCG-stimulated incubations in the presence or absence of MIX at times from 30 to 120 minutes. However, at earlier times (5 to 20 min), there were significant differences between control and hCG-stimulated values in the presence and absence of MIX ($p < 0.01$).

In contrast to the small differences in cAMP production, there was a marked difference in progesterone production by control and hCG-stimulated groups in the presence and absence of the phosphodiesterase inhibitor. In the absence of MIX, control values increased by 100% during 120 min incubation. During stimulation with hCG, the lag time for progesterone stimulation was 20 min and gradual increases were observed up to 4-fold stimulation over control values at 120 min. In the presence of MIX, control values were significantly increased over those observed in the absence of MIX. Stimulation of progesterone production by hCG during incubation in the presence of MIX was observed at the earliest time examined (5 min), in contrast to the 20 min time lag during stimulation in the absence of MIX. At all times during hormonal stimulation in the presence of MIX, progesterone production was higher than in the absence of the inhibitor.

A comparative study of progesterone levels was carried out over a wide range of hCG concentrations (0.2 to 200 pM) in the absence and presence of 0.2 mM or 0.5 mM MIX. As before, marked differences were found in basal and hCG-stimulated levels of progesterone in the presence or absence of MIX. Maximal steroid production was the same in the absence and presence of 0.2 mM MIX and was significantly decreased in the presence of 0.5 mM MIX. The inhibitory effect of high concentrations of MIX upon gonadotropin-stimulated steroidogenesis in the luteal cell is of interest in relation to earlier observations that large doses of theophylline or MIX can inhibit ACTH and gonadotropin-induced steroidogenesis in adrenal fasciculata cells and Leydig cells, respectively (3,10). This action of high concentrations of phosphodiesterase inhibitors has been attributed to inhibition of protein synthesis by the xanthine derivatives (10). During luteal cell incubations in the presence or absence of 0.2 mM MIX, dose-related increases were observed in intracellular receptor-bound and extracellular cAMP when measured over the hCG concentration range that elicited a full progesterone response ($10^{-13}$ to $10^{-11}$ M). Further increases in hormone concentration did not produce further rises in progesterone production despite additional occupancy of cAMP receptor protein (Fig. 2).

In contrast with studies on the Leydig cell during hormone action (6), a larger proportion of cAMP receptors in the luteal cell was occupied by endogenous cyclic AMP during dose-related increases in steroidogenesis. Receptor-bound cAMP increased from control levels of 480 to 1,500 femtomoles per $10^6$ cells during incubation with hCG concentrations that produced maximal steroidogenic responses (20 pM). Further increases in receptor-bound

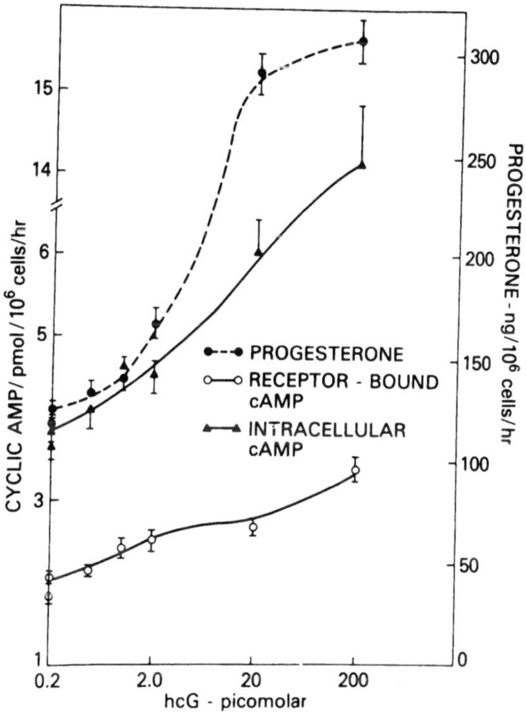


Figure 2. Comparison of intracellular and receptor bound cAMP with progesterone production in luteal cells incubated with hCG (0-200 pM) in the presence of 0.2 mM MTX.

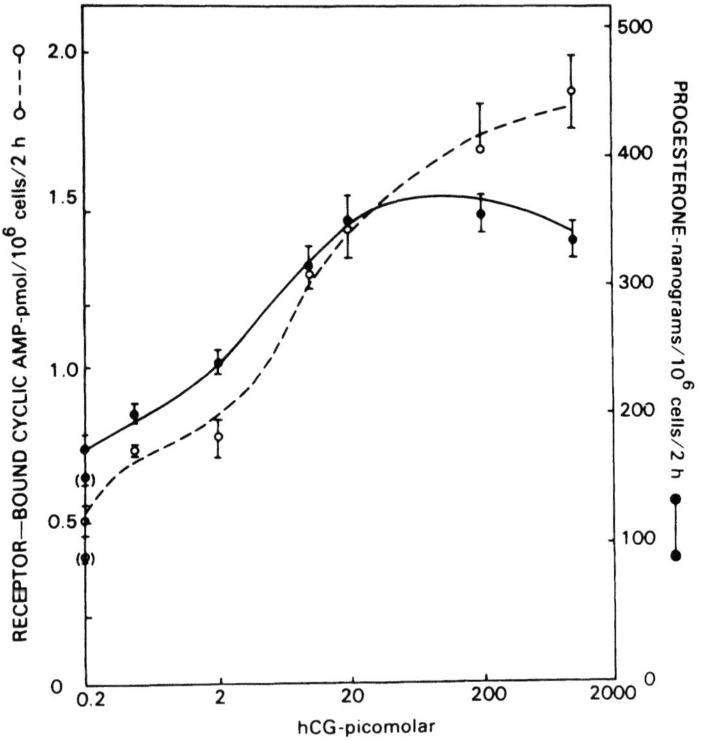


Figure 3. Comparison of receptor-bound cyclic AMP with progesterone production in luteal cells incubated with hCG (0-2000 pM) in the presence of 0.2 mM MIX.

cAMP were observed with higher hCG doses, and complete occupancy of nearly 2000 femtomoles of cAMP per $10^6$ cells occurred at $10^{-9}$M hCG. These results have indicated that whereas in the testis a 25% increase in receptor occupancy would produce the maximal steroid response (6), in the luteal cell a major proportion of receptors (about 75%) are occupied at hCG doses that produce a maximal progesterone response (Fig. 3).

The present observations have provided direct evidence that cAMP formation is increased during all phases of hCG action in dispersed rat luteal cells, in the absence and presence of a phosphodiesterase inhibitor. It is also evident that binding to the regulatory subunit of protein kinase occurs in a fashion commensurate with the role of the adenylate cyclase-protein kinase enzyme sequence in the mediation of gonadotropin action in the luteal cells.

References

1. Beall, R.J. and G. Sayers. Arch. Biochem. Biophys. 148 (1972)70.

2. Catt, K.J. and M.L. Dufau. Nature New Biology 244((973) 219.

3. Mendelson, C., M.L. Dufau and K.J. Catt. J. Biol. Chem. 250(1975)8818.

4. Conti, M., J. Harwood, M.L. Dufau and K.J. Catt. J. Biol. Chem. 252(1977)8869.

5. Ling, W.Y. and J.M. Marsh. Endocrinology 100(1977)1571.

6. Dufau, M.L., T. Tsuruhara, K.A. Horner, E. Podesta and K.J. Catt. Proc. Natl. Acad. Sci. USA /4(1977)3419.

7. Sala, G., M.L. Dufau and K.J. Catt. Clin. Res. 26(1978)312A.

8. Harper, J.F. and G. Brooker. J. Cyclic Nucleotide Res. 1(1975)207.

9. Abraham, G.E., R. Swerdloff, D. Tulchinsky and W. Odell. J. Clin. Invest. 32(1971)619.

10. Halkerston, I.D.K., M. Feinistein and O. Hechter. Proc. Soc. Exp. Biol. Med. 122(1966)896.

DISCUSSION - J. MARSH - CHAIRPERSON

R. KELLY: I would like to ask Dr. Niswender what amount of 30,000 G-pellet is required to measure the occupied receptors?

G.D. NISWENDER: We use the equivalent of 50-100 mg of wet weight of tissue.

R. Kelly: Do you pick up any differences in the free receptor concentration during periods of endogenous LH elevations?

G.D. NISWENDER: Yes, the greatest changes observed during the estrous cycle occurred in the concentration of unoccupied receptors. We also observed differences in the percentage of total receptors occupied by endogenous LH. Forty-eight hours after the preovulatory LH surge, the percentage of receptors occupied by endogenous hormone was about 8 times higher than during the mid-luteal phase of the cycle. This is not surprising since we have published other data which shows that in the living animal, the half-time of dissociation of hormone from the receptor in vivo is approximately 17 hours. So had we looked at a time closer to the preovulatory LH surge the percentage of receptors occupied by endogenous hormone would probably have been even higher than that.

F. LEDWITZ-RIGBY: Dr. Dufau, do you have any thoughts about a possible physiological role for all that extracellular cAMP?

M.L. DUFAU: Extracellular cyclic AMP rises in parallel with intracellular nucleotide in response to trophic hormone stimulation or non-hormonal stimuli such as choleragen. The mechanism of the extracellular increase is not known and could be due either to passive diffusion of the nucleotide or to an active secretory process. The levels of cyclic AMP accumulated in the extracellular compartment during maximal hormonal stimuli are quite substantial (about 160-fold over the intracellular levels) (M. L. Dufau, A. K. Horner, E. J. Podesta and K. J. Catt, Proc. Natl. Acad. Sci. U.S.A. 74, 3419 (1977)) and could influence other cells in the vicinity.

However, to date there is no evidence of cell to cell communication through extracellular cyclic AMP in mammalian tissues. However, there is an interesting recent study in which intracellular cyclic AMP was found to communicate hormonal stimuli (FSH or epinephrine) during co-culture of ovarian and muscle cells. In this case the flow of cyclic AMP between cells was shown to occur through gap junction (T. S. Lawrence, W. H. Beers and N. B. Gilula, Nature 272, 501 (1978).

I. ROTHCHILD: I was supposed to do the summary, and there is a piece of information missing from Dr. Dufau's paper. I thought it was my eyesight or my failing memory, but nobody in my neighborhood seems to know what animal this was done on.

M.L. DUFAU: I mentioned it was pseudopregnant rats.

B.F. RICE: Gordon, if you take the human corpus luteum and study hCG-LH receptors, the curves that you showed for the sheep are almost identical in terms of the amount of receptor sites and unbound sites that are present with one exception. When you study corpora lutea of pregnancy, you cannot demonstrate the binding of any labeled gonadotropin and the interpretation that we've always put on that is that all the receptor sites were bound. I presume the sheep doesn't have a placental luteotropin as such, but would you care to speculate about that?

G.D. NISWENDER: There have been many people, including our laboratory, who have searched exhaustively for the placental luteotropin in sheep, and other domestic animals. We have not been able to identify a placental "hCG-like" secretion. However, this does bring up an important point regarding prolactin receptors. We have tried to demonstrate prolactin receptors in the ovine corpus luteum but we were not successful. However, recently Patti Diekman in our laboratory has been treating ewes with ergocryptine for at least 5 days to reduce endogenous prolactin levels and she observed an increase in "specific binding" from 2% all the way up to 5-6%. Therefore, we are not certain as to whether or not there is really a prolactin receptor. Our interpretation of these data, at the present time, is that perhaps the reason we can't demonstrate prolactin receptors in ovine luteal tissue is because they are saturated by high levels of endogenous hormone. It is going to require considerable more effort to resolve this problem.

Unfortunately, we cannot elute prolactin from the receptor like we can elute hCG. We have tried high alkalinity and a number of other approaches but all have failed.

J. SAEZ: I have a comment to Dr. Dufau's paper. In the adrenal, it also has been reported that there is a discrepancy between the dose of ACTH able to increase steroidogenesis, and to increase cyclic AMP. Recently, we have used isolated human adrenal cells and we have found a good correlation between the dose of ACTH able to increase steroidogenesis and activation of protein kinase. For instance, the lowest dose able to stimulate these two parameters was $10^{-11}$ M, while to increase cAMP we required $10^{-9}$ M ACTH. I think you looked carefully at the activation of protein kinase. We found the same result that you have in the testes and in the adrenals.

In response to Dr. Ledwitz-Rigby's question, I think the big increase of cAMP, at least in the adrenals, is related in some way to protein and DNA synthesis.

M.L. DUFAU: We have no experience with the human adrenal, but using the rat adrenal we have demonstrated increases in cyclic AMP bound to receptor protein during stimulation of adrenal fasciculata cells with ACTH at all concentrations of the hormone that produce a dose-related response in corticosterone production ($10^{-12}$ to $10^{-9}$ M ACTH). Also, in the same studies we have shown during hormone action decreases in free cyclic AMP receptor sites, during binding studies with $^{3}$H cyclic AMP under non-exchange conditions. This change in free binding sites parallels the increase in occupied sites by endogenous cyclic AMP produced during hormone stimulation. (G. Sala, M. L. Dufau and K. J. Catt, Clin. Res. 26, 312A (1978).

T. LEE: I am pleased to find in Dr. Dufau's presentation, that the production of cyclic-AMP is correlated with binding and the concentration of hCG. I have some data to confirm that the binding of hCG and the hCG sensitive adenyl cyclase activity is the same for both...parameters, if you carry out both assays under exactly the same conditions. The $K_{ms}$ will not be the same, however, if you examine hCG binding and hCG sensitive adenyl cyclase under different conditions.

I also have a question for Dr. Dufau: as I remember, the cAMP production is not always correlated to hCG binding in her testis system, unlike the situation in the ovary. What do you think the explanation for this is?

M. DUFAU: My presentation today was not concerned with the correlation of hCG binding with cyclic AMP production, but with the role of cyclic AMP in hormone-induced steroidogenesis. However in response to your question, we have recently described the correlation between $^{125}$I-hCG binding to luteal cell receptors and cyclic AMP formation during hormone action (M. Conti, J. P. Harwood, M. L. Dufau and K. J. Catt, J. Biol. Chem. 252, 8869 (1977)). This relationship in the luteal cell was similar to our earlier observation on the Leydig cell, where stimulation of LH/hCG receptors with trophic hormone caused simultaneous increase in cyclic AMP and testosterone (K. J. Catt and M. L. Dufau, Advan. Exptl. Med. Biol. 36, 319 (1973); Nature New Biol. 242, 246 (1973)).

M. DIMINO: My question or comment is primarily directed toward Dr. McNatty. With regards to follicular predeterminants of luteinization, we've taken a somewhat different approach, and have been looking at changes in mitochondrial activity in follicular versus luteal tissue. We have shown that mitochondria from luteal tissue in addition to synthesizing pregnenolone, can also synthesize progesterone, and we feel that this ability to synthesize progesterone by luteal mitochondria occurs very early in luteinization of follicular tissue. We feel, too, that possibly, the mitochondrial steroidogenic activity may influence what steroids are produced by ovarian tissues. For no reason other than having two sites of progesterone synthesis occurring within the cell, you may be able to produce more progesterone. That may be too simplistic, but another concept is that progesterone synthesized by mitochondria and by endoplasmic reticulum (ER) are handled differently. The progesterone synthesized by ER would be readily available as the substrate for additional steroidogenesis. Whereas, that progesterone produced by mitochondria may be more easily secreted from the cell. Dr. McNatty, have you seen any other earlier changes at the cellular level which indicate that follicular cells are converting to luteal tissue?

K. McNATTY: There's very definite morphological evidence that granulosa cells luteinize or are undergoing the process of luteinization a long time before rupture occurs, which is particularly true in the human. In fact, we often make the mistake of calling late preovulatory follicles luteal tissue. And it's not until you look at these structures more carefully, that you can say that they actually are follicles.

J. HOLT: I would like to address a question to Dr. Keyes. I have been studying the uterus and cervix, and progesterone antagonizes the action of estrogen in these tissues. Since the corpus luteum is estrogen responsive, I wonder if you have any evidence that the luteal tissue also responds to progestin the way other tissues do.

L. KEYES: To my knowledge, progesterone receptors have not been found in the corpus luteum. Because of the high concentrations of progesterone in luteal tissue, it might be difficult to detect a cytoplasmic progestin receptor. I am not aware of any evidence at this point that the corpus luteum responds to progesterone.

M. SANYAL: This question is related to Dr. McNatty's paper. It is very difficult to identify or assess the quality of the follicle in the human, because, the follicle will often contain oocytes at different stages of atresia. This may explain why there is a high degree of variability of follicular steroidogenesis. Perhaps the data you have presented should be re-evaluated in terms of the quality of the ovum: the presence of the ovum, and whether it is a good ovum, which has the potential to mature (in terms of nuclear maturation) in vitro. Otherwise, the presentation was excellent.

K. McNATTY: We, in fact, assessed the stage of the ovum in our studies. I haven't discussed it, but we recovered the egg from all of the follicles, and looked at them.

M. SANYAL: Do you think they are all normal?

K. McNATTY: No, they were not all normal healthy oocytes. We found a very good correlation between viability and the ability of the oocyte to resume nuclear meiosis, in culture. Also, the health of the oocyte could be correlated with the androgen/estrogen levels in follicles. In fact, one has to do this, because it is easily overlooked when presenting data like this, since 99.9% of all the human follicles obtained at surgery are, in fact, atretic.

M. SANYAL: Yes, and also it depends upon the age of the patient at ovariectomy.

K. McNATTY: Yes, for sure!

A. SHAIKH: I have a question for Dr. Keyes. Nine years ago, we published a paper showing that luteal function can be prolonged in the rat by giving daily injections of LH. Subsequently, we published a paper to show that this luteotropic effect of LH was mediated through estrogen. We gave very small doses of estrogen to pseudopregnant rats, and we could prolong pseudopregnancy more than 22 days. In your studies, you have given rats testosterone to show the luteotropic effect of estrogen which is produced within the corpus luteum by aromatization of testosterone. Have you injected estrogen and if so, in what doses?

L. KEYES: Dr. Gibori and Dr. Rothchild have injected rats with estradiol alone and the effective dose is about 100 micrograms per day. This amount of estradiol must be injected to maintain normal intraluteal concentrations of estradiol.

A. SHAIKH: That's interesting in the light that in the pseudopregnant rat, a dose of 0.2 microgram of estradiol is luteolytic, whereas, a daily injection of 0.05 microgram can prolong pseudopregnancy.

S. DAY: Several years ago, Dr. Keyes, you and Dr. Nalbandov showed that in the rabbit with an ovary irradiated to destroy follicles progesterone secretion could be maintained by the injection of 3 micrograms of estradiol daily. We have administered estradiol in two different doses (3 micrograms per day or 30 micrograms per day) to pseudopregnant rabbits between Days 8 and 12 of pseudopregnancy and then, subjected them to a 2-day withdrawal of estradiol. Serum progesterone and LH-stimulable adenylate cyclase activity were then measured on Day 14. We found that after cessation of treatment with 3 micrograms of estradiol there was no effect on serum progesterone, or on LH stimulable cyclase activity. However, cessation of treatment with 30 micrograms per day resulted in a precipitous decline in progesterone as well as loss of LH-stimulable cyclase. My question is, have you tried any other intervals of time for replacement of the estradiol implant?

L. KEYES: No, we have not made a systematic study of this; we do not know how long estradiol replacement can be delayed, before progesterone secretion can no longer be stimulated. Since the corpora lutea begin to decrease in weight by 48 hours after withdrawal of estradiol, this may be the approximate time of irreversible regression.

S. DAY: Thank you.

J. WEISZ: Two brief comments and one question relating to McNatty's presentation which I have enjoyed immensely; those are a beautiful set of correlations, providing very pretty integrated pictures of the development of the follicle. First of all, if one thinks of the follicle as one determinant of the kind of corpus luteum that it's going to make, I wonder if we are right in starting to look at the follicle at the time of menstruation. If you recall the data of Dr. Griff Ross (Recent Progr. Hormone Res. 25:1, 1970) in which the FSH levels are plotted spanning 2 cycles, FSH rise begins as the last corpus luteum begins to regress, i.e., seven days before the next menstruation. There are also data that in the monkey the final growth spurt of the follicle that's going to ovulate during the next cycle begins in the previous cycle also about the time when the last cycle's corpus luteum begins to regress. It would seem that the menstruation is a very artificial division as far as the ovary is concerned. Menstruation is essentially a uterine event, not an ovarian one. That's my first comment.

My second comment relates to Dr. Rice's remark about the follicle's ability to put androstenedione. In our own studies of the human female treated with contraceptives, we have shown that the major C19 steroid in the ovarian vein blood of these women was androstenedione (J. Clin. Endocrinol. Metab. 37:254, 1973). Since these women were anovulatory, their ovaries contained only follicles and interstitial tissue but not corpora lutea. I know Dr. Rice would like the interstitial tissue to be the source of the androstenedione but I accept the follicle, at present, as the more probable source.

And now my question; you have made a correlation between the number of cells in corpus luteum and the number of granulosa cells.

You sort of imply that the population of cells in the corpus luteum is made up of granulosa cells. What happened to the theca cells? Is the concept of the theca granulosa and theca lutein cells to be abandoned? Are the theca cells there simply to replace the few granulosa cells that couldn't hang on but got ejected with the ovum?

K. McNATTY: Since I have been at this conference, I've listened to your questions with great interest, and you have great insight, and these are three very good points. The first point you made, namely, that collection of follicles at menstruation is artificial is quite true. Menstruation is merely a useful time to choose from a methodological point of view because it is easy to recognize. If one examines endocrine parameters in follicular fluid, the smallest follicle in which one can do this is 4 millimeters in diameter and has a volume of less than 50 microliters. It is very difficult to measure gonadotropins and sex steroids in that volume, as you can appreciate. Namely the largest follicle present at the time of menstruation is only 4 mm. But, I would agree that by the time of menstruation, if the large follicle is about 4 millimeters, then it is well on its way toward ovulation.

In regard to the second point, we have looked at steroid production by interstitial tissue, thecal tissue and granulosa cells. All three produce androstenedione, but the theca has the greatest capacity.

With regard to the human corpus luteum, there is a substantial theca-cell component. In vitro, human thecal tissue interacts with granulosa-lutein cells to produce substantially more progesterone than either cell-type can produce alone.

## CHANGES IN OVARIAN MITOCHONDRIA:

## EARLY INDICATORS OF FOLLICULAR LUTEINIZATION

Michael J. Dimino, Edna A. Elfont, and S. Kim Berman

Departments of Research and Laboratory Medicine
Sinai Hospital of Detroit
Detroit, Michigan 48235

The rapid growth of ovarian follicles occurring after gonadotropic stimulation requires ATP which is obtained primarily from the oxidative reactions in mitochondria. In addition, the rate limiting step of ovarian steroidogenesis appears to be associated with the conversion of cholesterol to pregnenolone by the cholesterol side chain cleavage enzyme complex that is located exclusively within mitochondria. Thus, two of the most important functions of the ovary--follicular growth and steroidogenesis--are dependent upon products produced by mitochondria. For the last few years, we have been studying the role of mitochondria in ovarian physiology and we have been especially interested in the differentiation of ovarian mitochondria that occurs with luteinization of follicular tissue.

Previously, we observed that both cholesterol conversion activity and cytochrome oxidase activity (a marker for the terminal part of the respiratory chain) in ovarian mitochondria increased after gonadotropic treatment of immature rats (1). In mature rats, however, hypophysectomy resulted in the loss of ovarian mitochondrial cholesterol conversion activity but did not change the specific cytochrome oxidase activity (2). These results indicated that the ability of ovarian mitochondria to convert cholesterol to pregnenolone requires gonadotropic stimulation of the ovarian cells, while other mitochondrial components, e.g. the respiratory chain, can be maintained in part without gonadotropic stimulation. However, as evidenced by our oxidative phosphorylation studies (1, 2), it still is likely that nonsteroidogenic mitochondrial functions are influenced to some degree by gonadotropic stimulation of the ovary.

To minimize the problem of studying mitochondria from samples containing several tissue types, we now are looking at mitochondria prepared from isolated follicles and corpora lutea of porcine ovaries. We have found that the ability of mitochondria to convert cholesterol to pregnenolone increased as the follicle enlarged, and that there were no apparent differences between granulosa and thecal mitochondria (3). However, cholesterol conversion activity was significantly greater in luteal mitochondria than in all follicular mitochondrial preparations. Highly purified follicular mitochondria had little ability to convert pregnenolone to progesterone, whereas mitochondria from corpora lutea showed a high specific progesterone synthesis rate. Thus, it appears that ovarian mitochondria undergo differentiation as the follicle grows and becomes luteinized.

To investigate the differentiation of ovarian mitochondria further, we incubated intact large follicles (8-10 mm diameter) for 18 and 24 h in medium 199 D (4) containing 0.05 μg LH/ml. As compared to those from unincubated follicles, mitochondria from follicles incubated for 18 h showed a two-fold increase in cholesterol conversion activity while mitochondria from 24 h-incubated follicles showed a five-fold increase in activity (5). In contrast to these increases in steroidogenic activity, specific cytochrome oxidase activity in follicular mitochondria decreased after incubation. However, specific cytochrome oxidase activity in luteal mitochondria was three times greater than that of follicular mitochondria. These results suggested that short-term incubation of follicles with LH stimulates the development of mitochondrial steroidogenesis but without initially affecting the respiratory chain.

We wanted to see if these changes in mitochondrial steroidogenesis occurring after incubation of follicles were related to luteinization. When intact follicles were incubated for 18 h with LH, there were 6.7 ± 0.3 pmol progesterone secreted/follicle. After an additional 6 h of incubation, progesterone secretion increased dramatically to 41.4 ± 7.1 pmol/follicle, suggesting that some luteinization of the follicular tissue had occurred.

Electron microscopic studies indicated that the morphology of the granulosa and theca changed after incubation of the follicles. In particular, the cristae of mitochondria from granulosa of unincubated follicles appeared to have begun the transformation from a lamellar to a tubular conformation and were surrounded by an electron-lucent matrix (Fig. 1). Also, the shapes of these mitochondria were irregular. In contrast, mitochondria from granulosa of 18 h-incubated follicles were rounded and more regular, and had matrices that were more electron-dense (Fig. 2). Also, their cristae appeared to be somewhat more tubulovesicular. Mitochondria from granulosa of 24 h-incubated follicles were similar in appearance to those from cells of 18 h-incubated follicles. The non-fibroblastic cells in the theca appeared morphologically to be steroidogenic and occupied larger areas after incubation of the follicles. The mitochondria of these cells resembled those of the granulosa.

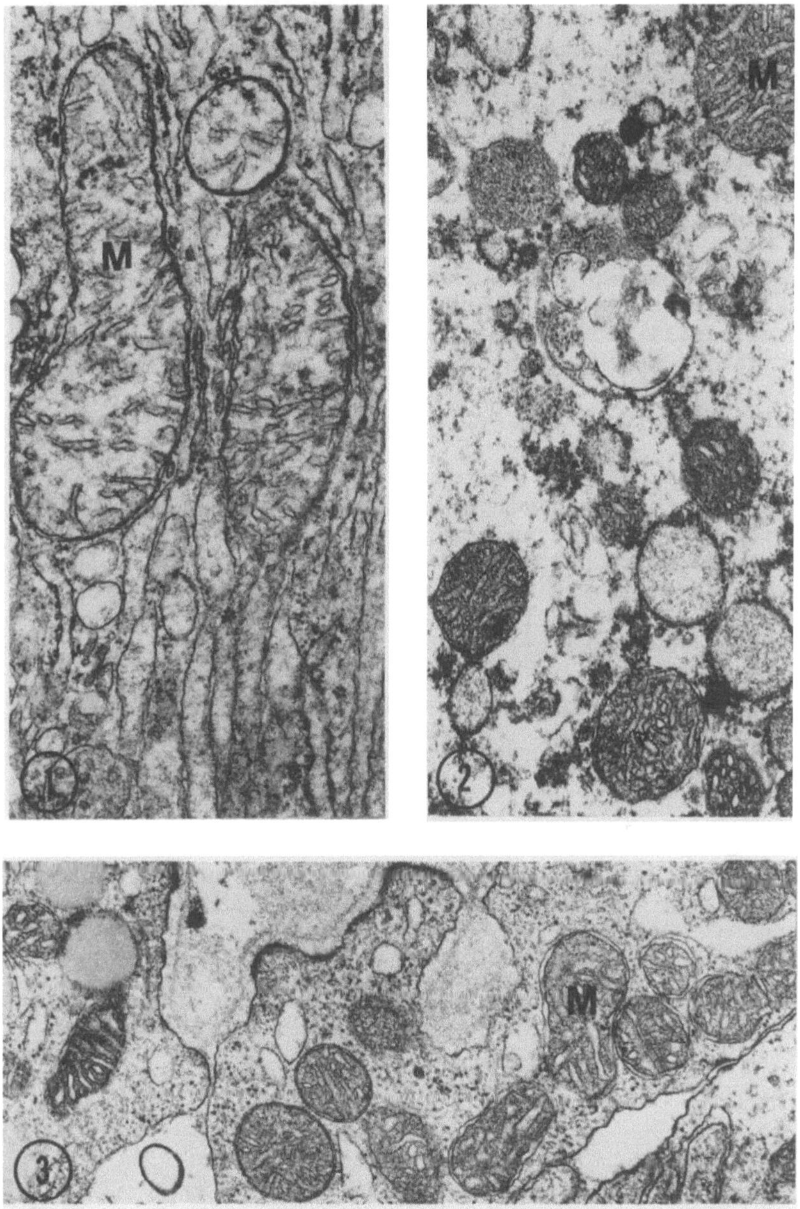


Figs. 1, 2, and 3. Mitochondria (M) of granulosa (X 31,500). 1) Unincubated follicle: mitochondria are irregular with somewhat tubular cristae and electron-lucent matrices; 2) 18 h-incubated follicle: mitochondria are rounded with tubulovesicular cristae and electron-dense matrices; 3) 24 h-incubated follicle: mitochondria are similar to those of 18 h-incubated follicle.

Mitochondrial and microsomal preparations were made from these follicles and their 3 β-hydroxysteroid dehydrogenase/isomerase (3 β-HSD) activities were determined as previously reported (3). Compared to mitochondria from unincubated follicles, mitochondria from incubated follicles, showed a two-fold increase in progesterone secretion rate after 18 h and an eleven-fold increase after 24 h (Table 1). In comparison, progesterone synthesis activity in microsomal preparations did not increase significantly after 18 h, and only two-fold after 24 h.

In these studies, 3 β-HSD activity by mitochondrial preparations was measured in the absence of exogenous NAD+ so that only intramitochondrial NAD+ was available for progesterone synthesis. Because pyridine nucleotides are compartmentalized within the cell, it is not likely that the observed increase in 3 β-HSD activity by mitochondria from incubated follicles was due to microsomal contamination (3).

The dramatic increase in progesterone secretion by intact follicles after 24 h incubation suggests that luteinization was initiated. In turn, this luteinization may be associated, at least in part, with changes in the steroidogenic capabilities of follicular mitochondria (Fig. 4). With dual sites for 3 β-HSD activity, luteal cells should have a greater capacity for progesterone production than do follicular cells. Also, it is possible that ovarian cells handle progesterone synthesized by mitochondria and endoplasmic reticulum in different ways. Progesterone produced by the endoplasmic reticulum should be readily available as a substrate for additional steroid transformation. In contrast, progesterone synthesized by

Table 1: Progesteone synthesis by mitochondria and microsomes from follicles incubated with LH

| Intact follicles incubated with LH for: | Progesterone synthesis rate for: mitochondria (nmol/min/mg protein) | microsomes |
|---|---|---|
| 0 h | 0.8 ± 0.1 | 8.8 ± 1.6 |
| 18 h | 1.9 ± 0.2** | 11.1 ± 2.0 |
| 24 h | 9.5 ± 1.4** | 16.1 ± 2.3* |

Numbers represent means ± S.E. of 6 observations. Intact follicles (8-10 mm) were incubated in medium 199 D (4) containing 0.05 μg LH/ml. Progesterone synthesis rates by mitochondria and microsomes were determined using [$^3$H]pregnenolone as substrate (3). NAD+ was added when assaying for microsomal activity but not for mitochondrial activity.

* P < 0.05 vs. 0 h
** P < 0.001 vs. 0 h

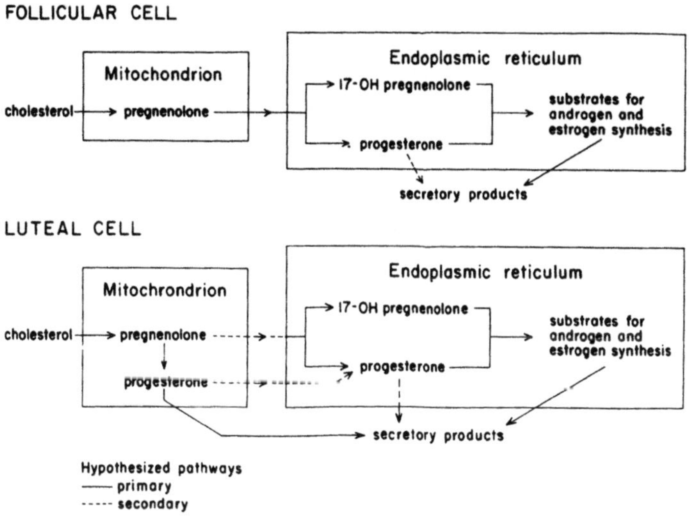


Fig. 4. Hypothesis of how mitochondrial steroidogenesis may influence what steroid products are secreted by ovarian tissues. Details of the hypothesis are presented in the text.

mitochondria may be secreted from the cell more readily since it would not be as available spatially to the enzyme systems of the endoplasmic reticulum.

The results of these studies and our other findings are summarized in Fig 5. It can be hypothesized that nonsteroidogenic components of ovarian mitochondria, e.g., the respiratory chain, increase as the ovarian tissue matures (1, 2). These nonsteroidogenic mitochondria appear to contain lamellar cristae. After gonadotropic stimulation, steroidogenic acitvity in follicular tissue develops with the mitochondria of these cells initiating pregnenolone synthesis (1, 2). As the follicular tissue continues to grow, there is an increase in the ability of follicular mitochondria to synthesize pregnenolone (3, 5). The mitochondria become larger and pleomorphic in shape, and appear to be going through a transitional stage. With the initiation of luteinization, mitochondria from follicular tissue develop the enzyme system required for progesterone synthesis. As luteinization progresses, components of the respiratory chain increase to meet the energy requirements for increased ATP and steroid syntheses (5). In luteinized tissue, mitochondria demonstrate characteristic tubulo-vesicular cristae which are surrounded by a matrix having a greater electron-density.

Thus, we feel that studies of mitochondria from incubated follicles may prove to be a unique means of monitoring the luteinization of follicular tissue, as well as some of the temporal effects of gonadotropins on ovarian steroidogenesis.

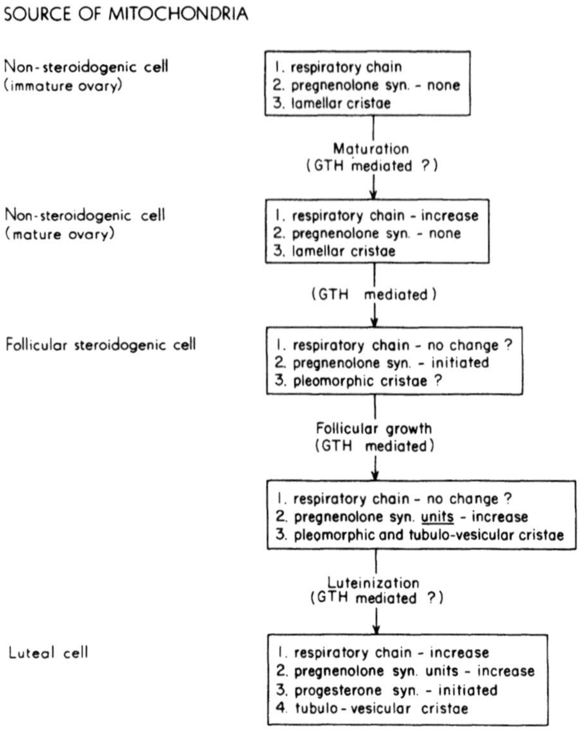


Fig. 5. Summary of changes occurring in mitochondria as ovarian tissue undergoes differentiation. (GTH = gonadotropin)

## ACKNOWLEDGMENTS

These studies were supported by National Institutes of Health Grant No. RR05641 and National Science Foundation Grant No. PCM-7618801. Ovine NIH-LH-S16 was a gift of the Endocrinology Study Section, National Institutes of Health.

## REFERENCES

(1) M.J. Dimino, D.M. Lloyd, E.A. Elfont, D.W. Peters, and R.J. Schubatis. Endocrinology 99(1976)1277.

(2) M.J. Dimino, Proc.Soc.Exp.Biol.Med. 156(1977)330.

(3) M.J. Dimino. Endocrinology 101(1977)1844.

(4) C.P. Channing and F. Ledwitz-Rigby. Methods Enzymol. 39(1975) 183.

(5) M.J. Dimino and S.K. Berman. Submitted for publication.

## PROSTAGLANDIN $F_2\alpha$ AND PROGESTERONE SYNTHESIS IN VITRO BY BOVINE CORPORA LUTEA OF EARLY PREGNANCY

J. Lukaszewska and W. Hansel

Physiology and Physical Biology

Cornell University, Ithaca, N.Y. 14853

### INTRODUCTION

The rapid regression of the bovine corpus luteum at the 17th-19th day of the estrous cycle requires the presence of the uterus and appears to be mediated by estrogen and one or more uterine luteolysins acting locally on the ipsilateral corpus luteum (1,2). Little information is available as to how the presence of a viable embryo in the uterus causes maintenance of the bovine corpus luteum. Del Campo et al. (3) suggested a local relationship between a gravid uterine horn and the adjacent ovary in the pregnant cow. Although the corpus luteum is necessary for maintenance of pregnancy in the cow, at least until Day 236 (4), the transition of the cyclic corpus luteum to the corpus luteum of pregnancy occurs before the embryo implants. Staples et al. (5) showed that very rapid elongation of the trophoblast occurs at about Day 15, just prior to the time of cyclic corpus luteum involution. Mesoderm formation also begins at Day 15, and the bi-nucleated giant cells of the trophoblast (6) first appear at this time (5). Greenstein et al. (7) suggested that these cells may play a role in stimulating maintenance of the corpus luteum of pregnancy. These facts suggest that the developing embryo either inactivates the normal uterine luteolytic mechanism, or produces a potent luteotrophin that overcomes this mechanism.

Numerous studies (8) show that corpora lutea of cyclic, pregnant and hysterectomized cattle all respond to LH and HCG. However, there does not appear to be any increase in radioimmunoassayable plasma LH either in the peripheral circulation after the 15th day of pregnancy (9) or in the uterine venous effluent of the gravid horn between the 17th and 18th day of pregnancy (Hansel and Lukaszewska, unpublished observations). Corpus luteum regression in the cyclic animal is not associated with a decline in plasma LH concentration, but administration of anti-bovine LH causes a significant reduction in corpus luteum weights and progesterone contents (1).

Although exogenous $PGF_2\alpha$ is luteolytic, and $PGF_2\alpha$ administered into the uterus appears to be preferentially transferred from the uterus to the ovarian artery in the cow, its role as a luteolysin in the normal animal is not well established (10). In vitro and in vivo studies indicate that the bovine ovary has the ability to convert arachidonic acid to PGF (10) and that luteal tissues specifically bind $PGF_2\alpha$ (11). Thus, the corpus luteum is able to respond to a uterine luteolysin and to produce its own "luteolytic" substance. These and the earlier observations of Armstrong and Black (12) that corpora lutea from cows in the late luteal stage did not respond to LH, suggested the need for a study in which comparisons were made of the abilities of luteal tissues of pregnant and nonpregnant animals to synthesize PGF and progesterone at the critical time for corpus luteum survival.

## MATERIALS AND METHODS

Eighteen pregnant and 17 nonpregnant (unmated) Holstein heifers were used. Pregnant animals were inseminated on the first day of estrus (Day 0). Corpora lutea were obtained by laparotomy 17 days after estrus from 8 pregnant and 7 nonpregnant and 18 days after estrus from 10 pregnant and 10 nonpregnant heifers. Pregnancy was confirmed by the presence of embryos in uterine flushings. Jugular vein blood samples were collected from all animals daily until the day before surgery, at which time the sampling frequency was increased to 6 h. Corpora lutea were dissected free of connective tissue, rinsed in saline, sliced and aliquots (120-200 mg) were incubated in a Dubnoff shaker in 3.5 ml of Kreb's-Ringer buffer for 2 h at 37 C in an atmosphere of $O_2$:$CO_2$ (95%:5%). Each corpus luteum provided tissue for unincubated control, incubated control, arachidonic acid-treated and LH-treated incubation flasks (5 each). The levels of LH and AA added were 3 μg and 30 μg/ml of incubation medium respectively. In addition, each corpus luteum provided tissue for the assay of lipoprotein lipase activity (300 mg) and arachidonic acid content (200 mg). The latter tissues were frozen immediately on dry ice and stored at -20 C until measurements were made.

The incubated tissue and media were frozen on dry ice and stored at -20 C until assayed. The extraction and radioimmunoassay of PGF were carried out as described by Jaffe et al. (13) and Auletta et al. (14), with minor modifications. The coefficient of variation between assays was 14.9 percent and the within assay coefficient was 10.3 percent.

The method described by Hixon et al. (15) was employed for progesterone radioimmunoassay. The coefficient of variation between and within assays was 16.5 and 8.4 percent, respectively. Lipoprotein lipase activity in the corpora lutea was determined as described by Shemesh et al. (76). Total lipids from luteal tissue were extracted according to the method of Folch et al. (17). Fatty acid esters were prepared as described by Morrison and Smith (18), and analyzed by gas liquid chromatography. Methyl arachidonate was identified according to its retention time. Yields of methyl arachidonate were determined by adding a known amount of n-heptadecanoic acid to the extracted lipids before methylation as an internal standard.

## RESULTS

Jugular plasma progesterone concentrations were clearly higher ($p<0.05$) in pregnant than in nonpregnant animals during Days 15-18 inclusive (Figure 1).

The in vitro studies of steroidogenic capabilities of the Day 17 and 18 corpora lutea indicated that corpora of pregnant cows had higher initial contents of progesterone and synthesized more progesterone during incubation. Loss of the ability to produce progesterone in the absence of added LH occurred between Days 17 and 18 in the corpora lutea of the nonpregnant cows but this response was maintained in the pregnant animals (Tables 1 and 2). However, the corpora of the nonpregnant animals retained the ability to respond to LH, since addition of LH to the incubation medium caused essentially the same mean rise in progesterone concentration in both groups on both days. Corpora from 2 of 10 Day 18 nonpregnant cows failed to synthesize progesterone in vitro, with or without added LH. Plasma progesterone levels in these two animals were extremely low (Figure 2), indicating that their corpora lutea were in an advanced state of regression. The preovulatory plasma LH peak had not occurred in either of these animals, but the preovulatory rise in plasma estradiol concentration had occurred in one animal.

Initial concentrations and synthesis of $PGF_{2}\alpha$ by bovine corpora lutea removed at Day 18 of the estrous cycle and of pregnancy are shown in Table 3. Corpora from both groups responded to additions of arachidonic acid with increased ($p<0.05$) $PGF_{2}\alpha$ synthesis. The mean concentration of $PGF_{2}\alpha$ was slightly but not significantly ($p>0.05$) higher in corpora lutea from nonpregnant than pregnant animals. LH did not stimulate $PGF_{2}\alpha$, although some corpora lutea responded to LH by a slight increase of $PGF_{2}\alpha$ during incubation. In the absence of added LH, the correlation between the concentration of $PGF_{2}\alpha$ and net progesterone synthesis was negative ($r = -0.77$, $p<0.01$) in the nonpregnant cows, but positive in the pregnant cows,

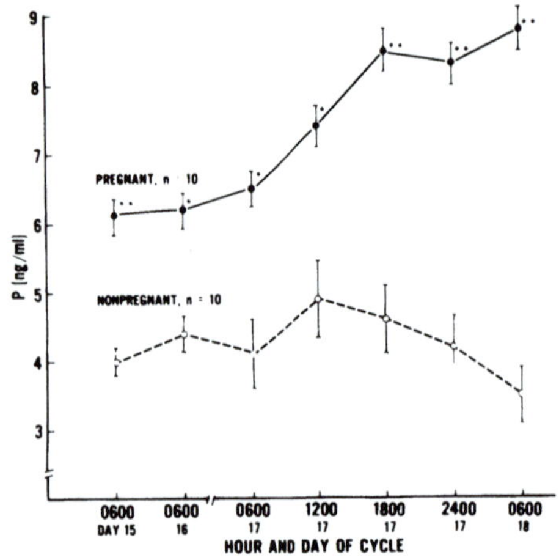


Fig. 1. A comparison of progesterone concentration in jugular plasma of pregnant and nonpregnant cows (mean ± S.E.M.). Different from nonpregnant; ** p<0.01; * p<0.05.

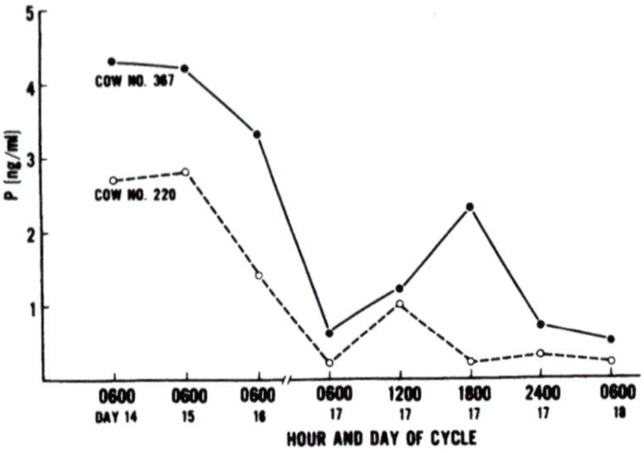


Fig. 2. Progesterone concentrations in jugular plasma of two nonpregnant cows whose corpora lutea did not respond to LH *in vitro*.

(r = 0.73), $p<0.05$). Similarly, when arachidonic acid was added, correlations obtained between tissue concentrations of $PGF_{2}\alpha$ and net progesterone synthesis in nonpregnant and pregnant animals were -0.70 ($p<0.05$) and 0.70 ($p<0.05$), respectively.

Lipoprotein lipase activities measured in the corpora lutea of the pregnant and nonpregnant cows are shown in Table 4. Lipoprotein lipase activity was higher in corpora lutea from pregnant cows than

Table 1. *In vitro* progesterone synthesis (mean ± S.E.M.) by bovine corpora lutea removed at Day 17.

| | Progesterone | | | | |
|---|---|---|---|---|---|
| | Total per/CL | Unincubated | Incubated | Incubated with LH | LH stimulated synthesis |
| | | μg/g | | | |
| Pregnant n = 8 | 235.4 ±48.9 | 51.7* ±5.4 | 63.3* ±6.4 | 76.9 ±6.7 | 13.6 ±2.4 |
| Nonpregnant n = 7 | 156.7 ±29.7 | 35.2 ±5.3 | 49.3 ±4.1 | 65.6 ±5.5 | 16.3 ±3.3 |

*Different from nonpregnant animals, $p<0.05$, Students t-test.

Table 2. In vitro progesterone synthesis (mean ± S.E.M.) by bovine corpora lutea removed at Day 18.

| | Progesterone | | | | |
|---|---|---|---|---|---|
| | Total per/CL | Unin- cubated | In- cubated | Incubated with LH | LH stimulated synthesis |
| | | μg/g | | | |
| Pregnant n = 10 | 233.2** ±31.8 | 48.9* ±9.0 | 66.5** ±9.8 | 81.1** ±8.8 | 14.6 ±3.3 |
| Non-pregnant n = 10 | 117.7 ±29.4 | 27.2 ±6.1 | 32.2 ±6.3 | 45.5 ±7.5 | 13.3 ±4.3 |

* Different from nonpregnant animals, p<0.05, Students t-test.
** Different from nonpregnant animals, p<0.01.

Table 3. In vitro prostaglandin $F_2\alpha$ synthesis (mean ± S.E.M.) by bovine corpora lutea removed at Day 18.

| | Prostaglandin F | | | | |
|---|---|---|---|---|---|
| | Total per/CL | Unin- cubated | In- cubated | Incubated with LH | Incubated with AA |
| | | ng/g | | | |
| Pregnant n = 10 | 119.6 ±37.9 | 22.1 ±4.6 | 22.5 ±4.5 | 22.1 ±4.7 | 82.4* ±14.0 |
| Non-pregnant n = 10 | 101.3 ±16.7 | 27.7 ±5.8 | 27.5 ±5.5 | 27.2 ±6.7 | 73.4* +6.9 |

* Different from incubated concentrations, p<0.05, Students t-test.

in those of nonpregnant animals, suggesting an association with maintenance of steroidogenic function of the corpus luteum. Arachidonic acid concentrations were high in both types of corpora lutea, 2.6±0.1 and 2.5±0.1 mg/g for pregnant and nonpregnant, respectively.

## DISCUSSION

These results suggest that complete loss of LH receptors by the bovine corpus luteum is a consequence rather than a cause of luteolysis. The possibility that some luteolytic factor(s) inhibits the ability of LH to function at some point subsequent to its

Table 4. Lipoprotein lipase activity in bovine corpora lutea (means ± S.E.M.).

| Group | Day 17 | Day 18 |
|---|---|---|
| | (nmole FFA /hr/100 mg acetone dried powder) | |
| Pregnant n = 11 | 396.3 ± 60.4* | 354.2 ± 61.4 |
| Nonpregnant n = 10 | 264.5 ± 45.9 | 262.3 ± 43.9 |

* Different from nonpregnant animals, $p<0.05$, Students t-test.

binding cannot be excluded. However, corpora lutea which retain their LH receptors can be "rescued", either by hysterectomy or by administration of LH or HCG (1), while corpora lutea which lose their receptors cannot. It is for those reasons that irreversible luteolysis occurs in the cow as late as two days after the first decline in plasma progesterone.

Prostaglandin $F_2\alpha$ has been suggested as a natural luteolysin in several species, including cattle. Evidence for and against this concept was recently summarized (10). $PGF_2\alpha$ receptors are present in the bovine corpus luteum of the cycle and of pregnancy and their number increases from Day 3 of the cycle onward (11, 19). On the other hand, the bovine corpus luteum contains appreciable amounts of $PGF_2\alpha$ and is able to convert arachidonic acid to $PGF_2\alpha$ in vivo and in vitro (10). Thus, the amount of $PGF_2\alpha$ found in the corpus may represent that produced within luteal cells, as well as that bound to luteal cell membranes.

Our data provide little evidence for a change in the ability of the corpus luteum to synthesize PGF that can be related to pregnancy. Corpora lutea from pregnant and nonpregnant animals contained equal amounts of arachidonic acid and PGF before and after incubation and responded to added arachidonic acid with similar increases in PGF synthesis.

In view of the luteolytic effects of exogenous $PGF_2\alpha$ in the cow, the negative association ($r = -0.77$, $p<0.05$) between PGF concentration and net progesterone synthesis might have been expected. However, the positive correlation noted between $PGF_2\alpha$ concentration and progesterone synthesis in the pregnant animal is of interest. This fact, along with higher activity of lipoprotein lipase found in the corpora lutea of pregnancy might imply a luteotrophic effect of PGF in the pregnant cow.

The fact that the corpora lutea of pregnancy reach a greater weight than those of nonpregnant animals at Day 18 ($5.1 \pm 0.4$ g vs. $4.1 \pm .4$g, $p<0.05$), contain and secrete more progesterone and are able to synthesize progesterone in vitro without exogenous luteotrophin, all suggest that pregnancy corpora are stimulated in vivo by an embryonic luteotrophic factor(s). The luteotrophic substance(s), which has not been isolated appears to differ from pituitary LH. In preliminary experiments (Hansel and Lukaszewska, unpublished observations), aqueous extracts of 18-day bovine embryos stimulated progesterone synthesis in vitro by dispersed bovine luteal cells.

## REFERENCES


(1) W. Hansel, P.W. Concannon and J.H. Lukaszewska. Biol. Reprod. 8(1973)222.
(2) D. Gengenbach, J. Hixon and W. Hansel. Biol. Reprod. 13(1975) 126.
(3) M.R. Del Campo, R.F. Rowe, L.R. French and O.J. Ginther. Biol. Reprod. 16(1977)580.
(4) T.Y. Tanabe, Pa. State Univ. Agr. Expt. Sta. Bul. No. 774. December, 1970.
(5) R.E. Staples, K. McEntee and W. Hansel. J. Dairy Sci. 44(1961) 2049.
(6) W.A. Wimsatt. Anat. Record 103(1949)148.
(7) J.S. Greenstein, R.W. Murray and R.C. Foley. Anat. Record. 132 (1958)321.
(8) W. Hansel, in: Ovarian Regulatory Mechanisms. J. Reprod. Fertil. Suppl. No. 1(1966)33.
(9) W.D. Oxender, H.D. Hafs and L.A. Edgerton. J. Anim. Sci. 35 (1972)51.
(10) W. Hansel, J. E. Fortune, in: Control of Ovulation, ed. D.B. Crighton, G.R. Foxcroft, W.B. Haynes, G.E. Lamming (Butterworth Inc., Woburn, Mass., 1978)p.237.
(11) Ch.V. Rao, V.L. Estergreen, F.R. Carman, G.E. Moss and K.A. Frandle. Fed. Proc. 37(1978)Abstr. 2684.
(12) D.T. Armstrong and D.L. Black. Endocrinology 78(1966)937.
(13) B.M. Jaffee, H.R. Behrman and C.W. Parker. J. Clin. Invest. 52(1973)398.
(14) F.J. Auletta, R.M. Zusman and B.V. Caldwell. Clin. Chem. 20 (1974)1580.
(15) J.E. Hixon, R. Nadaraja, R.J. Schechter and W. Hansel. Prostaglandins 4(1973)679.
(16) M. Shemesh, A. Bensadoun and W. Hansel. Proc. Soc. Exp. Biol. Med. 151(1976)667.
(17) J. Folch, M. Lees and G.H.S. Stanley. J. Biol. Chem. 226(1957) 497.
(18) W.R. Morrison and L.M. Smith. J. Lipid Res. 5(1964)600.

LUTEAL PHASE DEFICIENCIES IN PERIPUBERTAL RHESUS MONKEYS:

MECHANISTIC CONSIDERATIONS[1]

Thomas E. Nass[2], Donald J. Dierschke, James R. Clark[3], Patricia A. Meller, and Keith K. Schillo

Wisconsin Regional Primate Research Center, University of Wisconsin, Madison, Wisconsin 53706

## ABSTRACT

Short luteal phases in peripubertal monkeys were identified by abbreviated progesterone secretion. Evaluation of the hormonal profiles for LH, FSH, and estradiol revealed that depressed serum concentrations of FSH during the follicular phase and LH and estradiol during the luteal phase were associated with luteal defects. In addition, morphological differences in the corpora lutea of short and normal luteal phases were identified. A mechanism is presented whereby inadequate follicular phase FSH secretion results in abbreviated luteal function.

## INTRODUCTION

Deficient corpus luteum (CL) function, as characterized by serum progesterone ($P_4$) concentrations which are submaximal and show a premature decline, has been identified in cycling adult rhesus monkeys (1). Similarly, inadequate $P_4$ secretion has been reported in adult women and may cause infertility (2). In peri-

---

[1]Supported by the Wis. Alumni Res. Foundation and NIH Grant No. RR00167 to the Wis. Reg. Primate Res. Ctr. Publication No. 18-002 of the Wis. Reg. Primate Res. Ctr.

[2]NIMH Predoctoral Fellow (MH 07478).

[3]NICHHD Postdoctoral Fellow (HD 01816). Present address: Dept. of Animal Science, Texas Tech Univ., Lubbock, TX 79409.

pubertal girls, menstrual cycles in which the basal body temperature reflects an aberrant luteal condition are often predominant (3). Additionally, Foster (4) has shown that the initial post-menarchial ovulatory cycles in rhesus monkeys are associated with a high incidence of short luteal phases. We have independently confirmed these findings and have presented preliminary results identifying morphological differences in the CL of short (SLP) and normal luteal phases (NLP) in adolescent monkeys (5).

Investigation of the hormonal profiles associated with SLP in women and monkeys has revealed abnormalities in gonadotropin secretion which may be important to an understanding of the processes responsible for defective CL function. In this study we report additional abnormalities in gonadotropin secretion in peripubertal monkeys and suggest a possible mechanism whereby depressed preovulatory FSH secretion may result in deficient luteal function.

## MATERIALS AND METHODS

Eight rhesus monkeys (Macaca mulatta, age 3.5-5.0 years) were utilized for hormonal and laparoscopic evaluation. For comparisons of CL morphology, 27 menstrual cycles from an additional 8 monkeys of similar age were used. All animals were born at the Wis. Reg. Primate Res. Ctr. and housed under standard conditions (6). Six of the 8 animals in the hormone evaluation study had received anterior hypothalamic lesions as part of a related study. As previously reported (5), these lesions had no effect on the timing of puberty or the occurrence of SLP.

Twenty menstrual cycles were selected for hormone evaluation; 10 were classified as SLP (<11 days from LH surge to menses, range

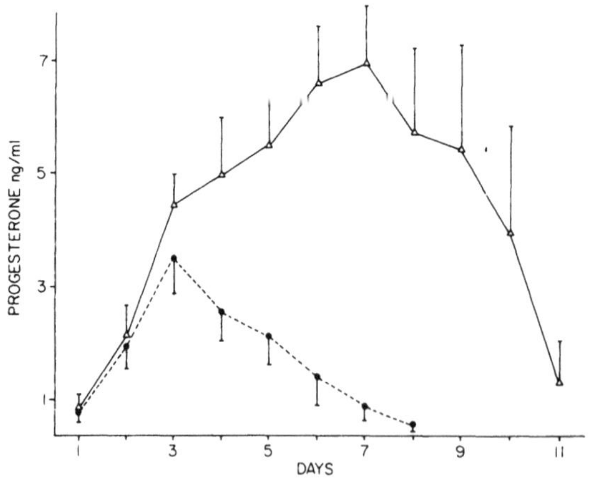


*Figure 1. Concentrations of $P_4$ in SLP (●---●) and NLP (Δ---Δ) cycles. ($\bar{x}$ ± SEM)*

6-10 days) while the 10 remaining cycles had NLP (>11 days, range 12-20 days). To obtain a more detailed hormonal pattern throughout the periovulatory and luteal portions of the menstrual cycle, blood samples were taken daily at these times (based on sex skin color (7) and laparoscopic observation), but samples were taken less frequently early in the cycle. Collection and storage of blood samples was as previously described (6). Laparoscopy (8) was utilized to evaluate follicular development and confirm ovulation. The ovaries of 2 monkeys having apparent SLP (based on morphological appearance) were removed for histological evaluation.

Serum concentrations of LH (9), FSH (10), and $P_4$ (6) were determined by radioimmunoassay (RIA) methods reported previously. Serum estradiol concentrations were determined by RIA utilizing antibody Lot $E_2$ (Holly Hill Biologicals, Hillsboro, Oregon). Methods and verification of the $E_2$ assay were as described by Hotchkiss et al. (11). In addition, pooled samples from the day of the LH surge and from 4 days before and after the surge were subjected to biological and radioimmunological assay to determine the ratio of immunological to biological activity (12).

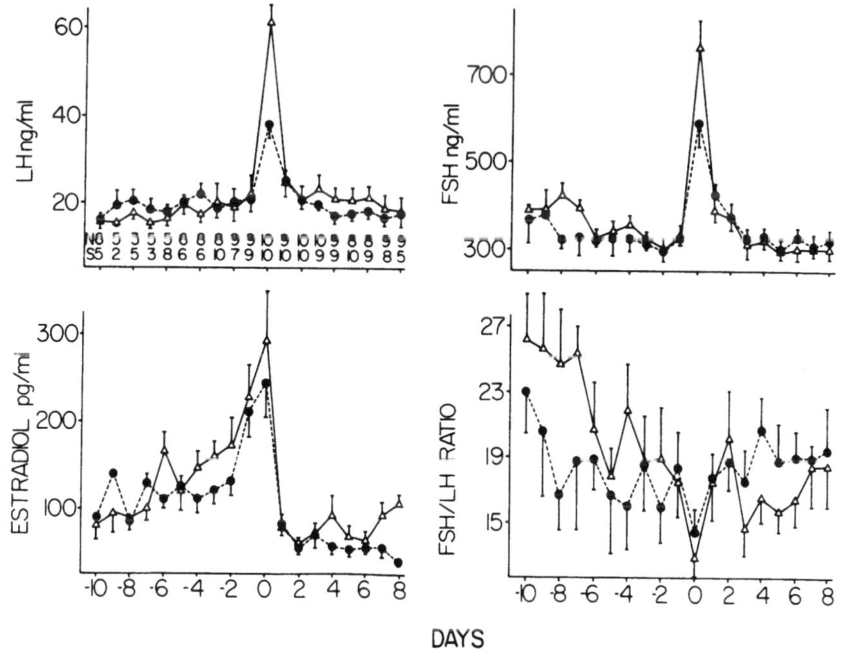


*Figure 2. Concentrations of LH, FSH, $E_2$ and ratios of FSH:LH in SLP and NLP cycles normalized to the day of the LH surge. Symbols are as in Fig. 1. Values below data points in top left panel represent the number of samples for each group in all graphs. ($\bar{x}$±SEM)*

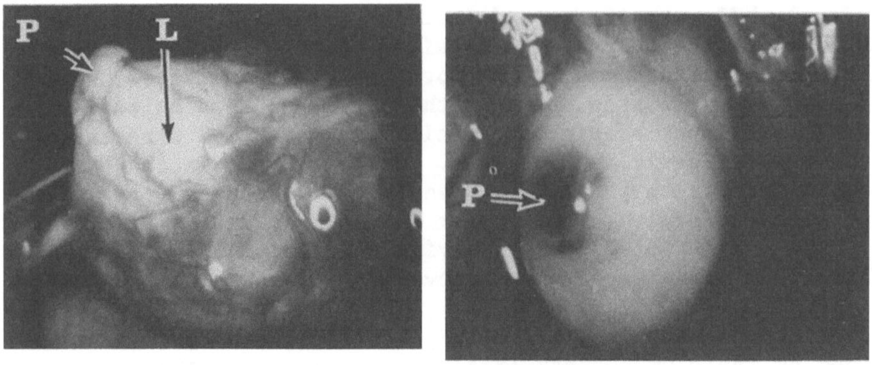


*Figure 3. Representative CL from an NLP (left) and SLP (right) cycle. P = papilla of CL. L = subsurface luteal tissue.*

The data were analyzed by a 2-way analysis of variance with unweighted means during the follicular and luteal phases, and a Student's t-test for the day of the LH, FSH, and $E_2$ surge.

## RESULTS

Serum concentrations of $P_4$ during SLP and NLP are illustrated in Fig. 1. In confirmation of previous reports (4,5), females in the SLP group had both abbreviated elevations of $P_4$ and lower peak concentrations when compared with the NLP group.

Figure 2 contrasts the serum concentrations of LH, FSH, $E_2$ and values for the FSH:LH ratio in SLP with those observed in NLP. Analysis of the data revealed statistically significant differences in serum concentrations of FSH during the follicular phase ($P < .05$), LH during the preovulatory surge ($P < .001$) and LH ($P < .01$) as well as $E_2$ ($P < .05$) during the luteal phase. Other values were not statistically different. FSH:LH ratios in the 2 groups were different for both the follicular ($P < .01$) and luteal ($P < .05$) portions of the menstrual cycle, but not on the day of the surge. In addition, the immunological/biological ratios for LH in the pooled samples were representative of adult values, in contrast to those observed in prepubertal monkeys (12).

By laparoscopic observation, the apparent size and appearance of the preovulatory follicles were not altered in animals with SLP compared with NLP. However, morphological differences in the resulting CL were noted (Fig. 3). In these representative examples the CL from the NLP had an abundance of highly vascular subsurface luteal tissue associated with the protruding papilla. By contrast, the CL from the SLP had a small vascular area with little subsurface luteal tissue visible. Of the 47 menstrual cycles evaluated, 16/20 CL from NLP had visible subsurface luteal tissue, whereas in SLP only 5/27 ovulations produced visible luteal tissue. Histological evaluation of the 2 ovaries which were removed during apparent SLP revealed CL of normal size and appearance.

## DISCUSSION

The regulation of CL function may be influenced by events occurring prior to and following ovulation. Thus, the depressed FSH concentrations observed during the follicular phase in SLP may have been inadequate for complete maturation of the preovulatory follicle. In particular, FSH secretion was low early in the follicular phase (days -8 and -7) in the SLP group and this in turn was followed by depressed $E_2$ concentrations on days -4 through -2. Although the apparent follicle size was similar in both SLP and NLP, the extent to which gross morphology of the follicle reflects its function is not known. Sherman and Korenman (13) have also reported significantly lower concentrations of FSH and $E_2$ in women and have attributed the accompanying luteal defects to these deficiencies. The depressed FSH secretion in SLP resulted in low FSH:LH ratios prior to ovulation, confirming previous findings in women with abbreviated luteal function (14).

Although the amount of subsurface luteal tissue is a subjective measure, the hypothesis that SLP results from inadequate FSH secretion and follicle development implies that morphological differences in the CL would exist and our observations support this contention. Though the 2 CL which were removed for histological study appeared to be normal, it is not known whether these CL would have maintained normal function had they remained _in vivo_.

Abnormal secretion of gonadotropins during the luteal phase may also contribute to SLP. Vande Wiele _et al_. (15) demonstrated that the human CL is dependent upon continuous exposure to small amounts of circulating LH. However, their data only compared CL exposed to continuous LH or none at all, thus it is questionable if the small differences in luteal LH secretion noted in our study were sufficient to account for SLP. Depressed secretion and biological activity of LH in adult monkeys experiencing inadequate luteal function was reported by Channing and Sakai (16); in agreement with their results, the biological activity of pooled samples from day 4 of the SLP cycle was depressed in our study. However, samples from the day of the LH surge did not differ in biological activity, in contrast to their immunological activity. Neill _et al_. (12) have previously reported changes in the immunological activity of LH on the day of the preovulatory surge in rhesus monkeys. Without additional analyses for biological activity and more frequent sampling on the day of the preovulatory surge, the differences in LH secretion observed in our study and their contribution to the early demise of the CL remain questionable.

In addition to FSH and LH, prolactin has also been implicated in the etiology of SLP in women. Corenblum et al. (17) were able to correlate galactorrhea with deficient CL function and reported that antiprolactin therapy restored NLP in these patients. However, prolactin has been associated with altered gonadotropin concentrations (18) and thus may exert its effect secondarily through FSH and LH secretion. The possible role of prolactin in luteal function and its site of action in rhesus monkeys remains to be determined.

Based on these results, we propose a mechanism whereby preovulatory events may determine luteal function in women and monkeys. The depressed FSH secretion throughout the follicular portion of the menstrual cycle, and particularly early in this phase, results in inadequate stimulation and maturation of the preovulatory follicle and reduced $E_2$ secretion. Ovulation of small or retarded follicles then gives rise to deficient CL (both structurally and physiologically) and short luteal phases.

## ACKNOWLEDGMENTS

The authors are indebted to Dr. J. D. Neill (Emory University, Atlanta) for performing immunological/biological activity determinations and Dr. V. Chandrashekar for performing the protein radioimmunoassays. We also wish to acknowledge the technical assistance of K. Rygasewicz and R. Abrams.

## REFERENCES

(1) J.W. Wilks, G.D. Hodgen and G.T. Ross. J. Clin. Endocrinol. Metab. 43 (1976) 1261.

(2) G. Jones, in: Progress in Infertility, eds. S.J. Behrman and R.W. Kistner (Little, Brown & Co., Boston, 1968) p. 299.

(3) G.K. Doring. J. Reprod. Fert., Suppl. 6 (1977) 77.

(4) D. L. Foster. Biol. Reprod. 17 (1977) 584.

(5) J.R. Clark, D.J. Dierschke and J.J. van der Werff ten Bosch. S.S.R. Abstract #97, 1976.

(6) J.R. Clark, D.J. Dierschke and R.C. Wolf. Biol. Reprod., in press.

(7) J.A. Czaja, J.A. Robinson, S.G. Eisele, G. Scheffler and R.W. Goy. J. Reprod. Fert. 49 (1977) 147.

(8) D.J. Dierschke and J.R. Clark. J. Med. Prim. 5 (1976) 100.

(9) P.T.K. Toivola, W.E. Bridson and J.A. Robinson. Endocrinology (1978), in press.

(10) G.D. Hodgen, J.W. Wilks, J.L. Vaitukaitis, H.C. Chen, H. Papkoff and G.T. Ross. Endocrinology 99 (1976) 137.

(11) J. Hotchkiss, L.E. Atkinson and E. Knobil. Endocrinology 89 (1971) 177.

(12) J.D. Neill, R.A. Dailey, R.C. Tsou and L.E. Reichert, Jr. Endocrinology 100 (1977) 856.

(13) B.M. Sherman and S.G. Korenman. J. Clin. Endocrinol. Metab. 38 (1974) 89.

(14) C.A. Strott, C.M. Cargille, G.T. Ross and M.B. Lipsett. J. Clin. Endocrinol. Metab. 30 (1970) 246.

(15) R.C. Vande Wiele, J. Bogumil, I. Dyrenfurth, M. Feru et al. Recent Prog. Horm. Res. 26 (1970) 63.

(16) C.P. Channing and C.N. Sakai. FASEB Abstract #383, 1978.

(17) B. Corenblum, N. Pairaudeau and A.B. Shewchuk. Obstet. and Gynecol. 47 (1976) 486.

(18) J. Meites and C.S. Nicoll. Ann. Rev. Psychol. 28 (1966) 57.

# CELLULAR MECHANISMS INVOLVED IN LUTEOLYSIS

Laurie G. Paavola

Department of Anatomy, Temple University School of Medicine, Philadelphia, Pennsylvania 19140

Regression of the mammalian corpus luteum is commonly thought to involve autophagy (cellular self-digestion) or heterophagy (removal of cells by macrophages), or both. However, few studies have focused on these mechanisms, and their actual importance in luteolysis thus remains uncertain. Moreover, studies dealing with these topics have often come to differing conclusions; e.g., van Lennep and Madden (1) reported that autophagy was probably only a minor component in the overall process of involution in humans, whereas it was proposed in a study on sheep that autophagy played a substantial role in the destruction of luteal tissue (2).

Similarly, little information is available on the role of macrophages in luteolysis. An involvement of phagocytes in luteolysis was suggested by earlier histochemical studies carried out at the light microscopic level, in which macrophages were reported to be particularly abundant in corpora lutea that were involuting (3,4). Further, macrophages containing debris-laden vacuoles, perhaps resulting from the phagocytosis of luteal cells or their remains, have been noted in fine structural studies of degenerating corpora lutea in several mammals (5,6). Although these findings suggest that macrophages may be involved in luteolysis, neither the suspected increase in numbers of these cells during involution, nor the phagocytosis of luteal cells has been documented.

The purpose of the present study therefore was to determine the mechanisms underlying the normal postpartum regression of corpora lutea in guinea pigs. To this end, the participation of macrophages in luteolysis was examined by conventional electron microscopy (EM). The role of autophagy in luteal destruction was investigated by the EM localization of two lysosomal enzymes, acid phosphatase (ACPase) and arylsulfatase. It was necessary to study the distribution of both enzymes since proof that a structure is involved in intracellular digestion rests on the demonstration that it contains more than one acid hydrolase (7). The corpora lutea used in this work were fixed in situ by vascular perfusion, a method that provides not only optimal preservation of fine structure, but also maintains enzyme activity at levels adequate to permit their cytochemical localization. Luteal tissue was examined during the period of peak steroid secretion, 20 to 33 days of gestation, to establish a normal baseline, and during late pregnancy and the postpartum period to elucidate changes related to involution.

## MATERIALS AND METHODS

Corpora lutea were obtained from Hartley guinea pigs at 25-33 and 60-68 (near term) days of gestation, on the day of parturition, and at days 5, 8, 10, 16, 20, 24 and 26 in the postpartum period. All tissue was fixed by vascular perfusion (8). Corpora lutea were prepared for morphological studies as previously detailed (8). The fixatives and incubation conditions used for the cytochemical procedures (ACPase, arylsulfatase) have been described elsewhere (9). The controls, especially important in cytochemistry, used to show specificity of the reaction were the following: deletion of substrate from the complete medium; heating of the tissue to inactivate enzymes, followed by incubation in a complete medium; or inclusion in a complete medium of an inhibitor specific for the enzyme. These controls proved to be negative.

## RESULTS

### Period of Maximum Progesterone Secretion

During the period of peak steroid secretion, the most striking localization of ACPase activity is to the well developed Golgi complex, where reaction product accumulates in several cisternae, filling the cisternal cavities (fig. 1). Located near the Golgi complex in these cells are smooth-surfaced elements that resemble the GERL (Golgi complex-endoplasmic reticulum-lysosomes) system described in other tissues (10). Since GERL is better developed in older luteal cells, it will be described with that stage. It should be noted, however, that in contrast to most cell types, GERL in luteal cells of this age is unreactive or shows low levels of enzyme activity after incubation for ACPase (fig. 1, inset).

Lysosomes, which are reactive for ACPase and arylsulfatase, occur in modest numbers in luteal cells of this stage, and are usually irregularly spheroid. Autophagic vacuoles are rarely observed. Also present in these cells are smaller bodies which contain catalase, and are therefore microperoxisomes (Paavola, unpublished observations).

## Regression

Since we have suggested previously that GERL cisternae play a role in luteolysis (8), they will be described in some detail. These distinctive smooth-surfaced cisternae occur near the inner (trans) face of the Golgi complex, and show specializations that distinguish them from other smooth membrane-bound organelles present in aging luteal cells;e.g., they may contain moderately dense material or a central dense line; they often display a "thickened" limiting membrane; and they generally are narrower in width (25 nm) than are tubules (50-60 nm) or cisternae (40-45 nm) of smooth ER. Although the origin of GERL remains uncertain, the occasional connections between these cisternae and ER suggests that GERL may be derived from ER (11). The GERL system in regressing luteal cells, in striking contrast to the time of peak steroid secretion described above, is hypertrophied and consistently displays intense ACPase (fig. 2, inset) and arylsulfatase activity. In addition, reaction product accumulates in small (coated and smooth) vesicles and larger vacuoles, which presumably represent developing lysosomes, that are attached to GERL.

Other marked changes typically noted in regressing luteal cells are the dramatic increase in the number of lysosomes, and the appearance and accumulation of characteristic double-walled autophagic vacuoles (fig. 2). After cytochemical incubation for either of the acid hydrolases, reaction product fills the space between the double walls of these vacuoles. Lysosomes also show both ACPase and arylsulfatase activity, and are often trapped in autophagic vacuoles. Often, several small acid hydrolase-positive vesicles cluster around and appear to fuse with larger reactive bodies (lysosomes). For further details on the fine structure and cytochemistry of these lytic bodies, see Paavola (8, 11).

In order to ascribe an autophagic function to a structure seen in a thin section, it must be shown that the structure is completely "sealed off" from the surrounding cytoplasm, and contains cellular organelles. From an examination of various structures in ninety consecutive serial thin sections, it was determined that the double-walled vacuoles described above are completely closed, roughly spherical, discrete entities, that contain other organelles. Some double-walled vacuoles show multiple compartments which are also sealed-off from the cytoplasm. It was further ascertained that GERL cisternae exist largely as cup-shaped structures. It thus seems clear that the double-walled bodies are autophagic in function

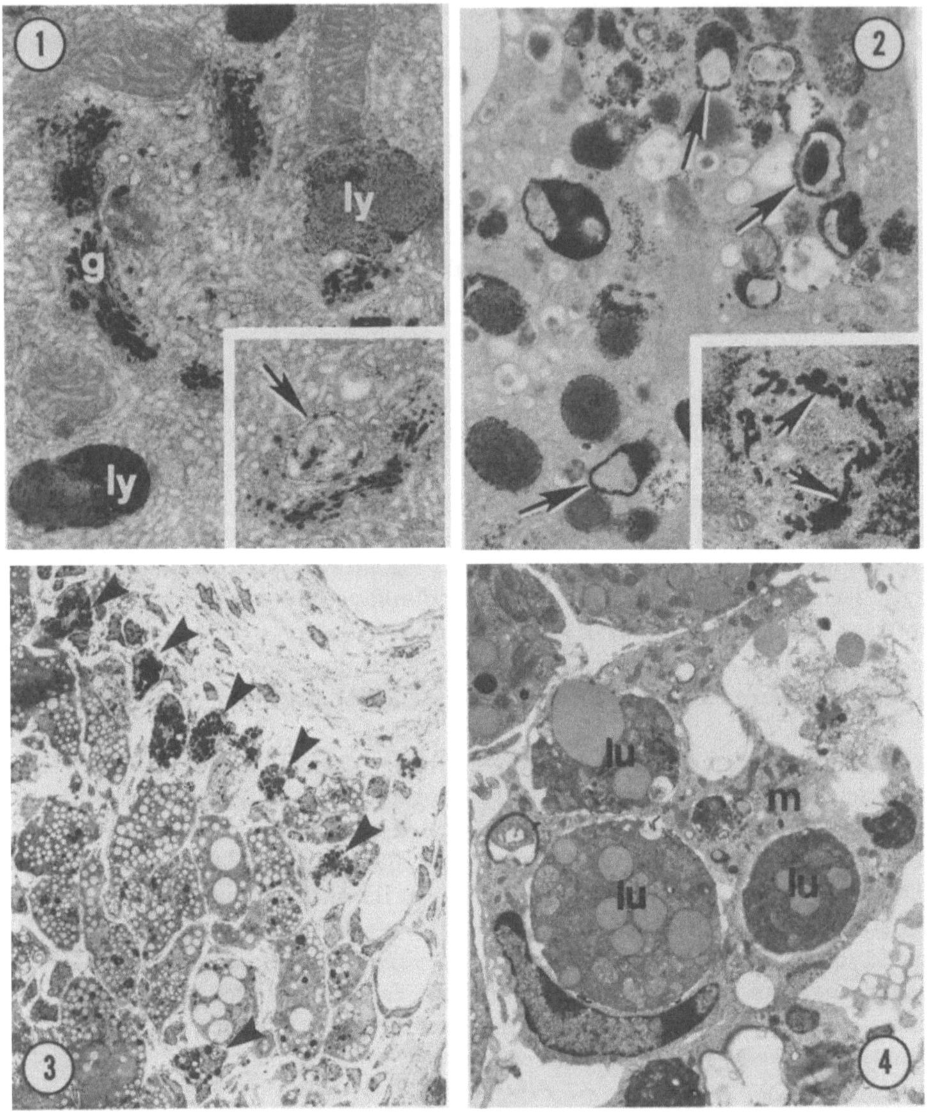


Fig. 1 Luteal cell at time of peak steroid secretion, reacted for ACPase. Golgi cisternae (g) and lysosomes (ly) are reactive. Inset shows low level of ACPase in GERL (arrow). X 21,000; inset, X 24,000. Fig. 2 Regressing luteal cell. ACPase-positive double-walled autophagic vacuoles (arrows) and lysosomes are abundant. Inset shows intensely reactive GERL of this stage. X 16,000; inset, X 20,000. Fig. 3 Regressing corpus luteum showing activated macrophages (arrowheads) along the edge. X 575. Fig. 4 Macrophage (m) displaying large heterophagic vacuoles that contain recognizable pieces of luteal cell cytoplasm (lu). X 4,500.

since they (a) are completely "closed" structures, (b) contain various cellular organelles, and (c) are reactive for more than one acid hydrolase.

In the present study, ACPase is not observed free in the cytoplasm of regressing luteal cells prior to their engulfment by macrophages.

Macrophages in Corpora Lutea of Pregnant and Postpartum Animals

Activated macrophages are conspicuous in involuting corpora lutea. As we have shown previously (12), they can be readily distinguished from luteal cells, even in low power electron micrographs (fig. 3).

Occasional luteal cells are ingested by macrophages in the pregnant animal. During the luteolytic period, however, phagocytosis of luteal cells is markedly increased. Macrophages are observed phagocytizing and digesting materials that range in size from small pieces of luteal cells to entire luteal cells (fig. 4). Luteal cell fragments often retain their normal morphology during the initial stages of ingestion. Generally, digestion begins at some point after phagocytosis is completed, and heterophagic vacuoles containing partially digested fragments of luteal cells are noted.

Activated macrophages are abundant along the perimeter of involuting corpora lutea (fig. 3), a finding that is consistent with the more or less symmetrical shrinkage which occurs as this tissue regresses. In the most advanced stages of luteolysis, virtually all luteal cells are enveloped by macrophages. Examination of material from fully regressed corpora lutea reveals that the scar consists of collagen, debris-laden macrophages, a few fibroblasts, and some cells that superficially resemble luteal cells. The identity of these latter cells remains uncertain. For other details regarding these cells, see Paavola (12).

## DISCUSSION

The data summarized here demonstrate that the normal postpartum regression of luteal tissue in guinea pigs involves both autophagy and heterophagy. The importance of autophagy as a mechanism in luteal cell involution is indicated by the striking increase in the number of lysosomes and the appearance of numerous autophagic vacuoles that occurs during luteolysis. Both are reactive for ACPase and arylsulfatase, establishing that they belong to the intracellular digestive apparatus, and thus probably provide the morphological bases for autolysis. The finding that autophagy plays a central role in the involution of guinea pig corpora lutea contrasts with that of van Lennep and Madden (1) in humans, and agrees with that of McClellan et al (2) in sheep.

A change characteristically associated with luteolysis is the remarkable development of GERL. This structure, which is rather modestly developed at the time of peak steroid secretion, becomes markedly hypertrophied near term, a condition that is maintained well into the postpartum period. This enlarged GERL system typically shows intense lytic enzyme activity, and functions in the packaging of acid hydrolases into lytic bodies in aging luteal cells, as it does in other cell types (10). Lysosomes appear to develop as focal protuberances of varying morphology that eventually separate from the parent GERL cisterna. In addition, larger lytic bodies seem to arise from the fusion of small vesicles that originated from GERL. Further, the double-walled autophagic vacuoles typically seen in regressing luteal cells of guinea pigs derive from GERL as follows: GERL cisternae sequester and wall-off small areas of cytoplasm, becoming in the process completely closed autophagic vacuoles that contain various cellular organelles. These vacuoles may then enlarge, perhaps sequestering additional bits of cytoplasm, or become structurally more complex, often developing multiple internal compartments. It seems likely that the acid hydrolases contained between the double walls of these vacuoles gain access to the interior of the vacuole through a disintegration or alteration of the inner limiting membrane. Release of lytic enzymes into the vacuole would initiate digestion of the contents, which results in the formation of residual bodies. Additional lytic enzymes may be gained by fusion of the vacuoles with lysosomes, or from lysosomes trapped in the vacuoles. For further discussion of the possible relationships between GERL, autophagic vacuoles, and lysosomes in regressing luteal cells, see Paavola (11).

Another conceivable mechanism for luteolysis, not supported by the present data, is the release of lysosomal enzymes directly into the cytoplasm. It has been suggested that as the corpus luteum ages, lysosomes become more fragile; thus, they are more susceptible to rupture (13). However, in the present study, ACPase activity was never observed free in the cytoplasm of luteal cells prior to their engulfment by macrophages, arguing against a generalized intracellular release of lytic enzymes.

The importance of heterophagy as a mechanism in the normal postpartum regression of luteal tissue in guinea pigs is emphasized by the abundance of macrophages that occur in regressing corpora lutea of this species. This finding contrasts with that of Bagwell et al (14), who reported that macrophages were absent from corpora lutea during luteolysin-induced regression in hysterectomized guinea pigs.

Activated macrophages, by means of processes or pseudopodia, make contact with, then engulf whole luteal cells as well as pieces of luteal cells. Once internalized in heterophagic vacuoles, luteal cells or their fragments are either digested or broken into small pieces, then digested. Continued degradation of these bits of luteal cell cytoplasm ultimately produces the distinctive large, dense residual bodies seen in these macrophages. Interestingly, activated macrophages phagocytize luteal cells that show no visible signs of degeneration as well as those that are obviously deteriorated. The avid phagocytosis of luteal cells or their remains by macrophages appears to bring about a reduction in the size of the corpus luteum, until little remains, except macrophages filled with luteal debris. Additional details on the role of macrophages in luteolysis appear in another report from this laboratory (12).

In conclusion, this study has demonstrated that cellular autophagy is a prominent mechanism underlying luteal cell involution, and has documented the previously suspected involvement of macrophages (i.e., heterophagy) in luteolysis. Complete regression of corpora lutea during normal postpartum luteolysis in guinea pigs thus depends on both autophagy and heterophagy.

(This work was supported by Research Grants HD 05897 and HD 09993 from the N.I.H., and by Biomedical Research Support Grant RR 05417 from the N.I.H.)

(1) E.W. van Lennep and L.M. Madden. Z. Zellforsch. 66(1965) 365.

(2) M.C. McClellan, J.H. Abel and G.D. Niswender. Biol. Reprod. 16 (1977) 499.

(3) P. Banon, D. Brandes and J.K. Frost. Acta Cytol. 8 (1964) 416.

(4) D. Bulmer. J. Anat. Lond. 98 (1964) 313.

(5) E.C. Adams and A.T. Hertig. J. Cell Biol. 41 (1969) 696.

(6) W.W. Leavitt, C.R. Basom, J.N. Bagwell and G.C. Blaha. Am. J. Anat. 136 (1973) 235.

(7) C. de Duve. in: Ciba Fdn. Symp. on Lysosomes, eds. A.V.S. de Reuck and M.P. Cameron (Churchill Ltd., London, 1963) p. 1.

(8) L.G. Paavola. Am. J. Anat. 150 (1977) 565.

(9) L.G. Paavola. J. Cell Biol. (II, in press).

(10) A.B. Novikoff. in: Lysosomes and Storage Diseases, eds. H.G. Hers and F. van Hoof (Academic Press, New York, 1973) p. 1.

(11) L.G. Paavola. J. Cell Biol. (III, in press).

(12) L.G. Paavola. Am. J. Anat. (in press).

(13) J.T. Dingle, M.F. Hay and R.M. Moor. J. Endocr. 40 (1968) 325.

(14) J.N. Bagwell, D.L. Davies and J.R. Ruby. Anat. Rec. 183 (1975) 229.

# ROLE OF PITUITARY LH AND PLACENTA IN LUTEAL PROGESTERONE PRODUCTION AND MAINTENANCE OF PREGNANCY IN THE RAT

H. G. Madhwa Raj,*† L. M. Talbert,† W. E. Easterling† and Ruth Chen Dy†.
Division of Reproductive Endocrinology and Fertility†
Departments of Obstetrics and Gynecology† and Pharmacology*
University of North Carolina School of Medicine
Chapel Hill, North Carolina 27514

## INTRODUCTION

It is now well-established that neutralization of circulating LH using specific antiserum leads to termination of pregnancy in the rat (1-3). In our extensive investigations, we demonstrated that LH antiserum is effective when administered from day 1-8 or any day between 8 and 11 (1,3). Neutralization of LH is effective even at midnight of day 11, but not at 10 a.m. on day 12, indicating a sharp 'cut off' point. The observed termination of pregnancy is accompanied by profuse vaginal bleeding 36 to 48 hrs after antiserum administration. Plasma progesterone levels drastically fall within a short period of neutralizing LH (4). This indicates that LH is essential for production of progesterone at this time of pregnancy. The effects of LH antiserum on pregnancy have been amply confirmed by other investigators (5-7). Morishige and Rothchild (5) extended these studies to earlier period of pregnancy and demonstrated that LH dependency sets in between day 6 and 8, simultaneously with the loss of dependency on pituitary prolactin. To investigate the mode of action of LH, Gibori et al utilized the LH antiserum treated rat model system (8). They demonstrated that testosterone implants can maintain pregnancy, in such rats. Further Gibori and Keyes have provided significant evidence that testosterone administered is converted to estrogen, which in turn maintains progesterone production in the pregnant rat. The rat corpus luteum contains high concentrations of estrogens during pregnancy (9, 10) and exogenous estrogen stimulates luteal growth and progesterone secretion in hypophysectomized-hysterectomized pregnant rats (11, 12). Intraovarian bursal estrogen implants also seem to maintain luteal functionality as judged by progesterone secretion, during days 10 and 11 of pregnancy (13). Thus at the present time it seems that the luteal progesterone production becomes dependent on pituitary LH and placental luteotropin by day 8 of pregnancy. The present series of investigations were undertaken to elucidate the contribution of the placenta and its potential to sustain steroidogenesis in the absense of LH.

## EXPERIMENTAL

### Studies During Peri-implantation Period

A specific and potent antiserum to purified ovine LH was prepared in rabbits using an immunization schedule described by us earlier (1). The antiserum was absorbed with normal lamb serum to remove contaminating serum specific antibodies. The absorbed antiserum showed a single precipitin band against purified sheep LH and cross reacted with rat pituitary extracts. A single subcutaneous injection of 0.5 ml of this antiserum when given on day 8 of pregnancy, induced vaginal bleeding within 48 hrs.

Pregnant rats were injected with 0.5 ml of LH antiserum daily from days 3-6 of pregnancy (day 1 = day of sperm finding). They were divided into two groups and one group also received 0.1 µg estradiol on day 4. The rats were autopsied on day 7 afternoon and incidence of implantation was observed. It can be seen from Table 1 that LH antiserum treatment inhibited implantation. However, estradiol supplementation permitted implantation in all the rats.

### Studies During Early Pregnancy

The above investigation indicated that luteal progesterone secretion is not dependent on LH between days 3-6. However,

Table 1. Effect of LH antiserum on early pregnancy in the rat:

| Gr. | daily treatment from d. 3-6 | day 7 nt/ns |
|---|---|---|
| 1 | 0.5 ml LH AS | 5/0 |
| II | 0.5 ml LH AS + 0.1 µg estradiol (on d. 4 only) | 5/5 |

nt: total number of rats.
ns: number of rats showing implantation sites.

Table 2. Effects of LH antiserum and dydrogesterone on pregnancy

| Daily Treatment, day 8 - 13 (number of animals) | ovarian weight mean ± S.E. | luteal weight mean ± S.E. | avg. no. of embryos |
|---|---|---|---|
| LH antiserum only | 64.0 ± 2.47 | 1.77 ± 0.25 | 0 vaginal bleeding observed. |
| LH antiserum + Dydrogesterone + Estradiol | 82.5 ± 5/8 | 2.76 ± 0.36 | 11; live. No bleeding. |

the same antiserum was capable of inducing abortion when given on day 8. Hence the potential of placenta to contribute to luteal progesterone secretion was tested in the following experiment: Adult Charles River female rats were bred with proven males and the day of sperm finding was considered as day 1. The rats were laparotomized on day 8 and only rats with normal complement of implantation sites were used for further experimentation. All the rats received 0.5 ml of LH antiserum daily between day 8 - 13. They were divided into two groups. One of the groups was supplemented daily with 4 mg dydrogesterone (6-dehydro-9-β-10-α- progesterone; trade name: Duphaston, Philips Roxane Laboratories, Inc.) and 0.1 μg estradiol as a single subcutaneous injection. The treatment was continued through day 13 and the rats were autopsied on day 14. At autopsy blood was collected and serum separated. The ovaries were removed, cleaned of extraneous tissue and quickly weighed. The corpora lutea were dissected in cold under a dissecting microscope and freed from follicles and interstitial tissue. The corpora lutea and the non-luteal tissue were separately weighed and frozen quickly on a block of dry ice. Serum and

Table 3: Effects of maintenance of pregnancy on luteal progesterone secretion in LH antiserum treated rats.

| Daily treatment between d. 8-11 | progesterone | | |
|---|---|---|---|
| | in serum ng/ml | in corpora lutea, mean ± S.E. | |
| | mean ± S.E. | ng/CL | ng/mg |
| LH antiserum only | 9.2 ± 2.6 | 3.8 ± 1.4 | 6.5 ± 1.6 |
| LH antiserum + dydrogesterone + estradiol | 72.0 ± 6.1 | 42.7 ± 13.7 | 114.2 ± 30.0 |

tissues were stored frozen at -40° c till analysed. Number and nature of embryos were noted at autopsy. Progesterone was measured in serum and ovarian tissue as an index of luteal function, using a specific antiserum obtained from Dr. G.E. Abraham. The extraction procedure as described by us previously (14) was utilized. It was ascertained that injected dydrogesterone does not interfere with progesterone assay. In the assay system used, less than 1% cross-reaction was observed with this synthetic progestin.

All the rats treated with LH antiserum alone, bled within 48 hrs and showed no implantation sites or live embryos at autopsy (Table 2). In contrast, rats supplemented with dydrogesterone and estradiol maintained pregnancy and live growing embryos were seen at autopsy. Both the total ovarian weight and weight of corpora lutea were significantly higher in this group given dydrogesterone.

It can be seen from Table 3 that luteal and serum progesterone levels drastically fell upon deprivation of LH. However, when pregnancy (and hence feto-placental unit) is maintained by supplementation with the synthetic steroid, luteal functionality also was maintained. This was reflected in the elevated serum progesterone levels, luteal content and concentration of progesterone (Table 3).

## DISCUSSION

In the present investigation an attempt has been made to maintain the feto-placental unit under conditions of LH deprivation, during the LH dependent phase of early pregnancy. This has been achieved by using dydrogesterone, a synthetic progestin closely related to retroprogesterone. This steroid was particularly suitable as 1.) it has been shown not to affect luteal progesterone production in women (15), unlike some synthetic progestins as Norgestrel, chlormadione acetate or medroxy progesterone acetate (16), and 2.) it cross-reacts to a negligible degree (less than 1%) in progesterone radioimmunoassay, thus permitting measurements of endogenous luteal progesterone secretion. Under these conditions, luteal functionality has been examined using progesterone production as an index. The results obtained clearly indicate that maintenance of pregnancy and feto-placental unit in an intact condition has a beneficial effect on luteal function, and progesterone production. It is well-established that the rat placenta produces a luteotropin which is prolactin-like in biologic action (17, see ref. 7 for review). The placenta is established with the occurrence of implantation between days 5 and 6 in the pregnant rat, and, by day 8, has already assumed the luteotropic function of pituitary prolactin as shown by Morishige and Rothchild (5). Even though during normal physiological progress of pregnancy the corpora lutea become dependent on pituitary LH, the placenta has the potential of maintaining luteal function, as demonstrated in the present experiments. As intra-luteal estrogen is being implicated in mediating progesterone secretion, it was of interest to measure ovarian estradiol concentrations in our current experimental model. Preliminary results indicate that LH antiserum treatment drastically reduces luteal estradiol concentration. However, concomitant with maintenance of feto-placental unit by dydrogesterone in LH antiserum treated rats, luteal estradiol concentration rises to high levels found in nontreated pregnant rats. The results seem to lend further support for involvement of estrogen in luteal progesterone production, in this species. The source of tropic support responsible for maintaining this high intra-luteal estrogen level in face of pituitary LH deprivation, is not known. A chorionic LH-like hormone (rCG) has been recently reported (18). As the placenta is intact in the present model system, the possibility that such an rCG might be maintaining luteal estradiol, cannot be excluded.

The observation that daily treatment with antiserum to LH from days 3-6 (Table 1) is not effective in terminating pregnancy (if estradiol is supplemented on day 4), is consistent with the demonstration that prolactin alone maintains progesterone secretion during this period (5, 19, 20). Indeed we have demonstrated that prolactin alone can, in the complete absence of LH, stimulate luteal progesterone secretion (14). However, the reasons for the onset of LH dependency and its mechanism are not clear at this time, and need to be investigated.

## REFERENCES

(1) H.G. Madhwa Raj and N.R. Moudgal. Endocrinology 86(1970)874.

(2) K.K. Loewit, S. Badawy and K.A. Laurence. Endocrinology 84(1969)244.

(3) H.G. Madhwa Raj, M.R. Sairam and N.R. Moudgal. Jour.Reprod. Fertil. 17(1968)335.

(4) N.R. Moudgal, H.R. Behrman and R.O. Greep. J.Endocrinol. 52(1972)413.

(5) W.K. Morishige and I. Rothchild. Endocrinology 95(1974)260.

(6) J. Akaka, E. O'Laughlin-Phillips and I. Rothchild. Endocrinology 100(1977) 1334.

(7) M.R. Sairam and C.H. Li in: Hormonal Proteins and Peptides, Vol. 1, ed. C.H. Li (Academic Press, New York, 1973) p. 101.

(8) G. Gibori, P.L. Keyes and J.S. Richards. Proc.9thAnn.Meet. Soc.Stud.Reprod. (1976) p. 39.

(9) H.B. Waynforth and D.M. Robertson. J.Endocrinol. 54(1972)79.

(10) D.J. Elbaum and P.L. Keyes. Endocrinology 99(1976)573.

(11) M. Takayama and G.S. Greenwald. Endocrinology 92(1973)1405.

(12) G. Gibori and P.L. Keyes. Endocrinology 102(1978)1176.

(13) G. Gibori, R. Rodway and I. Rothchild. Endocrinology 101(1977)1683.

(14) G.J. Macdonald, N.R. Moudgal, H.G. Madhwa Raj and R.O. Greep. Proc.Soc.Exp.Biol.Med. 144(1973)923.

(15) V. Larsson-Cohn, E.D.B. Johansson, L. Wide and C. Gemzell. J.Obst.Gynecol.British.Commwlth. 77(1970)840.

(16) E.D.B. Johansson. ActaEndocrinol. 68(1970)779.

(17) E.B. Astwood and R.O. Greep. Proc.Soc.Exp.Biol.Med. 38(1938)713.

(18) F.G. Haour and P. Sanchez. C.R.Acad.Sci. 282(1976)1183.

(19) J. Clemens, M. Sar and J. Meites. Proc.Soc.Exp.Biol.Med. 130(1969)628.

(20) E.S. Kisch and M.C. Schelesnyak. Jour.Reprod.Fertil. 15(1968)401.

# RELATIONSHIPS BETWEEN CHOLESTEROL SUPPLY AND LUTEAL MITOCHONDRIAL STEROID SYNTHESIS

M.E. Toaff, G.L. Flickinger, J.F. Strauss, III and S.J. Shattil, Depts. of OB/GYN and Medicine, University of Pennsylvania School of Medicine, Philadelphia, Pa.

LH acutely increases luteal progestin biosynthesis by stimulating cholesterol side-chain cleavage, a reaction catalyzed by a complex of mitochondrial enzymes containing cytochrome P-450 (1). The steroidogenic action of LH is inhibited by cycloheximide and puromycin, indicating a requirement for protein synthesis (2). It has been suggested that availability of cholesterol governs the rate of side-chain cleavage and that proteins formed under the influence of LH play a role in translocation of cholesterol to the mitochondrial enzyme complex (1,2). In the present study, relationships between cholesterol supply and *in vitro* steroid synthesis by rat luteal mitochondria were explored.

Luteinized ovaries were produced in 21-23 day old Sprague-Dawley rats by PMS-hCG treatment (3). Ovaries were removed from rats killed 3-4 days after hCG administration. In some experiments, animals were given 5 mg of cycloheximide or the 0.9% saline vehicle, i.p., 1 h before killing. Mitochondria-rich fractions were prepared from the ovaries, which were homogenized in the medium described by Robinson and Stevenson (4), by differential centrifugation.

Dispersions of cholesterol (C) and dipalmitoyl phosphatidylcholine (PL) were prepared in incubation buffer (4) without albumin in molar (C/PL) ratios of 0.5, 1.0 and 2.2 as described by Shattil *et al* (5). Incubation flasks contained 2.0 ml of mitochondrial suspension, 10 mM succinate and 1.5 ml of lipid dispersion or buffer. In certain studies, mitochondria were disrupted by sonication and were incubated in the presence of 1 mM NADPH rather than succinate. Zero time samples were collected prior to initiation of incubations at 37 C in a water bath under an atmosphere of air. When lipid contents of mitochondria were determined, incubations were terminated by cooling flasks to 4 C and immediately pelleting mitochondria by centrifugation. Supernatants were decanted and mitochondrial pellets were washed 3 times by suspension in 0.25 M sucrose and recentrifugation.

Mitochondrial lipids were extracted, fractionated by thin-layer chromatography, and cholesterol and phospholipids were determined by colorimetric assays as previously described (3). Mitochondrial cytochrome P-450 contents were measured as described by Omura and Sato (6). Pregnenolone and progesterone were quantitated by specific radioimmunoassays (7,8). Net steroid synthesis, expressed as the sum of pregnenolone and progesterone formed, was determined.

Lipid dispersions having different C/PL ratios were used as a means of altering availability of cholesterol for mitochondrial steroid synthesis. C/PL ratio, and not mass of sterol added to the incubation flask, determined accessability of exogenous substrate. We observed a direct relationship between cholesterol content of mitochondria which had been incubated with lipid dispersions and *in vitro* steroid synthesis ($r=0.92$, $p<0.001$). When mitochondria were incubated with dispersions having a C/PL ratio of 1 and 2.2, the resulting mitochondrial contents of cholesterol after 30 min were 2-fold and 3-fold greater, respectively, than mitochondrial cholesterol levels of control flasks which contained only buffer (control mitochondrial cholesterol contents: 37 nmol/mg protein). Steroid synthesis by mitochondria incubated for 30 min in the presence of dispersions having C/PL ratios of 1 and 2.2 was augmented 2-fold and 3-fold, respectively, above control levels (control mitochondrial steroid synthesis: 2.96 nmol/mg protein/30 min). Incubation of mitochondria in the presence of pure PL dispersions resulted in a 50% reduction in steroid synthesis due to sequestration of mitochondrial cholesterol.

The effects of the lipid dispersions on luteal mitochondrial cholesterol contents were due primarily to bulk transfer of sterol. Whereas mitochondrial cholesterol levels increased when the organelles were incubated in the presence of dispersions having a C/PL of 2.2, mitochondrial phospholipid contents were not significantly altered.

Mitochondria from ovaries of rats treated with cycloheximide 1 h before killing produced significantly less steroid from endogenous substrate ($p<0.001$) than mitochondria from ovaries of saline treated rats (cycloheximide-treated: 1.38 ± 0.15 nmol steroid formed/mg protein/20 min; saline-treated: 2.90 ± 0.12). Whereas mitochondria from ovaries of saline-treated rats increased steroid synthesis 2-fold when incubated in the presence of dispersions having a C/PL of 2.2, mitochondria from cycloheximide-treated animals did not respond to exogenous substrate with enhanced steroid production. However, the cytochrome P-450 levels of mitochondria from ovaries of cycloheximide and saline-treated rats were similar (cycloheximide-treated: 0.45 ± 0.10 nmol/mg protein; saline-treated: 0.43 ± 0.8). Furthermore, when mitochondrial architecture was disrupted by sonication, mitochondria from ovaries of cycloheximide-treated rats synthesized as much steroid as sonicated mitochondria from ovaries of saline-treated animals.

We conclude that the steroidogenic pool of cholesterol in isolated rat luteal mitochondria is of limited size. Sustained steroid synthesis _in vitro_ can only be maintained by supplying exogenous substrate. Besides the availability of cholesterol, which can clearly be influenced by the C/PL ratio of lipid dispersions, mitochondrial steroid synthesis is regulated by a cycloheximide-sensitive factor. Cycloheximide treatment results in diminished _in vitro_ steroid synthesis and prevents metabolism of exogenous sterol by intact mitochondria. These observations indicate that the locus of action of cycloheximide in inhibiting steroid synthesis is at the level of the mitochondria and that the cycloheximide-sensitive factor controls access of cholesterol to the side-chain cleavage system of intact organelles.

## REFERENCES

1. G.S. Boyd, J.R. Arthur, G.J. Beckett, J.I. Mason and W.H. Trzeciak. J. Steroid Biochem. 6 (1975) 427.

2. J.R. Arthur and G.S. Boyd. Eur. J. Biochem. 49 (1974) 117.

3. L.A. Schuler, G.L. Flickinger and J.F. Strauss, III. J. Endocrinol. in press, 1978.

4. J.R. Robinson and P.M. Stevenson. Eur. J. Biochem. 24 (1971) 18.

5. S.J. Shattil, R. Anaya-Galindo, J. Bennett, R.W. Colman and R.A. Cooper. J. Clin. Invest. 55 (1975) 636.

6. T. Omura and R. Sato. J. Biol. Chem. 239 (1964) 2370.

7. L.R. Johnson, A. Ruhmann-Wennhold and D.H. Nelson. Ann. N.Y. Acad. Sci. 212 (1973) 307.

8. G.O. De Villa, K. Roberts, W.G. Wiest, G. Mikhail and G.L. Flickinger. J. Clin. Endocrinol. Metab. 35 (1972) 458.

# A COMPARISON OF THE LH CONTROL OF PROGESTERONE SYNTHESIS IN SMALL AND LARGE CELLS FROM PREGNANT COW CORPUS LUTEUM

Jocelyne Ursely and Pierre Leymarie

Laboratoire de Biochimie - Faculté de Médecine

CAEN - 14033 - France

Like in many other mammalian species, the ovarian corpus luteum from the cow is mainly constituted by two different cell types (1). The large cells (diameter 30 to 50 u) and the small cells (diameter 15 to 20 u) (2).

The follicular origin of these two cell types is still the subject of controversy (1-4) and their respective contribution to progesterone synthesis is unknown. Histologic arguments have led Donaldson and Hansel (4) to attribute to the small cells, the main part in steroïd synthesis. For Priedkahns and Weber (2) however, progesterone would be produced by the large cells which contain numerous mitochondria and eventually stored in the small ones whose cytoplasm is rich in lipid droplets.

Recently the separation of large and small luteal cells has been successfully performed in the sow (5) and the steroïdogenic response to a single dose of LH has been proven to be markedly different in each cell type.

The present report concerns an "*in vitro*" study of cow corpus luteum small and large cells. Their respective ability to synthesize progesterone and to respond to LH has been investigated as well as the inhibition by estradiol of the LH stimulation.

Corpora lutea from 60 to 100 days pregnant cows were collected at slaughter and brought to the laboratory in ice cold saline.

In a typical experiment, about 2g of tissue, were enzymatically dissociated according to the technique of Gospodarowicz and Gospodarowicz (6) slightly modified. The cells were then separated by sedimentation at unit gravity according to an adaptation of the method developed by Miller and Philips (7). The sedimentation through a non linear gradient of Ficoll BSA in KRB, lasted 40 min at room temperature. The KRB contained half the usual concentration of $Ca^{++}$ and was continously gassed with $O_2$-$CO_2$ (95% - 5%). The cells were collected in fractions and centrifuged 5 min at 50 g After inspection under the microscope, the loose pellets, containing one cell type were pooled. About $10^6$ small cells (diameter 15-20 u) and $10^5$ large cells (diameter 30-50 u) were usually obtained. The poor recovery of large cells was apparently due to their relative fragility. The viability of large and small cells, evaluated after separation, by the Trypan Blue exclusion test (8) was 80% and 95% respectively.

Small cells were devoided of any contamination by large cells, but large cells were slighty contaminated by a variable amount of small cell agregates.

The "*in vitro*" incubation with or without LH were performed at 37°C under $O_2$-$CO_2$ (95% - 5%) immediately after separation. But cells could be kept at 4°C in the incubation medium (KRB pH 7.3 containing 4% BSA, 0.2% glucose, 0.1% Egg white trypsin inhibitor). Under these conditions, 93% of the small cells and 50% of the large ones were still alive 24 hours later.

Progesterone was measured by R.I.A. using an anti-serum raised against 3 carboxy-methyl-oxime progesterone-BSA. The steroïd content and "*de novo*" synthesis were expressed in pg of progesterone per cell.

Table 1 : Progesterone synthesis by large and small cells during the course of 2h incubations in the presence (+LH) or the absence (-LH) of added LH.

| | | Progesterone pg/cell (range) - LH | + LH | Stimulation ratio |
|---|---|---|---|---|
| Large cells (3 experiments) | 1st hour | 1.65 (1.3-2.4) | 4.42 (1.6-5.9) | 2.7 |
| | 2nd hour | 2.1 (1-3) | 4.86 (3-5.5) | 2.3 |
| Small cells (5 experiments) | 1st hour | 0.059 (0.03-0.08) | 2.17 (1.1-4.6) | 36.8 |
| | 2nd hour | 0.07 (0.02-0.09) | 1.63 (0.7-3) | 23 |

LH stimulation of progesterone synthesis

The endogenous content of the large cells (6.5 to 0.8 pg/cell) was about 10 times that of the small ones (0.03 to 0.09 pg/cell). Progesterone synthesis was measured for each cell type in the absence and in the presence of a saturating dose ($10^{-8}$ M) of LH during the course of two hour incubations. The results of five experiments are given on Table I.

In the absence of LH, large cells synthetized 10 to 30 times more progesterone than the small ones, whereas in the presence of LH, each cell type synthetized a comparable amount of progesterone, so that the factor of stimulation by LH was 10 fold higher for the small cells than for the large ones.

Progesterone synthesis at various LH concentrations was measured for the two cell types at the end of 2h incubations. The dose response curves are given on figure 1. Here again a striking difference can be observed between the two cell types. It concerns the minimum effective dose of LH as well as the minimum dose required to achieve complete stimulation. Both values were found several orders of magnitude higher in the large cells than in the small ones.

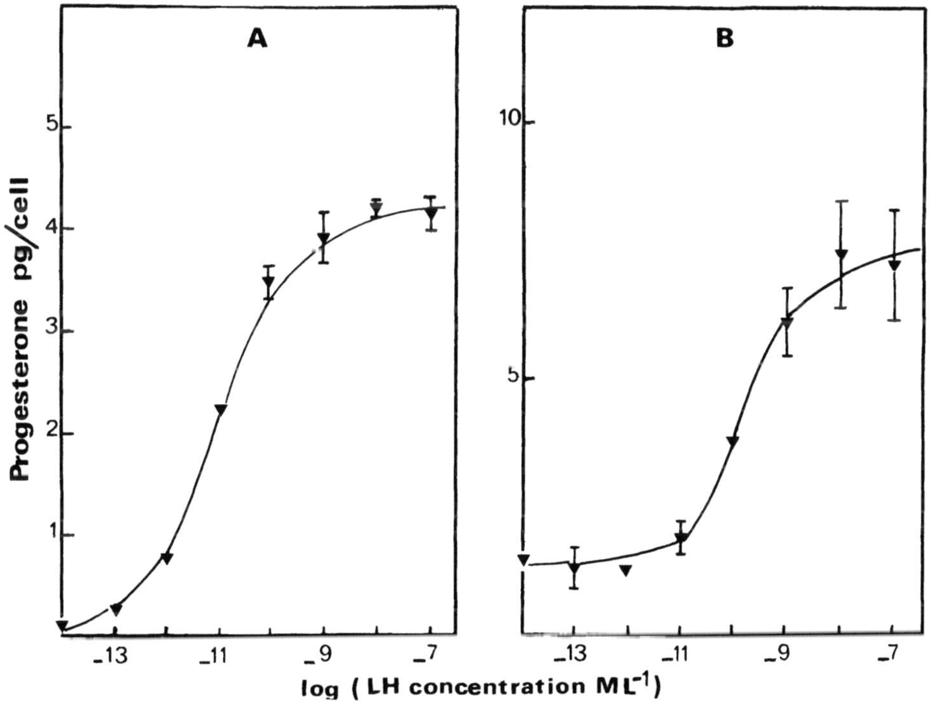


Figure 1 : Effects of various concentrations of bovine LH on progesterone synthesis by small (A) and large (B) luteal cells.

## Estradiol inhibition

In the cow, estradiol 17β has been reported to inhibit the LH stimulation of progesterone synthesis (9). We have also investigated in each cell type this effect of estradiol for various concentrations of the steroïd. The observed inhibition was found to be similar in both cell types. A slight inhibition was produced by steroïd concentration of $5.10^{-8}$ M and the total suppression of LH stimulation necessitated the use of $5.10^{-4}$ M concentration.

## CONCLUSION

Large and small cells from cow corpus luteum have been found to be able, "*in vitro*", to synthetize progesterone and to respond to hormonal stimulation.

Both of them are likely to be involved in progesterone production "*in vivo*" but their respective contribution is difficult to assess in the absence of data on the LH concentration at the cellular level.

"*In vitro*" isolation of homogenous suspensions of small cells could provide an improved model system for the study of the biochemical mechanisms underlying the hormonal regulation of progesterone synthesis in cow corpus luteum.

## ACKNOWLEGMENTS

We thank Mrs M. LEMON, I.N.R.A. Nouzilly France, for useful technical advices in the beginning of this study.

## REFERENCES

(1) H.W. Mossman and K.L. Duke. in: Handbook of Physiology, Section 7, Vol.II, part1, ed. S.R. Geiger (Whashington D.C. American Physiological Society, 1973) p. 389.

(2) J. Priedkahns and A.F. Weber. Z. Zellforsh. 91(1968)554.

(3) J. Priedkahns and A.F. Weber. Z. Zellforsh. 91(1968)574.

(4) L. Donaldson and W. Hansel. J. Dairy Sci. 48(1966)905.

(5) M. Lemon and M. Loir. J. Endocrinol. 72(1977)351.

(6) D. Gospodarowicz and F. Gospodarowicz. Endocrinol. 90(1972)1427.

(7) R.G. Miller and R.A. Philips. J. Cell. Physiol. 73(1969)191.

(8) J.R. Tennant. Transplantation. 2(1964)685.

(9) M.T. Williams and J.M. Marsh. Abstracts of the 59th Annual Meeting of The Endocrine Society (1977) p. 172.

# CYTOCHALASIN B INHIBITION OF LUTEINIZING HORMONE ACTION ON BOVINE LUTEAL CELLS

Marvin T. Williams and John M. Marsh

The Endocrine Laboratory and The Department of Biochemistry
University of Miami School of Medicine, Miami, Fl.33152

The mechanism of action of luteinizing hormone (LH)[1] in stimulating steroidogenesis in the corpus luteum has been studied for some time, and a concept of some of the major aspects of this mechanism has emerged (1). LH binds to receptors on the luteal cell surface and this binding brings about an activation of adenylate cyclase and an increase in the concentration of cyclic AMP. This cyclic AMP in turn binds to the regulatory subunit of a cyclic AMP-dependent protein kinase causing the release and activation of the catalytic subunit. The increase in protein kinase is believed to mediate most if not all of the effects of LH on the steroidogenic pathway, but its precise mechanism is not known at this time. The limiting step in the steroidogenic pathway appears to be between cholesterol and pregnenolone but there are several ways in which this step might be accelerated (1). One possibility is that LH and cyclic AMP might promote the transport of cholesterol into mitochondria, and there is some experimental support for this from studies on the mechanism of action of ACTH in stimulating adrenal corticosteroidogenesis (2,3). Furthermore, microfilaments, a contractile element of the cells, have been implicated in the ACTH stimulated transport of cholesterol into the mitochondria of Y-1 mouse adrenal tumor cells (4). We have attempted to investigate the possible role of microfilaments in the mechanism of action of LH in the luteal cell. Our approach has been to study the effect of an inhibitor of microfilament function, namely cytochalasin B.

---

[1] The following abbreviations are used: LH, luteinizing hormone; HEPES, N-2-hydroxy ethyl piperazine-N[1]-2-ethanesulfonic acid; cAMP, cyclic adenosine 3',5'-monophosphate; ACTH, adrenocorticotropic hormone.

Corpora lutea were obtained from cows 2 to 6 months pregnant and isolated luteal cells were prepared using collagenase (Worthington Type I) (5). Aliquots of 2 X $10^6$ cells were incubated at 37°C in air for 20 min in 0.5 ml of Krebs-Ringer salt solution with 25 mM HEPES (KR-HEPES) substituted for the bicarbonate or in 0.5 ml of KR-HEPES with or without glucose and various concentrations of cytochalasin B. The subsequent incubations were done in the presence of 1 μg/ml (where indicated). The test incubations were terminated by the addition of 55 μl of 100%. [8-$^3$H] Progesterone and [8-$^3$H] cyclic AMP were added as tracers to correct for losses during extraction and purification procedures. Progesterone was extracted from samples with diethyl ether and measured by RIA as previously described (5). Cyclic AMP was measured by the procedure described by Ling et al (6).

Table 1 shows the effect of $10^{-4}$M cytochalasin B on LH stimulated progesterone synthesis in isolated bovine luteal cells. LH (1 μg/ml) significantly stimulated progesterone synthesis over control ($p<0.001$). Moreover this LH stimulated synthesis is blocked by $10^{-4}$M cytochalasin B ($p<0.002$). Cytochalasin B did not affect control synthesis, indicating that this inhibition is not simply due to a poisoning effect and that it is directed toward the trophic hormone stimulation of steroidogenesis in this tissue.

TABLE I

Effect of cytochalasin B on progesterone synthesis in isolated bovine luteal cells

| Additions | Progesterone Synthesis (μg/ 2 X $10^6$ cells) | (N) | Significance |
|---|---|---|---|
| None | 0.344 ± 0.09 | (8) | |
| Cytochalasin B ($10^{-4}$M) | 0.304 ± 0.10 | (8) | |
| LH ( 1 μg/ml) | 2.130 ± 0.32 | (8) | $p<0.001$[1] |
| LH ( 1 μg/ml) + cytochalasin B ($10^{-4}$M) | 0.728 ± 0.17 | (8) | $p<0.002$[2] |
| LH (1 μg/ml) + cytochalasin B ($10^{-4}$) + glucose ( 50 mM) | 0.74 ± 0.17 | (5) | $p<0.002$[3] |

Approximately 2 X $10^6$ cells were incubated in 0.5 ml of KR-HEPES media at 37°C for 1 hour with the indicated test substances. Data is expressed as the mean ± standard error.

[1] - paired comparison with control
[2] - paired comparison between LH and LH + cytochalasin B
[3] - paired comparison between LH and LH + cytochalasin B + 50 mM glucose

Cytochalasin B is also known to competitively inhibit glucose transport into cells and this inhibition can be overcome by high concentrations of glucose (4). It was possible therefore that the observed inhibition of steroidogenesis in our luteal cell system was due to blockage of glucose transport. As can be seen in Table 1 we still observed an inhibition of steroidogenesis by cytochalasin B in the presence of 50 mM glucose. Thus it appears that the inhibitory effect of cytochalasin B is not due to an action on glucose transport.

The inhibition of LH stimulated progesterone synthesis by cytochalasin B is rapidly reversed when cells are washed and fresh medium containing LH is added (Fig. 1). Each data point is the mean of 2 separate experiments. LH stimulated the accumulation of progesterone over the control. Cytochalasin B at a concentration of $10^{-4}$M blocked the stimulatory effect of LH. However, this inhibition was readily overcome when the cells were washed (at 30 min) and fresh buffer containing LH was added.

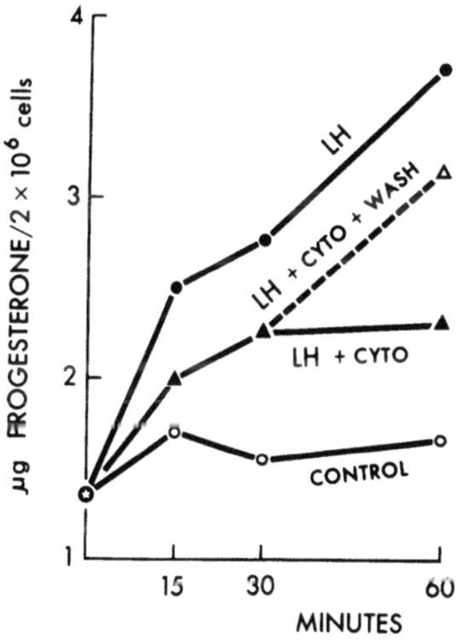


Fig. 1. The reversibility of cytochalasin B inhibition of progesterone synthesis. Incubations were carried out as described in Table I. Each data point represents the mean of 2 separate experiments.

Table 2 shows that the cytochalasin B inhibition of LH stimulated progesterone synthesis was dose related. We observed a partial but statistically significant inhibition of the LH stimulated progesterone synthesis at the lowest concentration of cytochalasin B tested ($10^{-6}$ M). The extent of the inhibition was increased with increasing concentrations of this inhibitor.

TABLE 2

Dose dependent inhibition of LH stimulated progesterone synthesis by cytochalasin B.

| Additions | Progesterone Synthesis ( μg/2 X $10^6$ cells) | (N) | Significance |
|---|---|---|---|
| None | 0.344 ± 0.09 | (8) | |
| LH (1 μg/ml) | 2.13 ± 0.32 | (8) | p<0.001[1] |
| LH (1 μg/ml) + $10^{-6}$M cytochalasin B | 1.49 ± 0.35 | (5) | p<0.01[2] |
| LH (1 μg/ml) + $10^{-5}$M cytochalasin B | 1.25 ± 0.36 | (4) | p<0.02[3] |
| LH (1 μg/ml) + $10^{-4}$M cytochalasin B | 0.728 ± 0.17 | (8) | p<0.002[4] |

Incubations were carried out as described in Table 1. Data is expressed as the mean ± the standard error.

[1] - paired comparison between LH and control

[2] - paired comparison between LH and LH + $10^{-6}$ M cytochalasin B

[3] - paired comparison between LH and LH + $10^{-5}$ M cytochalasin B

[4] - paired comparison between LH and LH + $10^{-4}$ M cytochalasin B

There is a large amount of evidence which shows that cAMP is a mediator of the LH stimulated progesterone synthesis in the corpus luteum, and we attempted to determine if cytochalasin B also blocked the effect of LH on cAMP accumulation.

As can be seen in Table 3, LH significantly stimulated the cAMP accumulation over control incubations ($p<0.05$). Moreover, this increase in cAMP caused by LH was not inhibited by $10^{-4}$ cytochalasin B. However in these same cells (although data not shown here) cytochalasin B virtually eliminated the LH stimulated progesterone synthesis. This suggests that cytochalasin B acted after the accumulation of cAMP.

Table 4 shows that dibutyryl cAMP at a concentration of 2 mM can mimic the effect of LH by significantly stimulating progesterone synthesis ($p<0.02$). Furthermore, the stimulatory effect of DBcAMP is also suppressed ($p<0.01$) by $10^{-4}$M cytochalasin B. These results therefore tend to confirm that the site of action of cytochalasin B is after cAMP accumulation.

TABLE 3

Effect of cytochalasin B on LH stimulated cAMP accumulation in isolated bovine luteal cells

| Addition | cAMP accumulation (p moles/2 X $10^6$ cell) | N | Significance |
|---|---|---|---|
| None | 15.9 ± 3.02 | (5) | |
| Cytochalasin B ($10^{-4}$M) | 19.8 ± 5.98 | (5) | |
| LH (1 μg/ml) | 930 ± 511 | (5) | $p < 0.05$ [1] |
| LH (1 μg/ml) + cytochalasin B($10^{-4}$M) | 1048 ± 409 | (5) | $p > 0.05$ [2] |

Incubations were carried out as described in Table 1. The data is expressed as the mean ± standard error.

[1] paired comparison between LH and control
[2] paired comparison between LH and LH + cytochalasin B

TABLE 4

Effect of cytochalasin B on DBcAMP stimulated progesterone synthesis in isolated bovine luteal cells

| Addition | progesterone synthesis (μg/ 2 X $10^6$ cells) | N | Significance |
|---|---|---|---|
| None | 0.344 ± 0.09 | (5) | |
| DBcAMP (2mM) | 1.310 ± 0.20 | (5) | $p<0.02$ [1] |
| DBcAMP (2mM) + cytochalasin B ($10^{-4}$M) | 0.537 ± 0.12 | (5) | $p<0.01$ [2] |

Isolated luteal cells were incubated as described in Table I. DBcAMP and cytochalasin B were used at the indicated concentrations. The data is expressed as the mean ± the standard error.

[1] paired comparison between DBcAMP and control
[2] paired comparison between DBcAMP and DBcAMP + cytochalasin B

Quite recently we have attempted to elucidate the mechanism of cytochalasin B action on steroidogenesis in the corpus luteum. In this attempt we have been guided by the work of Mrotek and Hall who worked with adrenal tumor cells (4). They reported that ACTH increased cholesterol transport into adrenal mitochondria as part of its effect on steroidogenesis. They also reported that cytochalasin B brought about its inhibition of the effect of ACTH by blocking this sterol transport. In our luteal cell system, however, we have not been able to demonstrate a stimulatory effect of LH on the transport of cholesterol into mitochondria and consequently no effect of cytochalasin B at this step. It may be that we have not hit upon the proper experimental conditions to demonstrate a stimulation of cholesterol transport with LH or that this aspect of the mechanism of ACTH action is not a prominent component of LH action.

## CONCLUSION

Our data shows that cytochalasin B can reversibly inhibit LH stimulated steroidogenesis in isolated bovine luteal cells and that this inhibition is not due to a block of glucose transport. We have also shown the site of the inhibition occurs after the accumulation of cyclic AMP. In view of the known effect of cytochalasin B on microfilaments, our data suggest that these components of the cytoskeleton may be involved in LH stimulated steroidogenesis in the bovine corpus luteum. Exactly how microfilaments may be involved in luteal cell steroidogenesis remains to be elucidated.

## Acknowledgements

The authors thank Ms. Adalgisa Roja and Mr. Jorge Cidre for their excellent technical assistance. We are also indebted to Dr. G. Niswender for the gift of progesterone antisera.

## References

(1) J.M. Marsh in: Advances in Cyclic Nucleotide Research, Vol.6 eds. P. Greengard and A.G. Robison (Raven Press, New York, 1975), p. 137.

(2) L.D. Garren, N.G. Gill, H. Masui and G.M. Walton in: Recent Progress in Hormone Research, Vol. 27, ed. E.B. Astwood (Academic Press, New York, 1971) p.433.

(3) D. Mahaffee, R.C. Reitz and L.R. Nez. J. Biol. Chem. 249 (1974) 227.

(4) J.J. Mrotek and P.F. Hall. Biochemistry 16:14(1977) 3177.

(5) M.T. Williams and J.M. Marsh. Abstracts of the 59th Annual Meeting of The Endocrine Society, Abst. No. 231 (1977)

(6) W.Y. Ling, J.M. Marsh, W.N. Spellacy, A.J. Thresher, and W.J. LeMaire. J. Clin. Endocrinol. Metab. 34(1974)479.

SESSION NUMBER: VII

TITLE: LUTEOLYSIS

PRINCIPAL SPEAKER: Harold Behrman

INVITED DISCUSSANT: John McCracken

PRESENTED ABSTRACTS:

15.) L.J. Dorflinger and H.R. Behrman. Effects of Cholera Toxin, LH and $PGF_{2\alpha}$ on Steroidogenesis in Luteal Cells.

16.) J.E. Hixon and W. Hansel. Effects of Prostaglandin $F_{2\alpha}$, Estradiol and Luteinizing Hormone in Dispersed Cell Preparations of Bovine Corpora Lutea.

CHAIRPERSON: John Wilks

# MECHANISMS OF $PGF_{2\alpha}$ ACTION IN FUNCTIONAL LUTEOLYSIS

H.R. Behrman, J.L. Luborsky-Moore, C.Y. Pang,
K. Wright and L.J. Dorflinger
Reproductive Biology Section, Departments of Obstetrics & Gynecology and Pharmacology
Yale University School of Medicine, New Haven, CT 06510

Regression of the corpus luteum is pivotal in regulating the estrous and menstrual cycle since it permits follicular development and the ovulatory surge of pituitary gonadotropins in the subsequent cycle. Thus, in mammalian species mechanisms have developed which cause regression of the corpus luteum, a response marked by a precipitious decline of circulating progesterone.

Recent progress in the study of luteal regression received its greatest impetus when Pharriss et al. reported (1) that $PGF_{2\alpha}$ caused luteolysis in the rat. Since that time substantial evidence has been accumulated to show that $PGF_{2\alpha}$ is probably the elusive uterine luteolytic hormone (2). It has been established that $PGF_{2\alpha}$ is present in uterine tissue, a block of $PGF_{2\alpha}$ synthesis prolongs luteal function, and direct administration of $PGF_{2\alpha}$ causes luteolysis in a wide variety of species (2).

The focus of this paper is on recent studies which show the sites and mechanisms of $PGF_{2\alpha}$ action in luteolysis. Emphasis is placed on studies of luteolysis in the rat, but when possible, substantiating or opposing data in other species will be discussed. The discussion is limited to functional luteolysis which is defined as loss of ability of the corpus luteum to secrete progesterone. In the rat this is also marked by elevated secretion of 20α-dihydroprogesterone (20α-OH-Prog). Morphological degeneration of the corpus luteum occurs later and involves a complex series of events such as leukocytic infiltration, cellular degeneration and eventual resorption of the gland.

## HORMONAL CONTROL OF THE CORPUS LUTEUM

The corpus luteum has several aspects of control similar to that of the adrenal cortex and the Leydig cells of the testes. Luteal, adrenal and Leydig cells are morphologically similar and they depend upon pituitary hormones for continued function. The corpus luteum is unique among steroidogenic tissues in that its function is transient. The nature of the hormone requirement of the corpus luteum varies among species, but LH seems to be a common component of the luteotropic hormone complex in all species. Other hormones which appear to support luteal function in conjunction with LH are prolactin in the rat, FSH in the hamster and estrogen in the rabbit. In the human, LH appears to be the predominant luteotropic hormone.

In the estrous cycle of the rat the corpus luteum is not functional and this is probably the basis for the short 4 or 5 day cycle. It becomes functional following mating or by direct mechanical stimulation of the cervix due to pituitary release of prolactin (3). The functional corpus luteum of the rat persists for about 12 days before it undergoes regression, but in pregnancy it persists for about 21 days. The nature of gonadotropin dependence appears to change during the life span of the functional corpus luteum of the rat. Thus, during the first 4 or 5 days prolactin is the major luteotropic signal but after 8 days LH appears to be the predominant luteotropin (4). It is of interest that $PGF_{2\alpha}$, except at heroic doses, does not cause luteal regression before day 5 in the pseudopregnant rat. A similar situation exists in the ewe (5), the cow (6), and the sow (7) in that $PGF_{2\alpha}$ becomes more potent as a luteolysin as the corpus luteum ages.

Several sites of action of $PGF_{2\alpha}$ in luteolysis have been proposed. It was suggested that $PGF_{2\alpha}$ may cause constriction of the ovarian vein (8) or release of a luteolytic level of LH (9). However, it appears that the site of $PGF_{2\alpha}$ action is directly upon the corpus luteum. This conclusion is based on several pieces of evidence. First, ovarian venous flow rate, measured directly, did not change 30 min (10), 6 hr or 12 hr (11) after $PGF_{2\alpha}$ treatment, although progesterone secretion was reduced. Second, $PGF_{2\alpha}$ is luteolytic following injection directly into the corpus luteum of the ewe (12). Third, $PGF_{2\alpha}$ is luteolytic in hypophysectomized pseudopregnant rats in which corpus luteum function is maintained with exogenous prolactin (13). Fourth, $PGF_{2\alpha}$ blocks LH-dependent progesterone secretion by luteal tissue <u>in vitro</u> (discussed later). Fifth, a local mechanism for delivery of uterine venous $PGF_{2\alpha}$ to the ovarian artery appears to exist.

## EFFECT OF $PGF_{2\alpha}$ ON CORPUS LUTEUM BLOOD FLOW

The site of action of $PGF_{2\alpha}$ within the corpus luteum is less clear. Some have suggested that $PGF_{2\alpha}$ reduces blood flow to the ovary (14) or to the corpus luteum (15) thereby causing luteal regression. This question has been examined in several species. From studies in the sheep (16), rabbit (17) and rat (18) a significant decrease in blood flow to the corpus luteum is associated with regression. However, evidence points to the conclusion that this is an effect rather than a cause of luteolysis. Blood flow to the corpus luteum was unchanged by intra-arterial infusion of $PGF_{2\alpha}$ at 250 μg/hr in the rabbit within the first 8 hr although it dropped to 71% of the control value at 24 hr (19). Plasma progesterone, however, was reduced within 2 hr and continued to fall thereafter (Table 1). Kirton (20) reported that in the rabbit high doses of $PGF_{2\alpha}$ reduce luteal blood flow and plasma progesterone levels, but at a minimal luteolytic dose no effect on luteal blood flow is seen. McCracken reported similar results in studies with the ewe (21). In the rat, $PGF_{2\alpha}$ (3 mg/kg; s.c.) had no measurable effect on luteal blood flow 1 hr or 24 hr after treatment, but 36 hr after $PGF_{2\alpha}$ treatment luteal blood flow was significantly reduced (22). In this same study, plasma progesterone was significantly reduced at all time periods after $PGF_{2\alpha}$ treatment (Table 2).

These data indicate that $PGF_{2\alpha}$ probably does not cause luteal regression by reducing blood flow to the corpus luteum since the reduction in luteal blood flow follows rather than precedes the decrease in progesterone secretion.

Table 1: Effect of $PGF_{2\alpha}$ Infusion (25 μg/hr) on Luteal Blood Flow and Plasma Progesterone in the Pregnant Rabbit.

| Infusion Time | Luteal Blood Flow | Plasma Progesterone |
|---|---|---|
| hr | ml/min/100g | % of Control |
| 0 | 1155 ± 212 | 100 |
| 0.5 | 1171 ± 196 | 100 |
| 8 | 1360 ± 245 | 39* |
| 24 | 689 ± 129* | 21* |

From Bruce & Hillier (19). *significantly different from 0 hr.

Table 2: Effect of $PGF_{2\alpha}$ (3mg/kg; s.c.) on Luteal Blood Flow and Plasma Progesterone in the Pseudopregnant Rat (Mean ± S.E.).

| Time | N | Luteal Blood Flow | Plasma Progesterone |
|---|---|---|---|
| hr | | ml/min/g | ng/ml |
| 0 | 6 | 4.7 ± 0.3 | 80.4 ± 4.1 |
| 1 | 5 | 3.5 ± 0.7 | 38.3 ± 3.0* |
| 24 | 5 | 3.2 ± 0.4 | 35.4 ± 5.2* |
| 36 | 5 | 2.3 ± 0.1* | 33.2 ± 4.8* |

From Pang & Behrman (22). *$p < .05$

## EFFECT OF $PGF_{2\alpha}$ ON LUTEAL GONADOTROPIN UPTAKE *IN VIVO*

One action of $PGF_{2\alpha}$ that has been shown *in vivo* but which does not occur *in vitro* is a block of specific uptake of radiolabelled human chorionic gonadotropin (hCG) (23) and prolactin in the rat (24). In Figure 1 the time course for inhibition of $^{125}$I-hCG uptake by luteal tissue following $PGF_{2\alpha}$ treatment *in vivo* is shown.

$^{125}$I-hCG uptake was reduced within 30 min and this effect was evident at 1, 2, 4, 8 and 24 hr after $PGF_{2\alpha}$ injection (23). This

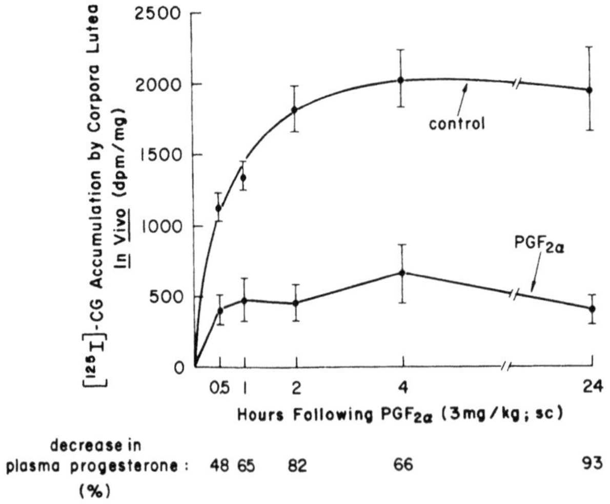


Fig 1. Effect of $PGF_{2\alpha}$ on $^{125}$I-chorionic gonadotropin (CG) uptake by rat luteal tissue. $^{125}$I-hCG was administered simultaneously with $PGF_{2\alpha}$ at time points of 0.5, 1 and 2 hr and 2 hr before sacrifice at 4 or 24 hr. From Behrman and Hichens (23).

response occurred in parallel with a significant reduction of plasma progesterone at all time periods. Inhibition of gonadotropin uptake by $PGF_{2\alpha}$ was judged not to be due to inhibition of blood flow to the corpus luteum (22). Moreover, inhibition of $^{125}$I-prolactin uptake by $PGF_{2\alpha}$ was elicited in hypophysectomized rats in which luteal function was maintained with exogenous prolactin (24). This observation indicates that dilution of radiolabelled hormone by endogenous hormone did not contribute to the decrease in radiolabelled hormone uptake produced by $PGF_{2\alpha}$. The early block of $^{125}$I-hCG uptake by $PGF_{2\alpha}$ also could not be accounted for by a loss in LH receptors in isolated luteal membranes (24). Surprisingly, $PGF_{2\alpha}$ has little effect on gonadotropin uptake by isolated luteal cells (33). At present, the basis for this action of $PGF_{2\alpha}$ *in vivo* is unexplained but it appears to be specific for $PGF_{2\alpha}$ and does not occur early in the luteal phase when the pseudopregnant rat is refractory to $PGF_{2\alpha}$ (H.R. Behrman, unpublished observations). It seems that $PGF_{2\alpha}$ reduces accessibility of the luteal cell *in situ* in a nonspecific manner to gonadotropins.

## EVIDENCE FOR A DIRECT LUTEAL CELL EFFECT OF $PGF_{2\alpha}$

One site of the luteolytic action of $PGF_{2\alpha}$ is directly on the luteal cell. In early studies $PGF_{2\alpha}$ was shown to reduce progesterone secretion by rat corpora lutea in prolonged organ culture (25) and within 6 hr in incubations of rabbit corpora lutea (26). $PGF_{2\alpha}$ also reduced LH-dependent progesterone secretion in organ cultures of hamster corpora lutea (27). In later studies, Henderson and McNatty (28) reported that $PGF_{2\alpha}$ reduced basal and LH + FSH stimulated progesterone secretion in long term cultures (9 days) of porcine granulosa-luteal cells. In bovine granulosa-luteal cells, $PGF_{2\alpha}$ reduced progesterone production when present in the culture media for several days (28). A similar finding was reported with cultures of human granulosa-luteal cells (29). Channing (30) also reported that $PGF_{2\alpha}$ directly inhibited LH-stimulated progesterone secretion in cultures of monkey granulosa-luteal cells. These data from *in vitro* studies indicate that $PGF_{2\alpha}$ may affect progesterone secretion by a direct action on the luteal cell in a variety of species.

## PROSTAGLANDIN RECEPTORS IN CORPORA LUTEA

$PGF_{2\alpha}$ receptors have been identified in corpora lutea of the human, ewe and cow (31). Binding of $PGF_{2\alpha}$ is specific and shows a dissociation constant ($K_d$) of about $10^{-7}$ M to $10^{-8}$ M. Recently, membrane receptors for $PGF_{2\alpha}$ have been demonstrated in rat corpora lutea (32). Such receptors are highly specific for $PGF_{2\alpha}$; the most active inhibitor which competed for binding against $PGF_{2\alpha}$, was $PGE_2$ with about 10% of the activity of $PGF_{2\alpha}$. The $K_d$ of the rat luteal $PGF_{2\alpha}$ receptor was 6.9nM with a binding capacity of 83 fM per mg wet weight of luteal tissue. Evidence that the receptor for $PGF_{2\alpha}$ is a cell surface receptor was obtained by conjugating $PGF_{2\alpha}$ to ferritin, incubating $PGF_{2\alpha}$-ferritin with luteal cells and detecting specific binding of $PGF_{2\alpha}$-ferritin by electron microscopy (32).

Although it has not been proven, it appears reasonable that receptor binding of $PGF_{2\alpha}$ mediates the luteolytic effect of $PGF_{2\alpha}$. Support for this suggestion arises from the specificity and high affinity of the receptor for $PGF_{2\alpha}$ similar to the specificity of the luteolytic action of $PGF_{2\alpha}$ and its synthetic analogues (31). Also, the concentration of $PGF_{2\alpha}$ found to modify progesterone secretion in dispersed rat luteal cells is similar to the $K_d$ for the $PGF_{2\alpha}$ receptor (33).

## RAPID ANTIGONADOTROPIC ACTIONS OF $PGF_{2\alpha}$

The first evidence that $PGF_{2\alpha}$ may induce luteolysis by inhibition of LH action on the corpus luteum arose from acute *in vivo* treatment of pituitary-intact and hypophysectomized pseudopregnant rats (10). In these studies intravenous $PGF_{2\alpha}$ reduced progesterone secretion within 30 min in the intact rat while simultaneous treatment with LH and $PGF_{2\alpha}$ prevented the effect of $PGF_{2\alpha}$. In hypophysectomized rats LH stimulated progesterone secretion almost 2 fold and simultaneous injection of $PGF_{2\alpha}$ inhibited this LH dependent effect. From these studies it was suggested that $PGF_{2\alpha}$ may acutely inhibit the action of LH and it was argued that this may occur through a loss in receptor integration for LH (Table 3) (10).

Direct *in vitro* evidence that $PGF_{2\alpha}$ inhibits LH dependent-progesterone secretion was obtained from organ culture experiments with hamster corpora lutea (27). In these studies $PGF_{2\alpha}$ directly inhibited stimulation of progesterone secretion produced by LH as well as the increase in progesterone secretion produced by co-incubation with homologous pituitary tissue. This effect occurred after 12 hr of incubation but earlier time points were not examined.

Table 3. Ovarian Progesterone Secretion in the Rat

| | Intact | Hypophysectomized |
|---|---|---|
| | μg/30 min | μg/30 min |
| control | 6.6 ± 0.7 | 3.7 ± 0.4 |
| LH (20 μg; iv) | 6.0 ± 1.5 | 6.7 ± 0.5 |
| $PGF_{2\alpha}$ (20 μg; iv) | 4.2 ± 0.7 | 4.8 ± 0.6 |
| LH + $PGF_{2\alpha}$ | 6.4 ± 0.8 | 4.8 ± 0.7 |

From Behrman *et al.* (10).

Table 4. Inhibition by $PGF_{2\alpha}$ of LH-stimulated Cyclic AMP Accumulation in Rat Corpora Lutea Explants.

| | cyclic AMP (pmoles/mg protein) |
|---|---|
| Control | 5.3 |
| LH (5 μg/ml) | 26.9 |
| $PGF_{2\alpha}$ ($4.2 \times 10^{-5}$ M) | 3.2 |
| LH + $PGF_{2\alpha}$ | 3.5 |

From Lahav *et al.* (35).

There is evidence that the corpus luteum becomes refractory to the action of LH particularly in older tissue. Herlitz *et al.* (34) showed that the ability of older rat corpora lutea to respond to LH *in vitro* with cyclic AMP production was much less than the response seen in younger corpora lutea. Recently Lahav *et al.* (35) showed that within intact rat corpora lutea, LH-stimulated cyclic AMP production was inhibited by $PGF_{2\alpha}$ within 2 hr of incubation (Table 4).

To further assess the effect of $PGF_{2\alpha}$ on progesterone secretion and on LH receptor function, Thomas *et al.* (33) cultured dispersed rat luteal cells with LH and $PGF_{2\alpha}$. In these studies the half maximal progesterone secretory response of rat luteal cells to LH was found to be about 50 ng/ml in 2 hr. $PGF_{2\alpha}$ ($1.1 \times 10^{-6}$M) inhibited progesterone secretion at doses of LH ranging from 1 ng/ml to 1 μg/ml (Figure 2). At the high dose of LH, $PGF_{2\alpha}$ was significantly less effective in inhibiting LH-dependent progesterone

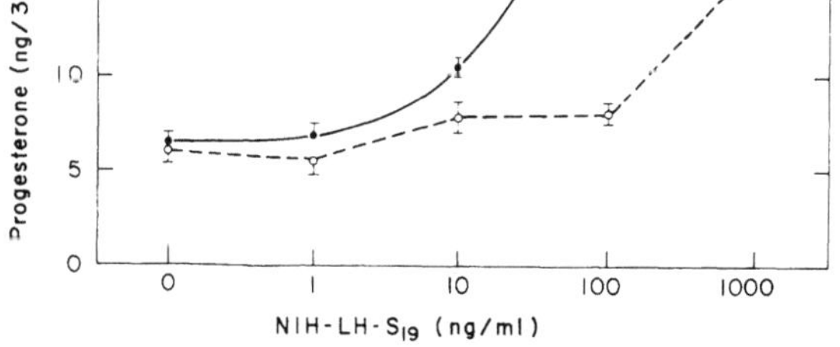


Fig 2. Dose-response of isolated luteal cells to LH and effect of $PGF_{2\alpha}$ (1.1 μm) on LH-dependent progesterone secretion. ●, LH; O, LH + $PGF_{2\alpha}$. From Thomas *et al.* (33).

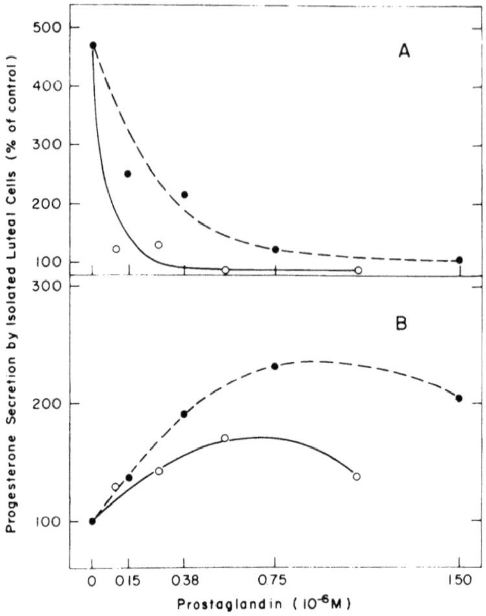


Fig 3. (A) Dose-response of isolated luteal cells to $PGF_{2\alpha}$(0) and $PGE_2$ (●) in the presence of LH at 50ng/ml. (B) Dose-response of isolated luteal cells to $PGF_{2\alpha}$ (0) and $PGE_2$ (●) in the absence of LH. From Thomas et al.(33).

secretion indicating that high levels of LH may reverse the anti-gonadotropic effect of $PGF_{2\alpha}$.

In cultures of rat luteal cells $PGF_{2\alpha}$ alone was found to stimulate progesterone secretion with a maximum effect seen at about $0.5 \times 10^{-6}$ M (33). However, maximal inhibition of LH-dependent progesterone secretion by $PGF_{2\alpha}$ was seen at $10^{-7}$ M (Fig 3). In this same study $PGE_2$ was more potent than $PGF_{2\alpha}$ as an agonist of LH with respect to stimulation of progesterone secretion, but this prostaglandin also inhibited LH-dependent progesterone secretion (Fig 3). It was estimated that $PGE_2$ had about 1/20 the activity of $PGF_{2\alpha}$ as an inhibitor of LH with respect to progesterone secretion (Fig 4).

Thomas et al.(33) reported that the acute effect of $PGF_{2\alpha}$ on LH-dependent progesterone secretion was reversed by the addition of dibutyryl cyclic AMP ($10^{-3}$M). These results and the finding that theophylline did not prevent the anti-LH effect of $PGF_{2\alpha}$ on progesterone secretion indicate that $PGF_{2\alpha}$ may inhibit LH-dependent adenylate cyclase activity. An alternate possibility considered was that $PGF_{2\alpha}$ may inhibit binding activity of LH receptors on intact luteal cells but direct studies showed no effect of $PGF_{2\alpha}$ on specific binding of $^{125}$I-hCG (33). Evidence for direct inhibition

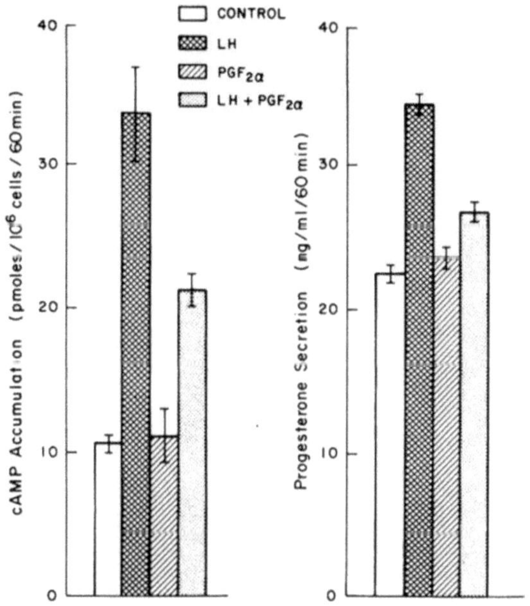


Fig 4. Effect of LH and $PGF_{2\alpha}$ on cyclic AMP accumulation and progesterone secretion in rat luteal cells cultured in the presence of theophylline (5mM). From Thomas et al. (33).

by $PGF_{2\alpha}$ of LH-dependent cyclic AMP accumulation in cultured luteal cells was demonstrated in incubations which contained theophylline (5 mM). These results are shown in Figure 4.

The results by Thomas et al. (33) indicate that $PGF_{2\alpha}$ inhibits activation of adenylate cyclase by the LH-receptor complex. In other studies (36) the possibility that $PGF_{2\alpha}$ may affect LH receptor distribution on the intact luteal cell was examined. Conjugates of LH and ferritin were prepared and specific binding of the hormone conjugate was examined by electron microscopy (Fig 5) in intact luteal cells incubated in the presence and absence of $PGF_{2\alpha}$. In these studies $PGF_{2\alpha}$ did not affect binding of ferritin-LH nor did $PGF_{2\alpha}$ appear to elicit major changes in the general (qualitative) distribution of LH receptors on the intact cell.

The mechanism of inhibition of LH dependent adenylate cyclase activity by $PGF_{2\alpha}$ is not known. However, the ionophore A-23187 was found to mimic this effect of $PGF_{2\alpha}$ on progesterone secretion and calcium was found to inhibit LH-dependent adenylate cyclase activity in isolated luteal membranes (37). $PGF_{2\alpha}$ is without effect however, on LH dependent adenylate cyclase activity in isolated membranes (37). It is possible that $PGF_{2\alpha}$ may modify calcium flux in the luteal cell membrane or modify intramembrane calcium availability

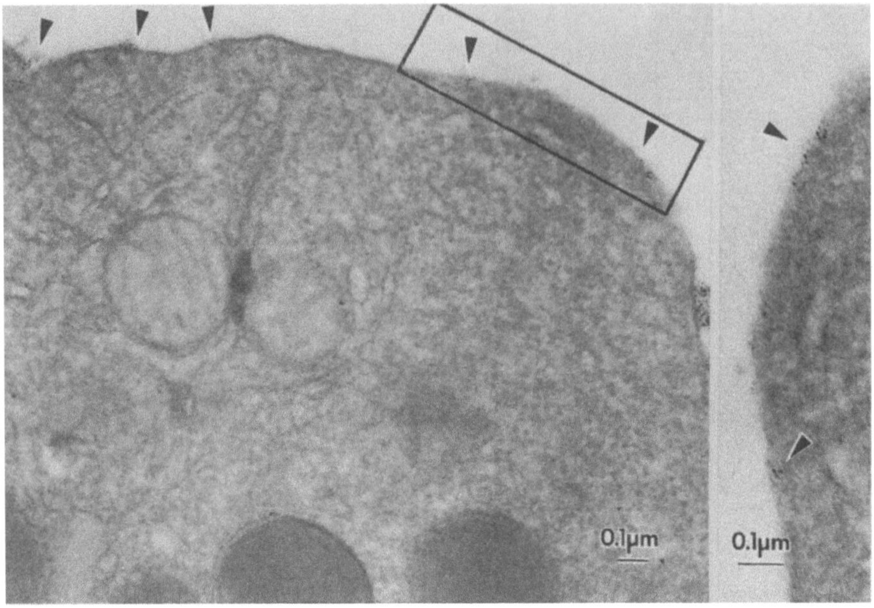


Fig 5. LH receptors labeled with ferritin-LH (↑) are distributed at intervals along the membranes of isolated luteal cells.

and thereby impair functional activity of the LH receptor complex. However, other mediators may be involved in the action of $PGF_{2\alpha}$. The antigonadotropic effect of $PGF_{2\alpha}$ appears not to be limited to the rat corpus luteum. For example, $PGF_{2\alpha}$ has been shown to antagonize LH stimulated progesterone secretion in luteal tissue from the hamster (27), pig (28), human (29) and the ewe (38). It is possible that this may be a general action of $PGF_{2\alpha}$ although $PGF_{2\alpha}$ was shown not to affect LH-stimulated testosterone secretion in dispersed Leydig cells from the rat (Thomas, J.P. and Behrman, H.R., unpublished observation).

The rapid antigonadotropic effects of $PGF_{2\alpha}$ are summarized in Figure 6. In vivo there is a rapid inhibition of radiolabelled hCG uptake by luteal LH receptors. The basis for this effect of $PGF_{2\alpha}$ is unknown but it directly parallels the decrease in progesterone levels in blood and it appears to be independent of blood flow to the corpus luteum. In vitro, there is a rapid but reversible inhibition of LH-dependent cyclic AMP accumulation by $PGF_{2\alpha}$ which appears to be due to inhibition of activation of adenylate cyclase by the LH-receptor complex. It appears not to be an effect of $PGF_{2\alpha}$ on enhanced cyclic AMP breakdown nor does $PGF_{2\alpha}$ appear to markedly alter LH receptor binding activity or distribution on the luteal cell. Inhibition of hormone dependent adenylate cyclase activity is not seen when $PGF_{2\alpha}$ and LH are exposed simultaneously to incubations

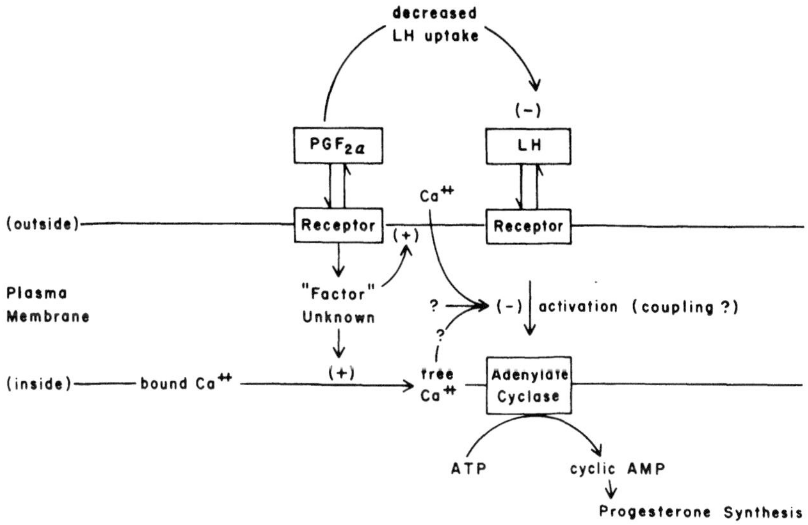


Fig 6. A model summarizing the rapid anti-LH effects of $PGF_{2\alpha}$ on the rat luteal cell _in vitro_ and _in vivo_.

of isolated luteal membranes and $PGF_{2\alpha}$ does not interfere with LH stimulated testosterone secretion by dispersed Leydig cells. These observations indicate that the effect of $PGF_{2\alpha}$ is not directly on adenylate cyclase. This raises the possibility that the effect of $PGF_{2\alpha}$ is mediated by some factor following binding of $PGF_{2\alpha}$ to its receptor on the luteal cell membrane. The similarity of effect of the calcium ionophore to the effect of $PGF_{2\alpha}$ on LH-dependent progesterone secretion by the intact luteal cell, and the direct inhibition by calcium of LH dependent adenylate cyclase activity in isolated luteal membranes, points to the possibility that a change in membrane calcium may mediate the rapid antigonadotropic effect of $PGF_{2\alpha}$.

## $PGF_{2\alpha}$-INDUCED LOSS OF LH RECEPTORS

_In vivo_, $PGF_{2\alpha}$ produces an acute drop in circulating progesterone levels seen within 30 min in the rat but no acute effect is seen on the binding activity of LH receptors in isolated luteal membranes from such animals (24). The time course for $PGF_{2\alpha}$-induced loss of LH receptors is shown in Figure 7. The first significant decrease in luteal membrane binding capacity occurs between 8 hr and 24 hr after $PGF_{2\alpha}$ treatment (39). It is of interest that functional luteolysis, marked by elevated $20\alpha$-OH-Prog secretion occurred within 8 hr after $PGF_{2\alpha}$ treatment. The loss of LH receptor binding activity produced by $PGF_{2\alpha}$ can be prevented by pretreatment with prolactin (39). Moreover, ergocryptine, which inhibits prolactin secretion,

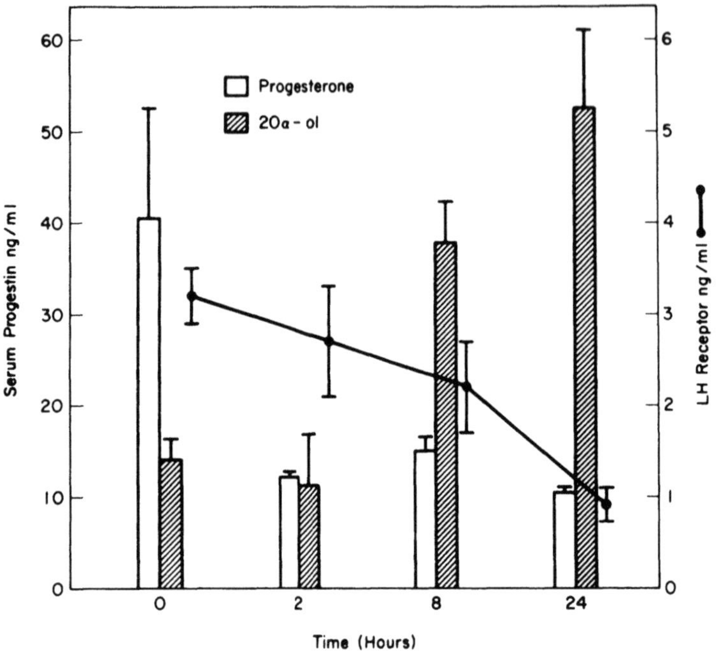


Fig 7. Time course for effect of $PGF_{2\alpha}$ on rat luteal LH receptor binding activity ( ● ) and on serum progesterone and 20α-dihydroprogesterone (20α - 01). From Grinwich et al. (39).

also produces a decrease in LH receptor binding activity with a time course similar to that of $PGF_{2\alpha}$ (39,40). The effect of ergocryptine was reversed by treatment of the rats with prolactin. In addition, hypophysectomy produces a loss of luteal LH receptor binding activity and this effect also is prevented by prolactin treatment (24,41). Taken together these data point to a role of prolactin in maintaining LH receptor binding activity in the rat and this may be the basis for the luteotropic activity of prolactin.

The long term effects of $PGF_{2\alpha}$ on functional luteolysis are summarized in Figure 8. It is presumed that the rapid effects of $PGF_{2\alpha}$ lead to eventual loss of LH receptors and this response may be due to binding by $PGF_{2\alpha}$ to specific receptors. This action in some unknown manner leads to a block in prolactin action. Early observations that $PGF_{2\alpha}$ produced a loss of luteal cholesterol esterase and cholesterol ester synthetase activity in the rat substantiate this conclusion (13). The activity of these enzymes is maintained by prolactin and $PGF_{2\alpha}$-induced loss of enzyme activity was interpreted as due to antagonism by $PGF_{2\alpha}$ of prolactin. Although $PGF_{2\alpha}$ *in vivo* inhibits prolactin uptake it is not known if a similar effect occurs in isolated luteal cells. The block of prolactin

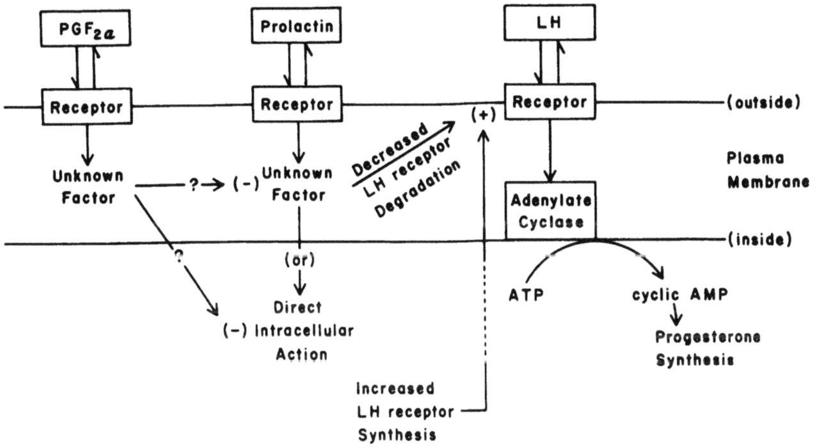


Fig 8. A model summarizing the sites and proposed mechanisms of long term luteolytic action of $PGF_{2\alpha}$.

action by $PGF_{2\alpha}$ results in loss of LH receptor binding activity possibly by increased degradation of LH receptors, by a decrease in synthesis of LH receptors, or both. In the absence of LH receptors corpus luteum function is prevented and as a consequence it is suggested that irreversible luteolysis occurs.

## SUMMARY AND CONCLUSIONS

The luteolytic action of $PGF_{2\alpha}$ appears to be due to loss of gonadotropin support of corpus luteum function. This "chemical hypophysectomy" produced by $PGF_{2\alpha}$ in the rat has several components:

1. A rapid block of gonadotropin uptake in vivo by an unexplained mechanism.

2. A rapid and direct block of adenylate cyclase activation by the LH receptor complex in isolated luteal cells.

3. An eventual loss of LH receptor binding activity produced by inhibition of prolactin action.

(Supported by grants from NIH-HD-10718 and the Ford Foundation 770-0534), AAUW postdoctoral fellowship (JLLM) and predoctoral traineeship USPHS GM-1113 (LJD).

## REFERENCES

1. B.B. Pharriss, L.F. Wyngarden and G.D. Gutknecht, in: Gonadotropins, ed. E. Rosenberg (Geron-X, Los Altos, 1968) p. 121.

2. E.W. Horton and N.L. Poyser. Physiol. Rev. 56 (1976) 595.
3. J.D. Neill, in: Handbook of Physiology, Section 7, Vol 4, part 2, ed. R.O. Greep and E.B. Astwood (Williams and Wilkins Co., Baltimore, 1974) p. 469.
4. W.K. Morishige and I. Rothchild. Endocrinology 95 (1974) 260.
5. H. Hearnshaw, B.J. Restall and A.R. Gleeson. J. Reprod. Fert. 32 (1973) 322.
6. L.E.A. Rowson, H.R. Tervit and A.J. Brand. J. Reprod Fert. 29 (1972) 145.
7. M.P.E. Moeljono, F.W. Bazer and W.W. Thatcher. Prostaglandins 11 (1976) 737.
8. B.B. Pharriss and L.J. Wyngarden. Proc. Soc. Exp. Biol. Med. 130 (1969) 92.
9. A. Labhsetwar. Prostaglandins 3 (1973) 729.
10. H.R. Behrman, K. Yoshinaga and R.O. Greep. Ann N.Y. Acad. Sci. 180 (1971) 426.
11. H.R. Behrman, K. Yoshinaga, H. Wyman and R.O. Greep. Am. J. Physiol. 221 (1971) 189.
12. D.T. Baird. J. Endocrinol. 62 (1974) 413.
13. H.R. Behrman, G.J. MacDonald and R.O. Greep. Lipids 6 (1971) 791.
14. G.W. Duncan and B.B. Pharriss. Fed. Proc. 29 (1970) 1232.
15. G.O. Thorburn and J.R.S. Hales. Proc. Austral. Physiol. Pharmacol. Soc. 3 (1970) 145.
16. G.D. Niswender, A.M. Akbar, M.A. Diekman, and T.M. Nett. Biol. Reprod. 9 (1973) 87.
17. M.J. Novy and M.J. Cook. Am. J. Obstet. Gynec. 117 (1973) 381.
18. C.Y. Pang and H.R. Behrman. Proc. Can. Fed. Biol. Soc. (1978).
19. N.W. Bruce and K. Hillier. Nature 249 (1974) 176.
20. K.T. Kirton, F.A. Kimball and S.E. Porteus, in: Advances in Prostaglandin and Thromboxane Research, Vol. 2, ed. B. Samuelsson and R. Paoletti (Raven Press, New York, 1976) p. 621.
21. J.A. McCracken, B. Barcikowski, J.C. Carlson, K. Green and B. Samuelsson. Adv. Biosciences 9 (1973) 599.
22. C.Y. Pang and H.R. Behrman. Proc. Soc. Study of Reprod. 1978.
23. H.R. Behrman and M. Hichens. Prostaglandins 12 (1976) 83.
24. H.R. Behrman, D.L. Grinwich, M. Hichens and G. J. MacDonald. Endocrinology (In Press) 1978.
25. L.M. Demers, H.R. Behrman and R.O. Greep. Adv. Biosciences 9 (1972) 701.
26. J.P. O'Grady, E.I. Kohorn, R.H. Glass, B.V. Caldwell, W.A. Brock and L. Speroff. J. Reprod. Fert. 30 (1972) 153.
27. H.R. Behrman, T.C. Ng and G.P. Orczyk, in: Gonadotropins and Gonadal Function, Ed. N.R. Moudgal (Academic Press, New York, 1974) p 332.
28. K.M. Henderson and K.P. McNatty. J. Endocrinol. 73 (1977) 71.
29. K.P. McNatty, K.M. Henderson and R.S. Sawers. J. Endocrinol. 67 (1975) 231.
30. C.P. Channing. Prostaglandins 2 (1972) 331.

31. W.S. Powell, S. Hammarstrom, V. Klyden and B. Samuelsson, in: Surface Membrane Receptors, ed. R.A. Bradshaw, W.A. Frazier, R.C. Merrell, D.I. Gottlieb, R.A. Hogue-Angeletti (Plenum Press, New York, 1976) p. 455.
32. J.L. Luborsky-Moore, K. Wright and H.R. Behrman. Proceedings of the Ovarian Workshop On Ovarian Follicular and Corpus Luteum Function. 1978.
33. J.P. Thomas, L.J. Dorflinger and H.R. Behrman. Proc. Nat. Acad. Sci. 75 (1978) 1344.
34. H. Herlitz, L. Hamberger, S. Rosberg and K. Ahren. Acta Endocr. 77 (1974) 737.
35. M. Lahav, A. Freud and H. Lindner. Biochem. Biophys. Res. Commun. 68 (1976) 1294.
36. J.L. Luborsky-Moore and H.R. Behrman. Fed. Proc. 37 (1978) 909.
37. L.J. Dorflinger and H.R. Behrman. Proc. Soc. Study Reprod. 1978.
38. M. Evrard, P. Leboulleux and C. Hermier. Prostaglandins (In Press) 1978.
39. D.L. Grinwich, M. Hichens and H.R. Behrman. Biol. Reprod. 14 (1976) 212.
40. J.A. Holt, J.S. Richards, A.R. Midgley and L.E. Reichert. Endocrinology 98 (1976) 1005.
41. J.S. Richards and J.J. Williams. Endocrinology 99 (1976) 1571.

---

DISCUSSION - CHAIRPERSON - J.W. WILKS

G.D. NISWENDER: Do you think there are major species differences in the effects of prostaglandins on the corpus luteum? We have incubated ovine luteal tissue with prostaglandins, with prostaglandins and estradiol, and with other agents; we have not seen a positive or a negative effect of prostaglandins on steroid output, even up to 12 hours of incubation. Secondly, if one assumes that the source of prostaglandins for the luteolytic mechanism is really the uterus, then are the concentrations of prostaglandin which you used in a normal range and similar to those found in uterine venous blood?

H.R. BEHRMAN: To answer your first question, inhibition of gonadotropin-dependent progesterone secretion has been demonstrated by our laboratory using hamster luteal explants (27) and dispersed rat luteal cells (numbers in parentheses refer to references in Dr. Behrman's paper). In the sow, Henderson and McNatty (28) reported that $PGF_{2\alpha}$ reduced basal and LH/FSH-dependent progesterone secretion by granulosa-luteal cells. A similar finding was reported by Channing (30) on the effect of $PGE_2$ in monkey granulosa-luteal cells. Ahren's laboratory has seen this effect of $PGF_{2\alpha}$ in isolated human luteal cells in culture. O'Grady et al., (26) reported that $PGF_{2\alpha}$ reduced progesterone secretion in rabbit luteal tissue in vitro. Evrard, et al., (38) reported that $PGF_{2\alpha}$ inhibited LH-dependent progesterone secretion in perifused ovine luteal tissue in vitro. In long term cultures of bovine and human granulosa-luteal cells $PGF_2\alpha$ reduced progesterone secretion as shown by McNatty et al., (29); on the other hand Hixon has found no rapid effect of PGF2α on LH-dependent progesterone secretion by isolated bovine luteal cells. Thus, the in vitro anti-LH effect of $PGF_2\alpha$ on progesterone secretion has been seen in several species but in some such as the cow and perhaps the ewe, this action of $PGF_2\alpha$ is less clear.

The second point you raised concerns the concentration of $PGF_2\alpha$ necessary to inhibit LH action on rat luteal cells in vitro. In our hands the concentration of $PGF_2\alpha$ which produces half maximal inhibition of LH-dependent progesterone secretion is about $10^{-7}$ M or less. This is in close agreement with the dissociation constant for the $PGF_2\alpha$ receptor we measured in rat luteal membranes. In incubated slices of rat luteal tissue 4.2 X $10^{-5}$ M $PGF_2\alpha$ was used (Lahav et al., 35) but this high level of $PGF_2\alpha$ may reflect problems of diffusion in that in vitro model.

G. GIBORI: You related the number of LH receptors with the luteolytic effect of prostaglandin. When luteolysis is induced with LH antiserum in the rat, progesterone secretion is drastically reduced while the number of receptors for LH remains elevated for 24-48 hours. In this case luteolysis is not correlated with the number of LH receptors.

H.R. BEHRMAN: Treatment of pregnant rats with LH antiserum has been shown to induce abortion and functional luteal regression judged from highly elevated 20α-dihydroprogesterone levels in plasma and reduced progesterone levels (Moudgal, et al., J. Endocrinol 52:413, 1972). However, in the pseudo-pregnant rat, within the same time interval after treatment with LH anti-serum, LH antiserum did not cause an elevation of serum 20α-dihydroprogesterone although progesterone levels were decreased (Behrman, et al., J. Endocrinol 42:419, 1972). Indeed, in this same report we found that luteal tissue removed from such animals 24 hr after LH antiserum treatment was more responsive to LH in vitro than control animals. Thus it is questioned whether LH antiserum induced luteal regression in the latter case or simply removed the tropic agent, namely LH. With regard to your comment that LH antiserum did not affect LH receptor binding activity yet reduced progesterone secretion, this observation is quite similar to the effect of $PGF_2\alpha$. LH receptor levels in luteal tissue were not reduced for up to 8 hours after $PGF_2\alpha$ treatment of rats in vitro although progesterone secretion was severely reduced (39). The rapid effect of $PGF_2\alpha$, as we discussed, appears to be due to block of the signal from the LH-receptor complex to activate adenylate cyclase which results in inhibition of progesterone secretion. Functional luteolysis, judged from elevated 20α-dihydroprogesterone levels, occurs due to a block by $PGF_{2\alpha}$ of prolactin action in rat corpora lutea through an unknown mechanism. It is, we believe, this latter effect of $PGF_{2\alpha}$ which causes the loss of LH receptors.

H.J. KIRCHICK: The concept of an action of $PGF_{2\alpha}$ via calcium transport is really interesting to me in light of the effects of prostaglandins on uterine contractions. Has your laboratory tested the actions of prostaglandins, using a chelating agent in culture, to see if you can block the effects of prostaglandins?

L.J. DORFLINGER: We have not examined the role of calcium in this process by using chelating agents but we have approached this question in other ways. First we examined the effects of LH and $PGF_{2\alpha}$ on luteal cells in calcium-free media. As with other steroid secretagogues, the stimulatory effect of LH requires some $Ca^{++}$. In these experiments, there appeared to be a blunting of the response to LH as well as a blunting of the effect of $PGF_{2\alpha}$. Rao's work (Biochem. Biophys. Res. Comm. 67:1242, 1975) on the $PGF_{2\alpha}$ receptor in luteal cells indicates that $Ca^{++}$ is necessary for the high-affinity binding component in membranes, therefore, this attenuation of the inhibitory action of $PGF_{2\alpha}$ may be due to its inability to bind to the cell. Secondly, the calcium ionophore (A 23187) mimicked the effect of $PGF_{2\alpha}$ in intact luteal cells. Thirdly, $Ca^{++}$ inhibited adenylate cyclase activity in isolated membranes and at 5 mM blocked LH stimulation completely. However, we find that $PGF_{2\alpha}$ is inactive in isolated membranes as an inhibitor of LH-dependent cAMP accumulation. This indicates that the effect of $PGF_{2\alpha}$ may be mediated by some factor in the intact cell. We do not know if this mediator is calcium in the intact cell, but this is our present working hypothesis.

K.J. CATT: I was intrigued by the close correspondence in the time course of the loss of LH receptors induced by $PGF_{2\alpha}$ with the similar time course one sees after giving LH or hCG to the same type of pseudopregnant animal. Do you have any information on the secretion of endogenous hormones after giving $PGF_{2\alpha}$? Can you exclude that there is a simultaneous release of endogenous LH which might be independently downregulating LH receptors, in addition to the demonstrated acute and direct actions of the $PGF_{2\alpha}$ on the luteal cell?

H.R. BEHRMAN: We have approached this question in two ways. First, we measured circulating levels of LH following treatment of the rat with a luteolytic dose of $PGF_{2\alpha}$ and no significant increases in LH levels were seen. Second, we hypophysectomized pseudopregnant rats and maintained luteal function with exogenous prolactin. In these animals $PGF_{2\alpha}$ produced the same luteolytic effect as in the pituitary-intact animals. Thus, we believe we have excluded the possibility that $PGF_{2\alpha}$ causes a loss of LH receptors in the corpus luteum by increasing LH secretion which subsequently causes homologous down regulation. We believe that $PGF_{2\alpha}$ directly causes heterologous down regulation of luteal LH receptors in the rat by an inhibition of prolactin maintenance of such receptors.

K.J. CATT: Is there an effect on the serum prolactin levels?

H.R. BEHRMAN: $PGF_{2\alpha}$ will cause an increase in prolactin secretion under some circumstances. However, in the studies discussed earlier with hypophysectomized animals we have controlled effects of $PGF_{2\alpha}$ on pituitary hormone secretion. In addition, we have been unsuccessful in down regulating luteal LH receptors with injections of large doses of prolactin, but we have induced loss of LH receptors by injections of hCG or LH. However, we found that homologous down regulation by treatment with LH or hCG is quite different from the heterologous down regulation of LH receptors produced by $PGF_{2\alpha}$. First the time course for this effect of $PGF_{2\alpha}$ is more rapid than that seen with hCG by several hours, but the time course for the $PGF_{2\alpha}$ action on down regulation of LH receptors closely approximates that seen following hypophysectomy. Second, we can prevent down regulation of LH receptors induced by $PGF_{2\alpha}$ with simultaneous injections of prolactin but we are unable to influence down regulation induced by hCG with prolactin. Third, inhibition of endogenous prostaglandin synthesis with indomethacin had no effect in our hands on down regulation induced with hCG. We were led from these observations to the conclusion that down regulation of LH receptors induced by $PGF_{2\alpha}$ occurs through a mechanism which is different from that involved in loss of LH receptors induced by LH or hCG.

I. ROTHCHILD: The explanation of how $PGF_{2\alpha}$ works through reducing prolactin, and in this way reducing LH receptors, sounds really fine. My question has to do with what else prolactin does in the regulation of progesterone secretion. In an LH-dependent rat the prostaglandin effect fits well with your explanation. The hypophysectomized rat bearing its autotransplanted pituitary in the kidney, however, is not LH-dependent (Lam and Rothchild, Endocrinology 101:1503, 1977). This independence from LH is not just a temporary effect; as long as four weeks after operation large doses of LH antiserum have absolutely no effect on the rate of progesterone secretion. Even though they are not LH-dependent, the number of LH receptors in rats with autotransplanted pituitaries are exactly the same as in intact animals. However, if you take the prolactin away these rats stop secreting progesterone. Would $PGF_{2\alpha}$ or other luteolytic prostaglandins have the same effect on progesterone secretion in these animals and would it have the same effect on LH receptors?

H.R. BEHRMAN: At present we really have no information on how prolactin acts to maintain progesterone secretion in the hypophysectomized rat. However, as discussed earlier in reply to Dr. Catt's question, $PGF_{2\alpha}$ produced the same luteolytic effect in the hypophysectomized-prolactin maintained rat as in the pituitary-intact rat. We have also found that LH receptors remain elevated in hypophysectomized rats treated with prolactin. In addition, ergocryptine, a prolactin secretion inhibitor, reduced rat luteal LH receptor levels (39). This "guilt by association" led us to postulate that prolactin is involved in maintenance of LH receptor function and $PGF_{2\alpha}$ interferes with this process. It is possible that prolactin may maintain the "coupling" of residual LH receptors to adenylate cyclase and thereby maintain function. I know of no situation where there is maintenance of luteal function in the absence of LH receptors in the fully differentiated corpus luteum.

In vitro $PGF_{2\alpha}$ seems to be equally effective in blocking LH-dependent progesterone secretion in luteal cells obtained from day 4 pseudopregnant rats (prolactin dependent) as in day 8 pseudopregnant rats (LH dependent). On the other hand, in vivo very high doses of $PGF_{2\alpha}$ are necessary to interrupt luteal function in the day 4 versus the day 8 pseudopregnant animal. This observation in conjunction with our observation that prolactin pretreatment can prevent the luteolytic effect of $PGF_{2\alpha}$ in vivo indicates that $PGF_{2\alpha}$ and prolactin may be antagonists with regard to luteal function. However, in vitro high levels of LH attenuate the inhibitory effect of $PGF_{2\alpha}$ on progesterone secretion. These observations do not prove, but are compatible with the possibility that prolactin may act to maintain coupling of the LH receptor complex to adenylate cyclase.

QUESTION: I have been looking at the effect of prostaglandins on dispersed rat Leydig cells; the only one that is not active in inhibiting the action of LH is prostaglandin $F_{2\alpha}$ while prostaglandin $E_1$, $E_2$, $A_1$, and $A_2$ all inhibit LH. Also, all of the active prostaglandins inhibit the action of dibutyryl cyclic AMP. Would you comment on this sex difference?

H.R. BEHRMAN: We have tested prostaglandin $F_{2\alpha}$ in Leydig cells, using the dispersion method of Catt and Dufau. Prostaglandin $F_{2\alpha}$ was totally inactive in the Leydig cell in inhibiting LH-dependent testosterone secretion at levels which were active in dispersed luteal cells. In our laboratory there is a definite sex difference with regard to $PGF_{2\alpha}$ action based on dose-response activity.

# PROSTAGLANDIN $F_{2\alpha}$ AND ITS 13-DEHYDRO ANALOGS:

# COMPARATIVE LUTEOLYTIC EFFECTS *IN VIVO*

J.A. McCracken, N. Einer-Jensen*
Worcester Foundation for Experimental Biology
J. Fried
Department of Chemistry, University of Chicago

## INTRODUCTION

In 1972 prostaglandin $F_{2\alpha}$ ($PGF_{2\alpha}$) was identified as a uterine luteolytic hormone in sheep (1) and there is good evidence to suggest that $PGF_{2\alpha}$ plays a similar physiological role in the guinea pig (2), in the pig (3; 4), in the cow (5; 6) and probably in other mammalian species (7). However in women and in non-human primates the evidence seems clear that the uterus is not essential for normal cyclic regression of the corpus luteum (8; 9). It would therefore appear that PGs of uterine origin are not the proximate cause of luteolysis in primates. In support of this conclusion it is reported that elevations of the primary metabolite of $PGF_{2\alpha}$ (15-keto-13:14-dihydro-$PGF_{2\alpha}$) do not occur in the peripheral plasma of women at the time of cyclic regression of the CL (10). This is in contrast to the cow and sheep where major elevations of the primary metabolite coincide with CL regression (6; 11). These metabolite studies thus serve to validate the role of $PGF_{2\alpha}$ as a uterine luteolytic hormone in the sheep and cow. Nevertheless the possibility that PGs originating within the ovary itself might initiate luteolysis locally in the primate has been suggested by several authors (12; 13; 14). Presumably the amount of $PGF_{2\alpha}$ so generated within the ovary would be so small that it would not yield sufficient primary metabolite to be detectable in peripheral plasma. The notion of a local origin and action of PGs in the primate ovary is given some credence by the finding that $PGF_{2\alpha}$ injected into the human CL appears to be luteolytic (15) whereas the injection of $PGF_{2\alpha}$ into the uterus is without effect (16). However there is some disagreement as to whether $PGF_{2\alpha}$

**Present address:* *Department of Physiology, University of Odense Odense, Denmark.*

levels rise in the human CL itself at the time of luteolysis (17; 18; 19). In any case the extra-luteal tissues of the ovary itself might still of course be a significant source of ovarian PG in the primate. Indeed a recent report indicates that estrogen treatment of monkeys increases $PGF_{2\alpha}$ concentration in blood draining the ovary (20).

Attempts have been made to demonstrate a luteolytic effect of administered PGs in both monkeys and women. In *M. mulatta*, 30 mg of $PGF_{2\alpha}$ given on days 22 and 23 apparently caused shortening of the cycle (21) while in *M. fascicularis* (22) but not *M. nemestrina* (23) it was necessary to pretreat with estradiol-17β ($E_2$-17β) to effect luteolysis with 15 mg of $PGF_{2\alpha}$. Several attempts have been made to induce luteolysis in women during the luteal phase of the cycle (24; 25; 26; 27; 28). However, no clear cut results emerged since some studies indicated a transient decline in progesterone and/or a shortening of the cycle while other studies reported no effects. It is possible in these studies in women that the amount of infused $PGF_{2\alpha}$ which actually reaches the ovary may be very small due to rapid metabolism of $PGF_{2\alpha}$. In addition the human CL like that of the pig (3) may be refractory in the early part of the luteal phase (28), perhaps resulting from low receptor binding of $PGF_{2\alpha}$ by human luteal cells (13).

More recently a variety of synthetic analogs have been used in an attempt to induce luteolysis in several species. The 15-methyl derivatives of $PGF_{2\alpha}$ were used by several groups (29; 30; 31) in a rational attempt to prolong activity. Unfortunately since these compounds are resistant to the 15-OH-PG-dehydrogenase (32) the powerful smooth muscle side effects of these analogs are also potentiated thus probably rendering them unsuitable for clinical use. However, a similar compound reported from Japan (33; 34), 16,16-dimethyl-trans-$\Delta^2$-$PGE_1$, apparently shows a selective increase in smooth muscle activity of the uterus but with little or no effect on blood pressure although mild gastrointestinal muscle stimulatory activity was present. This compound was recently reported to be effective in terminating early pregnancy in 86% of women given this compound in the form of repeated vaginal suppositories (35). Presumably this analog terminated pregnancy by causing intense uterine contractions. However its acceptability for clinical use remains unknown. The 16-aryloxy derivatives of $PGF_{2\alpha}$ have been used with some success in domestic animals (36; 37), however recent trials with this analog in monkeys (23; 38) and women (39) indicate that the side effects including increases in blood pressure and diarrhea reach unacceptable levels. $PGF_{2\alpha}$, 1-15 lactone has been administered to cycling and early pregnant rhesus monkeys (40). This analog was injected intramuscularly for two days and appeared to cause a reduction in progesterone levels in both cyclic and early pregnant monkeys. $PGF_{2\alpha}$ given in a similar regime appeared to be effective in cyclic monkeys but did not depress progesterone levels or terminate pregnancy in early pregnant monkeys. The smooth muscle side effects of this analog were not investigated in any detail but one out of three pregnant monkeys treated with this analog had diarrhea and a third trimester pregnant monkey showed increased uterine contractions after being given the analog.

We have previously reported on the luteolytic activity of the 13-dehydro analogs of $PGF_{2\alpha}$ in both sheep and monkeys (41; 42; 43). The present paper summarizes these previous findings and also describes studies on the effect of $PGF_{2\alpha}$ on the blood flow in the CL of the sheep. New results are reported on the luteolytic activity of the 13-dehydro analogs given by the oral and subcutaneous routes in the primate model.

STUDIES WITH LUTEOLYTIC 13-DEHYDRO PROSTAGLANDINS IN SHEEP

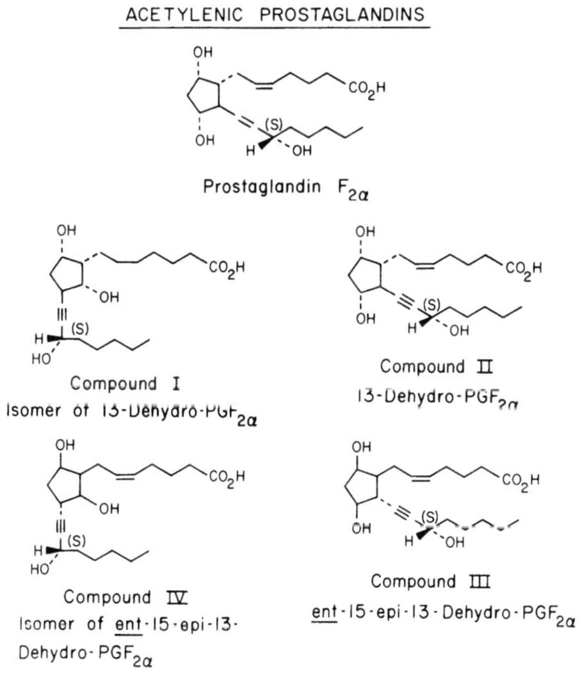


*Fig. 1 Examples of 13-dehydro (acetylenic)prostaglandin analogs. Further modifications included the addition of a fluorine atom in position 16 in compounds II & III.*

Following the identification of $PGF_{2\alpha}$ as a uterine luteolytic hormone in sheep (1) attempts were made to induce CL regression in normal cycling sheep by giving large amounts of $PGF_{2\alpha}$ subcutaneously or intravaginally as a single administration (44). This approach was only partially successful since only 2/3 of these animals showed evidence of luteolysis (44). It was observed that plasma levels of $PGF_{2\alpha}$ in these animals were elevated for less than 2 hours. It would appear that like other species the sheep has a remarkable capacity to metabolize $PGF_{2\alpha}$, probably mostly by the lungs (45). Thus the administered $PGF_{2\alpha}$ apparently did not persist long enough in plasma to have an effect on the corpus luteum in all subjects. In addition many of these animals showed evidence of severe side effects including diarrhea. The idea of devising a slow release method for administering $PGF_{2\alpha}$ was discarded because while this approach would probably be more efficient in terms of luteolysis, it would also probably increase the side effects observed with single administrations. It seemed more logical to seek luteolytic analogs of $PGF_{2\alpha}$ which would be resistant to metabolism and hopefully, at the same time, have reduced smooth muscle side effects.

Our initial experiments with synthetic PGs involved the use of the 7-oxa-analogs of $PGF_{1\alpha}$ (46; 47). The test system used was the ovarian autotransplant model (48; 49) which permits the infusion of PGs either directly into the arterial supply of the ovary or into

| Dose/hr/6 hr (μg) | Route | % Decrease Progesterone Secretion | Change in Blood Flow (↑↓) |
|---|---|---|---|
| 13-Dehydro-$PGF_{2\alpha}$: | | | |
| 25 μg | I.A. | > 90% | ↓ |
| 10 μg | I.A. | > 95% | ↓ |
| 5 μg | I.A. | > 90% | None |
| 2.5 μg | I.A. | > 90% | None |
| 50 μg | Sys. | ~ 30% | None |
| ent-13-Dehydro-15-epi-$PGF_{2\alpha}$: | | | |
| 25 μg | I.A. | > 95% | None |
| 10 μg | I.A. | > 90% | None |
| 5 μg | I.A. | ~ 50% | None |
| 50 μg | Sys. | ~ 30% | None |

*Fig. 2 Relative luteolytic potencies of 13-dehydro-$PGF_{2\alpha}$ and ent-13-dehydro-15-epi-$PGF_{2\alpha}$ in sheep ovarian transplant model. From (42).*

the systemic circulation to examine them respectively for direct luteolytic activity and for resistance to metabolism *in vivo*. However we found that the substitution of an oxygen atom in place of the carbon at position seven actually diminished the luteolytic properties of these analogs compared to the parent compound $PGF_{1\alpha}$ (47). Subsequently we examined a number of the acetylenic (13-dehydro) analogs of $PGF_{2\alpha}$ (50), examples of which are shown in *Fig. 1*. Two compounds had luteolytic activity (41; 42), one of which was equipotent to $PGF_{2\alpha}$ (13-dehydro-$PGF_{2\alpha}$) while the other had about 25% of the activity (ent-15-epi-13-dehydro-$PGF_{2\alpha}$). $PGF_{2\alpha}$ had no luteolytic activity in the ovarian transplant model when infused systemically (except at very high dose levels, >500 μg/hr). Both these analogs possessed weak systemic activity at 50 μg/hr (*Fig. 2*). This is in keeping with the known resistance of these

| Dose/hr/6 hr (μg) | Route | % Decrease in Progesterone secretion | Change in blood flow (↑↓) |
|---|---|---|---|
| | | **16-Fluoro-13-dehydro-$PGF_{2\alpha}$ (erythro)** | |
| 10 μg | I.A. | >95% | None |
| 5 μg | I.A. | >95% | None |
| 2 μg | I.A. | >95% | None |
| 1 μg | I.A. | >80%* | None |
| 100 μg | Sys. | >95% | ↑ |
| 50 μg | Sys. | >95% | ↑ |
| | | **16-Fluoro-13-dehydro-$PGF_{2\alpha}$ (threo)** | |
| 10 μg | I.A. | >95% | None |
| 5 μg | I.A. | >80% | None |
| 50 μg | Sys. | ∿15% | None |
| | | **Prostaglandin $F_{2\alpha}$** | |
| 2.5 μg | I.A. | >95% | None |
| 250 μg | Sys. | None | None |
| 500 μg | Sys. | ∿25% | None |

*Progesterone secretion rate was suppressed after 14 hours but rebounded at 24 hours (See Figure).

*Fig. 3* *Comparative luteolytic activities of 16-fluoro-13-dehydro-$PGF_{2\alpha}$ in erythro form (mixture of 15, 16 epimers) and threo form (mixture of 15,16 epimers) in sheep ovarian transplant. From (43).*

compounds to the 15-OH-PG-dehydrogenase *in vitro* (51). In addition these two analogs showed diminished uterine smooth muscle stimulating activity when infused into the arterial supply of the uterus *in vivo* (42). This suggested that the receptors governing the luteolytic effect on the one hand, and the smooth muscle effect on the other possessed different structural specificities. These findings therefore encouraged us to pursue the development of luteolytic (contraceptive) prostaglandin analogs which would be devoid of smooth muscle activity (side effects), but would be metabolically stable in the blood stream (resistant to the 15-OH-PG-dehydrogenase). When the above two analogs were tested in normal cycling sheep (41; 42) they both showed luteolytic activity in a low dose range of 0.25 to 1.0 mg/kg. The side effects previously noted with $PGF_{2\alpha}$ itself were either slight or absent in the case of these analogs, a finding in agreement with our smooth muscle tests *in vivo* (42).

In an attempt to improve further the luteolytic to smooth muscle ratio of the 13-dehydro analogs, fluorine was introduced into the molecule in position 16, i.e. on the carbon α to the 15-hydroxyl

| Dose/hr/6 hr (μg) | Route | % Decrease in Progesterone secretion | Change in blood flow (↑↓) |
|---|---|---|---|
| | | ent-15-epi-16-fluoro-13-dehydro-$PGF_{2\alpha}$ | |
| 5 μg | I.A. | None | None |
| 10 μg | I.A. | ∿30% | None |
| 20 μg | I.A. | ∿50% | ↑ |
| 40 μg | I.A. | >90% | None |
| 80 μg | I.A. | >95% | None |
| 100 μg | Sys. | ∿20% | None |
| 200 μg | Sys. | ∿30% | None |
| 300 μg | Sys. | >95%* | None |
| | | Prostaglandin $F_{2\alpha}$ | |
| 2.5 μg | I.A. | >95% | None |
| 250 μg | Sys. | None | None |
| 500 μg | Sys. | ∿25% | None |

*Progesterone secretion was completely suppressed after 12 hrs but rebounded to control values by 44 hours.

*Fig. 4* *Luteolytic activities of ent-15-epi-16-fluoro-13-dehydro-$PGF_{2\alpha}$ (erythro) in sheep ovarian transplant preparation. From (43).*

group in the PG molecule (42; 43). This was a direct analogy to work previously done in the steroid field (52) where the introduction of a fluorine atom adjacent to a hydroxyl group essential for biological activity led to greatly enhanced potency of the steroid. The prostaglandins would appear to be very suitable for such modification because of the existence of several hydroxyl groups and because of their broad range of activities.

The results of luteolytic potency tests in the ovarian transplant model with 16-fluoro derivatives of the two previously tested 13-dehydro compounds are shown in *Figs. 3* and *4*, specific examples of which are shown in *Fig. 5* (intraovarian) and in *Fig. 6* (systemic). The smooth muscle activity of the fluorinated compounds was tested in the sheep by infusing them into the uterine artery *in vivo* and measuring uterine activity via a fluid-filled balloon in the uterine lumen as described previously (53). These results are summarized in *Fig. 7*. An example of the smooth muscle activity (or more correctly the absence thereof) for ent-15-epi-16-fluoro-13-dehydro-$PGF_{2\alpha}$ (erythro) is shown in *Fig. 8*.

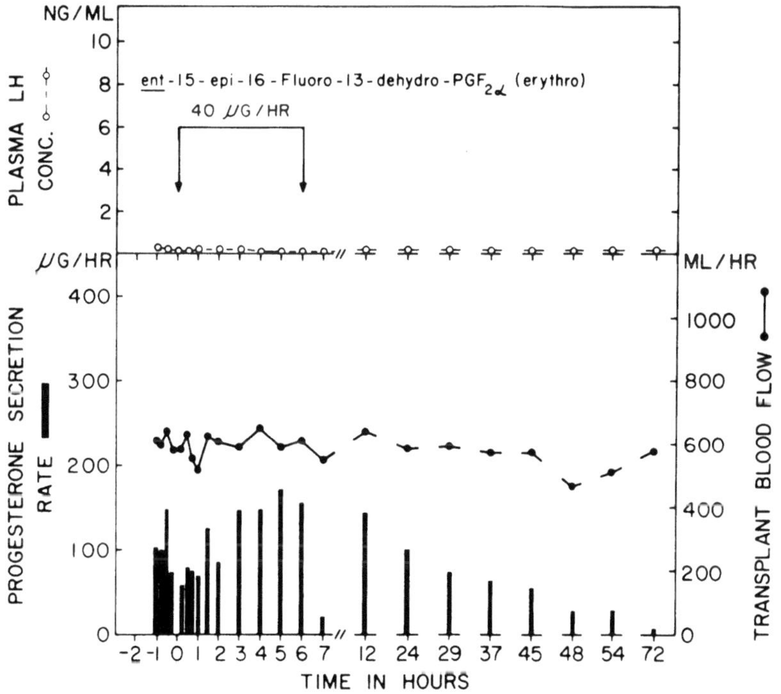


*Fig. 5* *Progesterone secretion and LH levels after the intra-arterial infusion of ent-15-epi-16-fluoro-13-dehydro-$PGF_{2\alpha}$ (erythro) at 40 μg/hr for six hours in sheep ovarian transplant. From (43).*

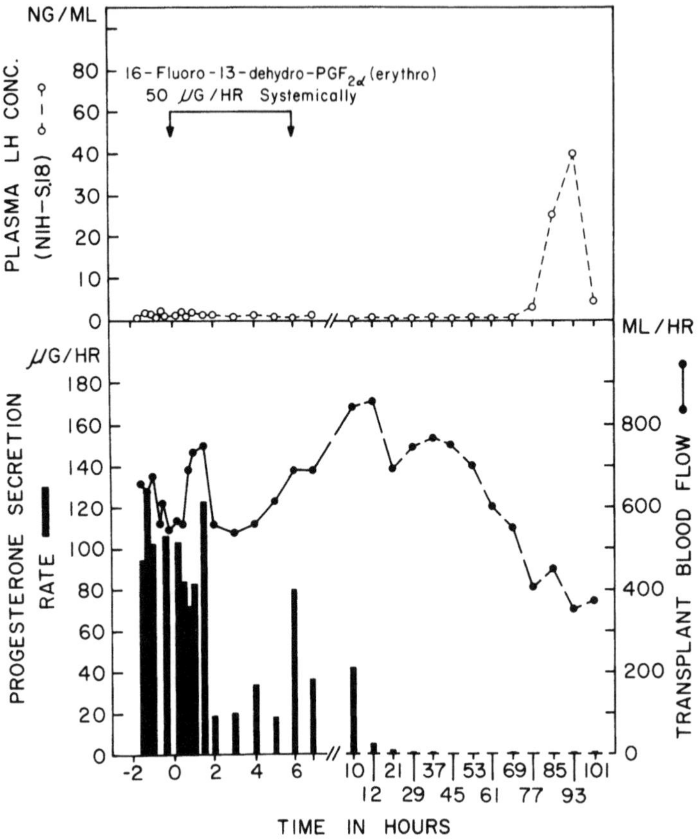


*Fig. 6* *Progesterone secretion and LH levels after the systemic infusion of 16-fluoro-13-dehydro-PGF$_{2\alpha}$ (mixture of 15,16 epimers, erythro) at 50 μg/hr for six hours. From (43).*

| | Potency (PGF$_{2\alpha}$=1) | |
|---|---|---|
| | Luteolysis | Smooth Muscle |
| 13-dehydro-PGF$_{2\alpha}$ | 1.0 | 0.4 |
| ent-15-epi-13-dehydro-PGF$_{2\alpha}$ | 0.5 | 0.01 |
| 16-fluoro-13-dehydro-PGF$_{2\alpha}$(erythro) | 5.0 | 0.2 |
| 16-fluoro-13-dehydro-PGF$_{2\alpha}$(threo) | 0.25 | 0.2-0.4 |
| ent-15-epi-16-fluoro-13-dehydro-PGF$_{2\alpha}$ (erythro) | 0.1 | 0.002 |

*Fig. 7* *Summary table of luteolytic and smooth muscle activities of 13-dehydro analogs relative to PGF$_{2\alpha}$ infused intra-arterially into the ovary or uterus of the sheep.*

In general we see an enhancement of luteolytic potency of the 16-fluoro derivative of the 13-dehydro analog in the erythro form (but not the threo form). Particularly striking is the luteolytic potency of this analog when given systemically in the sheep (*Fig. 6*). Presumably this is due in part to the increased luteolytic potency of the molecule *per se* and perhaps also in part to an increased resistance to metabolism in the circulation. The smooth muscle activity of this compound was essentially similar to its non-fluorinated counterpart.

With the 16-fluorinated ent-epi compound we see a reduction in luteolytic activity to about half of that of its non-fluorinated counterpart. However to compensate for this we see an order of magnitude reduction in smooth muscle activity with the fluoro compound which thus enhances the luteolytic to smooth muscle ratio. An unusual characteristic of the 16-fluoro-ent-epi compound is that luteolysis appears to occur long after the infusion with this analog has been completed (See *Fig. 5*). We presently have no explanation for this curious finding.

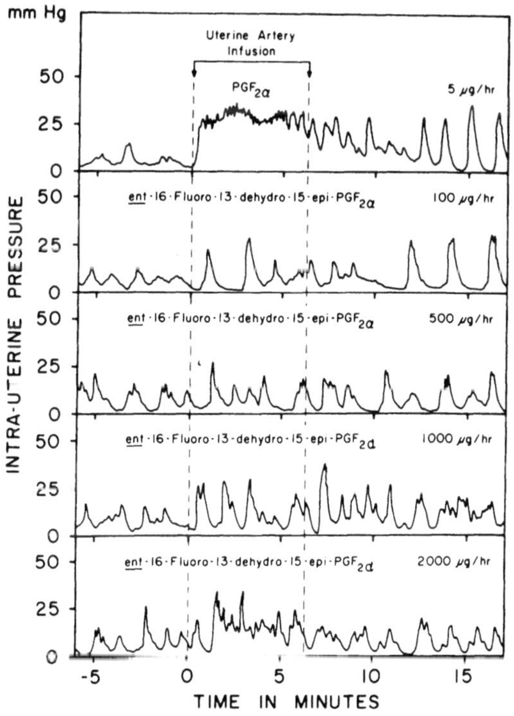


*Fig. 8* *Comparative effect of PGF$_{2\alpha}$ and ent-16-fluoro-13-dehydro-15-epi-PGF$_{2\alpha}$ on uterine smooth muscle activity in vivo in the sheep. From (43).*

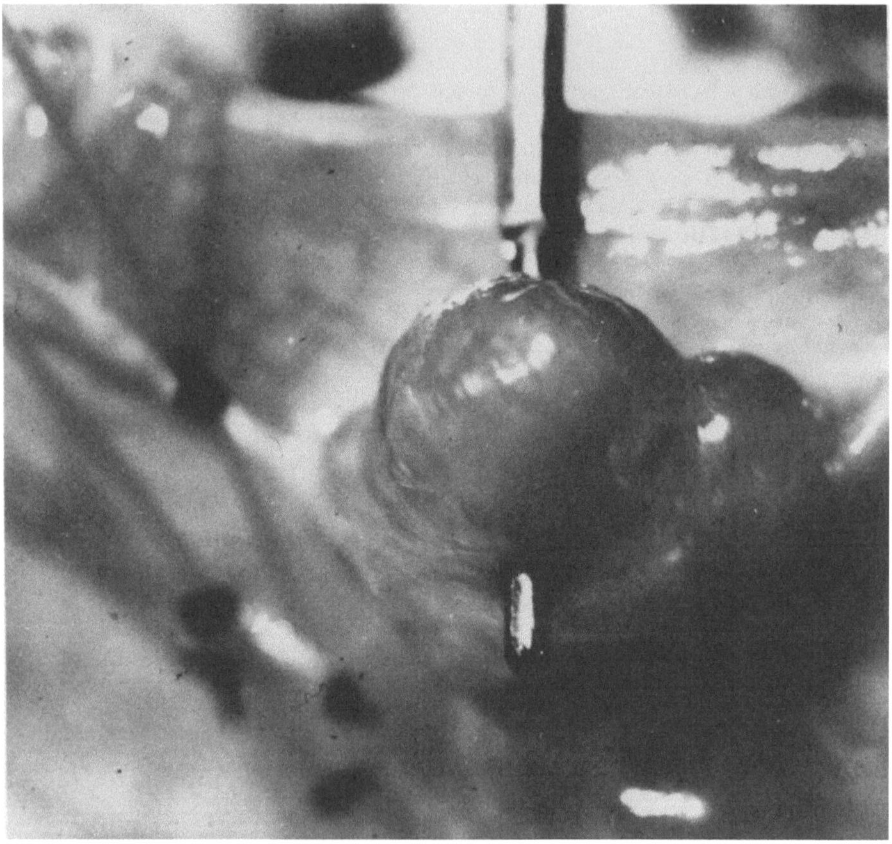

*Fig. 9* *Miniaturized Geiger-Müller probe in position in the center of an ovine corpus luteum. The catheters on the left are inserted in the uterine vein and in a branch of the ovarian artery (ramus uterinus). The probe in the CL is 1.5 mm in diameter. From (57).*

STUDIES ON THE EFFECT OF PGF$_{2\alpha}$ ON CAPILLARY BLOOD FLOW IN THE CL OF THE SHEEP

Although Pharris and Wyngarden (54) in their original hypothesis had suggested a veno or vascular constriction of the ovarian vessels as the mechanism of action of PGF$_{2\alpha}$, we were unable to observe (except with very high unphysiological doses) any reduction in total blood flow from the autotransplanted ovary of the sheep during PGF$_{2\alpha}$-induced luteal regression (49; 55). However the possibility still existed that PGF$_{2\alpha}$ might initiate luteolysis by a selective reduction in blood flow to the corpus luteum (CL) by redistributing blood through the other compartments of the ovary (55). In order to investigate this possibility capillary blood was measured in the CL of both the autotransplanted and the *in situ* ovary of the sheep (56; 57). A miniaturized Geiger-Müller probe was inserted through the center of the CL in each animal (*Fig. 9*). Periodically $^{85}$Krypton ($^{85}$Kr) (1 mC) dissolved in 0.4 ml saline was infused for 1 minute into a branch of the ovarian artery (*ramus uterinus*). The β-emission from $^{85}$Kr was picked up by the probe and registered on a detector. A print out was made every 10 seconds so that the washout curve (clearance rate) of $^{85}$Kr could be plotted (See *Fig. 10*). From

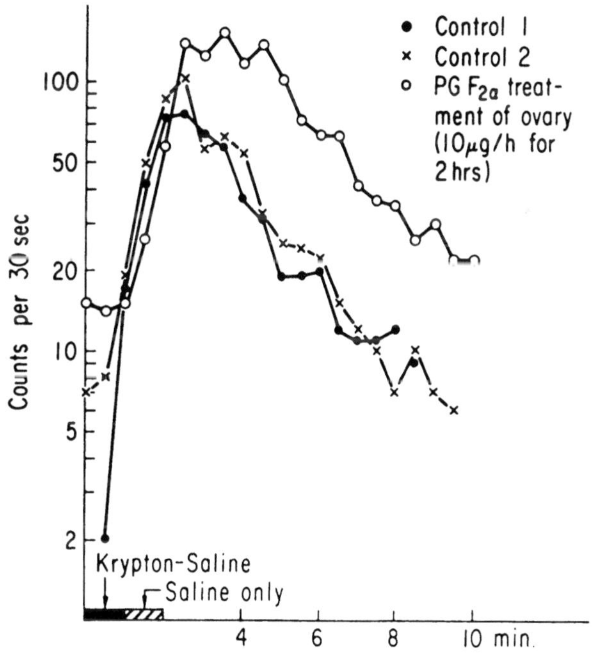


*Fig. 10* *Clearance rate of $^{85}$Krypton from CL of sheep before and 2 hours after treatment with PGF$_{2\alpha}$. The washout curve (clearance rate) of $^{85}$Krypton was not significantly different from controls 2 to 4 hours after treatment in six sheep. From (56; 57).*

the slope of this curve the 1/2 t could be obtained and hence the capillary flow of the CL could be accurately determined (58). After a control period of two hours an infusion of $PGF_{2\alpha}$ (10 μg/hr) was begun into a branch of the ovarian artery. Progesterone (P) concentration was measured in ovarian venous blood both before and after beginning the infusion of $PGF_{2\alpha}$ into either the *in situ* ovary or the autotransplanted ovary preparation. The onset of luteolysis was defined as the time when ovarian venous plasma progesterone concentration fell to 50% of mean control values (56). In the transplanted ovary preparation (n = 3) the mean time for luteolysis was 2.1 hr at which time CL capillary flow did not differ significantly from control flows (0.41 ml/g/min). In the *in situ* ovary (n = 3) the mean time to initiate luteolysis was 1.3 hr at which time capillary flow in the CL again did not differ from mean control flows (0.90 ml/g/min). The difference in flow between the CL in the transplanted ovary and the CL in the *in situ* ovary was considered to be due to their relative ages (> 60 days and < 10 days respectively). It was therefore concluded that the mechanism whereby $PGF_{2\alpha}$ initiates luteolysis is not dependent on a change in capillary flow through the CL, a finding which supports our results that luteolytic analogs which are devoid of smooth muscle activity are still luteolytic. Recent blood flow studies from other laboratories support the conclusion that CL regression is not initiated by a drop in blood flow through the CL (59; 60). As mentioned earlier there is also no drop in total flow through the ovary at the onset of luteolysis (func-

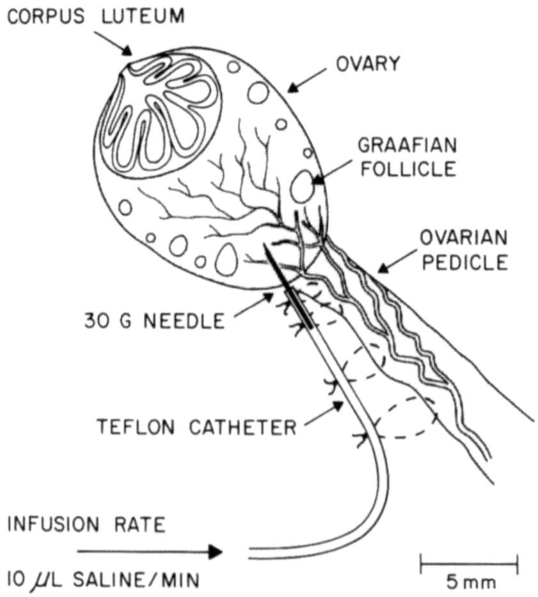


*Fig. 11 Diagram of technique for the intra-ovarian infusion of the primate ovary in vivo.*

tional) in the sheep ovary induced with physiological dose levels of $PGF_{2\alpha}$ (2.5 to 10 µg $PGF_{2\alpha}$/hr). However it should be noted that 24 to 48 hours after the onset of $PGF_{2\alpha}$-induced luteolysis in the sheep, total flow eventually declines probably due to structural changes within the regressing CL (structural luteolysis). This eventual decline in blood flow is normally seen in the *in situ* ovary of the sheep after naturally occurring luteolysis (61).

## STUDIES WITH 13-DEHYDRO ANALOGS IN A PRIMATE MODEL (*Macaca fascicularis*)

The crab-eating monkey (*Macaca fascicularis*) was selected for luteolytic studies because this species of monkey has a good temperament and is thus easier to handle. It also has very regular menstrual cycles and it does not show a pronounced reduction in ovarian cyclicity in the summer months which is often seen in other macaques.

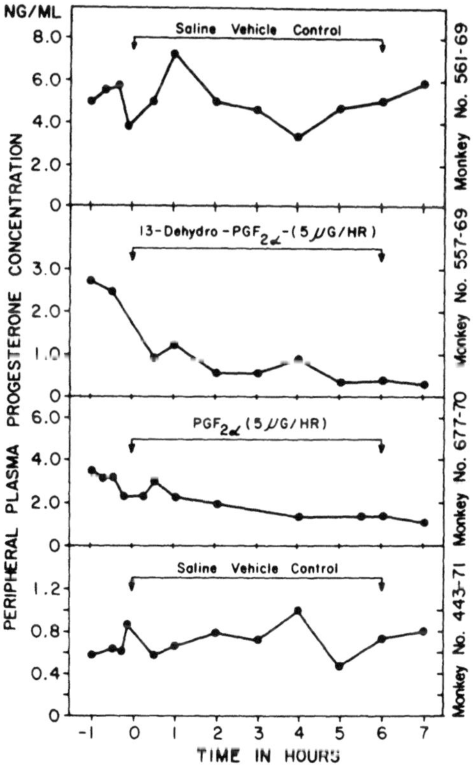


*Fig. 12* *Peripheral plasma progesterone levels in female monkeys (Macaca fascicularis) during the mid luteal phase following the intra-ovarian infusion of 13-dehydro-$PGF_{2\alpha}$ or $PGF_{2\alpha}$ alone (both at 5 µg/hr for six hours). From (43).*

Because the ovarian artery in *M. fascicularis* is very small it was not possible to infuse compounds into the ovary via an arterial cannula since the latter would probably occlude normal arterial blood flow to the ovary. A technique was devised to introduce $PGF_{2\alpha}$ and its analogs directly into the hilus of the ovary via a 30G needle (43). The method is shown diagrammatically in *Fig. 11*. Results using this technique are shown in *Fig. 12* where it can be seen that both $PGF_{2\alpha}$ and its 13-dehydro analog at 5 μg/hr for six hr caused a marked fall in peripheral plasma progesterone concentration during the mid-luteal phase of the cycle. Control infusions of vehicle alone (10 μl/min) were without effect. The 13-dehydro analog was also administered systemically (i.v.) to monkeys during the mid-luteal phase of the cycle (See example *Fig. 13*). Additional monkeys were also treated with the analog via the subcutaneous (s.c.) and oral routes. The results of these experiments are summarized in *Fig. 14*. It can be seen that at the higher doses given by these various routes, there was a rapid decline in peripheral plasma progesterone levels. The administration of saline controls by these same routes were without any immediate effect on progesterone levels although shortening of the cycle was observed in one or two animals perhaps due to the stress associated with chair confinement or perhaps even to a spontaneous short luteal phase (62). When $PGF_{2\alpha}$ itself was given at 1 mg/kg orally there was a transient fall in progesterone concentration in peripheral blood for 2 to 3 hr followed by a return to control levels (See *Fig. 15*). The ent-15-epi analog was also administered by various routes and also

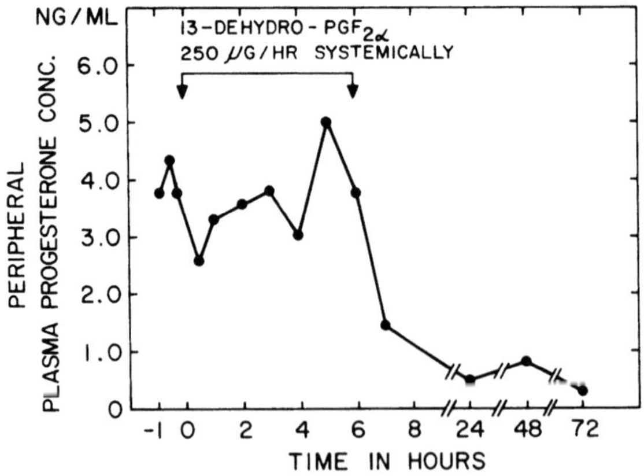


*Fig. 13* *Peripheral plasma progesterone concentration in a female monkey (Macaca fascicularis) on day 21 of cycle following the systemic infusion of 13-dehydro-$PGF_{2\alpha}$ at 100 μg/kg i.v. for six hours. From (43).*

appeared to cause a fall in progesterone levels at the higher dose levels (*Fig. 16*), although the fall was less precipitous and took longer to occur. A few experiments have been carried out with the 16-fluoro derivatives (erythro) of the 13-dehydro analog. Preliminary results suggest that this analog is somewhat more potent than its nonfluorinated counterpart. This is apparent when the 16-fluoro results are directly compared with the $PGF_{2\alpha}$ and 13-dehydro-$PGF_{2\alpha}$ each given orally at a dose level of 1 mg/kg (See *Fig. 15*).

| ROUTE | DOSE | LUTEOLYSIS |
|---|---|---|
| I.O. INFUSION | 2.5 μG/HR/6 HR | NEGATIVE |
| | 5.0 μG/HR/6 HR | POSITIVE |
| SYSTEMIC INFUSION | 25.0 μG/KG/HR/6 HR | NEGATIVE |
| | 75.0 μG/KG/HR/6 HR | POSITIVE |
| | 100.0 μG/KG/HR/6 HR | POSITIVE |
| S.C. INJECTION | 0.5 MG/KG | INCOMPLETE (50%) |
| | 1.0 MG/KG | POSITIVE |
| PER OS | 1.0 MG/KG | INCOMPLETE (80%) |
| | 2.0 MG/KG | POSITIVE |

*Fig. 14. Summary of luteolytic effects of 13-dehydro-$PGF_{2\alpha}$ in the monkey (Macaca fascicularis).*

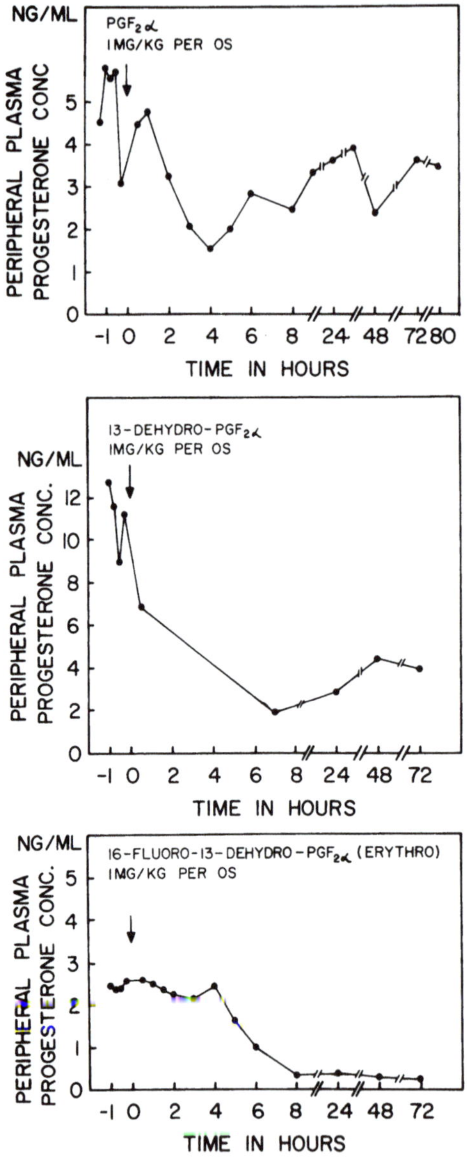


*Fig. 15* *Comparison of* $PGF_{2\alpha}$*, 13-dehydro-*$PGF_{2\alpha}$ *and 16-fluoro-13-dehydro-*$PGF_{2\alpha}$ *each given orally at 1 mg/kg on day 21 to three different monkeys (Macaca fascicularis).*

| ROUTE | DOSE | LUTEOLYSIS |
|---|---|---|
| I.O. INFUSION | 10.0 µG/HR/6 HR | INCOMPLETE (35%) |
| | 25.0 µG/HR/6 HR | INCOMPLETE (65%) |
| SYSTEMIC INFUSION (I.V.) | 100.0 µG/KG/HR/6 HR | NEGATIVE |
| | 250.0 µG/KG/HR/6 HR | POSITIVE |
| S.C. INJECTION | 0.5 MG/KG | INCOMPLETE (80%) |
| | 1.0 MG/KG | POSITIVE |
| | 2.0 MG/KG | POSITIVE |
| PER OS | 2.0 MG/KG | INCOMPLETE (50%) |
| | 3.0 MG/KG | POSITIVE |
| | 4.0 MG/KG | POSITIVE |

*Fig. 16 Summary of luteolytic effects of ent-15-epi-13-dehydro-$PGF_{2\alpha}$ in the monkey (Macaca fascicularis).*

Early attempts in the monkey to assess uterine smooth muscle effects of $PGF_{2\alpha}$ and its 13-dehydro derivatives, using a similar approach to that described above for the sheep, had to be abandoned because this method caused severe uterine hypertonus lasting many hours. A method previously described by Spilman and coworkers (63; 64) was therefore used to study the effect of PGs on uterine smooth muscle activity. This consisted of a small silastic balloon chronically implanted in the myometrium. The technique is shown diagrammatically in *Fig. 17*. The results are summarized in *Fig. 18*.

When the control dose of $PGF_{2\alpha}$ was increased to 100 µg i.v. emesis and wretching was consistently evoked in the test monkey. However no such effects were evoked with higher doses of the analogs when, in some cases, as much as 2000 µg were given i.v. These experiments carried out *in vivo* in the monkey so far compare very favorably with the results previously obtained in the sheep in terms of relative smooth muscle and luteolytic activity.

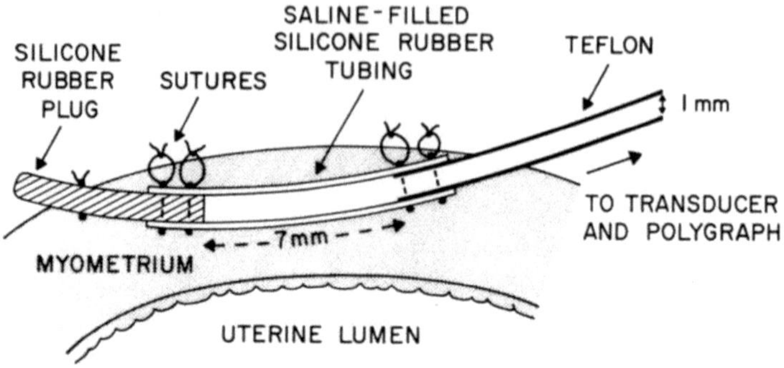


*Fig. 17* *Silicone rubber micro-balloon implanted in myometrium of monkey (Macaca fascicularis). Based on method of Spilman et al. (62; 63). Myometrial contractions were recorded on a polygraph following the intravenous injection of $PGF_{2\alpha}$ and compared to the effect of 13-dehydro analogs given by the same route.*

| Compound | Potency ($PGF_{2\alpha}$ = 1) |
|---|---|
| 13-dehydro-$PGF_{2\alpha}$ | 0.2 |
| ent-15-epi-13-dehydro-$PGF_{2\alpha}$ | 0.01 |
| 16-fluoro-13-dehydro-$PGF_{2\alpha}$ (erythro) | 0.5 |
| 16-fluoro-13-dehydro-$PGF_{2\alpha}$ (threo) | 0.25 |
| ent-15-epi-16-fluoro-13-dehydro-$PGF_{2\alpha}$ (erythro) | 0.02 |

*Fig. 18* *Comparative smooth muscle activity of the 13-dehydro analogs of $PGF_{2\alpha}$ in the uterus of the monkey (Macaca fascicularis).*

CONCLUSIONS:

The fact that the 13-dehydro analogs of $PGF_{2\alpha}$ have very weak smooth muscle stimulating activity but are potent luteolytic agents in both the sheep and the monkey, is compatible with our findings that $PGF_{2\alpha}$ does not initiate luteolysis by a vascular smooth muscle effect on the capillary bed of the CL. Furthermore these results suggest that the receptors governing the luteolytic effect on the one hand and the smooth muscle effects on the other, may possess different structural specificities. Thus the possibility is now opened up to develop analogs with even greater selectivity hence paving the way to the development of a once-a-month luteolytic contraceptive agent based on a synthetic PG analog.

ACKNOWLEDGEMENTS:

The work at the Worcester Foundation was supported by contract #HD-4-2801 (J.F.), in part by NIH grant #HD 08129 (J.McC.) and by the Population Council (N.E-J.). We wish to acknowledge the expert assistance of Stephen S. Hull, Jr., Larry C. Bovaird, Lawrence F. Underwood and Merrilyn E. Glew. Thanks are also due to David Watson and his staff at the Herschel Lab for the analysis of steroids in primate samples.

The studies involving primates were conducted at the New England Regional Primate Research Center, Southborough, MA which is supported by NIH grant #RR 00168 from the Division of Research Resources. Particular thanks are due to Drs. P. Sehgal and R. Rogers and to Mr. Jack Thompson for help with the primate study.

The work at the University of Chicago was supported by NIH grant AM 11499, contract HD-4-2801 and RCA AM 21846. Acknowledgement is also made to NSF (GP 33116) and NIH (GA 14999) for funds to purchase nmr equipment used in the synthetic work..

SUMMARY

*Prostaglandin $F_{2\alpha}$ ($PGF_{2\alpha}$) was identified as a luteolytic hormone in sheep (Nature, New Biol. 238, 129, 1972). Attempts to use $PGF_{2\alpha}$ for the pharmacological control of luteolysis in normal cycling sheep met with only partial success due to the rapid clearance of $PGF_{2\alpha}$ from the blood. In addition treated animals showed moderate to severe cardiovascular and gastrointestinal side effects. Accordingly, experiments were carried out to determine whether PG analogs might be more effective as pharmacological luteolytic agents. These compounds, which consisted of a number of the 13-dehydro analogs of $PGF_{2\alpha}$, were administered to both sheep and monkeys either directly into the ovary or into the systemic circulation to examine them respectively for direct luteolytic activity and for resistance to* *metabolism. In addition in both the sheep and the monkey smooth muscle activity of the analogs was determined by recording uterine contractions in vivo. Several 13-dehydro analogs including some 16-fluoro derivatives were shown to have luteolytic activity equal to, or in some cases greater than,* $PGF_{2\alpha}$ *itself. Furthermore most of these compounds showed a marked resistance to the 15-OH-PG-dehydrogenase enzyme in vivo as evidenced by their luteolytic activity when infused intravenously. In terms of uterine contractions, several luteolytic analogs showed markedly diminished smooth muscle activity, and in some cases, complete absence of activity. These results suggest that the receptors governing the luteolytic effect on the one hand, and the smooth muscle effect on the other, possess different structural specificities. Recent studies which we have carried out on the effect of* $PGF_{2\alpha}$ *on corpus luteum (CL) blood flow support this conclusion. CL capillary blood flow was continuously monitored by means of a miniaturized Geiger-Müller probe inserted through the center of the CL in both the in situ and the autotransplanted ovary of the sheep. Capillary blood flow was measured by the clearance rate of* $^{85}$*Krypton injected periodically into the ovarian artery before and during the induction of luteolysis with* $PGF_{2\alpha}$*. It was concluded that the initiation of luteolysis is not dependent on a smooth muscle effect of* $PGF_{2\alpha}$ *on the capillaries of the CL, a finding which supports the results with the synthetic analogs devoid of smooth muscle activity. More recently, in the primate model used (Macaca fascicularis) we have demonstrated that certain metabolically stable analogs are luteolytic when given intravenously, subcutaneously, or orally. These results demonstrate a rational approach to both drug synthesis and biological evaluation and suggest that a once-a-month contraceptive agent, based on a luteolytic analog of* $PGF_{2\alpha}$*, devoid of smooth muscle activity (side effects) and metabolically stable in the bloodstream may become a reality.*

REFERENCES

(1) J.A. McCracken, J.C. Carlson, M.E. Glew, J.R. Goding, D.T. Baird, K. Grēen and B. Samuelsson. Nature New Biol. 238(1972)129.

(2) N.L. Poyser. Adv.Prostaglandin and Thromboxane Research 2(1976)633.

(3) M.P. Eddy Moeljono, F.W. Bazer and W.W. Thatcher Prostaglandins 11(1976)737.

(4) A.R. Gleeson, G.D. Thorburn and R.I. Cox. Prostaglandins 5 (1974)521

(5) C.D. Nancarrow, J. Buckmaster, W. Chamley, R.I. Cox, I.A. Cumming, L. Cummins, J.P. Drinan, J.K. Findlay, J.R. Goding, B.J. Restall, W. Schneider and G.D. Thorburn. J.Reprod.Fert. 32(1973)320.

(6) H. Kindahl, L-E. Edquist, A. Bane and E. Granstrom. Acta Endocrinologica 82(1976)134.
(7) E.W. Horton and N.L. Poyser. Physiological Reviews 56 (1976) 595.
(8) E.L.G. Beavis, J.B. Brown and M.A. Smith. J.Obs.Gynaec. Brit.Cwlth. 76(1969)969.
(9) J.D. Neill, E.D.B. Johansson and E. Knobil. Endocrinology 84(1969)464.
(10) H. Kindahl, E. Granstrom, L.E. Edquist and P. Eneroth. Adv. Prostaglandin and Thromboxane Res. 2(1976)667.
(11) A.J. Peterson, H.R. Tervit, R.J. Fairclough, P.G. Havik and J.F. Smith. Prostaglandins 12(1976)551.
(12) J.A. McCracken. Prostaglandins in Fertility Control 2(1972) 234.
(13) W.S. Powell, S. Hammarström, B. Samuelsson and B. Sjoberg. The Lancet 1(1974)1120.
(14) K.M. Henderson and K.P. McNatty. Prostaglandins 9(1975) 779.
(15) R.C. Lyneham, A.R. Korda, D.A. Shutt, I.D. Smith and R.P. Shearman. Prostaglandins 9(1975)431.
(16) A.R. Korda, D.A. Shutt, I.D. Smith, R.D. Shearman and R.C. Lyneham. Prostaglandins 9(1975)443 and Prostaglandins 10(1975)727.
(17) D.A. Shutt, R.P. Shearman, R.C. Lyneham, A.H. Clarke, G.R. McMahon and P. Goh. Steroids 26(1975)299.
(18) J.R.G. Challis, A.A. Calder, S. Dilley, C.S. Forster, I. Hillier, D.J.S. Hunter, I.Z. MacKenzie and G.D. Thorburn. J.Endocrinol. 68(1977)401.
(19) I.A. Swanston, K.P. McNatty, and D.T. Baird. J.Endocr. 73 (1977)115.
(20) F.J. Auletta, H. Agins and A. Scommegna. Endocrinology, In Press (1978).
(21) K.T. Kirton, B.B. Pharriss and A.D. Forbes. Proc. Soc. Exptl. Biol. Med. 133(1970)314.
(22) A.A. Shaikh and E.L. Klaiber. Prostaglandins 6(1974)253.
(23) W. Russell. Prostaglandins 10(1975)163.
(24) F. Lehmann, F. Peters, M. Breckwoldt and G. Bettendorf. Prostaglandins 1(1972)269.
(25) K. Hillier, A. Dutton, C.S. Corker, A. Singer and M.P. Embrey. Brit. Med. J. 4(1972)333-336.
(26) W.J. Le Maire and A.G. Shapiro. Prostaglandins 1(1972)259.
(27) G.S. Jones and A.C. Wentz. Amer.J.Obstet.Gynecol. 114(1972) 393.
(28) W.S.M. Arrata and R.T. Chatterton. Amer.J.Obstet. and Gynec. 120(1974)954.
(29) C.H. Spilman and T.J. Roseman. Contraception 11(1975)409.
(30) P.F. Brenner, M.E. Lumkin, C. Ballard and D.R. Mishell, Jr. Contraception 11(1975)279.

(31) A. Leader, M. Bygdeman, P. Eneroth, J.N. Martin and N. Wiqvist. Advances in Prostaglandins and Thromboxane Research 2(1976)679.
(32) G. Bundy, F. Lincoln, N. Nelson, J. Pike and N. Schneider. Ann.N.Y.Acad.Sci. 180(1971)76.
(33) S. Takagi, H. Sakata, T. Yoshida, S. Nakazawa, K.T. Fujii, Y. Tominaga, Y. Iwasa, T. Ninagawa, T. Hiroshima, Y. Tomida, K. Itoh and R. Matsukawa. Prostaglandins 14(1977)791.
(34) K. Oshima, K. Matsumoto, T. Tsuda, K. Shibata and M. Hayashi. Prostaglandin 15(1978)473.
(35) Medicine Section, Time, April 17, 1978, p.55.
(36) D. Binder, J. Bowler, E.D. Brown, N.S. Crossley, J. Hutton, M. Senior, L. Slater, P. Wilkinson and N.C.A. Wright. Prostaglandins 6(1974)87.
(37) N.S. Crossley. Prostaglandins 10(1975)5.
(38) M. Dukes, W. Russell and A.L. Walpole. Nature 250(1974)330.
(39) A.I. Csapo and P. Mocsary. Prostaglandins 11(1976)155.
(40) C.H. Spilman, D.C. Beuving, A.D. Forbes and F.A. Kimball. Prostaglandins 14(1977)477.
(41) C.H. Lin, J. Fried, J.C. Carlson and J.A. McCracken. J. Steroid Biochem. 5(1974)400.
(42) J. Fried, M-S. Lee, B. Gaede, J.C. Sih, Y. Yoshikawa and J.A. McCracken. Advances in Prostaglandins and Thromboxane Res. 1(1976)183.
(43) J.A. McCracken, J. Fried, M-S. Lee, Y. Yoshikawa, and D.C. Mammato, in: Biochemical Aspects of Prostaglandins and Thromboxanes, ed. N. Kharasch and J. Fried (Academic Press, New York, 1977) p. 215.
(44) J.A. McCracken, B. Barcikowski, J.C. Carlson, K. Grėen and B. Samuelsson. Adv. Biosci. 9(1973)599.
(45) P.J. Piper, J.R. Vane and J.H. Wyllie. Nature 225(1970)600.
(46) J. Fried, M.M. Mehra and W.L. Kao. J. Am. Chem. Soc. 93(1971) 5594.
(47) J.C. Carlson, J.A. McCracken and J. Fried. Adv. Biosci. Suppl. 9(1972)101.
(48) J.R. Goding, J.A. McCracken and D.T. Baird. J. Endocrinol. 39 (1967)37.
(49) J.A. McCracken. Annals of the New York Acad. of Sciences 180 (1971)456.
(50) J. Fried and C.H. Lin. J. Med. Chem. 16(1973)429.
(51) J. Jarabak and S.S. Braithwaite. Arch. Biochem. Biophys. 177 (1976)245.
(52) J.Fried and A. Borman, in: Vitamins and Hormones, Vol. 16, ed. K.V. Thimann (Academic Press, New York, 1958) p.303.
(53) J.S. Roberts and J.A. McCracken. Biol. of Reprod. 15(1976)457.
(54) B.B. Pharriss and L.J. Wyngarden. Proc. Soc. exp. Biol. Med. 130(1969)92.
(55) J.A. McCracken, D.T. Baird and J.R. Goding. Rec. Prog. Horm. Res. 27(1971)537.

(56) N. Einer-Jensen and J.A. McCracken. Adv. Prostaglandin & Thromboxane Research 2(1976)901.
(57) N. Einer-Jensen and J.A. McCracken. Prostaglandins 13(1977) 763.
(58) N.A. Lassen and D.H. Ingvar. Experientia 17(1961)42.
(59) N.W. Bruce and K. Hillier. Nature 249(1974)176.
(60) P.O. Janson, I. Albrecht and K. Ahrén. Acta Endocrinologica 79(1975)337.
(61) P.E. Mattner and G.D. Thorburn. J. Reproduction and Fertility 19(1969)547.
(62) J.W. Wilks, G.D. Hodgen and G.T. Ross, in: Human Reproductive Medicine, Vol. 3, ed. E.S.E. Hafez (Elsevier/North Holland Biomedical Press, Amsterdam - New York, 1977)
(63) F.A. Kimball, K.T. Kirton, C.H. Spilman and L.J. Wyngarden. Biol. Reprod. 13(1975)482.
(64) C.H. Spilman, personal communication (1976).

DISCUSSION - CHAIRPERSON - J.W. WILKS

DR. GREENWALD (Kansas City) Do you have any preliminary information on the use of these compounds in the human? I assume that there are some studies being conducted in women.

DR. McCRACKEN: The 13-dehydro compounds synthesized by Dr. Fried have not yet been tested in women. Part of the rationale of administering these 13-dehydro analogs by various routes to monkeys is to determine how well these compounds are tolerated by the non-human primate and via what routes of administration they are most effective. This monkey work is in a sense a prelude to subsequent clinical trials in women. However more extensive testing with both nonpregnant and early pregnant monkeys will have to be done to show that these compounds are indeed safe in monkeys before being tried in humans. However as I report in the text of the paper several other PG analogs produced by ICI and by ONO have been tried on women. The ICI compounds appeared to be luteolytic in monkeys but apparently they were not very acceptable when tried on women. The side effects included quite marked increases in blood pressure together with some vomiting and diarrhea. We feel that the 13-dehydro analogs may offer a useful alternative in that they show very reduced smooth muscle activity in monkeys and sheep and thus we hope that they will have no side effects in women. In the case of the ONO analogs, a different approach has been taken in the sense that they have attempted to selectively enhance the smooth muscle activity of their analogs with the idea of inducing a menstrual-like bleeding by a direct stimulatory effect on the uterus. In a recent report (see text) they believe that some analogs have a specific smooth muscle stimulating effect on the uterus but a sparing effect on other smooth muscle in the body. In a clinical trial on women they report an 86% success rate in terminating early pregnancy. However I'm not sure how acceptable the approach is going to be in terms of routine clinical use because of the powerful smooth muscle effects on the uterus.

DR. CASTRACANE (San Antonio) Could you give us any information on the nature of any side effects which you have observed with your compounds in the monkey, particularly during systemic administration?

DR. McCRACKEN: The only side effect which we have seen with the 13-dehydro analogs given systemically is a rather marginal change in systemic blood pressure in the order of 5 to 10%. Apart from this we have not observed any diarrhea or vomiting at the dose levels we have employed. In the case of the <u>ent</u>-15-epi analogs the smooth muscle activity in both the sheep and the monkey is almost totally absent. On the other hand luteolytic activity is still present. This separation of the smooth muscle effect from the luteolytic effect is compatible with our parallel findings that luteolysis is not initiated via a smooth muscle effect on the capillary blood flow in the CL. The <u>ent</u>-15-epi compounds appear to show a very slow onset of a luteolytic effect, a finding for which we presently have no explanation.

DR. DUNN (University of Wyoming) How did you measure blood flow in the sheep with ovarian autotransplants?

DR. McCRACKEN: Blood flow from the transplanted ovary was measured by means of a wide bore cannula inserted into the jugular vein so that the tip came to rest opposite the internal orifice of the anastomosed utero-ovarian vein. Blood was collected during the occlusion of the jugular vein cranially and caudally to the tip of the cannula in the vein. The volume collected in unit time was recorded so that the secretion rate of progesterone by the transplanted ovary could be accurately determined.

DR. DUNN: I'm not sure how much faith I would put in this system as I have had some experience with this method in our laboratory.

DR. McCRACKEN: You can't get a much simpler method of blood flow measurement than the one I have described.

DR. DUNN: Well in the first place you have removed the ovary from its _in situ_ location and destroyed its original vascular connections. You then force the blood through a cannula, the diameter of which may vary. It seems to me that it is a crude system at best.

DR. McCRACKEN: I would accept that it could be called an empirical system, but I don't think that it could be called crude. In the proper hands it is accurate to within about 5% and you can't get better than that with any other system. I think that what you are really trying to say is that the blood flow figures which we obtain from the transplanted ovary are higher than the blood flow from the _in situ_ ovary in the sheep. This is, of course, correct but there are perfectly valid reasons for this. For example there is some blood contributed from the skin surrounding the preparation and in addition it is documented that when one ovary is removed (as we routinely do in this preparation) the remaining ovary undergoes substantial hypertrophy. We have never claimed that the blood flow from the ovarian transplant preparation is representative of the blood flow from the _in situ_ ovary. The main point is that we are assessing the luteolytic analogs by their ability to depress the secretion rate of progesterone from the transplanted ovary. While it is true that the higher blood flow from the ovarian transplant preparation will lower the concentration of progesterone in the venous effluent, the higher flow will not affect the calculation of progesterone secretion rates (SR = BF x conc.) The great advantage of the system is that one can monitor progesterone secretion rates in the conscious resting animal over long periods of time (up to 17 days or longer). In addition the same animal can be used repeatedly so that it can act as its own control.

## EFFECTS OF CHOLERA TOXIN, LH AND $PGF_{2\alpha}$ ON STEROIDOGENESIS IN LUTEAL CELLS

L.J. Dorflinger and H.R. Behrman

Reproductive Biology Section, Departments of Obstetrics, Gynecology, Pharmacology and Physiology
Yale University School of Medicine, New Haven, CT 06510

Recent studies using cultured rat luteal cells (1) showed that $PGF_{2\alpha}$ inhibits accumulation of cyclic AMP (cAMP) and progesterone secretion in response to LH, while having little effect on the binding of LH to its receptor. This inhibition of progesterone secretion was reversed by simultaneous addition of exogenous cyclic AMP but not by theophylline (1). It was concluded that $PGF_{2\alpha}$ prevents accumulation of cAMP stimulated by the LH-receptor complex. Secondary factors may mediate the action of $PGF_{2\alpha}$ in intact cells since it does not inhibit LH-stimulated adenylate cyclase activity in isolated luteal membranes (2).

Cholera toxin stimulates adenylate cyclase activity in a wide variety of cell types (3). The lack of cell specificity and the nature of adenylate cyclase activation (3) indicates that cholera toxin may stimulate adenylate cyclase activity in a manner different from that of LH. To provide information on the possible specificity of the inhibitory action of $PGF_{2\alpha}$, the effect of $PGF_{2\alpha}$ on cholera toxin stimulation of cAMP accumulation and progesterone secretion was compared to that produced by LH.

### MATERIALS AND METHODS

Immature (26-day-old) rats were superovulated by PMS-hCG regimen. Ovaries were removed 4-8 days after hCG injection and luteal cells dispersed enzymatically in Ca#-free media (1) followed by gentle mechanical dispersion to yield predominately single luteal cells (4). Cells were resuspended in 25 mM Hepes buffered media (Gibco 236) and cell numbers determined in a hemocytometer.

Cell viability was tested using the trypan blue exclusion test. Luteal cells routinely comprised 70-80% of the total cells as determined by their size and granular lipid inclusions as seen in the light microscope.

Cell incubations were carried out in 25 $cm^2$ Falcon Flasks as described previously (1) or in 12 x 75 mm pyrex test tubes under an atmosphere of 95% $O_2$/5% $CO_2$. Each flask or test tube contained 0.2 to 1.0 x $10^6$ cells in 2 ml media. Cells were preincubated in media alone for 1 to 2 hr at 37°C and media changed before addition of hormones or cholera toxin. Following treatment cell viability was determined and in all cases was greater than 90%.

Progesterone production by luteal cells was determined by RIA (5). When cAMP was measured, an aliquot of media was removed for progesterone assay and the remaining cells and media were treated with 0.5 ml 10% TCA. In test tube cultures, the contents were sonicated for 1 minute and cell debris pelleted by centrifugation @ 2200 rpm for 10 min. TCA was removed by ether extraction and the aqueous media evaporated under air @ 80°C. One ml of distilled $H_2O$ was added to each sample and aliquots assayed by RIA (1).

Hormones and Reagents: $PGF_{2\alpha}$ was kindly supplied by John Pike, Upjohn Co. (Kalamazoo, MI). Purified ovine LH (NIH-LH-S19) was a gift from NIAMDD; cholera toxin was purchased from Schwartz/Mann; 3-isobutyl-1-methylaxanthine (IBMX) and dibutyryl cyclic AMP (dbcAMP) were purchased from Sigma (St. Louis, MO).

## RESULTS AND DISCUSSION

Cholera toxin was found to be a potent stimulator of progesterone secretion in luteal cells with half the maximal stimulation at 2.5 ng/ml (Fig. 1). This is comparable to the 50% response observed in other steroidogenic systems (approximately 30 pM) (3). Maximum stimulation of progesterone secretion by cholera toxin occured at 100 ng/ml. In later experiments, doses up to 10 μg/ml produced no further stimulation. At all doses of cholera toxin tested, $PGF_{2\alpha}$ (1.2 μM) inhibited stimulation of progesterone secretion (Fig. 1).

A brief exposure to cholera toxin results in a rapid and irreversible binding to cells (3). Since binding is the first step in a sequence of reactions leading to progesterone synthesis, a decrease in binding of the toxin induced by $PGF_{2\alpha}$ would explain the antagonism of cholera toxin-stimulated progesterone secretion. Two experiments were designed to assess the effect of $PGF_{2\alpha}$ on binding and activation by toxin in these cells. In the first experiment, luteal cells were incubated for 90 minutes with cholera toxin (100 ng/ml) or with cholera toxin + $PGF_{2\alpha}$ (500 ng/ml). Cholera toxin stimulated progesterone secretion 2.0 fold above control levels

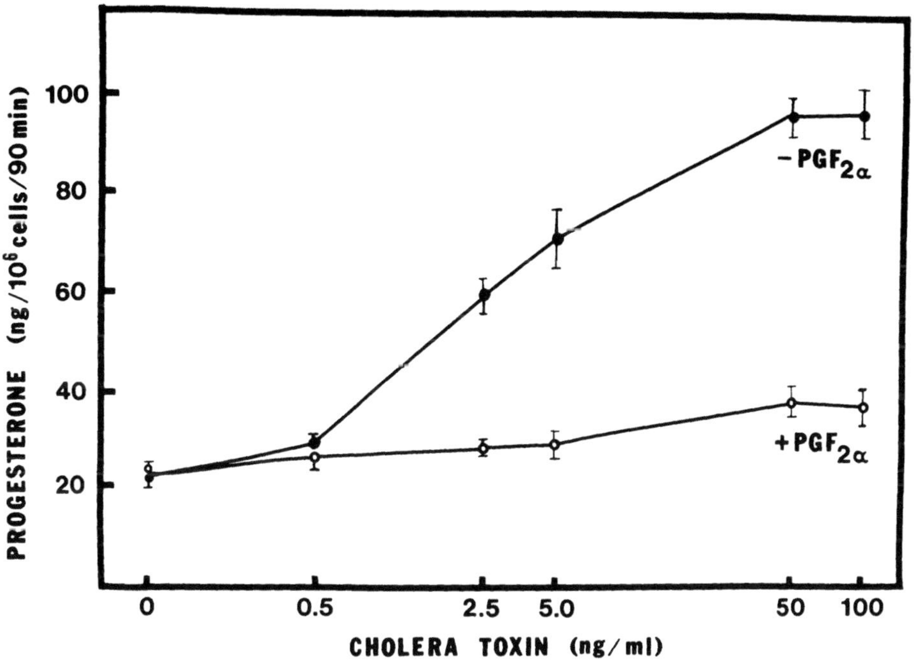


Fig. 1 Dose response of isolated luteal cells to cholera toxin in the presence and absence of $PGF_{2\alpha}$ (1.2 μM). Points are mean ± SEM of triplicate determinations. Statistical analysis: CT versus $PGF_{2\alpha}$ at 2.5, 5.0, 50, and 100 ng/ml of CT, $p < 0.001$.

($p < 0.01$). In the presence of $PGF_{2\alpha}$ there was a 95% inhibition ($p < .025$) of cholera toxin stimulated progesterone secretion. The cells were then washed and incubated in fresh media alone for an additional 90 minutes. Following the second incubation, cells exposed initially to either cholera toxin alone or cholera toxin + $PGF_{2\alpha}$ showed a similar increase in progesterone secretion over controls (4.3 fold; $p < .001$ versus 3.9 fold; $p < .001$, respectively). In the second experiment, cells were incubated in cholera toxin (100 ng/ml) for 60 minutes. $PGF_{2\alpha}$ was then added to determine if steroidogeneses in cells which were actively producing progesterone could be suppressed. This data is shown in Fig. 2. Progesterone secretion was linear between 30 and 90 minutes in cholera toxin treated flasks. After addition of $PGF_{2\alpha}$, secretion during the next 30 minutes was reduced by more than 80% ($p < .001$). This indicates that although binding of cholera toxin had taken place and the steroidogenic pathway activated, $PGF_{2\alpha}$ was capable of rapidly inhibiting further secretion. These results are similar to earlier findings (1) which show no major effect of $PGF_{2\alpha}$ on LH-binding to luteal cells.

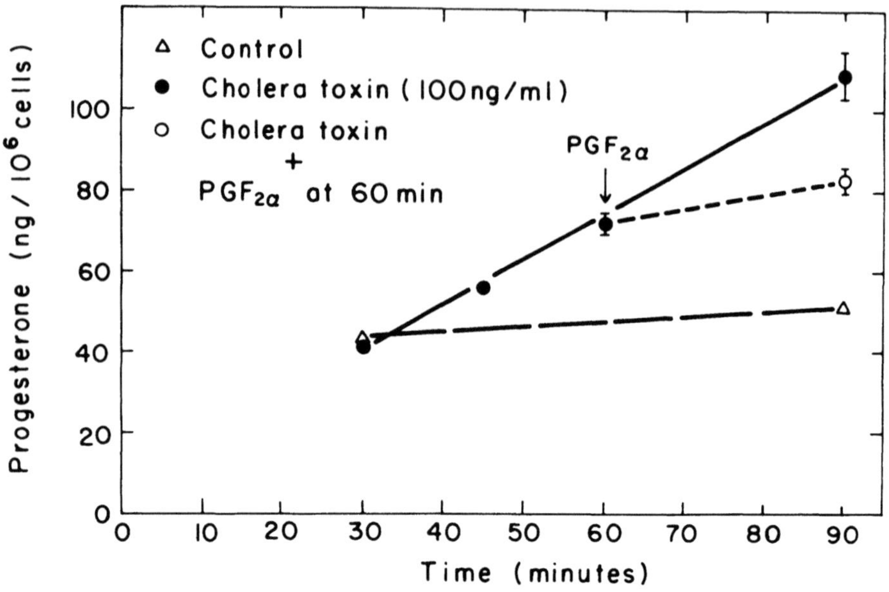


Fig. 2 Effect of adding $PGF_{2\alpha}$ 60 minutes after cholera toxin on subsequent progesterone secretion. Statistical analysis: CT (60 minutes) vs CT (90 minutes), $p < 0.005$; $CT_{90}$ vs CT + $PGF_{2\alpha\ 90}$, $p < 0.025$; $CT_{60}$ vs CT + $PGF_{2\alpha\ 90}$, NS.

The ability of dibutyryl cAMP to reverse the inhibition of cholera toxin stimulated progesterone secretion by $PGF_{2\alpha}$ was examined. In this study, $PGF_{2\alpha}$ inhibited cholera toxin stimulated progesterone secretion by 93% ($p < .025$) (Fig 3). Addition of dibutyryl cAMP to cells in the presence of cholera toxin + $PGF_{2\alpha}$ significantly reversed this inhibition. Just as the inhibition of the steroidogenic action of LH produced by $PGF_{2\alpha}$ appears to occur through a reduction of cAMP accumulation, the reversal by dbcAMP in these experiments would indicate a similar action in the case of $PGF_{2\alpha}$-induced inhibition of cholera toxin-stimulated progesterone secretion. The small component of inhibition remaining was not observed previously (1), but was reproducible and may be a consequence of a secondary effect of $PGF_{2\alpha}$.

The effect of cholera toxin (5 ng/ml) and LH (50 ng/ml) on cAMP accumulation is shown in Table 1. This dose of cholera toxin was considerably more potent as a stimulator of cAMP accumulation than

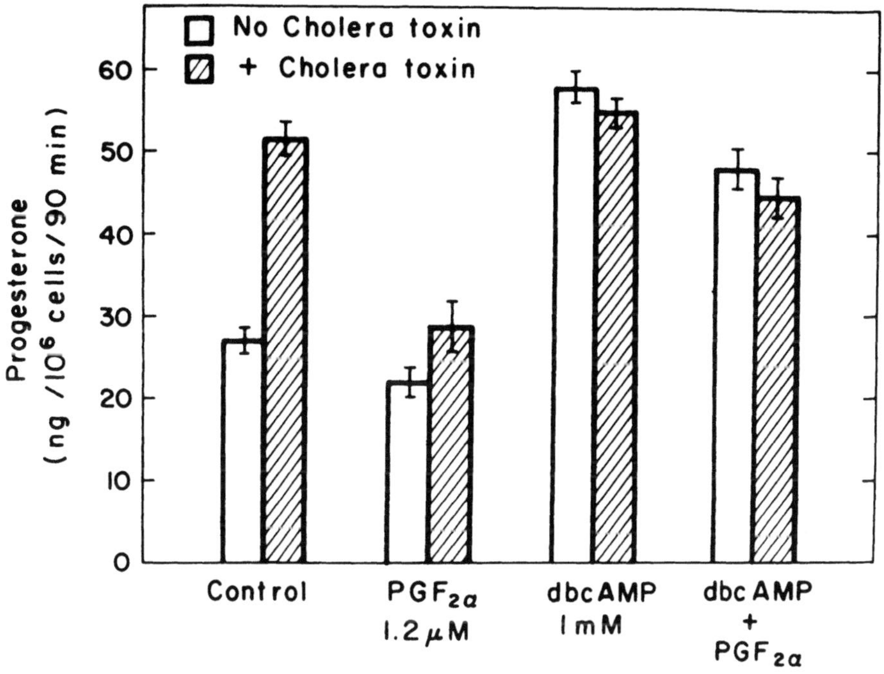


Fig. 3 Effect of addition of dbcAMP on $PGF_{2\alpha}$-induced inhibition of cholera toxin-dependent progesterone secretion. Values are mean $\pm$ SEM of triplicate determinations. Statistical analysis: Control vs CT, $p < .01$; Control vs CT + dbcAMP, $p < .005$; CT vs CT + $PGF_{2\alpha}$, $p < .001$; CT + $PGF_{2\alpha}$ vs CT + dbcAMP + $PGE_{2\alpha}$, $p < .025$; and CT + dbcAMP vs CT + dbcAMP + $PGF_{2\alpha}$, 0.05 $p < 0.10$.

was the dose of LH (3.8 versus 1.6 fold). $PGF_{2\alpha}$ reduced LH stimulation of cAMP accumulation by 75% ($p = .05$) but cholera toxin stimulation of cAMP accumulation was reduced by only 25%. When both LH and cholera toxin were present, cAMP accumulation was reduced by $PGF_{2\alpha}$ (-16%), but the difference was not significant. In contrast, progesterone secretion by both agonists, either alone or in combination, was inhibited by $PGF_{2\alpha}$. Even though progesterone secretion was completely inhibited, the level of cAMP accumulation in the cholera toxin + $PGF_{2\alpha}$ treatment group was significantly higher than that seen with LH treatment alone. This observation that $PGF_{2\alpha}$ completely blocked cholera toxin-dependent progesterone secretion while cholera toxin-dependent cAMP accumulation was only partially reduced, leads us to speculate that many of the adenylate cyclase moieties activated by cholera toxin may not be linked to progesterone biosynthesis. This would support the hypothesis that $PGF_{2\alpha}$ acts specifically to inhibit accumulation of cAMP linked to the progesterone biosynthetic pathway, independently or whether the stimulator is LH or cholera toxin.

Table 1. Effect of Cholera Toxin and $PGF_{2\alpha}$ on cAMP Accumulation and Progesterone Secretion in Luteal Cells in the Presence and Absence of LH.

| Addition | cAMP Accumulation pmol/$10^6$ cells | | Progesterone Secretion ng/$10^6$ cells | |
|---|---|---|---|---|
| | - LH | + LH | - LH | + LH |
| Control | 9.8±0.7 | 15.7±0.6 | 28.8±1.1 | 65.9±3.8 |
| $PGF_{2\alpha}$ | 8.2±1.3 | 11.3±0.3 | 23.6±0.4 | 38.4±1.4 |
| Cholera Toxin | 37.3±0.3 | 46.3±4.6 | 42.0±2.9 | 60.1±4.8 |
| Cholera Toxin + $PGF_{2\alpha}$ | 30.5±2.1 | 40.4±3.4 | 26.9±0.9 | 41.6±1.6 |

Statistical analysis: cAMP accumulation. Control vs LH and LH vs LH + $PGF_{2\alpha}$, $p < 0.005$; Control vs CT, $p < 0.001$; CT vs CT + $PGF_{2\alpha}$, $p < 0.05$. Progesterone secretion. Control vs LH, $p < 0.001$, LH vs LH + $PGF_{2\alpha}$, $p < 0.005$; Control vs CT, $p < 0.025$; CT vs CT + $PGF_{2\alpha}$, $p < 0.010$.

(Supported by grants from NIH-HD-10718 and Ford Foundation 770-0534 and a predoctoral traineeship USPHS GM-1113 (LJD).

## REFERENCES

1. J.P. Thomas, L.J. Dorflinger and H.R. Behrman. P.N.A.S. 75 (1978) 1344.
2. L.J. Dorflinger and H.R. Behrman. Abstract: Society for the Study of Reproduction, 1978.
3. D.M. Gill, in: Adv. in Cyclic Nuc. Res., Vol. 8, ed. P. Greengard and G.A. Robinson (Raven Press, New York, 1977) p. 85.
4. J.L. Luborsky-Moore, K. Wright, L.J. Dorflinger and H.R. Behrman, (in preparation).
5. G.P. Orczyk, M. Hichens, G. Arth and H.R. Behrman, in: Methods of Hormone Radioimmunoassay, eds. B.M. Jaffee and H.R. Behrman (Academic Press, New York), p. 347.

DISCUSSION - CHAIRPERSON - J.W. WILKS

J.M. MARSH: Did you measure the cyclic-AMP accumulation in the cells, the media, or both together?

L.J. DORFLINGER: We measured total cAMP. We are now beginning to measure and compare both intracellular and extracellular cAMP levels. In the case of the $PGF_{2\alpha}$-induced inhibition of cholera toxin-stimulated cAMP accumulation, the differential inhibition of one of these two compartments would indicate that there is a pool of cAMP within the cell whose synthesis and secretion is unaffected by $PGF_{2\alpha}$. This is one way of interpreting the data that we have on total cAMP levels.

J.M. MARSH: I think there probably is a pool of cyclic-AMP which is important for progesterone synthesis, and that the tiny changes which are observed are sufficient for stimulating steroidogenesis. I also think that cholera toxin produces pretty excessive amounts of cAMP. You may be blocking the pool of cAMP which is important for steroidogenesis, but cannot see the effect because it is masked by the larger amounts of cAMP in other pools.

L.J. DORFLINGER: Yes, I agree with you. The level of cAMP is generally higher in the presence of cholera toxin and the corresponding progesterone level lower than that seen with LH. As we showed, the change in the cAMP level with $PGF_{2\alpha}$ treatment was quite small but there was a complete block in progesterone secretion. Indeed, this is an indication that there is a pool(s) of cAMP which may not be coupled to progesterone biosynthesis.

M.L. DUFAU: Do you have any information about the composition of your cell preparation? Since you are using cells only just a few days after induction of pseudopregnancy it is possible that you are dealing with a mixed population of cells. It would be important to make sure that you are dealing with a homogenous population of cells since choleragen receptor sites have such a wide distribution and is possible that $PGF_{2\alpha}$ receptors are restricted to certain cell populations in the ovary.

Comment.

I noticed that all the differences reported during your presentation were based on the measurement of total cyclic AMP. In this case, most of the cyclic AMP that you are measuring is probably extracellular cyclic AMP, which is a reflection of what happens inside the cell, but may not have much to do with the fine control of metabolic events occurring during cell stimulation. In addition, the high levels of extracellular cyclic AMP usually found during stimulation of cells would mask the subtle changes that occur in the intracellular compartment during hormone action.

Since you were able to observe the differences by measuring total cyclic AMP, it is very likely that you would see more marked effects by measuring intracellular cyclic AMP levels.

In relationship to your findings, I think that it is appropriate to mention that we have recently reported functional compartmentalization of cyclic AMP responses in the Leydig cell. In those studies, we found that choleragen ($10^{-13}$ to $10^{-11}$M) increased cyclic AMP and occupied 75% of cyclic AMP receptor sites without producing changes in testosterone production. Only nearly all the cyclic AMP receptor sites were occupied when the expected changes in steroid production were observed. In contrast, during hCG stimulation small increases in cyclic AMP and minor receptor occupancy (25%) produced maximal steroidogenic responses. It is likely that this is the case as you propose in the luteal cell (M.L. Dufau, K .A. Horner, K. Hayashi, T. Tsuruhara, P.M. Conn and K.J. Catt, J. Biol. Chem. 253, 3721 (1978)).

L.J. DORFLINGER: We know that $PGF_{2\alpha}$ acts rapidly to inhibit the accumulation of cAMP. If all adenylate cyclase moieties were affected by $PGF_{2\alpha}$ then I would not expect to see extracellular accumulation. However, it appears that only a subpopulation of adenylate cyclases are affected and that a large portion of the cAMP that we measure is in the extracellular compartment. Therefore the subtle changes that we measure may indeed be more dramatic if we measure intracellular cAMP.

M. LAHAV: I wish to comment about the specificity of the $PGF_{2\alpha}$ affect with regard to the stimulant. When $PGE_2$ is used to stimulate cyclic-AMP accumulation in rat corpora lutea, $PGF_{2\alpha}$ does not reverse the effect. Similar results were found by McNatty in cultured human granulosa cells; the $PGE_2$ effect on progesterone synthesis was not abolished by $PGF_{2\alpha}$.

L.J. Dorflinger: We have also done that experiment, but the results were inconclusive.

A.R. MIDGLEY: I would like to add an obvious comment that perhaps not all of progesterone secretion is dependent upon cyclic-AMP stimulation. There's nothing very novel in this concept, but I am particularly reminded of it from a recent meeting that I attended at Keystone, Colorado. At the meeting Dr. Michael Gill presented evidence indicating that in activating adenylate cyclase a portion of cholera toxin not only enters cells, but also acts as an enzyme to ADP-ribosylate, and thereby inactivates the enzyme GTPase. This strengthens the possibility that, like cholera toxin, a portion of gonadotropins might enter cells and act enzymatically, perhaps to affect steps in the steroidogenic pathway that require GTP involvement. If this were true gonadotropins could give rise to effects beyond those arising through stimulation of adenylate cyclase. I don't think we should become too rigidly set on the requirement for a direct one-to-one relationship between a limited pool of cyclic-AMP and progesterone stimulation.

# EFFECTS OF PROSTAGLANDIN $F_2\alpha$, ESTRADIOL AND LUTEINIZING HORMONE IN DISPERSED CELL PREPARATIONS OF BOVINE CORPORA LUTEA

J. E. Hixon and W. Hansel

Depts. of Vet. Biosci., Univ. of Ill., Urbana, Ill., and

Physiology-Physical Biology, Cornell Univ., Ithaca, N.Y.

## INTRODUCTION

A luteolytic site of action for estradiol at the cellular level was suggested by the observation that administration of estradiol benzoate to cattle diminished the *in vitro* steroidogenic effects of LH (1). Inhibition of LH-induced progesterone synthesis was also observed after the addition of high concentrations of estradiol-17β ($E_2$) directly to dispersed cell preparations of bovine corpora lutea of pregnancy (2) and monkey corpora lutea (3).

Evidence that $PGF_2\alpha$ may exert its luteolytic effects at the cellular level was suggested by the observation that $PGF_2\alpha$ reduced the effects of LH in cultured granulosa cells of bovine origin (4, 5). These observations parallel evidence that $PGF_2\alpha$ inhibited LH induced cAMP production in the rat (6, 7).

In contrast to effects of $PGF_2\alpha$ observed with luteinized granulosa cell preparations, increased progesterone synthesis was observed when bovine luteal slices were incubated with $PGF_2\alpha$ (1,8). In view of this inconsistency, the *in vitro* effects of $PGF_2\alpha$ were reinvestigated with dispersed cell preparations of bovine corpora lutea removed at accurately determined stages of the estrous cycle. It was also of interest to determine if a luteolytic interaction between $PGF_2\alpha$ and $E_2$ occurred at the cellular level since an interaction of this nature has been demonstrated with uterine intact ewes (9), hysterectomized ewes (10) and intact cattle (Hixon and Hansel, unpublished observations).

## MATERIALS AND METHODS

Corpora lutea were surgically removed from normal heifers on days 12 and 13 of the estrous cycle and immediately placed in an excess of chilled Medium 199. Dispersed cell preparations were prepared by the method of Simmons et al. (11) modified by reducing the temperature for the dissociation to $33^{\circ}C$. Following completion of the dissociation, three to five aliquots were removed to a haemocytometer and the concentration of viable cells determined in each aliquot by counting those cells excluding Trypan Blue.

Incubations were conducted with $10^6$ viable cells/ml at $37^{\circ}C$ in a Dubnoff metabolic shaker with an atmosphere of 95% $O_2$, 5% $CO_2$. A suspension of cells was maintained and aliquots removed for progesterone determinations (14) represented medium plus cells. After a 30 minute preincubation period, initial aliquots were removed from all incubation tubes; treatments were then added and aliquots were removed at 2 h in the estradiol dose-response experiment or at 1, 2 and 4 h in all other experiments. Net progesterone accumulation was estimated as the difference between concentrations of progesterone at timed intervals following addition of treatments and that following preincubation.

Treatments consisted of the administration of LH (NIH-LH-B9), $PGF_2\alpha$-tromethamine salt (all doses of $PGF_2\alpha$ are expressed as free acid equivalents) and $E_2$. Stock solutions of LH and $PGF_2\alpha$ were prepared in Medium 199. Estradiol was dissolved in propylene glycol and diluted with Medium 199. Similar propylene glycol, Medium 199 mixtures were prepared as vehicles for all incubations where $E_2$ was included. The final concentration of propylene glycol in the incubation tube was 0.2 ul/ml of incubation. When multiple treatments were employed, each was mixed to deliver the final concentration simultaneously in a total volume of 100 μl.

Each experiment included corpora lutea from 5 animals with 5 replicates/treatment/animal. The data were subjected to analysis of variance using a split plot design (12). Heterogeneity of variance was observed (13) and therefore statistical analysis was performed after logarithmic transformation of the data. For this reason, results were graphed on a logarithmic scale with the corresponding antilogs (geometric means). Tukey's hsd procedure (13) was employed when multiple comparisons were required.

## RESULTS

The effects of 0,0.1,1,10,100 and 1000 ng/ml NIH-LH-B9 on net progesterone accumulation were determined with $10^6$ viable cells/ml (Fig. 1). Analysis of variance showed LH effects ($P<0.001$) and effects due to time ($P<0.001$). It is evident from Fig. 1 that 0.1

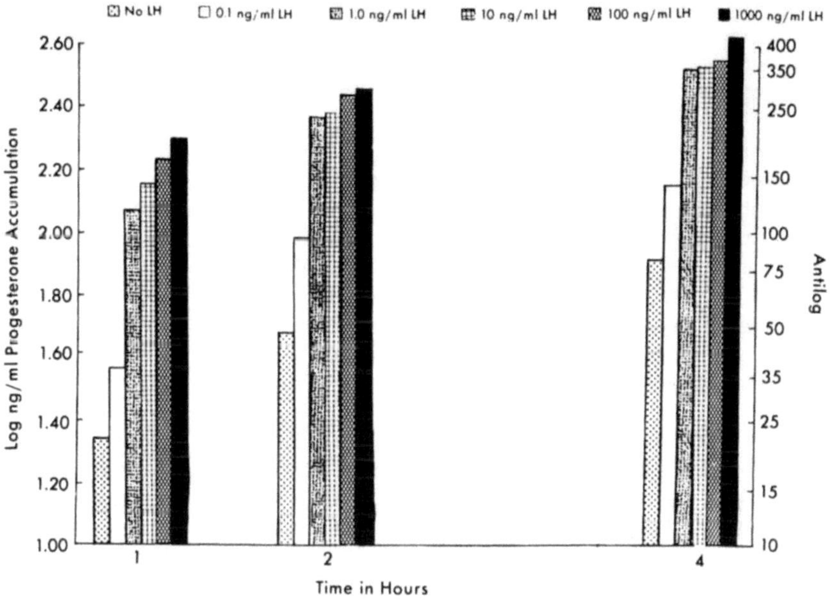


Figure 1. Log progesterone accumulation by $10^6$ viable bovine luteal cells/ml in response to the addition of 0, 0.1, 1.0, 10, 100 and 1000 ng LH/ml (NIH-LH-B9). S.E.M. = 0.08014.

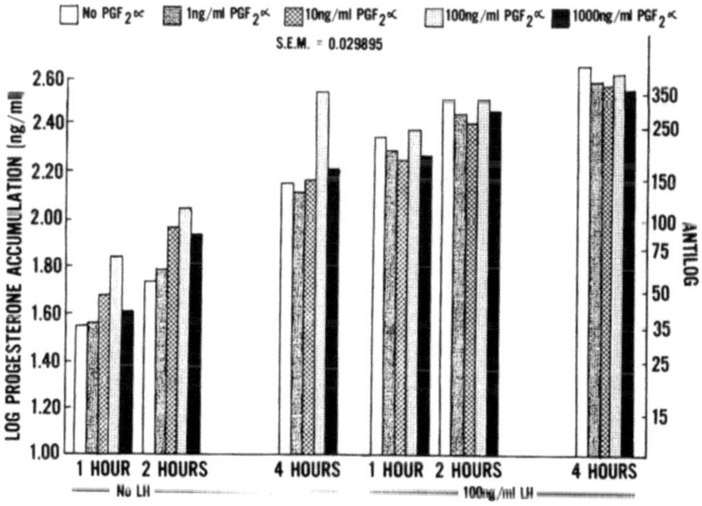


Figure 2. Log progesterone accumulation by $10^6$ viable bovine luteal cells/ml in response to the additions of 0, 1.0, 10, 100 and 1000 ng $PGF_2\alpha$/ml in the presence and absence of 100 ng LH/ml (NIH-LH-B9). S.E.M. = 0.02990.

ng/LH/ml increased progesterone accumulation ($P<0.01$) while a maximal response was obtained with 10 ng/LH/ml. The plateau in progesterone accumulation observed between 2 and 4 hours was typical of all incubations.

In a preliminary experiment, the effects of $PGF_2\alpha$ at concentrations between .001 and 1 ng/ml were determined. Since no effect of $PGF_2\alpha$ was observed at these concentrations, a second experiment was initiated in which net progesterone accumulation was determined after additions of 0,1,10,100 and 1000 ng $PGF_2\alpha$/ml in the presence and absence of 100 ng LH/ml (Fig. 2). While mean concentrations of progesterone were greater than controls after addition of 10, 100 or 1000 ng $PGF_2\alpha$, analysis of variance showed no $PGF_2\alpha$ effect ($P>0.1$) although LH increased progesterone concentrations ($P<0.01$).

The effects of 0,1,10,100 and 1000 ng $E_2$/ml on progesterone accumulation were determined in a 2 h incubation in the presence and absence of 100 ng LH/ml , (Fig. 3). It is apparent that $E_2$ was without effect on progesterone accumulation ($P>0.10$). On the other hand, LH increased progesterone accumulation without regard to $E_2$ treatment ($P<0.001$).

A possible luteolytic interaction between $PGF_2\alpha$ and estradiol was investigated in two, $2^3$ factorially designed experiments. In the first experiment, the effects of 100 pg $E_2$/ml and 100 ng $PGF_2\alpha$/ml were compared in the presence and absence of submaximal LH concentrations (1 ng/ml). These treatments were repeated in the second experiment but a maximal LH concentration of 10 ng/ml was employed. The results of these experiments are summarized as Figures 4 and 5. Similar effects of $E_2$, $PGF_2\alpha$ and LH were observed in both experiments. $E_2$ was without effect ($P>0.10$) while $PGF_2\alpha$ increased progesterone accumulation ($P<0.001$) in both experiments. LH in both concentrations increased progesterone accumulation ($P<0.001$) without regard to estradiol treatment. However, the nature of the $PGF_2\alpha$ by LH interaction differed in the two experiments and depended upon the concentration of LH employed. For example, the interaction was not significant ($P>0.15$) when submaximal LH concentrations were employed; i.e., $PGF_2\alpha$ has supplemental effects on progesterone accumulation in the presence of 1 ng/ml LH (Fig. 4). In the presence of maximal LH concentration (10 ng/ml), the LH by $PGF_2\alpha$ interaction was significant ($P<0.02$) suggesting that $PGF_2\alpha$ had no effect in the presence of LH (Fig. 5).

## DISCUSSION

Addition of LH to cell suspensions of bovine corpora lutea produced repeatable and significant increases in progesterone accumulation. The plateau in progesterone accumulation over time suggests

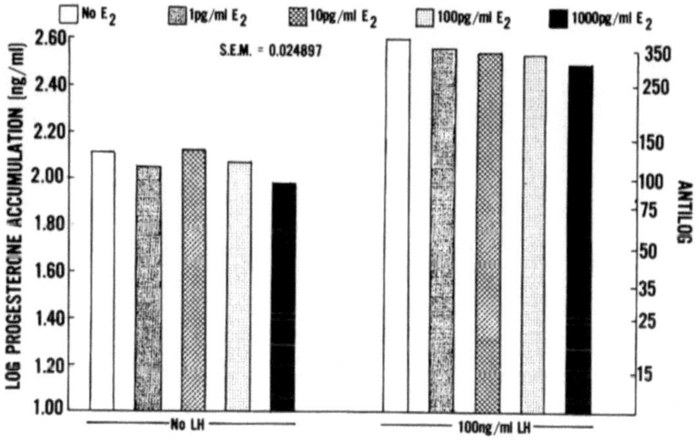


Figure 3. Log progesterone accumulation by $10^6$ viable bovine luteal cells/ml in response to 0, 1, 10, 100 and 1000 pg estradiol-17β/ml in the presence and absence of 100 ng/LH/ml (NIH-LH-B9) S.E.M. - 0.02490.

either substrate limiting kinetics or endproduct inhibition. It is not possible from these data to determine which alternative is correct, but retro-inhibition of $\Delta^5$-3β hydroxysteroid dehydrogenase activity has been reported for human corpora lutea (16) and subcellular preparations of human placentae (17).

$E_2$ was without effect on progesterone synthesis at the concentrations employed. This is in contrast to reports that very high concentrations of $E_2$ decreased the steroidogenic effects of LH in dispersed cell preparations of bovine (2) and primate (3) corpora lutea. There is no firm evidence that bovine corpora lutea are capable of estrogen synthesis, and it seems unlikely that concentrations higher than those we have used are within the physiological range. The results suggest that the luteolytic site of $E_2$ action is not at the luteal cell level.

The only observed effects of $PGF_{2}\alpha$ in dispersed cell preparations of bovine corpora lutea were to increase progesterone accumulation. This effect of $PGF_{2}\alpha$ is the same as was observed in incubations with sliced bovine corpora lutea (1,8). The observation that $PGF_{2}\alpha$ increased progesterone accumulation over that obtained at submaximal LH concentrations but not maximal LH concentrations provides circumstantial evidence that $PGF_{2}\alpha$ and LH share cAMP as an intracellular mediator. _In vitro_ effects of $PGF_{2}\alpha$ with primary cultures of bovine corpora lutea from pregnant cattle have also been attributed to cAMP (18).

It is difficult to rationalize the steroidogenic effects of $PGF_{2}\alpha$ observed with bovine corpora lutea with inhibition of LH-

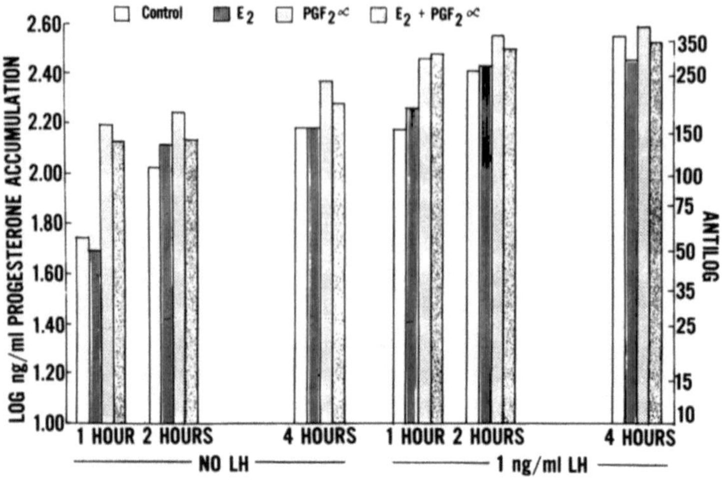


Figure 4. Log progesterone accumulation by $10^6$ viable bovine luteal cells/ml in response to addition of vehicles, 100 pg estradiol-17β ($E_2$)/ ml, 100 ng/ PGF$_2\alpha$/ml and 100 pg $E_2$/ml + 100 ng PGF$_2\alpha$/ml in the presence and absence of 1 ng LH/ml (NIH-LH-B9). S.E.M. = 0.03546.

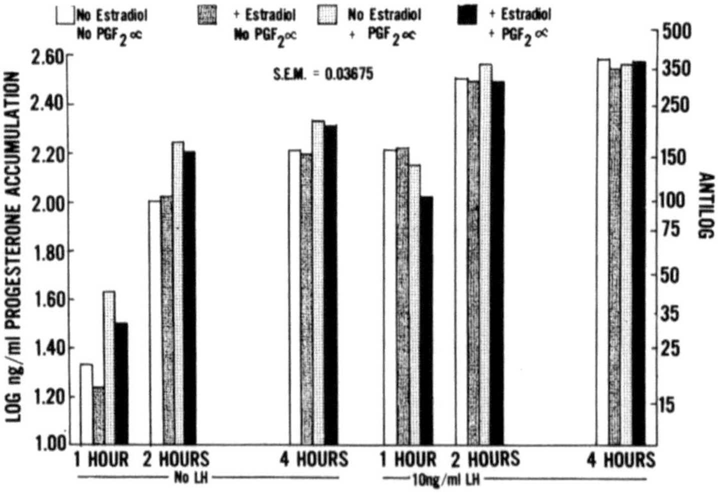


Figure 5. Log progesterone accumulation by $10^6$ viable bovine luteal cells/ml in response to addition of vehicles, 100 pg estradiol-17β ($E_2$)/ml, 100 ng PGF$_2\alpha$/ml and 100 pg $E_2$/ml + 100 ng PGF$_2\alpha$/ml in the presence and absence of 10 ng LH/ml (NIH-LH-B9). S.E.M. = 0.03675.

induced progesterone synthesis observed with granulosa cell cultures of bovine origin (4,5). Corpora lutea and luteinized granulosa cells differ in the presence and absence of cells of thecal origin which invade corpora lutea between days 4 and 10 of the estrous cycle (19). On this basis, the hypothesis that thecal cells play a determining role in the in vitro response to $PGF_2\alpha$ warrants investigation.

## REFERENCES

(1) W. Hansel, P.W. Concannon and J.H. Lukaszewska. Biol. Reprod. 8(1973)822.
(2) M.T. Williams, M.R. Clark and J.M. Marsh. Proced. 59th Annual Meeting of the Endocrine Society. June, 1977, p.172.
(3) R.L. Stouffer, W.E. Nixon and G.D. Hodgen. Endocrinology 101 (1977)1157.
(4) K.P. McNatty, K.M. Henderson and R.S. Sowers. J. Endocrinol. 67(1975)231.
(5) K.M. Henderson and K.P. McNatty. J. Endocrinol. 73(1977)71.
(6) M. Lahav, A. Freud and H.R. Lindner. Biochem. Biophys. Res. Commun. 68(1976)1294.
(7) J.P. Thomas, L.J. Dorflinger and H.R. Behrman. Proc. Natl. Acad. Sci. USA. 75(1978)1344.
(8) L. Speroff and P.W. Ramwell. J. Clin. Endocr. Metab. 30(1970) 345.
(9) J.E. Hixon, D. R. Gengenbach and W. Hansel. Biol. Reprod. 13 (1975)126.
(10) D.R. Gengenbach, J.E. Hixon and W. Hansel. Biol. Reprod. 16 (1977)571.
(11) K.R. Simmons, J.L. Caffrey, J.L. Phillips, J.H. Abel and G.D. Niswender. Proc. Soc. Exp. Biol. Med. 152(1976)366.
(12) J.L. Gill and H.D. Hafs. J. Anim. Sci. 33(1971)331.
(13) R.G.D. Steel and J.H. Torrie. Principles and Procedures of Statistics (McGraw Hill, New York, 1960).
(14) J. Apgar, D. Aspros, J.E. Hixon, R.R. Saatman and W. Hansel. J. Animal Sci. 41(1975)1120.
(15) B.R. Carr, G. Mikhail and G.L. Flickinger. J. Clinc. Endocr. 33(1971)358.
(16) B. Goldberg, G.E.S. Jones and H.I. Borkowf. J. Histochem. and Cytochem. 12(1964)880.
(17) F. Ferre, M. Breuiller and L. Cedrad. Steroids 26(1975)551.
(18) D. Gospodarowicz and F. Gospodarowicz. Endocrinology 96(1975) 458.
(19) L. Donaldson and W. Hansel. J. Dairy Sci. 48(1965)905.

DISCUSSION - CHAIRPERSON - J.W. WILKS

H.R. BEHRMAN: Do you preincubate your dispersed cells before you expose them to treatment?

J.E. HIXON: We expose them to a fifteen to thirty minute preincubation period.

H.R. BEHRMAN: We find with dispersed rat cells that there is an incredible dumping phenomenon that occurs when you first disperse them and place them in culture. You have a very high background noise which seems to be progesterone secretion which is independent of anything.

J.E. HIXON: I agree.

J.M. MARSH: We have used higher levels of estradiol than you have used, and see a very marked inhibition of the LH effect on progesterone production during a one hour incubation. Have you tried larger doses of estradiol?

J.E. HIXON: Larger doses of estradiol were used in preliminary experiments. One can argue about what is physiological and what's not physiological. We don't know whether these cells are making estradiol or not. If they are, I think one has the right to use much higher concentrations.

J.M. MARSH: We have done experiments similar to yours with $PGF_{2\alpha}$, but we did see a small significant stimulation. PGE was even more effective. These corpora lutea were taken from cows in pregnancy and workers from Sweden who have looked at human corpora lutea find that you have to select only certain corpora lutea of the menstrual cycle to see inhibition with $PGF_{2\alpha}$.

J.E. HIXON: I would agree that there is a definite difference between corpora lutea from pregnant and nonpregnant animals. Corpora lutea collected at day 7 or 10 of the cycle respond in a manner similar to those collected at day 12.

B.F. RICE: Nobody has ever shown that bovine luteal cells make estradiol, or any estrogen, or anything besides progesterone and the 20β-dihydroprogesterone.

K.P. McNATTY: Did you ever look at subnanogram levels of prostaglandins for their effects on progesterone synthesis in your dispersed cell populations?

J.E. HIXON: Yes, in initial experiments concentrations of $PGF_{2\alpha}$ between 1 and 1000 pg/ml were tested with no observed effect on progesterone accumulation.

## GENERAL DISCUSSION OF LUTEOLYSIS - CHAIRPERSON J.W. WILKS

H.D. GUTHRIE: I would like to present some data concerning luteolysis in the pig and sheep. Patek and Watson (Prostaglandins 12: 97, 1976) reported that in the pig luteal PGF synthesis in vitro was highest during the late luteal phase of the estrous cycle. They proposed that luteal $PGF_{2\alpha}$ synthesis might be involved in the process of luteolysis. We induced luteolysis in pigs by an intramuscular injection of 2.5 mg of Cloprostenol (synthetic $PGF_{2\alpha}$ analog) on Day 12 of the estrous cycle. Luteal and endometrial tissues for incubation were collected on Day 12 (no injection), 8 or 32 hr post-injection. Cloprostenol caused luteolysis with plasma progesterone and luteal progesterone synthesis in vitro reduced by 8 hr post-injection ($p < .01$). Luteal PGF synthesis in vitro was increased 15 fold by 8 hr post-injection compared to no injection ($p < .01$). Endometrial PGF synthesis in vitro was not significantly increased until 32 hr post-injection. We don't know if increased luteal PGF synthesis in vitro means that increased luteal $PGF_{2\alpha}$ in vivo is required for Cloprostenol induced luteolysis in the pig. In the ewe an injection of Cloprostenol causes both luteolysis and increased accumulation in vitro of luteal PGF. However, when indomethacin was injected along with Cloprostenol, luteolysis still occurred, but the increased accumulation in vitro of PGF by luteal tissue was completely blocked. Thus in the ewe, Cloprostenol induced luteolysis does not appear to be mediated through increased luteal $PGF_{2\alpha}$ synthesis.

J.W. WILKS: One of the biases I have developed as I read the literature is that while we can say that the luteolytic action of prostaglandin is mediated within the ovary, it's difficult to say that all the luteolytic actions of prostaglandin are mediated directly upon the luteal cell. Our discussion here has concerned direct actions upon the luteal cell, and I wonder if the speakers have any thoughts about the involvement of other ovarian cell types and other ovarian compartments, such as the follicle, in the luteolytic process? Could this explain the discrepancy between Dr. Behrman's data where several types of cells are present, as contrasted to Dr. Hixon's data in the cow where it is easier to solely harvest the luteal cell type?

J.E. HIXON: I would point out that in the cow, the corpus luteum has cells of both granulosa and thecal origin.

H.R. BEHRMAN: We believe that $PGF_{2\alpha}$ is acting directly on the luteal cell. This conclusion is based on several observations. First, we have demonstrated $PGF_{2\alpha}$ receptors directly on luteal cells with the aid of ultrastructural identification of $PGF_{2\alpha}$-ferritin on membranes of intact cells. Second, progesterone is derived primarily from luteal cells in response to LH and we have identified receptors for LH directly on the luteal cells by utilizing LH-ferritin conjugates. Third, the dispersed luteal cell model we use contains about 80% luteal cells by direct identification. These observations, however, do not preclude the possibility of cell-cell interactions which may be involved in functional luteal regression in vivo.

J.A. McCRACKEN: I have one comment about the possible role of other compartments of the ovary contributing to luteolysis: that is, one could interpret that estrogen from the follicles might play some role in maintaining or inducing the receptors for $PGF_{2\alpha}$ on the luteal cell itself.

B.F. RICE: In the case of the rat, my impression is, that the life span of the corpus luteum is like the telescope, with different sections being added as you go from the estrous cycle length to pseudopregnancy to the full pregnancy; and the control passes from one set of regulations to another. Does that happen in other species? Could $PGF_{2\alpha}$ turn on the corpus luteum in one phase of its life span and turn it off in another?

H.R. BEHRMAN: Your view of the corpus luteum and the analogy to a telescope is essentially the same as mine. In rodents such as the rat, the corpus luteum of the cycle is not functional whereas in species such as the sheep, cow, and primate it is. However, I view the functional corpus luteum in the pseudopregnant rat as a homologue of the gland seen in these other species. The hormonal signal in the rat which shifts the corpus luteum into a functional state is generally accepted to be prolactin which is released in response to cervical stimulation during mating. In the absence of pregnancy most species, except possibly for the dog, have developed a luteolytic process which allows an earlier return to ovulation than that seen with a successful pregnancy. Thus luteolysis plays an important role in species survival by permitting a greater number of ovulatory phases within the female's fertile life span.

A characteristic across species is the refractoriness that is shown to $PGF_{2\alpha}$ early in the luteal phase. For some species this period of refractoriness may extend for about 10 to 12 days (sow, human) but in the rat, cow, and ewe it is about 5 days. The basis for this refractoriness in the rat appears to be the antagonistic effect of prolactin to the luteolytic action of $PGF_{2\alpha}$. This response can be demonstrated with prolactin injections as late as day 9 in the rat pseudopregnant cycle. In early pregnancy, clearly luteal regression is prevented and the basis for continued function of the corpus luteum, in some species at least, appears to be due to production of placental luteotropic hormone(s). Whether the luteolytic signal ($PGF_{2\alpha}$) is present in pregnancy but its action blocked, or production of the luteolysin is inhibited, has not been established.

I see LH as a common luteotropic signal across most species and the role of this hormone appears to be acute modulation of progesterone synthesis. One theme from this discussion is that $PGF_{2\alpha}$ seems to block LH action in luteal progesterone production in several species. We have proposed that this may be the basis of $PGF_{2\alpha}$-induced luteolysis. I know of little sound evidence that $PGF_{2\alpha}$ plays a luteotropic role in any species at any stage of luteal development and function.

R.T. MOORE: Would you postulate what role the thromboxanes might play in the rapid component of luteolysis? Has anyone tried thromboxanes in their systems?

H.R. BEHRMAN: We haven't tried thromboxane but we have tried prostacyclin, and it was not as active as $PGE_2$ in stimulating progesterone secretion. The problem with doing such studies is the very short half-life of such materials in aqueous media of about 30 seconds which makes interpretation very difficult. There is evidence to show that $PGF_{2\alpha}$-induced luteolysis is not mediated through thromboxanes since inhibitors such as indomethacin and aspirin, which block the cyclooxygenase system do not prevent $F_{2\alpha}$-induced luteolysis.

P.L. KEYES: The uterus does not seem to play a role in regression of the human corpus luteum, however, there is evidence from Karsch and others that an intraluteal estrogen could be involved in regression of the primate corpus luteum. What is the current thinking about a possible role of intra-ovarian prostaglandins, or possibly intraluteal prostaglandins, in the human female?

J.A. McCRACKEN: There have been several studies on the concentrations of prostaglandins in the human corpus luteum, but there were conflicting results as to whether prostaglandins increased during luteal regression. Nonetheless, one could not rule out the possibility that other compartments of the ovary in the human might produce prostaglandins. Auletta has shown that infusions of estrogen will increase the concentration of $PGF_{2\alpha}$ in ovarian venous blood in the monkey, which is suggestive that estrogen can indeed stimulate prostaglandin synthesis in the primate ovary.

J.W. WILKS: Auletta et al. (Prostaglandins 11:724, 1976) have also demonstrated that estrogens are not luteolytic in rhesus monkeys when given in combination with indomethacin. Recently, Balmaceda et al. (Fertil. Steril. 23:238, 1978) have demonstrated that the intraluteal PGF:PGE ratio changes due to reduced PGE synthesis as the age of the corpus luteum increases in the rhesus monkey.

K.P. McNATTY: Dr. Behrman, would you care to speculate as to why the corpus luteum in various species is protected against prostaglandin early in its life, and how this protective nature is lost as the corpus luteum ages?

H.R. BEHRMAN: There have been reports in which prostaglandin receptor content has been measured during the cycle length. Dr. Rao has done some of these studies and we are doing similar studies with the rat. It is possible that the number of $PGF_{2\alpha}$ receptors may dictate the response of the corpus luteum to $PGF_{2\alpha}$. But whether this is the basis for the change in sensitivity to $PGF_{2\alpha}$ during the cycle is not clear yet. In the rat, prolactin levels appear to be the key. When prolactin levels are high, $PGF_{2\alpha}$ is only weakly active as a luteolytic agent.

P. BAGAVANDOSS: The mechanism of action of prostaglandin in causing luteolysis seems unclear; it doesn't affect LH binding and does not affect cholera toxin binding, and there is cAMP accumulation. Is it possible that prostaglandin can bind to the cAMP receptor--protein kinase?

H.R. BEHRMAN: I don't think $PGF_{2\alpha}$ has a major action on kinase. $PGF_{2\alpha}$ does produce a consistent and highly reproducible rapid block of LH-dependent adenylate cyclase activation and progesterone secretion. Hours later, there is eventual loss of LH receptors. It is possible that one action of $PGF_{2\alpha}$ may be on kinase but we believe this to be secondary to its action on adenylate cyclase. In addition, we believe that the action of $PGF_{2\alpha}$ may be determined by the presence of specific receptors for $PGF_{2\alpha}$ which we have identified in rat luteal cells. So we're looking at two different events, both of which result in the same end, no progesterone secretion. The bovine system appears to be different. There is a species commonality between hamsters, the rat, the sow, and the ewe, and the human in a block by $PGF_{2\alpha}$ of hormone-dependent progesterone secretion.

B.F. RICE: Has anybody ever observed any abnormalities in calcium homeostasis during prostaglandin treatment? Could this play any role in luteolysis?

H.R. BEHRMAN: There is a wide array of mechanisms for maintaining calcium homeostasis. Luteolysis occurs over a span of a few days. To upset calcium homeostasis, within such a short period, would be very difficult to show. Furthermore, $PGE_2$ was shown to be the "bone resorption factor" whereas it appears that $PGF_{2\alpha}$ is the "luteolytic factor".

I might add, that the block by prostaglandin of hormone action has been shown in other systems. For example, Dr. Sharpe showed such a response with prostaglandins on vasopressin in toad skin some years ago. I think that you can look at prostaglandins as cellular regulators whose actions are dependent upon the presence of specific receptors.

J.L. LUBORSKY-MOORE: In the preceeding discussions, the direct action of $PGF_{2\alpha}$ in blocking LH action on isolated rat luteal cells was described. However, prostaglandins have not been directly localized by morphological methods nor have $PGF_{2\alpha}$ binding sites (receptors) been demonstrated on rat luteal cells. We have studied the distribution of $PGF_{2\alpha}$ receptors on luteal cells with a ferritin-$PGF_{2\alpha}$ conjugate. The ferritin-$PGF_{2\alpha}$ conjugate was prepared with carbodiimide and separated from unconjugated $PGF_{2\alpha}$ by ultrafiltration, $PGF_{2\alpha}$ activity was assessed by radioimmunoassay, a membrane binding assay and by the ability of the conjugate to block LH stimulated progesterone secretion. Luteal cells from ovaries of gonadotropin primed rats were incubated in ferritin-$PGF_{2\alpha}$ ($10^{-6}$M) or ferritin-$PGF_{2\alpha}$ ($10^{-6}$M) + excess $PGF_{2\alpha}$ ($10^{-3}$M) as a control for nonspecific binding. We found that ferritin-$PGF_{2\alpha}$ was distributed at intervals along luteal cell surfaces while the number of ferritin molecules on control cells was significantly reduced. These findings were confirmed by binding of [$^3$H] $PGF_{2\alpha}$ to a membrane preparation from rat corpora lutea. Thus, we suggest that the action of $PGF_{2\alpha}$ is mediated by a specific cell surface receptor (cf., J.L. Luborsky-Moore et. al., "Demonstration of luteal cell membrane receptors for prostaglandin $F_{2\alpha}$ by ultrastructural and binding analysis").

# *IN VITRO* RELEASE OF PROGESTERONE AND PROSTAGLANDINS F AND E BY PORCINE LUTEAL AND ENDOMETRIAL TISSUE DURING INDUCED LUTEOLYSIS

H. D. Guthrie, C. E. Rexroad, Jr., and D. J. Bolt

U. S. Department of Agriculture, SEA
Beltsville, Maryland 20705

Prostaglandin $F_{2\alpha}$ ($PGF_{2\alpha}$) has been proposed as the uterine luteolysin in the ewe (1) and pig (2). The corpus luteum (CL) of the pig may also participate in its own demise by means of increased luteal PGF synthesis during luteolysis (3). Both endometrial and luteal PGF synthesis *in vitro* were higher during the late luteal phase of the cycle than at any other stage of the cycle, but the temporal relationship between endometrial and luteal PGF release *in vitro* during luteolysis was not established (3). We have found that an increase in endometrial PGF synthesis *in vitro* starts after day 12 of the cycle, preceding any increase in luteal PGF synthesis *in vitro* by approximately two days (4).

Because of the unpredictable timing of the exact onset of luteolysis in any particular pig and the possible transitory involvement of luteal PGF synthesis in luteolysis, luteal prostaglandin synthesis was investigated during induced luteolysis. In the present experiment, luteolysis was initiated by an injection of a synthetic $PGF_{2\alpha}$ analog (Cloprostenol, ICI 80996, ICI Americas). Luteal and endometrial tissues were incubated *in vitro* to determine if the initiation of luteolysis was followed by an increase in luteal PGF synthesis. An increase in luteal PGF synthesis coinciding with or followed by decreased luteal $P_4$ synthesis *in vitro* could be construed as evidence for an active role for the CL in its own demise. Various incubation treatments were imposed *in vitro* to evaluate to what degree the luteolytic effect of Cloprostenol *in vivo* could be potentiated or moderated *in vitro*.

## MATERIALS AND METHODS

Twenty-one mature, nulliparous crossbred pigs were assigned to one of three treatment groups on day 12 of their estrous cycles (day 0=first day of estrus). Seven pigs were not injected and were sacrificed on day 12. Two other groups of seven pigs each were sacrificed at 8 or 32 hr, respectively, after an i.m. injection of 2.5 mg of Cloprostenol on day 12. A blood sample was collected immediately before sacrifice and the plasma was frozen.

The ovaries and a 3 cm section of each uterine horn were transported to the laboratory in ice cold 0.9% NaCl solution. Luteal and endometrial tissues were minced with scissors while in ice cold incubation medium (Krebs Ringer bicarbonate buffer, pH 7.4, 5 mM glucose, 10 mM acetate and one-half concentration of calcium). Minced luteal and endometrial tissue (ca. 100 mg) were weighed into 25 ml Erlenmeyer flasks containing 5 ml of the incubation medium.

Each incubation treatment was represented by two flasks for each tissue from each pig. For incubation treatment 1, luteal tissue was incubated for 2 hr at 0° C in an atmosphere of air. Treatments 2 through 7 consisted of incubation at 37° C in an atmosphere of 95% $O_2$:5% $CO_2$ in a Dubnoff metabolic shaker as follows: 2) no additive, 3) 1 ug/ml of arachadonic acid, 4) $1.3 \times 10^{-4}$ M indomethacin, 5) 100 uM dibutyryl cyclic adenosine monophosphate (dbc-AMP), 6) indomethacin + dbc-AMP, and 7) 10 ug/ml of ovine NIH-S12 luteinizing hormone (LH). Incubation treatments 1 through 6 were also applied to endometrial tissue. After incubation the medium was frozen. Plasma $P_4$ and medium $P_4$, PGF and PGE from luteal incubations, and PGF and PGE from endometrial incubations were quantified by validated radioimmunoassays (5,6). PGE was quantified in medium from only incubation treatments 1, 2, 4 and 5. The hormone concentrations in the medium were analyzed as a split plot design. The whole unit analysis consisted of the three times of tissue recovery in relation to Cloprostenol treatment (0, 8 and 32 hr). The sub-unit analysis consisted of the incubation treatments.

## RESULTS

Both plasma $P_4$ and luteal $P_4$ release *in vitro* were significantly reduced at 8 hr after the Cloprostenol injection, demonstrating that Cloprostenol was an effective luteolytic agent in this experiment (Table 1). Luteal PGF and PGE release *in vitro* were increased significantly by 15 fold at 8 hr as compared to 0 hr. By 32 hr, the release of luteal PGF and PGE had decreasedto levels not significantly greater than at 0 hr. Endometrial PGF and PGE *in vitro* release did not increase significantly until 32 h after Cloprostenol treatment.

Incubation at 0° C significantly inhibited $P_4$ release into the medium by luteal tissue collected at 0 and 8 hr as compared to incubation treatments at 37° C (data not shown). By 32 hr, in vitro $P_4$ release at 37° C had declined to the extent that it was not significantly different than synthesis at 0° C.

The release of $P_4$ at 37° C was not significantly affected by the presence of dbc-AMP, arachadonic acid, LH or indomethacin in the medium during incubation.

TABLE 1. PLASMA PROGESTERONE ($P_4$) AND IN VITRO RELEASE OF $P_4$ AND PROSTAGLANDINS (PGF AND PGE) DURING INCUBATION AT 37°C OF TISSUES RECOVERED AFTER CLOPROSTENOL TREATMENT

| | Hours after Cloprostenol | | |
|---|---|---|---|
| Hormone concentration | 0 | 8 | 32 |
| Plasma P ng/ml | 37.2±2.7 [a] | 15.8± 2.8 [b] | 5.1± .6 [c] |
| Luteal P ng/ml | 23.2±3.2 [a] | 4.5± .6 [b] | 1.9± .5 [b] |
| Luteal PGF ng/100 mg | 17.3±2.4 [a] | 276.6±86.1 [b] | 63.7±10.3 [a] |
| Luteal PGE ng/100 mg | 3.8± .7 [a] | 56.6±19.9 [b] | 9.7± 3.6 [ab] |
| Endometrial PGF ng/100 mg | 14.2±2.4 [a] | 20.4± 4.1 [a] | 73.4±16.0 [b] |
| Endometrial PGE ng/100 mg | 7.8±2.4 [a] | 8.8± 2.6 [a] | 35.1± 7.6 [b] |

[abc]Means ± SE in a row with no superscript letter in common differ significantly by Duncan's Multiple Range Test (P<.01).

TABLE 2. EFFECT OF INCUBATION TREATMENT ON IN VITRO PGF RELEASE BY PORCINE LUTEAL TISSUE AFTER CLOPROSTENOL INJECTION IN VIVO

| | Hours after Cloprostenol[b] | | |
|---|---|---|---|
| | 0 | 8 | 32 |
| Incubation treatment[a] | PGF ng/100 mg of tissue (mean±SE) | | |
| 0°C | 2.0± .2[c] | 7.8± 2.6[c] | 4.2± .8[c] |
| 37°C | 17.3±2.4[d] | 276.6±86.1[c] | 63.7±10.3[d] |
| 37°C arachadonic acid | 15.4±1.9[d] | 244.9±93.1[c] | 67.4±13.7[d] |
| 37°C indomethacin | 7.8± .7[cd] | 30.0± 8.8[c] | 15.2± 1.9[c] |
| 37°C dbc-AMP | 39.6±3.6[e] | 277.9±87.3[c] | 127.0±20.9[e] |
| 37°C dbc-AMP + indomethacin | 19.5±1.7[d] | 52.0±13.6[c] | 32.2± 3.3[cd] |
| 37°C LH | 18.8±2.9[d] | 221.0±93.8[c] | 69.6± 7.8[d] |

[a]Main effect of incubation treatment significant, means for 0°, indomethacin and dbc-AMP+indomethacin were less than means for the rest of the incubation treatments by Duncan's Multiple Range Test (P<.01).
[b]Main effect hours after Cloprostenol significant, 8 hr was greater than 0 and 32 hr by Duncan's Multiple Range Test (P<.01).
[cde]Means in a column with no superscript letter in common differ significantly by Duncan's Multiple Range Test (P<.01).

TABLE 3. EFFECT OF INCUBATION TREATMENT ON IN VITRO PGE RELEASE BY PORCINE LUTEAL TISSUE AFTER CLOPROSTENOL INJECTION IN VIVO

| Incubation treatment [a] | Hours after Cloprostenol[b] 0 | 8 | 32 |
|---|---|---|---|
| | PGE ng/100 mg of tissue (mean±SE) | | |
| 0° C | .4± .1 | 5.8± 1.7 | 2.8±1.9 |
| 37° C | 3.8± .6 | 56.6±19.9 | 9.7±3.6 |
| 37° C indomethacin | .4± .1 | 7.0± 2.6 | 2.7±1.3 |
| 37° C dbc-AMP | 4.1±1.0 | 39.9±12.7 | 7.8±1.6 |

[a]Means for 0° C and indomethacin were less than means for 37° and dbc-AMP by Duncan's Multiple Range Test (P<.01).
[b]Mean for 8 hr was greater than mean for 0 hr (P<.01), but not greater than for 32 hr by Duncan's Multiple Range Test

TABLE 4. EFFECT OF INCUBATION TREATMENT ON IN VITRO PGF RELEASE BY PORCINE ENDOMETRICUM AFTER CLOPROSTENOL INNJECTION IN VIVO

| Incubation treatment [a] | Hours after Cloprostenol[b] 0 | 8 | 32 |
|---|---|---|---|
| | PGF ng/100 mg of tissue (mean±SE) | | |
| 0° C | 2.6± .5 [c] | 6.4±2.2 [c] | 10.9± 2.9 [c] |
| 37° C | 14.2±2.4 [cd] | 20.4±4.1 [cd] | 73.4±16.0 [c] |
| 37° C arachadonic acid | 16.1±2.1 [cd] | 21.5±4.0 [cd] | 66.6±11.4 [c] |
| 37° C indomethacin | 5.7±1.0 [c] | 7.7±1.9 [c] | 16.9± 2.7 [c] |
| 37° C dbc-AMP | 50.6±7.9 [e] | 56.8±7.4 [e] | 179.6±37.1 [d] |
| 37° C dbc-AMP+indomethacin | 23.5±3.9 [d] | 27.4±3.7 [d] | 48.7± 5.5 [c] |

[a]Main effect incubation significant; mean for dbc-AMP was greater than means for the rest; means for 0° C and indomethacin were less than the other means by Duncan's Multiple Range Test (P<.01).
[b]Except for incubation treatment 0° C, mean for 32 was greater than means for 0 and 8 hr for all other groups by Duncan's Multiple Range Test (P<.01).
[cde]Means in a column with no superscript letter in common differ significantly by Duncan's Multiple Range Test (P<.01).

TABLE 5. EFFECT OF INCUBATION TREATMENT ON *IN VITRO* PGE RELEASE BY PORCINE ENDOMETRIUM AFTER CLOPROSTENOL INJECTION *IN VIVO*

| Incubation treatment[a] | Hours after Cloprostenol[b] 0 | 8 | 32 |
|---|---|---|---|
| | PGE ng/100 mg of tissue (mean±SE) | | |
| 0° C | .7± .2 | 1.2± .5 | 4.6±1.5 |
| 37° C | 7.8±2.4 | 8.8±2.6 | 35.1±7.6 |
| 37° C indomethacin | .4± .1 | .6± .1 | 4.0±1.0 |
| 37° C dbc-AMP | 11.0±3.0 | 11.8±3.0 | 43.4±9.4 |

[a]Means for 0° C and indomethacin were less than means for 37° and dbc-AMP by Duncan's Multiple Range Test ($P<.01$).
[b]Mean for 32 hr was greater than means for 0 and 8 hr by Duncan's Multiple Range Test ($P<.01$).

The effects of incubation treatment on the release of PGF *in vitro* by luteal tissue from non-injected pigs and from pigs and 32 hr after the Cloprostenol injection are shown in Table 2. The presence of dbc-AMP in the medium during incubation increased the release of PGF at 0 and 32 hr relative to all other incubation treatments ($P<.01$). However, at 8 hr dbc-AMP did not cause any further increase in PGF release compared to no additive at 37° C. When indomethacin was present in the medium along with dbc-AMP, the effect of dbc-AMP was almost entirely canceled at 8 and 32 hr. In contrast, the presence of dbc-AMP had no significant effect on luteal PGE release as compared to no additive at 37° C (Table 3).

Prostaglandin F, but not PGE, release *in vitro* by endometrial tissue was significantly increased by the presence of dbc-AMP as compared to other incubation treatments (Tables 4&5). Incubation at 37° C in the presence of indomethacin greatly reduced PGF and PGE synthesis by endometrial tissue.

## DISCUSSION

The observation of a 15 fold increase in luteal PGF release *in vitro* 8 hr after the Cloprostenol injection *in vivo* supports the suggestion that luteal $PGF_2$ production in the pig may be involved in the process of luteolysis initiated by the uterus (3). A role for increased luteal PGE release *in vitro* has not been advanced. A general stimulation of the prostaglandin synthetase enzyme system probably caused endoperoxide formation which could result in a PGE spillover effect. The increased endometrial release of PGF and PGE at 32 hr was not likely due to a direct effect of Cloprostenol, but to the decline in plasma $P_4$.

Injections Cloprostenol into ewes also induced luteolysis and increased luteal PGF release *in vitro* (6). Luteolysis was induced in ewes injected with both Cloprostenol and indometha cin, but the increase in luteal PGF release *in vitro* was completely inhibited.

The increased luteal PGF release was not apparently required for Cloprostenol induced luteolysis. The possibility that Cloprostenol induced luteolysis and increased luteal PGF release in vitro can be disassociated by indomethacin treatment in vivo has not been tested in the pig.

The presence of the dbc-AMP in luteal and endometrial incubations increased the release of PGF but not PGE in vitro. The increased PGF release with both luteal and endometrial tissues was also observed in previous studies (8,12) in which incubations of tissue from mouse ovaries, rat testes and bovine CL have increased prostaglandin release in vitro in the presence of LH or dbc-AMP. Arachadonic acid as a prostaglandin precursor in luteal tissue did not appear to be limiting in the pig though it is in the cow (12).

Neither LH nor dbc-AMP affected luteal $P_4$ release in vitro in the present experiment, in contrast to the stimulator effect of LH or dbc-AMP on in vitro $P_4$ release by bovine luteal tissue (8). Other experiments utilizing porcine luteal tissue to assess LH in vitro have not agreed with respect to whether or not LH stimulated $P_4$ release (9,10,11). Luteal adenylate cyclase activity was stimulated in vitro by LH incubated with tissue collected from pigs on day 12 or 14 of the estrous cycle (8).

In conclusion, this experiment demonstrates that an injection of a $PGF_{2\alpha}$ analog, Cloprostenol, caused increased luteal PGF release in vitro, which supports the proposal that luteal $PGF_{2\alpha}$ synthesis is involved in the luteolytic process initiated by the uterus (3).

REFERENCES

(1) J.A. McCracken, J.C. Carlson, M.E. Glew, J.R. Goding, D.T. Baird, K. Green and B. Samuelsson. Nature New Biol. 238(1972)129.
(2) F.T. Bazer and W.W. Thatcher. Prostaglandins 14(1977)397.
(3) C.E. Patek and J. Watson. Prostaglandins 12(1976)97.
(4) H.D. Guthrie and C.E. Rexroad, Jr. Proc. ASAS (1978).
(5) H.D. Guthrie. J. Anim. Sci. 45(1977)1360.
(6) C.E. Rexroad, Jr. and H.D. Guthrie. (1978)this volume.
(7) K.M. Henderson and K.P. McNatty. Prostaglandins 9(1975)77
(8) J.M. Marsh. Biol. Reprod. 14(1976)30.
(9) G.W. Duncan, A.M. Bowerman, W.R. Hearn and R.M. Melampy. Proc. Soc. Expt. Biol. Med. 104(1960)17.
(10)B. Cook, C.C. Kaltenbach, H.W. Norton and A.V. Nalbandov. Endocrinology 81(1967)573.
(11)J. Watson and J.T.S. Leask. J. Endocrin. 64(1975)163.
(12)M. Shemesh and W. Hansel. Biol. Reprod. 13(1975)448.

# DEMONSTRATION OF LUTEAL CELL MEMBRANE RECEPTORS FOR PROSTAGLANDIN $F_{2\alpha}$ BY ULTRASTRUCTURAL AND BINDING ANALYSIS

J.L. Luborsky-Moore, K. Wright and H.R. Behrman

Reproductive Biology Section, Depts Ob/Gyn and Pharmacology
Yale Univ. School Medicine, New Haven, CT 06510

## INTRODUCTION

Prostaglandins have a ubiquitous distribution and may act on a variety of cell types (1). To date, there have been no studies in which prostaglandin binding sites were directly localized by morphological methods. Prostaglandin $F_{2\alpha}$ ($PGF_{2\alpha}$) has been shown to act directly on isolated rat luteal cells to block luteinizing hormone action (2) and is thought to be involved in luteolysis <u>in vivo</u> (3,4). However, specific binding sites for $PGF_{2\alpha}$ have not been demonstrated in rat luteal tissue. The purpose of this study was to determine the distribution and specificity of $PGF_{2\alpha}$ binding sites on cells and membranes isolated from rat luteal tissue by ultrastructural and binding analysis respectively.

## METHODS

Corpora lutea of immature, gonadotropin-primed (superovulated) rats (CD strain, Charles Rivers Laboratories) were removed 7 to 8 days after gonadotropin injection. The tissue was incubated with collagenase (2000 IU/g) and DNAase (3000IU/g) (Worthington Biochemicals) in a low calcium media (GIBCO, #138) for 1 hr at 37°C (1,5), rinsed, incubated in 2mM EDTA for 2 min, rinsed in MEM (GIBCO, #199) and cells mechanically dispersed by pipetting. Cell preparations which routinely contain 70-80% luteal cells were then maintained in media #199 at 37°C. Cell viability was 80-90% by the trypan blue exclusion test.

To prepare a ferritin-$PGF_{2\alpha}$ (FE-$PGF_{2\alpha}$) conjugate, ferritin (50mg), $PGF_{2\alpha}$ (2mg) and carbodiimide-HCl (2mg; Sigma Chemical Co.)

were incubated 4 hr at 22°C. The conjugate was dialyzed overnight against Tris buffer (pH 7.4) and free $PGF_{2\alpha}$ (mw 354) removed and the conjugate concentrated by ultrafiltration (AMICON: XMIOOA filter which excludes >100,000 mw). Free ferritin (mw 9 x $10^{+5}$) is not removed by ultrafiltration. The final yield of FE-$PGF_{2\alpha}$ was determined by specific radioimmunoassay (6) and competition with $[^3H]PGF_{2\alpha}$ in a membrane binding assay. Biological activity was assayed with isolated luteal cells. FE-$PGF_{2\alpha}$ concentration in this report refers to $PGF_{2\alpha}$ quantities determined by the membrane binding and radioimmunoassay.

Dispersed luteal cells were incubated with FE-$PGF_{2\alpha}$ (10μM), FE-$PGF_{2\alpha}$ (10μM) plus $PGF_{2\alpha}$ (10mM) or ferritin (50mg/ml) (to determine nonspecific binding) in media #199 for 90 min at 22°C. Following incubation, media was removed for progesterone radioimmunoassay (7). The cells were rinsed (2x) with fresh media and fixed in 2% glutaraldehyde (Polysciences), 0.5% tannic acid in cacodylate buffer (310 mOsM, pH 7.3, 1 hr, 22°C), rinsed, postfixed in 2% $OsO_4$ (1 hr, 22°C), pelleted, dehydrated and embedded in Epon 812. Thin sections were obtained with a Reichert OMU2 ultramicrotome and viewed in an Hitachi 12 electron microscope.

For binding studies, partially purified luteal cell membranes were isolated in 10mM Tris buffer (pH 7) containing 1mM $CaCl_2$, 250mM sucrose, 1mM dithiothreitol, 0.1% $NaN_3$ and 0.1% BSA. Membranes (6-10mg wet weight) were incubated with 1 to 100nM [5,6,8,9,11,12,14,-15-$^3$H]$PGF_{2\alpha}$ (New England Nuclear, SA=118-150 Ci/mmole) for 90 min at 22°C and membrane bound $PGF_{2\alpha}$ precipitated by centrifugation. Nonspecific binding was determined using 1μM $PGF_{2\alpha}$. Competing compounds were tested with 1 nM $PGF_{2\alpha}$ in a range of 10nM to 10μM.

## RESULTS

The final $PGF_{2\alpha}$ conjugate contained free ferritin and ferritin-$PGF_{2\alpha}$. Quantitation of $PGF_{2\alpha}$ concentration by specific radioimmuno-

Table 1: Effect of FE-$PGF_{2\alpha}$, $PGF_{2\alpha}$ and LH on Progesterone Secretion by Isolated Rat Luteal Cells (Mean ± S.E.)

| | No LH | *LH (50ng/ml) |
|---|---|---|
| | ng/3 x $10^4$ cells/hr | |
| Control | 15.1 ± 0.5 | 28.0 ± 1.2 |
| $PGF_{2\alpha}$ (500ng/ml) | 16.6 ± 1.1 | 21.2 ± 2.1 |
| FE-$PGF_{2\alpha}$ (500ng/ml) | 17.7 ± 0.5 | 23.6 ± 0.7 |

Control vs LH, $p < .001$, LH vs LH + $PGF_{2\alpha}$, $p < .01$, LH vs LH + FE-$PGF_{2\alpha}$, $p < .01$. *NIH-LH-S19.

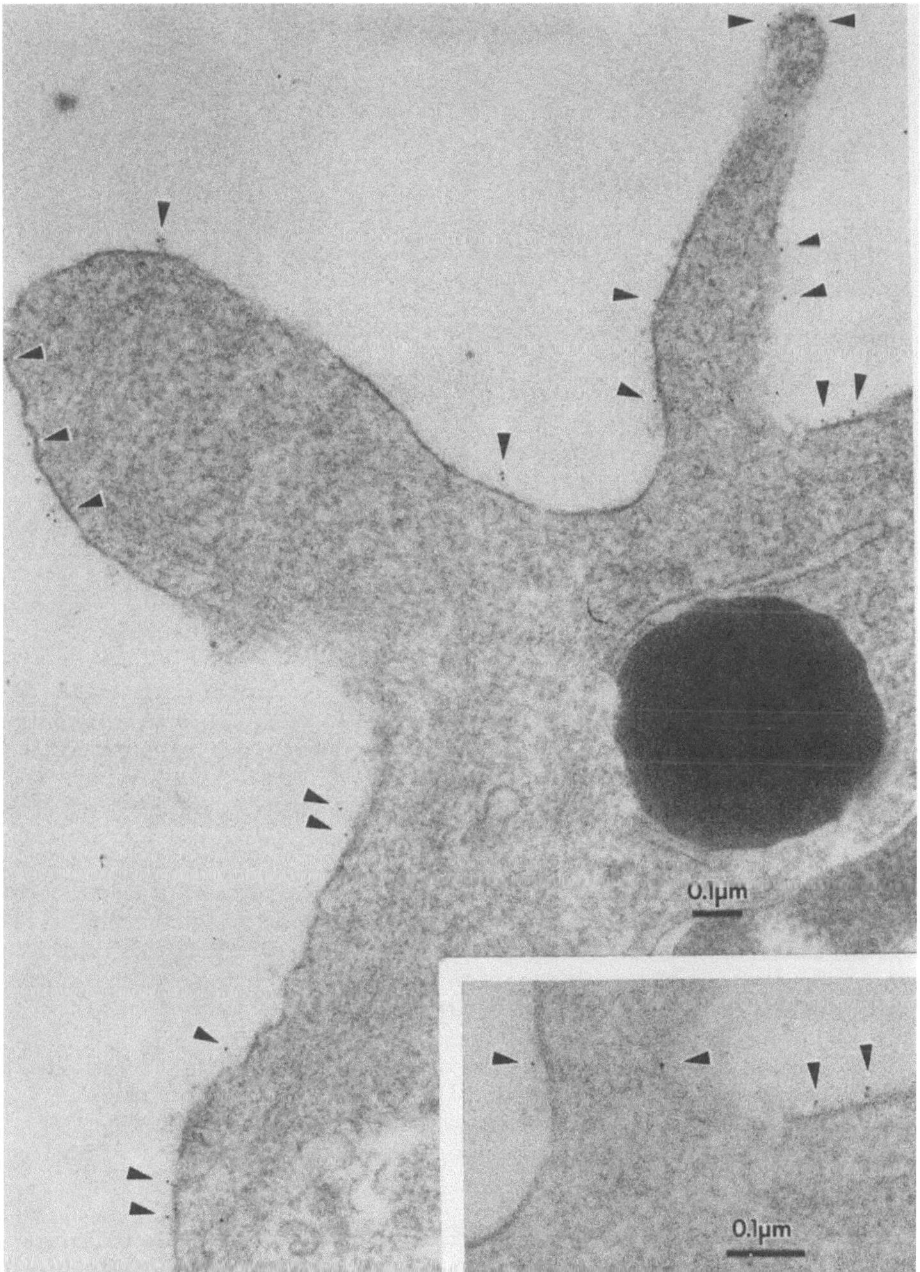


Fig. 1 $PGF_{2\alpha}$ binding sites labeled with FE-$PGF_{2\alpha}$ (↑) on membranes of isolated luteal cells. Ferritin is distributed at intervals along the cell surface. Unstained thin section. 75,000x. Inset: ferritin molecules shown at higher magnification. 120,000x.

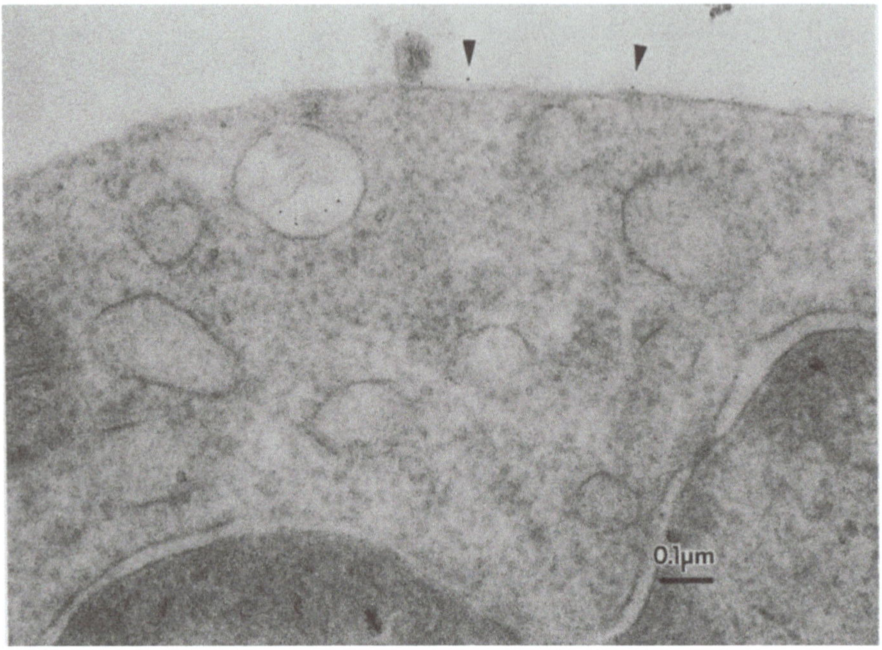


Fig. 2 Few ferritin molecules are seen on control luteal cells incubated with FE-PGF$_{2\alpha}$ plus excess (50 μg/ml) PGF$_{2\alpha}$. Some ferritin is occasionally seen in vacuoles in both control and experimental cells, possibly due to nonspecific uptake.75,000x.

assay and a membrane binding assay gave similar results with a final yield of 2μg of FE-PGF$_{2\alpha}$. The FE-PGF$_{2\alpha}$ was biologically active since 500ng/ml of FE-PGF$_{2\alpha}$ reduced LH stimulated progesterone secretion by isolated luteal cells (Table 1). Free ferritin did not show PGF$_{2\alpha}$ immunoreactivity, competition with [$^3$H]PGF$_{2\alpha}$ in the membrane binding assay, or biological activity on isolated luteal cells.

Following incubation of cells with FE-PGF$_{2\alpha}$, ferritin molecules (7nm in diameter) were seen primarily on luteal cell membranes (Fig 1). Luteal cells were identified by their cell size (10-30μm in diameter), granular lipid inclusions, smooth endoplasmic reticulum and numerous mitochondria.

Ferritin molecules were distributed at intervals along luteal cell surfaces mainly as single molecules but also in a few groups. Caps or large patches were not apparent. A few ferritin molecules were also seen on some non-luteal cells. Luteal cells incubated with PGF$_{2\alpha}$ plus FE-PGF$_{2\alpha}$, a test for nonspecific binding, showed a significant reduction of ferritin molecules (Fig 2). Ferritin was not found on cells treated with ferritin alone.

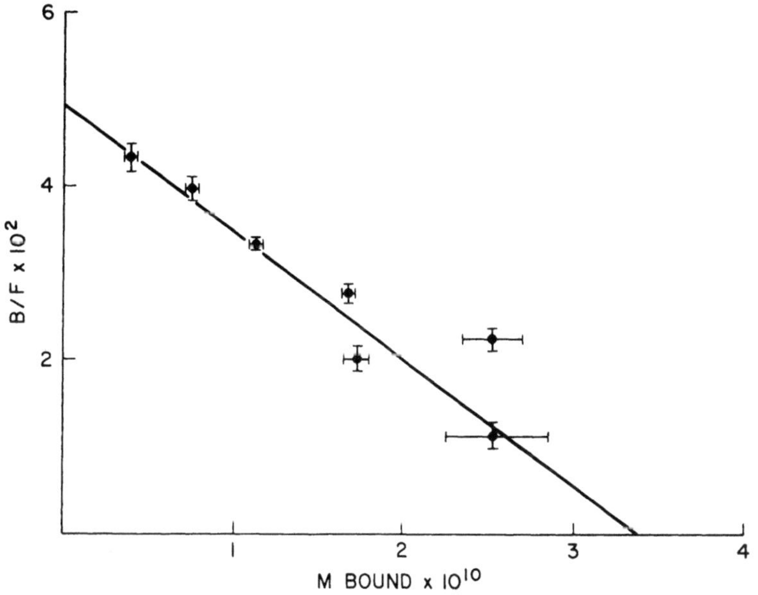


Fig. 3 Scatchard analysis of $PGF_{2\alpha}$ binding to membranes from rat corpora lutea at day 7. See text for details.

Binding of $[^3H]PGF_{2\alpha}$ to luteal cell membrane preparations was linear with respect to protein concentration over the range used. Maximum binding of $PGF_{2\alpha}$ was obtained after 90 min at 22°C in association studies. Fig 3 shows a typical Scatchard plot of data obtained from day 7 luteal membranes. Resolution of the curve revealed a high affinity, low capacity binding site with a $K_d$ = 6.9nM and a capacity of 8.3 fM/mg wet weight. Binding was highly specific; a small amount of competition was shown (mass/mass; $PGF_{2\alpha}$= 100%) by $PGF_{1\alpha}$ (9.2%), $PGE_1$, (0.1%), $PGE_2$ (1.8%), 6-keto-$PGF_{1\alpha}$ (0.1%), 15-keto-$PGF_{2\alpha}$ (0.5%) and thromboxane $B_2$ (1.0%).

## DISCUSSION

This report is the first direct demonstration of $PGF_{2\alpha}$ receptors on luteal cells by ultrastructural methods and the first report of $PGF_{2\alpha}$ binding to rat luteal cell membranes. Binding sites for $PGF_{2\alpha}$ were reported in the human, ewe, and cow ovary (3,8,9). The results of this report show that specific $PGF_{2\alpha}$ binding sites of high affinity and low capacity are found on membranes isolated from rat luteal tissue. The binding properties are similar to those found in studies of $PGF_{2\alpha}$ binding to bovine luteal membranes (10,11).

The FE-$PGF_{2\alpha}$ used in this study competes with $PGF_{2\alpha}$ for membrane binding sites and possesses biological activity similar to unconjugated $PGF_{2\alpha}$ used at approximately the same concentration. The luteal cell surface distribution of FE-$PGF_{2\alpha}$ is similar to ferritin-LH on luteal cells (12) and to ferritin-insulin on adipocytes (13), in that the binding sites are distributed at intervals along the cell surface (14). However, this distribution differs from the large patches and caps of ferritin-MSH on melanoma cells (14) and the small focal concentrations of ferritin-LDL associated with membrane thickenings on fibroblasts (15). These differences in distribution may be related to the different functions of the receptors. The results of this study suggest that the action of $PGF_{2\alpha}$ on luteal cells (2) is mediated by a specific receptor on the cell surface. (Supported by NIH-HD-10718-01 and Ford Foundation 770-0534, and an AAUW Postdoctoral fellowship (JLLM). Prostaglandin $F_{2\alpha}$ was a generous gift of Dr. John E. Pike, Upjohn Co., Kalamazoo, MI).

## REFERENCES

1. U. Zor and S.A. Lamprecht, in: Biochemical Actions of Hormones, Vol. IV, ed. G. Litwack (Academic Press, New York, 1977) p. 85.
2. J-P. Thomas, L.J. Dorflinger and H.R. Behrman. P.N.A.S. 75 (1978) 1344.
3. E.W. Horton and N.L. Poyser. Physiol. Rev. 56 (1976) 595.
4. A.P. Labhsetwar. Fed. Proc. 33 (1974) 61.
5. J.L. Luborsky-Moore, K. Wright, L.J. Dorflinger and H.R. Behrman. In preparation.
6. B.M. Jaffe and H.R. Behrman, in: Methods of Hormone Radioimmunoassay, eds. B.M. Jaffe and H.R. Behrman (Academic Press, New York, 1974) p. 19.
7. G.P. Orczyk, M. Hichens, G. Arth and H.R. Behrman, in: Methods of Hormone Radioimmunoassay, eds. B.M. Jaffee and H.R. Behrman.
8. W.S. Powell, S. Hammarstrom, U. Kyldén and B. Samuelsson, in: Surface Membrane Receptors, eds. R.A. Bradshaw, W.A. Frazier, R.C. Merrell, D.I. Gottlieb and R.A. Hogue-Agelletti (Plenum Press, New York, 1976) p. 455.
9. Ch. V. Rao, in: Advances in Prostaglandin and Thromboxane Research, Vol. 1, eds. B. Sammuelsson and R. Paoletti (Raven Press, New York, 1976) p. 247.
10. Ch. V. Rao. Biochem. Biophys. Res. Comm. 67 (1975) 1242.
11. S. Mitra and Ch. V. Rao. Arch. Biochem. Biophys. 185 (1978) 126.
12. J.L. Luborsky-Moore and H.R. Behrman. Fed. Proc. 37 (1978) 909.
13. L. Jarett and R.M. Smith. P.N.A.S. 72 (1975) 3526.
14. A.D. Pasquale, J.M. Varga, G. Moellmann and J. McGuire. Anal. Biochem. 84 (1978) 37.
15. R.G.W. Anderson, J.L. Goldstein and M.S. Brown. P.N.A.S. 73 (1976) 2434.

PROSTAGLANDIN $F_{2\alpha}$ AND PROGESTERONE RELEASE IN VITRO BY OVINE LUTEAL TISSUE DURING INDUCED LUTEOLYSIS

Caird E. Rexroad, Jr. and H. D. Guthrie

U. S. Department of Agriculture, SEA

Beltsville, Maryland 20705

Goding (1) has concluded that prostaglandin $F_{2\alpha}$ (PGF) is a naturally occurring luteolysin in the ewe and that its primary source for luteolysis is the uterus. Patek and Watson (2) have suggested that in the pig the corpora lutea may be involved in bringing about their own demise by PGF synthesis. They observed greater in vitro release of PGF from corpora lutea near the end of the estrous cycle than from corpora lutea during other parts of the cycle. Guthrie et al. (3) have shown that in pigs an injection of Cloprostenol (ICI 80996, a synthetic analogue of PGF) induced luteolysis and increased in vitro PGF release by luteal tissue by 8 hr after the injection. In the present study we investigated factors that affect in vitro synthesis of PGF by luteal tissue in ewes that were injected with Cloprostenol or PGF to induce luteal regression.

## MATERIALS AND METHODS

Parous ewes of 40-60 kg were used on day 10 or 11 of their estrous cycle. Ewes were checked for estrus twice daily by vascetomized rams. The day of estrus was designated day 0 of the cycle.

Experiment 1. A comparison was made of the effect of Cloprostenol and PGF on luteal PGF release in vitro. Twenty-four ewes were assigned at random to 3 groups and treated as follows: 1) control; 2) PGF, 15 mg injected i.m.; 3) Cloprostenol, 100 ug injected i.m. (table 1). Ewes were sacrificed at 4 hr after PGF or Cloprostenol and ovaries were removed immediately and transported to the laboratory in ice cold Krebs Ringer.

In the laboratory, corpora lutea (CL) were removed from the ovaries and weighed. CL were minced with scissors, and about 100 mg of luteal tissue were incubated in vitro for 2 hr with one of the following treatments: A) 0° C in room atmosphere, B) 37° C in 95%

$O_2$:5% $CO_2$ and C) 37° C in 95% $O_2$:5% $CO_2$ + $10^{-4}$ M indomethacin. Duplicate samples of each tissue for each treatment were incubated at 37° in 25 ml Erlenmeyer flasks containing 5 ml of Krebs Ringer solution with constant shaking. The dose of indomethacin was chosen because with pig luteal tissue $10^{-4}$ M indomethacin decreased the *in vitro* PGF release by more than 50% (see Guthrie *et al.*, this volume). After incubation, the media were assayed for PGF and progesterone by specific radioimmunoassays (4,5). For the PGF, six assays were run and duplicates of a low (360 pg) and high (2100 pg) laboratory standard had between assay coefficients of variability of 20 and 12%. The within-assay coefficients of variability were 11.8 and 6.8%. The laboratory standards were supernatants from homogenates of uterine myometrium from indomethacin and cylcoheximide treated ovariectomized ewes, respectively. When PGF was added to incubation media the regression of recovered on added PGF was 1.05 ($r = .989$ and $n = 13$). All samples were frozen and then assayed within the same week the incubation was performed. Cloprostenol when tested in the assay cross-reacted less than 1%.

Experiment 2. In experiment 1, addition of indomethacin to incubation media failed to reduce *in vitro* PGF release. Experiment 2 was to determine if the Cloprostenol-induced increase in *in vitro* luteal PGF release could be blocked with *in vivo* indomethacin. Sixteen ewes were assigned at random to 4 groups: 4) control, 5) indomethacin, injected 2 times with 2 ml of .1 M indomethacin 2 hr apart, 6) Cloprostenol, injected 1 time with 100 ug cloprostenol i.m. and 7) Cloprostenol plus indomethacin (table 2). The first indomethacin injection was concurrent with the Cloprostenol injection; the second was 2 hr later. Ewes were sacrificed 4 hr after the Cloprostenol injection. Tissue handling and *in vitro* treatments were as in experiment 1.

Experiment 3. Twenty-four ewes were assigned at random to 4 groups: 8) 0 ug, 9) 25 ug, 10) 37.5 ug and 11) 50 ug of Cloprostenol to determine the dose response curve for interval to estrus and decrease in plasma progesterone (table 3).

Experiment 4. Ewes were given a dose of Cloprostenol barely sufficient to induce luteal regression to determine if administration of indomethacin *in vivo* would inhibit luteal regression. Twenty ewes were assigned at random to two groups: 11) 37.5 ug Cloprostenol and 12) 37.5 ug Cloprostenol plus 4 ml of .1 M indomethacin at 0, 2 and 4 hr after the Cloprostenol injection. Ewes were bled from the jugular vein 0, 2, 4, 24 and 48 hr after the Cloprostenol injection (table 4), and plasma was assayed for progesterone. Ewes were checked for estrus by vasectomized rams.

## RESULTS AND DISCUSSION

Because PGF is a proposed *in vivo* luteolysin (1), the effect of PGF and Cloprostenol on *in vitro* PGF from luteal tissue was compared (table 1). Both PGF and Cloprostenol increased ($P<.05$) PGF release by 4 hr after *in vivo* administration. The difference between Cloprostenol and PGF on *in vitro* PGF release was not significant ($P>.05$).

TABLE 1. In vitro luteal PGF release after in vivo treatment with Cloprostenol or PGF in day 10 ewes

| In vivo treatment | In vitro treatment | PGF [a,b] (ng/g) | Progesterone [a,c] (ug/g) |
|---|---|---|---|
| 1) control | 0° | 13 ± 4 | 2.5 ± .2 |
| | 37° | 52 ± 9 | 26.1 ± 3.4 |
| | 37° + Indomethacin | 70 ± 8 | 25.5 ± 2.7 |
| 2) PGF | 0° | 11 ± 3 | 1.8 ± .2 |
| | 37° | 82 ± 24 | 12.1 ± 2.0 |
| | 37° + Indomethacin | 101 ± 14 | 10.8 ± 1.8 |
| 3) Cloprostenol | 0° | 23 ± 9 | 1.9 ± .2 |
| | 37° | 112 ± 18 | 8.3 ± .6 |
| | 37° + Indomethacin | 131 ± 29 | 8.6 ± .6 |

[a]Mean ± SE Accumulation in the media during a 2 hr incubation period; n=8 for all groups.
[b]Control vs PGF and cloprostenol, P<.05;
0° vs 37° and 37° + indomethacin, P<.0001.
[c]Control vs PGF and cloprostenol, P<.005;
0° vs 37° and 37° + indomethacin, P<.0001.

Progesterone release from tissue of control ewes was great (P<.005) than from tissue of either of the prostaglandin-treat groups.

In Experiment 1, in vitro indomethacin treatment did not redu (P>.05) PGF or change progesterone concentration in the media. experiment 2, ewes were given indomethacin in vivo to verify th the PGF released into the media was the result of synthesis rath than an alteration of membrane permeability. Experiment 2 was al conducted to establish that the induced increase in PGF could blocked by in vivo treatment.

In experiment 2, the increased in vitro PGF release followi Cloprostenol treatment was blocked by in vivo indomethacin treatme (interaction P<.05, table 2), but in vitro indomethacin failed reduce in vitro PGF release (P>.05). We conclude that the obser increase in in vitro PGF release was due to an effect of Clopro tenol on PGF synthesis by the CL, probably at some time prior incubation. In the pig (3), we have demonstrated that synthe occurs in vitro under identical circumstances. The failure of Pa and Watson (2) to observe increased PGF release in vitro by the of the ewe may have resulted from incubation conditions or me that did not support PGF synthesis by ovine luteal tissue.

The addition of PGF (1 ug/ml) to the medium did not aff progesterone release into the medium. This agrees with the obser tion (Table 2) that progesterone release was reduced by Cloproste (P<.001) but not by in vitro indomethacin (P>.05) which grea reduced prostaglandin release (P<.001). Indomethacin injections did not reduce the apparent luteolytic effect of Cloprostenol on progesterone release in vitro.

In experiment 3 (table 3), we found that the apparent minimum dose of Cloprostenol for shortening the return to estrus interval was about 37.5 ug. Some decline in the plasma concentrations of progesterone (Table 3) were observed for all doses of Cloprostenol by 36 hr after injection (Dose; P<.0001). Estrus expression then was probably an all or none effect, expressed when the CL had regressed below a threshold value for estrus suppression.

In experiment 4, ewes were treated with Cloprostenol (37.5 ug) or with Cloprostenol and indomethacin to determine if an indomethacin treatment that blocked in vivo synthesis of PGF would also block induced luteal regression. The rate (slope) of decline in plasma progesterone was similar in both groups (P>.05). Ewes in

TABLE 2. The release of PGF and progesterone by luteal tissue in vitro from ewes treated with Cloprostenol and indomethacin in vivo

| In vivo treatment | In vitro treatment | PGF [a,b] (ng/g) | Progesterone [a,c] (ug/g) |
|---|---|---|---|
| 4) control | 0° | 4 ± 1 | 2.6 ± .6 |
| | 37° | 40 ± 7 | 26.4 ± 4.2 |
| | 37° + Indomethacin | 26 ± 3 | 32.8 ± 7.7 |
| 5) Indomethacin | 0° | 5 ± 1 | 2.7 ± .4 |
| | 37° | 16 ± 2 | 39.1 ± 6.1 |
| | 37° + Indomethacin | 27 ± 3 | 31.8 ± 4.2 |
| 6) Cloprostenol | 0° | 4 ± 1 | 1.7 ± .4 |
| | 37° | 106 ± 21 | 10.0 ± 1.0 |
| | 37° + Indomethacin | 102 ± 40 | 8.6 ± 1.1 |
| 7) Cloprostenol + indomethacin | 0° | 4 ± 1 | 1.6 ± .4 |
| | 37° | 21 ± 5 | 11.6 ± 1.6 |
| | 37° + Indomethacin | 26 ± 4 | 10.7 ± 1.4 |

[a]Mean ± SE Accumulation in the media during 2 hr incubation; n=4 for all groups.

[b]Cloprostenol P<.05; in vivo indomethacin P<.05; interaction tion of Cloprostenol and in vivo indomethacin, P<.05; 0 vs all 37°, P<.01; interaction Cloprostenol and 0° vs all 37°, P<.01; interaction in vivo indomethacin and 0° vs all 37°, P<.01.

[c]Cloprostenol P<.001, 0° vs all 37°, P<.0001.

TABLE 3. Dose effect of cloprostenol on interval to estrus and plasma progesterone in day 10 ewes.

| Dose (ug) | Days to estrus[a,b] | Plasma progesterone (ng/ml) Hours after cloprostenol[a,c] 0 | 2 | 4 | 24 | 36 |
|---|---|---|---|---|---|---|
| 0 | 6.8 ± .7 | 2.6 ± .7 | 2.1 ± .6 | 2.2 ± .6 | 3.0 ± .9 | 2.0 ± .5 |
| 25 | 6.2 ± 1.2 | 3.0 ± 1.1 | 1.6 ± .3 | 1.0 ± .1 | 1.0 ± .3 | 1.3 ± .4 |
| 37.5 | 3.3 ± .9 | 3.5 ± .8 | 1.8 ± .3 | 1.6 ± .4 | .7 ± .2 | .8 ± .4 |
| 50 | 3.3 ± 1.2 | 2.7 ± .5 | 1.4 ± .1 | 1.1 ± .1 | .4 ± .1 | .6 ± .3 |

[a]Mean ± SE; n=6 for all groups.
[b]Dose P<.05; $r^2$ = .57 when fitted to quadratic and cubic components.
[c]Dose P<.0001; hours P<.0001; interaction P = <.01.

TABLE 4. Induced luteolysis in ewes treated with Cloprostenol and indomethacin on day 10 of the estrous cycle

| Treatment | Days to estrus[a,b] | Plasma progesterone (ng/ml) Hours after Cloprostenol[a,c] 0 | 2 | 4 | 24 | 48 |
|---|---|---|---|---|---|---|
| Cloprostenol | 4.6 ± .5 | 2.3 ± .2 | 1.7 ± .2 | 1.5 ± .1 | .8 ± .2 | 1.1 ± .3 |
| Cloprostenol + indomethacin | 5.1 ± .6 | 2.1 ± .2 | 1.5 ± .2 | 1.4 ± .1 | .6 ± .1 | .7 ± .2 |

[a]Mean ± SE; n=10 both groups
[b]P>.05.
[c]Hour P<.001; Cloprostenol P<.05.

both groups (table 4) returned to estrus at nearly the same time ($P>.05$). These data suggest that PGF synthesis by the CL or other tissue, is not needed fo r Cl prostenol induced luteolysis.

The results of these experiments demonstrated that an injection of Cloprostenol or PGF to induce luteolysis also caused an increase in the synthesis of PGF by luteal tissue *in vitro*. The mechanisms for the induced increase in PGF synthesis are not known, nor is it known whether increased PGF synthesis by luteal tissue occurs *in vivo* during induced luteolysis. Luteolysis occurred in ewes treated with Cloprostenol plus indomethacin, suggesting that increased luteal synthesis of PGF may not be required for luteolysis in the ewe.

## REFERENCES

1. J. R. Goding. J. Reprod. Fert. 38(1974)261-272.
2. C. E. Patek and J. Watson. Prostaglandins 12(1976)97-111.
3. H. D. Guthrie, C. E. Rexroad, Jr. and D. J. Bolt. (1978)in press.
4. C. E.Rexroad, Jr. Biol. Reprod. (1978)in press.
5. H. D. Guthrie, J. Anim. Sci. 45(1977):1360-1367.

SESSION NUMBER: VIII

TITLE: LUTEAL DESENSITIZATION

PRINCIPAL SPEAKER: Kevin Catt

INVITED DISCUSSANT: Lutz Birnbaumer

PRESENTED ABSTRACTS:

17.) U. Zor, B. Strulovici, S.A. Lamprecht, A. Oplatka and H.R. Lindner. Desensitization and Recovery of $PGE_2$ - Responsive Ovarian Adenylate Cyclase Modulated by Agents Affecting Cytoskeletal Function.

18.) F. Labrie, C. Auclair, L. Cusan, A. Lemay, A. Bélanger, P. Kelly, L. Ferland, G. Azadian-Boulanger and J.P. Raynaud. Inhibitory Effect of Treatment with LHRH or its Agonists on Ovarian Receptor Levels and Function.

CHAIRPERSON: Hans Lindner

# LUTEAL DESENSITIZATION: HORMONAL REGULATION OF LH RECEPTORS, ADENYLATE CYCLASE AND STEROIDOGENIC RESPONSES IN THE LUTEAL CELL

K. J. Catt, J. P. Harwood, N. D. Richert, P. M. Conn, M. Conti and M. L. Dufau
Endocrinology and Reproduction Research Branch, NICHD, National Institutes of Health, Bethesda, Maryland 20014

Regulation of peptide hormone receptors and target cell responses by increased concentrations of the homologous hormone has been demonstrated in a number of endocrine-dependent tissues (1-6). This process, described variously as desensitization, tachyphylaxis, refractoriness, tolerance, and down-regulation, is particularly marked in the testis and ovary following exposure to elevated gonadotropin levels (5-12). The finding that insulin (1) and growth hormone (2) are able to regulate their own receptor sites in target cells was followed by the demonstration of an inverse relationship between ligand concentration and receptor number for several other hormones, including catecholamines (3), thyrotropin releasing hormone (4) and gonadotropins (5-12). Although such negative regulation of receptor sites is now recognized to be a wide-spread property of peptide hormones, it should be noted that certain hormones, such as prolactin (13) and angiotensin (14) exert positive regulation upon their receptor sites, with an increase in binding capacity of the target cells in response to elevated hormone concentration.

The occurrence of hormone-induced receptor regulation in endocrine target cells raises a number of important questions concerning the mechanism, biological consequences, and physiological significance of this phenomenon. In several tissues in which receptor regulation has been demonstrated, there has not been a readily monitored target cell response with which to assess the effects of receptor loss. In others, it has been possible to analyze the effects of receptor regulation upon adenylate cyclase responses in target tissue homogenates. In the testis and ovary, the occurrence of gonadotropin-induced receptor regulation in steroidogenic target cells has provided an opportunity to analyze the biological responses of the affected target cells to hormonal stimulation during the desensitization process. This has been performed by evaluation of cyclic AMP and steroid responses during the successive phases of receptor occupancy, loss, and recovery in isolated target cells following exposure to high concentrations of gonadotropins in vivo.

The occurrence of gonadal desensitization at the level of adenylate cyclase in the ovary was first noted by Marsh et al (15), who found that the effect of LH upon cyclic AMP formation in rabbit Graafian follicles was reduced after the mating-induced endogenous LH surge, or following induction of ovulation by administration of hCG. This finding was later extended by Hunzicker-Dunn and Birnbaumer (16-18), who observed desensitization of follicular adenylate cyclase following spontaneous ovulation in the rat, and after injection of hCG in the rabbit. The latter effect was detected within minutes of gonadotropin administration, and persisted for up to two days following gonadotropin treatment.

Alterations in ovarian follicular sensitivity to hormone stimulation were also observed when pre-ovulatory rat follicles were exposed to high concentrations of gonadotropin or prostaglandin in vitro (19-21). In this case, the desensitization of adenylate cyclase responses was specific for the individual ligands, and became maximum within several hours of culture with the respective hormones. The effect of LH was slower than that of $PGE_2$, and was less readily reversible, being maintained during subsequent incubation in hormone-free medium or in the presence of anti-LH serum (20). In these studies, no loss of LH receptors was demonstrable during the phase of adenylate cyclase desensitization induced by exposure to gonadotropin in vitro.

Further analysis of the relationship between LH receptor regulation and ovarian target cell responses was performed following in vivo desensitization of luteinized ovarian tissue in PMSG/hCG-treated immature female rats (10,11). When the time course of receptor binding and adenylate cyclase activity in the luteinized ovary was followed for seven days after injection of hCG, it was apparent that after the first few hours, a close correlation existed between the number of free gonadotropin receptors and the hormonal responsiveness of the ovarian adenylate cyclase. After injection of high doses of hCG (2 or 20 μg), luteal receptor sites for LH (measured by $^{125}$I-hCG binding assay) fell by 90% or more within a few hours, remained low for four days, and recovered between five and seven days after hormone treatment. Treatment with lower doses of hCG (0.1 to 0.3 μg) caused partial (50%) loss of LH receptors from the luteinized ovary, with somewhat more rapid recovery of the receptor population (Fig. 1). These prominent changes in LH receptor concentration were not accompanied by any alteration in the binding affinity of the residual sites. In the same ovaries, adenylate cyclase responsiveness to LH was rapidly lost, and reappeared with the same time course as the recovery

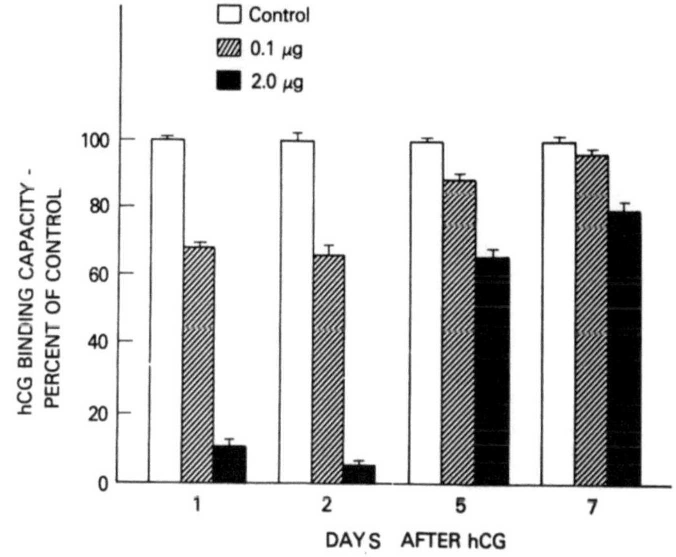


Fig. 1. Time course of LH receptor loss and recovery in luteinized ovaries of rats treated with 0.1 and 2.0 μg hCG.

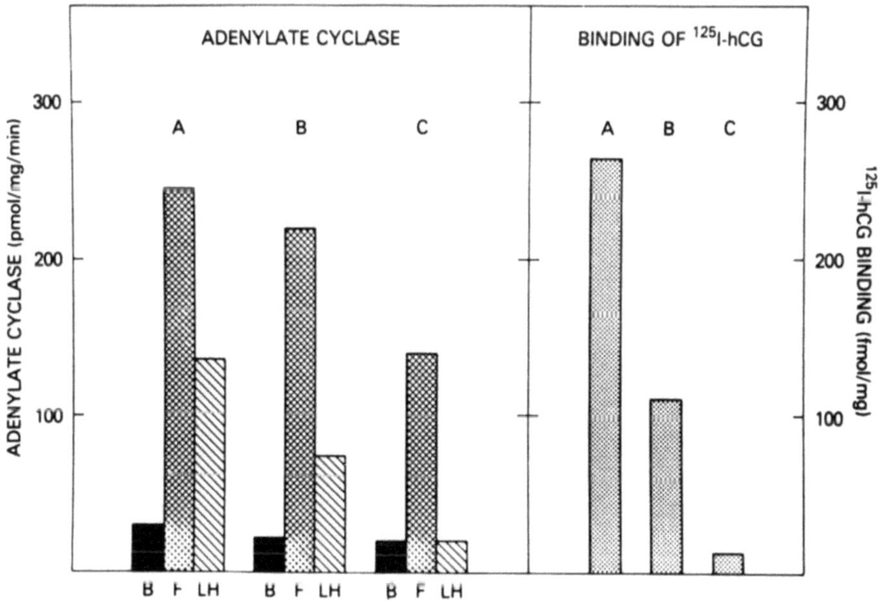


Fig. 2. Concomitant changes in luteal LH receptors and adenylate cyclase 24 hr after treatment with low and high doses of hCG. (A) Control. (B) 0.1 μg hCG. (C) 20 μg hCG.

of the LH receptor sites. In addition, there was a transient loss of the fluoride-stimulated adenylate cyclase response, during the first twenty-four hours following hCG treatment. The correlation between LH receptor number and adenylate cyclase responses is shown in Fig 2. In 24 hr. the intermediate (0.1 μg) and high (20 μg) doses of hCG produced partial or complete loss of receptor number and LH-stimulated enzyme activity. These changes in luteal receptor sites were shown to represent a true loss of receptors, following the initial phase of receptor occupancy due to binding of the administered dose of hCG. The receptor-bound hCG molecules were largely cleared from the ovary within twenty-four hours, and the sustained loss of receptor binding capacity measured in vitro during the subsequent several days was not due to receptor occupancy by the administered hormone. These initial studies showed that the prolonged loss of LH receptors after gonadotropin treatment was accompanied by a corresponding loss of hormonal activation of adenylate cyclase, though the basal and fluoride-stimulated enzyme activities were normal for several days after the initial acute desensitization phase. The disappearance of occupied receptor sites following gonadotropin treatment implies that each receptor is used only once during hormonal activation, and then processed or degraded. These findings also suggest that restoration of binding sites is dependent upon receptor synthesis, and that constant regeneration of receptors would be necessary to maintain hormonal responsiveness. In this regard, it should be noted that the so-called "spare" receptors observed in gonadal tissue could serve as a pool of receptor sites to replace those utilized after hormone occupancy and target cell activation.

In more detailed studies (10), administration of hCG to immature rats with luteinized ovaries was shown to cause a time- and dose-related loss of LH receptors, again with no change in binding affinity for hCG. Administration of quite small doses of hCG, producing serum concentrations of a few nanograms per ml, was followed by a relatively large decrease in ovarian LH receptors after 24 hours. Although it was possible to measure receptor occupancy during the first few hours after hormone treatment, by acid elution and radioimmunoassay of the bound hormone, this accounted for only a small proportion of the observed decrease in binding, and after twenty-four hours there was up to 90% loss of total receptor sites. In addition, the ovarian content of LH receptors was found to remain low for several days after the desensitizing dose of hCG had been cleared from the circulation. The finding that a significant loss of receptor sites could be detected following hCG doses as low as 30 ng/100 g of body weight indicated that receptor regulation could be expected to occur following physiological fluctuations in plasma gonadotropin concentration, as occurs during the endogenous LH surge prior to ovulation. In addition, relatively small degrees of receptor occupancy were followed by extensive loss of binding capacity, suggesting that an active or cooperative process might be involved in the mechanism of receptor loss. The loss of receptors was not due to sequestration within the plasma membrane, since analysis of detergent-solubilized ovaries during the desensitization process revealed no major difference in the number of sites measured in particulate or soluble receptor preparations. Also, no change in the sedimentation properties of the residual LH receptors was detectable following solubilization and analysis by density gradient centrifugation.

The initial phase of receptor occupancy following gonadotropin administration persisted in the ovary for 3 to 6 hours, during which there was no change in the total number of ovarian receptors (i.e., free plus occupied sites). Thereafter, the total receptor number began to decline rapidly, and reached a nadir at about twenty-four hours. The time course of receptor loss was similar whether the hCG was given by subcutaneous or intravenous injection, suggesting that receptor loss is not dependent on the continued presence of hCG in the circulation, just as the duration of receptor loss does not result from prolonged elevation of the circulating hCG concentrations. These studies demonstrated that the acute refractory state following hormone-receptor interaction in the ovary is followed by a more prolonged state of hormone insensitivity which depends upon the absence of LH receptor sites, and which recovers in proportion to the restoration of receptors in the luteal cell membrane.

The consequences of LH receptor regulation and adenylate cyclase desensitization upon target cell responses were next examined in isolated cells from the luteinized ovaries of PMSG/hCG treated rats (12). Such cells were prepared by collagenase dispersion of luteinized ovaries, from one to seven days after treatment with desensitizing doses of hCG (0.1 or 2 μg). The cells prepared from control luteinized ovaries showed highly sensitive cyclic AMP and progesterone responses to gonadotropin stimulation <u>in vitro</u>. The half-maximum progesterone response was elicited at about 1 pM hCG, and almost a 100-fold higher concentration of hCG was required for the corresponding half-maximum cyclic AMP response. Despite this difference, it should be noted that small increments in cyclic AMP production were detectable under appropriate conditions over the range of hCG concentrations that evoked the full steroidogenic response (22).

The luteal cells isolated from ovaries of rats given desensitizing doses of hCG exhibited marked decreases in LH receptors, and a corresponding reduction in cyclic AMP and progesterone formation during incubation with gonadotropin. Thus, the hCG-induced desensitization of adenylate cyclase previously observed in ovarian homogenates was evident in the intact cell system as a marked decline in cyclic AMP formation during hormone

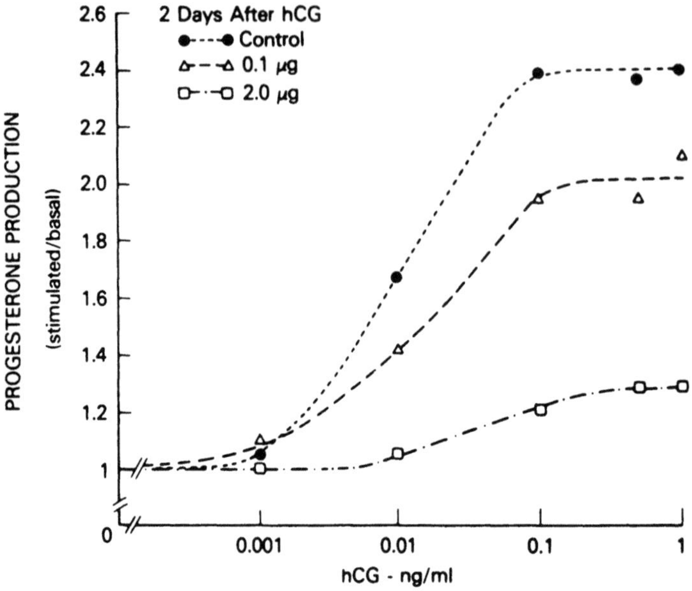


Fig. 3. Impairment of progesterone production in luteal cells of hCG-desensitized ovaries. Collagenase-dispersed cells were prepared from ovaries of primed rats 48 hr after administration of 0.1 μg or 2.0 μg hCG.

stimulation *in vitro*. The accompanying decreases in progesterone production during hormone stimulation of the desensitized cells were dose-related, and were characterized by loss of the maximum steroidogenic response to gonadotropin *in vitro* (Fig. 3). The latter impairment of progesterone formation was more extensive than might be expected from the fall in cyclic AMP formation, since as noted above, the full steroidogenic response can be elicited in normal cells with only a minor increase in cyclic AMP. Further studies revealed that the decrease in progesterone formation by desensitized luteal cells was not simply a consequence of the reduction in cyclic AMP formation, since the steroid responses of such cells to dibutyryl cyclic AMP or choleragen were also markedly depressed, whereas these agents stimulated progesterone production as effectively as hCG in control cells.

The impairment of the progesterone responses to the several stimuli of steroid biosynthesis, including those acting beyond the LH receptor site, indicated that the gonadotropin-induced decrease in LH receptors was accompanied by an additional lesion in the progesterone biosynthetic pathway. A similar defect in steroidogenesis has been recently noted in gonadotropin-desensitized Leydig cells (9, 23), in which biosynthetic lesions were detected

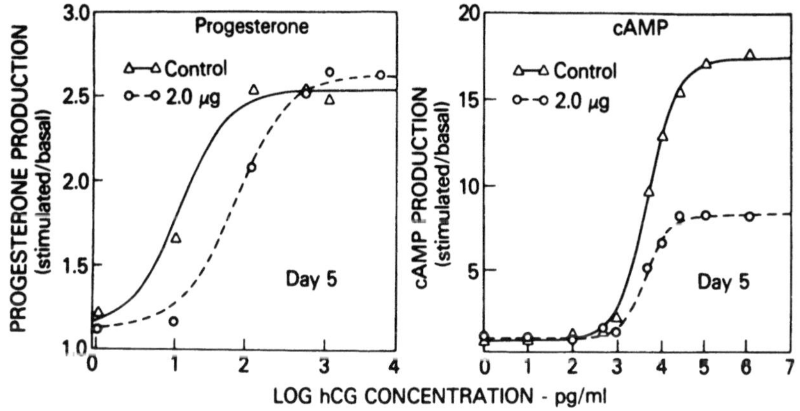


Fig. 4. Progesterone and cyclic AMP responses of luteal cells during recovery from hCG-induced desensitization. Cells were prepared from luteinized ovaries of primed rats 5 days after administration of 2.0 μg hCG.

at the 17-20 desmolase step, and also at a site prior to pregnenolone formation. In the luteal cell, the steroidogenic lesion following gonadotropin-induced desensitization has been localized to the early biosynthetic pathway prior to pregnenolone formation. This second lesion in receptor-depleted luteal cells, like that observed in the Leydig cell, recovered more rapidly than the LH receptor population. The occurrence of such steroidogenic defects may reflect the need for a continuous basal level of hormone-receptor interaction to exert a trophic action on the maintenance of the steroidogenic pathway. In addition, local feedback actions of gonadal steroids formed in response to the desensitizing dose of gonadotropin could exert an inhibitory effect upon progesterone biosynthesis in receptor-depleted luteal cells.

During recovery of ovarian luteal cells from the desensitized state, gonadotropin receptors, hormone-stimulated cyclic AMP production, and progesterone responses returned gradually to normal levels over about seven days. By the fifth day, the progesterone response had fully recovered, while the full complement of LH receptors had not yet returned. The recovery phase following receptor depletion was characterized by a shift in the steroid dose-response curve to gonadotropin stimulation, with reduced sensitivity to hCG during incubation in vitro. At this time, in the presence of fewer LH receptors and reduced cyclic AMP formation, maximum progesterone production could be evoked when the gonadotropin concentration was sufficiently increased in the incubation medium (Fig. 4). Thus, maximum steroid responses could be evoked in the presence of few receptors, but optimal sensitivity to gonadotropin required the full complement of LH receptors. These studies in dispersed ovarian luteal cells demonstrated that the hCG-induced decreases in LH receptors and hormone-sensitive adenylate cyclase activity result in a corresponding impairment of cyclic AMP production in the intact cell, and in more distal changes in the biological response as reflected by decreased progesterone production. A summary of the changes that occur in luteal cell receptors and responses to LH during the several phases of desensitization and recovery is shown in Table 1.

Table 1. Summary of Desensitization Processes in the Rat Ovary

I. Occupancy -- early phase, acute biological responses (1 min - 3 hr)
   (a) LH receptor number normal, minor occupancy
   (b) activation of adenylate cyclase
   (c) generation of cyclic AMP
   (d) production of progesterone

II. Occupancy -- late phase, primary refractory state (6 hr)
   (a) LH receptor number normal, decreased free sites
   (b) loss of adenylate cyclase responsiveness
   (c) loss of cyclic AMP response
   (d) loss of progesterone production

III. Receptor loss -- secondary refractory state (1-4 d)
   (a) loss of LH receptors
   (b) loss of LH stimulation of adenylate cyclase
   (c) loss of cyclic AMP response
   (d) loss of progesterone production

IV. Recovery -- of receptors and responses (3 to 10 d)
   (a) LH receptor number returns to normal
   (b) recovery of adenylate cyclase
   (c) generation of cyclic AMP
   (d) production of progesterone

The process by which homologous hormones cause a specific loss of receptor sites in their target cells has not yet been clarified, though a dependence on protein synthesis has been noted in several reports. Whether the decrease in receptor sites is caused by shedding of occupied sites from the cell surface, or by internalization of the hormone-receptor complexes may vary with the target cell concerned, but there is increasing evidence for the latter mechanism in a number of hormone-responsive tissues. In the ovary, the initial site of interaction of labeled gonadotropin with the luteal cell has been clearly located at the level of the plasma membrane by cell fractionation and morphological studies (24, 25). However, a significant proportion of the hormone-receptor complexes formed at the luteal cell membrane are later internalized, and subsequently degraded (26, 27). During this process, that fraction of the internalized complexes that is associated with particulate cellular structures retains the physico-chemical properties of the native hormone-receptor complex after solubilization with detergents and analysis by sucrose density gradient centrifugation (27). Presumably, the internalized complexes initially retain the structural properties of the membrane-bound receptors, and then undergo relatively rapid degradation to lower molecular weight forms. It was of interest to note that the internalization of labeled hormone occurred over the same time course as the loss of receptor sites induced by hormone treatment (27). Following the initial location of the majority of the bound radioactive hormone at the cell membrane after two to three hours, much of the bound hormone subsequently entered the cells, and a substantial degree of internalization was evident after seven hours. By twenty-four hours, when only a few percent of the originally bound tracer remained in the ovary, most of the radioactivity was intracellular. At this time, as noted above, the process of gonadotropin-induced receptor loss was almost complete, with a marked decrease in the absolute number of LH receptor sites in the luteal cell to about 10% of the control value.

Studies of this type have shown that a significant proportion of the hormone-receptor complexes undergoes endocytosis in the luteal cell, and this process probably accounts for much of the receptor loss induced by occupancy with the homologous hormone. These observations are also consistent with the possibility that LH receptor sites are occupied only once by the corresponding hormone, and are then processed and degraded, rather than vacated and re-utilized during subsequent hormone stimulation. It should be noted that these changes are not only induced by hCG, which has a relatively high affinity for LH receptors and a prolonged half-life in the circulation, but also by more rapidly metabolized gonadotropins such as ovine LH and endogenous LH secreted after stimulation with LHRH (28). Therefore, it is likely that physiological fluctuations in endogenous gonadotropin secretion during the reproductive cycle are sufficient to induce receptor regulation and desensitization of ovarian cell responses in vivo.

In the course of these studies, it was observed that the desensitization of adenylate cyclase induced by gonadotropin treatment was not completely specific. As noted earlier, the initial phase of desensitization was accompanied by a transient loss of the fluoride effect upon adenylate cyclase, suggesting the temporary formation of a refractory hormone-receptor-enzyme complex. To examine the effects of LH receptor regulation upon the properties of other ligands acting on the luteal cell, the actions of epinephrine on luteal cell responses were studied during gonadotropin-induced desensitization in the rat ovary. The ability of catecholamines to stimulate progesterone production in the bovine corpus luteum has been previously attributed to the presence of a β-adrenergic receptor system in the luteal cell (29). Also, stimulation of the β-adrenergic receptors in bovine luteal slices by isoproterenol was shown to increase cyclic AMP and progesterone synthesis by luteal slices in vitro (30). In the luteinized rat ovary, both epinephrine and LH were found to stimulate adenylate cyclase activity and progesterone production, and the combined effects of maximal stimulating doses of the two agents were not additive. The actions of epinephrine upon adenylate cyclase and progesterone production were inhibited by propranolol, but not by phentolamine. Also, neither of the antagonists had an effect upon the stimulation of adenylate cyclase and steroidogenesis produced by LH or hCG. These observations demonstrated that epinephrine acts upon the rat luteal cell via a β-adrenergic receptor, and that both LH and epinephrine act through independent receptors upon the same adenylate cyclase to stimulate progesterone production in the luteal cell.

When rats were pretreated with doses of hCG to induce luteal desensitization, the decreased ability of LH to stimulate adenylate cyclase activity was accompanied by a corresponding loss of the action of epinephrine on the enzyme. This was predominantly due to a decrease in the Vmax for cyclic AMP production, with relatively little change in the Km of the system. As in luteal slices, epinephrine was also shown to increase cyclic AMP and progesterone production in collagenase-dispersed rat luteal cells. However, when luteal cells were prepared from ovaries of animals treated with desensitizing doses of hCG, the decreased effect of LH upon progesterone production in vitro was accompanied by a corresponding loss of the response to epinephrine. Also, cyclic AMP production in isolated luteal cells during incubation with epinephrine was decreased in proportion to the loss of the LH effect after gonadotropin-induced luteal desensitization. Although the concentration of LH receptors in the desensitized ovaries was considerably reduced, no significant change in the number of β-adrenergic receptors was detected by binding studies with tritiated alprenolol.

These observations have demonstrated that the hCG-induced desensitization of LH receptors and responses in the ovarian luteal cell is accompanied by loss of the effects of epinephrine. However, in contrast to the marked loss of LH receptors, there appears to be little or no change in the ovarian content of β-adrenergic receptors during the phase of desensitization. The early loss of adenylate cyclase responses to LH and epinephrine during the phase of receptor occupancy, and the accompanying transient loss of the fluoride response, was again consistent with the formation of a refractory hormone-receptor-enzyme complex. The concomitant loss of LH and catecholamine responsiveness of adenylate cyclase, in the absence of a change in β-adrenergic receptors, suggests that the coupling of other receptors to adenylate cyclase is altered during gonadotropin-induced desensitization. These findings indicate that the process of desensitization is more complex than a simple loss of receptor sites and consequent failure of activation by the homologous hormone. It is possible that a common component in the coupling mechanism between hormone receptors and adenylate cyclase may undergo a relatively prolonged change following the acute phase of enzyme refractoriness and receptor regulation induced by occupancy of LH receptors by the homologous hormone.

In these studies, the luteal cell has been used to explore the relationships between LH receptor occupancy and the subsequent processes that are responsible for "desensitization" or refractoriness to further hormonal stimulation. The extent to which these changes can lead to altered gonadal function is illustrated by the marked regression of ovaries (28) and testes (31) observed in rats treated with potent LHRH agonists, and the natural LHRH molecule (32), presumably related to the marked loss of gonadotropin and prolactin receptors that results from such treatment. This LHRH-induced process is accompanied by post-receptor defects in the steroidogenic pathway, similar to those observed in the gonadotropin-treated rat (33), that accentuate the regression of steroid-dependent compartments of the testis and ovary. A similar process has been observed in the corpus luteum of pregnancy in the rhesus monkey, in which refractoriness of the luteal cells during exposure to mCG in early pregnancy has been attributed to impairment of the gonadotropin receptor/adenylate cyclase system and reduction of steroidogenic enzymes necessary for progesterone synthesis (34).

Of particular interest in the luteal cell is the relationship between desensitization and luteolysis. Although the mechanism of normal luteolysis remains unclear in most species, prostaglandin $F_{2\alpha}$ is regarded as the physiological mediator of luteolysis in the sheep (35) and has a marked luteolytic action in the rat (35). The latter effect has been shown to be accompanied by extensive loss of LH receptors during the 24 hours following treatment with $PGF_{2\alpha}$ (37), though the onset of functional luteolysis occurs even more rapidly, with decreased progesterone secretion within a few hours (38). The time course of the $PGF_{2\alpha}$-induced loss of LH receptors is similar to that produced by hCG treatment, raising the possibility that a common mechanism accounts for the two effects. It is also possible that the $PGF_{2\alpha}$ effect could result from a transient elevation of the plasma LH concentration, with secondary loss of LH receptors in the ovary.

It is well known that gonadotropins such as LH and hCG are able to induce luteolysis in the rabbit and rat ovary (39,40), and the possible relation between luteal desensitization and luteal regression is obviously an important issue. The extent to which gonadotropin treatment is followed by luteal regression is dependent upon the dose and duration of hormone treatment, but luteolysis has been observed after single doses of ovine LH (39). We have previously suggested that luteal desensitization may represent a form of transient or reversible luteolysis (6), and it is evident that the process of desensitization may be followed by true luteolysis when large doses of hormone are administered. If this occurs, the delayed recovery of LH receptors and luteal cell responses after large doses of hCG could result from the development of further luteal tissue following the desensitizing injection of gonadotropin.

On the other hand, the series of functional changes observed after low doses of hCG (e.g., 0-1 μg) are more likely to be reversible within the luteal cell, and to represent a biochemical sequence that can be equated with transient or reversible luteolysis. These changes (i.e., adenylate cyclase desensitization and LH receptor loss) when sufficiently extensive, as after large doses of hCG, could lead to irreversible regression of the luteal cells, following the same sequence of changes induced by luteolytic agents such as $PGF_{2\alpha}$. Thus, while the biochemical changes induced in luteinized ovaries by high doses of hCG can be regarded as a common feature of the initiation of luteolysis, it should be noted that low doses of gonadotropins cause a comparable series of changes in the testis, with reversible losses of Leydig cell receptors and responses. Thus, in the Leydig cell, the sequence of LH receptor loss and desensitization proceeds as in the luteal cell, with no evidence of target cell regression comparable to luteolysis. In the luteal cell, which has been regarded as programmed for regression, the initiation of this sequence by several stimuli appears likely to lead to luteolysis. The relationship of these hormone-induced changes to the normal control of luteal cell life span remains to be determined, but the marked sensitivity of the luteal cell to negative regulation makes it a valuable model for the study of hormonal regulation of target cell receptors and functional responses.

## Summary

Hormonal regulation of receptors and target cell responsiveness to subsequent hormone stimulation has been demonstrated in several endocrine tissues, and is particularly marked in the testis and ovary. In immature rats with ovaries luteinized by PMSG/hCG priming, further treatment with hCG caused a time- and dose-related loss of LH receptors which exceeded the degree of receptor occupancy by the administered hormone. The concomitant desensitization of adenylate cyclase also occurred in a dose-dependent manner after hCG treatment, with rapid loss of hormone (LH) responsiveness after high hCG doses and delayed recovery in unison with the return of LH receptors. However, the initial loss of adenylate cyclase responses preceded the decrease in total (occupied + free) LH receptors, being coincident with the early phase of receptor occupancy. Thus, gonadotropin-induced desensitization includes an acute process of enzyme refractoriness without decreased LH receptors and a more prolonged phase associated with LH receptor loss.

During the receptor occupancy and turnover that follows binding of LH or hCG, the β-adrenergic actions of catecholamines on the luteal cell were also reduced. However, in contrast to the marked loss of LH receptors, no change in β-adrenergic receptors was detected by binding studies with [$^{3}$H]-alprenolol. The early loss of adenylate cyclase responsiveness to LH and epinephrine during the phase of receptor occupancy, with the accompanying transient loss of the fluoride response, was again consistent with the formation of a refractory hormone-receptor-enzyme complex. The concomitant loss of LH and catecholamine responses, in the absence of a change in β-adrenergic receptors, suggests that the coupling of other ligands to adenylate cyclase is altered during gonadotropin-induced desensitization.

These changes in LH receptors and adenylate cyclase were accompanied by corresponding decreases in cyclic AMP and progesterone formation in isolated luteal cells. This decrease in steroidogenesis was not overcome by addition of dibutyryl cyclic AMP or during the action of choleragen, indicating that a second lesion must exist in desensitized luteal cells beyond the receptor-cyclase changes in the cell membrane. The second lesion may reflect the need for a continuous basal level of hormone-receptor interaction to maintain the steroidogenic pathway, or a local feedback effect of increased steroid synthesis following the desensitizing dose of hCG.

The mechanisms of LH receptor loss in ovarian luteal cells could involve shedding of occupied sites or internalization of the hormone-receptor complexes. Evidence for the latter process has been found in studies of several target cells, and has been demonstrated in corpus luteum cells during desensitization with labeled hCG. The internalized tracer behaved as the hormone-receptor complex on solubilization and sucrose gradient centrifugation of the particulate ovarian fraction. It is likely that internalization accounts for much of the receptor loss induced by occupancy with the homologous hormone. It is also probable that LH receptor sites are occupied only once, after which they are processed and degraded, rather than vacated and re-utilized during subsequent hormone stimulation.

## References


1. Gavin, J.R., J. Roth, D.M. Neville, P. deMeyts and D.N. Buell. Proc. Natl. Acad. Sci. USA. 71(1974)84.

2. Lesniak, M.A. and J. Roth. J. Biol. Chem. 251(1976)3720.

3. Mukherjee, C., M.G. Caron and R.J. Lefkowitz. Proc. Natl. Acad. Sci. USA 72(1975)1945.

4. Hinkle, P.M. and A.H. Tashjian, Jr. Biochemistry 14(1975)3845.

5. Hsueh, A.J.W. M.L. Dufau and K.J. Catt. Biochem. Biophys. Res. Commun. 72(1976)1145.

6. Conti, M., J.P. Harwood, A.J.W. Hsueh, M.L. Dufau and K.J. Catt. J. Biol. Chem. 251(1976)7729.

7. Sharpe. R.M. Nature 264(1976)644.

8. Hsueh, A.J.W., M.L. Dufau and K.J. Catt. Proc. Natl. Acad. Sci. USA 74(1977)592.

9. Richards, J.S., J.R. Ireland, M.C. Rao, G.A. Bernath, A.R. Midgley and L.E. Reichert. Endocrinology 99(1976)1562.

10. Tsuruhara, T., M.L. Dufau, S. Cigorraga and K.J. Catt. J. Biol. Chem. 252(1977)9002.

11. Conti, M., J.P. Harwood, M.L. Dufau and K.J. Catt. Mol. Pharmacol. 13(1977)1024.

12. Conti, M., J.P. Harwood, M.L. Dufau and K.J. Catt. J. Biol. Chem. 252(1977)8869.

13. Posner, B.O., P.A. Kelly and H.G. Freisen, Proc. Natl. Acad. Sci. USA 71(1974)1407.

14. Hauger, R.L., G. Aguilera and K.J. Catt. Nature 271(1978)176.

15. Marsh, J.M., T.M. Mills and W.J. LeMaire. Biochim. Biophys. Acta 304(1973)197.

16. Hunzicker-Dunn, M. and L. Birnbaumer. Endocrinology 99(1976)185.

17. Hunzicker-Dunn, M. and L. Birnbaumer. Endocrinology 99(1976)198.

18. Hunzicker-Dunn, M. and L. Birnbaumer. Endocrinology 99(1976)211.

19. Zor, U., S.A. Lamprecht, Z. Misulovin, Y. Koch and H.R. Lindner. Biochim. Biophys. Acta 428(1976)761.

20. Lamprecht, S.A., U. Zor, Y. Salomon, Y. Koch, K. Ahren and H.R. Lindner. J. Cyc. Nuc. Res. 3(1977)69.

21. Lindner, S.A., A. Amsterdam, Y. Salomon, A. Tsafriri, A. Nimrod, S.A. Lamprecht, U. Zor and Y. Koch. J. Reprod. Fert. 51 (1977)215.

22. Sala, G., M.L. Dufau and K.J. Catt. Clin. Res. 26(1978)312A.

23. Cigorraga, S., M.L. Dufau and K.J. Catt. J. Biol. Chem. 25 (1978)4297.

24. Rajaniemi, H.J., A.H. Hirshfield and A.R. Midgley, Endocrinology 95(1974)579.

25. Han, S.S., H.J. Rajaniemi, M.I. Cho, A.K. Hirshfield and A.R. Midgley. Endocrinology 95(1974)589.

26. Chen, T.T., H.J. Abel, M.C. McLellan, H.R. Sawyer, M.A. Dickman and G.D. Niswender. Cytobiologie 14(1977)412.

27. Conn, P.M., M. Conti, J.P. Harwood, M.L. Dufau and K.J. Catt. Nature (1978) in press.

28. Kledzik, G.S., L. Cusan, C. Auclair, P.A. Kelly and F. Labrie. Fertil. Steril. 29(1978)560.

29. Condon, W.A. and D.L. Black. Biol. Reprod. 17(11977)514.

30. Godkin, J.D., D.L. Black and R.T. Duby. Biol. Reprod. 17(1977) 514.

31. Auclair, C., P.A. Kelly, D.H. Coy, A.V. Schally and F. Labrie. Endocrinology 101(1977)1890.

32. Dufau, M.L., A.J. Hsueh, S. Cigorraga, A.J. Baukal and K.J. Catt. Int. J. Androl. Suppl. 2(1978)193.

33. Tsuruhara, T., M.L. Dufau, S. Cigorraga and K.J. Catt, J. Biol. Chem. 252(1977)9002.

34. Stouffer, R.L., W.E. Nixon and G.D. Hodgen. Biol. Reprod. 17 (1978)858.

35. Behrman, H.R. and B.V. Caldwell. in MTP International Review of Science, Vol. 8. Reproductive Physiology, ed. R.O. Greep (Butterworths, London, 1974) p. 63.

36. Pharris, B.B. and L.J. Wyngarden. Proc. Soc. Exp. Biol. Med. 130(1969)92.

37. Hichens, M., D.L. Grinwich and H.R. Behrman. Prostaglandins 7 (1974)449.

38. Grinwich, D.L., M. Hichens and H.R. Behrman. Biol. Reprod. 14 (1976)212.

39. Stormshak, F. and L.E. Casida. Endocrinology 75(1964)321.

40. Rothchild, I., Acta Endocrinol. 49(1965)107.

# INTERACTIONS AMONG ESTROGEN, PROLACTIN AND LUTEINIZING HORMONE AT THE LEVEL OF ADENYLYL CYCLASE IN THE CORPUS LUTEUM: FINDINGS AND PHYSIOLOGICAL CORRELATES

Sharon Day, Joel Abramowitz, Mary Hunzicker-Dunn* and Lutz Birnbaumer
Department of Cell Biology, Baylor College of Medicine, Houston, Texas 77030, and *Department of Biochemistry, Northwestern University Medical School, Chicago, Illinois 60611

LH-stimulable adenylyl cyclase activity has been observed to fluctuate in rat and rabbit corpora lutea (CL) of pregnancy or pseudopregnancy (1, 2) in a manner that parallels progesterone output by these tissues (3, 4). However, the physiological significance of some of these fluctuations remains elusive, since LH is not the primary luteotropin throughout all stages of pregnancy or pseudopregnancy in rats or rabbits (5-9). LH has been shown to affect luteal steroidogenesis acutely (10, 11). It is thought to do so by interacting with a membrane bound receptor which when coupled to the adenylyl cyclase system results in elevation of intracellular cAMP and increased progesterone output (12). Other hormones such as estrogen and prolactin also modify progesterone output by the corpus luteum but apparently do so by some means other than changing intracellular levels of cAMP.

Prolactin (PRL) is the hormone that "rescues" the rat corpus luteum of the cycle transforming it into the corpus luteum of pregnancy or pseudopregnancy (13). Prolactin has been shown to be required to maintain pregnancy or pseudopregnancy from day 1 through day 7 (14). *In vitro* studies suggest that prolactin alone will not stimulate luteal steroidogenesis (10) and its mode of action is obscure. It may induce LH/hCG receptors (15) and/or maintain LH/hCG receptors (16). It has also been shown to increase the levels of enzymes involved in the metabolism of cholesterol esters (17) and to prevent the induction of enzymes that catabolize progesterone (18, 19). It may act in other ways as well, since increased steroidogenesis can occur in the absence of measurable changes in LH/hCG receptor levels (20).

Estrogen is the primary luteotropin in the rabbit. In the absence of estrogen-supplying follicles, LH is unable to maintain progesterone output in vivo (9), and estrogen alone will maintain pregnancy in the hypophysectomized rabbit (8). This effect of estrogen is not unique to the rabbit, as indicated by the fact that estradiol alone will support steroidogenesis by the rat CL between days 9 and 12 of pregnancy (7, 16).

Still another aspect of corpus luteum regulation comes to light from studies performed in our laboratory. In the pseudopregnant rabbit maintained on exogenous estrogen, a lytic dose of hCG results in desensitization of the LH-responsive adenylyl cyclase system. Furthermore, the rate of desensitization of the rabbit CL cyclase varies with the stage of pregnancy. We have also shown that the rat CL of the cycle can be protected against desensitization on proestrus if estrogen is administered on diestrus 1, although the CL of pregnancy can be desensitized even during the period when estrogen is thought to be the primary luteotropin (21). These studies show that the ability to modify adenylyl cyclase activity changes in luteal tissue with varying hormonal states.

It appears, therefore, that fine regulation of the CL involves both steroid and protein hormones. How these hormones interact with the cyclase system and how regulation of desensitization of the cyclase system affects luteal steroidogenesis is unknown. The following questions come to mind: 1) Is a LH-stimulable adenylyl cyclase necessary for normal protesterone synthesis and secretion by the CL? 2) Since hormones other than LH affect steroidogenesis, do they modify the hormonal stimulability of the adenylyl cyclase system? 3) If adenylyl cyclase can be protected against desensitization, what hormones affect this protection? and 4) Of what physiological significance is this protection? The last two questions are of particular interest because the primate CL of pregnancy becomes refractory to LH stimulation after the chorionic gonadotropin peak, but recovers responsiveness at term, producing progesterone during lactation.

To gain further information on the interplay between the LH-stimulable adenylyl cyclase system and luteotrophic factors such as estrogen and prolactin, we now explored various conditions in both rats and rabbits in which the natural balance between these parameters was altered. Of special interest were situations arising from exposure of animals to high ("lytic") doses of hCG or to endogenous ovulatory (also "lytic") surges of LH, but which resulted in an apparent maintenance of functional CL. We shall discuss two such instances: a) treatment with hCG of the estrogen supported rabbit CL of pseudopregnancy; and b) exposure of the cycle CL in the estrogen treated rat to LH. In each case, we monitored the state of the adenylyl cyclase system, the physiological role of which we wished to explore, and poised it against tissue (CL) and serum progesterone levels, which we used as indicators of the state of activity of the CL under study.

## I. Experiments with Rabbits

### (a) Effect of Estrogen on Corpus Luteum Adenylyl Cyclase of Normal and hCG Treated rabbits.

Initial experiments (21) had shown that hCG treatment of pseudopregnant rabbits results in CL with a desensitized adenylyl cyclase system and that this desensitizing action could not be prevented by treatment with estrogen. Yet, estrogen is known to be a primary luteotrophic agent in this species. It became of interest therefore, to characterize in more detail the effects of hCG and estrogen both on CL steroid production and on CL adenylyl cyclase.

The treatment schedule to which rabbits (New Zealand Whites) were exposed is shown on Figure 1. Two doses of estradiol were used: 1.5 ug twice daily (E-3) and 15 ug twice daily (E-30). These doses were chosen because: a) E-3 is known to maintain rabbit CL in a functionally active state both after removal of the hypophysis (8) and after removal of estrogen-producing follicles by irradiation (9), and b) E-30 is known to prevent morphological regression of rabbit CL after a lytic dose of LH as assessed by weight maintenance (23). Oil was administered to control rabbits.

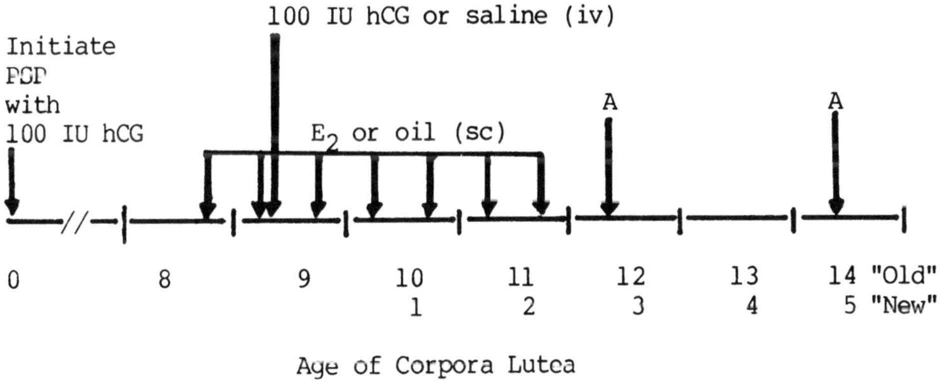


Figure 1. Treatment schedule for investigation of the effect of estrogen and its withdrawal on corpus luteum function and adenylyl cyclase activity in the presence and absence of desensitization. One group of 6 rabbits (3.5-4.5 Kg, having littered at least once) received a physiological dose of estradiol (1.5 ug/0.1 ml oil/12 h); another received a pharmacological dose (15.0 ug/0.1 ml oil/12 h); control animals received oil. Half of the animals in each group received a desensitizing dose of hCG (100 IU/0.5 ml saline iv).

Luteolytic doses of hCG (100 IU per 3.5-4.5 Kg rabbit, iv) not only affect CL but also induce ovulation, formation of new CL, and a transient reduction in intra-ovarian estrogen levels. To prevent existing CL from undergoing a period of estrogen withdrawal due to cessation of follicular steroidogenesis, estradiol treatments were started 16 h prior to hCG administration.

Thus the treatment schedule shown on Figure 1 afforded an exploration of the effect of hCG-induced desensitization in CL that had not been subjected to estradiol deficiency. Two experiments were performed. In experiment I, animals were sacrificed on day 12 of pseudopregnancy; i.e., 3 days after hCG injection and 3.5 days after initiation of estrogen treatment. In experiment II we explored the effect of discontinuing estrogen treatment on the evening of day 11 as seen on day 14. Daily bleedings were taken from the ear vein until the animals were sacrificed, at which time the LH- and isoproterenol (Iso)-stimulability of the adenylyl cyclase system was measured both in the old (original) CL and in new CL. Serum and luteal tissue progesterone levels were also determined using a radioimmunoassay based on antiserum # 337 provided by Dr. G. Niswender.

After hCG, even though ovarian estrogen levels were maintained with exogenous estradiol, serum progesterone levels fell within 24 h (Fig. 2). Neither physiological (E-3) nor pharmacological (E-30) doses of estradiol prevented functional luteolysis. As reported by others, estradiol alone did not alter serum progesterone levels (24).

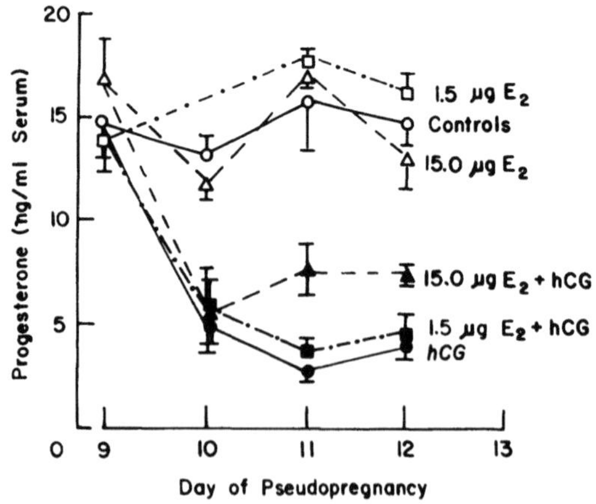


Figure 2. Progesterone content of rabbit serum as measured by RIA. Serum was collected from the marginal ear vein. Rabbits were treated as shown in Fig. 1. For days 9 and 12, n = 6, for days 10 and 11, n = 3. Values represent means ± standard errors.

From Table 1, it can be seen that hCG desensitized the hormonally responsive adenylyl cyclase system and that physiological doses of estradiol did not protect the CL from desensitization. Surprisingly, pharmacological doses of estradiol resulted in an inactivation of LH-reponsive adenylyl activity without affecting Iso-responsive activity. Since estradiol administration has been shown to elevate endogenous LH, and under certain conditions, to result in ovulation (26, 27, 28), the selective loss of LH-stimulable adenylyl cyclase activity could be due to endogenous LH. However, no new CL were observed in any of the animals subjected to E-30 only. The effect of E-30 on the adenylyl cyclase system was seen within 3 days of its administration and desensitizing doses of hCG had no further effect on the E-30 treated CL cyclase. In fact E-30 and hCG treated rabbits had CL with better LH-stimulable adenylyl cyclase than rabbits treated with hCG only. This indicated that even though E-30 treatment was deleterous *per se*, it affected protection against the hCG-induced desensitizing response. CL weight was maintained after hCG only in E-30 treated rabbits.

Since maintenance of physiological levels of estrogen (E-3 treatments) did not prevent functional luteolysis after hCG, as seen by the fact that serum progesterone levels fell, we conclude that conditions leading to desensitization of the LH-sensitive adenylyl cyclase system also lead to functional luteolysis by a mechanism unrelated to interruption of estrogen supply to the CL.

Table 1 Effect of Estrogen Treatment on Desensitization of LH-stimulable Adenylyl Cyclase Activity by hCG in Day 12 Pseudopregnant Rabbits (See Fig. 1 for Treatment Schedule).

| Treatments | CL Wt | Adenylyl Cyclase Activity in the Presence of | | |
|---|---|---|---|---|
| | | ----- | LH | Iso |
| | (mg) | (pmoles cAMP/min/mg protein) | | |
| Control | 20 ± 1¶ | 11.4 ± 1.4 | 74.0 ± 8.4¶ | 44.4 ± 5.8¶ |
| hCG | 9 ± 1* | 7.4 ± 0.8 | 10.5 ± 0.5* | 9.8 ± 3.2* |
| E-3 | 19 ± 2¶ | 10.8 ± 1.4 | 49.2 ± 8.6§# | 36.9 ± 3.5¶ |
| E-3 & hCG | 8 ± 1* | 5.9 ± 0.7 | 6.8 ± 0.1* | 8.3 ± 0.3* |
| E-30 | 24 ± 3$ | 8.4 ± 0.2 | 18.7 ± 4.0$ | 33.0 ± 0.9¶ |
| E-30 & hCG | 16 ± 2* | 11.1 ± 0.6 | 20.4 ± 1.6$ | 36.9 ± 1.9§ |

¶ vs *: $p < 0.01$; # vs *: $p < 0.01$; $ vs *: $p < 0.01$; ¶ vs $: $p < 0.01$; ¶ vs §: N.S. Values represent means ± standard error.

An additional puzzling finding appears to emerge from an initial analysis of tissue *vs* serum levels of progesterone. Thus, although further studies, especially with larger numbers of animals are necessary, we noticed that in rabbits treated with E-30 plus hCG, tissue levels of progesterone were significantly elevated over those found in CL of animals treated with hCG only (Table 2). However, inspite of this 10-fold difference in tissue levels, serum levels of progesterone did not differ, suggesting that estradiol while having a clear stimulatory effect on progesterone synthesis, may have also an inhibitory effect on progesterone secretion by the CL. We are currently exploring this phenomenon further.

Table 2 Effect of Estradiol and hCG on Tissue and Serum Progesterone in the Day 12 Pseudogrenant Rabbit (See Fig. 1 for Treatment Schedule).

| Treatments | n | Tissue P | Serum P |
|---|---|---|---|
| | | (pg/ug protein) | (ng/ml) |
| Control | 6 | 135 ± 15¶ | 14.9 ± 1.4¶ |
| hCG | 6 | 12 ± 1* | 4.0 ± 0.7* |
| E-3 | 6 | 172 ± 36¶ | 16.2 ± 1.1¶ |
| E-3 & hCG | 6 | 26 ± 13* | 4.4 ± 0.2* |
| E-30 | 6 | 184 ± 17¶ | 13.3 ± 1.7¶ |
| E-30 & hCG | 6 | 123 ± 10¶ | 6.9 ± 0.5* |

¶ vs *: $p < 0.01$; ¶ vs $: $p < 0.01$. Values represent means ± standard error.

(b) Effect of Withdrawal of Exogenous Estrogen on Luteal Steroidogenesis and Adenylyl Cyclase Activity.

Since we had found that pharmacologic doses of estrogen (E-30), but not physiological ones, resulted in an inactivation of LH-stimulable adenylyl cyclase in CL, we tested whether this effect was reversible by interrupting the estrogen treatment. Thus, rabbits were treated as in the previous study, but were subjected to a 2-day withdrawal by delaying the sacrifice until day 14 at which time

Table 3 The Effect of a 2-day Estrogen Withdrawal on Corpus Luteum Function in the Day 14 Pseudopregnant Rabbit (See Fig. 1).

| | Tissue P | | Serum P |
|---|---|---|---|
| Treatments | 5-day-old | 14-day-old | day 14 |
| | (pg/ug protein) | | (ng/ml) |
| Control | ----- | 140.4 ± 28.0¶ | 13.7 ± 3.3¶ |
| hCG | 198 ± 33 | 39.4 ± 7.0* | 3.6 ± 0.1* |
| E-3 | ----- | 151.8 ± 17.0¶ | 15.5 ± 1.2¶ |
| E-3 & hCG | 238 ± 21 | 55.5 ± 5.4* | 8.7 ± 0.9§ |
| E-30 | ----- | 23.4 ± 7.6* | 2.4 ± 1.2* |
| E-30 & hCG | 163 ± 18 | 43.6 ± 8.1* | 4.8 ± 1.4* |

¶ vs *: $p < 0.01$; ¶ vs §: not significant at 0.01 level (means ± SEM).

both the state of the adenylyl cyclase system and the levels of progesterone in serum and CL were determined. As seen in Table 3, withdrawal resulted in a precipitous decline of serum progesterone in rabbits exposed to E-30 treatment but not in rabbits exposed to E-3 treatment. Progesterone secretion was interrupted in E-30 treated CL in spite of maintenance of CL weight. Thus, withdrawal resulted in "functional luteolysis." Holt et al. (24) have reported a similar, rapid decline in serum progesterone after withdrawal of implants of estradiol. Tissue progesterone levels also fell after E-30 withdrawal and the adenylyl cyclase system remained unresponsive to LH (Tables 3, 4). These findings suggest that E-30 maintains CL artifically, since estrogen withdrawal from CL with hormonally stimulable adenylyl cyclase (E-3 treatment) affected neither progesterone output nor the responsiveness of the adenylyl cyclase system to LH or isoproterenol.

Table 4 The Effect of a 2-day Estrogen Withdrawal on Corpus Luteum Adenylyl Cyclase Activity in the Day 14 Pseudopregnant Rabbit (See Fig. 1 for Treatment Schedule).

| Treatments | CL Wt | Adenylyl Cyclase Activity in the Presence of ----- | LH | Iso |
|---|---|---|---|---|
| | (mg) | (pmoles cAMP/min/mg protein) | | |
| Control | 18 ± 2¶ | 11.6 ± 1.9 | 55.1 ± 3.7¶ | 35.5 ± 5.3¶ |
| hCG | 7 ± 1* | 12.7 ± 1.4 | 16.7 ± 2.0* | 19.1 ± 1.2* |
| E-3 | 21 ± 3¶ | 12.7 ± 0.1 | 74.7 ± 5.7¶ | 44.5 ± 6.2¶ |
| E-3 & hCG | 6 ± 1* | 11.3 ± 2.6 | 18.0 ± 5.3* | 19.6 ± 5.1* |
| E-30 | 18 ± 1¶ | 10.5 ± 1.2 | 19.5 ± 1.9* | 27.1 ± 3.6§ |
| E-30 & hCG | 18 ± 2¶ | 8.3 ± 0.5 | 21.2 ± 3.7* | 26.0 ± 0.2§ |

¶ vs *: $p < 0.01$; ¶ vs §: not significantly different at 0.01 level
Values represent means ± standard error.

It has been shown in rabbits (21) as well as in rats (22) that the luteolytic effect of hCG is irreversible in these species, i.e., that neither the adenylyl cyclase system recovers LH-stimulability nor the CL recovers the capacity to synthesize and secrete progesterone. Data presented here showing no return of LH-responsiveness (Table 4) or progesterone content (Table 3) as late as 5 days after hCG, are consistent with the irreversibility of its effect. This is not to say that serum progesterone levels will also remain low in animals treated with hCG, for new CL are formed as the old ones regress. Indeed, 5 days after hCG (Table 3) serum progesterone has clearly started to increase, as has the progesterone content in newly formed CL. Miller and Keyes (25) have shown that as early as 2 days after their formation, CL are capable of producing enough progesterone to significantly elevate blood levels of this steroid.

In summary, our studies in the rabbit have shown that once the hormonally stimulable adenylyl cyclase systen is desensitized in CL, their steroidogenic capacity is impared unless pharmacologic doses of estrogen are administered. This effect of hCG is not reversible. High doses of estradiol alone may alter the membrane bound adenylyl cyclase system through an as yet unknown mechanism. Once estradiol is withdrawn, luteal steroidogenesis collapses as though hCG had been administered. We conclude that both intraovarian estradiol and a hormonally responsive adenylyl cylcase are necessary for continued luteal steroid synthesis and secretion.

## II. Experiments With Rats

### (a) The Protective Effect of Prolactin in the Presence of Estradiol on the Corpus Luteum of the Rat Cycle

In the 4-day cycling rat, Krey and Everett (29) noted that an estradiol benzoate injection on diestrus 1 resulted in advancement of ovulation by one day followed by a pseudopregnancy characterized by "two sets" of CL. Their morphology suggested that both sets were functional. If so, then one set of CL (old set) had withstood the lytic effects of the ovulatory LH surge. Our laboratory has shown that while the cyclase system in CL of the cycle has normally lost all of its responsiveness to LH by the morning of proestrus, it retains and even increases its LH responsiveness if the rat is treated with estrogen on dietrus 1. The mechanism by which estrogen treatment results in protection of the cyclase system in CL and maintenance of its morphological integrity is not known. Since estrogen treatment is known to result in increased serum prolactin (30, 31), we investigated whether or not the effect of estrogen on CL is dependent on prolactin and whether the old CL that result after advancement of ovulation with estrogen treatment retain not only an LH-responsive adenylyl cyclase but also function, i.e. secrete progesterone. The possible involvement of prolactin was tested by first blocking prolactin release with CB-154 (2-Br-ergocryptine; 1 mg/0.2 ml sesame oil/12 h starting at 1000 h of diestrus 1) and then testing for reversal of the effect(s) of CB-154 by replacement therapy with exogenous ovine prolactin (200 ug/0.2 ml saline/12 h starting at 1000 h of diestrus 1). Advancement of ovulation due to estrogen (estradiol benzoate 20 ug/0.1 ml sesame oil at 1230 h of diestrus 1) was checked by counting ova. Functionality of the old CL was determined by measuring serum and tissue progesterone in old CL on the morning of proestrus, which due to ovulation advancement is the equivalent of estrus, a time at which new CL are neither synthesizing nor secreting progesterone. Effects on the basal, LH- and Iso-stimulable adenylyl cyclase system were determined as described by Birnbaumer *et al*. (32) in homogenates of old CL dissected under a dissecting microscope.

In homogenates of CL from control animals, little response to LH (1.5-fold increases over basal, Table 5) is seen when adenylyl cyclase is measured. This confirms our previous findings (21). Progesterone output on proestrus is low by our measurements also (Table 6). However, if estrogen is given both LH-stimulability of adenylyl cyclase in old CL and progesterone levels in serum are elevated (14:10 lighting schedule). Newly formed CL, however, are unresponsive with regard to LH-stimulable adenylyl cyclase activity and their progesterone levels are reduced compared to levels of progesterone found in the old CL (data not shown). Administration of 1 mg/12 h of CB-154, a dose known to inhibit prolactin secretion by the pituitary (13), did not effect the advancement of ovulation due to estrogen treatment, but blocked the maintenance of LH-stimulable adenylyl cyclase in old CL and resulted in reduced serum levels of progesterone, when compared to levels found in animals given estrogen alone. Treatment with prolactin resulted in elevated serum progesterone and in increased LH-stimulable adenylyl cyclase activity both in control animals and in estrogen plus CB-154 treated animals.

Table 5 The Protective Effect of Estrogen in the Presence of PRL on the LH-stimulable Adenylyl Cyclase System in the Rat Corpus Luteum of the Cycle.

| | | Adenylyl Cyclase Activities | |
|---|---|---|---|
| Treatments | Ovulations | LH | Iso |
| | | (-fold stimulation) | |
| Control | 0/5 | 1.5 ± 0.1¶ | 7.7 ± 0.6¶ |
| EB | 9/11 | 2.7 ± 0.2*$ | 7.5 ± 0.8¶ |
| EB & CB-154 | 7/8 | 1.3 ± 0.1§ | 9.6 ± 0.9¶ |
| CB-154 | 4/4 | 1.2 ± 0.1§ | 9.2 ± 1.1¶ |
| EB & CB-154 & PRL | 7/7 | 3.1 ± 0.6* | 8.0 ± 1.5¶ |
| PRL | 0/6 | 4.2 ± 0.5*# | 4.8 ± 0.6* |

¶ vs *: $p<0.01$; $ vs §: $p<0.01$; # vs §: $p<0.01$; # vs $: $p<0.01$; ¶ vs §: N.S. Values represent means ± standard error.

Although LH-stimulable adenylyl cyclase varied depending on the treatment, isoproterenol-stimulated cyclase remained relatively constant except in rats treated with prolactin alone, in which case the isoproterenol response was diminished to about 50% of that seen in all other groups. The constancy of isoproterenol-stimulable activity indicated that changes (loss) of LH-stimulable activity were specific for this hormone and its coupling process and not due to a general effect on the catalytic moiety of the system. The reason for the 50% decrease in isoproterenol-stimulable activity after treatment with prolactin only is not known.

However, this is also not due to a general change in the cyclase system since LH-stimulable adenylyl cyclase was actually increased under these conditions.

Table 6 The Physiological Correlates to the Protection of the LH-stimulable Adenylyl Cyclase System by Estrogen and PRL in the Rat Corpus Luteum of the Cycle.

| Treatments | n | Tissue P (pg/ug protein) | Serum P (ng/ml) |
|---|---|---|---|
| Control | 5 | 243 ± 25¶ | 5.6 ± 1.0¶ |
| EB | 13 | 609 ± 65*$ | 28.5 ± 4.3*$ |
| EB & CB-154 | 9 | 256 ± 50§ | 6.1 ± 1.4§ |
| CB-154 | 4 | 257 ± 66§ | 9.3 ± 2.2§ |
| EB & CB-154 & PRL | 7 | 496 ± 48* | 36.6 ± 4.3*$ |
| PRL | 6 | 448 ± 53* | 58.1 ± 5.5*# |

¶ vs *: $p < 0.01$; $ vs §: $p < 0.01$; # vs §: $p < 0.01$; # vs $: $p < 0.01$; ¶ vs §: N.S. Values represent means ± standard error.

These data demonstrate that prolactin in the presence of estradiol "rescues" the CL functionally and prevents desensitization by an endogenous LH surge. As in the rabbit, LH-stimulability of the adenylyl cyclase system correlates positively with increased progesterone production. The fact that prolactin administration resulted in increased LH-stimulable adenylyl cyclase activity and in increased progesterone output without effecting serum LH (13) lends further support to our hypothesis that LH-stimulable adenylyl cyclase is a necessary component for regulation of luteal steroidogenesis in the rat.

(b) Time Dependency of the "Rescue" of the Corpus Luteum by Prolactin.

Several groups (13, 33, 34) have shown that the CL of the rat cycle can be transformed into a CL of pregnancy or pseudopregnancy by prolactin, if administered at an early enough time. In addition to its trophic actions, prolactin has been shown to have lytic effects if administered at later times (35, 36). Using LH-stimulable adenylyl cyclase and serum progesterone as markers, we investigated how late prolactin treatment of the cycle CL could be initiated before it became no longer susceptible to "rescue". To this end, the procedure shown in Fig. 3 was followed. Prolactin treatment (200 ug/0.2 ml saline, sc), started as late as 1000 h on diestrus 2 (lighting 12:12), resulted in "rescue" of the CL of the cycle as evidenced by the elevated serum progesterone in 6 out of 8 animals (Fig. 4) and by elevated LH-responsiveness, also in

6 out of 8 animals. Interestingly, the longer the CL were exposed to prolactin the lower was their isoproterenol-stimulable adenylyl cyclase activity. By noon on diestrus 2, prolactin could no longer rescue the CL.

Previous studies have shown that in the absence of prolactin, adenylyl cyclase in CL of the cycle never achieves better than a 2-fold stimulation by LH (21); this occurs on diestrus 1 when serum progesterone is highest. Now we show that if prolactin is administered, a 4-fold stimulation is seen on proestrus. Thus, prolactin appears to be inducing rather than maintaining activity. This induction appears to take place, at least in part, within 24 h. What component of the LH-stimulable cyclase system is induced, is not known. Since prolactin induces LH-receptors in granulosa cells (15), it may be that this is one of the components induced in luteal cells upon "rescue". On the other hand, increased stimulability may also come about by induction of other

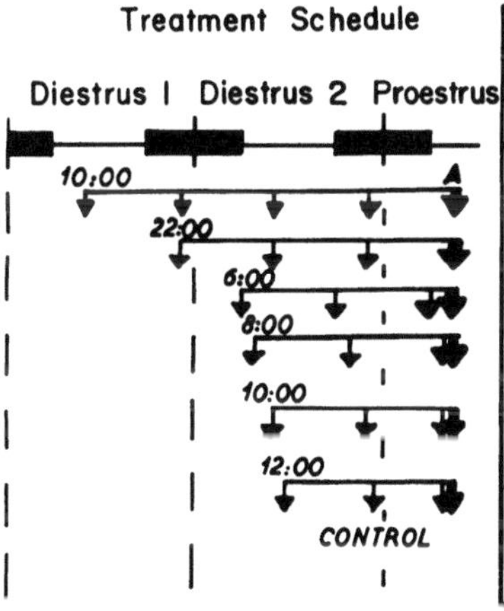


| Relative Stimulation of A.C. by: LH | Iso | n |
|---|---|---|
| 4.2 ± 0.5 | 4.8 ± 0.6 | (4) |
| 4.2 ± 0.6 | 5.8 ± 0.5 | (3) |
| 3.5 ± 0.2 | 6.0 ± 0.2 | (5) |
| 4.4 ± 0.6 | 7.7 ± 0.8 | (9) |
| 0.9 | 3.7 | (1) |
| 3.7 ± 0.2 | 8.0 ± 0.7 | (6) |
| 1.7 ± 0.1 | 8.2 ± 0.5 | (2) |
| 2.1 ± 0.2 | 8.7 ± 1.8 | (5) |
| 1.5 ± 0.1 | 7.7 ± 0.6 | (5) |

Figure 3. Treatment schedule for and effect on hormonal responsiveness of adenylyl cyclase activity (stimulated activity relative to basal activity) in 4-day cycling rats (Charles River, CD, outbred) subjected to prolactin treatment (200 ug/0.2 ml saline/12 h, sc) at successively later times in the cycle. Rats were sacrificeed at 0900 h on proestrus. Adenylyl cyclase activity was measured in homogenates of the CL of the cycle in the absence (basal) and presence of 10 ug/ml LH or $10^{-4}$ M isoproterenol , and hormonal responsiveness was calculated. The number of animals in each group is given in parentheses. Values represent means ± standard error.

components of the system, especially of elusive coupling factors related to nucleotide regulation of receptor and/or catalytic activity of the system. Binding studies are currently under way to clarify these aspects of prolactin action on CL.

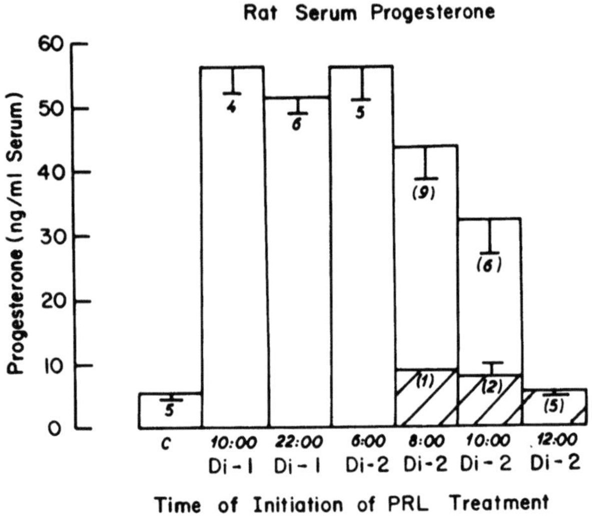


Figure 4. Serum progesterone levels on proestrus in control and prolactin treated rats. Hatched bars represent those animals in which LH responsiveness was not "rescued" by the treatment shown in Fig. 3. Values represent means ± standard error.

## Conclusions

Work in both rats and rabbits has shown that maintenance of an LH-stimulable adenylyl cyclase system is required for properly regulated progesterone output by the CL. In the rat, protection of the CL from LH-induced desensitization can be affected by prolactin in the presence of estradiol during the cycle; and prolactin alone enhances LH-stimulable adenylyl cyclase activity. In the rabbit, high estrogen levels suppress LH-stimulable adenylyl cyclase. This suppression may be due to a variety of factors such as altered prolactin secretion (37, 38), endogenous LH surge(s) (26, 27), or even a stress effect resulting in ACTH and glucocorticoid secretion. Our studies emphasize that the fine tuning of CL function involves an intricate interplay between LH, estrogen and prolactin, and that this interplay reflects itself in the properties of the hormone-responsive adenylyl cyclase system.

SUMMARY

Previously, we have shown that the luteolytic effect of LH is preceeded by desensitization of the adenylyl cyclase system (AC). In this chapter, we will present our current findings on the effects of estradiol (E) on uncoupling of the LH responsive AC in CL of pseudopregnant (PSP) rabbits and on the possible protection of rat CL of the cycle against the luteolytic and desensitizing action of LH by E and prolactin (PRL).

PSP rabbits treated with high E (15 ug E/12 h, 3.5 days) but not with low E (1.5 ug E/12 h, 3.5 days) lost LH-stimulable AC partially and selectively (no loss of isoproterenol [Iso] -stimulable AC) but maintained normal serum and luteal progesterone (P) levels. High E treatement did not protect against hCG-induced luteolysis but did block desensitization of AC beyond that caused by high E treatment alone. Although hCG plus high E treated PSP rabbits had low serum P their CL had normal P levels. Thus, a dissociation between steroid secretion and CL steroid levels was observed. Withdrawal of E after 3.5 days of treatment with low E had no detectable effects on AC activities or P parameters. Withdrawal after high E treatment resulted in functional luteolysis (low serum P), however both structure (2 days later) and Iso-stimulable AC activity remained. Lh-responsiveness did not return after withdrawal of high E. These studies show that high doses of E affect CL adversely even though this is not apparent until AC is measured or until E is withdrawn.

Cycling rats were treated with estradiol benzoate (EB, 20 ug) on D-1 to advance ovulation 1 day. This resulted in 4-day old CL that had been exposed to an ovulatory (luteolytic) level of LH and in newly induced 1-day old CL on the day of expected proestrus. The original CL were found to contain high levels of P and to have active LH- and Iso-stimulable AC activities while the 1-day old CL contained less P; LH-stimulability of their AC was low. Serum P was also elevated. Treatment of cycling rats with CB-154 and EB, still advanced ovulation, but old CL had low P content, low LH-stimulability and serum P was low as well. Iso-stimulable AC was as high as in active 4-day old CL. Replacement therapy with PRL resulted in "rescue" of CL (serum and luteal P were elevated) and maintenance of LH-stimulable AC. These studies suggest that both CL function and LH-stimulable AC may be protected against a desensitizing dose of LH by a combination of PRL and EB. An alternate possibility is that these CL are not yet sensitive to the luteolytic effects of LH. If so, they would be similar to 3-day old CL of PSP that are not yet sensitive to $PGF_{2\alpha}$.

Finally, we show that the "critical" periods for rescuing CL of the cycle, initiating PSP, and maintaining LH-stimulable AC with PRL coincide and that LH-stimulable AC activity and P output correlate.

Acknowledgements: This work was supported in part by grants from the National Institutes of Health (HD-09581 and HD-07495) and by the Clayton Foundation. CB-154 was generously provided by Sandoz Pharmaceuticals; hCG was a gift from Ayerst.

## REFERENCES

(1) M. Hunzicker-Dunn and L. Birnbaumer. Endocrinology 99(1976) 198.
(2) M. Hunzicker-Dunn and L. Birnbaumer. Endocrinology 99(1976) 185.
(3) W.K. Morishige, G.J. Pepe, and I. Rothchild. Endocrinology 92(1972)1527.
(4) J.R.G. Challis, I.J. Davies, and K.J. Ryan. Endocrinology 93 (1973)971.
(5) M. Takayama and G.S. Greenwald. Endocrinology 92(1973)1405.
(6) G. Gibori, E. Antczak, and I. Rothchild. Endocrinology 100 (1977)1483.
(7) G. Gibori, R. Rodway, and I. Rothchild. Endocrinology 101 (1977)1683.
(8) P. Rennie. Endocrinology 83(1968)323.
(9) P.L. Keyes and A.V. Nalbandov. Endocrinology 80(1967)938.
(10) D.H. Wu, W.G. Wiest, and A.C. Enders. Endocrinology 98(1967) 1378.
(11) H.G. Spies, J. Hilliard, and C.H. Sawyer. Endocrinology 83 (1968)354.
(12) J. Marsh, in: Advances in Cyclic Nucleotide Research, Vol. 6, eds. P. Greengard and G.A. Robison (Raven Press, New York, 1975) p. 137.
(13) M.S. Smith, B.K. McLean, and J.D. Neill. Endocrinology 98 (1976)1370.
(14) W.K. Morishige and I. Rothchild. Endocrinology 95(1974) 260.
(15) J.S. Richards and J.J. Williams. Endocrinology 99(1976) 1370.
(16) G. Gibori and J.S. Richards. Endocrinology 102(1978)767.
(17) H.R. Behrman, G.P. Orczyk, G.J. MacDonald, and R.O. Greep. Endocrinology 87(1970)1251.
(18) W.G. Wiest and W.R. Kidwell, in: The Gonads, ed. K.W. McKerns (Appleton, New York, 1969) p. 295.
(19) A. Zmigrod, H.R. Lindner, and S.A. Lamprecht. Acta Endocrinology (1972) 141.
(20) S.G. Hillier, A.J. Zelesnik, and G.T. Ross. Endocrinology 102(1978)937.
(21) M. Hunzicker-Dunn and L. Birnbaumer. Endocrinology 99(1976) 211.
(22) L. Birnaumer, S. Day, M. Hunzicker-Dunn, and J. Abramowitz, in: Ontogeny of Receptors and Mode of Action of Reproductive Hormones, eds. T.H. Hamilton, J.L. Clark, and W. Sadler

(Raven Press, New York, 1978) (in press).
(23) F. Stormshak and L.E. Casida. Endocrinology 77(1965)337.
(24) J.A. Holt, P.L. Keyes, J.M. Brown, and J.B. Miller. Endocrinology 97(1975)76.
(25) J.B. Miller and P.L. Keyes. Endocrinology 102(1978)31.
(26) S. Kanematsu, R.J. Scaramuzzi, J. Hilliard, and C.H. Sawyer. Endocrinology 95(1974)247.
(27) L. Dufy-Barbe, B. Dufy, and J.D. Vincent. Biol. Reprod. 18 (1978)118.
(28) C.H. Sawyer, J.W. Everett, and J.E. Markee. Proc. Soc. Exptl. Biol. Med. 74(1950)185.
(29) L.C. Krey and J.W. Everett. Endocrinology 93(1973)377.
(30) C.L. Chen and J. Meites. Endocrinology 86(1970)503.
(31) K. Ajika, L. Kurlich, C.P. Fawcett, S.M. McCann. Neuroendocrinology 9(1972)304.
(32) L. Birnbaumer, P.-C. Yang, M. Hunzicker-Dunn, J. Bockaert, and J.M. Duran. Endocrinology 99(1976)163.
(33) K.D. Dohler and W. Wuttke. Endocrinology 94(1974)1595.
(34) B.K. McLean and M.B. Nikitovitch-Winer. Endocrinology 93 (1973)316.
(35) P.V. Malven. Endocrinology 84(1969)1224.
(36) E. Billeter and E. Flickinger. Experientia 27(1971)464.
(37) S. Kanematsu, J. Hilliard, and C.H. Sawyer. Acta Endocrinol. 44(1963)467.
(38) S. Kanematsu and C.H. Sawyer. Endocrinology 72(1963)243.

# DESENSITIZATION AND RECOVERY OF $PGE_2$-RESPONSIVE OVARIAN ADENYLATE CYCLASE MODULATED BY AGENTS AFFECTING CYTOSKELETAL FUNCTION

U. Zor, Berta Strulovici, S.A. Lamprecht, A. Oplatka* and H.R. Lindner
Departments of Hormone Research and Polymer Research*

The Weizmann Institute of Science, Rehovot, Israel

There is reason to believe that hormone receptors are able to move freely within the fluid plasma membrane (1-3) and that such mobility is essential for effective coupling of the receptor to adenylate cyclase (4,5). The elements of the cytoskeleton seem able to influence this process (6).

Continued exposure of Graafian follicles to prostaglandin $E_2$ ($PGE_2$) results in the following sequence of events: (i) stimulation of cyclic AMP production, evident within seconds (7); (ii) progressive desensitization of the tissue to $PGE_2$, but not to other stimulants of adenylate cyclase, resulting in near-complete refractoriness within 3 h (8,9); and (iii), upon withdrawal of $PGE_2$ from the medium, a gradual recovery of responsiveness to $PGE_2$ (8,9), with complete resensitization after 4 h. We have recently shown that the first phase, viz. the response of the ovarian cyclase to $PGE_2$, is prevented by agents that impair the assembly or function of microtubules, but is unaffected by agents that disrupt the function of microfilaments (6). Here we report that the second phase, namely, desensitization to $PGE_2$, is prevented by agents believed to preserve the polymerized state of tubulin, such as $D_2O$ (10) and $Li^+$ (11), or by phalloidin, which is believed to stabilize actin filaments (12). Furthermore, the third phase, viz. recovery from $PGE_2$-induced refractoriness, is shown to be prevented by agents believed to interfere with the microfilament system, such as cytochalasin B, anti-actin serum or heavy meromyosin subfragment-1 (HMM S-1).

Three-month-old Wistar derived rats of the Departmental Colony were sacrificed between 8.00 - 13.00 h on the day of proestrus. Large intact Graafian follicles were isolated and cultured in

Eagle's medium as previously described (13). After culture with the test substance, the follicles were washed with saline and incubated with the $PGE_2$ for 30 min in Krebs-Ringer bicarbonate buffer containing 1 mg/ml glucose (KRBG) and 3-isobutyl-1-methylxanthine (IBMX; 0.1 mg/ml). The cyclic AMP content of the follicle was then assayed by a modification (7) of the Gilman method (14).

Follicles cultured for 6-20 h in a $PGE_2$-free medium responded to subsequent addition of $PGE_2$ to the medium within 30 min with a sharp rise in cyclic AMP accumulation (Figs. 1-3). Follicles cultured for 1-16 h in a medium enriched with $PGE_2$ (10 μg/ml) and subsequently challenged with fresh $PGE_2$ (10 μg/ml) as above showed a diminished response in terms of cyclic AMP accumulation (Figs. 1-3). Follicles cultured for various times (1-6 h) in a medium containing either colchicine (1 μg/ml) or anti-tubulin serum but no $PGE_2$ and subsequently incubated for 30 min with $PGE_2$ showed a gradually diminishing response. The time-course of this drug - or antiserum - induced desensitization resembles that of the development of refractoriness to $PGE_2$ (Fig. 1): after 3 h of incubation with colchicine, the $PGE_2$ effect was severely curtailed

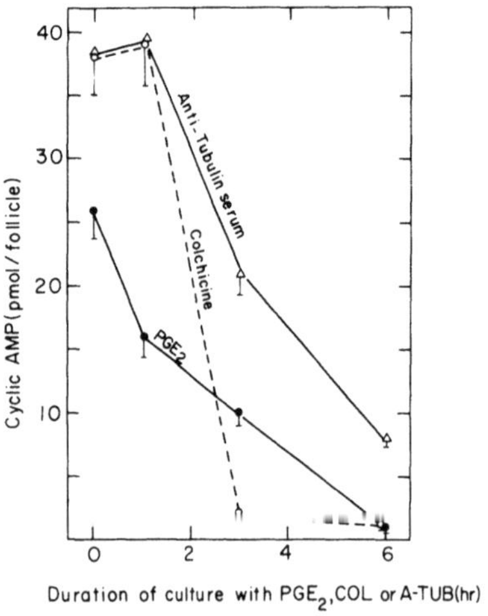


Fig. 1. Time-course of desensitization of Graafian follicles to $PGE_2$ and of inhibition of cyclic AMP response to $PGE_2$ by colchicine (COL) or anti-tubulin serum (A-TUB). $PGE_2$ (10 μg/ml), colchicine (1.0 μg/ml) or anti-tubulin serum were present in the medium during all (6 h) or none (0) of the 6 h culture period, or during the last 1-3 h as indicated. After rinsing, the follicles were incubated for 30 min in KRBG-IBMX, containing $PGE_2$ (10 μg/ml). Cyclic AMP was determined by a modification of the Gilman competitive protein binding assay. Vertical brackets, ± S.E.M. (n = 8). Arrow indicates basal cyclic AMP levels.

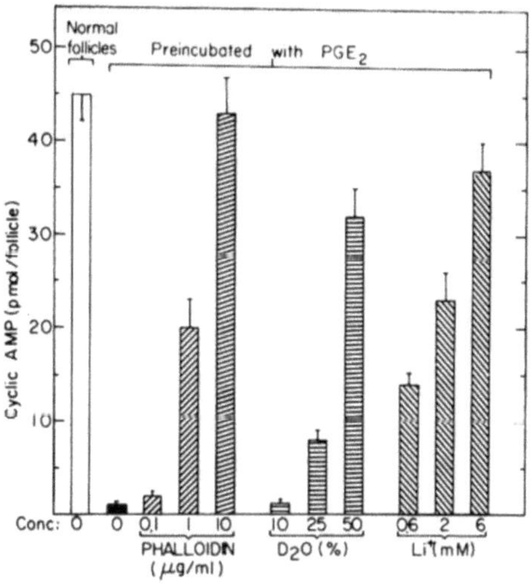


Fig. 2. Prevention by phalloidin, $D_2O$ and $Li^+$ of desensitization of cyclic AMP response to $PGE_2$ of rat Graafian follicles. The follicles were preincubated for 3 h with the drug indicated and cultured for a further 17 h with the drug together with $PGE_2$ (10 μg/ml). At the end of the culture period the follicles were rinsed, challenged with $PGE_2$ (10 μg/ml) in KRBG-IBMX and cyclic AMP accumulation was assayed as described in legend to Fig. 1 (n = 8). Normal follicles were cultured for 20 h in hormone-free medium.

(Fig. 1). However, concomitant addition of $D_2O$ (10-50%) or $Li^+$ (0.6 - 6 mM) or phalloidin (0.1 - 10 μg/ml) during culture periods of up to 20 h with $PGE_2$ prevented the hormonally induced refractoriness (Fig. 2). The degree of preservation of the response was directly related to the concentration of the stabilizer. When follicles refractory to $PGE_2$-stimulation were transferred to hormone-free medium, they fully regained responsiveness to $PGE_2$ in terms of cyclic AMP formation within 6 h (Fig. 3). This recovery of sensitivity was prevented by addition of cytochalasin B (3 μg/ml) or the myosin subfragment (80 μg/ml), or anti-actin serum to the medium. None of these three agents impaired the acute response of normal follicles to $PGE_2$. Furthermore, absorption of the anti-actin serum with exogenous actin (20 μg/ml) abolished the effect of the antiserum on restoration of the response to $PGE_2$. Anti-myosin serum was inactive in this respect (Fig. 3).

Development of refractoriness to the stimulatory action of protein hormones (8, 5-18), prostaglandins (8,9,16,18-22) and catecholamines (5,22,23) on cyclic AMP production has been shown to occur in a variety of tissues. The mechanisms responsible for hormonally induced refractoriness are poorly understood. Explan-

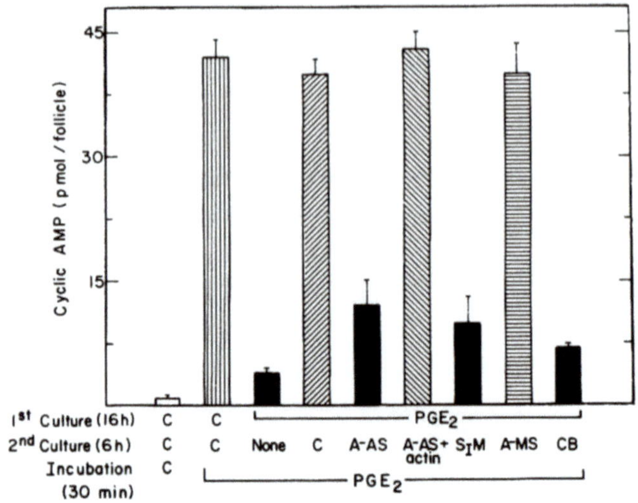


Fig. 3. Prevention of resensitization to $PGE_2$ of refractory Graafian follicles by anti-actin serum, HMM S-1 and cytochalasin B. The follicles were incubated with $PGE_2$ (10 µg/ml) for 16 h. The medium was then withdrawn, the follicles were rinsed with saline, incubated for another 6 h with hormone-free medium in the presence of the agents indicated and rinsed again. Cyclic AMP accumulation in response to $PGE_2$ (10 µg/ml) was then determined as described in legend to Fig. 1. C, control medium; None, 2nd culture period omitted; A-AS, anti-actin serum; actin (20 µg/ml); S1M, HMM S-1 (80 µg/ml); A-MS, anti-myosin serum; CB, cytochalasin B (3 µg/ml); (n = 12).

ations put forward include: (i) a decrease in the number or accessibility of receptor sites on the plasma membrane (24-26); (ii) uncoupling of the hormone-receptor complex from the catalytic moiety of adenylate cyclase (5,8,27-29); and (iii) induction of an inhibitor (30). It seems indeed that active protein synthesis is a prerequisite for induction of desensitization to PGE or LH, since actinomycin D, cycloheximide or puromycin prevent this process (8,20,31). Perhaps the inhibitor acts by masking the PGE-receptor or by uncoupling the hormone receptor complex/adenylate cyclase interaction. Alternatively, such a protein may be involved in the internalization of the occupied PGE-receptor complex as suggested by Chen *et al.* (32) and Amsterdam *et al.* (33) for the gonadotropin receptor complex, a process likely to involve the microtubule (MT) and microfilament (MF) system.

This report suggests that agents that stabilize tubulin or promote its polymerization, such as $D_2O$ (10), $Li^+$ (11) as well as stabilization of the polymerised state of actin by phalloidin (12), are able to prevent PGE-induced refractoriness, while treatments that disrupt the structure or function of microfilaments, such as cytochalasin B (34), anti-actin serum and the HMM S-1 (35), interfere with the recovery of the PGE response in the desensitized Graafian follicle.

Since we have shown earlier that the integrity of the microtubular system is essential for the response of follicular cells to $PGE_2$ (ref. 6; cf. Fig. 1), it is possible that desensitization involves a change in these elements of the cytoskeleton which is prevented by the stabilizing agents $Li^+$ and $D_2O$. The present results suggest that microfilaments, too, play a part in the desensitization process, since desensitization was prevented by phalloidin, although earlier results had shown that disruption of microfilaments by cytochalasin B does not impair the acute cyclic AMP response to $PGE_2$ (6). Thus, stabilization of one system may influence the stability of the second system, suggesting that a functional interrelationship exists between the MT and the MF systems (36). It is also possible that stabilization of the MT-MF system delays the translocation of hormone-receptor complex to the lysosomes, where receptor degradation appears to occur (32,33).

Gradual resensitization of the refractory tissues to hormonal stimuli generally occurs over a period of hours or days following hormone removal (8,9,23,25-27). This spontaneous recovery of the PGE response is prevented by treatment with agents that interfere with the MF system (Fig. 3), viz. anti-actin serum, HMM S-1 or cytochalasin B. If indeed loss of PG receptor sites leads to refractoriness, resensitization should involve newly-formed PGE receptors. The transmembrane movement of these receptors into the plasma membrane from the Golgi apparatus may require the participation of actin filaments. Likewise, the MT-MF system may influence lateral movement of the receptor within the plasma membrane, which may be essential for its effective coupling to the catalytic moicty of adenylate cyclase.

The observed effects of anti-tubulin and anti-actin sera on the intact follicle are rather surprising. It has been assumed that intact cells are largely impermeable to immunoglobulins. The possibility that the permeability of the cell membrane was altered by the culture procedure cannot be excluded. However, recent studies suggest that actin (37,38) and tubulin (38,39) are associated with the plasma membrane and may be exposed on the cell surface (40).

Evidently the initial step of hormone action, namely the interaction of the ligand with its specific receptor at the cell surface, though essential, is not sufficient for the activation of adenylate cyclase. In addition, a coupling system must be operative. It appears that the MT-MF system can affect this coupling mechanism. The functional integrity of cytoskeletal elements associated with the plasma membrane (or "exoskeleton") may thus play a role in the hormonal modulation of the cyclic AMP system. If so, changes in the organization (41) or distribution (42) of the components of the MT-MF in the membrane could account for the response - refractoriness - recovery sequence observed following exposure of cells to a hormone.

REFERENCES

(1) P. Cuatrecasas. Rec.Prog.Horm.Res. 31(1975)37.

(2) C. De Haën. J.Theor.Biol. 58(1976)383.

(3) E.J.M. Helmreich, H.P. Zenner, T. Pfeuffer and C.F. Cori. Current Topics in Cellular Regulation 10(1976)41.

(4) J. Orly and M. Schramm. Proc. Nat.Acad.Sci.USA 73(1976)4410.

(5) E.M. Ross and A.G. Gilman. Proc.Nat.Acad.Sci.UMA 74(1977)3715.

(6) U. Zor, B. Strulovici and H.R. Lindner. Biochem.Biophys.Res. Commun. 80(1978)983.

(7) S.A. Lamprecht, U. Zor, A. Tsafriri and H.R. Lindner. J.Endocrinol. 57(1973)217.

(8) S.A. Lamprecht, U. Zor, Y. Salomon, Y. Koch, K. Ahrén and H.R. Lindner. J.Cycl.Nucl.Res. 3(1977)69.

(9) U. Zor and S.A. Lamprecht. Biochemical Actions of Hormones IV(1977)85.

(10) J. Burgess and D.H. Northcote. J.Cell.Sci. 5(1969)433.

(11) B. Bhattacharyya and J. Wolff. Biochem.Biophys.Res.Commun. 73(1976)383.

(12) T.H. Wieland. Advances in Enzyme Regulations 15(1977)285.

(13) A. Tsafriri, H.R. Lindner, U. Zor and S.A. Lamprecht. J.Reprod. Fert. 31(1972)39.

(14) A.G. Gilman. Proc.Nat.Acad.Sci.USA 67(1970)305.

(15) S.J. Shuman, U. Zor, R. Chayoth and J.B. Field. J.Clin.Invest. 57(1976)1132.

(16) U. Zor, S.A. Lamprecht, Z. Misulovin, Y. Koch and H.R. Lindner. Biochim.Biophys.Acta 428(1976)761.

(17) J.M. Marsh, T.M. Mills and W.J. LeMaire. Biochim.Biophys.Acta 304(1973)197.

(18) M. Hunzicker-Dunn and L. Birnbaumer. Endocrinology 99(1976)185.

(19) T.J. Franklin and S.J. Foster. Nature (New Biol.) 246(1973)146.

(20) C.P. Ciosek, Jr., J.V. Fahey, Y. Ishikawa and D.S. Newcombe. J.Cycl.Nucl.Res. 1(1975)229.

(21) E. Remold-O'Donnell. J.Biol.Chem. 249(1974)3615.

(22) D.S. Newcombe, C.P. Ciosek, Jr., Y. Ishikawa and J.V. Fahey. Proc.Nat.Acad.Sci.USA 72(1975)3124.

(23) R.J. Lefkowtiz, C. Mukherjee, L.E. Limbird, M.G. Caron, L.T. Williams, R.W. Alexander, J.V. Mickey and R. Tate. Rec.Prog. Horm.Res. 32(1976)597.

(24) M.A . Lesniak and J. Roth. J.Biol.Chem. 251(1976)3720.

(25) P.M. Hinkle and A.H. Tashjian, Jr. Biochemistry 14(1975)3845.

(26) R.J. Lefkowitz, D. Mullikin, C.L. Wood, T.B. Gore and M. Chhabirani. J.Biol.Chem. 252(1977)5295.

(27) F.R. DeRubertis and P. Craven. J.Clin.Invest. 57(1976)435.

(28) J. Bockaert, M. Hunzicker-Dunn and L. Birnbaumer. J.Biol.Chem. 251(1976)2653.

(29) R. Simantov and L. Sachs. Proc.Nat.Acad.Sci.USA. 75(1978)1805.

(30) R-J. Ho and E.W. Sutherland. Adv.in Cycl.Nucl.Res. 5(1975)533.

(31) R.M. Sharpe. Biochem.Biophys.Res.Commun. 75(1977)711.

(32) T.T. Chen, J.H. Abel, Jr., M.C. McClellan, H.R. Sawyer, M.A. Diekman and G.D. Niswender. Cytobiol. 14(1977)412.

(33) A. Amsterdam, A. Nimrod, S.A. Lamprecht and H.R. Lindner (submitted).

(34) K.N. Wessells, S.B. Spooner, F.J. Ash, O.M. Bradley, A.M. Luduena, L.E. Taylor, T.J. Wrenn and M.K. Yamada. Science 171(1971)135.

(35) R.R. Weihing. J.Cell Biol. 75(1977)95.

(36) J. Cachon, M. Cachon, L.G. Tilney and M.S. Tilney. J.Cell Biol. 72(1977)314.

(37) B.H. Toh and G.C. Hard. Nature 269(1977)695.

(38) G. Gabbiani, C. Chaponnier, A. Zumbe and P. Vassali. Nature 269(1977)697.

(39) B. Bhattacharyya and J. Wolff. J.Biol.Chem. 250(1975)7639.

(40) M. Estridge. Nature 268(1977)60.

(41) T.T. Puck. Proc.Nat.Acad.Sci.USA. 74(1977)4491.

(42) G. Wickus, E. Gruenstein, P.W. Robbins and A. Rich. Proc.Nat. Acad.Sci.USA 72(1975)746.

# INHIBITORY EFFECTS OF TREATMENT WITH LHRH OR ITS AGONISTS ON OVARIAN RECEPTOR LEVELS AND FUNCTION

Fernand Labrie, Claude Auclair, Lionel Cusan, André Lemay, Alain Bélanger, Paul A. Kelly, Louise Ferland, Geneviève Azadian-Boulanger and Jean-Pierre Raynaud

Laboratory of Molecular Endocrinology, Le Centre Hospitalier de l'Université Laval, Quebec G1V 4G2, Canada; and Centre de Recherches Roussel-UCLAF, Romainville 93230, France

We have recently found that treatment of adult male rats with luteinizing hormone-releasing hormone (LHRH) or its agonists leads to a marked loss of testicular LH and prolactin receptor levels accompanied by decreased testis, seminal vesicle and prostate weight as well as lowered plasma testosterone concentration (1, 2). Since LHRH and its agonistic analogues have potent pre- and post-coital antifertility activity in the female rat (3-5) and rabbit (6), it occurred to us that these effects of LHRH and its analogues in the female could also be mediated by endogenous LH-induced desensitization of the ovary. Such a possibility was supported by the findings that treatment with LH or human chorionic gonadotropin (hCG) leads to a marked loss of LH receptors in the ovary accompanied by a reduced ability of LH to stimulate ovarian adenylyl cyclase activity (7).

The present report describes the potent inhibitory effects of acute treatment with LHRH and its agonist [D-Ala$^6$, des-Gly-NH$_2$$^{10}$] LHRH ethylamide on ovarian LH, FSH and prolactin receptor levels and plasma progesterone concentration in the pregnant and normal cycling rat. Following such dramatic effects of acute treatment with LHRH and its agonist, we will describe the effect of removal of endogenous LHRH by administration of an LHRH antiserum as well as the effect of long-term treatment with the LHRH agonist in the female rat on the same parameters. Data indicating the possibility of a luteolytic effect of LHRH administered during the early luteal phase in normal women will also be presented.

## RESULTS

### Effects in Pregnant Rats

As illustrated in Fig. 1, treatment with the potent LHRH agonist [D-Ala$^{6}$, des-Gly-$NH_2^{10}$]LHRH ethylamide (25 µg, s.c. four times a day, starting on day 7 of pregnancy) led to an almost complete loss of ovarian LH receptors measured on days 10, 12 and 18. The inhibitory effect of [D-Ala$^{6}$, des-Gly-$NH_2^{10}$]LHRH ethylamide on FSH receptor levels was already well established on day 8 and remained 70-75% inhibited on days 10 and 12. A similar pattern of inhibition was seen when LH and FSH receptor levels were expressed per 100 mg of ovary (8). Treatment with the LHRH agonist also led to a marked inhibition of ovarian prolactin receptor levels and to a 45 to 60% reduction of plasma progesterone levels at all time intervals studied. In this experiment, the post-implantation abortive effects of the LHRH agonist were illustrated by the almost complete resorption of foetuses already seen on day 12 and the complete absence of signs of pregnancy on day 18 in animals treated from days 7 to 12 (data not shown).

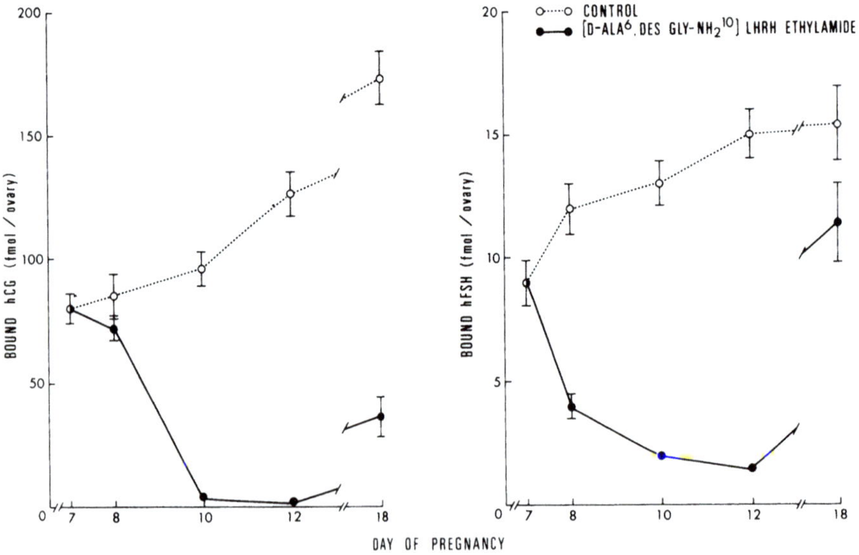


Fig. 1. Effect of treatment of pregnant rats with [D-Ala$^{6}$, des-Gly-$NH_2^{10}$]LHRH ethylamide on ovarian LH (left panel) and FSH (right panel) receptor levels. 25 µg [D-Ala$^{6}$, des-Gly-$NH_2^{10}$]LHRH ethylamide or saline were injected sc 4 times a day for various lengths of time starting on day 7 of pregnancy. The last injection was given 12h before sacrifice with the exception of groups on day 18 which were last injected on day 12 (8).

## Effects in Cycling Rats

As illustrated in Fig. 2, LH and FSH ovarian receptors are at their lowest level on estrus in intact 4-day cycling rats and increase progressively to a maximum on proestrus. A single injection of 25 μg of [D-Ala$^6$, des-Gly-$NH_2^{10}$]LHRH ethylamide on the morning of diestrus I led to a marked reduction of LH and FSH receptor levels measured on diestrus II and proestrus.

Since we have recently observed that a maximal inhibitory effect of [D-Leu$^6$, des-Gly-$NH_2^{10}$]LHRH ethylamide on testicular LH receptor levels was obtained with a single injection of only 40 ng of the peptide in the intact male rat (1), we next examined the effect of increasing doses of the LHRH agonist on ovarian LH and FSH receptor levels measured on the morning of expected proestrus. Single injection of as little as 8 ng of the LHRH agonist on diestrus I led to a significant reduction of ovarian LH receptors (30%). A near-maximal inhibition (60%) of ovarian LH receptors was seen at the dose of 40 ng, the inhibitory effect remaining of similar magnitude up to a dose of 25 μg (9).

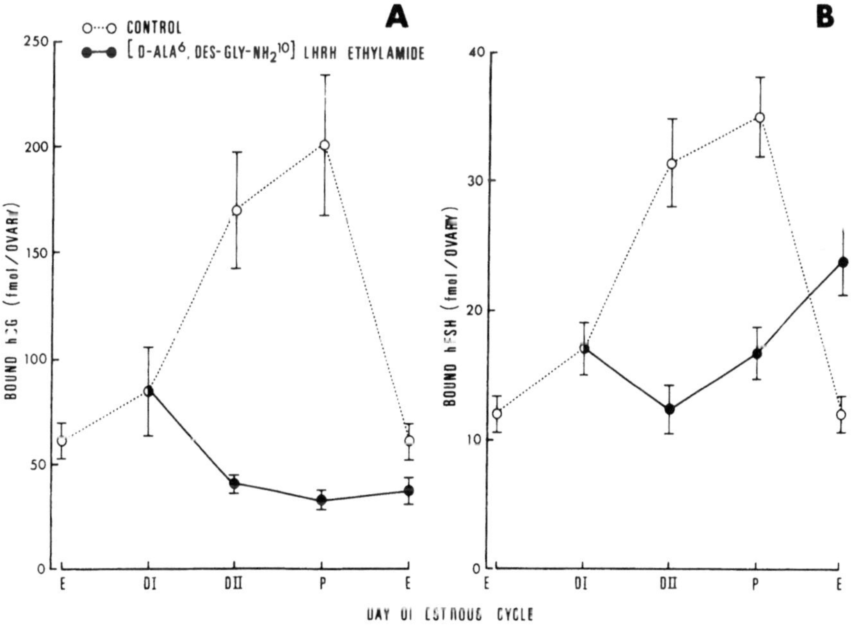


Fig. 2. Effect of a single injection of 25 μg [D-Ala$^6$, des-Gly-$NH_2^{10}$]LHRH ethylamide on ovarian LH/hCG (A) and FSH (B) receptor levels. Cycling female rats were injected on the morning of diestrus I and killed 1, 2 or 3 days later.

## Chronic treatment in the Female Rat

Following the dramatic inhibitory effects of short-time treatment with LHRH and its agonist on ovarian gonadotropin receptor levels and function, we studied the effect of long-term treatment with [D-Ala$^6$, des-Gly-NH$_2$$^{10}$]LHRH ethylamide on the same parameters. As illustrated in Fig. 3, daily treatment with the LHRH analogue led to a marked inhibition of ovarian LH, FSH and prolactin receptor levels. Similar patterns were obtained when receptor levels were expressed per ovary. It could also be seen in this experiment that plasma estradiol and progesterone levels were markedly reduced after treatment with the LHRH agonist. That the reduced steroidogenesis in animals treated with the LHRH agonist is not due to deficiency of gonadotropin secretion between the peaks of secretion induced by the LHRH agonist is well illustrated by the increased plasma LH and FSH levels measured in these animals (data not shown).

## Effect of Removal of Endogenous LHRH and Ovarian Gonadotropin Receptor Levels

This marked down-regulation of ovarian LH receptors which follows a surge of gonadotropin induced by LHRH or its agonists suggests that the spontaneous LH surge could play a physiological role in the control of ovarian receptors. Since administration of an

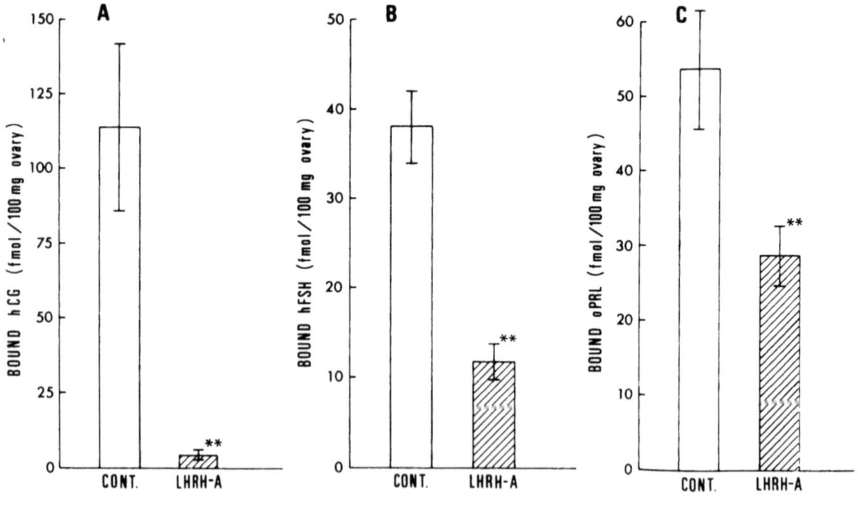


Fig. 3. Effect of chronic treatment (12 weeks) with [D-Ala$^6$, des-Gly-NH$_2$$^{10}$]LHRH ethylamide (5 μg daily) on ovarian LH (A), FSH (B) and prolactin (C) receptor levels in adult rats. The experiment was performed as described (8).

LHRH antiserum can efficiently block LH surges, treatment with the anti-LHRH serum (administered on the morning of proestrus) led to an almost complete prevention of the decrease of ovarian LH and FSH receptor levels normally found on the morning of estrus (data not shown).

### Possible Luteolytic Effect of LHRH in Normal Women

Administration of 250 µg of LHRH every 4h for 5 injections beginning at 0800h on one or two consecutive days starting between days 1 and 9 following the LH surge shortened the luteal phase from 1 to 4 days in 16 out of 17 treatment cycles. Fig. 4 represents a typical hormonal profile obtained in one woman treated for 1 day with 5 doses of 250 µg of LHRH. When LHRH was administered on day 7 of the luteal phase, an increase in serum LH and FSH concentrations was observed on the day of LHRH injection. This was followed by a precocious fall of serum progesterone while the beginning of menses was advanced by 3 days. A significant increase of serum estradiol can be noticed on the day of LHRH treatment.

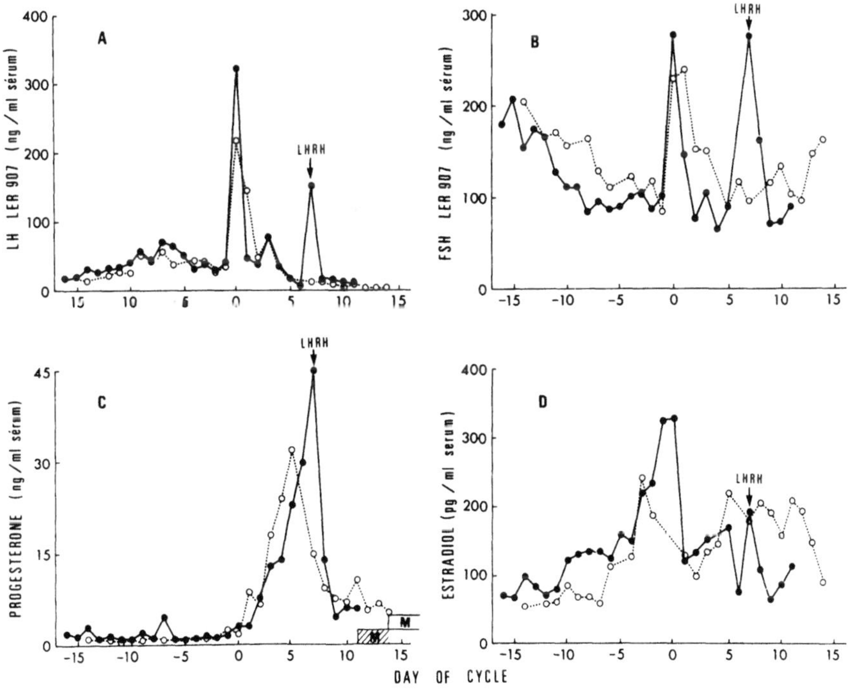


Fig. 4. Effect of 5 subcutaneous 250 µg doses of LHRH given every 4h on day 7 after the LH surge on serum LH (A), FSH (B), progesterone (C) and 17β-estradiol (D) and on time of menses in a normal woman (NL). Hormone levels were measured as described (1, 8, 10).

It could also be seen in this study that there exists a tendency for a better efficiency of treatment when LHRH is administered at a later stage of the luteal phase. In fact, in 10 cycles where LHRH was administered between days 1 to 5, the mean shortening of the luteal phase was 1.4 ± 0.2 days as compared to 3.3 ± 0.2 days for the cycles when the neurohormone was injected between days 6 and 9 following the LH surge ($p < 0.01$). Serum progesterone levels are also decreased to a greater extent when LHRH is administered after day 5 of the luteal phase. This appears to be analogous to the effect of LHRH treament on the time of apparition of menses. In fact, in 10 cycles where women were injected between days 1 and 5, serum progesterone, on average, decreased to 71.2 ± 6.0% of control while the mean serum progesterone levels were 43.9 ± 5.9% of control after LHRH administration between days 6 and 9 ($p < 0.01$).

## DISCUSSION

The present data clearly demonstrate that the pre- and postcoital antifertility effects of LHRH and its agonistic analogues in the rat are accompanied by a marked loss of ovarian gonadotropin receptor levels. The previous observation that injection of hCG or LH caused a desensitization of ovarian adenylyl cyclase activity which was closely paralleled by a loss of LH/hCG receptors (7) strongly suggests that the antifertility activity of LHRH or its analogues is due to desensitization of the ovary mediated by changes of endogenous gonadotropin release. As much as LH is known to be essential for normal ovarian function and maintenance of pregnancy up to day 12 in the rat, the present findings show that increased gonadotropin secretion can have deleterious effects. The luteolytic action of treatment with [D-Ala$^6$, des-Gly-$NH_2^{10}$]LHRH ethylamide during pregnancy is well illustrated by the 45 to 50% decrease of plasma progesterone levels in all treated groups. The present data also demonstrate that the single injection of relatively low doses of the LHRH agonist [D-Ala$^6$, des-Gly-$NH_2^{10}$]LHRH ethylamide on the morning of diestrus I in 4-day cycling rats leads to a marked loss of ovarian LH receptors. This receptor loss is accompanied by decreased plasma progesterone concentration and uterine weight measured on the morning of expected proestrus.

The present data clearly raise the possibility of a luteolytic effect of LHRH in normal women. The period of maximal sensitivity to this treatment appears to be on days 6-9 of the luteal phase. The effective dose was found to be 5 subcutaneous injections of LHRH at 4h intervals. The mean shortening of the luteal phase was 3.3 ± 0.2 days when LHRH was administered late (days 6-9) as compared to 1.4 ± 0.2 day when given on days 1-5. This more effective treatment on days 6-9 also correlates with a more precocious and marked decrease of serum progesterone after LHRH administration. Moreover, although the number of subjects is limited, the treatment with LH-RH seems to be more efficient when applied on 2 successive days as compared to one day. The findings obtained in normal cycling rats after suppression of the proestrus LH surge with the LHRH antiserum clearly indicate that physiological changes of endogenous LH release have an important role in the control of ovarian LH receptor levels. This could suggest that the ovarian failure observed at menopause is secondary to hypothalamic dysfunction. In fact, changes of the sensitivity of the hypothalamus to the negative feedback action of steroids could possibly lead to increased gonadotropin secretion and to secondary ovarian hypofunction.

The present data clearly indicate that ovarian LH receptors are highly sensitive to changes of circulating LH levels. They also indicate that the successful use of LHRH and its agonistic analogues for the treatment of infertility in both men and women has to take these findings into account. Moreover, the data summarized in this paper indicate the possible usefulness of LHRH and its agonists as contraceptives in the male and luteolytic agents in the female.

## REFERENCES

(1) C. Auclair, P.A. Kelly, D.H. Coy, A.V. Schally and F. Labrie, Endocrinology 101(1977)1890.

(2) G. Pelletier, L. Cusan, C. Auclair, P.A. Kelly, L. Désy and F. Labrie, Endocrinology (1978) in press.

(3) U.K. Banik and M.L. Givner. J. Reprod. Fertil. 44 (1975)87.

(4) R.R. Humphrey, B.L. Windsor, F.G. Bousley and R.A. Edgren. Contraception 14(1976)625.

(5) A. Corbin, C.W. Beattie, J. Yorkley and T.J. Toell. Endocr. Res. Commun. 2(1976)359.

(6) J. Hilliard, C.N. Pang and C.H. Sawyer. Fertil. Steril. 27 (1976)421.

(7) M. Hunzicker-Dunn and L. Birnbaumer. Endocrinology 99(1976)211.

(8) G.S. Kledzik, L. Cusan, C. Auclair, P.A. Kelly and F. Labrie. Fertil. Steril. 29(1978)560.

(9) G.S. Kledzik, L. Cusan, C. Auclair, P.A. Kelly and F. Labrie. Fertil. Steril. (in press).

(10)G.C. Olivier, J.E. Rioux and F. Labrie. Union Méd. Can. 104(1975 403.

## Discussion - H.R. Lindner, Chairperson

I. Rothchild: I am trying to stay out of asking questions, or discussing, but as an innocent bystander, who is being swamped by all the information about cyclic AMP, adenyl cyclase, protein kinase, etc., I can't get it through my head, why even this artificially induced pseudopregnant rat, which shows very strong correlation between the rate of progesterone secretion by the corpus luteum and the development and accumulation of corpus luteum LH/hCG receptors during the course of pseudopregnancy, should respond to the stimulus that is supposed to be maintaining progesterone secretion, not only with a loss of LH/hCG receptors, but with a cessation of progesterone secretion?

K. J. Catt: Your question is difficult to answer until we know more about the pattern of LH secretion and plasma gonadotropin levels during luteal cell development and desensitization. It is unlikely that our observations on receptor regulation by exogenous hormones are an exaggeration of a process that is proceeding during normal variations in plasma LH concentration. Even a small dose of hCG clearly occupies an adequate number of LH receptors for a sufficient time to induce membrane changes that lead to a major loss of receptor sites. This phenomenon also occurs after elevation of endogenous LH levels, as shown by Labrie and colleagues by treatment of male and female rats with potent LHRH analogues. We have observed the same changes after single injections of the natural LHRH molecule in the male rat, and it is clear that increased levels of endogenous LH also cause negative regulation of gonadal receptors. Yet, we all agree that LH is acting to maintain its target cells in the differentiated state, rather than to suppress their function. So I feel that we are observing an exaggeration of a normal process of receptor utilization during hormone action, and that occupancy of numerous receptors leads to a state of receptor depletion and secondary changes in luteal cell function.

You may recall that in the dose-response studies, we found that low doses of hCG caused a slight increase in LH receptors and cyclase responses, and larger doses produced desensitization and negative regulation of receptors.

The same changes can be seen in the Leydig cells of the testis, when very small doses of hCG have a positive effect on LH receptors and cell responses, whereas higher doses cause the negative actions observed in the ovary. Finally, we do not know enough about the relationship between plasma hormone concentration and receptor regulation. The receptor changes are probably not simply related to the degree of the plasma hormone change, but also to its constancy. It seems that a constant hormone level could be as important as a high level in this regard, and that the normal fluctuating pattern of hormone secretion may produce only minor variation in receptor concentration that are compensated by the normal process of receptor synthesis and replacement.

H. R. Lindner: I would agree that I think tonic maintenance of the corpus luteum by the pulsatile or episodic character of LH secretion precludes desensitization, and that we probably see desensitization physiologically only after the proestrus surge. But this still leaves me with the puzzle in relation to the primate corpus luteum of pregnancy, because I think already by the third week after conception, hCG levels in human blood would be high enough to cause desensitization if you go by any of the in vitro observations. And yet, the luteal placental shift only occurs about 4 weeks later. If corpus luteum function were shut off by desensitization, only temporarily, I think, pregnancy would probably be interrupted. So here this is a puzzle which I would like to have a solution during the general discussion.

G. Gibori: I think, Dr. Birnbaumer, that the reason you could not prevent the luteolytic effect of hCG in rabbit corpora lutea with estrogen, is that your so-called high dose of estrogen is not high enough. What is needed, is the maintenance of intraluteal estradiol levels, and injections of 15 ug twice daily cannot maintain the 6-8 pg of estradiol/mg of rat corpora lutea that the luteal tissue requires. We performed experiments where we needed to inject 100 ug/day to obtain physiological levels of intraluteal estradiol. Thus, it seems that your lack of protection against hCG might be due to insufficient estradiol and that if you would give higher levels you could possibly maintain progesterone synthesis.

L. Birnbaumer: While it is possible that progesterone synthesis might be maintained upon administration of still higher doses of estradiol, two aspects of our work lead me to believe that this would not lead to an actual improvement of the situation. One is that if we give even more estradiol, the cyclase system rather than improving would probably look worse.

G. Gibori: Did you try?

L. Birnbaumer: No, we did not try. The second reason for which I feel that higher doses of estradiol won't solve the problem, stems from our withdrawal experiment, where after having had the animals under continuous treatment with 15 ug estradiol/12 hours for 4 days (no hCG) we withdrew the steroid. We found that rabbits stopped making progesterone all together. This shows that such treatments affect animals adversely. As pointed out in our article, no such deleterious effect is seen with $1/10^{th}$ the estradiol dose.

G.J. Macdonald: Dr. Labrie, Dr. Craig Beattie and I have data related to your findings which suggest that caution should be exercised when considering the site of action of LHRH. We find that rats abort in a dose related manner by day 18, if they are hypophysectomized on day 12 of pregnancy and are given either 100 or 200 μg of LHRH b.i.d. on days 12 through 16. Day 1 is the day sperm are observed in the vaginal lavage. Serum progesterone values fall in response to the administered LHRH. Animals which totally abort have extremely low serum progesterone values. However, animals with viable sites have significantly reduced serum progesterone values when compared to hypophysectomized pregnant rats which received the sesame oil vehicle. This suggests that synthetic LHRH acts at sites other than upon the anterior pituitary.

F. Labrie: We think that the antifertility effects of LHRH and its agonists observed during pregnancy and the normal cycle are due to the release of endogenous gonadotropins. We have performed studies in hypophysectomized animals and could not find any effect of (D-Leu$^6$, des-Gly-$NH_2$$^{10}$) LHRH ethylamide on gonadotropin receptor levels. Although an extra pituitary effect of LHRH or its agonists is not completely excluded, it does not appear to play a significant role in the effects described at the ovarian level.

J. Sandow: I would like to mention that the desensitization observed after treatment with LHRH and its agonists follows very well the data obtained with hCG. It shows that physiological changes of LH secretion can desensitize the ovary.

A. Amsterdam: I would like to address myself to some points that were raised by the elegant presentation of Dr. Catt, and also Dr. Dufau about the clustering and internalization of receptors to LH in ovarian cells. I think that one should really pay attention to this phenomenon in order to understand some of the function of this receptor. We have used radioautography and immunofluorescent techniques to study this clustering and internalization of the receptor, and this work is presented in this book in a paper entitled "Lateral Mobility and Internalization of Hormone Receptors to Human Chorionic Gonadotropin in Cultural Rat Granulosa Cells" by Amsterdam et al.

G.J. Macdonald: I would like to respond to Dr. Birnbaumer's comment on physiological doses of estrogen by quickly reviewing studies from my laboratory which indicate that physiological doses of estrogen support luteal function or pregnancy. The pregnant rat, hypophysectomized and pituitary autografted before implantation will implant in response to 100 ng. of estradiol-17β. Estrogen must be administered daily for 9 days (equivalent to days 4 through 12 of normal pregnancy) and this allows pregnancy to proceed through day 20. Estrogen must be provided through day 12 for maximal fetal survival. We also know that the kidney bearing the autograft can be removed as early as day 8 equivalent. This suggests that pituitary prolactin is not required beyond day 8 of pregnancy. Ovariectomy as late as day 14, which would eliminate ovarian progesterone secretion, causes pregnancy failure. This work suggests a role in pregnancy for physiological amounts of estrogen.

K.J. Catt: In regard to the luteolytic effect of gonadotropins mentioned by Dr. Birnbaumer, there is no doubt that large doses of LH or hCG will cause luteolysis in the rat, as shown many years ago by Dr. Rothchild and others. The induction of irreversible luteolysis by the large doses of gonadotropin employed in the studies Lutz described is well-recognized, and this effect was discussed in our paper on luteal desensitization in 1976 (Conti et al, J. Biol. Chem. 251:7729-7731). Our concern in this presentation has also been with the effects of low doses of hCG, from 100 nanograms down to only 20 or 30 nanograms per animal. It is possible that the effect of these small doses of hCG represents the initiation of a process that can lead into luteolysis when higher doses of gonadotropin are given. What we observe with these low doses may be one of a spectrum of luteal cell responses that is initially reversible with only partial receptor loss, but leads into irreversible luteolysis with more substantial receptor loss after higher doses of gonadotropin.

H.R. Behrman: I have a question directed to Dr. Catt. In the low dose range of the dose-response studies that you have done, I was wondering if you have any information on whether there is stoichiometry in down regulation? That is, is there a one to one relationship between molecules of hormone bound and receptor moieties lost?

K.J. Catt: That is an interesting point. We believe that the degree of occupancy is relatively minor compared to the degree of receptor loss. Although the mechanism of this effect is difficult to prove, it seems likely that the process might be due to local internalization of receptors, probably after clustering, as suggested by Dr. Amsterdam's data. It is possible that internalization of the occupied receptors may be accompanied by concomitant endocytosis of adjacent unoccupied receptors, thus causing the discrepancy between occupancy and receptor loss. There is a much greater degree of receptor loss than measurable occupancy after low doses of hCG. For example, occupancy is barely measurable after low doses (100 ng) of hCG that caused quite substantial degrees of receptor loss.

D. Schomberg: I have a question for Dr. Labrie. What were the levels of luteinizing hormones which were produced in the rat following the injection of the analogue? I ask that because of the rapidity of the desensitization response. I would also like to get Dr. Lindner's response to this, because there was a recent paper from his laboratory (Nimrod et al, Biochem. Biophys. Res. Comm 78:977, 1978) showing that the level of hCG receptor binding in the rat on the day of proestrus had not fallen terribly significantly, by about midnight of the day of proestrus, and that was after the LH surge in the afternoon.

F. Labrie: The concentration of serum LH following injection of LHRH or its analogues varies between 300 and 600 ng equivalents LH-RP-1 per ml. This level of LH is that normally found at proestrus in the rat. The time-course of loss of ovarian LH receptors after the proestrus surge, the surge induced by LHRH or its agonists, or exogenous LH or hCG is the same: significant decrease between 6 and 12 hours and maximal loss of receptor 24 to 36 hours after the peak of circulating gonadotropins with a return to normal levels after 7 to 9 days.

H.R. Lindner: Regarding our results, I think we must distinguish now between down-regulation, a term I would restrict to the loss of receptor sites, and desensitization, which we think is an uncoupling between available receptors and the cyclase which is still stimulable by NaF or GMP. At early stages after stimulation, within 12 hours, we see a loss of LH stimulable cyclase, but we do not see a loss in receptor sites, and only later, you get this pronounced loss of receptor sites as Dr. Catt and others have described.

L. Nilsson: This is a question for Dr. Catt, and perhaps, Dr. Lindner. If you varied the mode of administration of the desensitizing dose of the hormone, could you separate the early and the late responses of uncoupling and receptor effects? If we injected intraperitoneally, would we get a long-standing desensitization, and if we have the hormone intravenously, would we get just a very short response?

K.J. Catt: In our experience the form or route of administration of the desensitizing gonadotropin has not made a lot of difference to the time course of LH receptor loss. The course of receptor regulation is much the same whether animals are treated by subcutaneous or intravenous injection of LH or hCG, with ovine LH (with its much shorter half-life in vivo) or with LHRH to elevate endogenous LH levels.

F. Labrie: Taking advantage of the time-course of receptor loss and desensitization, we have found that in the male or cycling female rat, a similar effect is found when the LHRH analogue is injected once or three times a day or twice a week. However, during pregnancy, the resistance to desensitization is higher and it is important to maintain serum LH at a high level for a relatively long period of time. During pregnancy, the same total dose can be up to hundred-fold more efficient if injected four times instead of once a day. The same appears to be true in the normal cycling woman with LHRH. In fact, a luteolytic effect was seen with 5 successive injections of LHRH while no effect was detected after 2 or 3 successive injections of the neurohormone.

M. Hunzicker-Dunn: I have two real quick questions. I don't mean to belabor the point of the permanency of desensitization, even though Dr. Catt has just commented on it, but when one does use an ovulatory desensitizing dose of hCG, one not only induces desensitization of the cyclase system in corpora lutea, but also one induces ovulation. This is true in rabbits and in pseudopregnant as well as in pregnant rats. So experiments using the whole ovary must be misleading.

In rabbits, we have looked up to 14 days following the injection of a desensitizing dose of hCG and have found the cyclase system in these corpora albicantria is still only minimally responsive to LH.

My second comment is on Dr. Lindner's question concerning the high levels of hCG in pregnancy that do not initially induce refractoriness in human corpora luteum. Lutz, Birnbaumer and I observed that in rabbits, corpora lutea obtained during the latter portion of pregnancy are partially protected from luteolytic desensitizing doses of hCG, whereas the corpus luteum in earlier pregnancy as well as in pseudopregnancy is not protected. This is not a permanent protection; LH responsiveness is maintained for a couple of days, and then one observes desensitization of the adenylyl cyclase to LH.

H.R. Lindner: It obviously must be protected, but what protects it?

M. Hunzicker-Dunn: We do not as yet understand the mechanism by which human or rabbit corpora lutea are protected against high doses of hCG.

R.A. Midgely: I want to begin with a few comments about dosages. It's very clear that in your rats a dose of hCG of 50 I.U. or greater is going to cause occupation of effectively all sites. This is not so for a dose of 5 I.U. Yet the latter dose is one that can cause ovulation of all follicles that are ready to ovulate. It strikes me that with dosages of hCG of 5 I.U. or less, one is at least approximating conditions that can occasionally be seen normally in vivo.

The issue has also been raised about estrogen concentrations; whether they're high or low. Here one has to mimic intraovarian concentrations, and intraovarian concentrations are enormously high. This can be particularly so when aromatization is occurring within the tissue compartment one is examining. Thus, if one wants to mimic the effects of intraovarian estradiol, one has to use extremely high concentrations.

I next want to consider the issues of pulsatile LH secretion. The possibility exists that these very low dose pulsatile events are extremely important. Each pulse could occupy a few receptor sites, cause clustering and internalization, and by this or some other mechanism, lead to loss of a few receptors. These repeated losses could in turn promote the continued synthesis and renewal of the receptor. If this were true, repeated, minor, down regulation respones to pulses of gonadotropins may be critically important to maintenance of the corpus luteum. I would like to ask how certain it is that the catecholamine beta-receptors are localized within the luteinized cell? I am wondering if it is absolutely certain that the luteinized cell itself is responding, or could the response be secondary to effects mediated by stromal connective tissue cells that contain catecholamine beta receptors? If it is the latter, if one is inducing down-regulation of LH receptors in luteal cells, and if this effect carries over to other cell types that happen to have an epinephrine receptor, than one is talking about cross-talk between cells. A similar phenomenon appears to occur between granulosa cells and theca cells.

K.J. Catt: We certainly feel that the occupancy and turnover process is also occurring at low hormone concentrations, and that the normal pulses of LH in the circulation could be initiating this type of mechanism. In regard to the β-adrenergic sites, though it's possible they are located in other cell types, there is no doubt about their occurrence in luteal cells. Catecholamines stimulate both cyclic AMP and progesterone production in isolated luteal cells, with similar maximum responses to LH, and there is no additivity between maximum effective doses of both stimuli.

L. Birnbaumer: Maybe I can address myself to that, as well. In adenylyl cyclase assays, there is no additivity between LH and catecholamine stimulation (Birnbaumer et al. (1976) Endocrinology 99, 163-184, c.f. Table 3B). Those experiments were carried out on superovulated ovaries. Other experiments with dissected corpora lutea from rats (pregnant) and rabbits (PSP and pregnant) have shown the same. Since dissected corpora lutea have very little extra corpora lutea tissue, the assumption that non-luteal cells rather than luteal cells, might be responsible for catecholamine-induced cyclase activity, and for catecholamine stimulated cAMP production, would require the additional assumption that the cyclase system in these other cells is 10-100 times more active (on a per mg protein basis) than that in luteal cells. There is no indication that this is so. Thus, I think it safe to accept that both LH and catecholamines act on a single luteal adenylyl cyclase.

I should like to come back to estradiol doses, once more. It is not that I am trying to criticize the general work done with high doses of estradiol, in fact, I am learning a lot from these types of experiments. It is just that when high doses of estradiol are used, they should be recognized as such and not mistaken as physiologic which they very possibly are not. Thus, in ovariectomized rats, Dr. Keyes' (Keyes (1973) Biol. Reprod. 8, 618-624) has clearly shown that 0.1 ug estradiol per day per rat maintains ectopic corpora lutea for 7 days since pregnancy is maintained for that time. Furthermore, Dr. Keyes showed us earlier that the affinity of rabbit estradiol receptors is just as high as that in any non-ovarian tissue. There is no apparent reason, therefore, why one should have to - as one has to - swamp the animal with estradiol to fulfill luteal requirements. One should also keep in mind that the fact that ovarian levels of estradiol are high, demonstrates neither that all the steroid is free and available at a correspondingly high concentration level - it might be trapped in secretory-type vesicles - nor that high levels of estradiol are required to physiologically maintain corpora lutea under physiologic conditions. Studies with ectopic corpora lutea in normal and hypophysectomized animals should allow us to clarify some of these issues.

MECHANISM AND SIGNIFICANCE OF LUTEAL DESENSITIZATION:

CHAIRMAN'S CONCLUDING REMARKS

H.R. Lindner

I have been charged with the task to sum up this wide-ranging discussion, to search for some take-home lessons and to identify issues that invite further investigation. We may first wonder why the Organizing Committee has seen fit to devote an entire session of this Workshop to luteal desensitization. The reason, I assume, was the belief that this pehonomenon is not merely a laboratory aretefact, but has some physiological relevance; and that in probing the molecular basis of the refractory state we may gain new insights into wider issues of peptide hormone action, the mode of transduction of the hormonal stimulus across the cell membrane and the organization of the cell surface. The decision may also have been prompted by a hunch that some of this information may eventually be put to practical use in the field of fertility regulation.

We have seen that continued exposure of ovarian cells to gonadotrophins or prostaglandin (PG) $E_2$ results in a characteristic sequence of events: (i) increased cyclic AMP formation; (ii) progressive desensitization of the adenylate cyclase (AC) system, attended or followed by "down-regulation" of receptor density on the cell surface; and (iii) upon removal of the hormone, gradual recovery of responsiveness.

Early indications for desensitization of an ovarian cyclase by gonadotrophins and prostaglandins came from studies on isolated Graafian follicles and cultured granulosa cells, and most of our own investigations have since centered on these two models [Lamprecht, S.A. *et al.*, J. Endocrinol. 57, 217 (1973); Zor, U. *et al.*, Biochim. Biophys. Acta 428, 761; J. Cyclic Nucleotide Res. 3, 69 (1977)]. Some of our conclusions were present at our previous Workshop (St. Cruz Workshop on Ovarian Follicular Development and Function, 1976, Ed. R. Midgley & W.A. Sadler, Raven Press, New York, 1978), and it may be instructive to compare these with what we learnt today on the corpus luteum.

In the follicle, we showed that hormone levels well above the minimal effective dose for cyclase activation were required for desensitization; that desensitization was specifically restricted to the homologous hormone, - e.g. follicles rendered refractory to LH remained fully responsive to FSH and $PGE_2$; that desensitization could be prevented or delayed by inhibitors of macromolecular synthesis; that AC in membrane preparations from LH-refractory follicles was still responsive to fluoride and p(NH)ppG; that desensitization was not due to increased degradation of cAMP by phosphodiesterase; and that desensitization of AC appears to precede loss of receptor from the cell surface when due allowance is made in the assay for occupied binding sites. To be sure, eventual loss of hormone receptors in desensitized cells - viz. down-regulation in its strict sense - has by now been well documented in several ovarian and testicular systems, but whether these changes are the cause or a sequel of desensitization remains to be firmly established. Indeed, it seems to me that many observations point to an uncoupling of the receptor from the cyclase as the basis of the early phase of desensitization.

Dr. Catt's beautiful lecture left us in no doubt that the desensitization mechanism is operative in the corpus luteum too, and that it affects steroidogenesis as well as cyclic AMP accumulation. In fact, - and this is a novel finding - luteal desensitization entailed a diminished steroidogenic response even to exogenous cAMP. Perhaps the maintenance of key steroid biosynthetic enzymes, such as cholesterol 20:22 lyase, depends on adequate tonic LH-support.

In contrast to our findings in the follicle, we learnt from Dr. Catt that even extremely low rates of receptor occupation can lead to marked down-regulation of the luteal LH-receptor, - a significant point that I hope to return to later. Again, while our data point to considerable specificity of the desensitization process with regard to LH, FSH and $PGE_2$ in the follicle, Dr. Catt presented evidence for an interesting exception to this generalization in the corpus luteum: LH was found to induce cross-desensitization to the action of β-adrenergic agonists. Dr. Midgley expressed the reservation that the catecholamine and LH receptors may be located on different cells. But perhaps we should also consider the alternative possibility that the same cell carries both receptors, in which case Dr. Catt's observation may have something in common with the phenomenon of co-patching and co-capping shown by H-2 and viral antigens [Schrader *et al.*, Proc. Natl. Acad. Sci. **72**, 5066 (1975); see also below].

Another feature of the luteal system brought to light by Dr. Catt's studies, which we failed to observe in the follicle, is a transient loss of fluoride-stimulable cyclase activity during the early phase of desensitization. To account for the initial loss of receptor function, Dr. Catt provided us with a provocative model of the cyclase molecule tumbling over, as it were, within the membrane, with formation of a refractory hormone-receptor-enzyme complex. If this model faithfully describes what is happening, one may suspect that such tumbling should involve cyto-skeletal elements. Dr. Amsterdam showed us that the process of down-regulation entails clustering and possibly capping of receptors followed by internalization of labelled hormone, probably receptor-bound, into lysozomes with subsequent partial degradation. Dr. Niswender's group has reported similar observations and Dr. Catt's electrophoretic data provided essential confirmation that the internalized readioactivity in fact represents largely hormone-receptor complex.

The data presented by Dr. Labrie - and similar observations recently reported by Dr. K. Yoshinaga (CIBA Symposium on Maternal Recognition of Pregnancy, London, May 1978), suggest that luteal desensitization can be achieved by endogenous LH when this is raised by administration of long acting GnRH analogues, and that this mechanism can be utilized for the interception of pregnancy in the rat. This gives the esoteric subject of desensitization a more practical dimension, suggesting possible applications in fertility control. In this connection we raised the question, what protects the primate corpus luteum from such desensitization by hCG, and I think we do not yet have a satisfactory explanation.

A number of other questions may be pertinent here. We were able to show that $PGF_{2\alpha}$ renders the AC of rat corpus luteum refractory to LH and abrogates LH-stimulated luteal cyclase activity within 15 min [Lahav, M., Freund, A. and Lindner, H.R. Biochem. Biophys. Res. Comm. **68**, 1291 (1976)]. Behrman, H.R. and Hichens, M. [Prostaglandins **12**, 83 (1976)] showed that this is followed within 1-2 h by a loss of LH-binding sites. Is this $PGF_{2\alpha}$-induced refractory state, which we believe to be the basis of cyclic luteolysis, mechanistically related to the desensitization of luteal cyclase of the homologous hormone which we discussed today?

Let us recall that desensitization of target cells to the action of drugs and hormones is a widespread - though by no means universal - phenomenon. It is shown also by peptide hormones not known to act via a cyclase mechanism, such as insulin. Does tachyphylaxis to drugs have anything in common with refractoriness to hormones? Conversely, what is the fundamental difference between cyclase systems that exhibit desensitization, such as the PG-stimulable AC of the ovarian follicle, and those which do not, such as the catecholamine-sensitive cyclase of turkey erythrocyte membrane [Hamski, E. and Levitzky, A. Life Sciences **22**, 53 (1978)]? Finally, desensitization of AC has been observed in crude membrane preparations from ovaries within 20 min of exposure to LH, in the presence of exogenous ATP and GTP (see presentations of Drs. L. Birnbaumer, M. Hunzicker-Dunn and Y. Salomon *et al.* at the 1976 and the present Workshops). In what respects, if any, is this phenomenon analogous to the desensitization we observe in isolated intact cells or *in vivo*, which takes hours to develop?

To understand the desensitization phenomenon, we must view it in the light of current concepts regarding the mechanism of hormone-stimulable AC action. These concepts, still tentative and partly derived from studies of catecholamine receptors [Orly, J. and Schramm, M. Proc. Natl. Acad. Sci. USA. **73**, 4410 (1976); Tolkovsky, A.M. and Levitzky, A. in Hormones and Cell Regulation, ed. J. Dumont and J. Nunez, p. 89. Elsevier, Amsterdam (1978)], tend to abandon the classical picture of a precoupled receptor-cyclase complex. Instead, it is postulated that receptors - and possibly cyclase molecules - are capable of indpendent movement within a fluid membrane. Elegant evidence based on experimental transfer of the LH-receptor to adrenal cortical cells has been presented by Dr. Dufau and her colleagues at this Workshop showing that receptors for gonadotrophins, too, are not permanently apposed to the catalytic moiety of the cyclase sytem, which can indeed interact with different hormone-receptors, and both Dr. Catt and Dr. Amsterdam showed us evidence for lateral mobility of the gonadotrophin receptor. We may further assume that a hormone-induced conformational change in the receptor permits its recognition, upon collision, by a binding site on the catalytic moiety of AC (Fig. 1); and that this receptor-AC interaction, accompanied by exchange of GTP for GDP at the nucleotide binding site of AC [Salomon, Y. and Rodbell, M. J. biol. Chem. **250**, 7245 (1975); Cassel, D. and Selinger, Z. Biochim. Biophys. Acta **452**, 538 (1976)] leads to a transition of the enzyme to its activated state. This activated state may persist after dissociation of the hormone-receptor complex [cf. Levitzky, A. (1978); loc. cit. ], but is abolished by the action of an intrinsic GTPase converting the bound nucleotide to GDP. These features have been incorporated in the hypothetical schema shown in Fig. 1.

I trust many aspects of this scheme will be obsolete by the time we meet for our next Workshop. Still, at this stage we may ask which if any of these putative steps is impaired during desensitization? And what determines whether a receptor molecule, after release from the cyclase, is available for another cycle of enzyme activation or becomes immobilized and eventually internalized (Fig. 2)?

This brings me to the question - is the cytoskeleton involved in receptor-AC coupling and in the process of desensitization? As you heard from Dr. Zor, our studies on Graafian follicles suggest

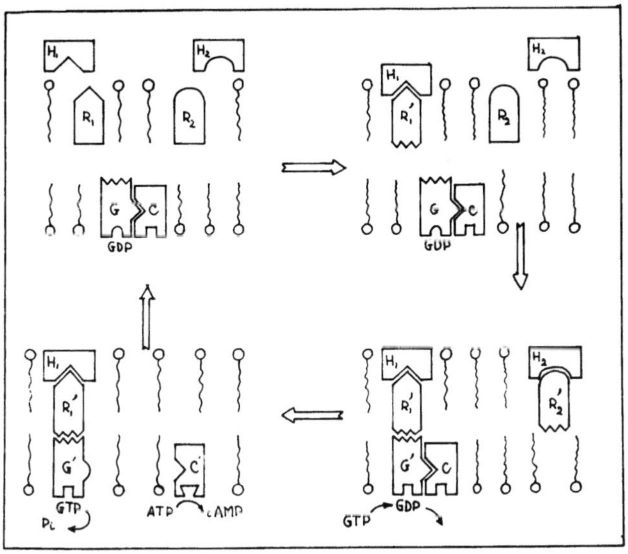


Fig. 1. Hypothetical schema of receptor/cyclase interaction.

Upper left: Mobile receptors $R_1$ and $R_2$ for hormones $H_1$ and $H_2$, respectively, facing extracellular space, are detached from adenylate cyclase (AC) comprising a catalytic (C) and a guanosine nucleotide-binding (G) domain facing the cytoplasm. (The possibility that R is a transmembrane protein and that part of R population is anchored in cytoskeleton is not considered in this scheme, - see text and Fig. 2.) Upper right: Interaction of $H_1$ and $R_1$ results in induced fit of $R_1$ with a recognition site on AC, here assumed to be located on the putative G subunit. ($H_2$ can induce similar complementariness between $R_2$ and G; see below.) Lower right: $H_1$-$R_1$ binds to AC, inducing a conformational change in nucleotide-binding site resulting in substitution of GTP for GDP and - Lower left - dissociation of G' from C, permitting in turn transition of the catalytic site to its active form (C') with formation of cAMP. Intrinsic GTPase activity of G' [Cassel, D. and Selinger, Z. Biochim. Biophys. Acta 452, 538 (1976)] now hydrolyses the bound GTP, G dissociates from $R_1$ and reassociates with C (upper left), quenching its activity. $R_1$ is available for rebinding hormone, unless it is immobilized by anchorage to submembranous MF-MT and removed by clustering and internalization (Fig. 2). Scheme based in part on collision coupling concept originally developed for adrenergic receptors [Tolkovsky, A.M. and Levitzky, A. In Hormones and Cell Regulation, Ed. Dumont, J. and Nunez, J., p. 89, Elsevier, Amsterdam (1978)], chosen here arbitrarily. Alternatively, the quarternary complex $H_1$-$R_1$-G'-C may have cyclase activity. (Salomon, R. and Lindner, H.R., unpublished.)

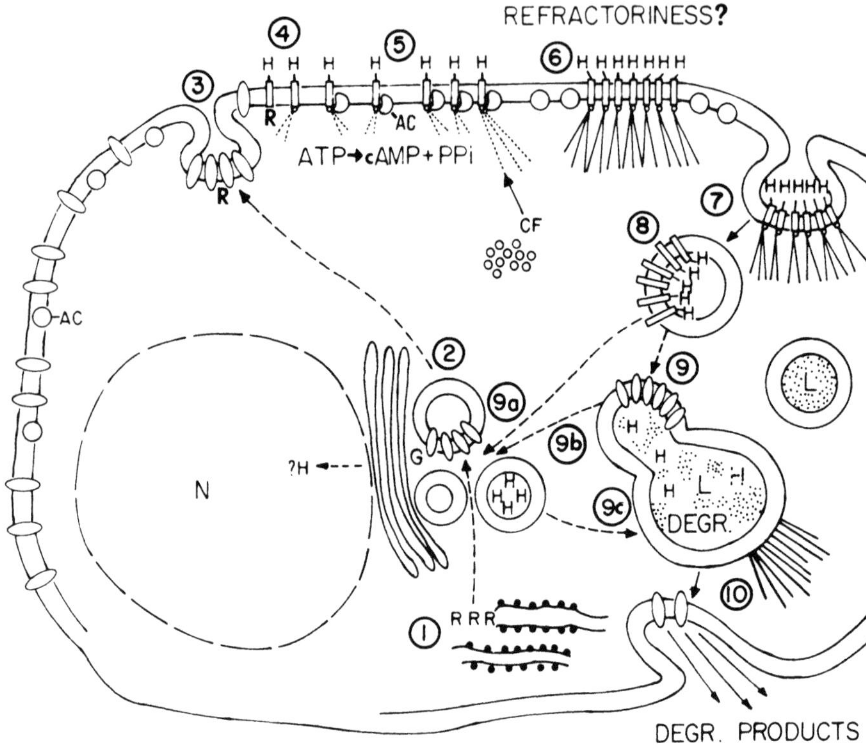


Fig. 2: Hypothetical model of gonadotropin receptor generation, function and turnover

1) Synthesis of receptor (R) or its precursor. 2) Glycosylation and insertion into membrane vesicle (?). 3) Fusion with cell membrane. 4) Hormone (H) binding and conformational change in receptor. 5) Coupling to adenylate cyclase (AC). 6) Clustering of receptors; anchorage to microfilaments and/or microtubules, possibly involving cytoplasmic adaptor molecule (CF)? 7 & 8) Endocytosis. 9 & 9a) Transfer of hormone receptor complex to lysosomes (L) or Golgi apparatus (G). Dissociation of the hormone receptor complex in lysosomes or Golgi complex. 9a & 9b) Partial reutilization of receptor? 9c) Degradation of hormone in lysosomes. 10) Discharge of lysosomal content by exocytosis. Penetration of hormone (?) or degradation products into nucleus (N). (Amsterdam, A. and Lindner, H. R., unpublished)

that this may be so. We have shown earlier [Zor *et al.*, Biochem. Biophys. Res. Comm. 80, 983 (1978)] that the acute response of follicles to LH can be prevented by prior exposure to agents believed to disrupt the microfilaments, such as cytochalasin B, but not by agents that impair the assembly or function of microtubules, such as colchicine, though the latter agent prevents the action of $PGE_2$. Dr. Zor now presented evidence from our laboratory that desensitization of follicles to $PGE_2$ can be prevented by stabilization of the MT-MF system, e.g. by agents believed to preserve the polymerized state of tubulin, such as by $Li^+$ and $D_2O^{13}$, or by phalloidin, which is believed to promote the polymerization of actin and to stabilize actin filaments. Conversely, recovery from $PGE_2$-induced refractoriness is prevented by cytochalasin B, anti-actin serum or heavy meromyosin subfragment 1, agents likely to interfere with the microfilament system, - possibly preventing the insertion of newly synthesized receptor into the membrane (Fig. 2). Thus it seems that the two major components of the cytoskeleton are involved, to varying degrees, with all three phases of the ovarian response to hormones. But we have been warned not to believe facts until we can prove them by theory; and we do not yet have a coherent theory that will account for all our observed facts.

There is, however, abundant, though equally indirect evidence, that the lateral movement of other cell membrane-associated proteins, antigens and lectin receptors is importantly influenced by elements of the cytoskeleton. Endocrinologists would be well-advised to closely follow developments in this exciting and rapidly expanding field [for review see Edelman, G.M. Science 192, 218 (1976)]. Of particular relevance to our discussion are indications that many receptor molecules appear to span the cell membrane [e.g. Tomita, M. and Marchesi, V.T. Proc. Natl. Acad. Sci. USA. 72, 2964 (1975); Tarrab-Hazdai, R., Geiger, B., Fuchs, S., and Amsterdam, A. Proc. Natl. Acad. Sci. USA. 75, 2497 (1978)] so that they may be able to interact with submembranous microfilaments and microtubules to form "surface modulating assemblies"; that lateral diffusion of receptors may be restricted to varying degrees by anchorage to these submembranous structures [Yahava, I. and Edelman, G.M. Proc. Natl. Acad. Sci. USA. 72, 1579 (1975); Schlessinger, J.*et al.*, Proc. Natl. Acad. Sci. USA. 74, 1110 (1977)], so that a given receptor population may be heterogeneous, comprising both free and immobile molecules, - the latter probably impaired in their ability to interact with adenylate cyclase. It appears that transition to the immobilized state can be effected by external agents such as cross-linking lectins, and that this "anchorage modulation" can be propagated by an MT-MF mediated process, so that occupation of a patch comprising as little as 4% of the cell surface by tetravalent Con-A can drastically reduce the mobility of Con-A receptors over the entire cell surface (Schlessinger et al., loc. cit.). These observations have obvious implications regarding Dr. Catt's findings that refractoriness in luteal cell can be induced by LH at extremely low occupancy rates.

Admittedly, thus far, our case for an implication of the MT-MF system in the ovarian cyclase response to hormone and the desensitization phenomenon is suggestive rather than conclusive; nor can we claim great novelty for the concept: some 22 years ago Sir Rudolph Peters [Nature London 177, 426 (1956)] proposed that there exists - and I quote - "a fluid anatomy in the geography of the cell, being some tenuous network for which Joe Needham proposed the term 'cytoskeleton', by the action of which the cell's enzymic activities are co-ordinated ........... and that this system possibly affects the patterns of diverse reactions which may be modified by the action of hormones on intact cells".

Details of observation forming the basis of some aspects of this discussion can be found in manuscripts at the end of this volume authored by Amsterdam, Kohen, Nimrod and Lindner; Zor and Amsterdam and Ezra, Lindner and Salomon.

# LUTROPIN INDUCED DESENSITIZATION OF RAT OVARIAN ADENYLATE CYCLASE: A GTP-DEPENDENT PROCESS

E. Ezra, H. R. Lindner and Y. Salomon

Department of Hormone Research

The Weizmann Institute of Science, Rehovot, Israel

Dr. Hunzicker-Dunn has pointed out that desensitization of pig ovarian adenylate cyclase (AC) requires - in addition to high LH concentrations - the presence of ATP. ATP could not be replaced by its non-phosphorylating analogue App(NH)p or by GTP (cf. Bockaert et al. 1). We would like to present evidence that desensitization of ovarian AC is also dependent on the presence of GTP.

Using purified rat ovarian plasma membranes (2) and a medium containing ATP (0.5 mM) and $Mg^{++}$ (5 mM), we found that the AC response to LH or GTP alone is weak (2-3-fold stimulation) but linear for at least 40 min (Figs. 1-3). When both agents were added together, AC activity increased 15 to 20-fold, but declined within 10-20 min to about 1/3 of its initial rate (Figs. 1-3). Addition of NaF (10 mM; Fig 2) or substitution of the phosphohydrolase-resistant analogue Gpp(NH)p for GTP (Fig. 1) prevented this decline.

The decline in AC activity in the presence of LH plus GTP is not due to destruction of the nucleotide, since it is neither prevented nor reversed by repeated addition of fresh GTP or Gpp(NH)p (data not shown).

Membrane preparations incubated with LH alone for up to 30 min remained fully responsive to GTP (Fig. 3), and conversely, membranes incubated for up to 30 min with GTP responded to subsequent addition of LH with a sharp rise in AC activity (data not shown), providing further evidence that the simultaneous presence of LH and GTP is essential for the development of the desensitized

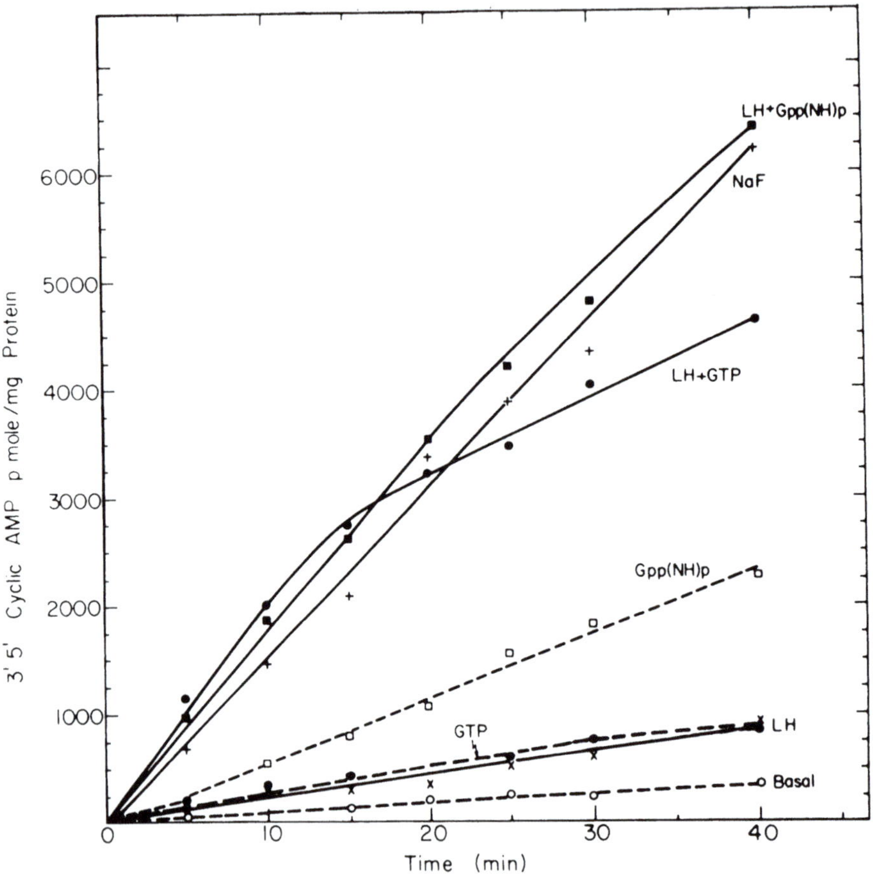


Fig. 1. Differential effects of GTP and Gpp(NH)p on the time course of LH-stimulated cyclase activity.
Adenylate cyclase activity was determined in the absence of any stimulant (basal activity) and in the presence of NaF (10 mM), GTP (10 μM), Gpp(NH)p (10 μM), LH (0.1 μM), LH + GTP, or LH + Gpp(NH)p. Assay conditions as described by Mintz et al. (2). Plasma membrane concentration was 60 μg protein/ml.

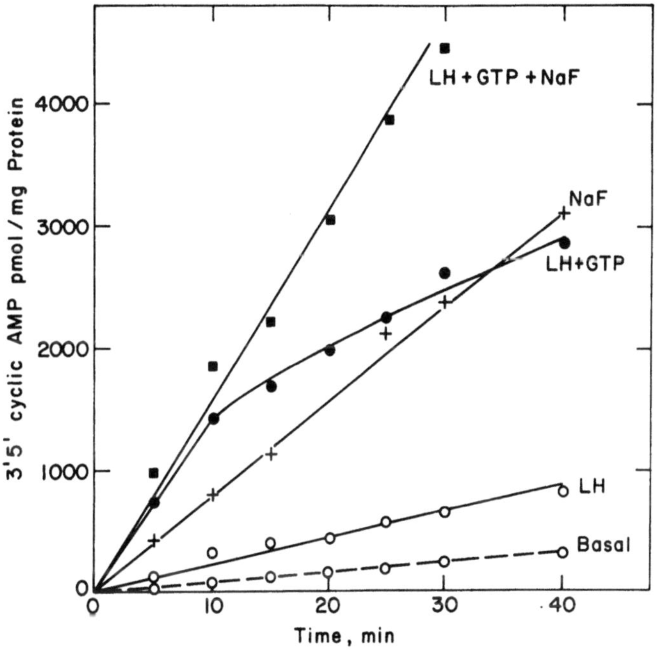


Fig. 2. Prevention by sodium fluoride of the GTP dependent decay of LH-stimulated adenylate cyclase activity.
Basal adenylate cyclase activity and the activity in the presence of NaF (10 mM), LH (0.1 μM), LH + GTP (10 μM) or LH + GTP + NaF were determined. Assay conditions and cAMP determination was as described by Mintz *et al.* (2). Plasma membrane concentration was 60 μg protein/ml.

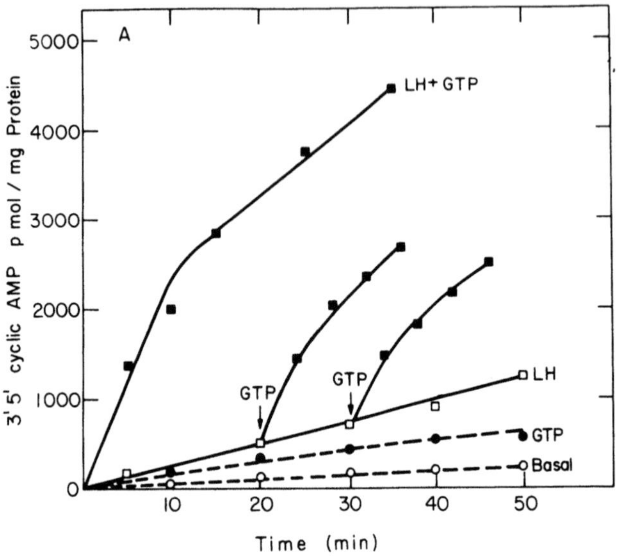


Fig. 3. Adenylate cyclase activity following sequential addition of LH and GTP.
Adenylate cyclase activity was determined under the following conditions: basal (0), LH (0.1 μM) (□), GTP (10 μM) (0.....0), GTP + LH (■——■). At times indicated (arrow) GTP (10 μM) was added to assay system containing LH only. Assay concentrations as for Fig. 1.

state. The addition of fresh plsma membranes to an assay system in which AC activity had already declined (50%) allowed full expression of the newly added enzyme (Fig. 4). This finding makes it unlikely that the decline in AC activity resulted from accumulation of an inhibitor.

The inability of Gpp(NH)p to substitute for GTP in the desensitizing mechanism is compatible with the idea put forward by Dr. Birnbaumer (1) that phosphorylation of an essential component of AC is responsible for desensitization, but suggests that the phosphate donor may be GTP rather than ATP. It should be pointed out that the ATP preparation used in our experiments was particularly low in GTP contamination (1,3) and this condition is crucial for demonstrating the GTP requirement. It is possible that the site of phosphorylation is the regulatory GTPase, recently shown by Cassel et al. (4) to be an integral component of AC. If phosphorylation is the basis of in vitro desensitization, one would expect the process to be reversible and thus of even greater regulatory significance. However, this remains to be established.

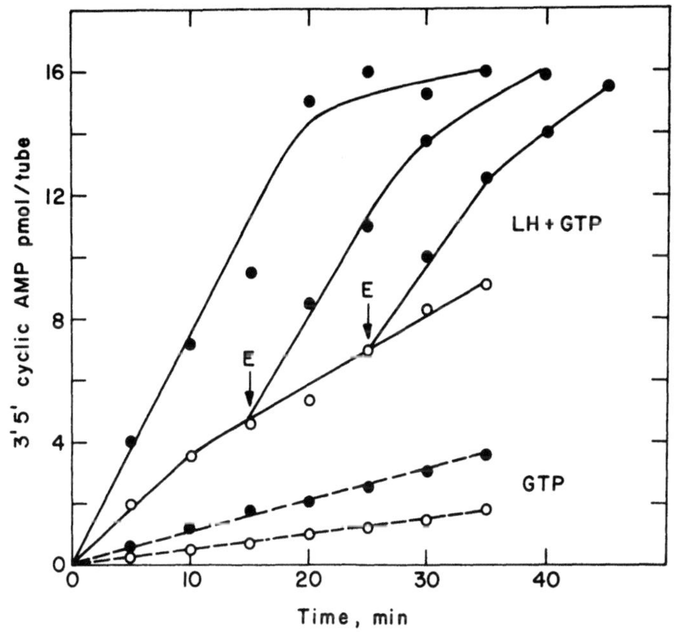


Fig. 4. Expression of fresh enzyme in desensitized LH-stimulable AC system.
Adenylate cyclase activity was determined at 40 μg (O) or 80 μg (●) membrane protein/ml in the presence of GTP (----) or LH plus GTP (——). At times indicated (arrow), fresh membrane preparation (40 μg protein/ml) was added to an assay system containing 40 μg membrane protein/ml. Other assay conditions were as described in legend to Fig. 1.

## REFERENCES

(1) J. Bockaert, M. Hunzicker-Dunn and L. Birnbaumer. J.Biol.Chem. 251(1976)2653.

(2) Y. Mintz, A. Amir, A. Amsterdam, H.R. Lindner and Y. Salomon. Mol.Cell Endocr. (1978)(In press).

(3) D. Cassel and Z. Selinger. Proc.Nat.Acad.Sci. 74(1977)3307.

(4) D. Cassel, H. Levkovitz and Z. Selinger. J.Cyclic Nucl.Res. 3(1977)393.

# hCG-INDUCED LOSS OF LH-hCG RECEPTOR AND DESENSITIZATION OF ADENYLATE CYCLASE

C.Y. Lee

Department of Laboratory Medicine and Pathology

University of Minnesota, Minneapolis, Minnesota

Experimental evidence has been accumulated to support the concept that gonadotropins are involved in the regulation of their own receptors (1-2). Exposure of ovarian follicles to hCG (LH) *in vivo* or *in vitro* results in a decline in the capacity of follicles to produce cyclic AMP in response to further hCG (LH) stimulation (3,4). Direct measurement of LH-stimulated adenylate cyclase activity in ovarian follicles and corpora lutea demonstrated that desensitization is due to decreased response of adenylate cyclase to subsequent gonadotropin stimulation (5). Furthermore, studies have shown that the desensitization of hCG-sensitive adenylate cyclase is closely related to the loss of hCG receptors (6,7). The present report demonstrates that hCG-induced loss of hCG (LH) receptors and desensitization of adenylate cyclase were both time and concentration-dependent and displayed a close correlation.

## MATERIALS AND METHODS

Immature female Holtzman rats were injected with 25 IU of PMSG at 22 days of age and 25 IU of hCG 56 h later. Desensitization was induced by subcutaneous injection of 0.5 to 50 IU of hCG when the ovaries were heavily luteinized (10-12 days after PMSG). Control groups (pseudopregnant rats without second hCG injection) received saline injections. Two to forty-six hours later, groups of 4-5 animals were sacrificed and the ovaries were removed.

Preparation of the membrane-enriched 2000xg pellets of heavily luteinized ovaries from PMSG and hCG primed rats and the procedures for hCG iodination and binding assay have been described before (8). The adenylate cyclase activity was determined according to the method described by Birnbaumer et al (9) with a minor modification (10).

## RESULTS AND DISCUSSION

### Loss of hCG Receptors and *In Vivo* Uptake of hCG

After injection of 50 IU hCG per rat, *in vitro* binding of human [$^{125}$I] iodo-CG to the ovarian membrane-enriched fractions was determined at various time intervals. The binding in hCG treated rats was expressed at a % of the binding of the control group. As illustrated in Fig. 1, hCG administration caused a decrease in available hCG receptors: 8% in 2 h, 61% in 5 h, 90% in 24 h and 86% in 46 h when a saturating dose of labeled hCG (100 ng/ml) was used for assay. Similar results were obtained when 4 ng/ml labeled hCG was employed (data not shown).

The uptake of human [$^{125}$I] iodo-CG *in vivo* by the ovary was studied at various time intervals following subcutaneous injection of labeled hormone. As shown in Fig. 1, uptake of labeled hCG by the ovary reached a peak in 6.5 h, decreased considerably in 24 h and the tracer was completely cleared from the ovary by 46 h. There is no inverse relationship between the binding of labeled hCG in vitro and the uptake of labeled hCG *in vivo.* These experiments suggest that the reduced binding capacity measured *in vitro* is not due to receptor occupancy. Similar results have been reported by Conti et al (6) in the pseudopregnant ovary, where the amount of hCG remaining in the ovary was measured after 20 IU hCG injection. The hCG binding capacity remained low for 4 days even though bound hCG is largely cleared from the ovary at 24 and 48 h. In hypophysectomized female rats, *in vivo* uptake and *in vitro* binding have been shown to decrease in a simultaneous manner following injection of labeled hCG together with saline or 5 IU unlabeled hCG (11).

### Desensitization of hCG-Sensitive Adenylate Cyclase

In the control group, as illustrated at the zero time in Fig. 2, epinephrine at 100 μM stimulated ovarian adenylate cyclase activity 3.6-fold relative to basal activity. hCG and NaF increased the enzyme activity 4.6- and 9.2-fold respectively. The desensitization of hCG-sensitive adenylate cyclase induced by 50 IU hCG injection is shown in Fig. 2. The enzyme activity declined 60% in 6.5 h and remained very low by 24 and 46 h, while epinephrine-and NaF-stimulated adenylate cyclase activities remained unchanged. The basal activity appeared slightly elevated at 2 and 6.5 h and

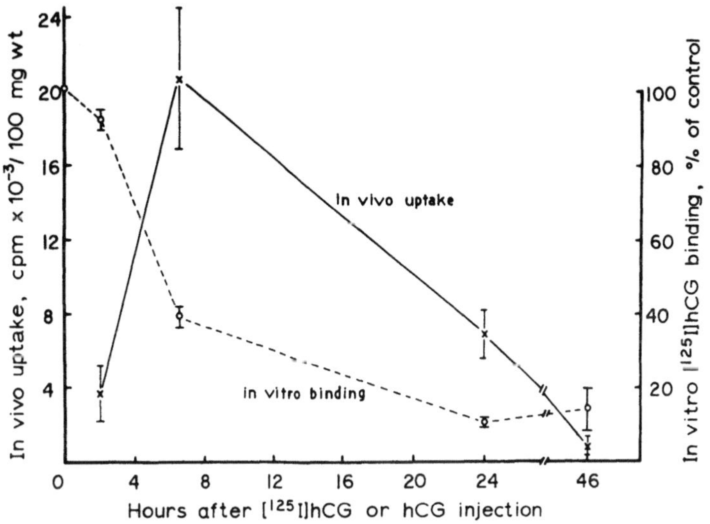


Fig. 1. In vitro binding and in vivo uptake of human [$^{125}$I] iodo-CG by the pseudopregnant rat ovary. The in vivo uptake was determined after subcutaneous injection of human [$^{125}$I] iodo-CG (2 μ Ci/rat). Each point represents mean ± SD from 5 animals. The in vitro binding after 50 IU hCG injection was measured using 100 ng/ml human [$^{125}$I] iodo-CG and was expressed as a % of control. In this and subsequent figures each point represents mean ± SD from a group of 4 animals.

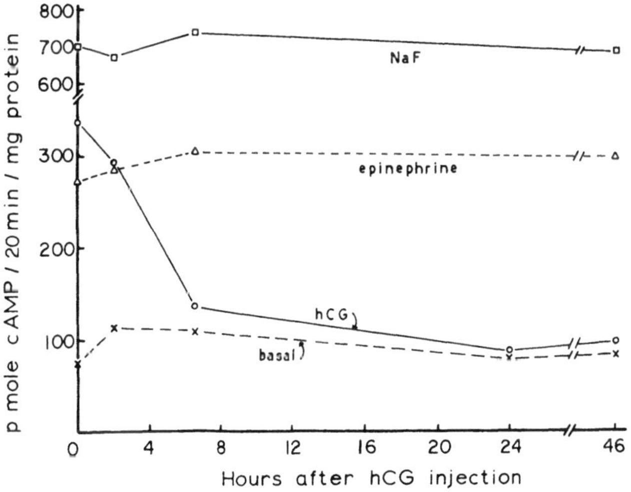


Fig. 2. Effects of hCG injection on basal and hormone-sensitive adenylate cyclase activities. Pseudopregnant rats were given a single injection of 50 IU hCG per rat. The concentrations used are: 10 μg/ml for hCG, 100 μM/ml for epinephrine and 10 mM for NaF.

returned to the pre-injection level by 24 and 46 h. At this period (24-46 h), ovarian adenylate cyclase remained unresponsive to hCG stimulation.

## Time Course and Dose Dependence

In order to examine the relationship between the hCG receptor loss and the decrease in hCG-sensitive adenylate cyclase activity, the time course after the hCG injection was followed. The decreases in hCG binding capacity (Fig. 1) was accompanied by a simultaneous decline in hCG-stimulated adenylate cyclase activity (Fig. 2). Both activities remained markedly suppressed 46 h after injection.

The dose-response curve for hCG-stimulated adenylate cyclase activity also directly parallels that for the binding capacity as demonstrated in Fig. 3. The binding activity 24 h after hCG injection decreased 23% by 0.5 IU, 47% by 1.5 IU, 58% by 5 IU, 67% by 15 IU, and 84% by 50 IU. The hCG-stimulated enzyme activity 24 h after hCG reduced 28% by 0.5 IU, 55% by 1.5 IU, 67% by 5 IU, 72% by 15 IU, and 79% by 50 IU. Pseudopregnant rats appeared sen-

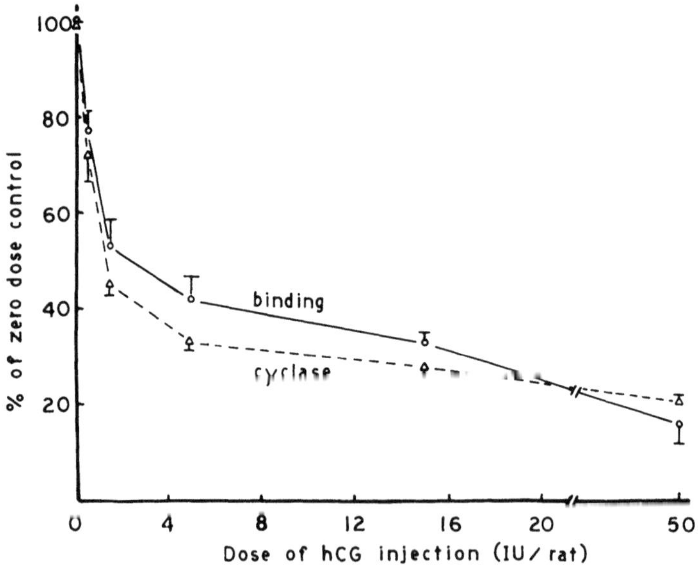


Fig. 3. Dose effect on human [$^{125}$I] iodo-CG binding and hCG-sensitive adenylate cyclase. Ovaries were obtained 24 h after a single subcutaneous injection of 0.5 to 50 IU hCG. The binding assay was performed as described in legend to Fig. 1 and the enzyme activity was determined in the presence of 10 μg/ml hCG.

sitive to hCG treatment. A small dose, as little as 0.5 IU hCG (Ayerst) or 50 ng hCG (CR-121) per rat, significantly reduced binding and adenylate cyclase activities 23 to 28%. These results clearly indicate that the desensitization of hormone-sensitive adenylate cyclase is associated with the decrease in the number of receptor sites. Conti et al (8) also reported a close relationship between receptor loss and refractoriness of adenylate cyclase during the time course of desensitization and regeneration of receptors.

Although the physiological role of LH (hCG)-induced desensitization is not clear at the present time, hCG-induced luteolysis has been shown to be associated with the hCG-induced desensitization of adenylate cyclase in the rat and rabbit corpora lutea (12). The desensitization of LH-and FSH-stimulated activities in ovarian follicles has been documented to be correlated with the decrease in steroidogenesis and to be related causally to the process of ovulation (3,5). It has been reported that in association with luteinization LH and hCG induce a loss in receptors for FSH and LH (2) and the loss of FSH receptors and FSH-sensitive adenylate cyclase activity (13). Thus, the desensitization process may play an important role in the regulation of the responsiveness of the ovary to gonadotropins by altering the gonadotropin receptor-adenylate cyclase system.

## Effects of hCG on Binding Affinity and Apparent Km for Adenylate Cyclase

A significant decrease in the number of the available receptor sites (33% decrease) was observed 24 h after treatment with 1 IU hCG per rat. There was no alteration in binding affinity. The concentrations of half-maximal binding obtained were 3.0 ng/ml ($6.5 \times 10^{-11}$M) and 3.2 ng/ml ($6.9 \times 10^{-11}$) for the treated and control animals respectively.

Ovarian adenylate cyclase was determined 24 h after treatment with 1 IU hCG. In both treated and control groups, maximal enzyme stimulation by hCG was reached at the concentration of 1.0 μg/ml. The apparent concentration giving half-maximal activation (Km) did not appear altered. The values were 50 ng/ml and 55 ng/ml for the control and treated groups respectively.

## SUMMARY

hCG-induced loss of hCG (LH) receptors and desensitization of adenylate cyclase in the pseudopregnant rat ovary have been studied. At various time intervals after hCG treatment, the decrease in hCG-stimulated adenylate cyclase activity was closely related to the reduction in available hCG receptors. The hCG-induced desensitization of adenylate cyclase and a fall in hCG binding activity were also dose-dependent and directly correlated. There were no changes in the affinity of receptors binding and in the apparent Km for adenylate cyclase activation. These findings indicate that the loss of hCG receptors is associated with desensitization of adenylate cyclase following the hCG injection.

## ACKNOWLEDGEMENT

This work was supported by NIH Research Career Development Award (HD 100191), NIH Grant (HD 10075) and the Graduate School of the University of Minnesota.

## REFERENCES

1. R.J. Ryan, L. Birnbaumer, C.Y. Lee, and M. Hunzicker-Dunn, International Review of Physiology. Reproductive Physiology II, Vol. 13, ed Roy O. Greep (University Park Press, Baltimore, 1977) p. 85.
2. J.S. Richards, and A.R. Midgley, Jr., Biol Reprod 14 (1976) 82.
3. J.M. Marsh, T.M. Mills, and W.J. LeMaire, Biochem Biophys Acta 304 (1973) 197.
4. S.A. Lamprecht, U. Zor, A. Tsafriri, and H.R. Lindner, J Endocrinol 57 (1973) 217.
5. M. Hunzicker-Dunn, and L. Birnbaumer, Endocrinology 99 (1976) 185.
6. M. Conti, J.P. Harwood, A.J.W. Hsueh, M.L. Dufau, and K.J. Catt, J Biol Chem 251 (1976) 7729.
7. C.Y. Lee, Prog 50th Annual Meeting Endocrine Society, Chicago, 1977, Abstract #223.
8. C.Y. Lee, and R.J. Ryan, Biochemistry 12 (1973) 4609.
9. L. Birnbaumer, P-C. Yang, M. Hunzicker-Dunn, J. Bockaert, and J.M. Duran, Endocrinology 99 (1976), 163.
10. C.Y. Lee, Endocrinology 101 (1977) 876.
11. M.C. Rao, J.S. Richards, A.R. Midgley, Jr., and L.E. Reichert, Jr., Endocrinology 101 (1977) 512.
12. M. Hunzicker-Dunn, and L. Birnbaumer, Endocrinology 99 (1976) 211.
13. C.Y. Lee, and H. Takahashi, Endocrinology 101 (1977) 869.

# EVIDENCE OF A GREATER ACTIVITY OF LH/HCH BINDING INHIBITOR PRESENT IN AQUEOUS EXTRACTS FROM OLD COMPARED TO YOUNG PORCINE CORPUS LUTEUM

Sandra Tucker, Lakshmi Kumari, Cornelia P. Channing

Department of Physiology, University of Maryland
School of Medicine, 660 West Redwood Street
Baltimore, Maryland 21201

## INTRODUCTION

Previously Yang, Saaman and Ward have shown the presence of an LH binding inhibitor in the 30,000 xg supernatant obtained from Parlow "pseudopregnant" and pregnant rat ovaries, which they have designated as LH receptor binding inhibitor or LHRBI. In addition, they have shown that LHRBI is a low molecular weight polypeptide of approximately 3,800 daltons (1). In 1977 Sakai, Channing and Engel demonstrated that crude extracts of porcine corpus luteum have the ability to inhibit the binding of $^{125}$I-hCG to porcine granulosa cells from large follicles (2). They observed that the inhibitory activity was localized in luteal tissue and absent in non-luteal ovarian tissue, granulosa cells, heart and lung. This inhibitory activity increased upon freezing of the ovaries and was present in the 100,000 xg supernatant of aqueous luteal extracts.

The question arose as to whether this LHRBI had a possible physiological role in regulation of luteal lifespan. In the intact nonpregnant pig the corpus luteum exists for a finite period of time, then regresses (see review by Hansel, Concannon and Lukaszewska (3)). As the corpus luteum ages it loses the ability to secrete progesterone in response to luteinizing hormone (4) and loses its sensitivity to LH in terms of ability to stimulate adenyl cyclase (5). It is possible that this loss in LH responsiveness is due to accumulation of an LHRBI. In order to test this hypothesis levels of LHRBI were measured in extracts of young, midcycle and old porcine corpora lutea. The inhibition of $^{125}$I-iodinated human chorionic gonadotrophin (hCG) binding to porcine granulosa cells was examined since hCG and LH share the same binding site on ovarian cells (6). Iodinated hCG is biologically active under conditions of our *in vitro* experiments and serves as a suitable tracer molecule (7).

## MATERIALS AND METHODS

Porcine ovaries were obtained from a local meat packing plant within 20 minutes of slaughter and were kept on ice until arrival at the laboratory about 1 hr. later. For these studies 4 batches of porcine ovaries were collected and frozen at -35°C or F for twelve to twenty-nine weeks. Prior to freezing, ovaries containing corpora lutea were separated arbitrarily into three pools on the basis of their color and gross morphology. Young corpora lutea included corpora hemmoragica and corpora lutea dark red in color, which were present probably from day 2 to 4 of the estrous cycle. Mid-cycle corpora lutea were pink in color and judged to be from pigs about day 5 to 12 of the estrous cycle. Old corpora lutea were pale pink and came from pigs about day 13-19 of the cycle (8).

All preparations of the tissue after thawing were carried out at 4° unless otherwise specified. All tissue homogenates were tested for their ability to inhibit binding of iodinated hCG to granulosa cells within one week of homogenization. The tissues (corpora lutea) were weighed in lots of about 40 g and homogenized for 1 min in a Waring Blender in 6 volumes of distilled water. The homogenized tissue was then filtered through three layers of cheese cloth, and the filtrate was centrifuged at 1000 g for 15 min; the supernatant was centrifuged for 15 min. at 30,000 g in Sorvall Model OTD Ultracentrifuge. In some instances the 30,000 g supernatants were frozen at -35°C or F and later assayed for protein by the Lowry technique (9) and for their ability to inhibit binding of iodinated hCG.

Highly purified hCG obtained from Dr. Robert Canfield (CR117; 14,000 IU/mg) was iodinated with 1 mCi of $^{125}$I (New England Nuclear) in aliquots of 10 μg with the chloramine-T technique (10) except that the ratio of hormone to chloramine-T was 1:1. The iodination was allowed to take place for 1 min. on ice and the reaction stopped with the addition of 10 μg of sodium metabisulfite. The mixture was poured on a 15 x 0.5 cm column of Bio-Gel and 1 ml fractions were collected. The peak fraction, which had a specific activity of 20-30 μCi/μg of hCG, was saved and used for these studies. This tracer was used within 2 weeks of iodination.

This procedure involved preincubating 1 x $10^6$ porcine granulosa cells obtained from large porcine follicles which were previously isolated, pooled and frozen with 100 μl of corpus luteum extract in a volume of 200 μl of 100 mM tris buffer pH 6.5 plus 0.1% bovine serum albumen for 20 minutes at 37° C. Subsequently 5 ng $^{125}$I-hCG was added, the mixture incubated at 37° C with shaking for 1 hour followed by centrifugation with the cell pellet washed twice and counted in an autogamma counter. Each count was corrected for nonspecific binding (< 5% total binding normalized per mg protein in one milliliter of ovine LH. Each aliquot of luteal extract was examined undiluted and in 1:5 dilution in 5 aliquots for its ability to inhibit binding of 5 ng $^{125}$I-hCG to porcine granulosa cells. Four batches of ovaries were examined in these studies.

## RESULTS

Addition of 100 μl of undiluted supernatant of homogenates of either young, midcycle and old corpus luteum to a total binding volume of 300 μl exerted complete inhibition of binding of $^{125}$I-hCG to large porcine granulosa cells. If the luteal extracts were diluted 1:5 the supernatant from midcycle and late luteal phase corpus luteum exerted a significantly greater ($p < 0.01$) ability to inhibit binding of $^{125}$I-hCG as compared to buffer alone. The extracts of old corpora lutea had a greater ability to inhibit binding of $^{125}$I-hCG compared to extracts of young corpora lutea (Fig. 1, Table 1, $p < 0.01$). Extracts of non-luteal tissue or lung did not show a comparable activity. When the data were standardized for mg protein the findings were similiar.

## DISCUSSION

The increase in LHRBI in luteal tissue observed as the porcine corpus luteum ages may explain the means whereby the porcine corpus luteum regulates its own lifespan. Increase in LHRBI with age of corpus lutea could inhibit the ability of circulating LH to stimulate progesterone secretion and maintain luteal function and initiate regression of the corpus luteum. In the pig and other animals prostaglandins have been suspected as playing a role in causing luteolysis, reviewed in (11). A plausible mechanism of prostaglandin action to cause luteolysis may be to increase LHRBI synthesis in the corpus luteum.

## SUMMARY

As porcine corpus luteum tissue ages there is an increase in amount of LH/hCG binding inhibitor. This inhibitor is absent in non-luteal ovarian tissue, heart and lung.

TABLE 1. Content of Luteinizing Hormone Binding Inhibitor (LHRBI) in extracts of young, midcycle and old porcine corpora lutea.

| Batch of Corpora lutea (weeks corpora lutea frozen) | Age of corpora lutea | Ability of extract of corpus luteum to inhibit binding of 5 ng $^{125}$I-hCG | | | |
|---|---|---|---|---|---|
| | | Extract assayed straight | | Extract assayed diluted 1 to 5 | |
| | | Binding % control | Binding % inhibition | Binding % control | Binding % inhibition |
| 1 (12) | young | 10 ± 0.5 | 90 | 69 ± 3.5* | 31 |
| | middle | 10 ± 0.1 | 90 | 43 ± 2.9* | 57 |
| | old | 1 ± 0.1 | >99 | 21 ± 5.7*+ | 79 |
| 2 (25) | young | <1 | >99 | 94 ± 2.9 | 6 |
| | middle | <1 | >99 | 40 ± 2.7* | 60 |
| | old | <1 | >99 | 28 ± 0.1*+ | 72 |
| 3 (29) | young | <1 | >99 | 100 ± 9.5 | 0 |
| | middle | <1 | >99 | 23 ±13.8* | 77 |
| | old | <1 | >99 | 20 ± 0.9*+ | 80 |
| 4 (12) | young | 32 ± 11* | 68 | | |
| | middle | 10 ± 2* | 90 | | |
| | old | 10 ± 2* | 90 | | |

Each batch of corpora lutea was examined in 5 aliquots in the binding assay and data are shown as the mean ± SEM percent of control binding (binding in the presence of buffer alone).

* $p < 0.01$ versus control (buffer alone)
\+ $p < 0.01$ versus extract of young corpus luteum

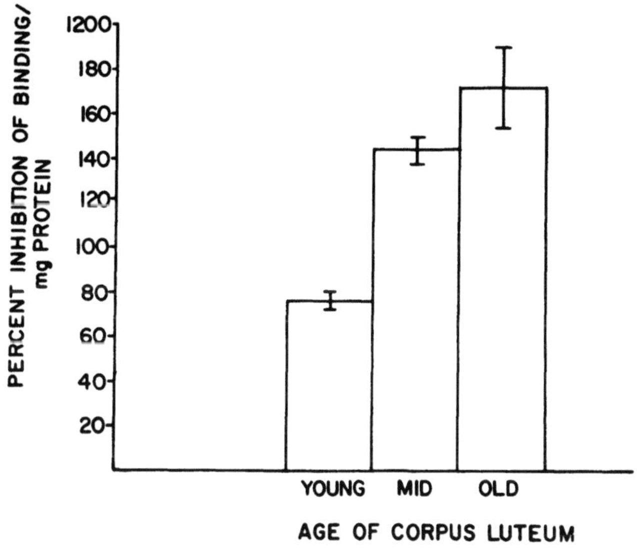


Figure 1. Ability of aqueous extracts of porcine corpus luteum to inhibit binding of 5 ng $^{125}$I-hCG to porcine granulosa cells. The values represent percent of control binding normalized per mg protein in one milliliter of crude extract. Each value represents the mean ± SEM of 5 replicate values.

## Acknowledgements

The support of NIH grant HD08834 and WHO contract number 77008 is gratefully acknowledged.

## REFERENCES

(1) K.P. Yang, N. Samaan and D.N. Ward. Endocrinology 98(1976) 233.

(2) C.N. Sakai, B. Engel and C.P. Channing. Soc. Exp. Biol. Med. 155(1976)155.

(3) W. Hansel, P.W. Cancannon and J.H. Lukaszewska. Biol. Reprod. 8(1973)222.

(4) L.L. Anderson and R.M. Melampy in: Reproduction in the female mammal. ed. G.E. Lamming and E.C. Amoroso pp. 285-316 Butterworth, London 1967.

(5) R.N. Anderson, F.L. Schwartz and L.C. Ulbers. Biol. Reprod. 10(1974)321.

(6) S. Kammerman, R.E. Canfield, J. Kolena and C.P. Channing. Endocrinology 91(1972)65.

(7) C.P. Channing and S. Kammerman. Endocrinology 92(1973)531.

(8) E.L. Akins and M.C. Marsisette. American Vet. Res. 29(1968) 1953.

(9) O.H. Lowry, N.J. Rosenbrough, A.L. Farr and R.J. Randall J. Biol. Chem. 193(1951)265.

(10) F.C. Greenwood, W.M. Hunter and J.S. Glover. Biochem. J. 89(1963)114.

(11) V.J. Goldberg and P.W. Ramwell. Physiol. Rev. 55(1975)325.

# SUPPRESSION OF OVARIAN FUNCTION BY LHRH AND ITS ANALOGUES IN PREGNANT RATS

Koji Yoshinaga

Laboratory of Human Reproduction and Reproductive Biology

Harvard Medical School, 45 Shattuck St., Boston, MA 02115

The chronic administration of large doses of LHRH and its analogues to rats from Days 1 through 7 of pregnancy delayed ovum implantation (1,2). The ovaries of pregnant rats treated with an LHRH analogue were significantly smaller than the ovaries of control intact rats; the corpora lutea were small and no large antral follicles were observed in the ovaries of the treated rats. In these rats serum levels of progesterone were significantly lower, and those of LH were significantly higher than those in control intact rats (2). The present studies were performed to determine (a) how soon after the initiation of the treatment is ovarian progesterone secretion affected by LHRH and its analogues, (b) what are the LH/hCG receptor levels in the ovaries of the rats treated with LHRH and its analogues, and (c) if chronic administration of LH would delay ovum implantation and reduce progesterone levels as was observed in the rats treated with LHRH and its analogues.

## Progestin levels in rats treated with LHRH and its analogues

Adult female rats, weighing 200-250 g, and showing regular 4-5 day cycles were used. On Day 1 of pregnancy (sperm positive day), the rats were assigned to four groups, each group consisting of at least 4 rats. Rats in Group I were left intact and served as controls. Rats in Groups II - IV were continuously infused with LHRH or its analogues by osmotic minipumps (Alzet) implanted sc between 1000 and 1400 h on Day 1. The infusion rates were as follows:

| Group | Peptide compound | Infusion rate(μg/24h) |
|---|---|---|
| II | LHRH | 100 |
| III | Analogue I (Des-Gly$^{10}$-LHRH-ethylamide) | 30 |
| IV | Analogue II [Des-Gly$^{10}$-(D-Ala$^{6}$)-LHRH-ethylamide] | 1 |

These doses were used because they were found sufficient to inhibit ovum implantation in 100% of the rats tested in our previous study on the dose-effect relationship (2).

Peripheral blood of these rats was collected under light ether anesthesia from a jugular vein every day between 1000 and 1400 h from Days 1 through 7 of pregnancy. On Day 8 peripheral blood, ovaries, and pituitaries were collected. Serum was separated from blood and frozen until progestin analysis. The ovaries were removed immediately after sacrifice, weighed and subjected to determination of LH/hCG receptor content. The uteri were examined for implantation swellings and the number of implantations was recorded. When no implantation swellings were observed, the uterine horns were flushed with saline and the flushings were examined for blastocysts.

Progestins were extracted from serum with diethylether. Progesterone and 20α-hydroxypregn-4-en-3-one (20α-OH-P) were separated by LH-20 column chromatography and were determined by radioimmunoassay (3). The serum concentration of progesterone in control pregnant rats increased rapidly from Day 1 to Day 5 of pregnancy. After reaching a peak on Day 5 it remained at a level above 60 ng/ml. This pattern of progesterone during early pregnancy is in agreemant with the results reported by Fajer and Barraclough (4) and Morishige *et al.* (5). As can be seen in Fig. 1, the serum concentration of progesterone in LHRH-treated rats and in Analogue I-treated rats increased from Day 1 to Day 3, but declined on Day 4 and remained low until Day 8. On the other hand, the serum concentration of progesterone in Analog II-treated rats increased from Day 1 to Day 4. It then declined to a low level on Day 5 and remained low until Day 8. Thus Analogue II was less potent than LHRH and Analogue I in reducing serum progesterone. At the doses used in the present study all three compounds were equipotent in inhibiting implantation. Since both progesterone and estrogen are required for implantation in the rat, it is likely that Analogue II is more potent than the other two compounds in suppressing estrogen secretion.

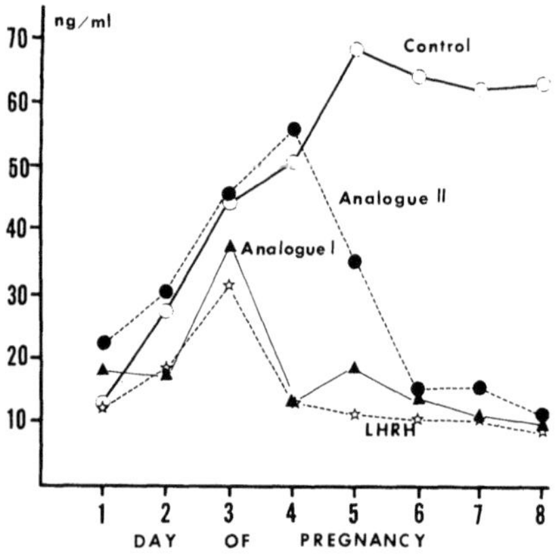


Fig. 1 Effect of LHRH and its analogues on serum concentration of progesterone in rats

The serum concentration of 20α-OH-P in control rats was high on Day 1 of pregnancy but gradually declined as pregnancy advanced. The pattern of 20α-OH-P during early pregnancy was influenced neither by LHRH nor by Analogues I and II. Ovum implantation was completely inhibited by LHRH and the analogues. Free blastocysts were flushed out of the uteri of all rats treated with these compounds except one rat treated with Analogue II.

Effect of LHRH and its analogues on LH/hCG receptor content in the ovarian tissue of early pregnant rats

It has been reported that treatment of male rats with LHRH analogues reduced testicular weight, LH/hCG and prolactin receptor levels and plasma androgen concentration (6,7). More recently, Kledzik *et al.* (8) reported that the daily dose of 100 µg of Analog II, when injected on Days 7 through 12 of pregnancy, completely terminated pregnancy. They found that this antifertility effect of Analogue II was accompanied by a decrease in plasma progesterone concentrations and complete inhibition of ovarian LH/hCG and FSH receptors. In the present study, we determined LH/hCG receptor levels in the ovarian tissues of the rats treated with LHRH and its analogues during the first week of pregnancy.

The LH/hCG receptor level in the ovarian tissue was determined according to the method reported by Auclair *et al.* (7) with minor modifications. The ovaries were homogenized in Tris-BSA buffer containing 0.25 M sucrose using a Tissumizer for 30 sec. The homogenate was centrifuged at 2,000 x *g* and the pellet was suspended in buffer using a Teflon-glass homogenizer. The pellet was resuspended in an aliquot of buffer so that each 200 µl contained 10 mg equivalent of wet ovarian tissue. An aliquot of 200 µl ovarian tissue suspension, 100 µl $^{125}$I-hCG (approximately 150,000 cpm, specific activity 77 µCi/µg, New England Nuclear), 200 µl buffer with or without 2 µg oLH (NIH-S-19) were mixed and incubated at 23$^{o}$C for 16 h. The reaction was stopped by adding 2 ml cold Tris-BSA buffer and the tubes were centrifuged at 2,000 x *g* for 30 min. The pellet was suspended in 2 ml cold buffer by vortexing. After centrifuging at 2,000 x *g* for 30 min the supernatant was aspirated out and the tubes were counted for radioactivity bound to the tissue.

The effect of LHRH and its analogues on ovarian LH/hCG receptor levels is shown in Fig. 2. In Fig. 2 the amounts of radioactivity bound to the membrane fraction in the absence of cold oLH are expressed as open circles and the amounts of radioactivity in the presence of cold oLH are expressed as closed circles. In control intact rats the amount of hCG bound to the tissue in the absence of cold oLH was approximately 3 times the amount of hCG bound to the tissue in the presence of cold hormone. In contrast to the high level of LH/hCG receptor in the control ovarian tissue, the receptor level in the ovaries of LHRH-treated rats was low, approximately one-ninth that in the control tissue. The receptor levels in the ovaries of rats treated with Analogues I and II were extremely

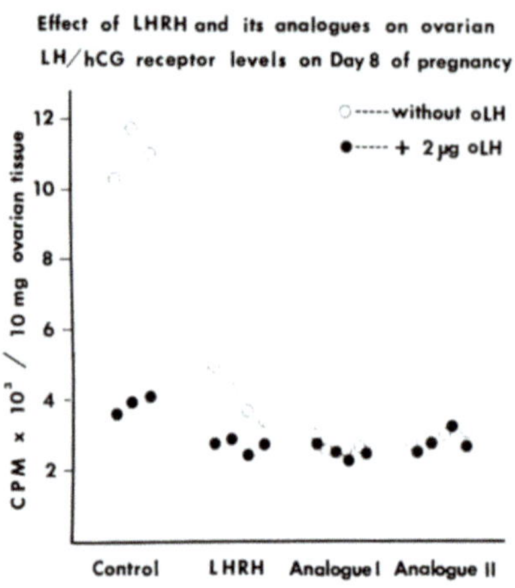


Fig. 2

low and there were no differences in hCG binding to the ovarian tissues whether or not they were incubated with cold hormone.

These results clearly show that the antifertility effect of LHRH and its analogues during the pre-implantation period is accompanied by a reduction of serum concentration of progesterone and inhibition of LH/hCG receptor in the ovarian tissue. The mode of action of Analogue II in inhibiting ovum implantation, therefore, appears to be the same as in terminating pregnancy during the post-implantation period of pregnancy.

Effect of LH on progestin secretion during early pregnancy.

In Analogue II-treated pregnant rats we showed that peripheral serum concentration of LH was significantly elevated and the pituitary content of LH was markedly decreased as compared with those in control intact pregnant rats (2). In contrast to the increased serum concentration of LH, the serum concentration of progesterone was suppressed in rats treated with LHRH and its analogues as shown in the above section of this paper. Since LH and hCG have been reported to terminate pregnancy (9) and to desensitize the gonads for LH by reducing the number of receptor sites (10,11), we investigated whether continuous infusion of LH to pregnant rats would inhibit ovum implantation in the same manner as did LHRH and its analogues.

Three groups of pregnant rats were used in this study. Ovine LH (NIH-S-19) was infused continuously by osmotic minipumps at the rate of 1 μg (Group V), 10 μg (Group VI) or 100 μg (Group VII) per 24 h from Day 1 through Day 8 of pregnancy. On Days 2,4 and 6 peripheral blood was collected from a jugular vein under light ether anesthesia and the rats were sacrificed on Day 8. At autopsy the ovaries and the pituitaries were weighed and the uterine horns were examined for implantation swellings. When no implantation sites were observed the uterine horns were flushed for blastocyst recovery.

Serum progesterone concentrations in these rats are listed in Table 1 and were compared with those values obtained in the rats treated with LHRH and its analogues. As can be seen in Table 1, serum concentration of progesterone in LH-treated rats was comparable or even higher than that in intact control rats. The serum

Table 1. Effects of LHRH, LHRH analogues and LH on peripheral serum concentration of progesterone and on ovarian and pituitary weight of pregnant rats.

| | | Serum concentration of progesterone (ng/ml) Day of pregnancy | | | | Organ weight on Day 8(mg) | |
|---|---|---|---|---|---|---|---|
| Group | Treatment | 2 | 4 | 6 | 8 | Ovaries | Pituitary |
| I | Control | 26.8±8.7* | 50.7±8.0 | 68.6±7.6 | 63.0±2.6 | 67.7±1.9 | 13.4±0.6 |
| II | LHRH | 18.4±3.2 | 13.3±2.4 | 10.2±1.4 | 8.6±2.5 | 47.6±2.1 | 12.1±0.3 |
| III | Analogue I | 16.9±3.3 | 13.3±2.6 | 13.8±2.6 | 9.1±1.9 | 60.5±3.1 | 12.4±0.8 |
| IV | Analogue II | 29.9±3.1 | 55.7±8.6 | 14.9±0.6 | 11.7±1.0 | 56.7±2.5 | 12.5±0.4 |
| V | LH ( 1 μg) | 17.0±1.5 | 71.6±5.2 | 74.5±3.1 | 75.2±14.3 | 69.9±4.4 | 12.8±0.9 |
| VI | LH ( 10 μg) | 13.9±1.5 | 62.8±7.3 | 64.6±6.4 | 63.1±5.9 | 72.9±5.1 | 12.9±0.5 |
| VII | LH (100 μg) | 21.0±1.7 | 51.4±8.0 | 76.7±7.3 | 99.0±10.2 | 144.0±18.0 | 21.1±0.9 |

* Mean ±SE

concentrations of 20α-OH-P were not different from intact control values in rats treated with lower doses of LH. However, 20α-OH-P concentrations remained elevated in rats treated with the highest dose of LH throughout the expeimental period.

Implantation of ova was not inhibited in rats treated with the low doses of LH. The uteri of the rats treated with 100 μg LH were soft and swollen due to water imbibition. Neither implantation sites nor free blastocysts were present in the uteri of 6 of 8 such treated rats. The uteri of the two remaining rats had 2-3 implantation sites. The ovaries contained large corpora lutea and large follicles. The ovaries and pituitaries in this group were heavier than those in the rats of any other group (Table 1). These results indicate that chronic administration of LH did not reproduce the results obtained by chronic administration of LHRH and its analogues. Interruption of pregnancy by a large dose of LH appears to be due to excessive secretion of estrogen and not due to reduction in progesterone secretion.

## SUMMARY

1. LHRH and its analogues, when infused continuously at the doses that inhibit ovum implantation, reduced serum concentration of progesterone in pregnant rats. The progesterone concentration fell between Days 3 and 4 of pregnancy in rats treated with LHRH and Analogue I (Des-Gly$^{10}$-LHRH-ethylamide), but fell between Days 4 and 5 in rats treated with Analogue II [Des-Gly$^{10}$-(D-Ala$^{6}$)-LHRH-ethylamide].

2. Ovarian LH/hCG receptor levels were markedly reduced in rats treated with LHRH and its analogues in the first week of pregnancy.

3. Continuous infusion of LH to pregnant rats did not reduce the serum concentration of progesterone even with the highest dose that inhibited implantation. Interruption of pregnancy by LH appears to be due to excessive secretion of estrogen as indicated by uterine swellings, presence of large follicles in the ovaries, and absence of free blastocysts in the uterine lumen. Thus, the antifertility effects of LHRH and its analogues were not reproduced by LH administration.

## ACKNOWLEDGMENT

The author thanks Dr. Masahiko Fujino of Takeda Chemical Industries, Osaka, Japan for the gift of LHRH and its analogues and NIAMDD for the gift of ovine LH (NIH-S-19). He also wishes to thank Mr. Robert L. Pilot and Mrs. Jaime Krenn for their excellent technical assistance, and Mrs. Stella Nieland for her help in the preparation of the manuscript.

## REFERENCES

(1) Y.C. Lin and K. Yoshinaga. Endocrinology 98 (1976) A-143.

(2) K. Yoshinaga and M. Fujino. In: The Maternal Recognition of Pregnancy. Ciba Foundation Symposium No. 64 (New Series) (1979) in press.

(3) J.J. Ford and K. Yoshinaga. Endocrinology 96 (1975) 329.

(4) A.B. Fajer and C.A. Barraclough. Endocrinology 81 (1967) 617.

(5) W.K. Morishige, G.J. Pepe and I. Rothchild. Endocrinology 92 (1973) 1527.

(6) C. Auclair, P.A. Kelly, F. Labrie, D.H. Coy and A.V. Schally. Biochem Biophys Res Comm 76 (1977) 855.

(7) C. Auclair, P.A. Kelly, D.H. Coy, A.V. Schally and F. Labrie. Endocrinology 101 (1977) 1890.

(8) G.S. Kledzik, L. Cusan, C. Auclair, P.A. Kelly and F. Labrie. Fertil. Steril 29 (1978) 560.

(9) W.H. Yang and M.C. Chang. Endocrinology 83 (1968) 217.

(10) S.A. Lamprecht, U. Zor, A. Tsafriri and H.R. Lindner. J. Endocr. 57 (1973) 217.

(11) A.J.W. Hsueh, M.L. Dufau and K.J. Catt. Biochem Biophys Res Comm 72 (1976) 1145.

SESSION NUMBER: IX

TITLE: CLINICAL APPLICATIONS IN THE AREA OF CONTRACEPTIVE DEVELOPMENT

PRINCIPAL SPEAKER: Kenneth Ryan

INVITED DISCUSSANT: William Sadler

PRESENTED ABSTRACTS: 19.) J. Quagliarello, R. Nachtigall, L.T. Goldsmith, J. Hochman, B.G. Steinetz, E.M. O'Byrne and G. Weiss. Serum Immunoreactive Relaxin Concentrations in Human Pregnancy, Labor and the Puerperium.

SUMMARY: Irving Rothchild

CHAIRPERSON: Sheldon Segal

# CLINICAL APPLICATIONS IN THE AREA OF CONTRACEPTIVE DEVELOPMENT

KENNETH J. RYAN

DEPT. OF OBS. & GYN., HARVARD MEDICAL SCHOOL

45 SHATTUCK STREET, BOSTON, MA 02115

"Leads" for new contraceptive development include possible direct manipulation of ovarian function and are an outgrowth of studies on basic gonadal physiology. Being able to perturb normal reproductive activity by altering some aspect of ovarian function also may provide evidence for a specific role for that function. The research thus serves the dual role of advancing basic knowledge while providing possible clues for new contraceptive modalities and human application. Present contraceptive methodology ranges from high efficacy with low (but marginally acceptable) risks of adverse effects as with the pill to the less effective techniques that carry the inherent risk of pregnancy when they fail as for most barrier techniques. With an effectiveness of 99% and a mortality risk one fifth that of a pregnancy, the pill is still a hard act to follow. It is unlikely that drug regulatory agencies will allow marketing of any new product that clearly does not surpass present methods. Contraceptive development involving the ovary clearly is not yet in so favorable a category.

## THEORETICAL CONSIDERATIONS

While effectiveness, safety, and patient acceptance are the ultimate standards against which any new contraceptive method will be judged, there are some caveats in development that might profitably be observed. These include:

## Functional Specificty of Action

The method should be expected to have a highly favorable advantage in acting predominantly, if not totally, on the target function. Inhibition of prostaglandin synthesis or action, administration of prostaglandins, and steroid biosynthesis inhibitors thus far all lack specificity to the reproductive system. It remains to be seen whether a functional specificity can be achieved. Immunological methods have promise from their inherent specificity of acting on a specific function.

## Target Organ Specificity

Preferential concentration of a contraceptive agent in a target organ can theoretically overcome some lack of specificity of action and/or can optimize a specific action. Methods involving a specific target cell receptor has the requisite potential for organ localization.

## Duration of Action and Reversibility

Techniques which have transient action have limitations unless related to ovulation and coitus. Prolonged duration of action and reversibility can influence clinical utility.

## Relevance to Human Physiology

Much research on ovarian function quite appropriately involves laboratory animal models and there is much in this which can be appropriately tested for relevance to primate biology. It is important to point out, however, that differences as well as similarities should be considered among animal species and an attractive contraceptive possibility for the laboratory rat or rabbit may be inapplicable to the human because of fundamental differences in ovarian function.

# SPECIFIC OVARIAN METHODOLOGIES

1. Ovulation anticipation would provide a natural insight into reproductive function present in essentially all animals, save the human. The chemical signals thus far identified; serum estradiol rise, the serum luteinizing hormone peak or some as yet unknown cyclic change need to be amplified or transduced for prac-

tical application. Thus far the basal body temperature and cervical mucus chemical or physical change fall far short of the goal. The inability to monitor the internal changes of the ovarian cycle remains a most fascinating challenge.

2. Immunization methods involving specific proteins of the ovum or follicle are theoretically attractive because of potential specificity but experience with the influenza vaccine and auto-immune mechanisms should inject a cautionary note into any chances of success.

3. Receptor regulation has the prospect of target organ localization especially with receptors for the gonadotropins. In this respect, animal models have been described which can result in loss of functional receptors and desensitization of the adenyl cyclase system of the corpus luteum (1,2). Inhibition of pregnancy has been achieved indirectly via an LH-RH agonist which results in overstimulation of gonadotropins with consequent desensitization (3). Specificity of action and target organ localization can be expected by this modality. On the other hand, regulation of steroid receptors in the ovary to the exclusion of other important tissue receptor sites seems more remote. In this regard, progesterone receptors in the uterus are a more attractive target.

4. Since prostaglandins have been demonstrated to be synthesized in the follicle (4,5,6), and important in the ovulatory process, attempts have been made to block ovulation with antagonists of prostaglandin synthesis (indomethacin) (7,8,9). There is, however, neither specificity of action nor specific target organ localization in this method. Indomethacin can block LH release and thereby inhibit ovulation (9) but can also inhibit gonadotropin induced ovulation in the rhesus and marmoset monkey (7,8). In the latter instances, the ovum is trapped and luteinization and hormone production are adequate. The toxic effects of the drug remain a problem.

5. Induction of luteolysis as an ovarian cycle regulator or indirect abortifacient remains an attractive possibility. The human corpus luteum functions differently than most mammals in having only an internal natural luteolytic process believed to be triggered by either local estrogens and/or prostaglandins. Three techniques have been explored: exogenous estrogens, exogenous prostaglandins and inhibitors of steroid synthesis aimed at the 3-β-ol dehydrogenase or cholesterol side chain cleavage sites.

a) Exogenous estrogens have been effective as post-coital agents and as luteolytic agents (10) but the high dose, attendent toxicity and need for repetitive use thus far limit wide applicability.

b) Administration of prostaglandin $F_{2\alpha}$ directly into the corpus luteum at a level of 1000 μg can provide a transient drop in progesterone and shortening of the luteal phase of the cycle (11). The lack of specificity of prostaglandins, however, when given systematically would restrict their use without significant improvement in delivery and localization of action. The human granulosa cell and corpus luteum synthesize prostaglandin $F_{2\alpha}$ (5,12). Some studies suggest a reciprocal relationship with progesterone production (12) while prostaglandin $E_2$ enhances progesterone synthesis (13). Activation of the endogenous system or blockage of corpus luteum maintenance mechanism invoked by pregnancy have not yet been achieved (14).

c) Steroid synthesis blockers have been considered as possible luteolytic agents. Aminoglutethimide which blocks cholesterol side chain cleavage in both the ovary and adrenal seemed to have a functional advantage for inhibition of the corpus luteum (15-18). Blockage of adrenal corticoid synthesis at high dosages in cancer cases has been overcome by simultaneous administration of dexamethasone or hydrocortisol (15). Finally, the ability of aminoglutethimide to block aromatase points up a further limitation in its specificity (16). Cyanoketones and other synthetic steroids have also been employed as inhibitors of the 3-β-ol steroid dehydrogenases but organ specificity and toxicity remain considerations for their use (19).

Recently a drug Danazol used for endometriosis and as part of a male contraceptive has received attention because of its lack of intrinsic hormone activity and supposed specificity in inhibiting gonadotropin release. Some animal work, however, indicates that it also acts as a steroid biosynthesis inhibitor in a wide variety of hormone producing tissues (20,21).

While no single agent has as yet been identified, the establishment of model screening systems for checking inhibitory actions and binding to specific P-450 enzymes might result in the development of a useful drug (20-22).

## SUMMARY

The ovary is not necessarily a prime locus for regulating reproductive function but in the process of understanding various aspects of ovarian function, opportunities have arisen to test both the role of specific functions in the reproductive process and the possible use of a given function as a site for contraceptive regulation. Better understanding of corpus luteum physiology from receptor site-gonadotropin interaction to natural luteolytic processes still hold promise for contraceptive development.

REFERENCES

(1) M. Conti, J.P. Harwood, A.J.W. Hsueh, M.L. Dufau, and K.J. Catt. J. Biol. Chem. 251(1976)7729.

(2) M. Hunzicker-Dunn, and L. Birnbaumer. Endocrinology 99(1976) 211.

(3) G.S. Kledzik, L. Cusan, C. Auclair, P.A. Kelly, and F. Labrie. Fertil. & Steril. 29(1978)560.

(4) J.M. Marsh, Norman S.T. Yang, and W.J. LeMaire. Prostaglandins 7(1974)269

(5) J.R.G. Challis, G.F. Erickson, and K.J. Ryan. Prostaglandins 7(1974)183.

(6) E.R. Plunkett, Y.S. Moon, J. Zamecnik, and D.T. Armstrong. Am. J. Obs. & Gyn. 123(1975)391.

(7) H. Mara, F. Barbosa, and E.M. Continto. Fertil. & Steril. 21 (1978)565.

(8) E.E. Wallach, A. de la Cruz, J. Hurt, K.H. Wright, and V.C. Stevens. Prostaglandins 9(1975)645.

(9) J.C. Carlson, B. Barcikowski, V. Cargill, and J.A. McCracken. J. Clin. Endocrinol. & Metab. 39(1974)399.

(10) Gore, B.Z., B.V. Caldwell, and L. Speroff. J. Clin. Endocrinol. & Metab. 36(1973)615.

(11) A.R. Korda, D.A. Shutt, I.D. Smith, R.P. Shearman, and R.C. Lyneham. Prostaglandins 9(1975)443.

(12) D.A. Shutt, A.H. Clarke, I.S. Fraser, P. Goh, G.R. McMahon, D.M. Saunders, and R.P. Shearman. J. Endocrinol. 71(1971)453.

(13) McNatty, K.P., K.M. Henderson, and R.S. Sawers. J. Endocrinol. 67(1975)231.

(14) Powell, W.S., S. Hammarstrom, B. Samuelsson, and B. Sjoberg. Lancet 1(1974)1120.

(15) R.J. Santen, S.A. Wells, R. Runic, C. Gupta, J. Kendall, E. B. Rudy, and E. Samojlik. J. Clin. Endocrinol. & Metab. 45 (1977)469.

(16) E. Samojlik, R.J. Santen, and S.A Wells. J. Clin. Endocrinol. & Metab. 45(1977)480.

(17) E.N. McIntosh, H.R. Holtrop, C. Alonso, and H.A. Salhanick. Endocrine Society, 1974 (Abstract).

(18) V.I. Uzgins, C.A. Whipple and H.A. Salhanick. Endocrinology 101(1977)89.

(19) A.S. Goldman and K. Sheth. Biochim. et Biophys. Acta 315(1973) 233.

(20) R.L. Barbieri, J.A. Canick, A. Makris, R.B. Todd, I.J. Davies, and K.J. Ryan. Fertil. & Steril. 28(1977)809.

(21) R.L. Barbieri, J.A. Canick, and K.J. Ryan. Endocrinology 101(1977)1676.

(22) D.B. Gower. J. Steroid Biochem. 5(1974)501.

# SERUM IMMUNOREACTIVE RELAXIN CONCENTRATIONS IN HUMAN PREGNANCY, LABOR AND THE PUERPERIUM

J. Quagliarello, R. Nachtigall, L.T. Goldsmith
J. Hochman, B.G. Steinetz, E.M. O'Byrne,G. Weiss
Depts. OB-GYN & Physiology, New York University
School of Medicine, New York, & Ciba-Geigy Corp.
Ardsley, New York

Relaxin is a peptide hormone of pregnancy which plays a role in pregnancy maintenance and parturition in several mammalian species (1). Immunoassayable relaxin is present in women during pregnancy. Relaxin is produced solely by the human corpus luteum of pregnancy. The placenta does not directly contribute to its production (2). Relaxin extracted from the human corpus luteum of pregnancy is active in the guinea pig pubic symphysis palpation assay. The biological activity paralleled the immunoactivity of these extracts (3). The purpose of this study is to determine secretion patterns of relaxin at various times during normal pregnancies.

## METHODS

Sera were obtained from 23 women during or after the luteal phase of conception cycles. The date of conception was determined by basal body temperature records. Eight of these patients had exogenously induced ovulations.

Sera were obtained serially from 16 women during the third trimester of normal pregnancies. Ten additional patients were sampled in the third trimester and again during labor. Samples from the first stage of labor were correlated with cervical dilation.

Daily serum samples from 13 lactating and 15 non-lactating women were obtained for three postpartum days. None of these women were treated with neuroleptics, narcotics, hormones or dopamine agonists. Groups of 12 women in the follicular phase of the menstrual cycle, the luteal phase of the menstrual cycle, during anovulatory cycles and taking combination oral contraceptive agents contributed serum samples, as did three hyperprolactinemic women.

Relaxin was measured by RIA using 125-I labelled polytyrosyl-relaxin(4) and rabbit anti-porcine relaxin serum R6-7/14/74 which has been used to measure relaxin in women (5).

## RESULTS

Immunoreactive relaxin was undetectable in the sera of all 51 nonpregnant women, whether they were in the follicular or luteal phases, taking oral contraceptives, were anovulatory or were hyperprolactinemic. Relaxin became detectable by the time of the missed menses in conception cycles.

Serum concentrations of relaxin were fairly constant in 16 third trimester patients (fig. 1). Two of these patients had levels more than twice that of any others. One had a spontaneous ovulation and one had a high dose clomiphene-induced ovulation. All pregnancies resulted in singleton births

No significant changes were noted between antepartum and intrapartum samples obtained in 10 patients (fig.2). Five of these patients were sampled within 4 days of the onset of labor.

The pattern of decline of circulating immunoreactive relaxin in lactating and nonlactating patients was similar for the first three postpartum days (fig.3).

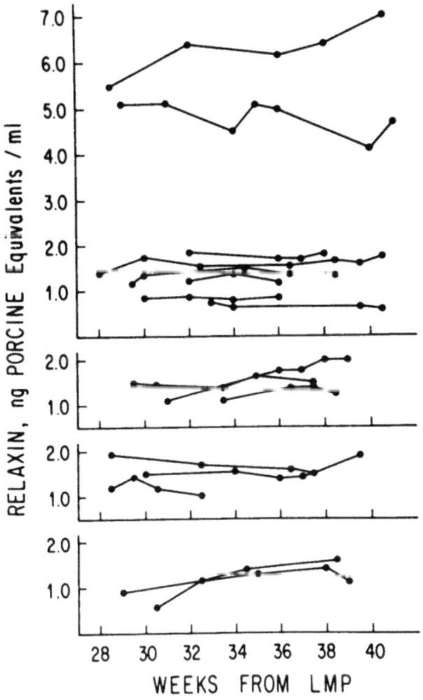


Fig.1. Serum Relaxin Concentrations in the third trimester of pregnancy

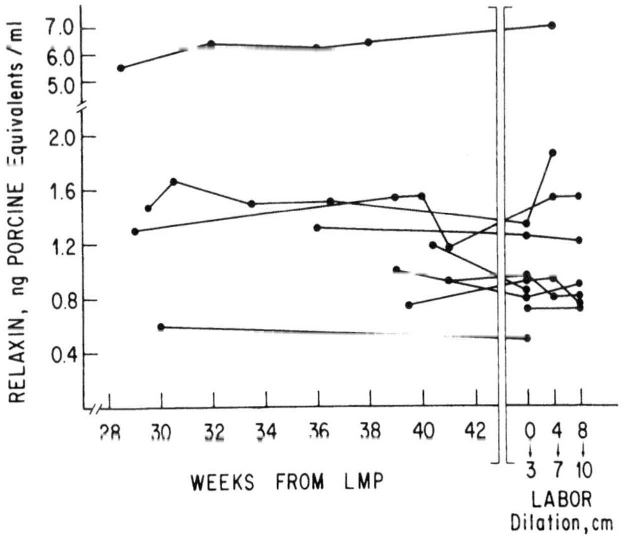


Fig.2. Antepartum and Intrapartum relaxin concentrations

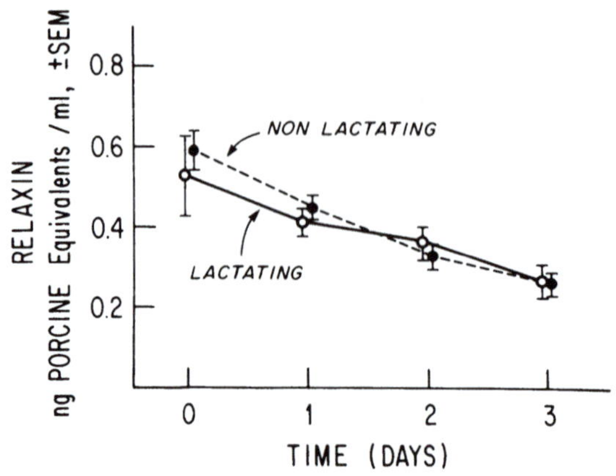


Fig. 3. Daily puerperal relaxin concentrations in 13 lactating and 15 nonlactating women.

## DISCUSSION

The detection of relaxin in early pregnancy at the time of the missed menses and the absence of relaxin in the nonpregnant state suggests that relaxin may be used as a specific pregnancy test. Since relaxin is a product of the corpus luteum only, it is the most specific indicator of luteal activity in pregnancy. The steroid hormones are also products of the placenta or fetus. Thus, relaxin is a substance that should prove most helpful in determining the exact role of the corpus luteum in early pregnancy maintenance. Relaxin inhibits uterine contractions in many species (1). We have found, in preliminary experiments, that relaxin-containing progesterone-free luteal extracts decrease the amplitude of spontaneous human myometrial contractions _in vitro_. Although the corpus luteum is necessary for early pregnancy maintenance, physiological doses of progesterone are inadequate to maintain pregnancies (6). _In vitro_ human myometrial contraction inhibition requires supraphysiological progesterone concentrations (7). Relaxin appearance in early pregnancy suggests a role, in concert with progesterone, in pregnancy maintenance.

The serum concentrations of immunoreactive relaxin vary little in late pregnancy and throughout spontaneous labor. They vary little in induced labors as well (8). We did not detect pre-labor surges in relaxin as were seen in other species (9) but our sampling procedures were not sufficient to detect short surges before the onset of spontaneous labor.

The similar decreases in serum relaxin concentrations during the postpartum period in lactating and nonlactating women suggests that the higher prolactin concentrations in nursing mothers are not luteotropic.

(1) B.G. Steinetz, G. Weiss, A. Segaloff, J.K. Mc Donald, E. O'Byrne, J. Hochman, B·Carriere and L. Goldsmith. Rec. Prog. Horm. Res. 34 (1978) in press.

(2) G. Weiss, E. M. O'Byrne, J. Hochman, B.G. Steinetz, L. Goldsmith and J.F. Flitcraft. Obstet. Gynecol. in press.

(3) E. M. O'Byrne, J.F. Flitcraft, W. K. Sawyer, J. Hochman, G. Weiss and B.G. Steinetz. Endocrinology. 102 (1978) 1641.

(4) O.D. Sherwood, K.R. Rosentreter and M.C. Birkhimer. Endocrinology. 96 (1975) 1106.

(5) E. M. O'Byrne and B.G. Steinetz. Proc. Soc. Exp. Biol. Med. 152 (1976) 272.

(6) A. I. Csapo and M. Polkkinen. Obstet. Gynecol. Survey. 33 (1978) 69.

(7) D. Kumar, T. Wagatsuma, W. J. Sullivan, Jr. and A. C. Barnes. Amer. J. Obstet. Gynecol.90 (1964)1355.

(8) J. Hochman, G. Weiss, B.G. Steinetz and E. M. O' Byrne. Amer. J. Obstet. Gynecol. 130 (1978)473.

(9) O.D. Sherwood, C.C. Chang, G. W. Be Vier, J. R. Diehl and P. J. Dziuk. Endocrinology. 98 (1976) 875.

Supported by United Cerebral Palsy Grant R-288-77.

DISCUSSION - S. SEGAL - CHAIRPERSON

J. HOLT: Have you looked at any granulosa cell tumors; is there a possibility of using relaxin as a marker for adnexal masses?

G. WEISS: We looked at about 25 blind tumors; the only tumors that had relaxin were two choriocarcinomas that were gestational. Two choriocarcinomas from postmenopausal women did not have relaxin.

H. LINDNER: Just to make this relevant, can you prevent or interrupt pregnancy by immunizing against relaxin in any species?

DR. WEISS: We've only studied this in humans and we're not about to get into that, quite yet.

L. NILSSON: I wonder, would you suggest that relaxin would be a good thing to measure to assess whether a prostaglandin analogue would be luteolytic or not in the human?

G. WEISS: I think in pregnancy, you have a problem of something to measure. You can't use progesterone measurements because this steroid comes from at least two sources. You don't know whether you're affecting the uterus in the pregnancy or whether you're affecting the corpus luteum. Since relaxin is only a luteal product, I think you can study the function of the corpus luteum by measuring relaxin.

K. RYAN: Gerson, you say it's only a luteal product, but have you excluded the placenta? In years gone by it had been described in the placenta with the old assays.

G. WEISS: It has previously been described in the placenta, but when we take out the corpus luteum, it drops and there's hardly any in the placenta.

# EFFECT OF REMOVAL OF CARBOHYDRATE RESIDUES UPON THE HALF LIFE AND IN VIVO BIOLOGICAL ACTIVITY OF HUMAN CHORIONIC GONDADOTROPIN

Satish K. Batta; M.A. Rabovsky; C.P. Channing and O.P. Bahl.* Department of Physiology, Univ. of Maryland Med. Sch., Baltimore., MD 21201 and *Div. of Cell and Molecular Biology, SUNY, Buffalo, N.Y. 14214

## Introduction

It has been previously shown (1, 2) that progressive removal of sialic acid from human chorionic gonadotropin (hCG) decreases the half life, as well as the *in vivo* biologic activity of the gonadotropin. The effect of removing sugar residues internal to sialic acid upon the half life and biologic activity of hCG has not yet been examined. *In vitro* studies, using isolated porcine (3) and monkey (4) granulosa cells, indicate that the successive removal of sugar residues internal to sialic acid from native hCG lead to a diminution in ability to stimulate cyclic AMP accumulation and progesterone secretion. The ability to inhibit binding of iodinated hCG to porcine granulosa cells was not appreciably altered provided the hCG derivative was added prior to the tracer molecule. In both monkey and pig granulosa cells the hCG derivatives with sugar residues internal to sialic acid removed, acted as competitive inhibitors of hCG action. In order to examine whether hCG derivatives have biologic activity *in vivo* and whether they have the ability to inhibit LH and hCG action *in vivo*, their possible interfering effect on ovulation in prepuberal female rats primed with pregnant mare serum gonadotropin was studied. In addition, the half life of the hCG derivatives was examined in adult female rats.

## Materials and Methods

hCG derivatives. The following hCG derivatives with carbohydrate residues enzymatically removed, as outlined previously (4, 5), were used in these studies: NG-hCG-(asialo-agalacto-hCG) 100% sialic acid and 60% galactose removed; NGA-hCG (asialo-agalacto-aglucosamino-hCG) 100% sialic acid, 60% galactose and 55% glucosamine removed; NGAM-hCG (asialo-agalacto-aglucosamino-amannoso-hCG) 100% sialic acid, 60% galactose, 55% glucosamine and 20% mannose removed). Purified native hCG with a potency of 12,000 IU/mg was used as a starting material. Adult cycling female rats (Sprague Dawley) and immature female rats were used in the study. The animals were maintained in environment controlled rooms with 14h light day and at 22 $\pm$ 0.5°C. The adult cycling females were examined daily for vaginal cytology. Those rats showing two consecutive four day cycles were used for half life studies of hCG and its derivatives: NG, NGA, and NGAM-HCG. The tissue distribution of these derivatives was also studied at the time of sacrifice.

## Half life and tissue distribution of hCG & its derivatives

Cycling female rats in diestrous were anaesthetized and an indwelling catheter was placed in the right atrium via the jugular vein. The indwelling catheter was used for both the injection of radiolabelled hCG and its derivatives and for blood sampling at various intervals. The blood volume withdrawn was replaced with an equal volume with physiologic saline (0.9% NaCl). After the injection of labelled hCG or its derivatives (1-5 X $10^6$ cpm of 125 [I] - hCG or hCG derivatives) blood samples were taken at 30 seconds, and every minute thereafter for the first five minutes, followed by every five minutes thereafter for a period of 1 hr. Between 1h and 4h, blood samples were taken every 30 minutes. $^{125}I$ was counted in each sample using a gamma counter with a detection efficiency of 70%. At the time of sacrifice various organs were removed and a portion of the tissue weighed on a torsion balance. The $^{125}I$ was counted in the tissue and the cpm/100 mg of wet tissue weight was calculated.

## Radio iodination of hCG and hCG derivatives.

Labelling of hCG and its derivatives with $^{125}I$ was performed by a modification (5) of the chloramine-T procedure employing Na $^{125}I$ in NaOH (7). Ten μg of hCG derivative in 0.05M phosphate buffer, pH 7.4 was kept on ice in a small glass tube, Na $^{125}I$ was added followed by addition of 10 μg of chloramine-T in 0.05M phosphate buffer. The contents were gently mixed and the iodination tube was kept on ice for 1 minute. The iodination reaction was stopped by addition of 10μg of sodium metabisulfite and applied to a biogel P50 column. The fraction containing the highest specific activity after the void volume was used for half life and distribution studies.

The biologic activity of the iodinated hCG and its derivatives was estimated by their ability to bind to porcine granulosa cells obtained from large follicles, as detailed previously (10). The assay consisted of incubating the labelled hCG or derivative in the presence and absence of excess unlabelled LH. A fraction was considered "active" if it exhibited >80% specific binding to the granulosa cells. Native hCG bound with 95% specific binding. The labelled hCG and its derivatives were used within one week of iodination.

Biologic activity in vivo

Immature, 25 day old Sprague Dawley female rats were primed to ovulate with 5U PMS given subcutaneously (8). After 24, 48, 50 and 52 hrs. rats were injected i.p. either with native hCG (10IU; 0.6 μg pure hCG equivalent) or hCG derivatives (4-7 μg NG-hCG, NGA-hCG, NGAM-hCG). The rats were sacrificed 73 hrs. after PMS injection. Control animals, which received saline, had an endogenous LH surge at 56 hrs. and ovulated before 70 hrs. (9) In each experiment, 5-10 rats were used. At the time of sacrifice, the ovarian ampullae were removed and placed between two microslides and examined under the microscope and the number of oocytes counted. In addition, the ovaries and uteri were removed, dissected free of fat and weighed. Statistical analysis was carried out using students "t" test or using a chi square test.

Results

Effect of removal of sugar moieties on the plasma half life of hCG

The half life of native hCG determined by the disappearance of $^{125}I$ in the plasma was 60-90 minutes (n=5). The half life of NG-hCG, NGA-hCG and NGAM-hCG (used within 3 days of iodination) was less than 2 minutes. (n=3, 4 & 2 respectively). (Figure 1)

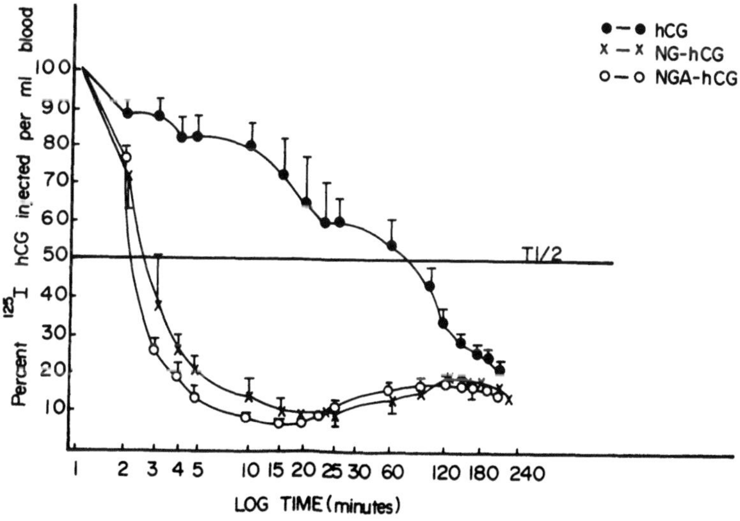


Figure 1. Half life of labelled hCG, NG-hCG and NGA-hCG in diestrous female rats.

Distribution of labelled hCG and its derivatives

At the end of each experiment carried out for half life studies of labelled hCG and its derivatives, the radio-activity present in each organ was normalized per 100 mg tissue as well as per 5 X $10^6$ cpm injected and is summarized in table I.

Table I

Uptake of $^{125}$I-hCG, NG-hCG, NGA-hCG & NGAM-hCG by various rat tissues 4 hrs. after administration.

| Rat Tissue | hCG (n=5) | NG-hCG (n=3) | NGA-hCG (n=4) | NGAM-hCG (n=1) |
|---|---|---|---|---|
| Thyroid | 69.8 | 67.6 | 193 | 34.5 |
| Corpus Luteum (C.L.) | 99.2 | 3.7 | 1.1 | 6.6 |
| Non C.L. ovary | 34.7 | 2.5 | 1.6 | 5.56 |
| Lung | 6.7 | <1.0 | <1.0 | <1.0 |
| Spleen | 3.4 | <1.0 | <1.0 | <1.0 |
| Stomach | <1.0 | 2.8 | 2.8 | <1.0 |
| Liver | <1.0 | 7.5 | <1.0 | <1.0 |
| Kidney | 8.6 | 7.0 | 3.7 | <1.0 |
| Urine | N.P. | 587 | 2090 | N.P. |
| Pituitary | <1.0 | <1.0 | <1.0 | 19.1 |

N.P. - not present

Results are expressed as the mean of ratio of cpm per 100 mg of tissue to the cpm per 0.1 ml whole blood.

The labelled native hCG was concentrated by corpora lutea and non-luteal portion of the ovary, in addition to the thyroid. Kidney, lung and spleen had more radioactivity than was present in 100 µl of blood at that time. The liver, kidney and corpora lutea had significant amounts of radioactivity in addition to thyroid after $^{125}$I-NG-hCG administration. The $^{125}$I-NGA-hCG was concentrated in thyroid and was probably excreted via kidney and stomach. The labelled NGAM-hCG was found only in pituitary and thyroid.

Biologic activity of hCG and its derivatives

PMS-injected, immature female rats ovulated with a rate of 9.2 1.2 (mean $\pm$ S.E.) ova per rat, however, when these animals were also administered hCG (0.6 µg/rat) at various intervals after PMS, surprisingly, there was not only a significant diminution in number of ova released per rat; but also the number of rats ovulating was decreased. There was no effect of NG-hCG, NGA-hCG, NGAM-hCG on either incidence of ovulation, number of oocytes per rat, or on the weight of ovaries and uteri from treated rats. (table II).

Table II

Effect of hCG and its various derivatives on incidence of ovulation, oocytes per rat and weight of uteri and ovaries in PMS primed prepuberal female rats.

| Group | Dose ug/ animal | Time (hr) after PMS | No. of rats used/ ovulated | Ova per rat (S.E.M.) | Uterine weight (mg, S.E.M.) | Ovarian weight (mg, S.E.M.) |
|---|---|---|---|---|---|---|
| Saline | - | 24, 48, 50, 52 | 5/5 | 9.2 ± 1.2 | 128.8 ± 5.8 | 40.3 ± 4.8 |
| hCG | 0.6 | - 11 - | 10/2[a] | 5.5[b] ± 1.5 | 159.7 ± 10.4 | 44.5 ± 4.4 |
| NG-hCG | 7.0 | - 11 - | 6/6 | 6.6 ± 1.3 | 131.8 ± 6.3 | 44.0 ± 2.4 |
| NGA-hCG | 5.0 | - 11 - | 6/5 | 8.4 ± 2.4 | 143.9 ± 1.9 | 45.9 ± 2.9 |
| NGAM-hCG | 4.0 | - 11 - | 6/5 | 7.6 ± 2.7 | 127.9 ± 4.0 | 46.7 ± 3.6 |

a = hCG VS Saline $p<0.05$ ($X^2$ test)

b = hCG VS Saline $p<0.05$ ('t' test)

In order to observe if hCG derivatives would compete with native hCG for ovarian receptor sites, another series of experiments was conducted. The endogenous surge of LH was blocked by injection of Nembutal (3 mg/100 g bw) during the "critical period", ie. 52h after PMS injection. After the Nembutal injection, hCG derivatives were given within one hour before the injection of native hCG as shown in the table III.

Table III

Effect of hCG and its various derivatives on incidence of ovulation in PMS primed prepuberal female rats.

| Group | Dose (μg/ animal) | Time after PMS (hr) | No. of rats used/ ovulated | Ova per rat (S.E.M.) |
|---|---|---|---|---|
| Saline | - | 52 | 5/5 | 7.0 ± 1.8 |
| hCG | 0.6 | 53 | 5/5 | 29.4 ± 8.3 |
| Nembutal | 3000 | 52 | 5/2 | 3.2 ± 1.2 |
| Nembutal<br>+ hCG | 3000<br>0.6 | 52<br>53 | 5/4 | 26.5 ± 4.0 |
| NC.hCG | 4.0 | 52 | 5/5 | 5.8 ± 1.4 |
| NG.hCG +<br>Nembutal<br>+ hCG | 4.0<br>3000<br>0.6 | 52<br>52<br>53 | 5/4 | 6.3 ± 1.3 |
| NGA-hCG | 0.6 | 52 | 5/5 | 11.0 ± 3.6 |
| NGA-hCG +<br>Nembutal<br>+ hCG | 0.6<br>3000<br>0.6 | 52<br>52<br>53 | 6/4 | 12.0 ± 4.4 |

The injection of 0.6 μg hCG to the PMS-primed rats increased the number of oocytes from 7.0 to 29.4. The injection of Nembutal alone reduced the number of oocytes per rat to 3.2. The inhibitory effect of Nembutal on ovulation could be reversed by a subsequent administration of native hCG (26.5 oocytes/rat). When NG-hCG and NGA-hCG were given at 52 hrs. post PMS, there was no effect on oocytes released per rat (5.8 ± 1.4 and 11.0 ± 3.6/ respectively VS 7.0 ± 1.8). When NG-hCG, NGA-hCG were given in concert with Nembutal and hCG, the stimulatory effect of hCG was not evident. The number of oocytes released per rat were similar to those obtained when NG-hCG and NGA-hCG was given alone to PMS primed rats.

## Discussion

Removal of sugar moieties from native hCG appears to reduce the effectiveness of hCG probably by reducing the half life. This corroborates the observations made by others (1, 2). This would appear to be the case, since desialylation of ceruloplasmin and other glycoproteins is known to cause increased clearance of these substances from plasma (11, 12). Removal of various sugar moieties appears to affect the accumulation of various derivatives. Concentration of native hCG by the ovarian tissue was similar to data reported in other species. (13, 14, 15, 16)

The diminution in the number of oocytes released per rat and the decreased number of rats ovulating by repeated injections of hCG may be due to the induced refractoriness of the ovarian follicle by the gonadotropin to suppress further action of endogenous LH (17). The finding that the hCG derivatives (NG-hCG and NGA-hCG) could inhibit the hCG stimulation of ovulation in Nembutal-blocked rats, which have no inherent activity when injected alone, may be explained by their ability to bind to the ovarian receptor and block the action of exogenous hCG, as has been shown in vitro (10). The fact that the hCG derivatives were unable to inhibit the ovulatory actions of the PMS-induced, endogenous LH surge could be due to poor timing of the administration of the derivatives; the endogenous LH surge could have started by the time the derivatives were administered. Since only 10% of the LH surge is required for the ovulation (18), exposure of the ovaries to a small concentration of endogenous LH before the administration of the derivatives could be enough to cause ovulation.

Summary

The effect of progressive enzymatic removal of various carbohydrates on the half life (T 1/2) and biologic activity in vivo of hCG was studied in rats. The cleavage of carbohydrates such as galactose, glucosamine and mannose, in addition to sialic acid, reduced the T 1/2 from 60-90 for native hCG to less than 3 minutes. The hCG derivatives had no inherent ability to cause ovulation, but in PMSG, Nembutal-treated immature rats, the hCG derivatives with carbohydrate residues internal to sialic acid removed blocked the stimulatory effect of native hCG upon ovulation.

## Acknowledgements

Thanks are due to Ms. Sandra Fowler for her expert technical assistance and to Dr. L. D. Anderson for labelling of gonadotropin. The investigation was funded by a WHO contract 77008 and by a grant from Lalor Foundation.

## References


1) Van Hall, E.V., J.L. Vaitukaitis, and G.T. Ross, Endocrinology 89 (1971) 11.

2) Tsuruhara, T., E.V. Van Hall, M.L. Dufau, and K.J. Catt, Endocrinology, 91 (1972) 463.

3) Channing, C.P., C.N. Sakai, and O.P. Bahl, Fed. Proc. 35 (1976) Ab. 3268.

4) Channing, C.P. and O.P. Bahl, Biol. Reprod. 17 (1978) 707.

5) Moyle, W., O.P. Bahl, and L. Marz, J. Biol. Chem. 250 (1975) 9163.

6) Kammerman, S. and R.E. Canfield, Endocrinology 90 (1972) 384.

7) Sakai, C.N., B. Engel, & C.P. Channing, Proc. Soc. Exp. Biol. Med. 155. (1977) 373.

8) Batta, S.K., F. Piva, and L. Martini, Adv. in Biosci. 9 (1972) 723.

9) Sasamoto, S., and T. Johke, Biol. Reprod. 13 (1975) 195.

10) Channing, C.P., C.N. Sakai, and O.P. Bahl, Endocrinology 1978 (in press).

11) Morell, A.G., R.A. Irvine, I. Sternlieb, I.H. Scheinberg and G.G. Ashwell, J. Biol. Chem. 243 (1968) 155.

12) Morell, A.G., G. Gregoriadis, I.H. Scheinberg, J. Hickman, and G.G. Ashwell, J. Biol. Chem. 246 (1971) 1461.

13) Lunenfeld, B. and A. Eshkol, Vitam. Horm. 25 (1967) 137.

14) Channing, C.P. and S. Kammerman, Endocrinology 93 (1973) 1035.

15) Mizejewski, G.J., W.H. Beierwaltes and J. Quinones, J. Nucl. Med. 13 (1972) 101.

16) Kammerman, S. and C.P. Channing, J. Clin. Endocrinol. Metab. 38 (1974) 463.

17) Nilsson, C., T. Hillensjö and C. Ekholm, Acta Endocrinol. (Kbh) 86 (1977) 384.

18) Turgeon, J. and C.A. Barraclough, Endocrinology 92 (1973) 755.

# A PROCEDURE FOR EVALUATING LUTEOLYTIC AGENTS IN PRIMATES

John W. Wilks

Fertility Research, The Upjohn Company, Kalamazoo, Michigan 49001

A method of human fertility control based on the regulation of corpus luteum function has been the objective of a continuing research effort by reproductive biologists. The vast array of physiologic control mechanisms governing the corpus luteum among the mammalian species presents a formidable obstacle in extrapolating concepts developed in animal research for clinical application in women. As an example the uterus clearly regulates the secretory activity and life span of the corpus luteum in infraprimate animals, but a role for the uterus in altering luteal function in primates has not been established. The importance of this biological distinction is illustrated by the successful application of prostaglandin $F_{2\alpha}$ (a physiologic luteolytic substance of uterine origin) to the regulation of estrous cycles in domestic animals (1), and the disappointing outcome of a clinical evaluation of this prostaglandin for control of the human corpus luteum (2).

Another consideration of biologic importance is the stimulus of human chorionic gonadotropin (hCG) upon the corpus luteum to offset luteolytic influences. While the physiologic factors controlling the life span of the corpus luteum during the nonfertile menstrual cycle are of interest, they may not be important in the design of post-conceptional agents with a luteolytic action. The function and life span of the human corpus luteum during early pregnancy is governed by the trophic and essential action of chorionic gonadotropin, and the tenacity of this stimulus may not be ignored. The significance of this biologic reality has been demonstrated by research experience. Oxymetholone (17β-hydroxy-2-[hydroxymethylene]-17-methyl-5α-androstan-3-one) elicited functional luteolysis when administered to women during nonfertile menstrual cycles (3), but did not inhibit the corpus luteum during early gestation (4) or in nonpregnant women given exogenous hCG (5). These results indicated that an early evaluation of luteolytic agents in nonpregnant women treated with hCG is important (5).

The physiology of the corpus luteum in the rhesus monkey has been studied in sufficient detail to reveal striking similarities with luteal function in the woman (6). A procedure employing the rhesus monkey for the evaluation of potential luteolytic agents was designed in anticipation that the data obtained would be of predictive value for the human.

## MATERIALS AND METHODS

Only those monkeys having an ovarian structure identifiable as a corpus luteum, at the time of laparoscopy during the second half of the menstrual cycle, were used; 190 menstrual cycles were studied in 101 rhesus monkeys. Blood samples were collected each morning from Day 20 of the menstrual cycle until the second day of menstrual bleeding or Day 31 from the previous menses. Intramuscular injections of hCG (Ayerst Laboratories, Inc.) were administered in 0.5 ml of saline on Days 20, 21 and 22 of the menstrual cycle at doses of 30, 60 and 90 I.U., respectively. Test compounds were administered intramuscularly or orally in an appropriate vehicle on Day 22 of the menstrual cycle, at the time of the last treatment with hCG. Three or more monkeys were used for the evaluation of each compound, or each dose of a compound. Control animals were given hCG injections and vehicle treatments. Control monkeys were examined occasionally during the operation of the testing procedure, and the results accumulated for comparison with treated animals. Serum progesterone concentrations in monkeys treated with test compounds were compared statistically (t test) to values obtained in control animals. To minimize between animal variation progesterone values obtained for each monkey on Day 23 of the menstrual cycle and thereafter (following the administration of the test compound) were expressed as a percentage of the mean serum progesterone concentration for Days 21 and 22 in the same animal; these percentages were also statistically compared to values calculated for control monkeys.

## THE hCG TREATMENT REGIMEN

The ultimate evaluation of a luteolytic agent is a demonstration of efficacy during early gestation in women. Preclinical selection of luteolytic agents could involve pregnant nonhuman primates, but this approach presents several practical problems. An antifertility action might be mediated at the uterus, placenta, or embryo and not directly upon the corpus luteum. The fertility rate among captive rhesus monkeys is approximately 30%, and large numbers of monkeys would have to be handled to select those which were pregnant. Simulation of the endocrine conditions of early gestation by the injection of hCG in monkeys was chosen as the experimental approach; a reduction of blood progesterone concentrations following administration of test compounds in this model may be ascribed to a direct action upon the corpus luteum.

Daily injections of 25 I.U. of hCG during the luteal phase of rhesus monkeys results in blood progesterone concentrations which mimic those of early gestation; larger doses do not elicit greater elevations in progesterone values (7). In a physiological setting serum concentrations of chorionic gonadotropin continuously increase throughout early gestation until a time when the corpus luteum is no longer essential for the maintenance of pregnancy, and the placenta assumes principal responsibility for progesterone secretion. It was considered important to develop a regimen of exogenous doses of hCG which would be of physiologic intensity and would closely mimic the endocrine status of early pregnancy.

Commercially available preparations of hCG were bioassayed against the Second International Standard for hCG. Preparations of hCG obtained from a pharmaceutical company consistently yielded bioassay potencies in agreement with the biological activity indicated on the label. Preparations of hCG obtained from several biochemical supply companies never achieved the biological potencies stated in the product specifications, and in the worst case varied from vial to vial within a lot.

Serum concentrations of chorionic gonadotropin were determined during early pregnancy in rhesus monkeys using a specific antiserum (8), and a standard for monkey chorionic gonadotropin (mCG) whose biological potency had been established by bioassay against the Second International Standard for hCG. Serum concentrations of hCG in nonpregnant rhesus monkeys following the injection of a preparation of hCG of confirmed biological activity, were determined by radioimmunoassay using the <u>same</u> antiserum employed for the measurement of mCG, but in this case the Second International Standard for hCG was employed as the reference standard. Using this approach doses of 30, 60, 90, 180 and 360 I.U. of hCG given on consecutive days were found to produce serum concentrations of hCG which bracketed the serum concentrations of mCG seen on the first five days that mCG is detectable in the peripheral circulation of pregnant rhesus monkeys during normal gestation (Figure 1).

Serum progesterone concentrations increase markedly in response to the initial secretion of mCG during early pregnancy, and then gradually decrease on subsequent days even though serum concentrations of mCG continue to increase (6). Exogenous hCG produces a similar pattern of serum progesterone--an initial, dramatic increase followed by declining values despite continued treatment (7). While

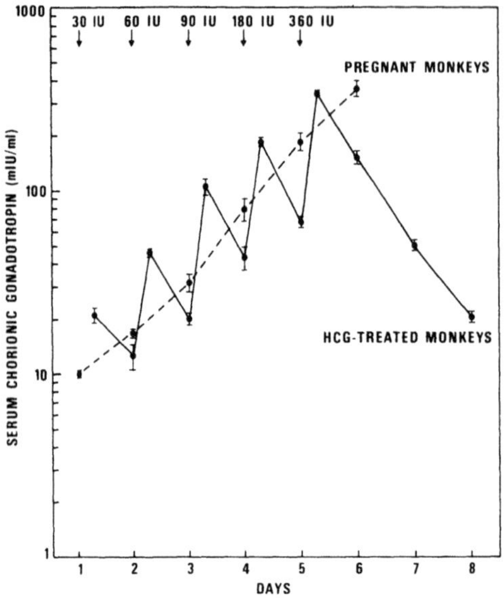


Figure 1. Serum concentrations of mCG (dotted line) in rhesus monkeys (n=7) standardized to the first day mCG was detectable in early pregnancy. Serum hCG concentrations (solid line) in nonpregnant monkeys (n=3) injected with the doses of hCG indicated at the top of the figure. hCG was determined daily and eight hours after each injection.

the use of hCG was considered essential in a testing procedure, continued administration of hCG alone results in a reduction in blood progesterone concentrations. Using the doses of hCG illustrated in Figure 1, it was learned that serum progesterone concentrations did not decline until after the fourth injection (180 I.U.) of hCG. If only three injections of hCG were given (30, 60 and 90 I.U.) progesterone levels were elevated and plateaued on Days 21, 22 and 23 of the menstrual cycle, and then declined in parallel with the decaying radioimmunoassayable serum concentrations of hCG. Any determination of functional luteolysis must be based on the demonstration of significant reductions in serum progesterone concentrations, and must be unambiuously attributed to the administration of the test compound, and not in part be related to the administration of hCG. Therefore only three injections of hCG were used, and the changes in blood progesterone within 24 hours of administering the test compound were emphasized in the assessment of luteolytic activity.

The time of initiating treatment with hCG for a correct simulation of early gestation would be following the first clear decline in luteal phase progesterone values. "Rescue" of the corpus luteum occurs approximately 11 days after the midcycle gonadotropin surges in the rhesus monkey (9). The time of midcycle gonadotropin surges during normal menstrual cycles in the rhesus monkey ranges from the 9th to the 13th day of the cycle, with a mean of 11 days; the duration of the menstrual cycle ranges from 22-30 days with a mean of 26 days (10). An examination of the responsiveness of the rhesus monkey corpus luteum to exogenous hCG at known times after the LH surge indicated that the response of the fully functional gland did not differ from that seen during early regression, but was significantly different from the responses of newly formed or premenstrual corpora lutea (11). The determination of an LH peak for each monkey would increase the complexity of the procedure, and it was desirable to maximize the use of those monkeys seen to have a corpus luteum by laparoscopy. Day 20 of the menstrual cycle was chosen as the time to initiate treatment; this time assured that the corpus luteum was fully mature or in the early phases of regression, and minimized the inclusion of premenstrual corpora lutea which would be minimally responsive to hCG. Of the 190 menstrual cycles studied, in only three cases did menstrual bleeding commence prior to the administration of the test compound, and in two additional cases the serum progesterone values in response to hCG and throughout the experimental interval were not consistent with the presence of a functional corpus luteum.

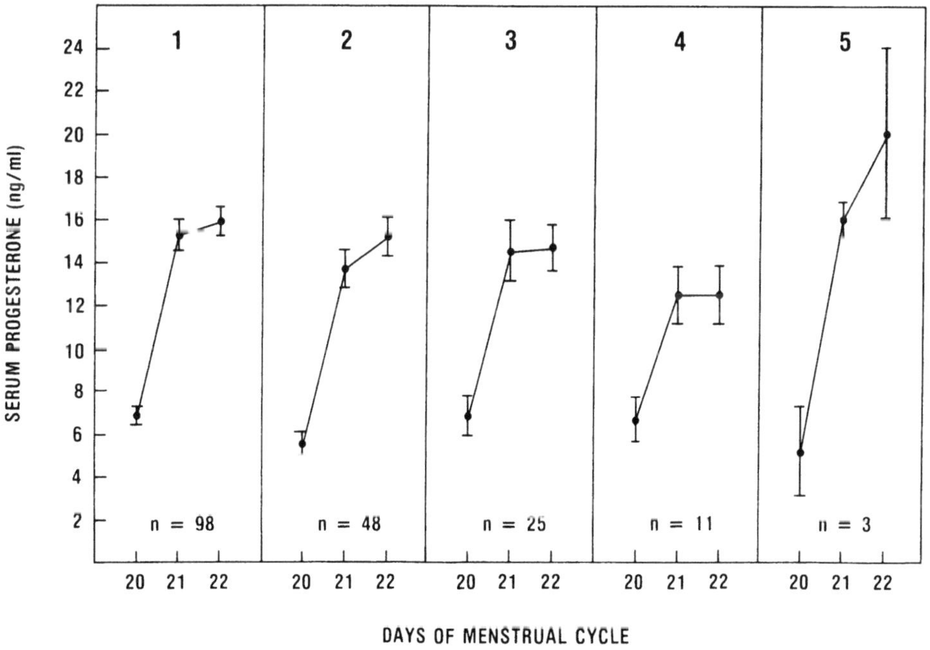


Figure 2. Serum progesterone concentrations prior to hCG injection (Day 20 values), and after injection of 30 and 60 I.U. of hCG on Days 20 and 21, respectively, in monkeys treated from one to five times.

One concern about the administration of hCG was the possibility of antibody formation. Attempts were made to locate antibodies in treated monkeys; none were detected by gel immunodiffusion or by examining sera for the binding of radioiodinated hCG. Some monkeys were subjected to the hCG treatment regimen up to as many as five times. If antibody formation was a problem, then the responsiveness of these monkeys to exogenous hCG, as judged by increments in serum progesterone concentrations, should have been diminished. Figure 2 illustrates the serum progesterone concentrations for monkeys following one, two, three, four and five exposures to the hCG treatment regimen. Statistical analysis revealed no significant differences between the serum progesterone concentrations during the first exposure to hCG and subsequent exposures. Additionally, the increment in serum progesterone in response to the first injection (30 I.U.) of hCG did not differ significantly between the first exposure and subsequent treatments. Conception rates did not differ between animals exposed to hCG and animals never exposed to hCG, again indicating that antibodies which might interfere with ovulatory secretion of LH or the nidatory secretion of mCG were probably not present. However, in other investigations in which the same commercial preparations of hCG were given to rhesus monkeys for a duration of 10 days, sera from some monkeys did bind radioiodinated hCG.

## CHARACTERISTICS OF OBSERVED RESPONSES

Five distinct patterns of serum progesterone concentrations have been observed in response to the administration of test compounds: 1) a sharp decline in serum progesterone concentrations within 24 hours to a value less than 20% of the mean serum progesterone concentration on Days 21 and 22 of the menstrual cycle; this fall in progesterone was sustained and values continued to decline to nearly undetectable levels; 2) a sharp decline in serum progesterone concentrations within 24 hours to a value less than 20% of the pretreatment level, but progesterone rebounded on subsequent days to concentrations not different from those in control animals; 3) a statistically significant decline in the serum progesterone concentration within 24 hours, however the concentration remained at a relatively high value (greater than 20% of the Day 21-22 mean); 4) concentrations and a pattern of serum progesterone which did not differ from that observed in control animals; and 5) a statistically significant increase in serum progesterone concentrations to values greater than those observed in control monkeys.

The responses to two prostaglandins evaluated by this method are illustrated in Figure 3. Panel A of the figure shows progesterone values for animals treated with (15S)-15-methyl prostaglandin $F_{2\alpha}$ methyl ester (15M-$PGF_{2\alpha}$; 12 milligrams/monkey). Prior to administration of 15M-$PGF_{2\alpha}$ serum progesterone values were higher than those observed in control monkeys (shaded area). Serum progesterone

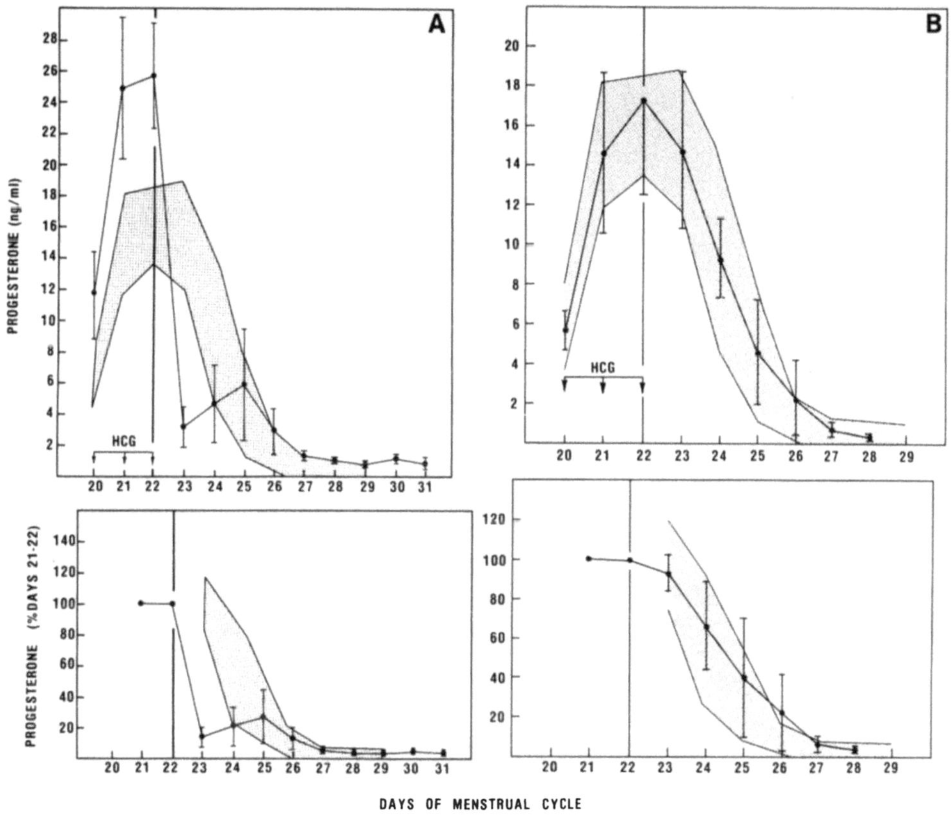


Figure 3. Progesterone values in monkeys (n=3) treated with (15S)-15-methyl $PGF_{2\alpha}$ methyl ester (Panel A), and with 2a,2b-dihomo-(15S)-15-methyl $PGF_{2\alpha}$ p-benzamidophenyl ester (Panel B) on Day 22 of the menstrual cycle. The shaded areas depict the 95% confidence intervals for values in control monkeys. The top portion of each panel illustrates the serum progesterone concentrations, and the bottom portion gives the same data as a percentage of the mean value from Days 21 and 22.

concentrations declined sharply within 24 hours after administering 15M-$PGF_{2\alpha}$, but then rebounded on subsequent days to values similar to those observed in control monkeys. The percentage values for the same data are shown in the bottom portion of Panel A, and again reflect a pattern on Day 23 which is divergent from that observed in control monkeys. 15M-$PGF_{2\alpha}$ was judged to provoke a significant inhibition of luteal function, but it produced diarrhea and appetite depression in each of the monkeys. Since surgical stress has been shown to cause luteolysis in nonpregnant rhesus monkeys (12), the question may be asked whether 15M-$PGF_{2\alpha}$ possesses intrinsic luteolytic activity or whether the reduction in serum progesterone concentrations is related to nonspecific stress associated with the observed side-effects. Panel B shows the results with another prostaglandin, 2a,2b-dihomo-(15S)-15-methyl $PGF_{2\alpha}$ p-benzamidophenyl ester (13 milligrams/monkey), which produced diarrhea and appetite depression in the test monkeys comparable to that observed with 15M-$PGF_{2\alpha}$. The serum progesterone concentrations after treatment with this second prostaglandin did not differ from those observed in control monkeys. These results suggest that the demonstrated luteolytic action of 15M-$PGF_{2\alpha}$ in the rhesus monkey is related to an inherent biologic activity of the molecule.

A steroidal agent, 17β-hydroxy-4,4,17α-trimethylandrost-5-eno [2,3-d]isoxazole, has been reported to be an inhibitor of 3β-hydroxysteroid dehydrogenase, and has been reported to terminate pregnancy in the rhesus monkey (13). Figure 4 shows the serum progesterone concentrations in monkeys given 500 milligrams of this steroid orally in the testing procedure. The reduction in serum progesterone after administration of this steroid was not statistically significant, and the pattern of values did not differ from that observed in control monkeys. When studies with this steroid were conducted in rhesus monkeys during early pregnancy (Figure 5), serum progesterone declined sharply within 24 hours of giving the agent; treatment was continued for five days at an oral dose of 500 milligrams daily. It is of interest that serum concentrations of mCG continued to increase

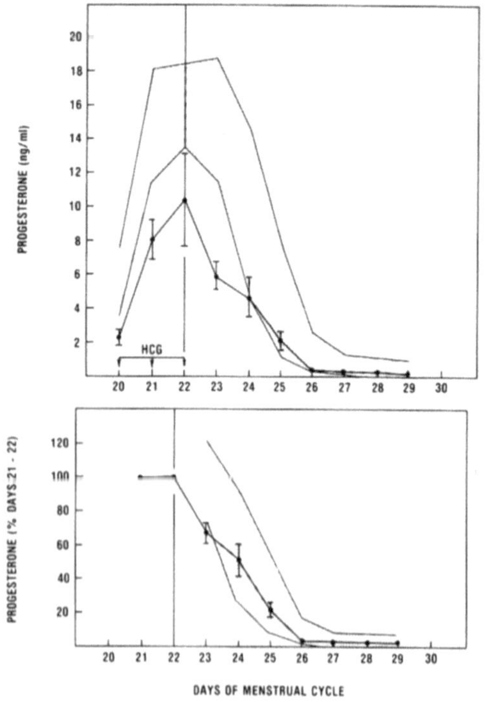


Figure 4. Serum progesterone concentrations in monkeys treated with 17β-hydroxy-4,4,17α-trimethylandrost-5-eno[2,3-d] isoxazole on Day 22 of the menstrual cycle.

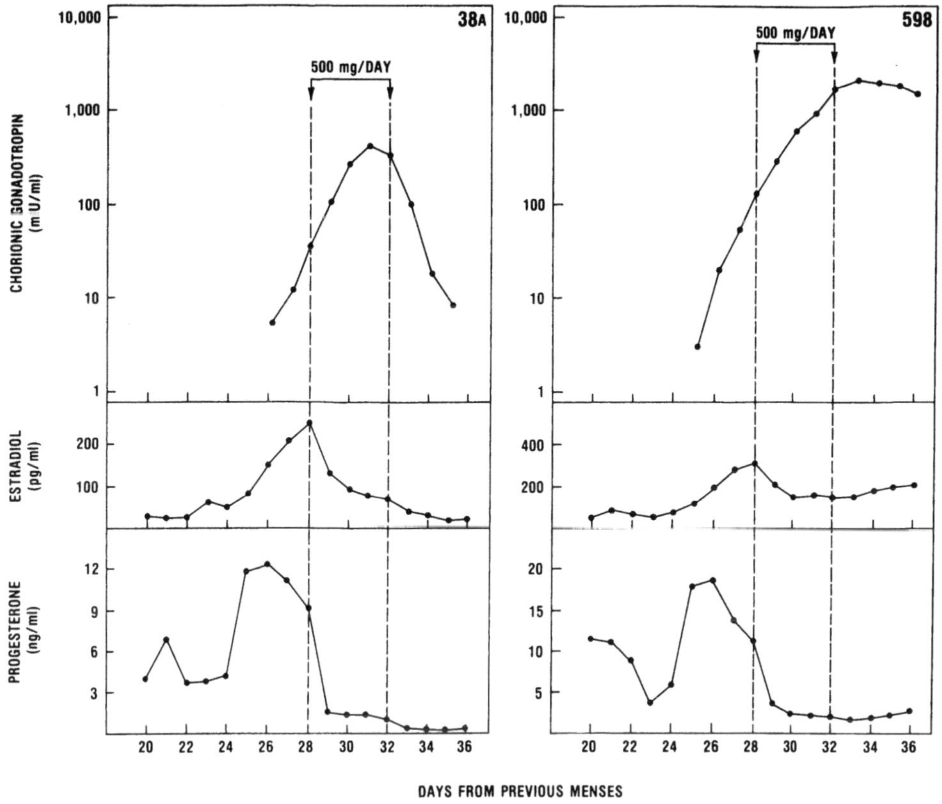


Figure 5. Serum concentrations of mCG, estradiol and progesterone in pregnant monkeys treated with 17β-hydroxy-4,4,17α-trimethylandrost-5-eno[2,3-d]isoxazole on Days 28 through 32 from the previous menses.

after serum progesterone fell; eventually serum mCG declined in monkey 38A and pregnancy was terminated, but pregnancy was maintained in monkey 598. The corpus luteum of early pregnancy is considered to be the primary source of blood estrogens (14); although serum estradiol declined following treatment with the steroid, the magnitude of this decline did not parallel that seen for progesterone. The developing placenta post-nidation may contribute to progesterone production and be more effectively inhibited by the steroid than the corpus luteum. This interpretation is consistent with the observation that the steroid terminates gestation in the rhesus monkey at lower doses during advanced gestation, when the corpus luteum is quiescent, than in early gestation when the corpus luteum is active (13).

## CONCLUSION

A procedure, based on physiologic concepts which are believed to be correct, has been described for the preclinical evaluation of luteolytic agents. Since a luteolytic agent for the woman has not yet been convincingly demonstrated in the literature, the predictive value of this procedure has not been established. When an effective luteolytic agent(s) is described, the method must be subjected to a critical appraisal. The procedure, however, will be useful in the investigation of the actions of chorionic gonadotropin, and for elucidating the mechanism of action of post-conceptional antifertility agents.

## REFERENCES

(1) J. W. Lauderdale, B. E. Seguin, J. N. Stellflug, J. R. Chenault, W. W. Thatcher, C. K. Vincent and A. F. Loyancano. J. Anim. Sci. 38(1974)964.

(2) W. J. LeMaire and A. G. Shapiro. Prostaglandins 1(1972)259.

(3) E. L. Klaiber, M. R. Henzl, C. W. Lloyd and E. J. Segre. J. Clin. Endocrinol. Metab. 36(1973)142.

(4) P. F. Brenner and D. R. Mishell. Contraception 11(1975)669.

(5) M. R. Henzl, E. J. Segre and R. M. Nakamura. Contraception 8(1973)515.

(6) E. Knobil. Biol. Reprod. 8(1973)246.

(7) J. D. Neill and E. Knobil. Endocrinology 90(1972)34.

(8) G. D. Hodgen, W. W. Tullner, J. L. Vaitukaitis, D. N. Ward and G. T. Ross. J. Clin. Endocrinol. Metab. 39(1974)457.

(9) L. E. Atkinson, J. Hotchkiss, G. R. Fritz, A. H. Surve, J. D. Neill and E. Knobil. Biol. Reprod. 12(1975)335.

(10) J. W. Wilks, G. D. Hodgen and G. T. Ross. J. Clin. Endocrinol. Metab. 43(1976)1261.

(11) J. W. Wilks, A. S. Noble, A. D. Forbes and K. K. Forbes. Abstracts, 10th Annual Meeting, Society for the Study of Reproduction (1977) p. 19.

(12) E. Knobil, J. D. Neill and E. D. B. Johansson. Endocrinology 82(1968)410.

(13) H. P. Schane, J. E. Creange and G. O. Potts. Fertil. Steril. 28(1977)301.

(14) W. T. K. Bosu and E. D. B. Johansson. Fertil. Steril. 25(1974)443.

# SUMMING UP

I. Rothchild

Department of Reproductive Biology
Case Western Reserve University School of Medicine
Cleveland, Ohio 44106

To paraphrase the stated purpose of the Workshop: it was to see where we stand today in our understanding of how the activities of the follicle and corpus luteum (CL) are regulated. My assignment, which really amounted to assessing the extent to which the Workshop fulfilled its purpose, was not easy. In fact, it was about the hardest job I've ever undertaken, and has made me feel very much like the soldier, who while on kitchen duty, was assigned the job of sorting the good from the bad potatoes, and peeling only the good ones. Several hours later, he was still sitting before the pile of potatoes, not a single one of which had been sorted, and when asked why he hadn't done his job, he replied: "I don't mind hard work, it's making all these decisions that's too much for me!" To have summarized in a sentence or two the contents of each paper presented at or submitted to the Workshop would have been child's play. To decide just what the Workshop did accomplish was not. The most important reasons for my difficulty probably arose from the way I look at things, but some also came from some of the current trends in biology and reproductive biology. This may become clearer as we go along.

Greenwald opened the Workshop with a review (1) of what he considered the most important problems in the field of ovarian physiology. For the most part I agree with his choices and his discussion of them. However, except for his 6th topic, "Intra-ovarian interactions of steroids and gonadotrophins", the Workshop touched on these subjects to only a minor degree or not at all, and, therefore, in my eyes at least, did not achieve its implied purpose, even though it was interesting and fun to attend. That it did accomplish something worthwhile should be apparent to anyone who attended it or who has read this published account of it.

This, plus the fact that it could, perhaps, have accomplished more, are the main themes of this Summing-Up.

I will begin with what I think is the most important problem in the field (it will not be too different, except in orientation, from Greenwald's), and will then take a look at the Workshop from the perspective of this problem. No apology is needed for what may seem to be an excessive use of the first person singular, because no one else is responsible for the evaluations expressed here but myself.

## THE MOST IMPORTANT UNSOLVED PROBLEM IN OVARIAN PHYSIOLOGY: INSTABILITY IN THE REGULATION OF FOLLICLE AND CL ACTIVITY

Two characteristics of mammalian ovarian activity are so striking that it seems trite even to remark on them. The first is that atresia is the fate of all but a very small minority of the follicles in almost all the mammals that have been studied. The other is that the CL is an ephemeral endocrine gland. The reason for being trite is to emphasize the point that any theory set up to explain how the follicle's activity is regulated must be based on the principle that the most characteristic element of the regulatory system is its instability. The theory must at the same time explain that it is still possible, within this essentially unstable regulatory system, for some follicles to grow, mature, ovulate and be transformed into functioning CL.

Any theory about how the CL is regulated must also be based on the principle that the system is essentially unstable, for otherwise the CL would not be ephemeral. Greenwald's discussion of the differences among antral follicles in relation to their stages of growth and development, the physiologic importance of the use of "pharmacologic" doses of hormones, the possibility that progesterone may play an important intraovarian regulatory role (see below), and steroid-gonadotrophin intraovarian interactions (1), in general, were particularly pertinent to this viewpoint about the requirements for a theory.

The need for such a theory is self evident. Although it still remains unsatisfied, it would be worthwhile to take a look at Falck's theory, because of its potentialities. Falck (2) is usually cited for his theory that only the combined action of the granulosa and theca interna cells was required for optimal estrogen secretion. He is not so often cited for his explanation of how a follicle grows and secretes estrogens (2). It is the best that has so far been formulated.

Hisaw's earlier review (3) had underlined the findings of Greep, van Dyke and Chow (4) that only the synergistic action of FSH and

LH could induce normal growth of follicles and estrogen secretion, and of Williams (5,6) that estrogens themselves might act as gonadotrophins , by preventing loss of responsiveness to gonadotrophins and atresia that follow hypophysectomy. Falck found that estrogens were secreted only when representative cells from each of the two main cell types of the follicle - granulosa cell and theca interna cells - were combined. Thus either granulosa or CL cells plus either theca or interstitial cells secreted estrogens, while each of these cells by itself, or combined with another cell from the same line (e.g., granulosa plus CL cells) did not do so. He then put these findings together with the points emphasized by Hisaw.

Thus: FSH stimulates the granulosa cells to grow; during this growth the cells produce a substance (he wondered whether it might be progesterone) which went to the theca cells and induced them to secrete estrogen in response to LH. The estrogens, besides going into the general circulation, also went back directly to the granulosa cells, and induced in them, an increased responsiveness to FSH, thus closing the circle. In this simple but brilliant synthesis, he explained the marked increase in response of the follicle to gonadotrophins when it acquires an antrum (antrum formation and the acquisition of a theca interna generally go together), the synergism between FSH and LH, and the requirement for both granulosa and theca cells for estrogen secretion. His theory also showed how these intrafollicular arrangements allowed the follicle to grow and secrete estrogens as a self-stimulating system, in the presence of constant levels of both gonadotrophins.

The theory is still basically sound and can be revised to take into account the newer information about the interplay between circulating and intrafollicular gonadotrophins and steroids (7-14), the physiologic roles of steroid and gonadotrophin receptors (15-20), and the possible regulatory effects of intrafollicular steroids (18,21-27).

But excellent as it still is, Falck's theory did not try to explain atresia, or the instability of the CL, or, since the CL is derived from the follicle which escaped atresia, that this follicle must be, in some way, different from all the others. Before looking at the possibilities for a new theory, let us look at the Workshop as a whole, in the perspective of this need for a new theory.

## THE ACCOMPLISHMENTS OF THE WORKSHOP: A ONE MAN'S BIRDS-EYE POINT OF VIEW

What did the Workshop accomplish? The zoological peculiarity of the heading is a crutch on which I lean to help me describe that which seems to best summarize the main accomplishments of the Workshop. The description has been sifted out from an enormous mass of detail.

The first thing worth noting was that the Workshop, in general, was more a time and place for digging in and establishing positions already won, than for scouting expeditions, or for probing thrusts into the enemy territory to uncover possible weak points that might yield to attack. Very little really new information came out of the meeting, and what there was was almost entirely on a minor level. It doesn't often happen at scientific meetings that the audience, as a man (or should I say, as a person?) rises to its feet, screaming with joy and enthusiasm over someone's contribution, but there wasn't even a chance of this happening at the Workshop.

The second interesting thing was that the majority of the contributions were concerned with problems either at the intraovarian, intrafollicular, intraluteal, intracellular, ultrastructural or macromolecular level (Table 1). This very interesting trend is partly responsible for my difficulty in getting a grasp of what was going on; I will comment on it further below.

With these two points in the background, I will try to answer the question of what the Workshop actually accomplished by making a few comments under several subject headings of my own choosing, but without trying to review or comment on each of the contributions; the latter are available to the reader and he (or she) can make his (or her) own comments and compare them with mine.

## Receptors and Ovarian Physiology

It's now very clear that practically all of what Falck described, allowing for the differences between his concept of where estrogens were made, and where they probably are made (see below), can be put in terms of the differences between granulosa and theca cells in their ability to generate and maintain FSH and LH receptors. The subject has been well reviewed (15,16,18,31). Richard's Workshop review (20), as well as several of the short unpresented papers (32-35), served to emphasize the specificity of LH receptor generation in the theca cells, of FSH receptor generation in the granulosa cells, and the gradual loss of FSH receptors and gradual acquisition of LH receptors by the granulosa cells, as the follicle within which they lie approaches ovulation.

We now know that hormones can stimulate generation of their own receptors (e.g., this is true of estrogen and FSH receptors),

Table 1. Subject matter of the Workshop classified according to approach to the problem[1]

| Subject Matter[2] | No. of Papers[3] | Primarily Intracellular | Primarily Classical Physiological | Combined Approach with emphasis on: Intracell. | Combined Approach with emphasis on: Physiol. |
|---|---|---|---|---|---|
| Follicle growth & atresia | 15 | 1 | 2 | 9 | 3 |
| Preovulation to ovulation | 9 | 2 | - | 6 | 1 |
| Follicular fluid factors, etc. | 12 | 1 | 4 | 4 | 1 |
| Oocyte maturation | 4 | - | - | 4 | - |
| Corpus luteum, in general | 15 | 2 | 5 | 6 | 2 |
| Luteolysis | 10 | 3 | 2 | 5 | - |
| Desensitization | 8 | 8 | - | - | - |
| General and miscellaneous | 12 | 2 | 4 | 6 | - |
| Totals | 83 | 19 | 17 | 40 | 7 |

[1] "Primarily intracellular" means that the investigators approached the problem with the primary object of uncovering the intracellular processes giving rise to or connected with it, such as EM, macromolecule processes, etc. The AC-cAMP-PK etc. system, looked at as the primary center of attention of the investigator, is also included here. "Primarily classical physiological" means the whole body approach, and can be best illustrated by, e.g., the Goodman and Hodgen paper (28) on the effect of CLectomy of pregnant monkeys on the interval to the next ovulation. "Combined Approach" means just that, but the investigators may have laid primary emphasis on an *in vitro* technique or problem, such as the conversion of androgens to estrogens by granulosa cells *in vitro* (29) or have used an *in vitro* technique to solve a classical approach problem, such as the Goebelsman *et al*. case report of a woman with an ovarian aromatase deficiency (30).

[2] My own classification of the Workshop's subject matter.

[3] For easy evaluation, only the short papers (presented and unpresented) have been considered here. However, the distribution among the principal papers and invited discussions was almost the same.

or the receptors for other hormones (e.g., estrogens plus FSH stimulate the generation of LH receptors in rat granulosa cells) (16,19,20) or can inhibit formation or maintenance of their own receptors (e.g., LH can decrease its own receptors) (16) (see also the comments on Desensitization, below, as well as, of course, the Workshop papers on that subject) or the receptors of other hormones (e.g., the loss of FSH receptors in granulosa cells associated with luteinization in response to LH) (36; see also Desensitization). These findings, and the work summarized by Richards (20) and others (17-19,31) in general, as well as in most specific details, have helped to establish that most, if not all the effects of gonadotrophins on the follicle are either mediated by specific receptors for each gonadotrophin or can at least be correlated with the presence of a specific receptor.

The most complete picture of the relationships between the gonadotrophins and their receptors, on the one hand, and estrogens and their receptors, on the other, within the follicle, has come from the work of Richards, Midgely, and their collaborators (20). The findings and the interpretations are interesting and provoking, and they can provide a modern translation, in the language of receptors, of the older story of intrafollicular gonadotrophin-estrogen relationships told by Falck (2). Nevertheless, one cannot help but feel a little uneasy about how much this picture represents the general mammalian one, since it is based on work done on immature rats, either intact or hypophysectomized, sometimes treated with enormous doses of estradiol, and in most cases with sequences of gonadotrophins that only roughly resemble how they are secreted in the intact, normal adult rat. This criticism should not diminish the importance of the work accomplished; its primary purpose is to emphasize how important it is to develop attacks, based on this work, to test the applicability of this picture of receptor-hormone interactions to most, or all mammals.

## Mechanism of Action of Protein Hormones

Since Sutherland and Rall's formulation of the theory that cAMP acts as the second messenger within the cell, through which protein hormones exert their effects (see Catt and Dufau (37) for one of several excellent reviews), it seems as though almost anyone with biochemical leanings, working in reproductive endocrinology, has tested the idea that the gonadotrophins act through the adenyl cyclase - cAMP - protein kinase - phosphorylated protein pathway to induce their specific effects in their target tissues. The findings have aroused confirmations of faith in the theory (31), sometimes skepticism (38), and in the end (?), if the Workshop is a typical example, a triumph of the belief in the one and only true faith (27,39-59).

I exaggerate a little, but perhaps this is permissable after hearing and reading so many people play the same tune, sometimes on different instruments, and sometimes to different accompaniments, but nevertheless, the same tune, throughout the Workshop. There is nothing intrinsically wrong with confirming the idea that the gonadotrophins work through cAMP as the second messenger. The danger is in thinking that they can act only in this way. Greengard's article on phosphorylated proteins (60) should be required reading for everyone, at every level in reproductive or other endocrinology, because the lesson in this thoughtful review is that cAMP is not the only way to activate a protein kinase to phosphorylate certain proteins, and furthermore, that even the phosphorylated proteins, important as they are, may not be the only means through which the action of the cell is expressed.

There's something about a good theory that, as far as I know, has never been explicitly stated.* The better a theory is, the more dangerous it becomes, because it blinds one to the possibility that there might be a still better theory. This should perhaps be remembered each time someone sets out to prove that something works through cAMP. The same principle can apply to the receptor theory in general, and to steroid hormone receptors, in particular (61).

## Where Are Estrogens Made?

The argument, which began over 50 years ago, about granulosa cells vs theca cells as the primary source of estrogens, has swung from the granulosa cell to the theca cell, and now, while it still continues (27), is slowly gathering momentum for the swing back to the granulosa cell as the ultimate, or most important site of estrogen production by the follicle. The evidence is most easily visible in the rat: the theca cells make androgens in response to LH and the androgens are converted to estrogens by the granulosa cells; the latter's aromatizing enzymes are stimulated or activated specifically by FSH (26,62-65). The pattern fits with that described by Falck, except for the site of estrogen production, and for the granulosa cell factor which he thought induced the theca cell to respond to LH. [The latter is probably not estrogen, since

---

* It would not surprise me, however, if someone had already said it. This usually happens with my best thoughts, and has prompted the following bit of nonsense:

I've often found, when I have written
Lines with which I'm deeply smitten,
That there was someone else who wrote
The lines that me so deeply smote!

estrogens seem to inhibit theca cell androgen production, in response to LH (66,67) as well as progesterone production (68). The possibility that it might be progesterone is mentioned below].

The Workshop brought out the findings that pig (27,29) and monkey granulosa cells (Channing, W-S open discussion) can also make estrogens from androgens *in vitro*. This is especially interesting in the case of the monkey, because *in vivo* findings suggested that the granulosa cells were not the source of the preovulatory estrogens (69). The pattern in the chicken may also be like the rat's since only the combination of granulosa and theca cells made estrogens well (70) and the theca cells seem to be mainly concerned with making androgens in response to gonadotrophins (71).

An interesting side light on these findings is the discrepancy between the obvious ability of granulosa cells to make estrogens, and the mass of ultrastructural and cytochemical evidence (e.g., 72,73) that they don't make steroids.

## Androgens and FSH Stimulate Progesterone Production in Granulosa Cells

The stimulation by androgens of progesterone production by rat granulosa cells, particularly in synergism with FSH, is a relatively new development in ovarian physiology (22-25), although, as Schomberg (27) pointed out, androgens have been known for a long time for their inhibitory actions on ovarian function. The stimulatory effect on progesterone production has also been found in pigs (74) not only in the granulosa cells of preantral follicles, but also in those from relatively mature follicles (27,29). The effect does not depend on luteinization of the granulosa cells, and the amount of progesterone produced nowhere even approaches that produced by the CL, or by luteinized granulosa cells maintained in tissue culture. Nevertheless, the effect can be very important, and the possible connections between it, on the one hand, and the inhibitory effects of androgens, the importance of the androgen/estrogen ratio in the induction of atresia (21), and the intrafollicular effects of progesterone (1,75-77), on the other, are mentioned below.

It is interesting that here also there seems to be a discrepancy between the obvious ability of granulosa cells to make progesterone and the cytochemical evidence that they don't contain the 3β hydroxysteroid dehydrogenase (3β olSDH). Does this mean that the granulosa cells make progesterone from some other precursor than pregneneolone, or that the enzyme findings are in error? [The heretical possibility *could* be considered that they make progesterone from testosterone, but the amounts made are far greater than the amount of testosterone required to stimulate progesterone production (25).

## Non-steroidal Regulatory Substances in Follicular Fluid

The Workshop sessions on this topic and on oocyte maturation were good opportunities for an overview of important and interesting subjects. The follicular fluid contains several substances of high molecular weight, probably secreted into it by the granulosa cells; they may all be part of the same molecule, but at least when seen as separate parts, they can inhibit completion of oocyte meiosis (31,51,53,78), prevent or diminish granulosa cell luteinization (53-55,79), depress the secretion of FSH (53,80-85), and depress the binding of FSH to granulosa cells (86).

Although all these effects are inhibitory, at least one may not be. In the pig, the follicular fluid of large follicles contains something which stimulates progesterone production by the granulosa cells of small follicles (54). The completion of oocyte meiosis may also involve both the reduction or loss of the oocyte maturation inhibitor, and the secretion into the follicular fluid of a factor that stimulates oocyte maturation (51,87).

## Oocyte Maturation

Oocyte maturation and how it is inhibited within the follicle, and released from this inhibition by either removal from the follicle, or by treatment of the follicle with LH, and the events going on within the oocyte as it completes meiosis, were well summarized from the intracellularist's viewpoint by Wassarman (52), and from a more classical point of view, by Tsafriri (51) and Schuetz (87). The phenomenon as a whole can't help but fall, eventually, into the hands of the molecular biologists, just as embryology already has fallen, since it is, essentially, a phase of embryology. Let us all hope, however, that its biologic significance, and the relation of oocyte maturation inhibition to follicular maturation, atresia, and ovulation will continue to be attractive to whole body physiologists, and, therefore, will continue to produce papers like that of Hunter et al. (88), showing the conditions under which it is possible to separate ovulation from oocyte maturation, or those coming from Thibault's laboratory [e.g. (89)] showing the conditions under which maturation itself and the ability to be fertilized can be separated from those which permit formation of the male pronucleus.

## Desensitization

The inability of certain agents, which otherwise can stimulate adenyl cyclase to form cAMP, to do so under certain conditions, usually a sudden increase in the dose of the agent, has been called (for reasons that still escape me) "desensitization". It occurs in other endocrine systems besides the reproductive one, as well as in non-endocrine systems. In reproductive endocrinology the story began with the loss of LH receptors associated with the surge of LH that results in ovulation and luteinization (16) and has since grown into almost a separate branch of the field. It occurs under very special conditions, e.g., in follicles or CL being maintained in an active state of secretion by small amounts of gonadotrophins, in response to treatment with a large dose of LH or hCG.

Although the Workshop contained only two 10 min. presented papers (42,43), two 15 min. so-called "invited discussant" papers (40,41), one 30 minute paper (39), and only six unpresented short papers (44-47,90,91) devoted to desensitization, it seemed to me to have contained many more.

Those interested in it have already looked at it in fine detail, and are now looking at it in the finest detail, and are rapidly reaching the point where they're talking more to one another than to the rest of us. The precise significance of desensitization, in relation to the facts that what induces it in the follicle is the ovulatory surge of LH, and that the accompanying changes in the follicle are an increasing rate of progesterone secretion and a decreasing rate of secretion of estrogens and androgens (92-96) remains, as they say in the trade, to be elucidated.

Actually, I find that, in spite of myself, the subject becomes more and more fascinating. It is especially interesting that the relationships among the inability of adenyl cyclase to be stimulated, the loss of LH receptors, (possible due to internalization), the compartmentalization of cAMP, the changes in the ultrastructural skeleton of the cell and its relation to the movements of macromolecules within the cytosol and to and from and within the cell membrane, in follicles desensitized in response to a surge of LH, seem to be the same as those which occur in a CL undergoing luteolysis in response to a similar surge of LH. Perhaps this similarity has to do with the use of one of the weirdest experimental animals - the immature rat subjected to superovulation by PMSG-hCG treatment - for so many of these studies, but that remains to be seen.

One of the more interesting of the contributions to the subject was the short paper of Hamberger *et al.* (47; unpresented, but described in the open discussion by Nilsson, one of the authors). They found that isolated granulosa cells of PMGS-hCG treated rats, *in vitro*, did *not* go through the desensitization phenomenon. The possibility, therefore, that desensitization may be a whole follicle phenomenon, and/or may depend on the presence of the theca cells was considered in the open discussion.

## THINGS CONSPICUOUS BY THEIR ABSENCE

Under the general heading of ovarian physiology, certain topics are usually taken to be so much a part of the subject, or are so interesting in themselves, that their absence from, or the small amount of time devoted to them at the Workshop is worth remarking on. The order of mention and amount of discussion bear no necessary relation to their order of importance.

### Controversy About the Corpus Luteum

The papers devoted to the CL in general [McNatty's paper (7) is not included under this heading, since it really belongs in what is labelled "Follicle Growth and Atresia", in Table 1] or particularly to luteolysis made no remarkable advances into unknown territory. The atmosphere surrounding these sessions, however, was distinguished in another way, namely, the absence of heated arguments about the CL. The silence on the subjects of, for example, what constitutes a luteotrophin, the possible existence of a universal luteotrophin, the luteotrophic effects of prolactin in mammals other than the rat, to mention only a few, was more than deafening, it was almost frightening, because it could imply that almost everyone thinks that room for argument no longer exists. It seemed to be taken for granted that LH was all a good CL, even a rat's, needed to make progesterone, and that all that was needed to put a stop to the CL's existence was a prostaglandin (PG).

I don't enjoy fighting and I'm not the kind of person who would argue endlessly with the umpire even after a bad decision, but I am disturbed by this absence of contention about the CL, because it smacks of a kind of intellectual apathy, an assumption that because the CL is an endocrine gland, it is like all other endocrine glands. Nothing could be further from the truth. There's no question that LH is a luteotrophin, and it may even be a universal one, but it certainly does not work by itself. Here is where one of the problems lies. The other is related to it, and is concerned with the luteotrophic action of prolactin.

The solution to the mystery of the mammalian CL may lie locked within the CL of the rat and other "short-cycle" mammals. Its CL looks like those of most other mammals but it does not behave like them. The short cycle mammals' CL evolved (out of the general mammalian CL) a special and peculiar insensitivity to the factors which in the other mammals induced the CL to secrete relatively large amounts of progesterone for periods of from about two weeks to two months, even in the absence of pregnancy. This adaptation allowed them to lose only about four days before ovulating again, instead of at least about 16, if a chance for coitus did not occur at ovulation. To short circuit this insensitivity, in case a chance for conception did occur, they also evolved a connection between copulation and the induction of prolactin secretion surges, and between the latter and the institution of a prolonged period of high progesterone secretion. Although prolactin does have luteotrophic effects in other mammals than the rat, mouse and hamster, these effects are on an altogether different level from its effect in the rat. If we could learn what prolactin does to a rat's CL, to enable it to raise and prolong its secretion of progesterone, we might understand why prolactin may not do (or may not have to do) the same thing in other mammals.

The ephermerality of the mammalian CL is the most important unsolved problem in CL physiology; the luteotrophic action of prolactin in the rat is the second most important unsolved problem. Neither was seriously attacked at the Workshop.

## Atresia

In spite of its importance, atresia was not considered as a subject in itself, and was touched on to only a small extent in both the presented and unpresented papers.

## Why Do Oogonia Stop Dividing?

No one even asked the question, yet how can it be separated from the problem of what induces an oogonium to begin meiosis I? Can one, in fact, even explain why and how meiosis I remains arrested in late prophase, without explaining why meiosis I even begins? And that means, why does the oogonium stop dividing mitotically?

## The Oocyte and the Inhibition of Luteinization

I would never expect Nalbandov and me to agree on anything even remotely related to the CL, and I'm still not sure we do, but at least I don't think the subject of the oocyte's possible inhibiting influence on granulosa cell luteinization is closed, and that the verdict is No. Of course, aside from the studies which originated in Nalbandov's laboratory (97-99), the evidence is against the oocyte having any influence on luteinization (100-102), but one can argue that the oocyte in question was not in exactly the same condition as one that would have remained within the normal follicular environment. The subject can still be argued (even if it involves the risk of finding myself agreeing with Nalbandov!), since there are at least theoretical grounds for postulating an inhibitory influence of the oocyte on granulosa cell luteinization.

## Pathways of Steroid Biosynthesis

Another resounding absence, especially when one considers what were major topics of discussion in the field 10-15 years ago, was the subject of steroid biosynthesis. For those of us who still have trouble thinking of systems below the level of whole tissues, this could be a small comfort, since it could mean that some day it might also be quiet on the mechanism-of-action-of-hormone front. But seriously: has the last word been said on the subject? Can we take the present-day concept of the $\Delta^5$ and $\Delta^4$ pathways from cholesterol to estrogens to be as solidly established as our knowledge of gross anatomy? Or is the relative silence about steroidogenesis only a symptom of the pauses between surges of discovery that are so typical of the way so many sciences develop?

## Ovary - Pituitary - CNS Relationships

The topic was not absent, but was definitely underplayed, and this had nothing to do with the selections of the program committee, since the same lack of interest could be seen among all the short papers submitted for presentation (Table 1). I do not have the same impression, however, from my efforts to review the literature on ovarian function, although there is an obvious trend in the direction already taken by the Workshop.

An interesting expression of this lack of interest in the external regulation of ovarian function was the almost complete silence about where follicular "inhibin" (the follicular fluid factor that suppresses FSH secretion) fits into the scheme of ovarian regulation. There's no question that it exists and is an interesting substance, but if it is true that some people have difficulty seeing where it belongs in the regulatory system (as someone remarked to me, "Why do we need it? Aren't estrogens enough?") why wasn't this discussed? Is the lack of interest in ovary-pituitary-CNS relationships at the Workshop a presage of things to come? I hope not, because a new theory of the regulation of the follicle must be concerned not only with how and why a follicle responds to constant levels of gonadotrophins, but with the events that occur through changes in the levels of gonadotrophins, particularly those preceding ovulation, and the widespread atresia that accompanies and follows ovulation.

## Why Granulosa Cells Don't Luteinize *In Vivo* Until After the LH Surge

The concentration of the factor in follicular fluid that inhibits luteinization of granulosa cells in tissue culture decreases with increase in size and maturity of the follicle to the point of non-detectability in the largest follicles (31), and now it seems that the large follicles also contain something that may even help granulosa cells to luteinize (54). If this is so, why don't the granulosa cells luteinize spontaneously within the follicle, well before the LH surge? And why wasn't this question raised at the Workshop? Is there something else, besides the follicular fluid factors and the LH surge, that controls luteinization?

## Luteinization

The absence of luteinization, i.e., the transition from follicle to CL, as a subject in itself at the Workshop can also be said to be remarkable. Its importance for an understanding of ovarian function is at least equal to that of how the follicle and the CL are regulated. It is probably the one event in the entire ovarian cycle that can tell us, if we knew why it occurs, why a follicle grows, or goes into atresia, or becomes a CL.

## Some Aspects of Practical Applications

Although a great many possibilities for the development of contraceptives, arising from or relating to the material presented at the Workshop, were discussed (103,104), no one mentioned two possibilities within the scope of the Workshop, and one peripheral to it. The first is that ovulation and CL formation can be separated from maturation of the oocyte, at least in pigs (88). The second is that, since the rupture of the follicle (i.e., ovulation) seems to depend specifically on PG's (18), it is possible for luteinization to be separated from ovulation through the use of PG inhibitors [see, e.g., (105) for the mouse, and for references to similar work in rats and rabbits; Ainsworth et al. (106) also showed that this can occur in pigs]. Both of these possible approaches have the potential of allowing otherwise normal, but non-fertile ovulation cycles to continue.

The third is to prevent implantation, without interfering with anything else, but that is a subject in itself!

## PROSPECTS FOR A NEW THEORY OF THE REGULATION OF FOLLICLE ACTIVITY

The Workshop did help, although not by design, to bring to our attention certain aspects of intrafollicular processes that must be essential elements of a new theory. These included: the interplay between estrogen, FSH and LH in the generation of estrogen, FSH and LH receptors in the granulosa cells and of LH receptors in the theca cells (20); the stimulation of granulosa cell growth by estrogens (7,20) and of granulosa cell secretory activity by FSH and estrogens (7); the stimulating effect of FSH on the conversion of androgens to estrogens by the granulosa cells (27,29); the production of androgens in the theca cells in response to LH (27,67); the synergism between FSH and androgens in the stimulation of progesterone production by follicular granulosa cells (27,29). The fact that theca cells also make progesterone, but at a much smaller rate than do granulosa cells (68) was also available, as was the less visible fact that estrogens can inhibit the production of androgens by theca cells in response to LH (67), a finding that was suspected many years ago (107,108).

The importance of the androgen/estrogen ratio within the follicle as a cause of atresia (21,109) came out of the above, through the generally opposite effects of androgens and estrogens on granulosa cell growth. The central problem as I see it, in the development of a new theory of how the follicle is regulated, embodying the principle of instability, is to fit the stimulating effects of androgens and FSH on the production of progesterone by the follicle granulosa cells with the other effects of estrogens and androgens, and with the effects of FSH and LH, and possibly of prolactin.

A typical example of what is involved is an old finding of Desclin (75) which suggested that progesterone inhibits the stimulation by LH of interstitial cell growth, in the absence of the pituitary, but does not prevent thecal "luteinization" in response to LH in the presence of the pituitary. The theory, in other words, will have to deal with all of what I've just mentioned plus at least the following questions:

Does progesterone stimulate or inhibit the production of thecal androgens?

Is progesterone the primary precursor for thecal androgens?

What is the precursor for the progesterone produced by granulosa cells in response to androgens and FSH?

Is progesterone the factor, produced by the granulosa cell in response to FSH, which sensitizes the theca cells to the stimulating effect of LH?

How do small amounts of estrogen facilitate and large amounts prevent the stimulating effect of androgens on progesterone production by follicular granulosa cells (25)?

How do large amounts of androgens prevent even this inhibiting effect of estrogens (25)?

Does progesterone actually decrease or prevent the follicle's response to gonadotrophins (18,76,77)? The granulosa cells' response to estrogens?

Does progesterone increase or decrease the rate of conversion of androgens to estrogens by granulosa cells?

If prolactin acts in general as an antigonadotrophin [e.g. (110)] and progesterone acts to diminish the follicle's response to gonadotrophins, how does the inhibiting effect of prolactin on follicular progesterone production (111) affect follicle growth or atresia?

Why is normal granulosa cell luteinization (in the rat) accompanied by degeneration of the theca cells (93)?

Is there a difference, in hormonal background, between follicles which undergo atresia accompanied by thecal "luteinization" and those in which the theca cells also disappear?

Is thecal "luteinization" always an effect of atresia of the granulosa cells, or can it also cause granulosa cell atresia?

Do androgens, estrogens or progesterone stimulate the secretion of the FSH binding inhibitor into the follicular fluid (86)?

Is the coordination of the events associated with ovulation (e.g., follicle rupture, luteinization, oocyte maturation) a function primarily of the stimulating effects of estrogens (88), or of the inhibiting effects of progesterone or prolactin?

The job of answering these questions and trying to fit the answers, together with the other information, into a theory, was even more difficult than the job of writing this Summing-Up, in spite of what I said in its first paragraph. The puzzle is both fascinating and terribly important, and whoever solves it, I have a hunch, will not only solve the mystery of how the follicle is regulated, but also a major part of why the CL is ephemeral.

## THINGS TO BE CONCERNED ABOUT?

The trend, which was so much a part of the Workshop, toward looking at intraovarian, intrafollicular and intraluteal processes for the causes of follicle growth, and of CL growth and regression, is beyond question a good sign. Especially in the case of the follicle, it is obvious that the explanation for atresia must lie within the organization of the follicle, because all of them live in the same external environment.

What may be disturbing about the trend, however, is that it may not have developed by design, but as part of a general trend toward studying all aspects of reproduction (and other aspects of biology) at the molecular level. The bandwagon phenomenon has always been part of the history of endocrinology. The steady increase, during the last 5-plus years, in the number of studies, for example, about receptors and about how hormones work within the cell, worthwhile as they are, suggests that today's bandwagon is molecular endocrinology. This is a little bit like being primarily concerned with the physical properties of the cement that allow it to bind bricks to one another, when what one really wants to know is how the building functions. Does it really make that much difference what kind of cement is used as long as it holds the bricks together?

Granted that we need molecularologists, or even if we didn't, that there's no legal way to eradicate them, the solution to the puzzle about how the ovary is regulated will probably not be solved by them. The viewpoint needed must be one which, even if it was concerned with tiny details, can rise above them, and most molecularologists are not able to do this. I'm worried that if the trend continues the number of whole body physiologists left in the field will fall below a critical mass!

I'm also concerned that the rat continues to be the most

frequently used animal for studies of reproduction. Over 50% of the Workshop papers dealt with the rat, 25% with the pig, and the remaining 25% were spread over nine other species (Table 2). The rat is a lovely animal and it lends itself especially well to studies of the CL, because it is so easy to manipulate, and because the differences between it and other mammals are what's most important. But it was only after many years of working with it that I realized that its greatest value, as far as the CL is concerned, is in its difference from other mammals. The majority of my colleagues who work with rats, and especially those who work with the PMSG-hCG treated immature one, I'm afraid, are using it with no more rationale than that it is ready to hand, easy to use, and can serve as a convenient "model". Of what?

## FUTURE WORKSHOP(S)

In spite of anything said in this Summing-Up that might be taken as critical, this Workshop was fun to attend. Such meetings usually are, since even if one does not learn anything startlingly new, it is reassuring to know that one has not grown startlingly ignorant.

But it was, in format, more of a mini-meeting of the Endocrine Society or of the Society for the Study of Reproduction than a workshop; it was a workshop only in the sense of being devoted exclusively to follicular maturation and CL function. There's nothing wrong with this except the name, for a workshop connotes a time and place for unlimited and informal exchange of information, something that was not possible at this meeting. As such, the idea of a workshop is probably inconsistent with the idea of open attendance, because the smaller the group, down to a certain point, the more readily can information be exchanged. The problem for future organizing committees to solve, and it won't be easy, will be to retain as much as possible of the freedom of exchange of information of a true workshop, without imposing limits on who can attend and discuss.

## SUMMARY OF THE SUMMING-UP

The Workshop has been looked at, in a highly subjective way, from the perspective of the need for a new theory of the regulation of the follicle and CL, embodying the principle of instability, since most follicles become atretic and the CL is an ephemeral endocrine gland. Some Workshop topics either bore on this need, or were prominent in themselves, and so were commented on; these included: the role of receptors in ovarian physiology; mechanism of action of protein hormones; where estrogens are made; the

Table 2. Distribution of Workshop articles, regardless of length of oral presentation, according to species studied, in relation to subject matter.

| Species | Follicle growth and atresia | Pre-ovulation to ovulation | Follicular fluid: Source | Follicular fluid: Subject | Oocyte maturation | Corpus Luteum in general | Luteolysis | Desensitization | General and miscell. | All subjects |
|---|---|---|---|---|---|---|---|---|---|---|
| Mouse | - | - | - | - | 1 | - | - | - | 1 | 2 |
| Rat | 5 | 3 | - | 4 | 4 | 7 | 2 | 2 | 5 | 32 |
| PMS-hCG rat | - | - | - | 1 | - | 2 | 2 | 7 | 2 | 14 |
| Hamster | - | 1 | 1 | 1 | - | - | - | - | 1 | 4 |
| Guinea pig | - | - | - | - | - | - | 1 | - | 1 | 2 |
| Rabbit | 2 | 2 | - | - | - | 1 | - | 2 | - | 7 |
| Sheep | 1 | - | - | - | - | - | 3 | - | - | 4 |
| Cow | 1 | - | 3 | 2 | - | 2 | 1 | - | - | 9 |
| Pig | 4 | 3 | 8 | 4 | 1 | 2 | 2 | 1 | 1 | 26 |
| Human | 2 | 1 | - | - | - | 1 | - | - | 3 | 7 |
| Monkey | 1 | - | - | 1 | - | 1 | 3 | - | - | 6 |
| Chicken | 2 | - | - | - | - | - | - | - | - | 2 |

The total number of studies in this table does not correspond with that in Table 1, since in several cases an animal might have been used in a single study for both a source of follicular fluid and as a subject, and in others more than one species might have been studied in a single article. Some of the review articles also covered more than one species, of course. This summary is not meant to be exact, but to be a rough account of which species are being studied.

stimulation by androgens and FSH of progesterone production by follicular granulosa cells; the non-steroid regulators in follicular fluid; oocyte maturation; and desensitization. Some things were either underemphasized or absent and so these were also commented on: controversy about the CL; atresia as a subject in itself; luteinization as a subject in itself; why oogonia stop dividing; whether the oocyte inhibits luteinization; pathways of biosynthesis; ovary-pituitary-CNS relationships.

The prospects for a new theory were looked at in the light of these and other observations and a list of questions offered, most of them having to do with atresia and the role of progesterone in intrafollicular relationships. The trend toward the molecular level, and the continued use of the rat for most studies of reproduction, were commented on as things to be concerned about, and the Summing-Up, except for this Summary, concluded with some thoughts about future workshops.

## ACKNOWLEDGEMENTS

I would like to express my thanks here to the University of Leiden for the opportunity to study there for a year, during which the effort to write this Summing-Up must be included as a part of the opportunity. I am also grateful to the Organizing Committee of the Workshop and to Dr. William Sadler and the NIH, for making it financially possible, at least in part, to take part in the Workshop. Finally, as ever, my thanks to Ms. Rosa Garnett for her preparation of the final typescript.

## REFERENCES

(The Workshop articles referred to in this list are designated by author(s), and the following abbreviations: WS-PP: principal paper; WS-ID: invited discussion paper; WS-SO: short paper orally presented; WS-S: short paper not orally presented; WS-CR: chairman's remarks.)

1. G.S. Greenwald. WS-PP.
2. B. Falck. Acta Physiol.Scand. 47(1959)Suppl. #163.
3. F.L. Hisaw. Physiol.Rev. 27(1947)95.
4. R.O. Greep, H.B. van Dyke and B. Chow. Endocrinology 30(1942) 635.
5. P.C. Williams. Nature 145(1940)388.
6. P.C. Williams. J.Endocr. 4(1945)125;127;131.
7. K.P. McNatty. WS-ID.
8. K.P. McNatty and R.S. Sawers. J.Endocr. 66(1975)391.
9. K.P. McNatty, W.M. Hunter, A.S. McNeilly and R.S. Sawers. J. Endocr. 64(1975)555.

10. K.P. McNatty, D.T. Baird, A. Bolton, P. Chambers, C.S. Corker and H. McLean. J.Endocr. 71(1976)77.
11. C. Hagen, K.P. McNatty and A.S. McNeilly. J.Endocr. 69(1976) 33.
12. P. Kemeter, H. Salzer, G. Breitenecker and F. Friedrich. Acta Endocrinol. 80(1975)686.
13. M.K. Sanyal, M.J. Berger, I.E. Thompson, M.L. Taymor and H.W. Horne, Jr. J.Clin.Endocrinol.Metab. 38(1974)828.
14. H. Eiler and A.V. Nalbandov. Endocrinology 100(1977)331.
15. A.R. Midgley, Jr. Adv.Exper.Biol.Med. 36(1973)365.
16. J.S. Richards and A.R. Midgley, Jr. Biol.Reprod. 14(1976)82.
17. R.J. Ryan and C.Y. Lee. Biol.Reprod. 14(1976)16.
18. H.R. Lindner, A. Amsterdam, Y. Solomon, A. Tsafriri, S.A. Lamprecht, U. Zor, and Y. Koch. J.Reprod.Fert. 51(1977)215.
19. J.S. Richards, J.S. Ireland, M.C. Rao, G.A. Bernath, A.R. Midgley, Jr. and L.E. Reichert, Jr. Endocrinology 99(1976)1562.
20. J.S. Richards. WS-PP.
21. J.-P. Louvet, S.M. Harmon, J.R. Schreiber and G.T. Ross. Endocrinology 97(1975)366.
22. D.T. Armstrong and J.H. Dorrington. Endocrinology 99(1976)1411.
23. A. Nimrod and H.R. Lindner. Mol.Cell.Endocrinol. 5(1976)315.
24. A.W. Lucky, J.R. Schreiber, S.G. Hillier, J.D. Schulman and G.T. Ross. Endocrinology 100(1977)128.
25. S.G. Hillier, R.A. Knazek and G.T. Ross. Endocrinology 100 (1977)1539.
26. J.E. Fortune and D.T. Armstrong. Endocrinology 102(1978)227.
27. D.W. Schomberg. WS-PP.
28. A.L. Goodman and G.D. Hodgen. WS-S.
29. L.D. Anderson, F.W. Schaerf and C.P. Channing. WS-SO.
30. U. Goebelsman, V. Davajan, J.H. Mestman and E. Z. Stanczyk. WS-S.
31. C.P. Channing and A. Tsafriri. Metabolism 26(1977)413.
32. R. Farookhi. WS-S.
33. M. Sar, W.E. Stumpf and P. Petrusz. WS-S.
34. R.S. Carson, J.K. Findlay and H.G. Burger. WS-S.
35. J.Th.J. Uilenbroek and J.S. Richards. WS-S.
36. C.Y. Lee and R.J. Ryan. Endocrinology 101(1977)869.
37. K.J. Catt and M.L. Dufau. Biol.Reprod. 14(1976)1.
38. J.M. Marsh. Biol.Reprod. 14(1976)30.
39. K.J. Catt, J.P. Harwood, N.D. Richert, P.M. Conn, M. Conti and M.F. Dufau. WS-PP.
40. L. Birnbaumer. WS-ID.
41. M. Hunzicker-Dunn, L. Birnbaumer, D. Derda, R. Jungmann and J. Bockaert. WS-ID.
42. F. Labrie, C. Auclair, L. Cuson, A. Lemay, A. Bélanger, P.A. Kelly, L. Ferland, G. Azadian-Boulanger and J.-P. Raynaud. WS-SO.
43. U. Zor, B. Strulovici, S.A. Lamprecht, A. Oplatka and H.R. Lindner. WS-SO.

44. M. Hunzicker-Dunn, M.-A. Wang and R. Jungmann. WS-S.
45. H. Rajaniemi, K. Jääskelainen and T. Hyvänen. WS-S.
46. C.Y. Lee. WS-S.
47. L. Hamberger, L. Nilsson, K. Nordenström and A. Sjögren. WS-S.
48. S. Azhar and K.M.J. Menon. WS-S.
49. M.L. Dufau, K. Hayashi, G. Sala, A. Baukal and K.J. Catt. WS-SO.
50. A.C. Kripner and C.P. Channing. WS-S.
51. A. Tsafriri. WS-ID.
52. P.M. Wassarman, R.M. Schultz, G.E. LeTourneau, M.J. LaMarca, W.J. Josefowicz and J.D. Bleil. WS-PP.
53. C.P. Channing. WS-PP.
54. F. Ledwitz-Rigby and B.W. Rigby. WS-ID.
55. A. Amsterdam, M. Shemesh and Y. Salomon. WS-S.
56. G.B. Sala, M.L. Dufau and K.J. Catt. WS-SO.
57. H.R. Behrman, J.L. Luborsky-Moore, C.Y. Pang, K. Wright and L.J. Dorflinger. WS-PP.
58. H.D. Guthrie, C.E. Rexroad and D.J. Bolt. WS-S.
59. L.J. Dorflinger and H.R. Behrman. WS-SO.
60. P. Greengard. Science 199(1978)146.
61. P.J. Sheridan. Life Sci. 17(1975)497.
62. Y.S. Moon, J.H. Dorrington and D.T. Armstrong. Endocrinology 97(1975)244.
63. J.A. Dorrington, Y.S. Moon and D.T. Armstrong. Endocrinology 97(1975)1328.
64. D.T. Armstrong and H. Papkoff. Endocrinology 99(1976)1144.
65. J.E. Fortune and D.T. Armstrong. Endocrinology 100(1977)1341.
66. P.C.K. Leung, A.K. Goff, T.G. Kennedy and D.T. Armstrong. Biol.Reprod. (1978)in press.
67. P.C.K. Leung, B.K. Tsang and D.T. Armstrong. WS-S.
68. J.E. Fortune and W. Hansel. WS-SO.
69. C.P. Channing and S.P. Coudert. Endocrinology 98(1976)590.
70. E.S.-R. Huang, K.J. Kao and A.V. Nalbandov. WS-S.
71. E.S.-R. Huang and A.V. Nalbandov. WS-SO.
72. S.S. Guraya. Physiol.Rev. 51(1971)785.
73. B. Flerko, F. Hajos and G. Setalo. Acta morph.acad.sci.hong. 15(167)163.
74. D.W. Schomberg, R.L. Stouffer and L. Tyrey. Biochem.Biophys. Res.Comm. 68(1976)77.
75. L. Desclin. Ann.d'endo. 10(1949)1.
76. D.L. Hess and J.A. Resko. Endocrinology 92(1973)446.
77. G.S. Greenwald. J.Endocr. 73(1977)151.
78. T. Hillensjö, A.S. Kripner, S.H. Pomerantz and C.P. Channing. WS-SO.
79. F. Ledwitz-Rigby, B.W. Rigby, V.L. Goy, M. Stetson, J. Young and C.P. Channing. J.Endocr. 74(1977)175.
80. M.L. Marder, C.P. Channing and N.B. Schwartz. Endocrinology 101(1977)1639.
81. S.C. Chappel, T. Ascott and H.G. Spies. WS-SO.

82. J.R. Lorenzen and N.B. Schwartz. WS-SO.
83. D. Shandar, L.D. Anderson, C.A. Barraclough and C.P. Channing. WS-S.
84. C.P. Channing, L.D. Anderson and G.D. Hodgen. WS-S.
85. K.F. Miller, J.K. Critser and O.J. Ginther. WS-S.
86. N.C. Darga and L.E. Reichert, Jr. WS-SO.
87. A.W. Schuetz. WS-CR.
88. R.H.F. Hunter, B. Cook and T.G. Baker. Nature 260(1976)156.
89. C. Thibault and M. Gerard. Ann.biol.anim.biochim.biophys. 13(suppl.)(1973)145.
90. J.P. Harwood, N.D. Richert, M.L. Dufau and K.J. Catt. WS-S.
91. K.K. Sen, S. Azhar and K.M.J. Menon. WS-S.
92. M.F. Hay and R.M. Moor. J.Reprod.Fert. 45(175)583.
93. K. Ahren, L. Bjersing, L. Hamberger, T. Hillensjö and A. Sjögren. WS-S.
94. S.K. Saidapur and G.S. Greenwald. WS-S.
95. C. D'Amato, J. Bahr, B. Stockert, F. Calvo and D. Kesler. WS-S.
96. C. Readhead, M.H. Kaufman, A.W. Schuetz and G.E. Abraham. WS-SO.
97. M.A. El-Fouly, B. Cook, M. Nekola and A.V. Nalbandov. Endocrinology 87(1970)288.
98. M. Nekola and A.V. Nalbandov. Biol.Reprod. 4(1971)154.
99. S. Stokolosowa and A.V. Nalbandov. Endocrinology 91(1972)25.
100. C.P. Channing and A. Tsafriri. J.Reprod.Fert. 50(1977)103.
101. R.H.F. Hunter and T.G. Baker. J.Reprod.Fert. 43(1974)193.
102. S.V. Nicosia and G. Mikhail. Fert.Steril. 26(1975)427.
103. K.J. Ryan. WS-PP.
104. W.A. Sadler. WS-ID.
105. P. Neal, T.G. Baker, K.P. McNatty and R.J. Scaramuzzi. J.Endocr. 65(1975)19.
106. L. Ainsworth and G.J. Marcus. WS-S.
107. S. Croes-Buth, S.E. de Jongh and F.J.A. Paesi. Acta physiol. pharm.neerl. 9(1960)303.
108. F.J.A. Paesi and J.C. de Wit. Acta physiol.pharm.neerl. 3(1953)71.
109. J.-P. Louvet, F. Bayard, A. Pujol, C. Boulard Toulouse. Ann.d'endo. 37(1976)487.
110. O. Riddle. J.Natl.CancerInst. 31(1963)1039.
111. K.P. McNatty, P. Neal and T.G. Baker. J.Reprod.Fert. 47(1976)155.

## Participants

Guy E. Abraham, M.D.
Torrance, California

Joel Abramowitz, Ph.D.
Dept. Cell Biology
Baylor College of Medicine
Houston, TX 77030

Kurt Ahren (Cancelled)
Goteburg, Sweden

Louis Ainsworth, Ph.D.
Animal Research Institute
Agric. Canada Research Branch
Ottawa, Ont. Canada KIA OC6

J. Steven Alexander, B.S.
Dept. Anatomy
Georgetown University
Washington, D.C. 20007

Max S. Amoss, Ph.D.
College Station, Texas

Abraham Amsterdam, Ph.D.
Rehovot, Israel

Richard N. Andersen, Ph.D.
Division of Reproductive Medicine
Univ. of Tennessee
Center for Health Sciences
800 Madison Avenue
Memphis, TN 38163

Larry D. Anderson, Ph.D.
660 W. Redwood St.
Univ. of Maryland Medical School
Baltimore, MD 21201

Professor D. Aravantinos, M.D.
c/o Dr. Diamantis
10 Brookside Road
Canton, Massachusetts 02021

Shawky Z. A. Badaway, M.D.
SUNY, Upstate Medical Center
750 E. Adams Street
Syracuse, NY 13210

P. Bagavandoss
Ann Arbor, Michigan

Dr. Janice Bahr
Animal Genetics Laboratory
University of Illinois
Urbana, IL 61801

Kenneth Barker, Ph.D.
Omaha, Nebraska

Marylynn Barkley, Ph.D.
Dept. of Animal Science
Univ. of California at Davis
Davis, CA 95616

Anna-Lisa Barofsky, Ph.D.
14912 McIntosh Court
Silver Spring, Maryland 20904

Satish K. Batta, D.V.M., D.Sc.
Room 5037
660 W. Redwood Street
Baltimore, Maryland 21201

Harold R. Behrman, Ph.D.
New Haven, Connecticut

Frank Bellino, Ph.D.
Buffalo, New York

E. Willem Bergink
Oss, Netherlands

George Betz, M.D., Ph.D.
Denver, Colorado

Frederick J. Bex, Ph.D.
Philadelphia, Pennsylvania

Gabriel Bialy, Ph.D.
Bethesda, Maryland

Lutz Birnbaumer, Ph.D.
Houston, Texas

Barry I. Bluestein
Boston, Massachusetts

Katryna Bogovich, Ph.D.
Ann Arbor, Michigan

Douglas Bolt, Ph.D.
USDA-BARC
Reproduction Lab
Beltsville, MD 20705

Charles O. Boyd, B.A.
Frazer, Pennsylvania

Robert M. Brenner, Ph.D.
Oregon Primate Research Center
505 N.W. 185th Street
Beaverton, OR 97005

Angela M.H. Brodie, Ph.D.
Bethesda, Maryland

Harry J. Brodie, Ph.D.
Bethesda, Maryland

Roy L. Butcher, Ph.D.
Morgantown, West Virginia

Elise Camelio
Cambridge, Massachusetts

Kenneth L. Campbell, Ph.D.
Ann Arbor, Michigan

Rocco Carsia
Trenton, New Jersey

V. Daniel Castracane, Ph.D.
8848 W. Commerce (P.O. Box 28147)
San Antonio, TX 78284

Kevin J. Catt, Ph.D., M.D.
Bethesda, Maryland

Grace M. Centola, Ph.D.
Department of Anatomy
Univ. of Maryland Dental School
666 W. Baltimore Street
Baltimore, Maryland 21201

Gagan Chainy, Ph.D.
Miami, Florida

Cornelia P. Channing, Ph.D.
Baltimore, Maryland

Scott Chappel, Ph.D.
Beaverton, Oregon

Swee H. Cheah
Urbana, Illinois

Thomas T. Chen, Ph.D.
Dept. of Physiology and Biophysics
Colorado State University
Ft. Collins, Colorado 80523

Jasti B. Choudary, Ph.D.
Dept. Biological Research
Searle Laboratories
P.O. Box 5110
Chicago, Illinois 60680

Marjorie Hearns Christie, Ph.D.
Frenchtown, New Jersey

Martin R. Clark, Ph.D.
Miami, Florida

Del Collins, Ph.D.
1495 E. Rock Springs
Atlanta, GA 30306

Kathryn G. Connell, M.S.
New York, New York

Julio M. Cortes, M.D.
3236 N. Miami Avenue
Miami, FL 33137

Nina C. Darga, D.V.M.
2348 Bermuda Run Court
Decatur, GA 30035

Sharon Day, Ph.D.
Houston, Texas

Nava Dekel, Ph.D.
New York, New York

Mark Diekman, Ph.D.
2628 Hunt Street
Ames, Iowa 50010

Donald J. Dierschke, Ph.D.
University of Wisconsin
Wisconsin Regional Primate
Research Center
1223 Capitol Court
Madison, WI 53706

Michael J. Dimino, Ph.D.
Dept. of Research
Sinai Hospital of Detroit
6767 West Outer Drive
Detroit, MI 48235

Gere diZerega, M.D.
Los Angeles, California

Laneta Dorflinger, B.S.
New Haven, Connecticut

Bruce R. Downey, D.V.M.
16 Muir Park Road
Senneville, P.2
Canada H9X1V2

Maria L. Dufau, M.D., Ph.D.
Bethesda, Maryland

Joyce Duncan
Ann Arbor, Michigan

Thomas G. Dunn, Ph.D.
Div. of Animal Sci.
Univ. of Wyoming
Box 3354 Univ. Station
Laramie, WY 80071

Edna Elfont, Ph.D.
Detroit, Michigan

David Bruce Endres, Ph.D.
Dept. Physiology and Biophysics
Colorado State University
Fort Collins, Colorado 80523

Ludeman A. Eng
Miami, Florida

John J. Eppig, Ph.D.
Bar Harbor, Maine

Greg Erickson, Ph.D.
Dept. Reproductive Medicine C-105
UCSD School of Medicine
La Jolla, CA 92093

Riaz Farookhi, Ph.D.
Ann Arbor, Michigan

Michael Fields, Ph.D.
Gainesville, Florida

Phillip A. Fields, Ph.D.
Gainesville, Florida

Jock Findlay, Ph.D.
Babraham, Cambridge, U.K.

Tony Fitz
1316B Antone
Bryan, Texas 77801

George L. Flickinger, Ph.D.
Philadelphia, Pennsylvania

J. Joe Ford, Ph.D.
U.S. Meat Animal Research Center
Clay Center, NE 68933

Joanne E. Fortune, Ph.D.
4 Fox Hollow Road
Ithaca, NY 14850

Douglas L. Foster, Ph.D.
Ann Arbor, Michigan

Guela Gibori, Ph.D.
Skokie, Illinois

Alan K. Goff, Ph.D.
London, Ontario, Canada

Laura T. Goldsmith
16 West 16th Street
New York, N.Y. 10011

Arnold L. Goodman, Ph.D.
Bethesda, Maryland

Neil Goodman, M.D.
9000 S.W. 87th Court
Miami, FL 33176

Robert Goodman, Ph.D.
Ann Arbor, Michigan

Wayne L. Gordon, M.S.
St. Joseph, Illinois

Gilbert S. Greenwald, Ph.D.
Department of Physiology
Univ. of Kansas Medical Center
Kansas City, KS 66103

Roy O. Greep, Ph.D.
45 Shattuck Street
Boston, Mass 02115

H. David Guthrie, Ph.D.
USDA Sea APGI Reproduction Lab
Beltsville, MD 20705

James M. Hammond, M.D.
Div. of Endocrinology
M.S. Hershey Medical Center
Hershey, PA 17033

A. F. Haney, M.D.
Box 2971 D.U.M.C.
Department of OB-GYN
Durham, NC 27710

Rajaniemi Hannu, Ph.D.
Department of Anatomy
Univ. of Oulu, Kajaanintie 52A
Oulu 22, Finland

William Hansel, Ph.D.
816 Veterinary Research Tower
Cornell University
Ithaca, NY 14853

France Haour, Ph.D.
Lyon, France

Mineko Fukushima Hariu, M.D.
Dept. of Obstetrics & Gynecology
Akita University
School of Medicine
Akita, Japan

James Harwood, Ph.D.
Bethesda, Maryland

Keith M. Henderson, Ph.D.
London, Ontario, Canada

Torbjörn Hillensjö, M.D., Ph.D.
Dept. of Physiology
University of Göteborg
Fack
S-400 33 Göteborg
Sweden

Stephen G. Hillier, Ph.D.
Bethesda, Maryland

James E. Hixon, Ph.D.
University of Illinois
Veterinary Medicine Bldg.
Urbana, Ill 61801

Valerie S. Hobbs, B.S.C.
Laramie, Wyoming

Gary D. Hodgen, Ph.D.
Bethesda, Maryland

John A. Holt, Ph.D.
Chicago Lying-In Hospital
University of Chicago
5841 Maryland Avenue
Chicago, Illinois 60637

J.G.M. Horbach, Ph.D.
Netherlands

W. F. Howard, M.D.
13227 Hill Forest
San Antonio, Texas 78230

Aaron Hsueh, Ph.D.
La Jolla, California

Eileen Su-Rong Huang, Ph.D.
R-10
700 Morreene Road
Durham, N.C. 27705

C. Huckins, Ph.D.
Houston, Texas

Mary Hunzicker-Dunn, Ph.D.
Chicago, Illinois

Barry Jacobs, M.D.
Department of OB-GYN
Univ. of Texas School of Medicine
Houston, Texas 77030

Robert B. Jaffe, M.D.
Dept. OB-GYN & Rep. Soi.
Univ. of California,
San Francisco
3rd & Parnassus
San Francisco, CA 94143

Sandra Kammerman, M.D.
New York, New York

Edward R. Katz, M.D.
Scottsdale, Arizona

Yair Katz, Ph.D.
Monell Chemical Senses Center
3500 Market
Philadelphia, Pa. 19104

Duane H. Keisler, B.S.
Star Route 2, Box 43 C
Beeville, TX 78102

Paul Kelly, Ph.D.
Quebec, Canada

June Kendall, Ph.D.
524 Burrill Hall
Department of Physiology
University of Illinois
Urbana, Illinois 61801

Edward J. Keogh, M.D., Ph.D.
Perth, Australia

Landis P. Keyes, Ph.D.
Ann Arbor, Michigan

Iqbal Khan, M.Sc.
Göteborg, Sweden

Howard J. Kirchick, Ph.D.
Ann Arbor, Michigan

O. Ray Kling, Ph.D.
P.O. Box 26901
Dept. OB/GYN
University of Oklahoma
Oklahoma City, OK 73190

Andrew R. LaBarbera, Ph.D.
Rochester, Minnesota

Dr. Fernand Labrie, Directeur
Groupe du Conseil de Récherches
Médicales en Endocrinologie
Molcculaire,
Centre Hospitalier de
l'Université Laval,
2705 boulevard Laurier,
Quebec, G1V 4G2, Canada

Michal Lahav
Haifa, Israel

Lynn H. Larkin, Ph.D.
Gainesville, Florida

Florence Ledwitz-Rigby, Ph.D.
Dept. of Biological Science
Northern Illinois University
DeKalb, Ill 60115

Ching y Lee, M.D., Ph.D.
Box 198 Mayo
Dept. Lab. Med. and Pathology
University of Minnesota
Minneapolis, Minn. 55455

Sandra J. Legan, Ph.D.
Ann Arbor, Michigan

William LeMaire, M.D.
Miami, Florida

Andre Lemay, M.D., Ph.D.
2223 Onesime Lagnon
Sulery, Quebec, Canada G1T 271

Peter C.K. Leung
London, Ontario, Canada

P. Leymarie, M.D.
Caen, France

Bernard Lieberman, M.D.
Great Neck, New York

H.R. Lindner, Ph.D.
Department of Hormone Research
Weizmann Institute of Science
Rehovot, Israel

Harry Lipner, Ph.D.
Dept. Biological Science
Florida State University
Tallahassee, FL 32306

Keith Joseph Lookingland
Baltimore, Maryland

Janice R. Lorenzen, Ph.D.
Evanston, Illinois

Judith Lee Luborsky-Moore, Ph.D.
Reprod. Biol. Section
Dept. OB/GYN
333 Cedar Street
New Haven, CT 06510

Janina H. Lukaszewska, Ph.D.
Ithaca, New York

Gordon J. MacDonald, Ph.D.
Piscataway, New Jersey

Damodar K. Mahajan, Ph.D.
Cleveland, Ohio

V. B. Mahesh, Ph.D.
Augusta, Georgia

Richard P. Marrs, M.D.
323 26th Street
Manhattan Beach, CA 90266

John Marsh, Ph.D.
Miami, Florida

John R. Marshall, M.D.
OB-GYN, Harbor General Hospital
Torrance, CA 90509

Pierre Matton, Ph.D.
University of Sherbrooke
Biology Department
Sherbrooke, Quebec,
CANADA

John A. McCracken, Ph.D.
Shrewsbury, Massachusetts

Ken McNatty, M.D.
Boston, Massachusetts

Carol McRae, M.S.
Tallahassee, Florida

Rodney Mead, Ph.D.
Dept. of Biological Sciences
University of Idaho
Moscow, Idaho 83843

K. M. J. Menon, Ph.D.
Ann Arbor, Michigan

Priv. Doz. Dr. W. Mestwerdt
Universitaets-Frauenklinik
Josef-Schneider-Strabe 4
8700 Wuerzburg
Germany

Sandra D. Michael, Ph.D.
Dept. Biological Sciences
SUNY
Binghamton, NY 13901

A. Rees Midgley, M.D.
Ann Arbor, Michigan

Josephine B. Miller, Ph.D.
Rochester, Minnesota

Kurt F. Miller, B.S.
1655 Linden Drive
Madison, Wisconsin

Thomas Mills, Ph.D.
Dept. of Endocrinology
Medical College of Georgia
Augusta, Georgia 30902

Isao Miyakawa, M.D.
Los Angeles, California

Ruth T. Moore, Ph.D.
R 3044 Kresge 11
Dept. Internal Medicine
University of Michigan
Ann Arbor, MI 48109

Teresa Morales, Ph.D.
Miami, Florida

Bruce D. Murphy, Ph.D.
Saskatoon, Saskatchewan,
Canada

Thomas Nass, M.S.
Madison, Wisconsin

M. Neves-E-Castro, M.D.
Organon Research
Houterbrugweg 22
Geldrop 5664-EV
Holland

Lars Nilsson, M.D.
Hammarkroken 146
S-424 36 Angered,
Sweden

Gordon Niswender, Ph.D.
Fort Collins, Colorado

Reid Norman, Ph.D.
Beaverton, Oregon

Ehard F. Nutting, Ph.D.
Box 5110
Searle Laboratories
Chicago, IL 60680

Nathan O. Okia, Ph.D.
Ann Arbor, Michigan

Juraj Osterman, M.D.
Hershey, Pennsylvania

James Overpeck
Washington, D.C.

Augustine Oyake, M.D.
Montreal, Canada

Laurie G. Paavola, Ph.D.
Dept. of Anatomy
Temple Univ. School of Med.
3420 N. Broad Street
Philadelphia, PA 19140

C. Y. Pang
New Haven, Connecticut

Harold Papkoff, Ph.D.
San Francisco, California

Rube Pardo, M.S.
Gainesville, Florida

Edward D. Plotka, Ph.D.
Marshfield, Wisconsin

Seymour Pomerantz, Ph.D.
Baltimore, Maryland

John Quagliarello, M.D.
25 Waterside Plaza - DD
New York Univ. Medical Center
New York, New York 10010

Mike Quasney, B.S.
Ann Arbor, Michigan

Madhwa Raj, Ph.D.
Chapel Hill, North Carolina

Judith A. Ramaley, Ph.D.
Department of Physiology
Univ. of Nebraska Medical Center
Omaha, Nebraska 68105

Naomi Rance
Baltimore, Maryland 21201

Ch. V. Rao, Ph.D.
Jeffersonville, Indiana

Dr. Carol Readhead
Division of Biology 156-29
California Institute of Technology
Pasadena, California 91125

Robert W. Rebar, M.D.
Dept. of Reproductive Medicine (T-002)
Univ. of California, San Diego School of Medicine
La Jolla, California 92093

Leo E. Reichert, Ph.D.
Dept. of Biochemistry
Emory Univ. School of Med.
Atlanta, Ga. 30322

John A. Resko, Ph.D.
Beaverton, Oregon

Nasser Rezai, M.D.
222 Old Line Avenue
Laurel, MD 20810

Bernard F. Rice, M.D.
Kansas City, Missouri

Joanne S. Richards, Ph.D.
Ann Arbor, Michigan

B. W. Rigby, Ph.D.
DeKalb, Illinois

Griff T. Ross, M.D.
Bethesda, Maryland

Michael S. Roth, M.D.
Miami, Florida

Irving Rothchild, M.D., Ph.D.
Cleveland, Ohio

Subir Roy, Ph.D.
470 Lakeview Road
University of Southern California
Pasadena, CA 91105

Michael E. Rush, M.S.
Biology Unit 1
Florida State University
Tallahassee, FL 32306

Kathleen D. Ryan, Ph.D.
Ann Arbor, Michigan

Kenneth J. Ryan, M.D.
Boston, Massachusetts

Robert J. Ryan, M.D.
Rochester, Minnesota

William A. Sadler, Ph.D.
Bethesda, Maryland

Jose Saez, M.D.
Hospital Debrousse
29 Rue Soeur Bouvier
69322 LYON Cedex 1
France

Srinivas K. Saidapur, Ph.D.
Univ. of Kansas Med. Ctr.
Bldg 37, MRRC 416
39th and Rainbow Blvd.
Kansas City, Kansas 66103

M.R. Sairam, Ph.D.
Montreal, Canada

Teruyo Sakakura, Ph.D.
Aichi Cancer Center Research Institute
Tashiro-cho, Chikusa-ku
Nagoya, 464 Japan

Michel Sanders
Ann Arbor, Michigan

J. Sandow, M.D.
Frankfurt, Germany

Mrinal K. Sanyal, Ph.D.
Research Triangle Park, N.C.

Frederick W. Schaerf
Baltimore, Maryland

David Schomberg, Ph.D.
Box 3323
Duke Univ. Medical Center
Durham, NC 27710

Judy Schoonmaker
Ann Arbor, Michigan

Allen W. Schuetz, Ph.D.
615 N. Wolfe Street
Johns Hopkins University
School of Hygiene
Baltimore, MD 21205

Richard M. Schultz, Ph.D.
Boston, Massachusetts

Neena B. Schwartz, Ph.D.
Dept. of Biological Science
Northwestern University
Evanston, Ill. 60201

Alison Schwartz-Kripner, M.P.H.
Baltimore, Maryland

Meinhardt Schweditsch, M.D.
1620 McEldery Street
Baltimore, Maryland 21205

Sheldon J. Segal, Ph.D.
New York, New York

Christodulos Sengos, M.D.
18 Trikalon T.T. 607
Athens, Greece

Abubakar A. Shaikh, Ph.D., D.V.M.
P.O. Box 28147
Southwest Foundation
San Antonio, TX 78284

Ruth Shalgi, Ph.D.
New York, New York

Arthur G. Shapiro, M.D.
Dept. of OB-GYN
Univ. of Miami Sch. of Medicine
P.O. Box 016960
Miami, FL 33101

O. David Sherwood, Ph.D.
Dept. of Physiology and Biophysics
Room 524
Burrill Hall
Urbana, Illinois 61801

Kunio Shiota
Dept. of Veterinary Physiology
Faculty of Agriculture
University of Tokyo
Bunkyo-ku, 113 Tokyo
Japan

Samuel A. Sholl, Ph.D.
Madison, Wisconsin

Maryann Simonelli, B.S.
Gainesville, Florida

Charles W. Simpson, M.D.
London, Ontario, Canada

Karl Slotta, Ph.D.
Miami, Florida

Maja Slotta, Ph.D.
Miami, Florida

Patrick Sluss, M.S.
1509 Crestmore Place
Fort Collins, Colorado

Michael F. Smith, M.S.
Beeville, Texas

Michael C. Snabes
Ann Arbor, Michigan

Jimmy L. Spearow
Davis, California

Harold G. Spies, Ph.D.
Beaverton, Oregon

Frank Stanczyk, Ph.D.
21721 Roscoe Blvd., Apt 37
Canoga Park, CA 91304

Bernard G. Steinetz, Ph.D.
336 Longbow Drive
Franklin Lakes, N.J. 07417

Sarah Lipford Stone, Ph.D.
7602 Wildwood Drive
Takoma Park, MD 20012

Richard L. Stouffer, Ph.D.
Tucson, Arizona

Jerome F. Strauss, III, M.D., Ph.D.
Philadelphia, Pennsylvania

Joan Sulewski, M.D.
Dept. of OB-GYN
Milton S. Hershey Medical Ctr.
Hershey, Pa. 17033

Kalyan Sundaram, Ph.D.
New York, New York

R. James Swanson, Ph.D.
Dept. Biological Sciences
Old Dominion University
Norfolk, VA 23508

H. Taavon, M.D.
930 N. Charles Street
Baltimore, MD 21201

Kazuyoshi Taya, D.V.M.
Rainbow Towers
Apt 1101, 3838 Rainbow Blvd.
Kansas City, Kansas 66103

Rosemarie B. Thau, Ph.D.
Population Council
Rockefeller University
York Ave. & 66th Street
New York, N.Y. 10021

G. D. Thorburn, M.D.
Australia

Alexander Tsafriri, Ph.D.
Rehovot, Israel

Jan Uilenbroek, Ph.D.
Ann Arbor, Michigan

Judith Vaitukaitis, M.D.
Boston, Massachusetts

Johannes Veldhuis, M.D.
Dept. of Medicine
Endocrinology Division
The Milton S. Hershey
Medical Center
Hershey, PA 17033

Nicholas Vorys, M.D.
Columbus, Ohio

Michael W. Vernon, M.S.
Gainesville, Florida

Paul M. Wassarman, Ph.D.
Boston, Massachusetts

Michael Waters, Ph.D.
Physiology Department
Univ. of Queensland
Brisbane, Queensland
Australia

Frank T.G. Webb, Ph.D.
Human Reproduction Unit
World Health Organization
1211 Geneva 27
Switzerland

Gerson Weiss, M.D.
New York Univ. Medical Center
566 First Avenue - Suite 2C
New York, NY 10016

Judith Weisz, Ph.D.
Div. of Reproductive Biology
Dept. of OB-GYN
Milton S. Hershey Med. Ctr.
Hershey, Pa. 17033

Klaus U. Weithmann, Ph.D.
Frankfurt, Germany

Anne Colston Wentz, M.D.
3411 Waynoka
Memphis, Tenn. 38111

Margaret Whelan
Miami, Florida

William J. Whelan, D.Sc.
Miami, Florida

Walter G. Wiest, Ph.D.
Dept. of OB-GYN
Washington Univ. Sch. of Med.
4911 Barnes Hospital Plaza
St. Louis, MO 63110

John W. Wilks, Ph.D.
Fertility Research
The Upjohn Company
Kalamazoo, MI 49001

A. Trent Williams
Tallahassee, Florida

Julie J. Williams
Ann Arbor, Michigan

Marvin T. Williams, Ph.D.
Miami, Florida

Robert Williams, Ph.D.
Bethesda, Maryland

Karen H. Wright, M.S.
Dept. OB-GYN
Pennsylvania Hospital
8th & Spruce
Philadelphia, PA 19107

Kristina Wright, Ph.D.
Dept. of OB/GYN
Yale School of Medicine
333 Cedar St.
New Haven, CT 06510

Chung H. Wu, M.D.
3400 Spruce Street
Department of OB-GYN
Univ. of Pennsylvania
Philadelphia, PA 19104

Masaki Yanagishita, M.D.
NICHD, ERRB
NIH, Bld 10 Rm 12N206
Bethesda, MD 20014

Koji Yoshinaga, Ph.D.
LHRRB Harvard Medical School
45 Shattuck Street
Boston, Mass. 02115

Edward V. YoungLai, Ph.D.
McMaster Univ. Medical Centre
Dept. of OB-GYN
1200 Main St. West,
Hamilton, Ontario Canada L8S 4J9

Kathy Khe Ching Yuh, Ph.D.
2481 Towner Blvd.
Ann Arbor, MI 48104

Stephen J. Zimniski
Boston, Massachusetts

Lawrence C. Zoller, Ph.D.
Hershey, Pennsylvania

Uriel Zor, Ph.D.
Rehovot, Israel

INDEX

MIX
Papier aus verantwortungsvollen Quellen
Paper from responsible sources
FSC® C105338

If you have any concerns about our products,
you can contact us on
**ProductSafety@springernature.com**

In case Publisher is established outside the EU,
the EU authorized representative is:
**Springer Nature Customer Service Center GmbH**
**Europaplatz 3, 69115 Heidelberg, Germany**

Printed by Libri Plureos GmbH
in Hamburg, Germany